Progress in Optical Science and Photonics

Volume 32

Series Editors

Indexed by Scopus

The purpose of the series Progress in Optical Science and Photonics is to provide a forum to disseminate the latest research findings in various areas of Optics and its applications. The intended audience are physicists, electrical and electronic engineers, applied mathematicians, biomedical engineers, and advanced graduate students.

Prasanth Ravindran · Deepa K. G. · Adersh Asok ·
Durga Shankar · Al Jumlat Ahmed
Editors

Photon to Power

Harvesting the Sun

Springer

Editors
Prasanth Ravindran
Department of Green Energy Technology
Pondicherry University
Pondicherry, India

Deepa K. G.
Department of Physics
University of Kerala
Thiruvananthapuram, Kerala, India

Adersh Asok
Centre for Sustainable Energy Technologies
CSIR-National Institute for Interdisciplinary Science and Technology (NIIST)
Thiruvananthapuram, Kerala, India

Academy of Scientific and Innovative Research (AcSIR)
Ghaziabad, India

Durga Shankar
Department of Physics, Chemistry and Biology (IFM)
Linköping University
Linköping, Sweden

Al Jumlat Ahmed
CSIRO
Clayton, VIC, Australia

University of Wollongong
Wollongong, Australia

ISSN 2363-5096 ISSN 2363-510X (electronic)
Progress in Optical Science and Photonics
ISBN 978-981-96-5913-5 ISBN 978-981-96-5914-2 (eBook)
https://doi.org/10.1007/978-981-96-5914-2

This Springer imprint is published by the registered company Springer Nature Singapore Pte Ltd.
The registered company address is: 152 Beach Road, #21-01/04 Gateway East, Singapore 189721, Singapore

Preface

The quest for sustainable and efficient energy sources has driven extensive research and technological advancements across multiple disciplines. Understanding the fundamental processes that convert photons into usable energy becomes paramount as the world transitions toward clean energy solutions. *Photon to Power* is an interdisciplinary exploration of the scientific principles, technological developments, and real-world applications of solar energy conversion.

This book is designed to serve as a comprehensive resource for researchers, engineers, academicians, and students engaged in renewable energy. It bridges the gap between fundamental physics, materials science, and engineering applications, offering insight into photovoltaics, solar thermal systems, photochemical energy conversion, and emerging hybrid technologies. Each chapter is meticulously crafted to provide a balanced blend of theoretical concepts, experimental methodologies, and practical implementations.

The discussion extends beyond conventional solar cells to advanced materials, innovative characterization techniques, and integrated energy storage solutions. Special emphasis is placed on nanomaterials, tandem, multi-junction solar cells, photonic enhancement strategies, and solar-driven hydrogen production. Additionally, computational modeling and artificial intelligence-driven approaches for optimizing solar energy systems are explored.

This book is the culmination of collaborative efforts from experts across diverse domains, reflecting the dynamic nature of solar energy research. By bringing together state-of-the-art developments and prospects, Photon to Power aims to inspire further innovations and provide a valuable reference for those shaping the next generation of energy solutions.

The book is structured into five parts, each dedicated to a specific aspect of solar energy conversion:

Part One: Conversion of Solar to Photovoltaic Energy This part delves into the fundamental principles of photovoltaic technology, exploring the working mechanisms of solar cells, advancements in materials, and efficiency enhancement strategies.

Part Two: Conversion of Solar to Thermal Energy Solar thermal energy is crucial in power generation and industrial applications. This part discusses the principles of solar thermal conversion, concentrating solar power (CSP) systems, and novel solar-to-thermal solutions that enhance efficiency and sustainability.

Part Three: Conversion of Solar Energy to Electrochemical Energy Electrochemical energy storage and conversion, mainly through solar-driven water splitting, are vital for energy sustainability. This part explores advancements in photocatalysis, solar fuel generation, and electrochemical solar energy harvesting systems.

Part Four: Advancements Beyond Conventional Solar Conversion Techniques Beyond traditional photovoltaic and thermal systems, new approaches continue to emerge. This part highlights cutting-edge research in innovative technologies that push the boundaries of solar energy utilization.

Part Five: Solar Optics Efficient solar energy harvesting is intricately linked to optical design. This part covers the role of solar optics in maximizing light absorption, including light trapping techniques, anti-reflective coatings, and the application of nanophotonics to enhance solar conversion efficiency.

Through this structured approach, *Photon to Power* provides a holistic understanding of solar energy conversion, bridging the gap between fundamental principles and practical applications. We hope that this book will serve as a valuable resource for advancing research and innovation in sustainable energy technologies.

We sincerely thank all contributors, reviewers, and researchers whose invaluable efforts have enriched this volume. We sincerely thank research scholars Adhithya S. Thampi (NIIST), Sangeetha M. S. (NIIST), and Aravind S. P. (PU) for their invaluable support in consolidating and tracking the review process. Your dedication and assistance have been instrumental in ensuring a smooth and organized workflow.

We hope that this book serves as a guiding light for students, professionals, and enthusiasts striving to harness the full potential of photon-to-energy conversion.

Pondicherry, India	Prasanth Ravindran
Thiruvananthapuram, India	Deepa K. G.
Thiruvananthapuram/Ghaziabad, India	Adersh Asok
Linköping, Sweden	Durga Shankar
Clayton/Wollongong, Australia	Al Jumlat Ahmed

Contents

Conversion of Solar to Photovoltaic Energy

Fundamentals of Photovoltaic Technologies

V. J. Anjana Nair, S. Heera, and Deepa K. G.

Abstract Carbon emissions are the greatest threat faced by Earth today, with the burning of fossil fuels being the primary contributor to CO_2 emissions worldwide. Currently, it is estimated that renewable and clean energy sources contribute roughly 16% of global energy demand. This analysis underscores the urgent need to find alternative energy sources to meet the ever-growing global demand, which is a top priority for both developed and developing nations. Solar photovoltaic (PV) technology is one such solution, and it is nearly maintenance-free. Solar cells, or photovoltaic cells, convert sunlight directly into electricity. They have a wide range of applications, such as supplying electricity to remote areas, powering spacecraft, and being used in military, automotive, and building-integrated PV systems. This chapter covers the fundamentals of converting solar light energy into electrical energy in photovoltaic devices. It also explores the evolution of solar cells through various generations and provides an overview of the market status of the PV industry. Additionally, a brief introduction is given to emerging fields like bifacial photovoltaics, floatovoltaics, and agrivoltaics.

Keywords Photovoltaic · Generations · Shockley–Queisser limit · Parameters · Efficiency · Solar panel · Floatovoltaics · Agrivoltaics

1 Introduction

Harnessing sunlight to generate electricity is the key feature of photovoltaic (PV) technology, a sustainable and renewable method for clean energy generation. PV technology is based on the principle of the photovoltaic effect, discovered in 1839 by French physicist Edmond Becquerel. He observed an increase in electrical current when silver-coated platinum electrodes immersed in an electrolyte were exposed to light. The effect is more pronounced when the electrodes are coated with light-sensitive materials, and the wavelength of the light source also influences the results.

V. J. Anjana Nair · S. Heera · Deepa K. G. (✉)
Department of Physics, University of Kerala, Thiruvananthapuram, Kerala, India
e-mail: deepachaithanya@gmail.com

P. Ravindran et al. (eds.), *Photon to Power*, Progress in Optical Science and Photonics 32, https://doi.org/10.1007/978-981-96-5914-2_1

This discovery spurred further research in the field, which continues today, with the goal of improving renewable energy generation.

Solar cells are devices made from light-sensitive materials capable of generating electricity upon exposure to light, and they are a fundamental component of PV technology. In the early stages of PV research, the primary focus was on developing PV-active materials. However, current research is centered around designing device structures that improve light absorption and minimize energy losses. Materials used for solar energy generation should exhibit high light absorption followed by efficient electron generation. These materials, generally referred to as semiconductors, include oxides, selenides, sulfides, and others. The devices are engineered using various strategies, such as thin-film absorbers, p–n junctions, and tandem cells.

As solar cell efficiency is greatly influenced by material quality, highly pure materials were used, which obviously increased the production cost. This sets the technological applications only for use in space devices where power generation by sunlight is highly valuable. PV technology should be commercialized for a well-pronounced global transition to clean energy. Further research focused on this aspect, reduced the material and production cost of the technology, and made it accessible to the common public.

Today's PV systems include solar panels, mounting structures, batteries, and inverters. Solar panels comprise several individual silicon solar cells, either connected in series or in parallel, depending on the desired output power. The abundance of Si and the use of single and polycrystalline Si wafers have reduced the cost and aided in the commercialization of PV systems. Solar panels should be mounted at optimal places to capture maximum solar irradiance. The direct current (DC) electricity produced by the solar cells can be stored in batteries or fed to inverters to convert into alternating current (AC) electricity, which can be directly used for household and commercial appliances.

Years of research have witnessed many PV technological advancements that significantly increased their efficiency, durability, and cost-effectiveness. Material innovations, together with innovative designs like thin-film solar cells, perovskite solar cells, and tandem solar cell designs, accelerated the performance and lowered manufacturing costs. The introduction of grid integration technologies and advancements in energy storage solutions promotes energy generation and makes solar energy more viable as a primary power source and for income generation.

1.1 Energy from Sun

The primary source of energy for our planet is the Sun, delivering large amounts of energy either as heat or as light. Photovoltaic cells make use of this energy to generate electricity. The photoconversion process relies significantly on solar irradiance and the solar cell's ability to convert light to electricity.

Solar irradiance refers to the power per unit area received from the Sun at a given area on Earth. It is measured in watts per square meter (W/m^2). Various factors create non-uniformity in solar irradiance across the Earth's surface, including weather conditions, geographical location, the angle at which sunlight strikes the surface and the time of day.

Direct and diffuse irradiance are the two primary components of solar irradiance. The light that travels from the Sun to the Earth's surface directly without being scattered or absorbed by the atmosphere is the direct irradiance. In contrast, the scattered light arriving at the surface from all directions is termed as diffuse light. The total solar irradiance at a location is considered as the combination of the direct and diffuse components.

The Sun emits energy across a broad spectrum of electromagnetic radiation, from ultraviolet (UV) and visible light to infrared (IR) radiation. The total energy emitted by the Sun is distributed across these wavelengths, but on reaching the Earth's atmosphere, scattering by the atmospheric constituents, like water and oxygen molecules, ozone, CO_2, etc., occurs, resulting in spectral attenuation [1]. The atmospheric attenuation of the solar spectrum is estimated by "Air mass", $n_{\text{Air mass}}$, defined by

$$n_{\text{Air mass}} = \frac{\text{Optical path length to sun}}{\text{Optical path length if sun directly overhead}},$$

$$n_{\text{Air mass}} = \operatorname{cosec}\gamma_s,$$

where γ_s is the angle of Sun elevation.

The standard solar spectrum is considered as the Air Mass 1.5 (AM 1.5) where a standard thickness of Earth atmosphere attenuates the extraterrestrial solar spectrum. Considering the Sun's elevation angle at 42°, AM 1.5 spectrum attenuates the solar irradiance to 900 Wm^2, normalized to 1000 Wm^2. The solar cells should directly face the Sun without any shadowing for efficient light collection. Innovations like solar tracking were also implemented for better light absorption but were not commercialized due to economic concerns.

1.2 *Need for Sustainable Energy*

In recent decades, the increasing urbanization and population growth have surged in the global energy demands. The higher energy demands and environmental and economic challenges have emphasized the need to switch to sustainable energy solutions. Meeting the current energy criteria from renewable energy sources without compromising the energy needs of future generations or the climate can be termed a sustainable energy source. Only a sustainable energy source can address the global issues of resource depletion and climate change. Though traditional energy sources, like fossil fuels, had geared human development, they came up with significant environmental impacts such as greenhouse gas emissions, climate change, air pollution, and habitat destruction.

Fossil fuels are the conventional energy sources, including oil, coal, and natural gas. They need to be extracted underneath the Earth and can often lead to environmental degradation, mainly deforestation, soil erosion, pollution, and loss of biodiversity. Also, burning fossil fuels releases carbon dioxide (CO_2), a greenhouse gas contributing to global warming and climate change. The limited fossil fuel supply will also create a situation of energy insecurity that can only be accomplished by renewable energy sources. Renewable energy sources have minimal environmental impact and do not release harmful pollutants into the atmosphere. The rise in global temperatures leads to an increase in sea levels at an alarming rate, frequent and intense weather events, and disruptions to ecosystems [2].

By reducing reliance on fossil fuels and increasing the share of clean energy in the global energy mix, we can significantly reduce greenhouse gas emissions and limit global warming to safer levels. Considering the geographic conditions, at least one of the energy sources, such as solar, wind, hydro, and geothermal sources, can be harnessed from any part of the world, making renewable energy feasible for all. The abundance and the capability of locally harnessing from almost all parts of the world promote the application of renewable energy sources and can address the issues of energy poverty. The environmental impact of the installation and disposal of renewable energy systems is far smaller than what is created with fossil fuels. Since most sustainable energy sources are in their infancy, innovative approaches need to be carried out to exploit maximum efficiency from renewable resources, creating further opportunities for research. The quality of life can be improved by deploying renewable energy systems in remote areas, and economic growth, jobs, and support for local businesses can be achieved [3]. More than a distant idea, renewable power generation is an urgent necessity globally to meet the rising energy demands.

1.3 Photovoltaic Effect

The twofold nature of light, known as the dual nature of light, where light can behave both as a particle and wave, is a key concept in quantum mechanics and features the photoelectric and photovoltaic effect. The nineteenth-century discovery of photovoltaic effect can be well explained by the particle nature of light. The photons, quanta of light gets absorbed by photosensitive material, causes the formation of electron-hole pairs which is the vital mechanism behind photovoltaic cells.

Every material is composed of a very large number of atoms interacting with each other by bonding. These interactions result in splitting atomic energy levels, creating a resultant continuum or bands of energy levels with nearly similar energies. The filling of the individual atomic orbitals determines the filling of bands. The energy separation between the highest occupied band (valence band, VB) and the lowest unoccupied band (conduction band, CB), known as bandgap, classifies the material into metals, semiconductors and insulators [1].

Metals are materials with partially or completely filled VB overlapped with the CB. The band overlapping enables valence electrons to easily occupy the available empty states with similar energies in the CB, generating free electrons. These free electrons are responsible for the electrical and thermal conductivity in metals. The existence of an energy gap creates a difficulty in transferring valence electrons to the CB, affecting their intrinsic properties.

In semiconductors, the VB is completely filled and has a bandgap of 0.5–3 eV. Since the valence band electrons are employed in bonding, a sufficient energy greater than or equal to the bandgap is essential for exciting the valence electrons to the CB. Whereas an increased bandgap above 3 eV is incapable of promoting the valence electrons to the CB, such materials are termed as insulators.

As mentioned earlier, a sufficient energy, greater than or equal to the bandgap energy, can excite the valence electrons to the CB. The CB electrons are free to move around, enabling thermal and electrical conduction. Different forms of energies like optical, thermal, electrical energy, etc., can be used for valence electron excitation to the CB. When the valence electrons get sufficient energy, they become capable of breaking the bonding and exciting the freed electrons to the CB, leaving vacant sites in the VB known as holes. For each material, the excitation energy will be different depending on the bandgap of the material. For wide bandgap materials, UV or blue-green light is required, whereas near IR or red wavelength is used up by materials with a bandgap of approximately 1.5 eV.

The photovoltaic effect is a phenomenon that results in voltage or electricity being generated inside a photovoltaic cell when exposed to sunlight [4]. This effect enables solar panels to generate electrical energy from sunlight. A solar cell or photovoltaic cell is a device that converts light energy into electrical energy through the photovoltaic effect. The term "photovoltaic" comes from the Greek "phos", which means light, and "volt", which is the unit of electromotive force. They are basically p–n junction diodes; when exposed to sunlight, they generate a voltage or electric current. The mechanism of the photovoltaic effect is shown in Fig. 1.

If a suitable wavelength of light is incident on a photovoltaic cell, the material absorbs the radiation creating electron hole pairs. The presence of a p-n junction creates a potential difference and separates the charge carriers according to their polarity, then ended up in generation of photovoltage [5].

1.4 Basics of Solar Cell

The fundamental principle of solar cells is the photovoltaic effect, generating electrical energy from optical energy. Solar cells, categorized as two terminal devices, function as diodes in the absence of light and are capable of power generation upon illumination. The illumination wavelength, intensity, illumination area, working temperature, and many other factors influence the photovoltage generation in solar cells [6].

This section discusses the structure of solar cell, different materials employed in solar cell fabrication, mainly light absorbers, charge transport materials and contacts,

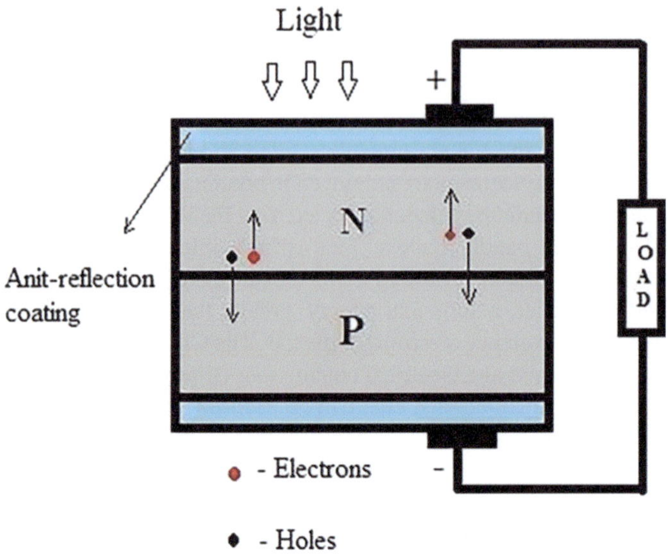

Fig. 1 Photovoltaic effect

solar cell structure, different types of solar cells and the major factors influencing the performance of solar cells.

1.4.1 Solar Cell Structure

A typical solar cell consists of several layers that work together to convert light into electricity. Each layer functions uniquely to improve the overall device performance. A typical structure of solar cells is shown in Fig. 2. While the specific design can vary, the following are the basic components of most solar cells:

Fig. 2 Schematic of solar cell structure

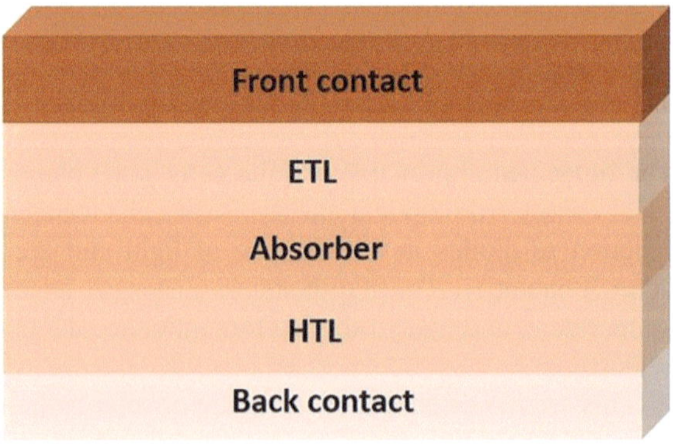

1. Anti-reflection Coating: The top layer of a solar cell is usually coated with an anti-reflective material to reduce the amount of sunlight that is reflected off the surface. This layer ensures that as much light to enter the cell, enhancing its efficiency.
2. Electron Transport Layer (ETL): The primary function is to facilitate the efficient movement of electrons from the photoactive layer (such as the perovskite or organic material) to the electrode, typically the cathode. The ETL helps reduce recombination losses, ensuring that electrons do not recombine with holes before being collected [7].
3. Absorber: Absorbers are the fundamental component in solar cells where photo-conversion occurs. These materials absorb the incident light and generate electron–hole pairs for current generation.
4. Hole Transport Layer (HTL): HTL serves the opposite function to the ETL, focusing on the transport of holes from the photoactive layer to the anode. The HTL is designed to facilitate the smooth movement of holes while blocking electrons to prevent recombination. Like the ETL, HTL is made from materials with good hole mobility. By promoting efficient hole collection, the HTL helps in improving the charge separation and collection efficiency, which is crucial for maximizing the power conversion efficiency of the solar cell [8].
5. Passivation Layers: Passivation layers are thin coatings applied to the surfaces of the active material in solar cells, especially on materials like perovskites or silicon, to reduce surface defects and minimize recombination losses. These layers help to "passivate" or neutralize the surface states that can trap charge carriers, thereby improving the efficiency of charge collection and reducing energy losses. Passivation is particularly important for materials prone to degradation or having high surface recombination rates. The passivation layer may be composed of materials like silicon nitride (SiN_x), aluminum oxide (Al_2O_3), or organic molecules, depending on the type of solar cell and the specific materials used. By reducing surface defects and improving the stability of the material, passivation layers contribute to enhancing both the longevity and efficiency of solar cells [9, 10].
6. Contacts: The solar cell has metal contacts on both the front and back sides. These contacts allow the generated electricity (electrons) to flow out of the solar cell and into an external circuit. Transparent conductors are preferred front contacts, whereas light-reflective back contacts improve light absorption by the absorber and provide structural support and electrical conductivity.

The basic operation of any solar cell includes:

- Generation of electron–hole pairs/excitons by the absorption of sunlight.
- Separation of oppositely charged carriers.
- Extraction of carriers to an external circuit.

1.4.2 Materials Employed in Solar Cell Fabrication

The absorber material, charge transport layers, and contacts uniquely enhance the solar cell performance. The absorber material captures light followed by the generation of electron–hole pairs. While the charge transport layers ensure efficient movement of the generated charge carriers, they are extracted as electric current by the contacts. These materials should be carefully chosen and integrated to design highly efficient solar cells.

Absorber Materials

Solar cell performance greatly depends on the materials employed in their fabrication. Different types of materials are used in a device, each with specific functions. In solar cells, the absorbers are primarily semiconductors with properties between metals and insulators. For enhanced electrical conductivity, the semiconductors are usually doped n-type or p-type, adding extra charge carriers by including impurity atoms. Upon stacking n-type material upon a p-type material, a p–n junction is formed with a built-in electric field, which is responsible for separating the electrons and holes generated by light absorption. This field drives the electrons toward the n-side and the holes toward the p-side, creating an electric current [1].

1. The absorber quality and characteristics determine the photo absorption capability of the device, hence should be carefully chosen based on the solar cell application. To be effective, the absorber material must have specific properties.
2. Optimum Bandgap: For the efficient absorption of solar energy, the optimum bandgap of absorber materials is centered around 1.4 eV. This bandgap range favors the absorption of visible light, constituting most of the solar spectrum.
3. Absorption Coefficient: A high absorption coefficient, greater than 105 cm^{-1}, ensures that the material can absorb a large fraction of incident light, reducing the transmittance of useful visible light. This reduces the requirement for higher material quantity, promotes lightweight device fabrication, and reduces material costs.
4. Charge Carrier Lifetime: The material should allow large carrier diffusion paths for the generated charge carriers, allowing them to be collected at the electrodes before recombining. Better crystallization and reduced grain boundaries reduce the chances of carrier recombination, hence demanding better quality of the materials with the least impurity incorporation [11].
5. Stability: The material should be durable under environmental stress, such as UV radiation, temperature fluctuations, and humidity. The absorber material stability improves the lifetime of the solar cell making it more economic.

Charge Transport Layers

The free charge carriers, electrons, and holes, generated in the absorber material, should be collected immediately before recombining. If the recombination rate is comparable with the rate of carrier generation, the solar cell performance will decline. When the absorber layer is sandwiched between charge transport layers, the charges are efficiently collected and transported to the electrodes. They act as intermediaries, facilitating the movement of free charges to the electrodes, where they can be extracted as electrical current. They effectively reduce recombination losses and provide a pathway for charge flow. This enhances the overall device efficiency. The primary feature of charge transport layers is their selective charge transport, allowing only one type of charge carrier to pass through while blocking the other to reduce recombination losses. The materials should have high electrical conductivity, enabling effortless flow of carriers to the electrodes. Proper band alignment between the absorber material and the charge transport layers ensures efficient charge extraction without any losses. The charge transport layers should be chosen appropriately for each absorber material so that bands are well aligned without creating a potential barrier for maximum carrier extraction.

Types of Charge Transport Layers

Electron Transport Layer (ETL)

The ETL is responsible for extracting and transporting generated free electrons from the absorber material to the cathode (negative electrode). ETL materials must have a higher electron affinity than the absorber material to enable efficient electron flow. Titanium Dioxide (TiO_2), Zinc Oxide (ZnO), and Phenyl-C61-butyric acid methyl ester (PCBM) are the commonly used ETLs in solar cells.

Hole Transport Layer (HTL)

Contrary to ETLs, HTL collects and transports positively charged holes from the absorber material to the anode. For effective hole transport, HTL should ensure higher hole mobility than the absorber material. In general, the conductivity of HTLs is lower than that of ETLs mainly because of the higher mobility of electrons. Various organic and inorganic materials are employed as HTLs in solar cells, including Spiro-OMeTAD, Poly(3,4ethylenedioxythiophene):poly(styrenesulfonate) (PEDOT:PSS), CuSCN, NiO_x, CuI, Cu_xO, and CuS.

Contacts (Electrodes)

The contacts (or electrodes) are vital in collecting the charge carriers from the charge transport layers and transferring them to an external load. They ensure that the electrical energy generated by a solar cell is effectively captured and used. Typically made of metal, the properties of the contacts must be compatible with the surrounding materials to ensure optimal performance. Highly conductive materials with lower contact resistance between the absorber, exhibiting excellent stability should be chosen as contacts.

Types of Contacts

Front Contact (Anode)

The front contact should ideally be transparent to allow maximum sunlight to reach the absorber material while extracting charges. Fluorine-doped Tin Oxide (FTO) and Indium Tin Oxide (ITO) are commonly used as front contacts, mainly in thin-film solar cells. Apart from conductive oxides, very thin layers of metals (silver, gold, etc.) or grid-like metal contacts are used, ensuring transparency and minimized contact shading, allowing more light to reach the absorber material [12].

Back Contact (Cathode)

Back contact serves the dual purpose of carrier collection and back surface reflection, where the unused light is bounced back into the absorber material for further collection and improves the overall efficiency. Thick metal layers, like aluminum for silicon-based cells, are the commonly used back contacts [13].

1.4.3 Applications of Solar Cells

Solar cells are used in a wide range of applications, from small portable devices to large-scale solar power plants. Some of the most common uses include:

Residential Solar Panels

As researches continue to improve the efficiency and affordability of solar cells, the technology has popularized with a rise in residential solar power installations. Rooftop solar panels are of interest because of their easy installation and low maintenance. These systems can be used to power homes and, in some cases, store excess energy in batteries for later use.

Commercial and Industrial Solar Power

Solar panels are also used in commercial and industrial applications to power businesses, factories, and warehouses. Large-scale solar installations produce enough energy to meet industrial energy needs, reducing dependence on the grid and offering long-term savings.

Solar-Powered Vehicles

Solar cells are being integrated into electric vehicles (EVs), reducing their dependence on fossil fuels, embarked as a revolutionary approach for eliminating greenhouse gas emissions. While this technology is still in the early stages, solar-powered cars, buses, and airplanes are being tested and developed.

Off-Grid Power Systems

In many developing countries, primitiveness in rural areas is predominantly created by the lack of access to conventional power sources. This condition can be eradicated by solar cell technology, which enables providing off-grid electricity in remote areas, thereby improving living conditions and providing vital services among the underdeveloped.

Space Applications

Solar cells are the primary source of energy for spacecraft and satellites. Since space missions often occur in environments without access to external power sources, solar panels convert sunlight into electricity, providing power for onboard systems and instruments.

1.5 External Solar Cell Parameters

The peak power P_{max}, the short-circuit current I_{sc}, the open-circuit voltage V_{oc}, and the fill factor FF are the key aspects used to characterize the performance of solar cells. The illuminated I-V characteristics are used to calculate these parameters. These factors can be used to calculate conversion efficiency. An I-V curve is shown in Fig. 3.

1.5.1 Standard Test Conditions

It is important to perform the measurements under standard test conditions (STC) for a reliable measurement of the I-V characteristics. The spectrum should resemble the AM1.5 spectrum and the total measured irradiance on the solar cell is 1000 W/ m^2. Since the temperature depends on the performance of the solar cell, it needs

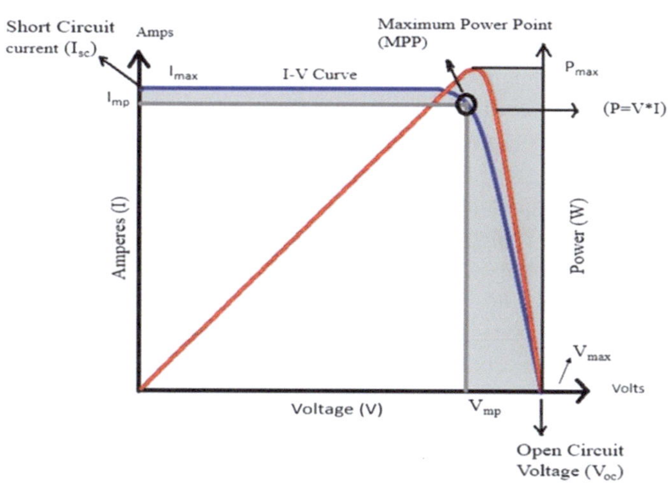

Fig. 3 I-V curve of the solar cell

to be kept at a temperature of 25 °C. On a clear day, these parameters match the spectrum and irradiance of sunlight incident on a 37° slanted surface facing the Sun, which is at an angle of 41.81° above the horizon. The actual world, however, hardly ever encounters these conditions. Many manufacturers' flash testing uses STC-based performance metrics. However, lamp spectrum, cell temperature, environment temperature, and irradiation variances result from manufacturers' frequent failure to follow STC standards during the production process [14].

1.5.2 Short-Circuit Current Density

When the electrodes of the solar cell are short-circuited, the current that flows through the external circuit is called short-circuit current, I_{sc}. It is determined by the collection and generation of photocarriers. In an ideal solar cell with efficient loss mechanisms, the short-circuit current and the light-generated current are the same. Factors that affect the I_{SC} include the solar cell area, the number of photons, the intensity of falling light, spectrum, absorption and reflection, and collection. It also depends on the life of a minority carrier at the base, and surface passivation [15]. The area of the solar cell affects I_{sc}. To avoid this dependence, short-circuit current density is used to describe the maximum current obtained from the solar cell. In an ideal PV cell, with moderate resistive loss, I_{sc} and the light-generated current are identical. The greatest current density that crystalline silicon solar cells can produce under an AM1.5 spectrum is 46 mA/cm^2. While commercial solar cells have a J_{sc} over 35 mA/cm^2, experimental c-Si solar cells have a measured J_{sc} above 42 mA/cm^2.

1.5.3 Open-Circuit Voltage

The voltage at which no current flows through the external circuit is called open-circuit voltage. It is regarded as the maximum voltage that can be delivered by a photovoltaic cell. The forward bias voltage corresponds to V_{oc} at which the photocurrent density gets compensated by the dark current density. If the solar cell is at the open-circuit state, the free carriers generated under steady illumination will accumulate at the anode and cathode, forming a potential difference which cancels out the built-in potential. Under this condition, carrier generation and recombination exactly balance each other, and no net current exists at any point inside the device. The system is now in a quasi-equilibrium state, and the potential difference formed between the two electrodes is the open-circuit voltage (V_{OC}), which represents the maximum voltage a solar cell can provide to an external circuit. In particular, the dependence of V_{OC} on four important factors, namely temperature, light intensity, work function of the electrode and material microstructure based on the model of density of states [16]. Under normal AM1.5 circumstances, the V_{oc} of laboratory crystalline silicon solar cells can reach 720 mV, although the V_{oc} of commercial solar cells is usually greater than 600 mV.

1.5.4 Fill Factor

The power of the solar cell is zero at both operating points corresponding to I_{sc} and V_{oc}. The fill factor (FF) determines the maximum power from a solar cell which is in conjunction with V_{oc} and I_{sc}. Fill factor is given by the relation below.

$$\mathrm{FF} = \frac{I_{\mathrm{MP}} V_{\mathrm{MP}}}{I_{\mathrm{SC}} V_{\mathrm{OC}}}.$$

The fill factor (FF) is a crucial parameter in determining the efficiency of a solar PV module. It should be maximum and approaching unity on the I-V curve, ensuring maximum efficiency. FF is the ratio of maximum power to the product of open-circuit voltage (V_{OC}) and short-circuit current (I_{SC}). The empirical relationship between FF and V_{oc} indicates that FF is dependent only on V_{OC}. Maximum Power Point Tracking (MPPT) is used to track maximum power points on the I-V curve and generate peak power. The FF and efficiency of the solar PV system are also affected by changes in series and shunt resistance. The maximum power point is always present at the point of intersection from the voltage and current axes. Optimizing the fill factor ensures higher overall system efficiency. The highest value of FF (=1) indicates the maximum power value, and achieving unity FF is crucial for high efficiency. Optimizing the fill factor value leads to more squareness in the I-V characteristic curve of the solar PV module, resulting in maximum system efficiency [1].

A device's fill factor indicates how well its series and shunt resistances are balanced. Shunt resistance is resistance brought on by external connections, whereas series resistance is the resistance of the cell's internal components. Efficiency is decreased when current flow through the solar cell is reduced by increased series resistance. Solar cells should have better contact design and low material resistance to lower series resistance. Power losses brought on by electron and hole recombination through other paths, including defects in the device, are referred to as shunt resistance. Increased shunting lowers the current flowing through the cell, and in severe situations, low shunt resistance may result in a short circuit of the device. Therefore, it is essential to increase shunt resistance in solar cells, as low resistance indicates a manufacturing defect rather than a material property.

It is crucial to raise shunt resistance and decrease series resistance in order to optimize a solar cell's fill factor. By adjusting the cell's design, this can be accomplished by altering the cell's size and shape, the thickness of the active or transport layers, the device layers, and the doping profile and concentration of the solar cell. By increasing the current pathways across the active layer, surface treatments like as texturing, roughening, and doping can raise shunt resistance. By increasing the number of points of contact between the cell and contacts, surface treatments can also lower series resistance. By reducing current-carrying channels and blocking minority carriers, charge carrier layers such as ETLs and HTLs can increase shunt resistance. Furthermore, making sure that the device and external connectors are connected properly might enhance the fill factor by applying conductive paint at contact points [17].

1.5.5 Efficiency

The efficiency (sometimes the power conversion efficiency, η) is defined as the ratio of output electrical power at maximum power point on the I-V curve to the incident light power—typically by a standard AM1.5G simulated solar spectrum [18]. Thus, the efficiency can be written as,

$$\eta = \frac{P_{\text{max}}}{P_{\text{in}}} = \frac{I_{\text{MP}}V_{\text{MP}}}{P_{\text{in}}} = \frac{I_{\text{Sc}}V_{\text{OC}}\text{FF}}{P_{\text{in}}}.$$

The conversion efficiency of a photovoltaic (PV) cell is the percentage of solar energy that is converted into usable electricity. Improving this conversion efficiency is crucial for making PV technologies cost-competitive with conventional sources of energy. However, not all sunlight reaches a PV cell, as most is lost. Multiple factors in solar cell design limit a cell's ability to convert sunlight.

The materials used for solar cell fabrication should be of the best quality, with the least impurity presence and fewer defects. This can ensure lower carrier loss by recombinations at defective sites, leading to higher efficiency. Wavelength plays a role in determining efficiency. Sunlight has a wide range of wavelengths, from ultraviolet to infrared. When light hits a solar cell's surface, some photons are reflected, while others pass through. Some energy is turned into heat, while the remainder is used to produce charge carriers and electric current. Recombination is another factor that limits efficiency. Direct recombination occurs when electrons and holes recombine, canceling out their contributions to the electrical current. Indirect recombination occurs when electrons or holes encounter impurities that make it easier for them to recombine and release energy as heat. Temperature is crucial for solar cells to function optimally. Higher temperatures can cause semiconductor properties to shift, resulting in increased current but larger voltage decrease. Proper thermal management improves efficiency and lifetime [4, 18]. Solar cells are more efficient when exposed to high-intensity light, such as direct sunlight, compared to indirect or diffuse light [53]. Reflection adversely affects efficiency. By minimizing light reflection on the cell's surface, it can be compensated. Untreated silicon reflects over 30% of incident light, while anti-reflection coatings and textured surfaces help decrease reflection. High-efficiency cells will appear dark blue or black [19].

A crystalline silicon solar cell's typical exterior characteristics are as follows: FF between 0.75 and 0.80, V_{oc} up to 0.65 V, and $J_{sc} = 35$ mA/cm^2. The conversion efficiency is between 17 and 18%.

2 Shockley–Queisser Limit

The detailed balance limit, also referred to as the Shockley–Queisser limit, is the maximum light-to-electric power conversion efficiency of a single junction solar cell for a specific illumination spectrum is shown in Fig. 4. William Shockley and Hans-Joachim Queisser initially proposed in 1961, based on detailed balance considerations, the computation of the maximum conversion efficiency of a p–n junction solar cell under solar illumination, where a black body with a surface temperature of $T_s = 6000$ K was used to approximate the Sun spectrum.

Radiative recombination, which establishes the upper bound for the minority carrier lifetime, is thought to be the only recombination pathway in an ideal solar cell that cannot be reduced to zero. It was assumed that photons with energies below the bandgap would not interact with the solar cell, whereas photons with energies above the bandgap would be transformed into electron–hole pairs with a 100% quantum efficiency. At a cell temperature of $T_c = 300$ K, Shockley and Queisser determined the efficiency limit for a single junction solar cell.

While the AM 1.5G spectrum includes scattered light from the atmosphere and is relevant for PV systems without light concentration, the AM 1.5D spectrum is relevant for solar conversion systems based on light concentration with mirrors or lenses because it only contains direct sunlight. According to ASTM G173-03, the Shockley–Queisser limit, which is the theoretical maximum light-to-electric power conversion efficiency for single junction solar cells, is computed for the AM 1.5G

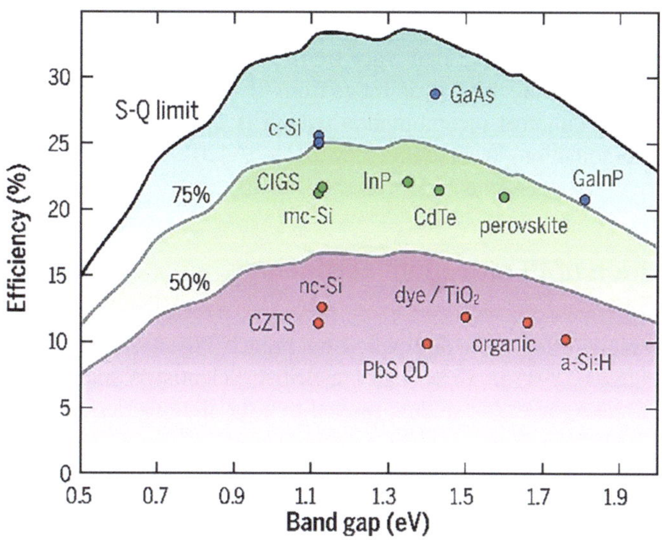

Fig. 4 Shockley–Queisser limit of various solar cells (Reference: Di Vece, Marcel. (2019). Using Nanoparticles as a Bottom-up Approach to Increase Solar Cell Efficiency. Powder and Particle 36. 72–87. https://doi.org/10.14356/kona.2019005.)

spectral irradiance. According to Shockley and Queisser's detailed balance considerations, the maximum light-to-electric power conversion efficiency for AM 1.5G illumination is 33.16%, necessitating a semiconductor bandgap of 1.34 eV (928 nm). The efficiency curve exhibits two local maxima of 32.85% at 1.15 eV (1080 nm) and 33.16% at 1.34 eV (928 nm) and is not entirely smooth because of atmospheric absorption bands at about 900, 1150, 1400, 1800, and 2500 nm. The highest conversion efficiencies are attained for bandgaps ($J_{max} = 30$–45 mA/cm^2), where the light absorber does not absorb about half of the photons. A perfect solar cell's fill factor is above 90% for bandgap wavelengths smaller than 800 nm and decreases linearly as the bandgap wavelength increases.

The fill factor is decreasing from above 85% to below 60%, particularly at low bandgap energies below 1 eV, which results in a lower electron and hole conversion efficiency into electric energy. It should be mentioned that the J_{sc}, V_{oc}, and η are the highest values that can be obtained for a specific bandgap energy, and they can only be attained when the solar cell is run at the detailed balance limit. On the other hand, when the cell is run at conversion efficiencies lower than the detailed balance limit, the fill factor may increase beyond the reported values. That might sound odd at first, but it makes clear when you remember that the fill factor is only a comparison of the voltage and current density at the greatest power point in relation to the open- and short-circuit values, respectively. Therefore, a solar cell with a certain bandgap operating at V_{oc} and/or a J_{sc} below the detailed balance limit may have a fill factor that is higher. This occurs particularly when the V_{mpp} and/or J_{mpp} are not significantly different from the V_{oc} and J_{sc}, respectively. Nevertheless, in this scenario, the total conversion efficiency η will continue to fall below the detailed balance limit.

For a solar cell bandgap between 0.93 and 1.61 eV, conversion efficiencies above 30% are achievable. The V_{oc} may vary from 0.69 to 1.31 V, or 74 to 82% of the bandgap potential, while the J_{sc} can range from 51.7 to 25.2 mA/cm^2. At a bandgap of 1.52 eV, non-radiative recombination, which is 10^6 times more powerful than radiative recombination, lowers efficiency to 21.7% [20].

3 Evolution of Photovoltaic Technology

The photovoltaic effect was discovered by French Physicist Alexandre-Edmond Becquerel (1820–1891) in 1839. Later, a British Scientist William Grylls Adams (1836–1915) along with his student proved that the illumination of selenium bars produces a flow of electricity. In 1883, American inventor Charles Fritts (1850–1903), discovered the first working photovoltaic cell by sandwiching selenium between an iron plate and a semi-transparent gold top layer. Even though the cell has very low efficiency (<1%), it was considered a milestone in the history of solar cells. He was also considered as the forerunner in the field of solar panels. In the mid-1880s, he connected several selenium modules together and placed them on the rooftop.

During subsequent years, photovoltaic effects were observed in copper–copper oxide thin-film structures in lead sulfide and thallium sulfide. The early cells contain a semi-transparent layer of metal deposited on the top of a semiconductor which creates an asymmetric electronic junction necessary for photovoltaic action. So, they were considered as thin-film Schottky barrier devices. In 1914, Goldman and Brodsky relate this effect to rectifying action (i.e., the existence of a barrier to current flow at one of the semiconductor–metal interfaces). After 16 years, Walter Schottky and Neville Mott developed the theory of metal–semiconductor barrier layers.

It is not the photovoltaic property, but the photoconductivity of the selenium that creates curiosity in researchers. Since the intensity of the incident light is proportional to the current produced and is related to the wavelength, it is considered that photoconductive materials were an ideal candidate for photographic light meters. Until 1950, there was no further progress in the field of photovoltaics. The development of high-quality silicon wafers and their application in new solid-state electronics made advances in photovoltaic devices.

Followed by the discovery of silicon wafers, manufacturers developed p–n junction in silicon. The exposure of inherent n-type silicon wafers in boron trichloride gas develops a p-type skin. A part of this skin is etched off to give access to the n-type layer beneath. The photovoltaic behavior of the p–n junction shows much better-rectifying action than Schottky barriers.

In 1954, the first silicon solar cell with an efficiency 6% was reported by Chapin, Fuller, and Pearson. This efficiency didn't show any significant rise in the following years. Due to this low efficiency and high production cost, solar cells were considered inefficient in power generation for several decades [21]. During the 1950s and 60s, silicon solar cells found applications in space such as in satellites. Due to its reliability and low weight, its high manufacturing cost became irrelevant in this field.

Cadmium sulfide p–n junction was developed with an efficiency of 6% in 1954. In succeeding years, theoretical work was stimulated for gallium arsenide, indium phosphide, and cadmium telluride which indicates higher efficiency. After all, silicon remained the foremost photovoltaic material and has applications in the microelectronics industry.

The crisis in energy supply experienced by the oil-dependent western world in the 1970s stimulates the interest to found alternative sources of energy and funding for research and development in those areas. This was a period during which a range of strategies for producing photovoltaics with cheap manufacturing costs and high efficiency were explored. The cost was reduced by introducing photoelectrochemical junction and replacing crystalline silicon with polycrystalline silicon, amorphous silicon, other thin-film materials, and organic conductors. The efficiency of the cells was increased by introducing tandem and other multiple bandgap designs. Nevertheless, none of these led to a widespread commercial application.

During the 1990s, a realization of the need for secure sources of electricity drives the development in the area of photovoltaics. At the end of the decade, photovoltaic production expanded at a rate of 15–25% per annum, driving the reduction in cost [22].

4 Generations of Solar Cell

All the four generations of solar cells are comprehended in Fig. 5.

4.1 First-Generation Solar Cell (Wafer-Based)

The first-generation solar cell was formed on silicon wafers. They were considered the most promising and commonly used technology due to their high efficiency.

The silicon wafer-based solar cells are classified into two:

- Single/monocrystalline silicon solar cell.
- Poly/multicrystalline silicon solar cell.

Since silicon is abundant and non-toxic, both single crystal and multicrystalline were widely popular. Monocrystalline solar cells are fabricated from a homogeneous continuous crystal structure [23]. The silicon single crystals were manufactured by the Czochralski process. In this process, Si crystals were sliced from big ingots. The large single crystals thus produced require precise processing. However, they are also the most expensive type of solar cell to manufacture, and the efficiencies of

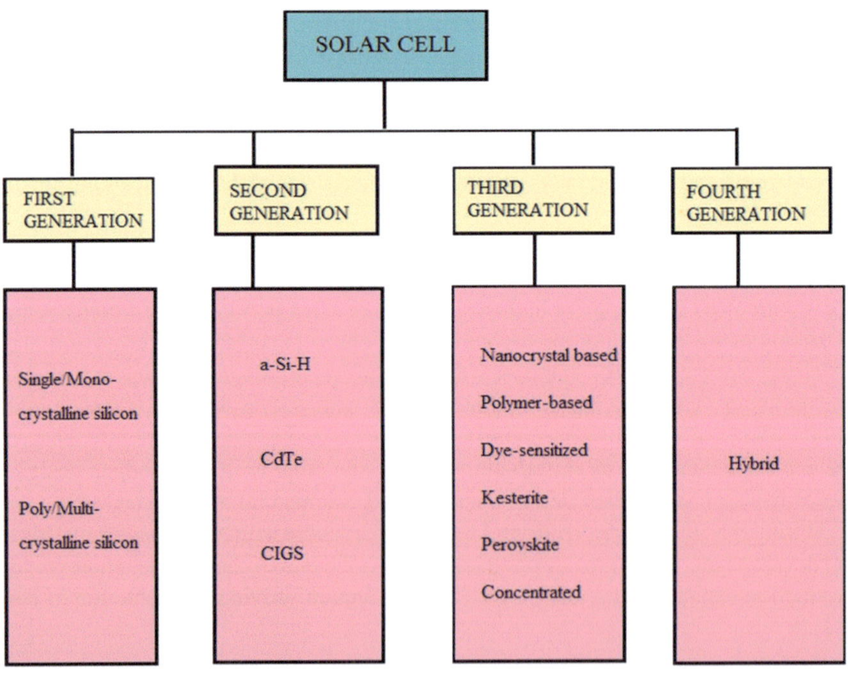

Fig. 5 Generations of solar cell

these cells are ~ 27.3% [24]. In addition, they are most space-efficient and have high durability.

Polycrystalline solar cells are made of materials with a polycrystalline structure. It is the most commercially used solar cell. It is manufactured by a variety of methods such as casting and ribbon growth and doesn't undergo any cutting process similar to that in the case of the monocrystalline solar cell. Though the processing of multicrystalline solar cells was economical, its efficiency is ~ 13–16% [25] which is comparatively lower than that of single-crystalline silicon solar cells.

4.2 Second-Generation Solar Cell (Thin Film)

The second-generation solar cells are more economical in comparison to that of first-generation silicon wafer solar cells. In first-generation solar cell, the light-absorbing material (i.e., silicon wafer) has a thickness of ~ 350 μm, whereas in second-generation solar cell, the thickness of light-absorbing material is in the order of 1 μm.

They are classified into three different types:

- Amorphous silicon (a-Si).
- Cadmium telluride (CdTe).
- Copper indium gallium di-selenide (CIGS).

The word "amorphous" in connection with solar cell means "not highly ordered". It lacks a definite arrangement of atoms in lattice (non-crystalline structure). The cell is environmentally friendly, since it doesn't use any toxic materials such as cadmium or lead, but its efficiency is low. They are fabricated at a low processing temperature, which enables its use in a variety of low-cost polymers and other flexible substrates that require less energy for processing; therefore, an a-Si solar cell is comparatively cheaper and widely applicable. Amorphous silicon was once expected to become a major contributor in the field of photovoltaics but lost its market due to the strong competition from conventional crystalline silicon cells and other thin-film technologies such as CdTe and CIGS cells. The efficiency of the cell is ~ 10.2% [24].

Among the thin-film solar cells, CdTe is one of the leading photovoltaic candidates. It is cheaper than other conventional solar cells made of crystalline silicon in multi-Kilowatt systems. CdTe has a bandgap of ~ 1.5 eV as well as a high optical absorption coefficient over 5 × 105/cm and chemical stability. Being a direct bandgap crystalline compound, the absorption of light is easier and efficiency is comparatively high (~ 21%) [24]. The Cadmium Telluride based solar cell was synthesized from polycrystalline material and glass was chosen as a substrate. However, there are various environmental issues associated with cadmium present in the solar cell. It is a heavy metal and a potentially toxic agent.

CIGS was chosen as a potential photovoltaic material. It has an attractive bandgap ~ 1.0 eV and a high optical absorption coefficient. Its preparation method is quite inexpensive. The cell components were available in adequate quantities. The manufacturing deployment and decommissioning of the technology fall within acceptable environmental constraints [26, 27] The highest reported efficiency of cells is ~ 23.35% [24].

4.3 Third-Generation Solar Cell

Third-generation solar cells are considered to be cost-effective and facilitate easy fabrication. These types of cells don't have a conventional p–n junction. They are classified as follows:

- Nanocrystal-based solar cell.
- Polymer-based solar cell.
- Dye-sensitized solar cell.
- Kesterite solar cell.
- Perovskite solar cell.
- Concentrated solar cell.

The nanocrystal-based solar cells are usually known as quantum dot solar cells. They were semiconducting materials of II-VI, III-V, or IV-VI materials which are in the nanocrystal range. Due to their size, quantum mechanical effects play a major role in the performance of solar cells. By adjusting the dot's size, the bandgap of the solar cell can be tuned across a wide range of energy levels. The preparation method of quantum dot solar cells was simple [28].

Generally, polymer-based solar cells are flexible due to the polymer substrate. In the cell, thin functional layers are serially coated on polymer foil or ribbon which functions as a combination of the donor (polymer) and acceptor (fullerene). The different types of absorbing materials include organic materials like conjugate or conducting polymer. The highest efficiency achieved by the polymer thin-film cell is ~ 19.2% [24].

The dye-sensitized solar cells are basically photoelectrochemical cells. They were highly flexible, transparent, and low-cost thin-film solar cells. It consists of four components—a semiconductor electrode (n-type TiO_2 and p-type NiO), a dye sensitizer, a redox mediator, and a counter electrode (C or Pt). DSSC had achieved wide attention due to simple conventional processing methods like printing techniques. The photosensitization is caused by the optically active dyes present on the porous TiO_2. The most concerning limitation of DSSC was its stability issue. The exposure of dye toward UV and IR radiations is detrimental to the lifetime and stability of the cell. The highest recorded efficiency of DSSC was ~ 12% [24].

The basic two synthetic compounds in Kesterite solar cells were copper zinc tin sulfide (Cu_2ZnSnS_4) (CZTS) and copper zinc tin selenide ($Cu_2ZnSnSe_4$) (CZTSe). Even though the electronic properties of CZTS and CZTSe were similar to CdTe and CIGS, they didn't contain any toxic or rare-earth elements like cadmium and indium, respectively. The highest recorded efficiency of thin-film CZTS is ~ 10%

[29] and that of CZTSe was ~ 13.45% [24]. Dominant surface recombination, high series resistance, and low minority carrier lifetime limit the performance of these cells.

The light-absorbing component of perovskite solar cells is a perovskite-structured compound, which is typically a hybrid organic–inorganic lead or tin halide-based material. Perovskite was a class of elements that can be represented by formulae ABX_3, where A and B were cations of different sizes and X was a halogen such as I-, Br-, and Cl-. Perovskite solar cells possess several advantages over conventional silicon solar cells and thin-film solar cells. They have high efficiencies and low production costs. The highest recorded efficiency of thin-film perovskite cells is ~ 25.2% [24]. The current issues with the perovskite cell are its stability and durability. The material degrades over time and causes an overall drop in its efficiency [30].

Solar cell research and development has recently focused on concentrated photovoltaics. Concentrated cells work on the concept of collecting a significant amount of solar energy through a tiny area over the PV solar cell. To focus sunlight, this technology necessitates the use of big mirrors and lens combinations. The converged solar radiation produces heat energy. This heat energy is then channeled through a heat engine that is controlled by a power generator with integrated CPVs, which have demonstrated their potential in the solar realm. High cell efficiencies, absence of any moving components, no thermal mass, a quick response time, and the ability to scale to a variety of sizes are all advantages of concentrated photovoltaics [29, 31].

4.4 Fourth-Generation Solar Cells

Being the latest generation in solar cells, the fourth-generation solar cell utilizes the combination of low-cost flexible polymer thin film with inorganic nanostructures like metal nanoparticles and oxides. So, it is also known as "inorganics-in-organics" solar cell sometimes, and it uses carbon nanotubes, graphene, and its derivatives. The architecture of fourth-generation solar cells provides stability and convenience for solution processing. Hence, the fourth-generation solar cell is cheap to manufacture. The charge dissociation and charge transport within the cells get improved by the inclusion of the nanomaterials. So they are often referred as nanophotovoltaics. Recently, scientists incorporated graphene (G) into the cell since it is a material of the highest technological and scientific expectations.

Due to the efficient flexibility and transparency, G and its derivative can be used as transparent conducting electrodes. The efficiency of the PSCs gets increased from 3.68 to 3.98% when using G, and its derivative is used as transparent conducting electrodes instead of ITO [6]. Even though the increment is marginal, it can be concluded that G has more potential than ITO.

The active layer of organic perovskite solar cells in bulk heterojunction (BHJ) has a large interfacial area with a donor–acceptor interpenetrating network at the nanoscale. This structure is the reason behind the efficiency of BHJ solar cells. The charge carrier collection is due to the generation of a donor–acceptor interpenetrating network and the large heterojunction interface causes the efficient dissociation of excitons. Due to the large surface area which helps in the formation of an interpenetrating donor–acceptor network, G can be used as an active layer in the solar cell.

The oxidized form of G, i.e., GO can be used as a hole transporting layer since its work function is lower than that of G. So, it provides better hole injection [32].

5 New Developments

5.1 Bifacial Photovoltaics

In bifacial photovoltaic technology, light is transmitted from both sides of the solar cell via the photoelectric effect which is different monofacial photovoltaics (shown in Fig. 6). It replaces the back surface field (BSF) at the rear end of the cells with back contacts and an anti-reflection coating. Photons that are more energetic than the bandgap give up part of their energy to electrons when they come into contact with sunlight, creating electron–hole pairs. Attracted by the internal electric field, carriers close to the depletion area diffuse to the substrate and emitter. An electromotive force is created between the front and rear contacts as a result of electrons flowing into the N-type semiconductor and holes flowing into the P-type semiconductor. After connecting both sides of the PV cells, electrons go via the external load [33]. Novel silicon solar cell technologies such as heterojunction with an intrinsic thin layer (HIT), interdigitated back contact (IBC), passivated emitter rear contact (PERC), passivated emitter rear locally diffused (PERL), and passivated emitter rear totally diffused (PERT) have made bifacial solar cell operation possible [34].

They were first introduced in 1960 and have a number of benefits over monofacial cells, including the ability to capture photons from both sides, use a more affordable substrate, produce more power, and lower cell temperature. It is possible for bifacial cells to function at low temperatures. Reflected light from the implanted surface illuminates the rear side of the PV cell; a higher reflectivity of the surface is necessary for improved power production. Tilt angle, module elevation, system design, and module position are some of the variables that affect rear side estimate. The ground's reflected power, or albedo, is a component of the output power from the rear. The incident light that enters the bifacial module is made up of isotropic radiations, diffuse reflections, and beams. The International Technology Roadmap for Photovoltaics (ITR) projects that by 2024, bifacial solar modules would account for more than 15% of the global market.

Monocrystalline and multicrystalline silicon were initially used to create bifacial solar cells, which may now be produced using a variety of substrates, including CIGS, CdTe, GaAs, and Ge types. Although silicon has been used as a substrate in the majority of research publications, bifacial solar cells are capable of handling this expensive material with ease. It is possible to shrink the silicon wafer's thickness to 58 μm, which can increase productivity and save expenses. Experiments indicate that

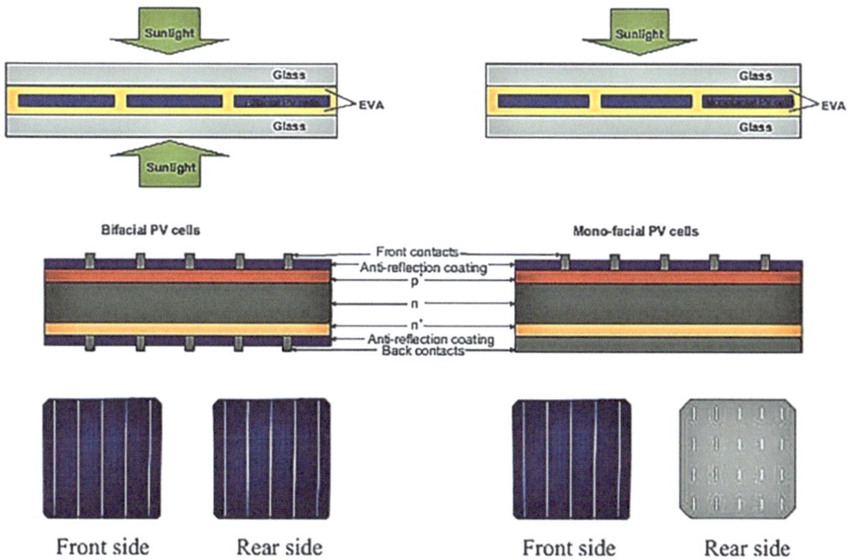

Fig. 6 Difference between monofacial and bifacial photovoltaics (Reference: Lee, J., & Park, H. (2020). A systematic literature review of the bifacial photovoltaic module and its applications. *Renewable and Sustainable Energy Reviews, 116*, 12–28. https://doi.org/10.1016/j.rser.2020.109437)

bulk recombination and impact reduction enhance efficiency when the cell thickness is decreased from 240 to 140 μm, but further study is required. They have the unique ability to increase output power by absorbing radiation from both the front and the back. It is possible to make solar cell cane flexible, lightweight, and suitable for semi-transparent applications such thin-film-based PV cells and dye-sensitized PV cells (DSSCs). When utilized as a counter electrode in bifacial dye-sensitized solar cells, graphene oxide can detect 80% transparency in the 300–1000 nm range and exhibit efficiency comparable to that of traditional dye-sensitized PV cells (DSSCs) based on platinum. Although there are currently no experimental investigations on bifacial solar cells, perovskite PV cells are being studied in monofacial solar cells. With the goal of becoming extensively commercialized in future, bifacial solar cells that use these applications are now being researched.

There are several uses for bifacial solar cells (PVs), such as in space applications and building windows. Their transparency in the infrared band, which raises light energy by 11–41% and lowers the equilibrium operating temperature, makes them very helpful in low-Earth orbits. Bifacial photovoltaic cells can boost power output by 10–20% when deployed in space missions. Their performance, however, deteriorates when applied to spacecraft subjected to fluxes, trapped particles, or solar protons. Windows also utilize bifacial photovoltaics to generate power for charging heavy-duty gadgets. This application can employ the majority of transparent dye-sensitized solar cells; however, the technology is still in its infancy and needs more study and efficiency improvement. Compared to monofacial PV cells, bifacial PV cells are less costly, yet yield more electricity.

Research on bifacial solar cells will continue to yield improvements in cell design, materials, and PV cell efficiency. They are especially well-suited for energy collecting in a variety of settings, including deserts with snowy or reflecting surfaces. Adding solar power systems to buildings so they can have both practical and aesthetic uses is known as building-integrated photovoltaics, or BIPV. By transferring energy between the front and rear surfaces, BIPV may improve aesthetics and bring it closer to the point of consumption. Another use for bifacial solar cells is the construction of floating solar farms, which make the most of unused waterways and accessible land. Unused water bodies may be efficiently used by reflecting sunlight onto their highly reflective surface, which is then absorbed by the opposite side of the bifacial solar cells. Combining solar energy with agricultural uses is another usage for the solar cells. Researchers and developers must use new technologies, such as improved rear contact designs, light-trapping structures, novel materials, and alternative cell designs, to increase the efficiency of bifacial PV cells. Bifacial PV cells can thus play a major role in the creation of environmentally friendly, energy-efficient houses [35].

Although bifacial solar cells have demonstrated higher efficiency, they still have problems with scalability, durability, and installation complexity. They raise expenses since they need tracking and accurate positioning systems. It is challenging to anticipate energy production accurately because of location-specific characteristics like ground reflectance and the albedo effect. Additionally, bifacial panels are susceptible to soiling and shade, which lowers the amount of energy captured on the back side. Improvements in materials, manufacturing processes, installation systems, energy modeling tools, and site selection are required to overcome these obstacles and boost the performance and economic feasibility of bifacial solar cells [36].

5.2 *Floating Photovoltaics (Floatovoltaics)*

Around the world, PV systems are becoming more and more popular as a means of producing electricity. However, because of their poor efficiency and large land needs, many countries—particularly those with dense metropolitan areas—face difficulties in finding room for more panels. This may be addressed by installing floating photovoltaic systems (FPVSs) on bodies of water such as rivers, canals, dam reservoirs, and ponds to free up land for housing and agricultural activities. A pontoon, mooring and anchoring equipment, photovoltaic panels, and electrical cables and connections make up an FPVS which is shown in Fig. 7. While the anchoring and mooring system secures the floating structure, pontoons float on their own to allow for maintenance access. It is also possible for other electrical parts, including central inverters and combination boxes, to float on water. Electricity is delivered to the land via connectors and wires, which can be fed to the grid or stored in batteries.

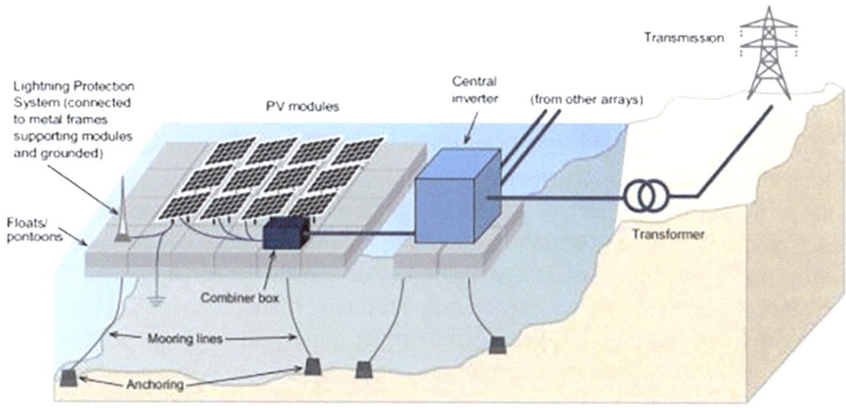

Fig. 7 Components of floating photovoltaics (Reference: Smith, J., & Lee, M. (2021). Solar PV energy: From material to use, and the most commonly used techniques to maximize the power output of PV systems: A focus on solar trackers and floating solar panels. *Renewable Energy Reviews, 123*(4), 45–67. https://doi.org/10.1016/j.rser.2021.101231)

In 2007, the first FPV system (FPVS) with a 20-kW capacity was constructed in Aichi, Japan. Since then, miniature FPVSs have been used for study and demonstration in nations including the US, France, Italy, Korea, and Spain. The first commercial installation was a 175 kWp FPVS at the Far Niente Winery in California in 2008. The FPV market is still expanding at an exponential rate after reaching its first megawatt capacity in 2009. The overall installed capacity of FPVSs surpassed that of overland PV systems in 2000, reaching 1.314 GWp by the end of 2018. With more than 950 MWp of installed capacity—nearly 73% of the global total—China has emerged as the market leader.

Since it can be installed in hydroelectric dam reservoirs, wastewater treatment ponds, or other water bodies, they offer shade, reducing evaporation and preserving water stored. FPVS can minimize water losses by 15–25,000 m^3 per MWp of installed capacity. The shade also helps reduce algae blooming, improving water quality. The water's inherent reflectivity enhances solar radiation incidence, increasing PV energy production. Floating PV panels can operate at higher efficiency than terrestrial panels due to the natural cooling action of water. The lower operating temperature and higher humidity above water result in increased efficiency, up to 12 and 20% in summer [37].

The tracking systems employ single-axis or dual-axis trackers to optimize energy output, fixed floating solar systems are stable and appropriate for calm conditions. Integrated floating solar systems offer a varied energy solution by combining FPV with additional renewable energy sources. While modular systems provide flexibility and scalability for different water bodies, offshore floating solar systems are appropriate for locations with limited land availability. Energy generation and structural support are provided by building-integrated floating solar (BIPV), which incorporates floating solar panels into infrastructure. Concentrated floating solar is more efficient in areas with strong solar radiation because it employs mirrors or lenses to direct sunlight onto a receiver. A special chance to install solar systems on existing infrastructure while lowering land use and minimizing water evaporation is provided by floating solar on canals and waterways [38–41].

By obstructing sunlight, floating photovoltaic (PV) arrays can have a substantial negative influence on aquatic habitats, impacting plant life and photosynthesis. Long-term impacts on fish habitats, biodiversity, and water quality may result from this. Chemicals may seep into the water from the materials used in PV modules or floating platforms, which might have an impact on aquatic life and water quality. The growth of some algae species may also be promoted by large floating solar arrays, leading to toxic algal blooms that degrade water quality and endanger aquatic life. Because platforms, wiring, and panels may deteriorate over time and cause pollution or structural collapse, the long-term viability of floating solar infrastructure is also an issue. Careful site selection, the use of sustainable materials, ongoing monitoring, and adaptive management are examples of mitigation techniques. By staying away from environmentally sensitive areas and using eco-friendly materials, FPV systems can minimize environmental risks and ensure the health and well-being of aquatic ecosystems [42].

5.3 Agrivoltaics (APV)

Agrivoltaics (APV) simultaneously resolves two important needs, food production and renewable energy generation. Agricultural land is used alongside with power generation which is shown in Fig. 8. Sustainable use of space can be achieved by using the same land for different purposes. Since food and power are produced together, the rising demand for food and clean energy can be tackled efficiently [43]. This section discusses the principles, benefits, challenges, and future of agrivoltaic systems.

1. Fixed Agrivoltaic Systems: In such systems, solar panels are placed at a set height and angle, offering shade to plants. This helps to reduce water usage and protect crops from bad weather.
2. Tracking Agrivoltaic Systems: Panels are designed to follow the Sun, enhancing energy capture. They are mounted on tracking systems. Agrivoltaics makes better use of land and lessens environmental impacts, suitable for large farms and small family-run operations.

Simultaneous usage of land for different purposes improves the land utility providing extra revenue to the cultivators. The energy produced can power farm operations or be sold to the grid, providing extra income for farmers. A study shows that agrivoltaics could meet up to 20% of global energy needs by using just 0.5% of agricultural land, highlighting its scalability [44]. With rapid urbanization and limited land availability, vast land areas are getting converted for energy generation,

Fig. 8 Power generation in agricultural fields (Reference: Martinez, P., & Wang, T. (2023). Environmental life cycle assessment of a stilted and vertical bifacial crop-based agrivoltaic multi land use system and comparison with a mono land use of agricultural land. *Renewable and Sustainable Energy Reviews, 158*, 112543. https://doi.org/10.1016/j.rser.2023.112543. Agrivoltaic systems came in two different configurations)

which threatens food production. This situation can be turned down by technique of agrivoltaic systems where the land can be more efficiently used. Power generation without compromising agricultural activities eliminates natural habitat loss, and land degradation. Areas with high population density will be highly benefited, as the limited available land gives both food and electricity.

Agrivoltaic systems can boost agricultural productivity by providing partial shade to crops, reducing heat stress in dry areas [45]. In hot climates, crops like vegetables, fruits, and grains benefit from this shade, which lowers evaporation and conserves water. They also help maintain a stable microclimate, protecting crops from extreme weather events [46]. Research indicates that crops grown under solar panels generally grow better and yield more than those grown in full sunlight. A study in Japan showed that tomatoes, lettuce, and spinach grew more effectively in the shade from agrivoltaic systems [47]. This increase in productivity and lower water use makes agrivoltaics appealing for farmers in dry areas.

Agrivoltaics helps with climate change adaptation and mitigation. For adaptation, the shade from solar panels protects crops from extreme temperatures, making them more resilient to climate stress, which is crucial as climate change causes more heatwaves, droughts, and floods that threaten food production globally.

Several challenges need to be tackled for efficient commercialization of the technology, mainly includes high initial investments and technical and operational challenges [48].

1. High Initial Investment

The initial cost of setting up agrivoltaic systems can be expensive, especially for small farmers. Costs include solar panels, mounting systems, and electrical connections. Besides possibility of extra income generation and long-term savings, the higher installation costs will be a hurdle for the common cultivators, which detach the farmers from the technology.

2. Land and Resource Management

The companionship should be carefully planned as some plants will be benefited while others are harmed from the shadow created. So skilled professionals are made available for proper planning and designing of the agricultural fields for expected results.

3. Operational Challenges

The maintenance costs and efforts will be more than the usual roof top systems because of the higher chances of panel encounter with dust and birds. Extra labor will be needed for proper cleaning and maintenance of the panels which will further be a barrier for the farmers, mostly economically. As this is more like an experimental approach for the farmers, many cultivators refuse to experiment their yields. So adequate awareness should be given by the officials [4].

6 Deployment and Market Status of Solar Cell

In 2022, the Paris Agreement has seen a significant expansion in climate commitments, with 10 countries revising their Nationally Determined Contributions (NDCs) to reduce emissions. However, only 25 of the 193 countries that have signed the agreement have set net-zero emissions targets. In response, around 80 new renewable energy policies were announced in 2022, mostly in the form of fiscal/financial incentives. This totaled 454 policies across sectors, with 94 countries having either policies or targets in at least one end-use sector by year's end. Only three countries (Spain, Portugal, and Türkiye) had targets or policies in all four end-use sectors. The Paris Agreement's climate policies need to translate into concrete enabling policies and mandates in all sectors. As of the end of 2022, 52 countries have implemented policies supporting renewable energy adoption in the buildings sector, including incentives for renewable technologies like rooftop solar and solar water heaters, and mandates to ban fossil fuel use for heating in new and existing buildings. In the agriculture sector, 14 countries had targets and policies for renewables as of the end of 2022, with seven of these countries announcing them during the year. All of the policies comprised either investments, subsidies, grants, or tax deductions, mostly for solar-powered irrigation and agrivoltaics [49].

In the last ten years, the production capacity for solar PV worldwide has transitioned from Europe, Japan, and the USA to China, where more than $50 billion has been invested in new PV supply capacity, resulting in the creation of over 300,000 manufacturing jobs. China accounts for over 80% of all manufacturing phases of solar panels, which is more than double its worldwide PV demand. The nation hosts the leading 10 suppliers globally of solar PV manufacturing equipment. China has lowered solar PV costs globally, aiding clean energy transitions. Nevertheless, the geographical concentration within global supply chains poses possible challenges that governments must tackle [50].

Major market participants are concentrating on addressing the drawbacks of solar panels and reducing the costs of PV modules, while heightened investment in renewable energy sources is driving market expansion. The market is divided into thin-film solar cells (CdTe, a-Si, CI(G)S), crystalline solar cells (multi-Si, mono-Si), and installation types for residential, commercial, and utility scales. The market growth is also fueled by investments in renewable energy sources.

The crystalline solar cell segment leads the global market, representing more than fifty percent of total solar cell revenue thanks to its enhanced efficiency. C-Si solar cells consist of very pure silicon wafers and transform solar energy into electrical energy. The demand for renewable energy and the growing global population drives the market. The restricted supply of fossil fuels and stringent governmental policies on carbon emissions generate a significant demand for efficient and affordable renewable energy options such as solar power. Crystalline silicon solar cells exhibit greater efficiency when compared to amorphous silicon and non-silicon solar cells. Advancements in technology and higher government investment in renewable energy initiatives are anticipated to create growth prospects in the market.

The global solar cell market is primarily led by the utility-scale segment. Utility-scale solar refers to large-scale solar power plants. A solar energy facility that is sufficiently large to be considered "utility scale" when compared to rooftop solar installations. Utility scale pertains to electrical facilities or devices, where the functioning of a single unit would significantly affect the performance of a utility. It is producing dependable, clean energy with a consistent fuel cost for many years. Solar power facilities can be designed to align environmental conservation with rising energy needs and climate objectives. Utility-scale solar can be combined with energy storage to handle evening energy surges, deliver backup power, and additional functions. The existence of the previously mentioned broad range of uses for solar cell panels in different industrial and enhanced applications is likely to offer many opportunities for the market.

Solar PV technology became popular in the early 2000s, as support initiatives for home systems were launched in leading countries. In the last twenty years, solar PV systems have gained global popularity, significantly contributing to the transition to renewable energy. Yearly on-grid installations achieved GW milestones of 1 GW in 2004, 10 GW in 2010, 100 GW in 2018, and 226 GW in 2022. Having achieved a compound annual growth rate of 37% since the year 2000, solar PV has emerged as the most rapidly expanding energy technology. The market's volume in 2022 is 38% greater than in 2021, propelled by the worldwide energy crisis and the increasing need for clean and affordable power.

Japan and Germany were trailblazers in solar PV advancement, spearheading efforts through the creation of a Renewables Portfolio Standard (RPS) legislation and a feed-in tariff (FIT) initiative. By 2004, they had functioned at a capacity of 1 GW, signaling the "golden age" of solar PV in Europe. Europe maintained a dominant role in the global market, with yearly installation shares between 66 and 85%. Germany continued its growth path, while other European nations such as Italy, Spain, France, and Belgium exceeded the 1 GW milestone. Italy's 9.2 GW link in 2011 is recognized as the top in Europe. The European solar market declined from 21.9 GW in 2011 to 7 GW in 2016, allowing the Asia–Pacific region to take on a significant role in the worldwide solar scene.

The Asia–Pacific (APAC) area has witnessed a notable rise in solar PV installations, with China, India, and Japan as the primary economies fueling this expansion. In 2013, APAC launched 21.6 GW of solar PV, accounting for 51% of the worldwide total of 37.9 GW. In 2022, APAC reinforced its status as a worldwide frontrunner, obtaining 59% of the yearly installed capacity. Nonetheless, China's major influence on solar installations resulted in an increase in its global market share from 34 to 42% in just one year. The remainder of APAC saw only slight advancement, increasing from 39 GW in 2021 to 40 GW in 2022. India's slight increase of 3.2 GW did not offset the decline seen in Australia and South Korea, each of which faced a major decrease of almost 4 GW in their yearly market.

Europe has seen a notable increase in installed capacity, making it the second-largest market for yearly solar PV deployment. With 47.1 GW installed in 2022, or 21% of the worldwide market, the solar industry's performance resulted in a 14.1 GW annual increase. With an 8.4 GW contribution, Spain has surpassed Germany's 7.4 GW, making it the new solar leader. Additionally, Poland, the Netherlands, and Italy outperformed the year before.

With the USA and Brazil at the forefront, solar PV has grown steadily throughout the Americas throughout the 2000s. With 38 GW built, solar PV's yearly share in the Americas grew from 8% in 2011 to 17% in 2022. Nonetheless, the continent's worldwide market share has decreased, suggesting that the technology is being adopted more slowly than in APAC and Europe. This was the first dip since 2017 as the Americas' share fell from 19.6% in 2021 to 16.7% in 2022. This drop is due to the US, the largest regional market, contracting. In 2022, the US market's annual size dropped from 23.4 to 21.9 GW.

Largely due to multi-gigawatt projects started through government bids in the Arabian Peninsula, the Middle East and Africa (MEA) solar sector has grown gradually. With 7.1 GW of solar power built in 2022, the MEA area accounted for 3% of the worldwide market. With the addition of more than 1 GW of new capacity, Israel has emerged as the MEA's main solar market. In 2022, South Africa installed 1.3 GW of solar capacity, maintaining its position. Outside of South Africa, on-grid photovoltaic development is difficult in the Middle East and North Africa (MENA) area, which is still bigger than Sub-Saharan nations [51].

The size of the worldwide solar cell market is estimated to be USD 149.45 billion in 2024, USD 175.15 billion in 2025, and USD 730.74 billion by 2034, with a high compound annual growth rate (CAGR) of 17.2% from 2024 to 2034. In 2024, the Asia Pacific solar cell market was valued at USD 73.23 billion, and it is projected to expand at the fastest compound annual growth rate (CAGR) of 17.32% during the forecast [52].

7 Summary

The demand for photovoltaics has increased drastically as a result of the rapid industrial growth and depletion of fossil fuel supply globally. With the increase in the technological developments and an increasing emphasis on decarbonization worldwide, solar energy becomes more popular. The development of hybrid solar systems, improvements in energy storage integration and the use of building-integrated photovoltaics in urban infrastructure are some of the major themes propelling the sector. Today, nearly 97% of all cells are produced using c-Si technology. Thin-film modules are less expensive to manufacture compared to crystalline cells and possess reasonably good conversion efficiencies. New concepts such as floatovoltaics and agrivoltaics will further aid in the widespread acceptance of photovoltaic technology. It is expected that solar PV will contribute to 21% of the overall energy sector CO_2 emissions reductions by 2050.

Acknowledgement Heera S acknowledge DST, Government of India (IF200250) for providing fellowship through the Inspire scheme. Authors also thank KSCSTE, Government of Kerala, India for the funding through SRS scheme(KSCSTE/268/2024/SRSPS).

References

1. J. Nelson, *The Physics of Solar Cells*, 2nd edn. (CRC Press, 2014)
2. S. Kuşkaya, F. Bilgili, E. Muğaloğlu, K. Khan, M.E. Hoque, N. Toguç, The role of solar energy usage in environmental sustainability: fresh evidence through time-frequency analyses. Environ. Sci. Pollut. Res. **30**(24), 68452–68467 (2023)
3. C.R.J. Kumar, M.A. Majid, Renewable energy for sustainable development in India: current status, future prospects, challenges, employment, and investment opportunities. Renew. Sustain. Energy Rev. **131**, 109976 (2020)
4. N.F. Voudoukis, Photovoltaic technology and innovative solar cells. Renew. Sustain. Energy Rev. **91**, 830–843 (2018)
5. W. Shockley, H.J. Queisser, Detailed balance limit of efficiency of p-n junction solar cells. J. Appl. Phys. **32**(3), 510–519 (1961)
6. A.S. Al-Ezzi, M.N.M. Ansari, Photovoltaic solar cells: a review. Renew. Sustain. Energy Rev. **118**, 109548 (2020)

7. K. Mahmood, S. Sarwar, M.T. Mehran, Current status of electron transport layers in perovskite solar cells: materials and properties. Materials **14**(8), 2029 (2021)
8. Y. Zhao, S. Zhang, A brief review of hole transporting materials commonly used in perovskite solar cells. J. Mater. Sci. Mater. Electron. **40**, 2712–2729 (2021)
9. S.W. Glunz, F. Feldmann, SiO_2 surface passivation layers—a key technology for silicon solar cells. Prog. Photovoltaics Res. Appl. **18**(6), 395–405 (2010)
10. D.I. Kim, J.W. Lee, R.H. Jeong, J.H. Boo, A high-efficiency and stable perovskite solar cell fabricated in ambient air using a polyaniline passivation layer. Sci. Rep. **12**, 697 (2022)
11. D. Kiermasch, P. Rieder, K. Tvingstedt, A. Baumann, V. Dyakonov, Improved charge carrier lifetime in planar perovskite solar cells by bromine doping. Nat. Commun. **8**, 14769 (2017)
12. J. Van Deelen, Y. Tezsevin, M. Barink, Multi-material front contact for 19% thin film solar cells. Sol. Energy Mater. Sol. Cells **163**, 109–115 (2017)
13. L. Gouillart, A. Cattoni, J. Goffard, F. Donsanti, G. Patriarche, M. Jubault, N. Naghavi, S. Collin, Development of reflective back contacts for high-efficiency ultrathin Cu(In,Ga)Se_2 solar cells. Sol. Energy Mater. Sol. Cells **210**, 110477 (2020)
14. Delft University of Technology. Solar energy (2023). https://ocw.tudelft.nl/courses/solar-energy/
15. S. Rühle, Tabulated values of the Shockley-Queisser limit for single junction solar cells. Sol. Energy **130**, 139–147 (2016). https://doi.org/10.1016/j.solener.2016.02.015
16. L. Szabó, *The History of Using Solar Energy* (2017). https://www.academia.edu/33396080/The_History_of_Using_Solar_Energy
17. M.A. Green, E.D. Dunlop, M. Yoshita, N. Kopidakis, K. Bothe, G. Siefer, D. Hinken, M. Rauer, J. Hohl-Ebinger, X. Hao, Solar cell efficiency tables (version 64). Prog. Photovoltaics Res. Appl. **32**(7), 425–441 (2024). https://doi.org/10.1002/pip.3831
18. Encyclopaedia Britannica: Solar cell (2024). https://www.britannica.com/technology/solar-cell
19. Electrical4U: Solar cell: working principle & construction (diagrams included) (2024). https://www.electrical4u.com/solar-cell/
20. S. Sharma, K.K. Jain, A. Sharma, Solar cells: in research and applications—a review. Mater. Sci. Appl. **6**(12), 1145–1155 (2015). https://doi.org/10.4236/msa.2015.612113
21. R.A. Deshpande, Advances in solar cell technology: an overview. J. Sci. Res. **65**(2), 72–75 (2021). https://doi.org/10.37398/jsr.2021.650214
22. K. Raut, A review on comparative studies of diverse generation in solar cells (2018). https://www.academia.edu/73297963/A_Review_on_Comparative_Studies_of_Diverse_Generation_in_SolarCell
23. S. Bhattacharya, S. John, Beyond 30% conversion efficiency in silicon solar cells: a numerical demonstration. Nat. Commun. **11**(1), 2175 (2020)
24. International Journal of Scientific Development and Research, Analysis of the effect of fill factor on the efficiency of solar PV system for improved design of MPPT (n.d.). https://www.ijsdr.org/papers/IJSDR1902043.pdf
25. V.V. Tyagi, N.A.A. Rahim, N.A. Rahim, J.A.L. Selvaraj, Progress in solar PV technology: research and achievement. Renew. Sustain. Energy Rev. **20**, 441–448 (2013)
26. J.A. Luceño-Sánchez, A.M. Díez-Pascual, R. Peña Capilla, Materials for photovoltaics: state of art and recent developments. Int. J. Mol. Sci. **20**(4), 976 (2019). https://doi.org/10.3390/ijms20040976
27. REN21: Global Status Report 2023 [PDF] (2019). https://www.ren21.net/wp-content/uploads/2019/05/GSR2023_GlobalOverview_Full_Report_with_endnotes_web.pdf
28. International Energy Agency: Executive Summary—Solar PV Global Supply Chains—Analysis (n.d.). https://www.iea.org/reports/solar-pv-global-supply-chains/executive-summary
29. Precedence Research: Solar cell market size to surpass USD 730.74 billion by 2034 (n.d.). https://www.precedenceresearch.com/solar-cell-market
30. Allied Market Research: Solar cell market statistical analysis and growth forecast—2031 (n.d.). https://www.alliedmarketresearch.com/solar-cell-market-A08602
31. International Solar Alliance: Solar energy trends and statistics report (n.d.). https://isolaralliance.org/uploads/docs/a85d84a7a402f9209262f855b273da.pdf

32. Niclas, Standard test conditions (STC): definition and problems. Manage risks and maximize ROI for your PV and energy storage projects (2023). https://sinovoltaics.com/learning-center/quality/standard-test-conditions-stc-definition-and-problems/
33. J. Smith, L. Zhang, A comprehensive review and outlook of bifacial photovoltaic (bPV) technology. Renew. Sustain. Energy Rev. **133**, 100–115 (2021). https://doi.org/10.1016/j.rser.2021.100115
34. R. Jones, M. Patel, A systematic literature review of the bifacial photovoltaic module and its applications. Renew. Energy **150**, 50–65 (2020). https://doi.org/10.1016/j.renene.2020.01.110
35. D. Lee, H. Kim, A review of photovoltaic cell generations and simplified overview of bifacial photovoltaic cell technology. Energy Rep. **5**(3), 200–215 (2019). https://doi.org/10.1016/j.egyr.2019.04.002
36. Y. Chen, Z. Zhang, Durability of bifacial solar cells: a review. Sol. Energy Mater. Sol. Cells **214**, 122–137 (2022). https://doi.org/10.1016/j.solmat.2022.112206
37. J. Smith, H. Lee, Solar PV energy: from material to use, and the most commonly used techniques to maximize the power output of PV systems: a focus on solar trackers and floating solar panels. Renew. Energy **150**(2), 123–145 (2021). https://doi.org/10.1016/j.renene.2020.12.025
38. T. Xu, X. Zhang, Floating solar photovoltaic systems: a review of technologies, applications, and environmental impacts. Renew. Sustain. Energy Rev. **101**, 1–18 (2019). https://doi.org/10.1016/j.rser.2018.11.021
39. H. Tsai, C. Li, Solar tracking systems and floating photovoltaic systems: a comprehensive review. Renew. Energy **156**, 1135–1150 (2020). https://doi.org/10.1016/j.renene.2020.03.080
40. X. Yuan, Y. Li, Y. Li, Building-integrated floating photovoltaics: a new sustainable solution for urban energy systems. Energy Build. **261**, 111912 (2022). https://doi.org/10.1016/j.enbuild.2022.111912
41. X. Lu, H. Li, Concentrated floating solar photovoltaics: a comprehensive review on emerging technologies and applications. Sol. Energy Mater. Sol. Cells **253**, 111975 (2023). https://doi.org/10.1016/j.solmat.2022.111975
42. A. Brown, K. Johnson, Potential environmental impacts of floating solar photovoltaic systems. Environ. Sci. Technol. **54**(7), 456–478 (2020). https://doi.org/10.1021/es2045768
43. Y. Zhang, X. Zhao, W. Zhang, Agrivoltaics: a promising pathway for sustainable land use. Environ. Sci. Technol. **54**(10), 6650–6662 (2020)
44. S. Hegedus, X. Lu, Agrivoltaics: combining food and solar energy production. Nat. Commun. **11**(1), 1–11 (2020)
45. C. Dupraz, M. Aguiton, F. Guerin, Agrivoltaics: synergies and trade-offs between agriculture and photovoltaics. Renew. Sustain. Energy Rev. **15**(4), 1999–2008 (2011)
46. T. Kato, S. Matsumura, K. Kondo, T. Akiyama, Agrivoltaic systems: a review of the current state of research. Nat. Sustain. **2**(5), 378–391 (2019)
47. Y. Liu, J. Zhang, **The** effect of agrivoltaics on crop yield and energy production: a comparative study. Appl. Energy **267**, 114927 (2020)
48. H. Takahashi, T. Saito, The future of agrivoltaics: potential impacts and challenges in the global food-energy nexus. Renew. Sustain. Energy Rev. **137**, 110549 (2021)
49. D.D. Rooij, Short circuit current. Manage risks and maximize ROI for your PV and energy storage projects (2023). https://sinovoltaics.com/learning-center/basics/short-circuit-current/
50. S. Izawa, Open-circuit voltage in organic solar cells, in *Organic Solar Cells* (2020), pp. 195–216. https://doi.org/10.1007/978-981-15-9113-6_8
51. M. O'Kane, *Fill Factor of Solar Cells* (Ossila, n.d.). https://www.ossila.com/pages/fill-factor-solar-cells
52. U.S. Department of Energy: Solar performance and efficiency (n.d.). https://www.energy.gov/eere/solar/solar-performance-and-efficiency
53. Z. Li, J. Yang, P.A. Nejad Dezfuli, Study on the influence of light intensity on the performance of solar cell. J. Renew. Sustain. Energy **13**(4), 045302 (2021)

Silicon-Based Technologies for Solar to Photovoltaic Conversion

Rameeja T. Abdul Rasheed and Aldrin Antony

Abstract Governments worldwide are committed to tripling global renewable energy capacity by 2030 to limit global temperature rise to 1.5 °C above pre-industrial levels. Solar energy, powered by silicon solar cells, plays a critical role in this transition with silicon (Si)-wafer-based technology holding 97% of the market share. Achieving this ambitious goal for renewable energy generation requires significant advancements in efficiency and cost-effective production. The efficiency of silicon solar cells has evolved dramatically, from the 6% of the first cell developed at Bell Labs in 1954 to the record-breaking 27.3% achieved in a back-contacted architecture (HBC) by LONGi in 2024. This chapter describes the development of silicon solar cell architectures, detailing the variations, design improvements, and technological breakthroughs that have shaped the industry. The small changes in cell architecture, passivation, etc., were found to significantly impact the efficiency and production scalability. Beyond lab-scale innovations, how these technologies transition to mass production is discussed. Each architecture, including Aluminium Back Surface Field (Al-BSF), Passivated Emitter and Rear Cell (PERC), Silicon Heterojunction (SHJ), Interdigitated Back Contact (IBC), Dopant-Free Asymmetric Heterocontact Solar Cells (DASH), Carrier-Selective Solar Cells (CSCs), Tunnel Oxide Passivated Contact (TOPCon), and Polycrystalline Oxide Layer (POLO), is explored in-depth, with a focus on how design adjustments ensure industrial feasibility. The challenges and solutions to making these cells viable for large-scale deployment are discussed, highlighting the processes required, while maintaining high efficiency. The chapter gives a comprehensive insight into the enhancement in cell performance with the developments in cell architectures and in facilitating their integration into the global renewable energy landscape.

R. T. Abdul Rasheed · A. Antony (✉)
Department of Physics, Cochin University of Science and Technology, Kochi, Kerala, India
e-mail: *aldrin@cusat.ac.in

A. Antony
Centre of Excellence in Advanced Materials, Cochin University of Science and Technology, Kochi, Kerala, India

P. Ravindran et al. (eds.), *Photon to Power*, Progress in Optical Science and Photonics 32, https://doi.org/10.1007/978-981-96-5914-2_2

Keywords Silicon solar cell · High-efficiency cells · Photovoltaics · Solar cell architectures · PERC · Silicon Heterojunction · IBC solar cells · TOPCon · POLO · Carrier-Selective Solar Cells · Solar cell mass production

1 Introduction

Solar energy has emerged as a cornerstone of the global transition towards sustainable and renewable energy sources. As global concerns about climate change and the demand for energy security intensify, solar power will have the lead role in this transition to renewable energy. In terms of energy generation, solar power produced 1294 TWh of electricity in 2022, representing a 25.6% increase from the previous year. This growth contributed to a 29.1% renewable energy share of the 29,031 TWh global electricity generation. In terms of renewable capacity addition, the world added an impressive 473 gigawatts (GW) of renewable energy capacity in 2023, marking a notable 14% increase from 2022—the largest annual growth since 2000. This progress is encouraging and demonstrates a strong global commitment to expand the renewable energy sources. At COP 28, governments committed to triple the global renewable energy capacity by 2030. However, to meet this ambitious goal, renewable capacity of at least 11 terawatts (TW) is essential by 2030. If the current rate of expansion continues, this could lead to only 9.7 TW by 2030. This highlights both the remarkable strides we've made and the urgent need to intensify our efforts to ensure a sustainable future [1] (Fig. 1).

The global fuel usage in this decade is on the verge of witnessing a record peak in oil, natural gas, and coal consumption, and after which this demand for these fossil fuels is expected to decline, paving the way for a surge in renewable energy

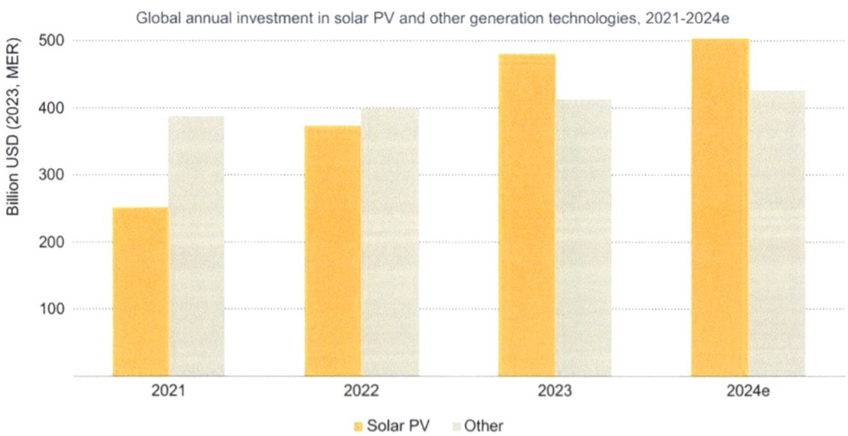

Fig. 1 Global annual investment in solar PV and other generation technologies, 2021–2024e. Reprinted from World Energy Investment 2024, IEA, Paris [2]

growth. Solar energy is set to play a key role in this transition. The Stated Policies Scenario (STEPS)—which projects energy trends based on current policies—indicates a substantial expansion in global solar PV deployment, growing from 220 GW in 2022 to 500 GW by 2030. Despite this substantial growth in solar energy capacity, the utilization rate is expected to remain below 40% through 2030, highlighting a significant opportunity to further accelerate solar deployment worldwide [3]. Currently, China leads the world with an annual output of 500 GW, accounting for 80% of the global market. Vietnam and India follow, each holding 5% and 3% of the market share, respectively [3].

The shift towards clean energy is further emphasized by the evolving investment ratio: in 2015, the investment ratio between clean power and unabated fossil fuel power was approximately 2:1; by 2024, this ratio is projected to increase to 10:1. In the United States, investment in clean energy is anticipated to exceed USD 300 billion in 2024, while the European Union is expected to invest USD 370 billion. China is leading this global push, set to spend USD 680 billion in 2024, supported by its large domestic market and extensive development in industries such as solar cells, lithium batteries, and electric vehicle manufacturing [2].

There is ample opportunity for the growth in solar PV generation and deployment, as it holds the potential to help achieve the net-zero emissions (NZE) goal by 2030. The substantial increase in solar PV capacity has spurred the creation of extensive resources and production centres for silicon modules. This growth, signals that silicon photovoltaics are on the brink of a massive boom in the coming years. Consequently, any research aimed at reducing costs, boosting efficiency, or enhancing production methods for silicon-based solar cells will have a direct and transformative impact on the solar PV market.

2 Brief History of the Development of Si Photovoltaic Cells

The true beginning of the modern silicon solar cell era is often credited to Russell Ohl's invention in 1941 at Bell Labs, USA. He created the first photovoltaic cell using silicon rods with natural potential barriers in recrystallized melts formed through impurity segregation during the melting process. Ohl filed a patent for this innovation, titled "Light Sensitive Electric Device Using Silicon" [4]. In this work, they explained that "one end of the rod developed a negative potential when illuminated or heated, and when supplied with a negative potential, it showed low resistance to current flow across the barrier." This observation led to the terminology of "n-type" for the negative end, while the opposite end was designated as "p-type." Later, with advancement in understanding of the semiconductor mechanisms, it was understood that crystal lattices with donor impurities act as n-type material, while crystal lattices with acceptor impurities behave as p-type material [5]. In 1954, the first 6% efficient silicon solar cells utilizing a dopant diffused wrap-around p–n junction with dual rear contact were developed at the Bell laboratory, USA by Chapin et al. [6, 7]. Although the initial efficiencies generated significant enthusiasm, this excitement

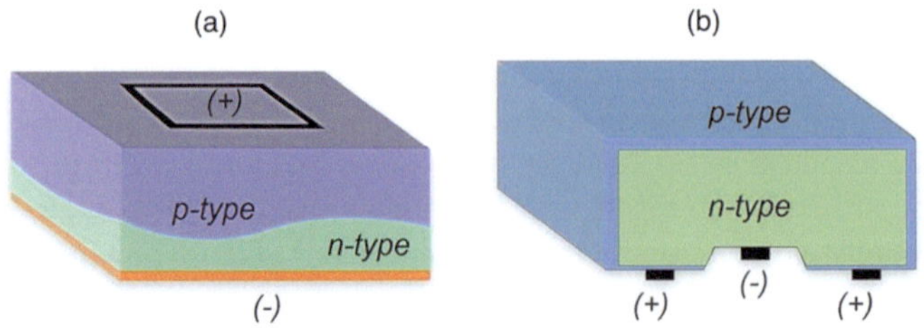

Fig. 2 Schematic representation of **a** an early solar cell developed by Russell Ohl, featuring a recrystallized melt junction that formed the basis for the p–n junction in crystalline silicon, **b** the first 6% efficient silicon solar cells utilizing a dopant diffused wrap-around p–n junction with dual rear contact were developed at the Bell laboratory, USA, by Chapin, Fuller, and Pearson

soon faded due to the prohibitive manufacturing costs and the underdeveloped state of the silicon production industry at the time. In 1955, the semiconductor division of Hoffman Electronics developed a commercial solar cell with 2% efficiency, priced at $1500/W. Just a few years later, in 1958, Vanguard I, NASA's the first solar-powered satellite, was launched, equipped with a 0.1 W, 100 cm^2 silicon solar panel [8]. Over the following years, the efficiencies improved and by 1959, Hoffman Electronics created a 10% efficient commercial silicon solar cell using grid contacts. The first commercial communication satellite Telstar1 built by Bell labs was launched by NASA in 1962. This satellite was powered by 3600 solar cells, generating over 14 W power. Between 1954 and 1970, significant advancements were made in solar cell technology to meet the demands of space missions. Solar cells were regarded as an ideal standalone power source for space applications, leading to intense research focused on adapting these cells to withstand the harsh radiation and other challenges of space environments. Innovations during this period included the development of grid metallization on the front side, the use of 10 Ω-cm p-type substrates for enhanced resistance to radiation in space, heavily phosphorus-diffused emitters to prevent shunting from metallization, and the application of SiO_x anti-reflection coatings to improve light coupling into the cell [9] (Fig. 2).

3 C-Si-Based Solar Cell Technologies

The evolution of silicon solar cell technology represents a confluence of fundamental research and applied engineering, culminating in the efficient, scalable devices we see today. In the subsequent sections, we delve into the key milestones that have shaped this field, offering a detailed exploration of the architectures, mechanisms, and processes that underpin modern silicon photovoltaics.

The section begins with the Aluminium Back Surface Field (Al-BSF), the first matured industrial technology that introduced basic passivation and set the stage for further advancements. From there, we transition to the Passivated Emitter and Rear

Cell (PERC), which revolutionized solar cell efficiency by addressing carrier recombination losses with practical, cost-effective modifications. A detailed examination of commercial PERC structures follows, highlighting how this technology has been adapted for large-scale manufacturing while retaining its core principles.

Building on this foundation, we investigate the development of Tunnel Oxide Passivated Contact (TOPCon) and Poly-Si on Oxide (POLO) technologies. These configurations represent a paradigm shift in carrier-selective contacts, offering enhanced passivation and conductivity. How these technologies translate from laboratory research to industrial production is critically analysed, with an emphasis on their scalability and compatibility with the existing fabrication lines.

Next section presents the evolution of Silicon Heterojunction (SHJ) solar cells, which have set new benchmarks in efficiency. By dissecting the design principles and unique advantages of SHJ cells, we shed light on their role in defining the current landscape of photovoltaic technology.

The latest efficiency records in each of the cell architectures are compared, discussing the interplay between material properties, interface engineering, and advanced processing techniques. Special attention will be given to the commercialization strategies and industrially viable processes that have enabled these high-performance designs to reach the market.

3.1 Aluminium-Back Surface Field (Al-BSF) Solar Cells

In the 1970s, a simple, cost-effective, and efficient method was developed to create a robust metal contact along with a p^+ diffused layer on the rear side of silicon known as Al-BSF (Aluminium Back Surface Field). This technology proved to be highly useful and compatible for forming rear contacts with p-type wafers. Owing to its ease of implementation, all cell configurations during this period utilized p-type wafers with n-type emitters diffused using phosphorus atoms (Fig. 3).

Al-BSF solar cells are made of p-type crystalline silicon and the rear contact is achieved through screen-printing of Al paste followed by the firing at temperatures 600 °C to 800 °C. This annealing process creates a p^+ region at the rear side as a result of alloying of Al with silicon just beneath the aluminium contact [10, 11]. The n^+-p-p^+ doping profile of the solar cell enabled favourable band alignment for the collection of photogenerated holes while selectively preventing electron recombination at the base, providing both selectivity and passivation at the rear, which helped reduce the loss of photogenerated current. Screen-printed Al-BSF contacts remained a top choice for many years, influencing the evolution of later solar cell structures to continue using p-type substrates with Al-BSF. However, the industry recently transitioned to n-type wafers because of its high lifetime and to overcome limitations such as light-induced degradation (LID) and light and elevated temperature-induced degradation (LETiD) in p-type solar cells. Until 2017, Al-BSF solar cells dominated the commercial market, holding a share of over 90% (Fig. 4).

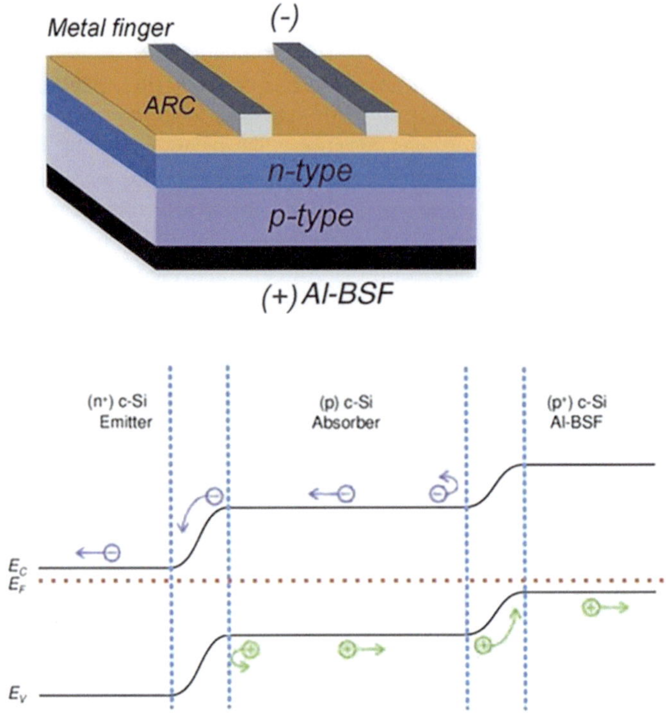

Fig. 3 Schematic representation of conventional Al-BSF structure and corresponding band diagram

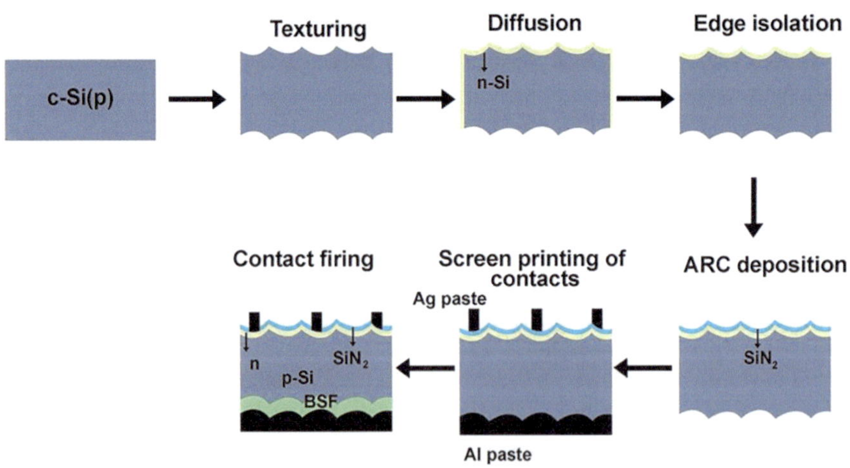

Fig. 4 Manufacturing steps for the production of Al-BSF solar cells

The fabrication of Al-BSF solar cells typically involves the following steps [12, 13].

1. *Wafer Preparation*: p-type (boron or gallium-doped) Czochralski (Cz)-grown single-crystalline or multi-crystalline wafers of resistivity 1–5 Ω cm are treated in alkaline solution (KOH) for saw damage removal, followed by texturing to enhance light coupling into the cell reducing the reflection losses. Wafers are either isotropically etched in acidic solution of hydrofluoric acid and nitric acid or anisotropically etched in an alkaline solution of KOH/NaOH and isopropyl alcohol.
2. *Emitter Formation*: This is accomplished by diffusing phosphorus atoms into the p-type wafer with gas sources like phosphoryl chloride ($POCl_3$) or phosphine (PH_3) in a high-temperature quartz furnace at temperatures 800 to 1100 °C to create the n^+ layer. Industrial solar cell process uses a two-stage diffusion process where $POCl_3$ gas is used and is known as *pocal diffusion*. N_2 or Ar is used as the carrier gas, and oxygen is also used during the first-stage diffusion process at around 800 °C. This *pre-deposition* process creates a layer of phosphosilicate glass (PSG) on the surface with a shallow diffusion of P atoms. This phosphorous rich PSG acts as the source of diffusion of P atoms at a higher temperature 1000–1100 °C, which is known as *drive-in* process which is tuned to get the desired junction depth. After this drive-in process, the phosphosilicate glass (PSG) is etched away using hydrofluoric acid (HF). The sheet resistance of the n^+ layer is can be varied in the range 40–100 Ω/cm and the emitter depth is usually kept less than 0.5 microns. The resistance of this emitter layer decides the busbar configuration. The higher resistance layers have the advantage of low parasitic light absorption and needs multiple busbars for the carrier collection.
3. *Edge Isolation*: The emitter diffusion might have occurred on the front and edges of the silicon wafer since wafers are normally stacked as coin-cells during the diffusion process. The edges of the wafer are isolated using laser scribing, chemical or plasma etching to prevent electrical shorting between the front and rear sides of the wafer.
4. *Antireflective Coating (ARC)*: A layer of hydrogenated silicon nitride (SiN_x:H), with a thickness of 70–80 nm, is the typical ARC layer used for the n-type emitter layer which is usually deposited using plasma-enhanced chemical vapour deposition (PECVD) or MWPECVD technique. This layer serves as both an anti-reflective coating and also offers excellent passivation to the n-doped layer. Multilayer ARCs are also used for enhanced light trapping.
5. *Front Contact Deposition*: Front contacts are made by screen-printing silver (Ag) paste in a grid pattern with fingers and busbars, with a thickness ~ 25 microns.
6. *Rear Contact Deposition*: Aluminium paste, at a coverage of 7–10 mg/cm^2, is screen printed over the entire rear to establish the rear contact.
7. *Co-firing*: To enable actual contact formation from the deposited paste, the cell is heated in a furnace at 850 °C in a process called co-firing, enabling both the front and rear contacts to make electrical contact with respective surfaces. During this process, aluminium atoms diffuse into silicon at the rear side, forming a p^+ back

surface field (BSF). At the front side, the grits in the silver paint penetrates the SiN_x layer and makes ohmic contact with the emitter layer. This firing process is done using either RTP furnaces in lab or belt furnace in the industry.
8. *Testing and Characterization*: The fabricated Cells are tested for their Current–Voltage characteristics (I-V) and sent for module preparation.

Despite the ease of manufacturing and widespread industry adoption, the efficiency of Al-BSF configurations remained ~ 20%. Recombination at the front and rear surfaces, caused by the large semiconductor/metal contact area, was the significant drawback. To reduce production costs and, consequently, the cost per watt in solar energy, it became necessary to explore methods for improving cell efficiency. Later structures addressed numerous factors affecting the open-circuit voltage, leading to the evolution of higher-efficiency designs.

3.2 Evolution of Passivated Emitter and Rear Cell (PERC) Solar Cells

In 1984, Green and his team at UNSW developed a silicon solar cell with an impressive 18% efficiency [14]. This was achieved by taking a new approach to address the recombination at the semiconductor–metal interface at the front contacts. A Metal–Insulator–Semiconductor (MIS) contact was employed to prevent direct contact between the two interfaces, and to provide passivation at the emitter surface [10, 15]. They focused on a completely new approach to achieve high efficiencies in silicon-based solar cells which concentrated on minimizing recombination losses at the emitter, both in areas covered by the metal grid and in uncovered regions, with the goal of achieving high open-circuit voltages. This approach led to the development of several key solar cell configurations, including MIS, MINP, PESC, PERC, PERL, PERT, and PERF. The key features and significant step-by-step evolution of these configurations are explained in the following sections.

3.2.1 Metal–Insulator–NP Junction (MINP)

The concept of a metal–insulator–semiconductor (MIS) contact in the solar cell emitter was first introduced by Godfrey and Green in 1976. This design involved contacting the emitter with metal through a thin layer of high-quality tunnel oxide (SiO_2) that enabled quantum mechanical tunnelling of the charge carriers while providing passivation to the emitter surface, and significantly reduced surface recombination velocities. This structure was soon modified to include a lightly doped n layer to give better results. Green coined the acronym "MINP" for Metal–Insulator–NP junction cell for the resulting device [11, 16]. This was achieved through improved passivation of the top surface using a 5 nm layer of SiO_2, combined with a low grid coverage of just 4.6%. The metallization was accomplished using a 5-µm stack of

Ti/Pd/Ag. Additionally, they employed a double-layer anti-reflection coating (ARC) made from MgF_2 and ZnS. This device marked the first of 18 world records set by this group over the following 30 years [16].

The MINP concept concentrated on several crucial aspects of solar cell design, establishing a foundation for the development of high-efficiency solar cells. The different improvements implemented in MINP are the following [17] (Fig. 5):

Front Contact: A Ti/Pd/Ag contact structure was used, with vacuum-deposited Ti and Pd, along with electroplated Ag, to form a thick contact that lowered series resistance arising from the front contacts.

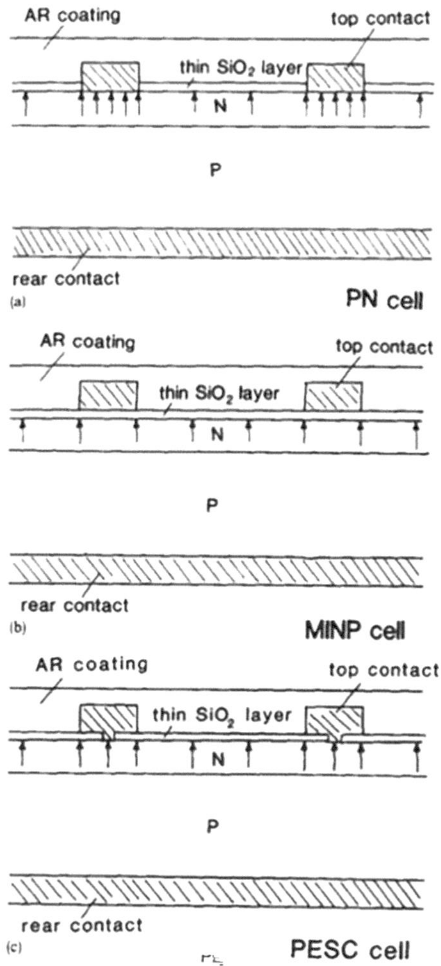

Fig. 5 Schematic of **a** NP junction cell, **b** metal–insulator–NP junction (MINP) cell, and **c** passivated emitter solar cell (PESC). Reprinted with permission from [Appl Phys Lett, vol. 44, M. A. Green, A. W. Blakers, J. Shi, E. M. Keller, and S. R. Wenham, "19.1% efficient silicon solar cell," no. 12, pp. 1163–1164, Jun. 1984, https://doi.org/10.1063/1.94678] with the permission of AIP Publishing [17]

Reduced Metal/Semiconductor Contact Regions: The metal/semiconductor contact area was minimized to below 4.6%, effectively reducing recombination losses at these critical interfaces.

Antireflective Coating (ARC): A double-layer anti-reflective coating composed of ZnS/MgF_2 was employed, reducing the surface reflection to below 3% and thereby enhancing light absorption into the cell.

Passivating Thermal Oxide on Emitter Surface: A 5 nm layer of high-quality trichloroethane (TCA) grown thermal oxide was used to passivate the emitter surface, significantly reducing surface recombination and recombination losses at the semiconductor/metal interface. The thin oxide layer was designed to allow tunnelling of charge carriers while avoiding resistive losses.

Substrate Resistivity: The resistivity of the substrate wafers was carefully selected to be as low as 0.1–0.2 Ω cm, optimizing the material for efficient charge carrier transport.

3.2.2 Passivated Emitter Solar Cell (PESC)

An improvement of efficiency to 19.1% with open-circuit voltage of 653 mV was achieved by Green et al [17] following minor modifications to the MINP structure:

Reduction in Series Resistance: In the PESC structure, fine openings in the passivating oxide layer, located directly beneath the metallization, allowed for controlled metal/semiconductor contact, effectively reducing series resistance. The different improvements implemented in PESC are the following:

Reduced Metal/Semiconductor Contact Regions: The metal/semiconductor contact area was further reduced to 3–3.5% range, reducing the shading losses and contact resistance at the metal/semiconductor junction.

An efficiency of 20% was achieved in this structure by introducing photolithographically defined micro-grooves on the front surface, which significantly reduced reflection losses.

3.2.3 Passivated Emitter and Rear Cell (PERC)

The improvement in open-circuit voltage (V_{oc}) due to reduced recombination at the emitter surface as described in earliest session made it apparent that the primary factor limiting further increases in efficiency was the recombination losses occurring at the semiconductor/metal interface on the rear surface [15]. To address this issue, the following modifications were introduced:

Rear Surface Oxide Passivation: A thin dielectric layer of SiO_2 was introduced at the interface between silicon and aluminium on the rear side. An array of openings was created in this dielectric layer, allowing the metal to contact the silicon surface in a controlled manner. Although this scheme aimed to passivate the rear surface, it was anticipated that the minority charge recombination at this region might increase. This was because the aluminium layer beneath the oxide would only weakly deplete the silicon, leading to less effective prevention of recombination. In addition to improving passivation, the SiO_2/Al stack served as a superior optical reflector compared to aluminium alone, achieving an impressive 97% reflection of incident light versus 88% for just the aluminium layer. This stack, referred to as "dielectrically displaced rear reflectors," significantly enhanced the reflection of light back into the cell by inserting a lower refractive index dielectric layer between the high refractive index silicon and the aluminium. This innovation was crucial, as silicon's indirect bandgap and low absorption coefficient meant that multiple passes of light through the silicon layer were necessary to effectively enhance its spectral response [18, 19].

Inverted Pyramid Structures on Emitter for Light Trapping: Inverted structures facilitated light trapping on the surface while simultaneously working in conjunction with the rear reflector to create a highly efficient light-trapping scheme. This combination maximized the amount of light absorbed within the solar cell, further enhancing its overall efficiency.

Heavily Doped Regions Below the Metal Contacts in Emitter Surface: The n^+ diffused regions beneath the metal grid allowed for reduced recombination by creating a favourable n^+/n contact for the photogenerated carriers.

Enhanced Cleaning Techniques: Chlorine-based cleaning and processing techniques were adopted to ensure growth of high-quality thermal oxide and increased effective minority carrier lifetimes.

3.2.4 Passivated Emitter Rear Locally Diffused (PERL)

Improvement at the rear contact, without compromising the passivation, was achieved by introducing the following modifications in the PERC structure.

Locally Diffused Boron: The key modification in this structure, compared to its predecessor, was the localized diffusion of boron through specifically opened points of dimension 10 μm × 10 μm in the rear oxide layer, forming p^+ regions as shown in Fig. 6b. Boron diffusion was achieved by the diffusion using solid sources, boron tribromide, or spin-on dopant films. This localized diffused region enhanced the electrical performance of the solar cell by providing a p/p^+ doping profile at the rear contact. In 1999, Zhao et al. [20] reported a milestone efficiency record of 25% with the PERL structure [21, 22].

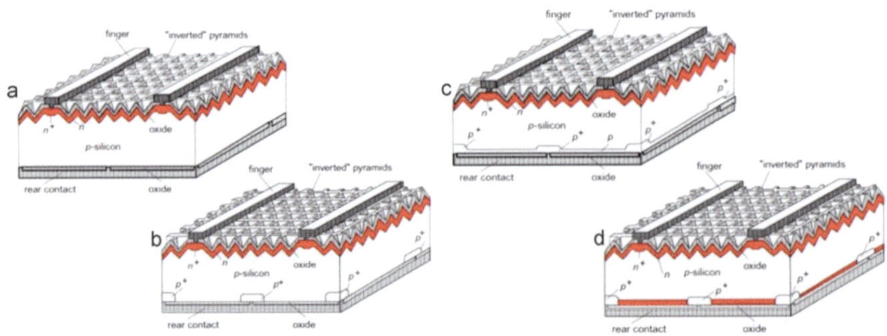

Fig. 6 **a** Simple PERC cell (Passivated Emitter Rear Contacted), **b** PERL cell (Passivated Emitter Rear Locally doped), **c** PERT cell (Passivated Emitter Rear Totally diffused), and **d** PERF cell (Passivated Emitter Rear Floating-junction). Reprinted from [Solar Energy Materials and Solar Cells, vol. 143, M. A. Green, "The Passivated Emitter and Rear Cell (PERC): From conception to mass production", pp. 190–197, Jul. 2015, https://doi.org/10.1016/j.solmat.2015.06.055] with permission from Elsevier [18]

3.2.5 Passivated Emitter Rear Totally Diffused (PERT)

In the PERT structure, a *p*-type region is diffused across the entire rear surface of the cell, with an additional localized heavily doped *p*+ region applied directly beneath the local openings [20].

Light p Doping at the Rear Side: The additional step introduced in this structure compared to PERL was the boron diffusion to form a lightly doped *p-region* along the entire rear surface of the cell. This lightly doped back surface field was introduced to reduce minority carrier recombination.

The cell structure now features a symmetrical design, with both the emitter and rear surfaces incorporating oxide passivation as shown in Fig. 6c. The top side has n-type diffusion, while the rear side has p-type diffusion, with heavily doped regions directly beneath the metal contacts, ensuring efficient carrier collection and reduced recombination at both interfaces.

3.2.6 Passivated Emitter Rear Floating Junction (PERF)

As the complexity of the rear surface increased, an injection-dependent recombination mechanism emerged, which led to a high ideality factor at the Maximum Power Point (MPP) and, consequently, resulted in lower fill factor values for the solar cells.

Introduction of a Rear Floating Junction: An electrostatically induced n-type layer, also known as a "floating" junction, was introduced to shift the region of high ideality factor away from the MPP [9]. This approach helped in maintaining better device performance. Later, the floating junction was replaced by a phosphorus-diffused n-type layer, further enhancing efficiency by providing improved control over recombination as shown in Fig. 6d.

3.2.7 Fabrication of Commercial PERC Solar Cells

The commercialization of PERC took a span of 25 years after its success in the laboratory. The path to commercialization involved incorporation of the fabrication process into existing Al-BSF manufacturing lines with minimal additional steps. Large-scale commercialization is only feasible when the processes involved are cost-effective. Typical commercial PERC cells featured a front surface design similar to that of Al-BSF, facilitating easy integration. Aluminium oxide (Al_2O_3), with its fixed negative charge ($Q_{fix} = 10^{13}$ cm^{-2}) deposited through spatial Atomic Layer Deposition (spatial ALD) or plasma-enhanced chemical vapour deposition (PECVD) enabling field-effect passivation through hole selectivity, was employed for rear-side passivation. This layer, combined with a silicon nitride (SiN_x) capping layer, deposited through PECVD, which protect the fixed charge during the firing process, formed the rear surface stack. A crucial feature of PERC was the local openings at the rear, which were initially created using lithography in labs. In industry, laser ablation replaced photolithography due to its low cost, simplicity and scalability. The rear contact then followed the Al-BSF cell steps of aluminium paste screen-printing and sintering [23].

In the 2000s, the industrial PERC concept (i-PERC) emerged, utilizing a laser-ablated rear passivation layer and Laser Contact Opening (LCO) to create openings in the passivation oxide before metallization. Additionally, the concept of Laser Fired Contacts (LFC) involved firing the laser through the rear passivation stack and aluminium to dope the silicon wafer, forming silicon-aluminium alloy doped regions [24–26]. Different manufacturing lines utilized rear stacks with either Al_2O_3/SiN_x, SiO_2/SiN_x or $SiON_x/SiN_x$ for passivation. These techniques became the foundational steps for fabricating PERC cells in the industry, collectively known as the i-PERC concept. The equipment manufacturing company, Centrotherm commercialized the first PERC solar modules, "Centaurus PERC," which were produced using a modified sequence on the existing Al-BSF sequence on 156 mm × 156 mm wafers and achieved efficiencies exceeding 19% [27]. During this period, other manufacturers began adopting PERC technology as they realized its potential to boost cell efficiency with relatively minor changes to existing production lines [28].

PERC technology rapidly gained traction in the industry from 2014 onwards, becoming the dominant technology in monocrystalline silicon solar cells. Companies like Trina Solar, Solar World, Jinko Solar, and LERRI Solar pushed the efficiency boundaries, with Trina Solar achieving a champion efficiency of 24.5% for a 210 mm p-type monocrystalline PERC module in 2022 [29]. Throughout this period, improvements in laser contact opening (LCO) techniques, rear passivation materials, and the overall manufacturing process further enhanced PERC's performance and cost-effectiveness. Bifacial PERC, which enables sunlight absorption from both the top and rear sides with aluminium grid contact at the rear, was developed. Bifaciality increases module output by ~ 25%. The technology is named as PERC+. Solar World has started mass production of these cells on p-type wafers, achieving 21.5% efficiency on the front side and 16.7% on the rear side [30, 31]. With the Al-BSF concept phased out in 2023, PERC technology emerged as the most dominant cell technology. PERC p-type monocrystalline silicon with a passivating layer of Al_2O_3 and a SiN_x capping layer held a 64% market share in 2023 [32].

3.3 Tunnel Oxide Passivated Contacts (TOPCon) Solar Cells

This solar cell configuration utilizes n-doped crystalline silicon wafer with n+ doped polysilicon interface (nc-Si/n^+poly-Si) with a thin tunnel passivating layer of SiO_2 in between them. Polysilicon emitters were explored for bipolar transistors and BiCMOS devices for forming shallow junctions enabling upscaling of lateral device configurations [33]. In further enhancement of this architecture, contacts employing polysilicon structures such as n + poly-Si/insulator/p-Si, known as SIS emitters (semiconductor–insulator–semiconductor), were developed in 1978. A thin 2 nm layer of wet oxide was deposited between the crystalline and polycrystalline silicon layers, where quantum mechanical tunnelling enabled charge transport in these structures [34]. The idea of using polysilicon-based contacts for solar cells was proposed by Overstraeten in 1981 [28]. He suggested that a polysilicon layer could be placed beneath the metal grid to reduce surface recombination velocity under the front contacts. In 1985, Yablonovitch predicted that an ideal solar cell with a double heterojunction configuration could be created by placing a narrow bandgap active layer between two wide bandgap layers with opposite doping. He also noted that precise lattice matching with silicon isn't necessary for these layers to work effectively; instead, the electronic quality of the wide bandgap materials must support efficient quasi-Fermi level splitting in the active layer. Yablonovitch and his team investigated the use of SIPOS (semi-insulating polysilicon), a combination of microcrystalline silicon and SiO_x, for creating heterocontacts. They fabricated a double heterojunction with an n^+ SIPOS layer on p-type FZ silicon, achieving open-circuit voltages as high as 720 mV. This high voltage was attributed to the excellent surface passivation provided by the heterocontact, which reduced the reverse saturation current density to around 10 fA/cm^2 [35, 36]. Green, during his work with MIS contacts on solar cells, also explored using "degenerately doped polycrystalline silicon as the metal layer in MINP devices" and obtained encouraging open-circuit voltages of 660–665 mV [37]. Although polysilicon contact-based solar cells exhibited high open-circuit voltages, they couldn't provide higher short-circuit current density (J_{sc}) or fill factor (FF). Further research into polysilicon-based contacts for solar cells was limited and did not receive extensive exploration.

Polysilicon contacts did not make a significant return until 2013, when Feldmann et al. at Fraunhofer ISE demonstrated their high potential. They utilized a poly-Si/SiO_2 layer as a passivating rear contact on an n-type wafer with a boron-diffused emitter, achieving a high efficiency of 23% and an open-circuit voltage of 698 mV [38]. This publication coined the term "TOPCon"—Tunnel Oxide Passivated Contacts—and marked the beginning of a new device structure for solar cells that is now increasingly dominating the solar market. In this architecture (Fig. 7), n-type wafers are used, in contrast to the p-type wafers that were prevalent in all previous architectures. A tunnel oxide (SiO_2) with a thickness of 1–2 nm is grown at the rear side. A phosphorus-doped, silicon-rich a-SiC_x layer is deposited using plasma-enhanced chemical vapour deposition (PECVD) and then annealed at high temperatures of 800–900 °C to form a 20 nm n^+-doped polysilicon layer. This is followed by 400 °C Remote Plasma Hydrogen Passivation (RPHP) for 30 min, which enhances the layer's performance through passivation with atomic hydrogen. LPCVD-grown n^+ poly silicon can also be used. This highly doped poly-Si layer functions as a passivating carrier-selective layer, allowing the selective transport of electrons while blocking the passage of holes by inducing band bending in the crystalline silicon absorber. The emitter is formed by boron diffusion, with local p++ boron doping beneath the Ti/Pd/Ag grid contacts [38, 39]. The rear contact is formed using silver paste. This cell achieved a record efficiency of 25.8% for a front and back-contacted c-Si solar cell with an open-circuit voltage (V_{oc}) of 724 mV, short-circuit current density (J_{sc}) of 42.9 mA/cm^2 and fill factor (FF) of 83.1% [40]. The TOPCon configuration employs a similar light-trapping strategy as PERC solar cells. Random pyramids, along with a double-layer anti-reflection coating of SiN_x/MgF_2, are applied to reduce optical losses. A p^+ emitter, combined with a local p^{++} region beneath the metal contacts, is used to minimize recombination at the front contacts. Additionally, a thin layer of Al_2O_3 is applied to passivate the surface and decrease surface recombination velocity at the emitter.

One disadvantage of the PERC configuration was that its local rear contacts led to crowding of charge carriers at the contact points, resulting in a 3D carrier transport mechanism. In contrast, TOPCon contacts the entire rear surface with highly doped poly-Si through a tunnel oxide layer, enabling a simpler 1D carrier transport mechanism, enabling high fill factors (FF) [41]. This carrier transport mechanism enables high efficiency of the cell, independent of the wafer resistivity [39].

The tunnel oxide at the rear contact is essential for passivating the rear surface, which reduces the reverse saturation current (J_{0b}) at the base that results in higher open-circuit voltages (V_{oc}). Besides passivation, this layer also act as a diffusion barrier during processing, preventing the dopants from diffusing from poly-Si layer to c-Si bulk. The advantage of this layer is that it can be grown using straightforward methods, such as Nitric acid oxidation of silicon (NAOS), UV-O_3 oxidation, DI-O_3 oxidation, plasma oxidation or thermal diffusion [42, 43]. The application of TOPCon passivating contacts resulted in record-breaking minority carrier lifetimes on n-type FZ wafers, reaching 0.5 s, with a minority carrier diffusion length of 2.5 cm [44]. The thickness of the SiO_2 layer is a critical factor that influences carrier transport. As the tunnel oxide thickness increases, its passivation quality improves, but its resistivity rises. Therefore, it is essential to keep the thickness of the oxide between 1–2 nm to allow quantum mechanical tunnelling through this layer. In

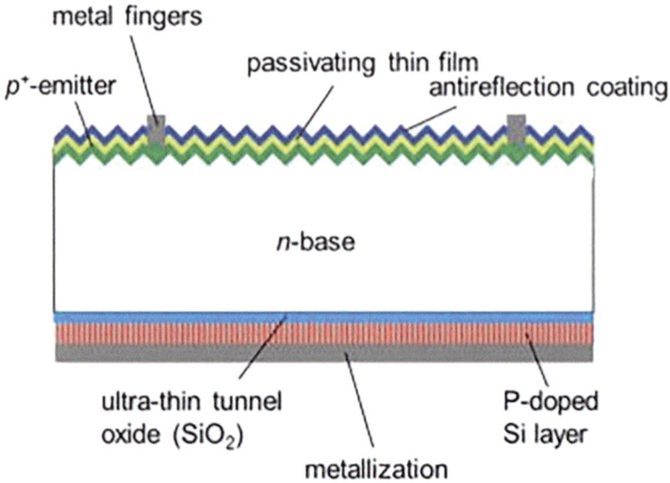

Fig. 7 Schematic of the TOPCon solar cell structure with the passivated rear contact. Reprinted from [Solar Energy Materials and Solar Cells, vol. 120, no. PART A, F. Feldmann, M. Bivour, C. Reichel, M. Hermle, and S. W. Glunz, "Passivated rear contacts for high-efficiency n-type Si solar cells providing high interface passivation quality and excellent transport characteristics" pp. 270–274, 2014, https://doi.org/10.1016/j.solmat.2013.09.017] with permission from Elsevier [38]

POLO (polycrystalline silicon on oxide) structures, thicker oxide layers (2–4 nm) are also used. In these cases, carrier transport is facilitated by direct current flow through pinholes in the SiO_2 [45]. The SiO_2 layer thickness dictates the type of carrier transport mechanism: in thick oxide layers, pinholes created during high-temperature post-anneal treatments (900–1050 °C) enable charge transport, whereas in thin oxides (1–2 nm), a high-temperature anneal step is unnecessary, and quantum mechanical tunnelling allows the charge transport [46].

The polysilicon layer can be coated directly through LPCVD (Low-Pressure Chemical Vapour Deposition) or by first depositing amorphous silicon (a-Si) and then annealing it at very high temperatures to convert it into polysilicon. LPCVD is a well-established industrial technique, widely recognized for its reliability and scalability in manufacturing processes. The advantage of LPCVD is that it produces high-quality, pinhole-free polysilicon layers with good thickness distribution at relatively lower deposition temperatures (600 °C).It can also include the option of in situ doping of polysilicon layers with uniform doping profile [47]. The disadvantage of this method is its low throughput, and it also creates a wrap-around layer on the sides and front of the cell, which needs to be removed by an additional chemical etching step. Tubular plasma-enhanced chemical vapour deposition (PECVD) is the favoured method for initially coating amorphous silicon, as it provides higher deposition rates, a more one-sided layer, and also enables in situ dopant diffusion [48]. The dopant deposition can be executed in situ, during the a-Si:H deposition or ex situ, after the a-Si:H deposition. A high-temperature anneal (800–900 °C) is required in such cases for dopant activation and conversion of deposited a-Si:H to poly-Si.

The SiO_2/poly-Si stack is particularly attractive from an industrial perspective because it is resistant to high temperatures (up to 800 °C) without affecting the passivation [38]. The shift to n-type wafers from p-type wafers was a positive development, as n-type wafers did not experience light-induced degradation (LID), a common problem in p-type silicon wafers caused by the formation of boron-oxygen (B-O) complexes. Although manufacturing lines for p-type substrate-based solar cells were well-established, the benefits of the TOPCon structure justified the transition. However, research on developing p-type TOPCon is ongoing. Passivating contacts can be formed on lightly doped p-type wafers, which helps mitigate the effects of light-induced degradation (LID). P-type TOPCon cells featuring rear passivated contact of SiOx and B-doped Si layer with n-type TOPCon contacts as the front emitter have demonstrated high open-circuit voltages (V_{oc}) of 694 mV and fill factors (FF) of 81%[41]. Another innovative configuration using the TOPCon concept was implemented with p-type wafers, known as Tunnel Oxide Passivated Contact with Rear Emitter (TOPCoRE). In all the configurations we've discussed so far, the p–n junction was located on the front side of the cell. In this configuration, the p–n junction is placed on the rear side of the cell. An n-type TOPCon contact deposited on p-type wafers creates the p–n junction at the rear. This configuration omits boron diffusion at the front, with boron diffusion occurring only just beneath the front contacts to reduce recombination at the grid contacts. Such cells are now referred to as back-junction (BJ) cells, as opposed to the Front Junction (FJ) cells we've discussed previously. The front side of the cell is passivated with Al_2O_3, and this configuration, denoted as p-BJ (p-type wafer with back junction), achieved remarkable efficiency of 26% on 1 Ω cm p-type wafers, surpassing conventional n-FJ cells (n-type, front junction). The cell recorded an impressive open-circuit voltage (V_{oc}) of 732.3 mV, a short-circuit current density (J_{sc}) of 42.05 mA/cm^2, and an excellent fill factor (FF) of 84.3% [49]. This configuration set a record for efficiency among front and back-contacted (FBC) silicon cells without a lateral conducting transport layer on the front side.

3.3.1 Fabrication of Commercial TOPCon Solar Cells

The TOPCon structure can be implemented in PERC manufacturing lines with only a few modifications, as both process flows involve metal contact firing, front surface passivation, and anti-reflection coating. When the TOPCon configuration transitioned to industrial manufacturing, it utilized a bifacial structure on n-type Cz wafers, incorporating a thermal oxide SiO_x layer and polysilicon deposited by LPCVD or PECVD. The industrial cell featured a boron-diffused emitter with Al_2O_3 passivation on the front side, and both sides were coated with an SiN_x:H anti-reflection layer, with screen-printed grid contacts applied on both the front and rear (Fig. 8).

The first transition of this technology to an industrial cell configuration was achieved by Stodolny and Lenes of ECN Solar Energy and Tempress Systems in the Netherlands. They fabricated a 20.72% efficient bifacial n-type TOPCon solar cell on 6-inch Cz-grown n-type wafers with a resistivity of 5 Ω cm. The cell featured

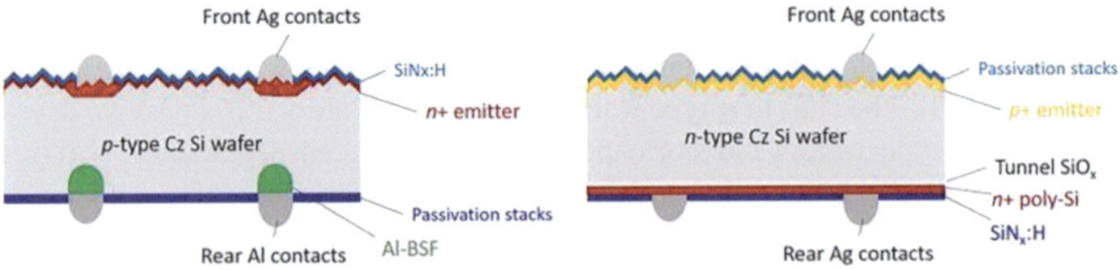

Fig. 8 Cross-sectional schematic of Passivated Emitter and Rear Cell (PERC) (left) and industrial-tunnel oxide passivated contact (i-TOPCon) (right). Reprinted from [Progress in Photovoltaics: Research and Applications, vol. 31, Y. Chen et al., "Technology evolution of the photovoltaic industry: Learning from history and recent progress", no. 12, pp. 1194–1204, Dec. 2023, https://doi.org/10.1002/pip.3626] with permission from Wiley Materials [50]

a typical TOPCon structure with LPCVD-grown polysilicon and an additional layer of SiNx:H on the rear side for anti-reflection and hydrogen passivation. Both sides of the cell were grid-contacted using screen-printed metal pastes. The emitter was passivated with ALD-grown Al_2O_3, and the technology was named Passivated Emitter Rear Polysilicon (PERPoly) [51]. Later, an efficiency improvement to 23% was achieved on M2 (244.3 cm^2) Cz n-type wafers with a resistivity of 1–5 Ω cm. This was accomplished by Nandakumar et al. [52] at SERIS, Singapore, using a SiO_x/poly-Si layer grown via inline PECVD (Fig. 9; Table 1).

The State Key Laboratory of Photovoltaic Science and Technology at Trina Solar, China, made significant advancements by developing industrial Tunnel Oxide Passivated Contacts (i-TOPCon) bifacial solar cells on M4 (161.7 mm) Cz n-type silicon

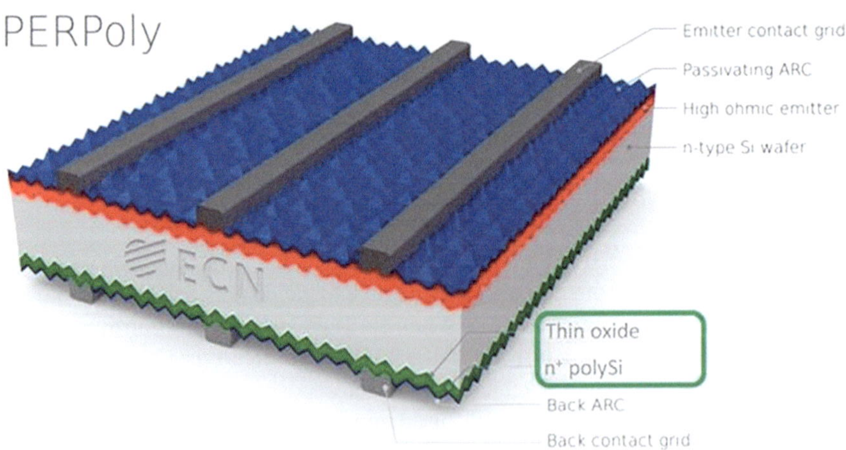

Fig. 9 Schematic drawing of the bifacial n-type solar cell design, featuring n-poly-Si/SiOx contacts, named Passivated Emitter and Rear Polysilicon (PERPoly). Reprinted from [Solar Energy Materials and Solar Cells, vol. 158, M. K. Stodolny et al., "n-Type polysilicon passivating contact for industrial bifacial n-type solar cells," pp. 24–28, Dec. 2016, https://doi.org/10.1016/j.solmat.2016.06.034] with permission from Elsevier [51]

Table 1 Process flow of PERC and i-TOPCon cells in Trina's manufacturing lines

	PERC	i-TOPCon
1	Saw-damage etch and texturing	Saw-damage etch and texturing
2	Phosphorus diffusion	Boron diffusion
3	Laser doping of n^{++}	Inline rear borosilicate glass (BSG) removal
4	Thermal oxidation	Thermal or plasma assisted tunnel oxide
5	Inline rear phosphosilicate glass (PSG) removal and cleaning	Poly-Si (i) by LPCVD or poly-Si (n) by PECVD
6	Thermal oxidation in tube	Phosphorus diffusion or annealing
7	Deposition of rear passivation	PSG removal and cleaning
8	Front SiN_x deposition in tube	Deposition of rear passivation
9	Rear laser ablation for BSF	Deposition of front passivation
10	Rear Ag and Al printing	Rear Ag printing
11	Front Ag printing	Front Ag printing
12	Firing	Firing
13	Hydrogenation	Hydrogenation
14	I–V measurement	I–V measurement

Reprinted from [Progress in Photovoltaics: Research and Applications, vol. 31, Y. Chen et al., "Technology evolution of the photovoltaic industry: Learning from history and recent progress", no. 12, pp. 1194–1204, Dec. 2023, https://doi.org/10.1002/pip.3626] with permission from Wiley Materials [50]

Abbreviations: BSF, back surface field; i-TOPCon, industrial-tunnel oxide passivated contact; PERC, Passivated Emitter and Rear Cell

wafers with resistivity ranging from 0.2 to 2 Ωcm. The i-TOPCon manufacturing line was developed from an existing Al-BSF manufacturing line with a few modifications, including the addition of an alkaline texturing wet bench and an LPCVD tool [53]. By 2018, Trina Solar had commenced mass production of n-type bifacial TOPCon modules. By 2022, Trina Solar achieved an efficiency of 25.42% on 210 mm wafers, with a batch average efficiency of 25.04% for 1000 pieces [50]. In the same year, Trina Solar took a significant step forward by launching a new 210 mm i-TOPCon cell factory with a capacity of 8 GW, marking a major expansion in their TOPCon production capabilities. By the end of 2022, Jolywood and Jinko Solar had achieved impressive efficiencies of 26.1% and 26.4%, respectively, on their TOPCon structures, marking the significant milestone of surpassing the 26% efficiency threshold in the TOPCon configuration [54]. As the push for higher efficiencies and cost-friendly mass production continued, several key challenges were addressed to further optimize the TOPCon structure. Despite being projected as a stack that could withstand high temperatures, the SiO_2/poly-Si stack showed some degradation during the metal contact firing process due to oxide breakage. To overcome this challenge, thicker polysilicon layers, around 200 nm, were used, but this solution came with trade-offs, including increased parasitic absorption of the IR spectrum at the rear end and higher manufacturing costs due to the additional material. Methods that enhanced the metal electrical contact without relying on such high temperatures were also explored.

One such method, Laser-Enhanced Contact Optimization (LECO), has garnered recent interest for this purpose. LECO is a non-destructive technique used to enhance the electrical contact between the metal layer and the underlying layer in underfired contacts, enabling the use of low-doped emitters. It is typically carried out using high intensity laser pulse on the contacts after the last firing process for screen-printed contacts in PERC cells, ensuring better contact and improved performance [55]. In 2022, Fellmeth et al. at Fraunhofer ISE explored the LECO technique in industrial TOPCon cells to reduce the contact firing temperature, which enabled lower contact resistance leading to higher open-circuit voltages (V_{oc}) and simultaneous improvements in fill factor (FF) values [56]. As efficiencies soared above 26%, recombination at the emitter and grid on the front surface became a dominant loss mechanism. To reduce recombination at the semiconductor/metal interface on the front side, highly doped regions just beneath the metal contacts showed promise [57]. This approach, known as the selective emitter, was a proven mechanism of high-low doping to lower majority carrier concentration at the contacts. It had already been employed in PERC cells, where n+ doped regions beneath metal contacts enabled very low emitter reverse saturation current density (J_{0e}). However, the drawback of this method was the number of photolithographic steps involved. A new technique, Laser-Doped Selective Emitter (LDSE), was developed, where a laser is used to diffuse in deposited boron dopant to create p+ regions on the emitter, effectively reducing recombination without the need for complex photolithography [58]. Laser-Doped Selective Emitter (LDSE) and Laser-Enhanced Contact Optimization (LECO) approaches have significantly contributed to pushing industrial TOPCon cell efficiencies beyond 25%. For instance, solar cells exceeding 25% efficiency on 210 mm wafers, manufactured by Trina Solar, incorporated PECVD-grown poly-Si layers and an industrial laser-doped selective boron emitter, which further boosted the efficiency by 0.2% absolute [59]. By 2023, Jolywood and Jinko Solar achieved remarkable efficiencies of 26.7% and 26.89%, respectively, on 330 cm^2 cells [60].

3.4 Polycrystalline Silicon on Oxide (POLO) Solar Cells

Polycrystalline silicon on oxide (POLO) contacts are another version of the TOPCon contact. This configuration uses a similar passivating stack of doped polycrystalline silicon layer with a SiO_x layer on a crystalline silicon (c-Si) wafer. The development of these POLO contacts occurred in parallel at the Institute for Solar Energy Research in Hamelin (ISFH) while Fraunhofer ISE was developing TOPCon cells. POLO contacts were developed following the work of Gan and Swanson on poly-Si/SiOx/c-Si structures, which achieved very low emitter reverse saturation current densities (J_{0e}) of less than 20 fA/cm^2 and contact resistances of 0.1 m$\Omega \cdot cm^2$ [61]. In 2014, Römer et al. began studying n^+ and p^+ POLO contacts with SiO_2 oxide thicknesses ranging from 1.4 to 3.6 nm and achieved J_0 values as low as 5 fA/cm^2. The

ISFH team conducted extensive studies on polysilicon-based contacts for both polarities. In 2015, Römer et al. investigated poly-Si/c-Si junctions with ion-implanted poly-Si layers. The n-doped poly-Si junction and p-doped poly-Si junction achieved excellent surface passivation values, with J_0 of 1 fA/cm^2 and 4.4 fA/cm^2, respectively [62]. This sparked research into the carrier transport mechanism across the junction, as tunnelling alone couldn't explain the very low recombination current and junction resistance observed in p^+ poly-Si/c-Si junctions. In the case of n^+ poly-Si/SiO_x/c-Si interfaces, the highly doped poly-Si induces band bending in the c-Si, leading to a band bending at the c-Si/SiO_x interface, which enables field-effect passivation. Additionally, the tunnelling model in n^+ poly-Si/c-Si structures so far utilized the difference in tunnelling barrier heights. With a conduction band offset of 3.2 eV and a valence band offset of 4.7 eV between Si and SiO_2, a larger barrier is created for holes, while enabling electron transport. This explains the low junction recombination current and low junction resistance at these junctions. In contrast, for p^+ poly-Si/SiO_x/c-Si junctions, larger recombination currents and junction resistances are expected given the band offsets. However, the observed values were surprisingly low, with a J_0 of 4.4 fA/cm^2. This can't be explained by the quantum mechanical tunnelling model. To address this, Peibst et al. proposed a model based on current flow through pinholes that form during thermal annealing of the junctions, which was previously utilized by BJT community in poly-Si emitters [63]. This theory was supported by experimental evidence from TEM images, which revealed pinholes, oxide breakage, and local thickness reductions in the oxide layer. The mechanism was then explained using a model that described current flow between c-Si and poly-Si through regions with thinner oxide and pinholes. The model also considered doped c-Si regions formed by dopants diffusing from the poly-Si layer through these openings. Overall, the mechanism results from reduced minority carrier recombination in areas with pinholes or reduced oxide thickness [63] The development of POLO-IBC cells primarily focused on p-type wafers, as it was believed that leveraging the existing industry infrastructure for p-PERC mass production would make it economically viable to develop cell structures that align with this setup (Fig. 10).

With the highest efficiencies reported in back-contact structures and Kaneka's SHJ-IBC achieving 26.67% record efficiency on back-contacted cells, ISFH quickly began implementing both polarities of selective POLO stacks in IBC structures. In 2017, Haase et al. fabricated an IBC solar cell with POLO contacts of both n^+ and p^+ polarities [65]. The cell, fabricated on a FZ p-type wafer with a resistivity of 1.3 Ωcm, featured a 2.2 nm thermally grown SiO_x layer followed by an LPCVD-grown poly-Si layer. Patterned photoresist was used to selectively implant boron, with the rest of the rear was covered in photoresist. The process was then repeated for phosphorus implantation, while the boron-implanted regions remained covered. The regions where n^+ and p^+ directly contacted were removed using KOH-based etching, as they served as high recombination centres. To passivate the region between the two contacts, a 20 nm Al_2O_3 layer was deposited. A PECVD SiO_x layer was then used to passivate the entire rear region. Aluminium contacts were patterned onto each polarity contact. The emitter was created by ion implanting phosphorus, while the front side featured a classic textured surface for light trapping, along with an Al_2O_3/

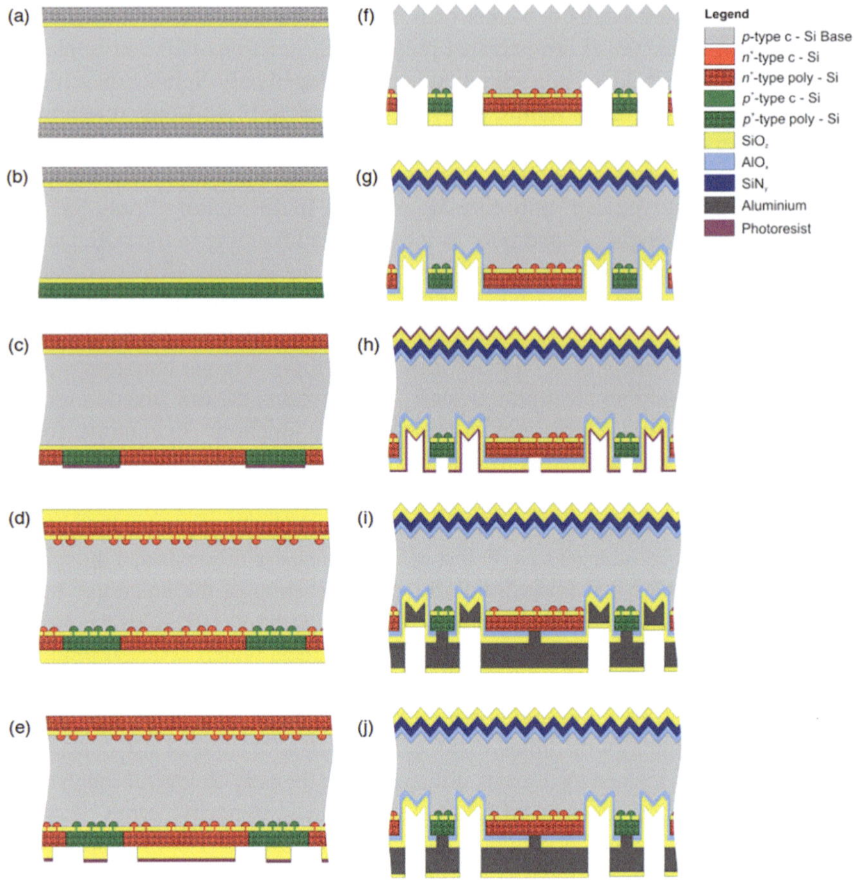

Fig. 10 Process sequence of POLO-IBC solar cells. Reprinted with permission from [Jpn J Appl Phys, vol. 56, F. Haase et al., "Interdigitated back contact solar cells with polycrystalline silicon on oxide passivating contacts for both polarities," no. 8S2, p. 08MB15, Aug. 2017, https://doi.org/10.7567/JJAP.56.08MB15] © The Japan Society of Applied Physics. Reproduced with permission from IOP Publishing limited. All Rights Reserved [64]

SiN_x/SiO_x stack for passivation and anti-reflection. This cell achieved an efficiency of 25% for designated area of 3.97 cm^2 with a remarkable open-circuit voltage (V_{oc}) of 723 mV. Further enhancement in efficiency was anticipated, as significant recombination was identified at the Al_2O_3 passivated regions on the front side and at the rear between the two contacts. Reducing recombination in these areas was seen as a key to reach higher-efficiency levels [65].

Haase et al. modified the fabrication process of the POLO-IBC cell by introducing an additional thermally grown SiO_x layer between the aluminium contact and poly-Si to function as a reflector, enhancing the cell's rear reflection. For metallization, a non-fire-through paste was utilized, necessitating the manual creation of openings in the SiO_x reflector layer to establish Al/Poly-Si contacts [45]. The openings were created using Layer Selective Laser Ablation (LASA) in a damage-free manner, utilizing a pulsed UV laser (Nd:YO4) to form openings in the SiO_x reflector layer, allowing contact between the aluminium and doped poly-Si without damaging the underlying poly-Si. The ion implant barriers were designed to maintain a 30 μm separation between the n-type and p-type contacts, resulting in an intrinsic poly-Si layer between the p and n contacts, thus creating a poly-Si p-i-n junction at the rear. The fabrication process flow was named the $POLO^2$-IBC process. The n-POLO and p-POLO contacts of this cell achieved exceptionally low reverse saturation current densities of 2 fA/cm^2 and 6 fA/cm^2, respectively. With a designated area of 3.986 cm^2, the cell reached a record efficiency of 26.1% with p -type c-Si, featuring an excellent open-circuit voltage (V_{oc}) of 726.6 mV, a short-circuit current density (J_{sc}) of 42.62 mA/cm^2, and a fill factor (FF) of 84.28%. A detailed analysis of the losses in this configuration revealed that reducing the series resistance and improving surface passivation could potentially yield efficiencies of up to 28% on 1.3 Ωcm p type wafers [66].

3.4.1 Fabrication of Industrial POLO Structures

Once the record lab efficiency for IBC POLO structures was achieved on p-type wafers, the next objective was to efficiently transfer these passivating contacts to industrially viable structures. As previously mentioned, the developers of the POLO structure concentrated on achieving high efficiencies on p-type wafers due to the existing PERC technology on manufacturing lines. They aimed to align the POLO-IBC processes with the current equipment setup to facilitate easier integration into manufacturing.

The "$POLO^2$-IBC" structure refers to IBC cells where both polarities are formed using poly-Si contacts. Alternatively, the "POLO-IBC" cell design was introduced, where the n-type contact remains as an n^+ doped poly-Si contact, while the p^+ contact is achieved using an Al-BSF layer. This approach shortens the manufacturing process and makes it more similar to the PERC process flow. The manufacturing of POLO-IBC differs from that of $POLO^2$-IBC in several ways.

The industrial POLO-IBC cells were realized on KOH textured Cz wafers with a resistivity of 1.5 Ωcm. The SiO_2 passivating layer was grown through DI-O_3 oxidation instead of thermal oxidation. In $POLO^2$-IBC, the deposition of poly-Si was done using LPCVD of a-Si, followed by high-temperature annealing and doping via ion implantation through patterned photoresists. In contrast, POLO-IBC structures employ in situ doped poly-Si layers that are laser-patterned. For $POLO^2$-IBC, contacts were formed using aluminium through laser-patterned SiO_2 layers. In the POLO-IBC process, Al contacts for p-type and Ag contacts for n-type were screen-printed and co-fired. AlO_x and SiN_y layers were coated on both the front and rear sides for passivation. Initially, industrial type POLO-IBC prototypes were fabricated as 2 × 2 cm cells on M2-sized wafer which achieved efficiencies of 21.8% in the proof-of-concept batch [67]. The screen-printing of aluminium contacts led to

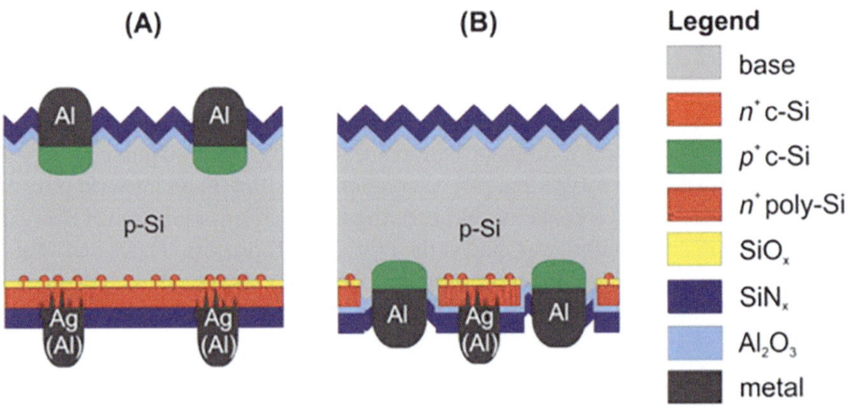

Fig. 11 Schematic representation of (A) the POLO back-junction (BJ) cell and (B) the POLO interdigitated back-contact (IBC) cell. Reprinted from [Progress in Photovoltaics: Research and Applications, vol. 31, no. 4,R. Peibst et al., "On the chances and challenges of combining electron-collecting nPOLO and hole-collecting Al-p + contacts in highly efficient p-type c-Si solar cells," pp. 327–340, Apr. 2023, https://doi.org/10.1002/pip.3545] with permission from Wiley Materials [69]

high recombination at the aluminium contact sites, with reverse saturation current density reaching up to 2250 fA/cm^2. By optimizing the contact fraction of Al-p$^+$ contacts, the efficiency was further improved to 23%, achieving a good open-circuit voltage of 716 mV [68]. By additionally optimizing the contact widths for the Al-p+ contacts and increasing the thickness, the reverse saturation current density was significantly lowered to 600 fA/cm^2, marking a substantial improvement. This reduction contributed to an enhanced cell efficiency of 23.7% for a designated area of 4 cm^2, highlighting the cumulative impact of these advancements on overall performance [69] (Fig. 11).

The POLO contacts were explored in a top-and-bottom-contacted cell configuration on a p-type wafer, similar to the TOPCon p-BJ (p-type wafer, back junction) structure that achieved a record efficiency of 26% [49]. This structure utilized POLO contacts for both p and n polarities at the top and bottom contacts of the cell. The POLO-BJ cell was fabricated on a p-type Cz wafer with a resistivity of 1 Ωcm. The p+ contact is achieved using the Al-p$^+$ technique, while the n$^+$ contact is formed with n-doped polysilicon. The junction formation occurs at the rear of the cell, classifying it as a back-junction cell (POLO-BJ). Unlike the TOPCon p-BJ, the POLO-BJ is bifacial and features grid metallization on both sides. A passivation stack of AlO_x/SiN_y is applied to passivate both the front and rear surfaces. Aluminium paste is screen-printed on the front side, and silver paste is screen-printed on the rear side, followed by co-firing. The POLO-BJ cell has achieved a highest efficiency of 22.9% with a notable open-circuit voltage (V_{oc}) of 714 mV, short-circuit current density (J_{sc}) of 39.6 mA/cm^2 and fill factor (FF) of 80.9% on a large area of 244 cm^2 [70]. This structure achieved higher efficiency in 2023 by employing a superior quality AlO_x/SiN_y passivation layer deposited using an industrial LF-PECVD tool. Even when the costly RCA clean was replaced with a more economical HCl, HF, O_3 cleaning process, the passivation quality of this stack remained impressive, achieving a low reverse saturation current density of 4.3 fA/cm^2 on textured surfaces—comparable to that of the more expensive ALD AlO_x/PECVD SiN_y stack. An efficiency of 24.2% was achieved on p-type Cz textured M2-sized, Ga-doped wafers, with an open-circuit voltage (V_{oc}) of 725 mV, a short-circuit current density (J_{sc}) of 40.2 mA/cm^2, and a fill factor (FF) of 83% [71].

The POLO-IBC and POLO-BJ structures were developed on p-type Cz substrates with minimal deviation from the PERC process flow to facilitate smooth integration into existing industrial settings. These innovative POLO structures have demonstrated promising efficiencies on industrial prototype cells and are anticipated to make a significant impact in the commercial market soon, thanks to their excellent cell parameters and compatibility with current manufacturing technologies.

3.5 Silicon Heterojunction Solar Cells (SHJ)

The strategy of separating the metal contact from the silicon surface is employed in many solar cell architectures. In PERC, TOPCon, and POLO designs, a thin SiO_2 layer provides this "decoupling," effectively passivating the silicon surface by reducing dangling bonds and minimizing interface recombination. In contrast, the SHJ architecture, developed by Sanyo in Japan in the 1980s, utilizes a heterojunction between hydrogenated amorphous silicon and crystalline silicon (a-Si:H/c-Si) for effective passivation of the c-Si surface. One requirement for effective passivation of the c-Si/a-Si:H interface is the presence of an atomically sharp interface without any epitaxial growth. a-Si:H was an excellent choice for passivating the c-Si surface because it shares the same short-range order as c-Si and forms covalent bonds with it. Additionally, any remaining dangling bonds are effectively passivated by atomic hydrogen species released from the hydrogenated a-Si (a-Si:H) [72]. This high degree of passivation in SHJ solar cells allows V_{oc} values to surpass 700 mV with ease, offering better V_{oc} values than all other solar cell configurations.

The early concept of a-Si/c-Si junctions were explored by Matsurra et al. in the 1980s, but their application did not extend to that of solar cells [73]. In 1992, Tanaka et al. from the Functional Materials Research Centre at Sanyo Electric (now Panasonic), Japan, published their work on a solar cell employing a p-type a-Si/intrinsic a-Si/n-type c-Si/n-type a-Si junction with a transparent conducting oxide layer (TCO) at the emitter, achieving a power conversion efficiency (PCE) of 18.1% for the first time using processes conducted at temperatures below 200 °C. They introduced the term "Artificially Constructed Junction-Heterojunction with Intrinsic Thin Layer (ACJ-HIT®)" for this innovative structure [74]. The inclusion of the thin (5–10 nm) intrinsic a-Si layer on the front side enhanced surface passivation and lowered the reverse saturation current density, resulting in an improved open- circuit voltage (V_{oc}) that was 30 mV higher compared to a junction without the intrinsic a-Si layer. This structure was developed by modifying earlier designs that lacked both the intrinsic a-Si layer and the BSF n-type a-Si layer. Further enhancements to this structure were achieved by introducing a HIT configuration at the rear side, inserting a thin intrinsic a-Si:H layer between the c-Si and n-type a-Si:H layers at the rear. This enhanced structure achieved an efficiency of 20% for a 1 cm^2 cell fabricated on a Cz n-type wafer in 1994 [75]. Large-area cells of 100 cm^2 with this configuration achieved an efficiency of 17.3%, leading Sanyo to start commercial production of HIT modules in 1997. This efficiency further improved to an impressive 20.1% with open-circuit voltage (V_{oc}) over 700 mV by the year 2000 [16, 76]. These modules, named HIT Power 21™, offered a module efficiency of 15.2% and an electrical output of 180 W [76]. With a symmetrical structure on front and rear sides, bifacial structure (Fig. 12) became the new standard for HIT cells, featuring intrinsic a-Si for passivation on both sides, doped a-Si of each polarity on either side, a TCO layer on both sides, and screen-printed metal grids on both surfaces. In the same year, a new record efficiency of 20.7% was achieved by Sanyo for a large area of 100.5 cm^2 using the bifacial HIT structure, with an impressive open-circuit voltage (V_{oc}) of 719 mV [77].

The fabrication process of HIT solar cells is relatively straightforward. N-type Si wafers are cleaned and textured, followed by the deposition of an intrinsic amorphous silicon layer (i a-Si:H) and doped a-Si:H layers via plasma-enhanced chemical vapour deposition (PECVD) at low temperatures (below 200 °C). The i a-Si:H layer passivates the c-Si surface, while the doped a-Si:H layers provide carrier selectivity. The stack of intrinsic a-Si:H and doped a-Si:H is approximately 10 nm thick. This is followed by the sputtering of transparent conducting oxides (TCO), such as ITO, to a thickness of 70–80 nm to enable lateral conductivity while providing good transparency. Finally, silver paste is screen-printed on both sides to form a grid contact structure.

The HIT structures demonstrated remarkable open-circuit voltages, primarily due to the excellent passivation at the a-Si:H/c-Si interface. Sanyo quickly recognized

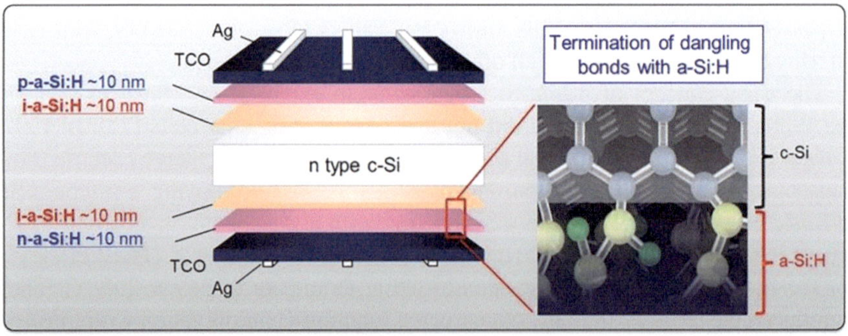

Fig. 12 Schematic representation of HIT® solar cell. Reprinted with permission from [ECS Journal of Solid State Science and Technology, vol. 10, no. 2, M. Taguchi, "Review—Development History of High Efficiency Silicon Heterojunction Solar Cell: From Discovery to Practical Use," p. 025002, Feb. 2021, https://doi.org/10.1149/2162-8777/abdfb6] [72]

the potential of the bifacial configuration and began producing bifacial HIT cells, branded as HIT Power Double™. These bifacial modules generated 8.1% more power than the HIT Power 21™ modules [76]. Improvements in contact grid geometry and the conductivity of the TCO enabled further increases in efficiency. By 2009, a 23% efficiency cell was achieved on wafers with a thickness of 200 μm. The focus then shifted to achieving high performance on thin wafers (less than 100 μm). Thin wafers were pursued not only for cost reduction but also to enhance V_{oc} by collecting photogenerated carriers with minimal loss. Continuous efforts in this direction led to a high V_{oc} of 750 mV and a PCE of 24.7% for cells of area 101.8 cm^2 on 98 μm thick wafers in 2014 [78]. It is worthwhile to note that throughout this efficiency climb, the basic structure of the HIT solar cell remained unchanged. Improvements were made in surface cleaning, the deposition process of a-Si:H to enhance its quality, grid structure optimization, and increased TCO conductivity to achieve higher efficiencies. Unlike other solar cell structures, which were developed in institutional laboratories and later transferred to industry for conversion into industrial designs, SHJ-HIT solar cells evolved directly within the industry on Cz wafers of area 100 cm^2. They quickly attained a practical structure, bypassing the need for extensive efforts in industrializing or converting lab-based designs into scalable industrial structures.

The shadowing loss from front contacts posed a significant challenge to attain efficiency beyond 25%. The next approach involved experimenting with back-contacted architectures, which eliminate this shadowing loss to achieve higher J_{sc}. The interdigitated back-contacted heterojunction solar cell (IBC-SHJ) by Sanyo (Panasonic) was fabricated on n-type, 150 μm thick Cz wafers with an area of 154 cm^2. A SiN_x passivation layer was applied to the front side, which featured a textured surface for enhanced light trapping. The rear side contained both polarity contacts, where the n-contact utilized an (i a-Si:H/ n a-Si:H) stack and the p-contact used an (i a-Si:H/ p a-Si:H) stack. The removal of the front-side a-Si:H layer improved the EQE by eliminating parasitic absorption from the a-Si:H layer, and the removal of the front grid eliminated optical shadowing losses, resulting in a 2.3 mA/cm^2 increment in the J_{sc} value. In 2014, this configuration led to a 25.6% efficient cell with a V_{oc} of 750 mV, a J_{sc} of 41.8 mA/cm^2, and a fill factor (FF) of 83.2% [79].

Kaneka, Japan, in 2015, adopted a different approach to surpass the 25% efficiency mark. In addition to introducing electroplated Cu metallization scheme to reduce costs, they analysed the minority carrier lifetime in front- and back-contacted SHJ solar cells and discovered that the a-Si:H/c-Si interface contained recombination centres, which contributed to efficiency losses by reduction in minority carrier lifetime. By optimizing the fabrication process to improve the interface and reduce the losses caused by lower minority carrier lifetime, they achieved a front and back-contacted SHJ cell on a 151.9 cm^2 Cz n-type wafer, with an efficiency of 25.1%, V_{oc} of 738 mV, J_{sc} of 40.8 mA/cm^2, and fill factor (FF) of 83.5% [80] (Fig. 13).

Kaneka switched to the fabrication of back-contacted architectures and broke the efficiency record for silicon solar cells twice by surpassing the 26% efficiency mark by 2017. A 26.3% efficiency was achieved on a 165 μm wafer with a V_{oc} of 744 mV, J_{sc} of 42.3 mA/cm^2, and FF of 83.8%, followed by the 26.67% efficient cell fabricated on n-type Cz wafers with a thickness of 200 μm with designated area

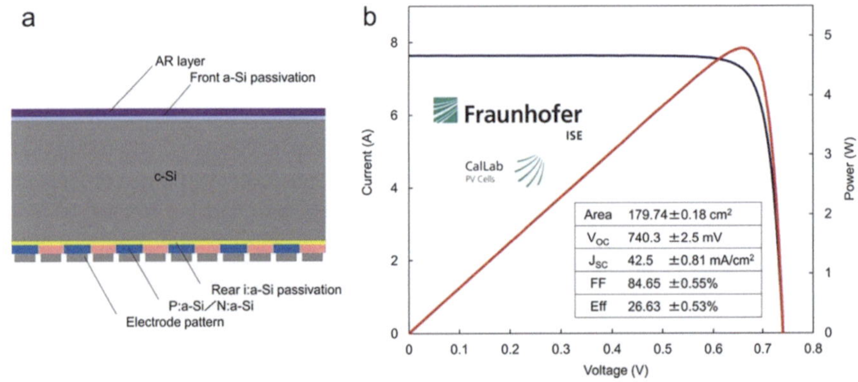

Fig. 13 Schematic representation of the record efficiency HJ-IBC Solar cell by Kaneka. Reprinted with permission from [Solar Energy Materials and Solar Cells, vol. 173,K. Yoshikawa et al., "Exceeding conversion efficiency of 26% by heterojunction interdigitated back contact solar cell with thin film Si technology", pp. 37–42, Dec. 2017, https://doi.org/10.1016/J.SOLMAT.2017.06.024] [81]

of 180 cm^2 attaining V_{oc} of 740 mV, J_{sc} of 42.5 mA/cm^2, and FF of 84.6%. The cell structure is simple, with a textured front side coated with a layer of intrinsic a-Si:H for passivation, followed by an anti-reflection coating (ARC). The rear side contains both polarity contacts, with the n-contact stack consisting of intrinsic a-Si:H and n-type a-Si:H, while the p-contact is made from an intrinsic a-Si:H and p-type a-Si:H stack, along with patterned electrodes. This marked a milestone achievement as the first cell to surpass the 26% efficiency mark [81, 82].

Hanergy in China developed a front and back-contacted SHJ solar cell with several innovative approaches in 2019. One notable feature of this design was the utilization of a rear junction, where the p–n junction was located at the rear of the cell instead of the conventional front-side configuration [83]. We have previously discussed similar structures in TOPCon solar cells. In this setup, the p-type contact stack was deposited on the rear side of the n-type wafer, while the n-type majority carrier contact was placed on the front side. This configuration offers several advantages: it allows the use of less conductive TCOs on the front side, as the majority carrier transport occurs within the wafer. As a result, the focus could shift more towards optimizing the optical properties and transparency of the TCO, enhancing light collection [84]. Additionally, optimizing the n-type a-Si:H contact is less complex than the p-type a-Si:H contact, and placing it at the emitter reduces power losses caused by 2D transport at the emitter [85]. The second feature was the introduction of a 0.5–1 nm thick a-Si:H layer with a high hydrogen content and a large microstructure factor, which was deposited between the n-type c-Si and the usual intrinsic a-Si:H layers using RF-PECVD. This layer, referred to as the buffer layer, was designed to improve the passivation at the interface. The third feature was the use of n-a-Si:H deposited with μc-SiOx:H as the majority carrier contact at the emitter. Alloying μc-Si:H with oxygen increases the bandgap, which in turn reduces the extinction coefficient (k) of the layer, thereby minimizing parasitic absorption losses at the emitter [84]. This cell, fabricated on an n-type Cz wafer with an area of 244.5 cm^2 (M2 wafer), achieved an efficiency of 25.11% with a V_{oc} of 747 mV, J_{sc} of 39.5 mA/cm^2 and a fill factor (FF) of 84.98%.

This cell structure highlights the ongoing research and efforts to replace doped a-Si:H in the SHJ structure with doped nanocrystalline silicon (nc-Si:H) for improved passivation and carrier selectivity. nc-Si:H is formed by embedding nanocrystallites within an a-Si:H matrix. With better crystallinity, it is expected to exhibit enhanced conductivity. Additionally, its higher bandgap of 2 eV, compared to a-Si's 1.7 eV, helps reduce parasitic absorption in the UV region, leading to an improved J_{sc}. nc-Si:H also demonstrates higher doping efficiency than a-Si:H. Typically, doped nc-Si:H is deposited through RF-PECVD process on a thin 5 nm of intrinsic a-Si:H layer [86]. The challenge with creating p-type nc-Si:H lies in the fact that boron hinders crystalline growth, making it more difficult to develop p-type nc-Si:H. Despite these challenges, various pre-deposition treatments and fine-tuning of deposition conditions have proven effective in obtaining p-type nc-Si:H films. To reduce constraints on this already complex boron-doped nc-Si:H, it is simpler to place the p-type contact at the rear, forming a rear junction SHJ, as discussed earlier (Fig. 14).

The next major breakthrough came when LONGi announced high-efficiency SHJ cells utilizing nanocrystalline hole-selective layers. In 2021 and 2022, they developed new Front and back-contacted (FBC) cell structures achieving efficiencies of 26.3%, 26.74%, and 26.81% by incorporating nanocrystalline-silicon layers and further optimizing the TCO layers and rear reflective stack for enhanced optical performance [87]. These cells were fabricated in a rear junction configuration, with the p-type contact stack deposited at the bottom. n-type Cz wafers, with an area of 274 cm^2 (M6 wafer) and a thickness of 130 μm, were used to produce high-efficiency cells. RF-PECVD (13.6 MHz) was employed to deposit the intrinsic a-Si:H layer, while VHF (40.68 MHz) PECVD was used for the deposition of front-side n-type nc-SiO_x:H and rear-side p-type nc-Si:H layers. To prevent epitaxial growth at the c-Si/a-Si:H interface, a thin layer of wet oxide was applied through chemical methods. Pre-deposition treatment utilizing CO_2 plasma was performed before the deposition of p-type nc-Si:H at the rear. Improved TCO layers were grown using a reactive plasma deposition technique, rather than sputtering. A 1% CeO_2 doped In_2O_3 layer was used as the TCO on the front emitter side, while the conventional ITO (10% SnO_2 in In_2O_3) was applied at the rear. Silver electrodes were screen printed, and a 150 nm layer of MgF_2 was used as an anti-reflective coating (ARC) on the front side, similar to PERC high-efficiency designs. For the rear reflection, an MgF_2/Ag stack was applied, contributing to the champion cell efficiency of 26.81% in a monofacial cell configuration [87]. This cell holds the record for the highest efficiency SHJ cell with a front and back-contacted configuration with V_{oc} of 751.4 mV, J_{sc} of 41.45 mA/cm^2 and a fill factor (FF) of 86.07% (Fig. 15).

The development of SHJ cells has primarily focused on n-type wafers due to their superior minority carrier lifetimes, which result from lower SRH recombination rates. This advantage allows n-type wafers to overcome the challenges faced by p-type

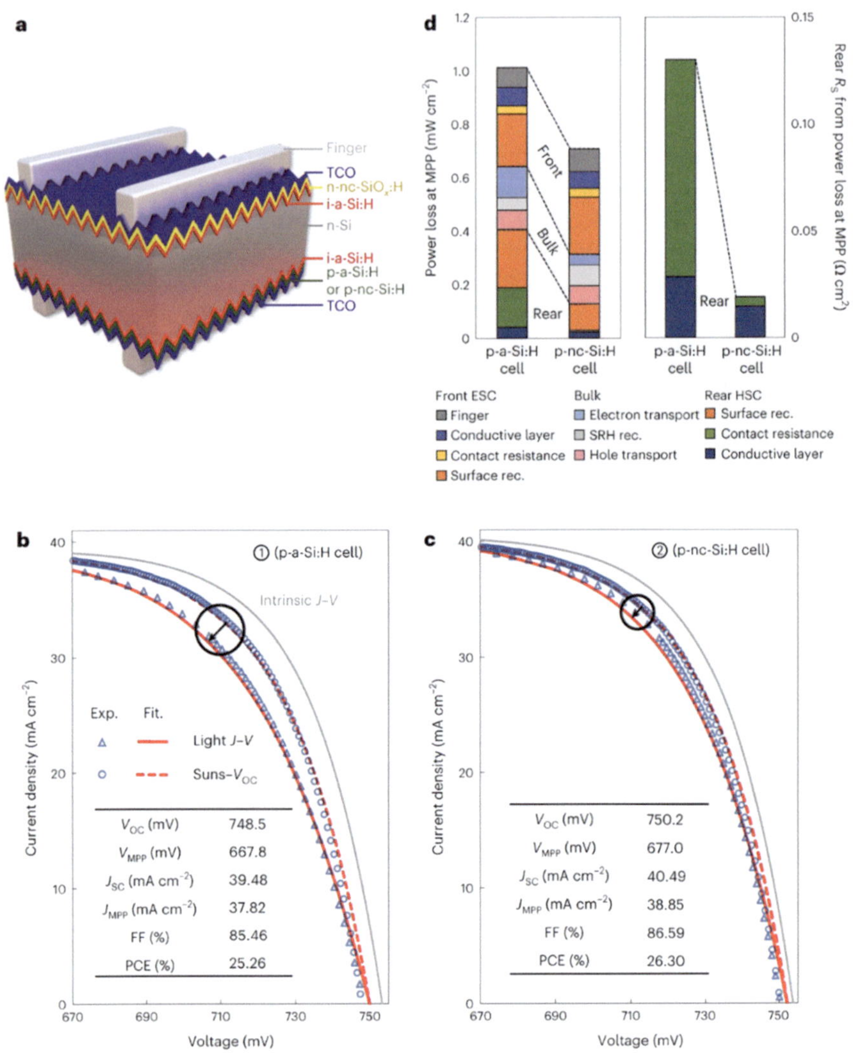

Fig. 14 Schematic representation of SHJ solar cells employing nanocrystalline-silicon (nc-Si:H) layers along with a-Si:H layers. Reprinted with permission from [Nat Energy, vol. 8, no. 8,H. Lin et al., "Silicon heterojunction solar cells with up to 26.81% efficiency achieved by electrically optimized nanocrystalline-silicon hole contact layers," pp. 789–799, Aug. 2023, https://doi.org/10.1038/s41560-023-01255-2] [87]

wafers, particularly light-induced degradation caused by the formation of boron-oxygen complexes (B-O) from boron doping. New architectures, such as TOPCon, have also been developed using n-type wafers for these reasons. Nevertheless, there have been advancements in SHJ cells using p-type wafers, although their efficiencies

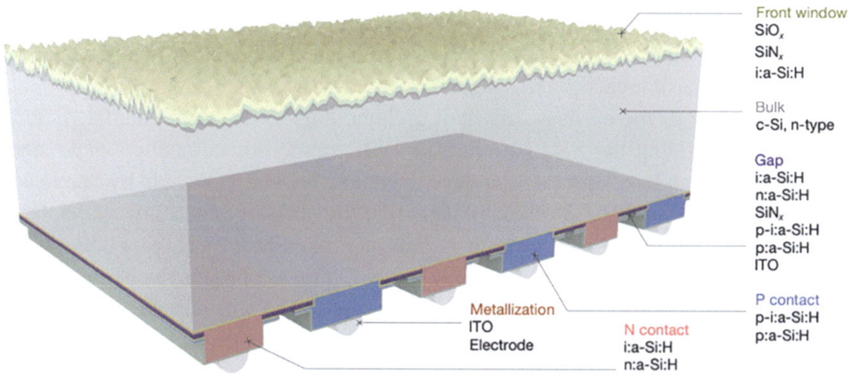

Fig. 15 Schematic diagram of the highest efficiency (27.3%) cell by LONGi. Reprinted with permission from [Nature, H. Wu et al., "Silicon heterojunction back-contact solar cells by laser patterning", Oct. 2024, https://doi.org/10.1038/s41586-024-08110-8] with permission from Springer Nature Limited [92]

tend to lag behind n-type counterparts by at least 0.5% [88]. High-intensity illuminated annealing, also known as light soaking, of cells fabricated on boron-doped wafers has proven effective in reducing the light-induced degradation (LID) caused by the formation of boron-oxygen (B-O) complexes in the cells. Recently, with the mass production of gallium-doped Cz wafers, further research has been conducted on this structure. In 2022, SHJ cell fabricated on a lightly doped p-type wafer achieved an efficiency of 24.47% [89]. In April 2024, LONGi demonstrated a 26.6% efficiency for a front and back-contacted, front junction SHJ cell on gallium-doped p-type M6 Cz wafers with a thickness of 150 μm. To enhance carrier lifetime and suppress SRH recombination losses, LONGi employed effective gettering through phosphorus diffusion gettering (PDG) to minimize metallic impurities in the p-type wafers. The cell configuration featured intrinsic a-Si:H passivation layers on both sides, with a front junction formed from n-type nc-SiOx:H and a p-contact made from p-doped nc-Si:H. The front TCO used cerium-doped indium oxide, similar to the n-type champion cell. These improvements resulted in a 26.6% efficiency on a 274.1 cm^2 wafer, with a V_{oc} of 751.3 mV, J_{sc} of 41.3 mA/cm^2, and a fill factor (FF) of 85.6% [90].

LONGi achieved record-breaking efficiencies for Heterojunction Back-Contacted (HBC) solar cells, reaching 27.09% in January and 27.3% in May 2024 [91]. The world record efficiency of 27.3% was achieved on a designated area of 243 cm^2 using cells fabricated on 175-μm-thick textured n-type wafers with a resistivity of 1.3–1.8 Ωcm. These advancements were made possible by employing dense a-Si:H passivating contacts with reduced hydrogen content, which enhanced conductivity and minimized porosity. The manufacturing process incorporated an industrially viable laser patterning technique using pulsed green (532 nm) and UV (355 nm) picosecond laser, carefully designed to prevent laser-induced damage. Additionally, modifications to the wet chemical process effectively mitigated undercut issues between the c-Si and SiN_x mask layers [92].

SHJ was one of the earliest solar cell architectures to be commercialized. With Sanyo industrializing the SHJ structure in 1997, numerous companies have since adopted and expanded its production once the patent for HIT cells held by Sanyo and Panasonic expired in 2010. Notable players like Meyer Burger in Switzerland, REC Solar headquartered in Norway which is currently taken over by Reliance Industries Ltd, Gold Stone Solar in China, and Huasun in China are now actively engaged in the mass production of SHJ modules, contributing to the widespread adoption and ongoing innovation of this architecture [93]. With its advantages of a simple structure, enhanced passivation leading to high V_{oc} values, low-temperature processing below 200 °C, and easy bifaciality, this architecture quickly evolved within the industry. It has now reached record efficiencies beyond 27%. The recent incorporation of nanocrystalline-silicon-based contacts has further boosted the potential of SHJ cells by enhancing conductivity and reducing contact resistance. Future cost reductions could be achieved by focusing on Cu metallization techniques, developing indium-free TCOs, as seen in some of the champion cells, and exploring the use of lower-cost p-type wafer substrates. Additionally, SHJ architectures are being actively researched as bottom cells for perovskite-silicon tandem structures, opening new possibilities for efficiency improvements.

3.6 Dopant-Free Asymmetric Silicon Heterocontacted (DASH) Solar Cells

Dopant-Free Asymmetric Heterocontacted (DASH) Solar Cells utilize the concept of carrier selectivity to separate photogenerated electron–hole pairs in the absorber layer, owing to which this configuration is often referred to as "Carrier-Selective Solar Cells" (CSCs). This concept was already applied in SHJ solar cells, where heterojunctions formed by p-type a-Si:H acts as a hole-selective layer, selectively blocking electrons, and n-type a-Si:H serves as an electron-selective layer, selectively blocking holes. The intrinsic a-Si:H layer acts as the passivating layer, effectively reducing the defect density at the c-Si interface. Similarly, TOPCon and POLO architectures employ highly doped polysilicon layers as selective contacts, inducing band bending in the crystalline silicon absorber. These contacts are created through doping, which involves using gases like diborane (B_2H_6) or trimethyl borane for p-type doping and phosphine (PH_3) for n-type doping. However, SHJ structures require fabrication techniques like LPCVD and PECVD. To reduce costs, toxic gas usage, and the number of processing steps, wide bandgap materials such as transition metal oxides (TMOs) and alkali metal fluorides (AMFs) are now being explored as replacements for hole and electron-selective contacts. Dopant-free contact stacks are found to perform the dual function of providing passivation and carrier selectivity at the interface. These materials can be deposited using low-cost techniques such as thermal evaporation, Atomic Layer Deposition (ALD), magnetron sputtering, spin-coating, and spray pyrolysis, making them attractive alternatives for reducing both the cost and complexity of the fabrication process in DASH solar cells. The wide bandgap materials used in the DASH configuration help eliminate parasitic losses on the front side, which are commonly experienced in SHJ solar cells. This reduction in losses offers the potential for higher current densities, providing optimism for improved overall performance in DASH solar cells.

3.6.1 Hole-Selective Contacts Using Transition Metal Oxides

Typically, materials that form heterojunctions with c-Si will have a low ΔE_v (valence band offset) and a high ΔE_c (conduction band offset) to function as hole-selective layers (HSL), while those with a low ΔE_c and a high ΔE_v act as electron-selective layers (ESL). Transition metal oxides with wide bandgaps and suitable work functions are well-suited for use in the DASH configuration to achieve terminal selectivity. Similar to p-doped layers in the SHJ configuration, p-type TMOs such as NiO and Cu_2O in their non-stoichiometric forms were investigated as hole-selective layers due to their favourable band alignment with silicon, providing good hole selectivity [94–96]. In these materials, the ΔE_v (valence band offset) is low and ΔE_c (conduction band offset) is high, making them a selective-membrane enabling the transfer of holes, while blocking electrons. What was even more surprising was the hole-selective behaviour of n-type transition metal oxides like MoO_x, V_2O_x, and WO_x. Despite the n-type nature of these oxides, their high work functions allow them to induce a p^+ inversion layer on the n-type c-Si surface through an upward band bending, allowing these materials to selectively pass holes from the c-Si absorber layer to the external electrical contact [97–99]. This charge transport occurs either through Trap-Assisted-Tunnelling (TAT) or Band-to-Band tunnelling (B2B) mechanism at the n-Si/TMO interface. In TAT, holes in the p-type inversion region of the n-type c-Si absorber layer are captured by defect states within the bandgap of the TMO material and subsequently transferred to the conduction band of the TMO layer. The abundance of oxygen vacancies in films deposited by methods such as thermal evaporation facilitates the presence of these defect states within the bandgap. The hole transport in MoO_{3-x}, V_2O_{5-x}, and WO_{3-x} is explained through this mechanism [97]. In cases where the defect state density is low and the work function of the TMO is very high, the band-to-band tunnelling (B2B) mechanism dominates, with holes from the p-type inversion region of n-type Si being directly transported to the conduction band of the TMO. The work function of MoO_{3-x} ranges from 4.5 eV to 7 eV. MoO_{3-x} films with higher work function values are found to induce a stronger inversion layer at the MoO_{3-x}/n-Si interface, improving hole selectivity. The density of oxygen defect states, and consequently the work function of these materials, is often influenced by deposition techniques, external treatment processes and environmental conditions. For instance, UV-O_3-exposed MoO_{3-x} films have been found to exhibit an increased work function value of 6.6 eV [100]. The work function, and consequently the band bending at the interface, determines the type of charge transfer mechanism that occurs at the Si/TMO interface.

TMOs such as NiO, Cu_2O, CrO_3, MoO_3, V_2O_5, WO_3, and ReO_3 have been explored as hole transport materials. Among these, the most extensively studied TMO is MoO_3. In 2014, Battaglia et al., at University of California, Berkeley, demonstrated a 14.3% efficient solar cell utilizing a 15 nm MoO_3 layer as the HSL, replacing the p-doped a-Si:H layer in the SHJ configuration [101]. By inserting PECVD-deposited intrinsic a-Si:H layers for passivation, the cell parameters were further improved, resulting in an increase in current density of 2.4 mA/cm^2, an improved J_{sc} of 39.4 mA/cm^2, and a V_{oc} of 711 mV—surpassing the 700 mV mark—with an overall efficiency of 18.8% [98]. In 2015, Geissbühler et al., at EPFL, Switzerland, presented a MoO_3-based solar cell featuring Cu-plated front electrodes, passivation using intrinsic a-Si:H layers, a 7 nm MoO_3 layer as the HSL, and an n-type a-Si:H/ITO/Ag stack as the rear contact, achieving an efficiency of 22.5%, a V_{oc} of 725.4 mV, a J_{sc} of 38.6 mA/cm^2, and a fill factor (FF) of 80.36%, demonstrating the potential of dopant-free structures to achieve efficiencies comparable to commercial counterparts. In 2019, Cho et al. studied the thickness-dependent variation of MoO_3 films on solar cell parameters. A 6 nm thick MoO_3 layer was found to be optimal in minimizing parasitic absorption loss due to defect states in the MoO_3 film. Reducing the thickness from 10 to 6 nm led to an improvement in cell performance, resulting in an efficiency of 19.3%, a V_{oc} of 724 mV, a J_{sc} of 36 mA/cm^2, and a fill factor (FF) of 74.1%. This study also revealed the presence of a SiO_x layer at the intrinsic a-Si:H/MoO_3 interface [102]. DASH solar cells, fabricated using a very thin 4 nm MoO_3 as the HSL with 6 nm intrinsic a-Si:H passivation layers and a HIT rear structure for the ESL on textured n-type c-Si wafers, achieved a peak efficiency of 23.5% in 2020, with an excellent fill factor of 81.8% [103]. The cell also demonstrated high stability after 11 months of air storage, showing only a minimal loss of 1 mV in V_{oc}. By 2022, cell configurations utilizing MoO_3 achieved an efficiency of 23.83%, with a V_{oc} exceeding 721 mV and a fill factor (FF) of 82.18%, through the use of plasma treatments involving boron radicals on intrinsic a-Si:H layers [104]. This indicates that MoO_3-based DASH cells have strong potential for attaining excellent efficiencies in mass production and commercialization. All the above architectures used MoO_3 as a replacement for the p-doped a-Si:H layer while retaining the intrinsic a-Si:H passivation layer from the HIT structure, with the added advantage of eliminating parasitic absorption. Additionally, this approach provides a method to implement hole-selective layers without relying on toxic gases or complex PECVD setups, thereby simplifying the overall cell fabrication process (Fig. 16).

Gerling et al. at Universitat Politècnica de Catalunya, Spain, conducted extensive studies comparing the performance of MoO_3, WO_3, and V_2O_5 in solar cell configurations without a-Si:H layers. n-type Si wafers of 280 μm thickness and 1.5 Ω·cm resistivity were used to fabricate CSCs. The front carrier-selective contact was formed using a TMO/ITO structure followed by an Ag grid, while the rear contact was achieved by depositing a passivating back reflector and a Ti/Al stack with 15 nm Ti and 1 μm Al, which was laser fired—similar to the PERC process—to achieve lower contact resistivity. The cell fabricated using V_2O_5 as the HSL demonstrated

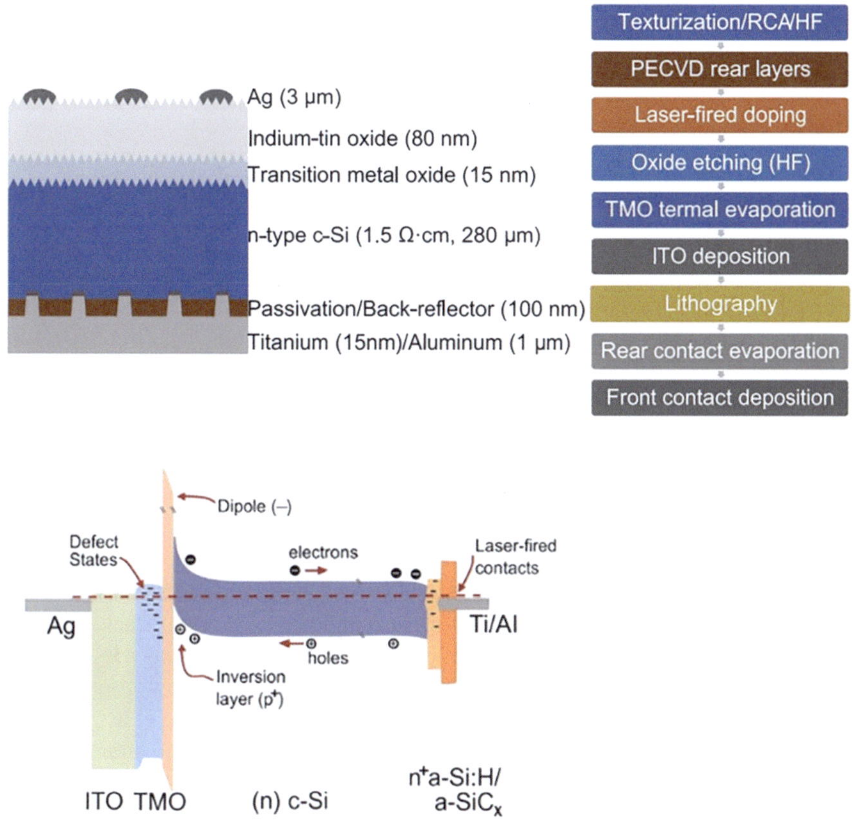

Fig. 16 Schematic representation of a Dopant-Free Carrier-Selective Solar Cell employing hole-selective transition metal oxide (TMO) layers, the corresponding process flow and the band diagram. Reprinted from [Solar Energy Materials and Solar Cells, vol. 145, L. G. Gerling et al., "Transition metal oxides as hole-selective contacts in silicon heterojunctions solar cells," pp. 109–115, Feb. 2016, https://doi.org/10.1016/j.solmat.2015.08.028] with permission from Elsevier [105]

superior performance, achieving an efficiency of 15.7% and a V_{oc} of 606 mV, while MoO_3 and WO_3 based cells achieved efficiencies of 13.6% and 12.5%, respectively. This highlighted the potential of V_2O_5 as an effective HSL layer. The cell parameters also indicates that the passivation provided by TMO layers is significantly lower than that provided by a-Si:H layers due to lattice mismatch at the interface with c-Si [105]. The promising performance of V_2O_5 as an HSL ignited further research into configurations utilizing V_2O_5 as the HSL. In 2016, Weiliang Wu et al. at Sun Yat-Sen University, China, employed TMOs (MoO_x, V_2O_x, WO_x) and Cs_2CO_3 as HSL and ESL in an all-back-contacted c-Si solar cell. The cell was entirely a-Si:H -free and featured a front-side passivation and anti-reflection stack of SiO_2 and SiN_x. The champion cell in the study, using V_2O_5 as the HSL, achieved an efficiency of 16.59%, a V_{oc} of 610 mV, a J_{sc} of 38.85 mA/cm^2, and a fill factor (FF) of 70%. These structures were fabricated using a simple hard mask deposition process, eliminating the complexity of lithographic patterning, which signalled the ease of fabricating DASH solar cells in IBC structures [106].

3.6.2 Electron-Selective Contacts Using Transition Metal Oxides Alkali Metal Halides

Materials with a low conduction band offset and a large valence band offset with crystalline silicon, function effectively as electron-selective layers (ESLs). Transition metal oxides and alkali metal halides such as TiO_2, MgO, SnO_2, ZnO TaO_x, LiF, MgF_2, and CaF_2 are commonly used as electron-selective materials. Besides these, alkali carbonates such as Cs_2CO_3, $CaCO_3$, are also explored as electron-selective materials [107]. Among these, the most extensively explored TMO is TiO_2. TiO_2 deposited through modified CVD was first explored as an ESL by Avasti et al. at Princeton University, USA. The TiO_2/n-Si junction was found to selectively conduct electrons while effectively blocking hole transport, resulting in a device with efficiency of 7.1% [108]. Further studies conducted by Cho et al. at K. U. Leuven, Belgium, on ALD-deposited TiO_2 were found to result in enhanced passivation and selectivity. When combined with a low work function Ca/Al metal stack, a device employing a rear a-Si:H/TiO_2 ESL with an 8 nm a-Si:H layer and a 2 nm TiO_x stack yielded an efficiency of 18.2% [109]. Combining TiO_x with low work function capping materials has proven to enhance cell efficiencies. Bullock et al., from the University of California, Berkeley, demonstrated a highly efficient solar cell with a 23% efficiency by incorporating an ESL stack consisting of TiO_2/LiF/Al on n-type silicon, showcasing the potential of TiO_2 configuration with low work function capping materials [110].

Another extensively studied ESL material is LiF, an alkali metal fluoride with low work function of nearly 2.9 eV and a wide bandgap of > 10 eV. LiF has a high dipole moment, and when combined with Al, the LiF/Al stack effectively reduces the work function of Al and prevents Fermi level pinning at the n-type c-Si/Al junction. This stack has been successfully implemented in both organic LEDs and solar cells [111, 112]. In 2011, Liang Fang et al., at Korea Advanced institute of Science and technology, Korea, demonstrated an a-Si rear Schottky junction solar cell that implemented a LiF/Al electrode with low series resistance, with the LiF thickness optimized up to 2 nm [113]. In 2016, James Bullock et al. at the University of California, Berkeley, demonstrated a partial rear LiF/Al ESL stack on n-type c-Si with a p-doped emitter, an Al_2O_3 passivation layer, and a SiN_x anti-reflection coating (ARC) at the top. This cell achieved efficiencies exceeding 20%, demonstrating the effectiveness of the LiF/Al stack in replacing highly n-doped regions in the cell architecture [114]. They also demonstrated a fully dopant-free silicon solar cell using MoO_x as the HSL and LiF/Al as the ESL stack, with passivating intrinsic a-Si:H layers on both sides. This cell achieved an impressive efficiency of 19.42%, with a V_{oc} of 716 mV, a J_{sc} of 37.07 mA/cm^2, and a fill factor (FF) of 73.15% [115]. This dopant-free architecture provided clear proof that metal oxides and alkali metal fluorides can replace doped a-Si:H layers. Further research is being conducted to replace the intrinsic a-Si:H with other wide bandgap passivating layers such as SiO_x, HfO_x, Al_2O_3, and Ga_2O_3. Materials like Al_2O_3 and SiN_x provide a fixed charge density on their surface, enabling field-effect passivation by depleting the junction of one type of carriers. Corona-charged dielectric layers are one method used to achieve this field-effect passivation on materials that lack these fixed charges. Bonilla et al., at the University of Oxford, United Kingdom, have made significant progress in this area by achieving excellent passivation on SiO_2 layers, with surface recombination velocities as low as 0.65 cm/s on n-type silicon wafers [116]. Corona-charged dielectric layers can enhance dopant-free architectures by providing field-effect passivation (FEP) along with carrier selectivity, improving cell parameters. This approach enables the complete replacement of a-Si:H layers in silicon cell architectures, further simplifying the design while maintaining high performance.

Dopant-free architectures have shown promising results at the laboratory scale, demonstrating high efficiencies and simplified cell designs. However, further improvements in the stability of the materials used and successful large-scale upscaling onto large-area cells are essential to make this approach industrially viable. With continued research and advancements in material durability and production techniques, we can expect industrial adoption of these configurations by 2030, offering a more efficient, cheaper, and streamlined alternative to traditional silicon solar cell technologies.

4 Conclusion

The evolution of silicon solar cell technologies has been remarkable, with each new architecture contributing to significant efficiency gains and cost reductions. Starting with Al-BSF cells, which laid the foundation for industrial solar cell production, the introduction of PERC cells marked a major leap forward by enhancing rear-side passivation and boosting efficiency. TOPCon and POLO technologies have further advanced this by introducing tunnel oxide and carrier-selective passivating contacts, enabling even higher efficiency and improved stability.

Silicon Heterojunction (SHJ) solar cells have pushed the boundaries of performance by combining crystalline silicon with thin layers of amorphous silicon, achieving some of the highest efficiencies in commercial solar cells today. The introduction of dopant-free architectures, which replace doped a-Si:H layers with metal oxides and alkali metal fluorides, offers a compelling alternative that simplifies cell design while maintaining high performance.

The future of silicon solar cells looks promising, with several key developments on the horizon. Continued optimization of passivation techniques, especially in POLO and SHJ cells, is expected to further reduce recombination losses. Industrial adoption of dopant-free architectures could bring about a new era of cost-effective and scalable solar cell manufacturing, reducing reliance on toxic gases and complex doping processes.

As the solar industry moves towards mass production of high-efficiency cells, we can expect technologies like TOPCon and POLO to dominate the market by 2030, with dopant-free architectures likely to gain traction as stability and scalability improve. By combining these innovations, silicon solar cells will continue to play a pivotal role in the global transition to renewable energy, driving further efficiency improvements and cost reductions to meet the growing demand for clean energy solutions.

References

1. IRENA: *Renewable Energy Statistics 2024* (International Renewable Energy Agency, Abu Dhabi, 2024) [Online]. Available: https://www.irena.org/-/media/Files/IRENA/Agency/Publication/2024/Jul/IRENA_Renewable_Energy_Statistics_2024.pdf?rev=c1fd271e788d42d189fa1b1f625cf521. Accessed 11 Aug 2024
2. IEA: *World Energy Investment 2024* (IEA, Paris, 2024). https://www.iea.org/reports/world-energy-investment-2024. Licence: CC BY 4.0
3. IEA: *World Energy Outlook 2023* (IEA, Paris, 2023). https://www.iea.org/reports/world-energy-outlook-2023. Licence: CC BY 4.0 (report); CC BY NC SA 4.0 (Annex A)
4. O.H.L.S. Russell, *Light-Sensitive Electric Device Including Silicon* (1948) [Online]. Available: https://lens.org/135-426-191-683-784
5. E.F. Kingsbury, R.S. Ohl, Photoelectric properties of ionically bombarded silicon. Bell Syst. Tech. J. **31**(4), 802–815 (1952). https://doi.org/10.1002/j.1538-7305.1952.tb01407.x
6. D.M. Chapin, C.S. Fuller, G.L. Pearson, A new silicon p-n junction photocell for converting solar radiation into electrical power. J. Appl. Phys. **25**(5), 676–677 (1954). https://doi.org/10.1063/1.1721711
7. D.M. Chapin et al., *Solar Energy Converting Apparatus* (1957) [Online]. Available: https://lens.org/026-898-384-753-471
8. A. Luque, S. Hegedus, *Handbook of Photovoltaic Science and Engineering* (Wiley, 2003)
9. S.R. Wenham, M.A. Green, Silicon solar cells. Prog. Photovoltaics Res. Appl. **4**(1), 3–33 (1996). https://doi.org/10.1002/(SICI)1099-159X(199601/02)4:1%3c3::AID-PIP117%3e3.0.CO;2-S
10. M.A. Green, R.B. Godfrey, MIS solar cell—general theory and new experimental results for silicon. Appl. Phys. Lett. **29**(9), 610–612 (1976). https://doi.org/10.1063/1.89161
11. A. Blakers, N. Zin, K.R. McIntosh, K. Fong, High efficiency silicon solar cells. Energy Procedia 1–10 (2013). https://doi.org/10.1016/j.egypro.2013.05.033
12. V.R. Mehta, B.L. Sopori, N.M. Ravindra, Screen-printed contacts for crystalline silicon solar cells: an overview. Emerg. Mater. Res. **11**(3), 284–302 (2022). https://doi.org/10.1680/jemmr.22.00021
13. A. Smets, K. Jäger, O. Isabella, R. van Swaaij, M. Zeman, *Solar Energy: The Physics and Engineering of Photovoltaic Conversion, Technologies and Systems* (Bloomsbury, USA, 2016) [Online]. Available: https://books.google.co.in/books?id=as3JEAAAQBAJ
14. A.W. Blakers et al., 18-percent efficient terrestrial silicon solar cells. IEEE Electron Device Lett. **5**(1), 12–13 (1984). https://doi.org/10.1109/EDL.1984.25813
15. M.A. Green, A.W. Blakers, J. Zhao, A.M. Milne, A. Wang, X. Dai, Characterization of 23-Percent Efficient Silicon Solar Cells (1990)
16. M.A. Green, *Silicon Solar Cells to Power the Future 2022 Millennium Technology Prize Lecture* (2022) [Online]. Available: https://millenniumprize.org/events/
17. M.A. Green, A.W. Blakers, J. Shi, E.M. Keller, S.R. Wenham, 19.1% efficient silicon solar cell. Appl. Phys. Lett. **44**(12), 1163–1164 (1984). https://doi.org/10.1063/1.94678

18. M.A. Green, The passivated emitter and rear cell (PERC): from conception to mass production. Sol. Energy Mater. Sol. Cells **143**, 190–197 (2015). https://doi.org/10.1016/j.solmat.2015.06.055
19. A.W. Blakers, A. Wang, A.M. Milne, J. Zhao, M.A. Green, 22.8% efficient silicon solar cell. Appl. Phys. Lett. **55**, 1363–1365 (1989)
20. J. Zhao, A. Wang, M.A. Green, 24·5% efficiency silicon PERT cells on MCZ substrates and 24·7% efficiency PERL cells on FZ substrates. Prog. Photovoltaics Res. Appl. **7**(6), 471–474 (1999). https://doi.org/10.1002/(SICI)1099-159X(199911/12)7:6%3c471::AID-PIP298%3e3.0.CO;2-7
21. M.A. Green, The path to 25% silicon solar cell efficiency: history of silicon cell evolution. Prog. Photovoltaics Res. Appl. **17**(3), 183–189 (2009). https://doi.org/10.1002/pip.892
22. J. Zhao, A. Wang, P. Altermatt, M.A. Green, Twenty-four percent efficient silicon solar cells with double layer antireflection coatings and reduced resistance loss. Appl. Phys. Lett. **66**(26), 3636–3638 (1995). https://doi.org/10.1063/1.114124
23. A. Blakers, Development of the PERC solar cell. IEEE J Photovolt **9**(3), 629–635 (2019). https://doi.org/10.1109/JPHOTOV.2019.2899460
24. E. Schneiderlöchner, R. Preu, R. Lüdemann, S.W. Glunz, Laser-fired rear contacts for crystalline silicon solar cells. Prog. Photovoltaics Res. Appl. **10**(1), 29–34 (2002). https://doi.org/10.1002/pip.422
25. R. Preu, S. W. Glunz, S. Schaefer, R. Luedemann, W. [Fraunhofer-I.fuer S. E. (ISE) "Wettling Freiburg im Breisgau (Germany)]," and W. [Forschungszentrum K. G. T. und U. (Germany)] Pfleging, *Laser Ablation—A New Low-Cost Approach for Passivated Rear Contact Formation in Crystalline Silicon Solar Cell Technology* (Germany, 2000)
26. G. Beaucarne, G. Agostinelli, J. Szlufcick, P. Choulat, G. Beaucarne, *Local Contact Structures for Industrial PERC-Type Solar Cells* [Online]. Available: https://www.researchgate.net/publication/284670341
27. K.A. Münzer et al., Rear side passivated and locally contacted solar cells with laser diffused selective emitter. Energy Procedia 1–9 (2012). https://doi.org/10.1016/j.egypro.2012.02.001
28. R.J. Van Overstraeten, Summary Advances in Silicon Solar Cell Processing
29. *210 PERC Cell Efficiency Achieves 24.5%, Trina Solar Breaks World Record for the 24th Time* [Online]. Available: https://www.trinasolar.com/en-glb/resources/newsroom/en210-perc-cell-efficiency-achieves-245-trina-solar-breaks-world-record-24th-time. Accessed 29 Aug 2024
30. T. Dullweber et al., Present status and future perspectives of bifacial PERC+ solar cells and modules. Jpn. J. Appl. Phys. **57**(8S3), 08RA01 (2018). https://doi.org/10.7567/JJAP.57.08RA01
31. T. Dullweber, J. Schmidt, Industrial silicon solar cells applying the passivated emitter and rear cell (PERC) concept-a review. IEEE J. Photovolt. **6**(5), 1366–1381 (2016). https://doi.org/10.1109/JPHOTOV.2016.2571627
32. Photovoltaics Equipment International Technology Roadmap for Photovoltaics (ITRPV) 2023 Results
33. I.R.C. Post, P. Ashbum, G.R. Wolstenholme, *Poly Silicon Emitters for Bipolar Transistors: A Review and Re-evaluation of Theory and Experiment* (1992)
34. H.C. de Graaff, J.G. de Groot, The SIS tunnel emitter, in *1978 International Electron Devices Meeting* (IRE, 1978), pp. 333–335. https://doi.org/10.1109/IEDM.1978.189421
35. E. Yablonovitch, T. Gmitter, R.M. Swanson, Y.H. Kwark, A 720 mV open circuit voltage SiO_x:c-Si:SiO_x double heterostructure solar cell. Appl. Phys. Lett. **47**(11), 1211–1213 (1985). https://doi.org/10.1063/1.96331
36. Y.H. Kwark, R.M. Swanson, *SIPOS Heterojunction Contacts to Silicon* (United States, 1986) [Online]. Available: https://www.osti.gov/biblio/5905985
37. M.A. Green, A.W. Blakers, *Advantages of Metal-Insulator-Semiconductor Structures for Silicon Solar Cells* (1983)
38. F. Feldmann, M. Bivour, C. Reichel, M. Hermle, S.W. Glunz, Passivated rear contacts for high-efficiency n-type Si solar cells providing high interface passivation quality and excellent

transport characteristics. Sol. Energy Mater. Sol. Cells **120**(PART A), 270–274 (2014). https://doi.org/10.1016/j.solmat.2013.09.017
39. A. Richter, J. Benick, F. Feldmann, A. Fell, M. Hermle, S.W. Glunz, N-type Si solar cells with passivating electron contact: identifying sources for efficiency limitations by wafer thickness and resistivity variation. Sol. Energy Mater. Sol. Cells **173**, 96–105 (2017). https://doi.org/10.1016/j.solmat.2017.05.042
40. A. Richter et al., *Both Sides Contacted Silicon Solar Cells: Options for Approaching 26% Efficiency*
41. F. Feldmann, M. Bivour, C. Reichel, H. Steinkemper, M. Hermle, S.W. Glunz, Tunnel oxide passivated contacts as an alternative to partial rear contacts. Sol. Energy Mater. Sol. Cells **131**, 46–50 (2014). https://doi.org/10.1016/j.solmat.2014.06.015
42. A. Moldovan, F. Feldmann, M. Zimmer, J. Rentsch, J. Benick, M. Hermle, Tunnel oxide passivated carrier-selective contacts based on ultra-thin SiO_2 layers. Sol. Energy Mater. Sol. Cells **142**, 123–127 (2015). https://doi.org/10.1016/j.solmat.2015.06.048
43. A. Moldovan et al., Simple cleaning and conditioning of silicon surfaces with UV/ozone sources. Energy Procedia 834–844 (2014). https://doi.org/10.1016/j.egypro.2014.08.067
44. B. Steinhauser, T. Niewelt, A. Richter, R. Eberle, M.C. Schubert, Extraordinarily high minority charge carrier lifetime observed in crystalline silicon. Solar RRL **5**(11) (2021). https://doi.org/10.1002/solr.202100605
45. F. Haase et al., Laser contact openings for local poly-Si-metal contacts enabling 26.1%-efficient POLO-IBC solar cells. Sol. Energy Mater. Sol. Cells **186**, 184–193 (2018). https://doi.org/10.1016/j.solmat.2018.06.020
46. S.W. Glunz et al., Silicon-based passivating contacts: the TOPCon route. Prog. Photovoltaics Res. Appl. **31**(4), 341–359 (2023). https://doi.org/10.1002/pip.3522
47. B. Kafle, B.S. Goraya, S. Mack, F. Feldmann, S. Nold, J. Rentsch, TOPCon—Technology options for cost efficient industrial manufacturing. Sol. Energy Mater. Sol. Cells **227** (2021). https://doi.org/10.1016/j.solmat.2021.111100
48. W. van Sark, B. Hoex, A. Reinders, P.J. Verlinden, N.J. Ekins-Daukes, *Photovoltaic Solar Energy: From Fundamentals to Applications, Volume 2* (IEEE Press, Wiley, 2024) [Online]. Available: https://books.google.co.in/books?id=W68OEQAAQBAJ
49. A. Richter et al., Design rules for high-efficiency both-sides-contacted silicon solar cells with balanced charge carrier transport and recombination losses. Nat. Energy **6**(4), 429–438 (2021). https://doi.org/10.1038/s41560-021-00805-w
50. Y. Chen et al., Technology evolution of the photovoltaic industry: learning from history and recent progress. Prog. Photovoltaics Res. Appl. **31**(12), 1194–1204 (2023). https://doi.org/10.1002/pip.3626
51. M.K. Stodolny et al., N-type polysilicon passivating contact for industrial bifacial n-type solar cells. Sol. Energy Mater. Sol. Cells **158**, 24–28 (2016). https://doi.org/10.1016/j.solmat.2016.06.034
52. N. Nandakumar et al., Investigation of 23% monoPoly screen-printed silicon solar cells with an industrial rear passivated contact, in *2019 IEEE 46th Photovoltaic Specialists Conference (PVSC)* (IEEE, 2019), pp. 1463–1465. https://doi.org/10.1109/PVSC40753.2019.8980861
53. Y. Chen et al., Mass production of industrial tunnel oxide passivated contacts (i-TOPCon) silicon solar cells with average efficiency over 23% and modules over 345 W. Prog. Photovoltaics Res. Appl. **27**(10), 827–834 (2019). https://doi.org/10.1002/pip.3180
54. W. Shen, Y. Zhao, F. Liu, *Highlights of Mainstream Solar Cell Efficiencies in 2022* (Higher Education Press Limited Company, 2022). https://doi.org/10.1007/s11708-023-0871-y
55. S. Groser, E. Krassowski, S. Swatek, H. Zhao, C. Hagendorf, Microscale contact formation by laser enhanced contact optimization. IEEE J. Photovolt. **12**(1), 26–30 (2022). https://doi.org/10.1109/JPHOTOV.2021.3129362
56. T. Fellmeth et al., Laser-enhanced contact optimization on iTOPCon solar cells. Prog. Photovoltaics Res. Appl. **30**(12), 1393–1399 (2022). https://doi.org/10.1002/pip.3598
57. M. Meßmer et al., Stack diffusion process for cost- and energy-efficient boron emitter formation. IEEE J. Photovolt. **12**(6), 1393–1399 (2022). https://doi.org/10.1109/JPHOTOV.2022.3214437

58. W. Lin et al., Green-laser-doped selective emitters with separate BBr3 diffusion processes for high-efficiency n-type silicon solar cells. Sol. Energy Mater. Sol. Cells **210** (2020). https://doi.org/10.1016/j.solmat.2020.110462
59. Y. Chen et al., 690 WP N-Type i-TOPCon modules in mass production with >25% efficiency solar cells based on large-area 210 mm wafers, in *2023 IEEE 50th Photovoltaic Specialists Conference (PVSC)* (IEEE, 2023), pp. 1–1. https://doi.org/10.1109/PVSC48320.2023.10359700
60. W. Shen, Y. Zhao, F. Liu, Highlights of mainstream solar cell efficiencies in 2023. Front. Energy (2024). https://doi.org/10.1007/s11708-024-0937-5
61. J.Y. Gan, R.M. Swanson, Polysilicon emitters for silicon concentrator solar cells, in *IEEE Conference on Photovoltaic Specialists* (IEEE), pp. 245–250. https://doi.org/10.1109/PVSC.1990.111625
62. U. Römer et al., Ion implantation for poly-Si passivated back-junction back-contacted solar cells. IEEE J. Photovolt. **5**(2), 507–514 (2015). https://doi.org/10.1109/JPHOTOV.2014.2382975
63. R. Peibst et al., Working principle of carrier selective poly-Si/c-Si junctions: is tunnelling the whole story? Sol. Energy Mater. Sol. Cells **158**, 60–67 (2016). https://doi.org/10.1016/j.solmat.2016.05.045
64. F. Haase et al., Interdigitated back contact solar cells with polycrystalline silicon on oxide passivating contacts for both polarities. Jpn. J. Appl. Phys. **56**(8S2), 08MB15 (2017). https://doi.org/10.7567/JJAP.56.08MB15
65. F. Haase et al., Interdigitated back contact solar cells with polycrystalline silicon on oxide passivating contacts for both polarities. Jpn. J. Appl. Phys. **56**(8) (2017). https://doi.org/10.7567/JJAP.56.08MB15
66. C. Hollemann, F. Haase, S. Schäfer, J. Krügener, R. Brendel, R. Peibst, 26.1%-efficient POLO-IBC cells: quantification of electrical and optical loss mechanisms. Prog. Photovoltaics Res. Appl. **27**(11), 950–958 (2019). https://doi.org/10.1002/pip.3098
67. F. Haase, C. Hollemann, S. Schafer, J. Krugener, R. Brendel, R. Peibst, Transferring the record p-type Si POLO-IBC cell technology towards an industrial level, in *2019 IEEE 46th Photovoltaic Specialists Conference (PVSC)* (IEEE, 2019), pp. 2200–2206. https://doi.org/10.1109/PVSC40753.2019.8980960
68. F. Haase, B. Min, C. Hollemann, J. Krügener, R. Brendel, R. Peibst, Fully screen-printed silicon solar cells with local Al-p+ and n-type POLO interdigitated back contacts with a VOC of 716 mV and an efficiency of 23%. Prog. Photovoltaics Res. Appl. **29**(5), 516–523 (2021). https://doi.org/10.1002/pip.3399
69. R. Peibst et al., On the chances and challenges of combining electron-collecting nPOLO and hole-collecting Al-p+ contacts in highly efficient p-type c-Si solar cells. Prog. Photovoltaics Res. Appl. **31**(4), 327–340 (2023). https://doi.org/10.1002/pip.3545
70. B. Min et al., Approaching 23% with p-Type Back Junction Solar Cells Featuring Screen-Printed Al Front Grid and Passivating Rear Contacts Preprint-38th European Photovoltaic Solar Energy Conference and Exhibition Approaching 23% with p-type Back Junction Solar Cells Featuring Screen-Printed Al Front Grid and Passivating Rear Contacts [Online]. Available: https://www.researchgate.net/publication/354504060
71. B. Min et al., 24.2% efficient POLO back junction solar cell with an AlO_x/SiN_y dielectric stack from an industrial-scale direct plasma-enhanced chemical vapor deposition system. Prog. Photovoltaics Res. Appl. (2024). https://doi.org/10.1002/pip.3828
72. M. Taguchi, Review—development history of high efficiency silicon heterojunction solar cell: from discovery to practical use. ECS J. Solid State Sci. Technol. **10**(2), 025002 (2021). https://doi.org/10.1149/2162-8777/abdfb6
73. H. Matsuura, T. Okuno, H. Okushi, K. Tanaka, Electrical properties of n-amorphous/p-crystalline silicon heterojunctions. J. Appl. Phys. **55**(4), 1012–1019 (1984). https://doi.org/10.1063/1.333193
74. M. Tanaka et al., Development of new a-Si/c-Si heterojunction solar cells: ACJ-HIT (artificially constructed junction-heterojunction with intrinsic thin-layer). Jpn. J. Appl. Phys. **31**(11R), 3518 (1992). https://doi.org/10.1143/JJAP.31.3518

75. T. Sawada et al., High-efficiency a-Si/c-Si heterojunction solar cell, in Proceedings of 1994 IEEE 1st World Conference on Photovoltaic Energy Conversion—WCPEC (A Joint Conference of PVSC, PVSEC and PSEC) (IEEE, 1994), pp. 1219–1226. https://doi.org/10.1109/WCPEC.1994.519952
76. M. Taguchii et al., *Special Issue HIT TM Cells*High Efficiency Crystalline Si Cells with Novel Structure*
77. H. Sakata et al., 20.7% highest efficiency large area (100.5 cm^2) HIT/sup TM/cell, in *Conference Record of the Twenty-Eighth IEEE Photovoltaic Specialists Conference—2000 (Cat. No.00CH37036)* (IEEE, 2000), pp. 7–12. https://doi.org/10.1109/PVSC.2000.915742
78. M. Taguchi et al., 24.7% record efficiency HIT solar cell on thin silicon wafer. IEEE J. Photovolt. **4**(1), 96–99 (2014). https://doi.org/10.1109/JPHOTOV.2013.2282737
79. K. Masuko et al., Achievement of more than 25% conversion efficiency with crystalline silicon heterojunction solar cell. IEEE J. Photovolt. **4**(6), 1433–1435 (2014). https://doi.org/10.1109/JPHOTOV.2014.2352151
80. D. Adachi, J.L. Hernández, K. Yamamoto, Impact of carrier recombination on fill factor for large area heterojunction crystalline silicon solar cell with 25.1% efficiency. Appl. Phys. Lett. **107**(23) (2015). https://doi.org/10.1063/1.4937224
81. K. Yoshikawa et al., Exceeding conversion efficiency of 26% by heterojunction interdigitated back contact solar cell with thin film Si technology. Sol. Energy Mater. Sol. Cells **173**, 37–42 (2017). https://doi.org/10.1016/j.solmat.2017.06.024
82. K. Yoshikawa et al., Silicon heterojunction solar cell with interdigitated back contacts for a photoconversion efficiency over 26%. Nat. Energy 2(5) (2017). https://doi.org/10.1038/nenergy.2017.32
83. X. Ru et al., 25.11% efficiency silicon heterojunction solar cell with low deposition rate intrinsic amorphous silicon buffer layers. Sol. Energy Mater. Sol. Cells **215** (2020). https://doi.org/10.1016/j.solmat.2020.110643
84. J. Haschke, O. Dupré, M. Boccard, C. Ballif, Silicon heterojunction solar cells: recent technological development and practical aspects—from lab to industry. Sol. Energy Mater. Sol. Cells **187**, 140–153 (2018). https://doi.org/10.1016/j.solmat.2018.07.018
85. M. Bivour, H. Steinkemper, J. Jeurink, S. Schroer, M. Hermle, Rear emitter silicon heterojunction solar cells: fewer restrictions on the optoelectrical properties of front side TCOs. Energy Procedia 229–234 (2014). https://doi.org/10.1016/j.egypro.2014.08.035
86. M. Sharma, J. Panigrahi, V.K. Komarala, Nanocrystalline silicon thin film growth and application for silicon heterojunction solar cells: a short review. Nanoscale Adv. (2021). https://doi.org/10.1039/d0na00791a
87. H. Lin et al., Silicon heterojunction solar cells with up to 26.81% efficiency achieved by electrically optimized nanocrystalline-silicon hole contact layers. Nat. Energy **8**(8), 789–799 (2023). https://doi.org/10.1038/s41560-023-01255-2
88. Q. Wang et al., Efficient heterojunction solar cells on p-type crystal silicon wafers, Appl. Phys. Lett. **96**(1) (2010). https://doi.org/10.1063/1.3284650
89. A. Danel, N. Chaugier, J. Veirman, R. Varache, M. Albaric, E. Pihan, Closing the gap between n- and p-type silicon heterojunction solar cells: 24.47% efficiency on lightly doped Ga wafers. Prog. Photovoltaics Res. Appl. **31**(12), 1235–1244 (2023). https://doi.org/10.1002/pip.3635
90. X. Ru et al., Silicon heterojunction solar cells achieving 26.6% efficiency on commercial-size p-type silicon wafer. Joule **8**(4), 1092–1104 (2024). https://doi.org/10.1016/j.joule.2024.01.015
91. *LONGi Sets New World-Record for Silicon Solar Cell Efficiency, Launching 2nd Generation Ultra-Efficient BC-Based Module* [Online]. Available: https://www.longi.com/en/news/longi-hi-mo9-bc-world-record/. Accessed 21 Sep 2024
92. H. Wu et al., Silicon heterojunction back-contact solar cells by laser patterning. Nature (2024). https://doi.org/10.1038/s41586-024-08110-8
93. Z. Sun et al., Toward efficiency limits of crystalline silicon solar cells: recent progress in high-efficiency silicon heterojunction solar cells. Adv. Energy Mater. (2022). https://doi.org/10.1002/aenm.202200015

94. M. Xue et al., Contact selectivity engineering in a 2 μm thick ultrathin c-Si solar cell using transition-metal oxides achieving an efficiency of 10.8%. ACS Appl. Mater. Interfaces **9**(48), 41863–41870 (2017). https://doi.org/10.1021/acsami.7b12886
95. H. Imran, T.M. Abdolkader, N.Z. Butt, Carrier-selective NiO/Si and TiO_2/Si contacts for silicon heterojunction solar cells. IEEE Trans. Electron Devices **63**(9), 3584–3590 (2016). https://doi.org/10.1109/TED.2016.2585523
96. M. Nayak, S. Mandal, A. Pandey, S. Mudgal, S. Singh, V.K. Komarala, Nickel oxide hole-selective heterocontact for silicon solar cells: role of SiOx interlayer on device performance. Solar RRL **3**(11) (2019). https://doi.org/10.1002/solr.201900261
97. L.G. Gerling, C. Voz, R. Alcubilla, J. Puigdollers, Origin of passivation in hole-selective transition metal oxides for crystalline silicon heterojunction solar cells. J. Mater. Res. **32**(2), 260–268 (2017). https://doi.org/10.1557/jmr.2016.453
98. C. Battaglia et al., Silicon heterojunction solar cell with passivated hole selective MoO_x contact. Appl. Phys. Lett. **104**(11) (2014). https://doi.org/10.1063/1.4868880
99. J. Bullock et al., Stable dopant-free asymmetric heterocontact silicon solar cells with efficiencies above 20%. ACS Energy Lett. **3**(3), 508–513 (2018). https://doi.org/10.1021/acsenergylett.7b01279
100. C. Battaglia et al., *Hole Selective MoO_x Contact for Silicon Solar Cells* [Online]. Available: http://pubs.acs.org
101. C. Battaglia et al., Hole selective MoO_x contact for silicon solar cells. Nano Lett. **14**(2), 967–971 (2014). https://doi.org/10.1021/nl404389u
102. J. Cho et al., Interface analysis and intrinsic thermal stability of MoO_x based hole-selective contacts for silicon heterojunction solar cells. Sol. Energy Mater. Sol. Cells **201** (2019). https://doi.org/10.1016/j.solmat.2019.110074
103. J. Dréon et al., 23.5%-efficient silicon heterojunction silicon solar cell using molybdenum oxide as hole-selective contact. Nano Energy **70**, 104495 (2020). https://doi.org/10.1016/j.nanoen.2020.104495
104. L. Cao et al., Achieving 23.83% conversion efficiency in silicon heterojunction solar cell with ultra-thin MoO_x hole collector layer via tailoring (i) a-Si:H/MoO_x interface. Prog. Photovoltaics Res. Appl. **31**(12), 1245–1254 (2023). https://doi.org/10.1002/pip.3638
105. L.G. Gerling et al., Transition metal oxides as hole-selective contacts in silicon heterojunctions solar cells. Sol. Energy Mater. Sol. Cells **145**, 109–115 (2016). https://doi.org/10.1016/j.solmat.2015.08.028
106. W. Wu et al., Dopant-free back contact silicon heterojunction solar cells employing transition metal oxide emitters. Phys. Status Solidi Rapid Res. Lett. (RRL) **10**(9), 662–667 (2016). https://doi.org/10.1002/pssr.201600254
107. S. Acharyya et al., Dopant-free materials for carrier-selective passivating contact solar cells: a review. Surf. Interfaces **28**, 101687 (2022). https://doi.org/10.1016/j.surfin.2021.101687
108. S. Avasthi, W.E. McClain, G. Man, A. Kahn, J. Schwartz, J.C. Sturm, Hole-blocking titanium-oxide/silicon heterojunction and its application to photovoltaics. Appl. Phys. Lett. **102**(20) (2013). https://doi.org/10.1063/1.4803446
109. J. Cho et al., Passivating electron-selective contacts for silicon solar cells based on an a-Si:H/TiO_x stack and a low work function metal. Prog. Photovoltaics Res. Appl. **26**(10), 835–845 (2018). https://doi.org/10.1002/pip.3023
110. J. Bullock et al., Dopant-free partial rear contacts enabling 23% silicon solar cells. Adv. Energy Mater. **9**(9) (2019). https://doi.org/10.1002/aenm.201803367
111. C.J. Brabec, S.E. Shaheen, C. Winder, N.S. Sariciftci, P. Denk, Effect of LiF/metal electrodes on the performance of plastic solar cells. Appl. Phys. Lett. **80**(7), 1288–1290 (2002). https://doi.org/10.1063/1.1446988
112. S.E. Shaheen et al., Bright blue organic light-emitting diode with improved color purity using a LiF/Al cathode. J. Appl. Phys. **84**(4), 2324–2327 (1998). https://doi.org/10.1063/1.368299
113. L. Fang, S.J. Baik, S. Lim, S. Yoo, K.S. Lim, Amorphous Si rear Schottky junction solar cell with a LiF/Al back electrode. IEEE Trans. Electron Devices **58**(9), 3048–3052 (2011). https://doi.org/10.1109/TED.2011.2160267

114. J. Bullock et al., Lithium fluoride based electron contacts for high efficiency n-type crystalline silicon solar cells. Adv. Energy Mater. **6**(14) (2016). https://doi.org/10.1002/aenm.201600241
115. J. Bullock et al., Efficient silicon solar cells with dopant-free asymmetric heterocontacts. Nat. Energy **1**(3), 15031 (2016). https://doi.org/10.1038/nenergy.2015.31
116. R.S. Bonilla, C. Reichel, M. Hermle, P. Hamer, P.R. Wilshaw, Long term stability of c-Si surface passivation using corona charged SiO_2. Appl. Surf. Sci. **412**, 657–667 (2017). https://doi.org/10.1016/j.apsusc.2017.03.204

Thin Film Solar Cells: An Overview of Materials and Fabrication Methods

Arya Mohan, S. Heera, V. J. Anjana Nair, and Deepa K. G.

Abstract Thin film solar cells have shown its dominancy over crystalline silicon solar cells in terms of cost, flexibility and ease of fabrication. As manufacturing processes improve, these solar cells are becoming more cost-effective, appealing to both residential and commercial consumers seeking affordable solar options. The technology enabled the creation of printed ultrathin solar cells on flexible fabric, which can be integrated onto the surface of any shape or structure. The global market for thin film solar cells continues to grow at rate of 10.74% expecting the market to reach US$ 41.0 Billion by 2032. Cu(In,Ga)Se_2, CdTe, a-Silicon, and GaAs are the most established and commonly used materials in thin film solar cells, with Cu(In,Ga)Se_2 leading the market, achieving a module efficiency of approximately 20%. The book chapter provides an overview of the development and current status of Cu(In,Ga)Se_2, CdTe, a-Silicon, and GaAs-based solar cells. It also outlines the typical solar cell structure and the commonly used fabrication methods for each material, offering a comprehensive understanding of the existing technology for each type.

Keywords Thin film · Flexibility · Conversion efficiency · Vacuum techniques · Solution processing · Screen printing

1 Introduction

Thin film solar cells (TFSCs) were developed in the 1970s as second-generation solar cells with the goal of reducing production costs and enabling versatile fabrication techniques. The absorber layer in TFSCs ranges in thickness from nanometres to a few micrometres, whereas silicon (Si) requires hundreds of micrometres to absorb enough photons. Thin film technology also paved the way for the development of roll-to-roll flexible solar cell technologies. Although silicon (Si) possesses excellent long-term stability and high efficiency, traditional crystalline silicon (c-Si) solar cells have high production costs, primarily due to the cost of wafers, the high thermal budget over a

A. Mohan · S. Heera · V. J. Anjana Nair · Deepa K. G. (✉)
Department of Physics, University of Kerala, Thiruvananthapuram, Kerala, India
e-mail: deepachaithanya@gmail.com

P. Ravindran et al. (eds.), *Photon to Power*, Progress in Optical Science and Photonics 32, https://doi.org/10.1007/978-981-96-5914-2_3

prolonged period, and the complex manufacturing processes. Compared to that thin film solar cells utilize one or more thin layers of photovoltaic materials over a glass, metal, or plastic substrate. In addition, thin film offers the feasibility of versatile deposition methods on a variety of substrates.

Recently the concept of ultrathin solar cells has been introduced in which the thickness of absorber layer could be reduced as low as 100 nm by increasing the effective optical path length [1]. This could be achieved using effective light-trapping techniques. Ultrathin solar cells have already achieved efficiencies greater than 15%, highlighting the need for further development in this area to make them scalable, cost-effective, and industrially viable. The major thin film solar cell technologies that are widely commercialized include copper indium gallium selenide (CIGS), cadmium telluride (CdTe), amorphous silicon (a-Si), and gallium arsenide (GaAs). In these technologies, there is less material wastage of the absorber compared to first-generation solar cells. Major features of the four TFSCs are summarized in Table 1 as reported in the solar cell efficiency table, version 64 [2].

All four materials share a direct bandgap, which allows the use of thin layers. Furthermore, all four technologies can be utilized in building-integrated photovoltaics [3]. CIGS solar cells are highly efficient and easy to integrate, making them a popular choice. The stability and efficiency of CIGS cells are comparable to those of Si solar cells. Among thin film solar cells (TFSC), CdTe has proven to be the most scalable and reproducible material to date. However, the toxicity of cadmium in CdTe solar cells makes the processing and disposal of CdTe modules difficult. The particular advantages of a-Si solar cells are their low production cost and the flexibility, compared to its crystalline Si counterparts. However, the Staebler-Wronski effect, which is degradation caused by illumination, poses a significant obstacle to their practical use [4]. For low-cost terrestrial applications a-Si:H solar modules are an ideal match because of their minimal material cost and reliable manufacturing process. In the meanwhile, GaAs-based solar cell has exhibited the highest power conversion efficiency (PCE) among thin film solar cells. GaAs is highly radiation resistant which makes is it suitable for extra-terrestrial domain. The manufacturing methods of GaAs solar cell are very expensive, which limits their terrestrial market.

Table 1 Properties of CIGS, a-Si, CdTe, and GaAs TFSCs

	CIGS	a-Si	CdTe	GaAs
Bandgap (eV)	1.02–1.69	1.75	1.43	1.42
Absorption coefficient (cm^{-1})	10^5	$> 10^5$	$> 10^4$	$> 10^4$
Sufficient thickness (μm)	~ 1–2	~ 1	~ 5	~ 1–3
Disadvantage	Rare elements (In)	Efficiency	Toxicity	Rare elements High cost
Best efficiency (lab) (%)	23.35	14	22.1	29.1

Green et al. [2]

The following sections describe the structure, properties and fabrication methods of some of the widely used thin films in solar cell application.

2 Cu(In,Ga)Se_2 (CIGS)

Chalcopyrite semiconductors are an excellent pick among all the investigated and accessible materials since they can be used to create thin film solar cells with high efficiency and impressive cost-effectiveness [5]. Polycrystalline copper indium gallium selenide (CIGS) material has become one of the most promising photovoltaic materials within the global community in recent years since it reached a conversion efficiency of 23.35% on glass substrates at the laboratory level [6].

High optical absorption coefficient, optimum bandgap, and superior efficiency make CIGS stand out among the various available materials [7]. Its excellent outdoor stability, radiation hardness, and reliable long-term performance make it suitable for terrestrial applications. In addition, these solar cells have proven to be exceptionally stable at high temperatures, which makes them an excellent option for aircraft applications where reducing volume and weight is crucial [8]. Additionally, its potential for monolithic integration results in lower production costs due to a lesser number of processing steps. One of the issues with CIGS solar cells is the scarcity of In and Ga, which can lead to an increase in module costs. However, with the help of improved light-trapping techniques, the thickness of CIGS-based cells could be reduced to sub-micrometre levels.

Together with a configurable bandgap, CIGS is a good semiconductor to be used in tandem solar cells. Cell efficiency increases when CIGS is used with organic and inorganic solar cells because the combination produces varying bandgap energies that absorb broader regions of the solar spectrum, which can be converted into electricity [9]. Apart from its relatively higher manufacturing cost, CIGS outperforms cadmium telluride (CdTe) and hydrogenated amorphous silicon (a-Si:H) solar cells in terms of performance [10]. With an efficiency comparable to polycrystalline silicon technology, industrial CIGS technology may soon be able to compete with single-crystalline silicon technology. The ternary compound semiconductor CIGS crystallizes with the $CuXY_2$ (X = In, Ga Al, and Y = S, Se) tetragonal chalcopyrite structure. It possesses an absorption coefficient of 10^5 cm^{-1}. As a direct bandgap semiconductor, CIGS lessens the demand for long minority carrier diffusion lengths. The superior absorption features with p-type characteristics of CIGS make it a promising absorber in photovoltaic applications [11]. The bandgap of CIGS can be tailored by partially substituting indium with gallium in the basic $CuInSe_2$ (CIS) crystal structure. A bandgap ranging from 1.02 eV to 1.69 eV can be obtained by altering the ratio of gallium to indium. CIGS-based solar cells possess numerous advantages over CIS-based solar cells including the modification of optical bandgap, the improvement in V_{oc} resulting from Ga incorporation, and the easiness of creating a single-phase CIGS absorber rather than a similar CIS absorber [12].

In terms of flexible solar cells, CIGS offers (one among) the highest and mature technology for commercial applications. The efficiency outperforms alternatives such as DSSCs and organic solar cells. Flexible solar cells offer novel prospects for building-integrated photovoltaics. In general, the flexible solar cells have received significant attention due to four key requirements: high efficiency, low-cost manufacture, high specific power, and high throughput [6]. The common substrates employed in flexible solar cells are polyimide, polyethylene naphthalate (PEN), polyethylene terephthalate (PET), ceramics, and metal foils like titanium (Ti), aluminium (Al), molybdenum (Mo), and stainless steel (SS). Flexible CIGS solar cells have made remarkable development, especially when polyimide substrates are used [12]. It has been claimed that a high efficiency of 22.2% on a polyimide sheet hints that solar cells on flexible substrates could potentially be as efficient as those on rigid substrates [13].

2.1 *Typical Cell Structure*

A schematic of the typical CIGS solar cell structure of SLG/Mo/p-CIGS/n-CdS/i-ZnO/AZO/ARC/metal-grid is shown in Fig. 1. The commonly used substrate is soda-lime glass (SLG) due to its low cost, chemical inertness and sodium content that boosts the electrical characteristics of CIGS. Also, it can withstand the high-temperature reactive environment that CIGS demands to become a good-quality absorber layer. Flexible substrates like stainless steel foils, and polyimide films are also used. Molybdenum (Mo) metal is usually DC sputtered onto the SLG, serving as a back contact. In addition, Mo functions as a reflector sending back the unabsorbed light into the absorber. Also, the sodium from the SLG substrate diffuses to the CIGS absorber layer through the Mo contact, increasing the carrier concentration and grain boundary passivation. The film deposition usually follows a substrate structure from back contact to the front as the formation of CIGS requires an elevated temperature that might affect the CIGS-CdS interface in the other hand.

The CIGS absorber layer which is about 1–3 microns thick is deposited using physical vapour deposition techniques. The majority of the solar spectrum is CIGS

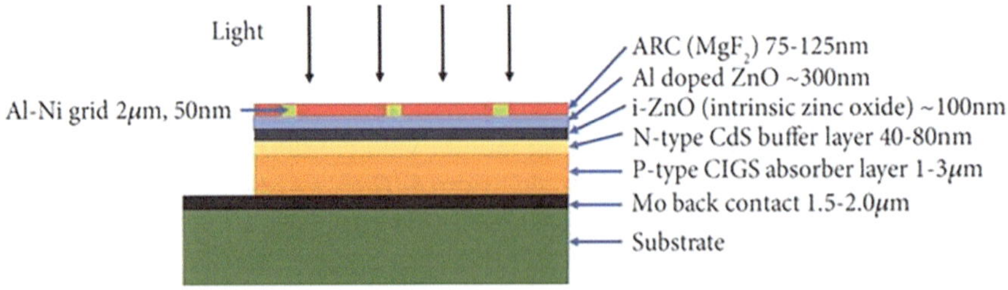

Fig. 1 Schematic of CIGS solar cell. Reproduced from Ong et al. [14]

absorbed within 1 μm, requiring only a thin layer of 2.0–2.5 μm for the device. Such that the raw material utilization can be minimized. The n-type cadmium sulphide (CdS) buffer layer on top of the absorber fabricated via chemical bath deposition forms the major p–n junction with CIGS. Other non-conventional buffer layers like Zn (O, S, OH) that can be deposited via vacuum techniques are also used. In this way, we can lessen the environmental risks associated with the usage of toxic Cadmium and also, the improved transmission with these layers permits higher current. In a recent method, another buffer layer of Mg-doped ZnO (50 nm) was deposited via the ALD technique on top of the 50 nm thick Cd-free buffer layer Zn (O, S, OH) to improve the cell performance even more. Also, boron-doped ZnO can be fabricated via the MOCVD technique, which serves as TCO [15]. A high resistive intrinsic zinc oxide layer (i-ZnO) on top of CdS is followed by a conductive Al-doped ZnO layer (AZO), both fabricated via sputtering. By using the i-ZnO layer, shunting channels/paths caused by the diffusion of Al from the AZO to the absorber can be avoided. The AZO is used for the lateral collection of photogenerated carriers into the current collecting Ni/Al metal grids [6]. Aluminium and MgF2 are used as electrode and antireflection coating, respectively.

The n-CdS layer allows light till 2.4 eV into the absorber layer following which the creation of electron–hole pairs takes place. As a result of the built-in electric field across the CIGS-CdS interface, electrons are swept away from the absorber layer to the n-buffer layer and are collected by the n-type electrode. In the same manner, the Mo rear contact collects all the holes within the diffusion length region that have been swept away from the n-CdS layer to the p-CIGS layer. A further mechanism that reflects electrons towards the p–n junction, where they are eventually collected by the n-type electrode, is the back surface field (BSF) in the CIGS layer formed as a result of the Ga gradient at the Mo back contact. In addition, the minority carrier recombination near the CIGS-Mo interface is decreased by the BSF [12].

High-efficiency solar cells should have a low Ga atomic ratio [Ga/(Ga + In) ~ 0.26] and a (Cu)/(In + Ga) ratio of 0.88–0.92 in the CIGS layer. Higher Ga ratios (> 4 0.3) lead to more defects in the film. Likewise, the absorber having one atomic fraction (about 0.1 at%) of sodium (Na) enhances the efficiency of CIGS solar cells. The advantageous role of Na prompted the study of other alkali metals like Li, Na, K, Rb, and Cs. Tiwari et al. used a "post-deposition potassium fluoride (KF) treatment" to incorporate potassium after CIGS deposition [16]. This resulted in flexible CIGS devices on alkali-free substrates with up to 20.4% efficiency [17]. However, the Cs treatment of a CIGS photo absorber layer yielded the best recorded photovoltaic efficiency to date [11].

2.2 Fabrication

There exist various vacuum and non-vacuum deposition techniques by which the CIGS layer can be coated on flexible and rigid substrates. Selecting the right deposition technique is an important aspect affecting the device's performance. The vacuum techniques employed include co-evaporation, sputtering, electron beam deposition, molecular beam epitaxy, pulsed laser deposition, CVD, and MOCVD. Over the past decade, low-cost techniques for non-vacuum deposition have emerged. Solution-based methods for developing CIGS films involve either depositing CIGS directly onto the substrate surface via electrodeposition or chemical bath deposition, or treating the surface with liquid precursors via screen printing, spray pyrolysis, spin coating, and dip coating. Although some of these methods use toxic reagents, they have the potential to reduce manufacturing costs by reducing the need for large initial expenditures in production machines [18]. However, due to the improved crystallinity of the CIGS layer, CIGS-based solar cells manufactured using vacuum techniques often have better cell efficiencies than those made using non-vacuum techniques.

Even though there are numerous deposition processes, only a few are widely used in lab-scale and industrial manufacturing. The lab-scale approach prioritizes precise monitoring of the film composition and cell efficiency while industrial production adheres to low costs, efficiency, high throughput, reproducibility, and process tolerance. The first technique commonly employed is co-evaporation of Cu, In, Ga, and Se elemental sources. The next method usually opted for in industry is reactive sputtering as this technique allows production in large areas [5]. Although sputtering is preferred for industrial applications, solar cells made from sputtered CIGS have lower efficiency compared to evaporation. This is likely due to the lack of Ga grading in sputtered CIGS absorbers. Still, the modules made using both techniques perform likewise [12].

Thermal evaporation is one of the simplest physical vapour deposition techniques by which the source material is heated till it evaporates in the vacuum atmosphere and eventually condenses as a thin layer on the substrate. There are many types of co-evaporation processes. Growth processes are typically referred to as single-stage, two-stage, and three-step processes based on the number of stages involved. Single-step co-evaporation involves simultaneously evaporating four elemental sources (Cu, In, Ga, and Se) to deposit CIGS in a single step. This method is easier to control, faster, and less expensive than the three-step co-evaporation process. Industrial scale application of the one-stage procedure is much simpler. In the study of Karaagac et al., CIGS thin films were grown by single-stage thermal evaporation by using a stoichiometric CIGS powder, and a single-phase CIGS film was successfully generated when the material was post-annealed at a temperature of at least 250 °C. Following the annealing process, there was a significant shift in the surface composition, which was brought about by the movement of Cu and Ga atoms to the bulk, creating the ordered CIGS structure [19].

In a two-stage process, Cu-rich growth is employed in the first stage, while Cu-poor growth is attained in the second stage via decreasing the Cu flux. This can be accomplished in three ways. A certain approach involves depositing an $(In,Ga)_2Se_3$ precursor at 350 °C substrate temperature before supplying Cu and Se components. Another implementation involves depositing Cu, In, Ga, and Se in the Cu-rich stage before turning off the Cu flux. The third procedure involves supplying Cu, In, Ga, and Se first, then decreasing the Cu flux [20].

Very often the complex multistage co-evaporation process results in high-efficiency CIGS TFSCs (>20%). In the three-stage co-evaporation process, elements such as In, Ga, and Se are evaporated at a temperature of 400 °C in the first stage. As shown in Fig. 2, step two involves raising the temperature to ~ 550 °C and then evaporating Cu and Se until the CIGS film becomes Cu-rich. From the second stage onwards, the substrate temperature is retained the same. Ultimately, In, Ga, and Se are evaporated in the third step, and the deposition is stopped when the required thickness and composition are attained. The large grain size of CIGS and an optimized Ga gradient formation are often credited with this method's effectiveness. In a study carried out by Salomé et al., a comparison of CIGS TFSCs fabricated via single-stage and three-stage co-evaporation processes was made. The three-stage sample was found to display the usual Ga notch profile, whereas the one-stage sample showed a nearly linear Ga gradient. The J-V characteristics of the devices were comparable, with the three-stage process exhibiting a marginally higher median electrical performance (17% Vs. 16.3% for the one-stage sample). It was inferred that the three-stage technique's notch profile and the larger grain size relative to the one-stage process could result in somewhat greater solar cell efficiency [21].

Sputtering involves bombarding a target with an ionized gas, usually argon, to deposit the desired substance. A repeated collision process transfers energy among the target's atoms, and atoms that gain enough momentum to break their bonds with nearby atoms are released from the target [22]. In general, sputtering has the advantage of being scalable to vast regions while maintaining homogeneity and simple control over processes. The high deposition rate and enhanced adhesion are added advantages. Numerous multi-component sputtering processes are being used to deposit CIGS. In the two-stage deposition method, metals are typically deposited from metal mixture targets or alloys, then selenized with vapour from a solid Se source or H_2Se gas. To enhance solar cell efficiency, different precursor layer structures have been investigated including In/CuGa/Mo structure, CuGa/In/Mo structure, multi-stack structure, and thin CuGa/In/CuGa/Mo structure [23]. Alternatively, metals can be co-sputtered sputtered from In and CuGa alloy targets, then annealing the metal precursor stack in a chalcogen-containing environment. Despite high-efficiency solar cells being developed through sputtering ternary target precursors, the annealing process remains complex and difficult to regulate. In addition, the usage of H_2Se, which contains highly hazardous chemicals, necessitates very costly

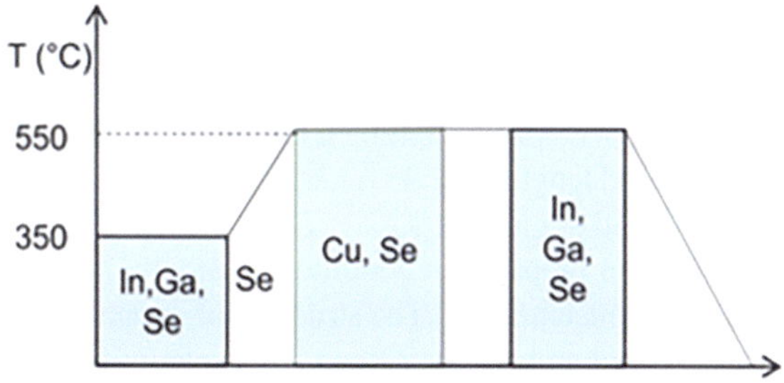

Fig. 2 Conventional three-stage co-evaporation process

annealing equipment. These factors provide significant impediments to its promotion and implementation in the industry at large [24].

Quaternary sputtering offers a possible alternative to traditional deposition processes for producing CIGS thin films. This process uses a single sputtering target with all four elements to deposit the CIGS film. Quaternary sputtering has numerous advantages over conventional deposition processes, including superior uniformity over vast regions, reduced costs, elevated material utilization, and less dependence on hazardous selenization procedures. Sputtering of quaternary targets pressed at elevated temperatures can stop the phase separation of CIGS. As a result, the film surface turns out to be smoother and the composition becomes more homogeneous [24]. No techniques other than sputtering can replicate such uniformity of the large-area deposition. By adjusting the elemental flux ratios during deposition in the thermal co-evaporation method, multistage processing, and grading can be easily attained. However, with quaternary sputtering, varying the elemental ratio with a single sputtering target is quite difficult. Using several targets can enable bandgap grading in quaternary sputtering [21]. Despite many advantages, there are a few downsides. Currently, devices produced using quaternary sputtering without further selenization have a maximum efficiency of about 16%, and achieving bandgap grading to match the performance of the best-evaporated devices is quite challenging [25].

Acciarri et al., introduced a hybrid sputtering-evaporation (HSE) method which combines the features of both techniques allowing for precise stoichiometry control and easy adaptation to industrial processes. Here the metal precursors are first sputtered onto suitable graphite plates held on a circulating transfer cylinder, which are then thermally evaporated in a Se atmosphere to obtain the final film. Including an evaporation phase allows for metal ratio in-depth profiles, much like the three-stage produced CIGS absorbers. Solar cells were constructed employing HSE-grown CIGS films on various substrates and the efficiency through this hybrid approach has been consistently high [26].

3 Amorphous Silicon (a-Si)

a-Si is a very fascinating substance whose intrinsic disorder allows for efficient solar absorption [27]. Compared to crystalline silicon, a-Si:H has a significant visible light absorption coefficient of around 10^5–10^6 cm^{-1}. It takes only one micrometre of thickness to absorb visible light. Therefore, a-Si:H can be used in solar cells, particularly when a thin coating is needed [28]. The a-Si bond length is 1% longer than the c-Si bond length (= 2:35 Å), while the tetrahedral bond angle varies by 10% from the c-Si bond angle (=109.47°). The bandgap of a-Si:H is 1:7–1:9 eV, which varies according on preparation conditions, namely the deposition temperature Ts, or the hydrogen concentration. The structure of crystalline silicon, amorphous silicon and hydrogenated amorphous silicon are shown in the Fig. 3. Innovative manufacturing techniques have been developed to lower the production costs.

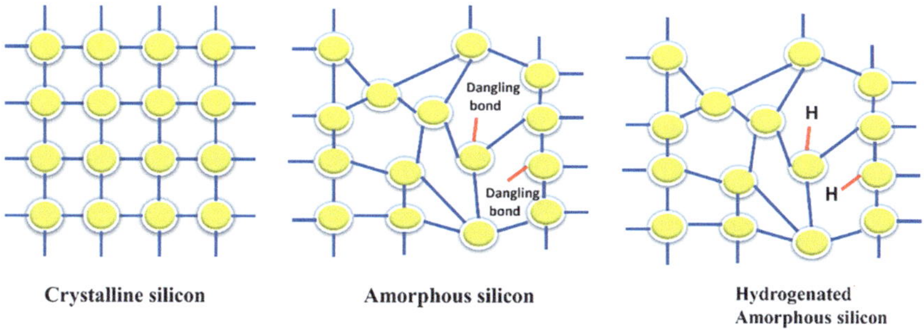

Fig. 3 Structures of crystalline, amorphous and hydrogenated amorphous silicon

The amorphous silicon has short-range order. Dangling bonds, or unsatisfied valences on an immobilized silicon atom, occur at $10^{19}/cm^3$. These dangling bonds function as recombination hotspots, significantly lowering the carrier lifespan. This is the exact opposite of attempting to separate charge carriers in a solar cell. Even worse, defects pin the Fermi energy level, preventing amorphous silicon from being doped into p-type or n-type. It is a semiconductor of minimal value, having lost all other useful properties for the synthesis in electronic gadgets. Dissociated silane gas (SiH_4) was produced on heated substrates using the glow discharge technique, which involves applying a voltage between electrodes at both ends of a glass tube to create a plasma. Adding 10% hydrogen during deposition significantly decreases defect density to $10^{16}/cm^3$.

Hydrogen atoms join with a dangling bond in amorphous silicon, creating hydrogenated amorphous silicon (a-Si:H). With reduced defect density, hydrogenated amorphous silicon may be doped and employed in junction devices [29]. The two main characteristics that set hydrogenated amorphous silicon (a-Si:H) apart from crystalline silicon are these high bonded hydrogen concentration (usually about 10% in device-quality a-Si:H) and the absence of long-range order. The material is suitable for (opto-) electronic devices because hydrogen atoms passivate the majority of the unsaturated Si dangling bonds.

For photon energies over the band-gap energy, the disorder results in large optical absorption coefficients because it relaxes the momentum conservation constraint. Disorder, however, results in a large density of localized states (defect and band-tail states) in the energy gap, which impedes carrier transit and increases recombination losses. Furthermore, during doping, band tails and defects obstruct the Fermi level from fully shifting from the bandgap centre towards the band edges. In fact, for strongly doped n- and p-type a-Si:H, respectively, the Fermi level cannot be pushed closer to the delocalized band states than 0.2 eV and 0.3 eV, respectively. The presence of band tails in a-Si:H solar cells reduce the practical open-circuit voltage (V_{oc}) to levels that are less than those predicted by the band-gap energy in theory. When exposed to light, more defects in the bandgap are formed, which serve as recombination sites for photogenerated carriers and degrade the electrical characteristics of a-Si:H. Solar cell efficiency and a-Si:H film photoconductivity both drop during light-soaking, but only to a limited extent until saturation. The sample temperature and light intensity have a significant impact on the deterioration behaviour and final saturation levels.

Despite these limitations, a-Si:H is a "high potential" material to consider for large-scale solar power applications: Low process temperatures and large-area deposition are compatible with the plasma-enhanced chemical vapour deposition (PECVD) deposition technique. This makes it possible to employ a wide range of low-cost substrate materials. It is simple to dope a-Si:H by introducing gases containing phosphorus or boron during the deposition process. It is possible to adjust the optical bandgap, which is normally about 1.7 eV, by alloying germanium towards lower photon energy and carbon or oxygen towards higher photon energies. Furthermore, by altering the hydrogen content as a result of certain deposition conditions and techniques, the bandgap of a-Si:H itself may also be somewhat changed [30].

Increasing hydrogen dilution is thought to promote hydrogen coverage on the growing surface. The impinging SiH_3 species diffuse across a longer distance before being absorbed into the film, leading to a better structure. Excessive hydrogen dilution can produce microcrystalline material. Hydrogen functions as an etchant as well, and it might remove weak bonds on the developing surface.

The highest-grade material is produced at a hydrogen dilution slightly below the amorphous-to-microcrystalline transition. Microcrystalline inclusions in high-quality materials may lead to enhanced stability through recombination, as these bonds are more robust and difficult to break. It has been hypothesized that the majority of the hydrogen is integrated in the microcrystallite areas based on hydrogen effusion investigations and the remainder of the material has a low hydrogen concentration. The light-induced deterioration is thought to be connected to the hydrogen in the alloy. As a result, the majority of the material, which has greater order and less hydrogen, is more stable when exposed to light.

The general view is that the energy released during the recombination of extra electron–hole pairs produced by illumination breaks a weak bond, resulting in the formation of two dangling bond defects. An adjacent Si–H bond allows a hydrogen atom to come in and prevent the two nearby broken bonds from recombining. Further bond swapping may cause the defect pair to become even further apart. During the process of annealing, the hydrogen atoms return to their initial locations. The initial values can be recovered by annealing for one hour at temperatures of about 150 °C [27].

3.1 *Typical Cell Structure*

As seen in Fig. 4, the fundamental layout of a-Si:H solar cells is the p–i–n structure, which consists of a relatively thick intrinsic layer placed between thin p-type and n-type doped layers. The single-junction solar cell is often built on an FTO coated glass substrate. The p layer is usually a boron-doped hydrogenated amorphous silicon

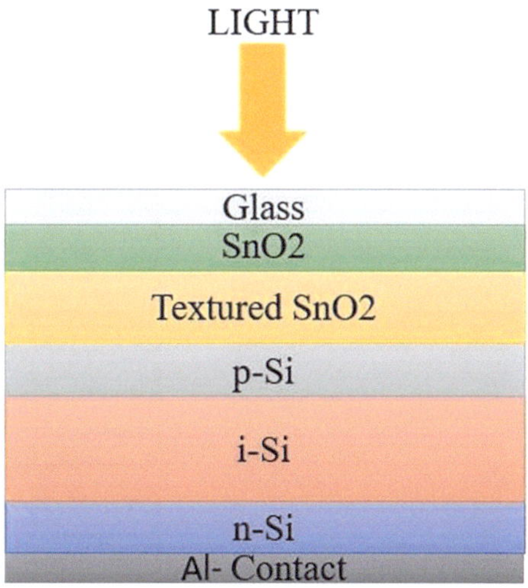

Fig. 4 Layout of a:Si solar cell in substrate configuration. Reproduced from Sivaraj et al. [32]

carbon alloy, whereas the n-layer (typically 20–30 nm thick) is either phosphorus-doped microcrystalline a-Si:H or phosphorus-doped a-Si:II [31].

The intrinsic (i) portion of the solar cell is where the majority of light absorption and carrier photogeneration occurs in such a p–i–n cell. With a typical thickness of 250–500 nm, the *i* layer aids in the electrical transport by creating an electric field that separates and gathers the charge. 0.1–1 ppm of boron levels are typical in the *i* layer [31]. On account of their short lifetime in heavily doped a-Si:H, there is only a little or no contribution to the photocurrent by the electrons and holes produced in the doped layers. The intrinsic layer's material quality and the electric field's distribution and strength are in accountable for charge carrier collection, which ultimately determines the performance of the solar cell. The two ways in which the defects impact the charge carrier collection are (i) they serve as recombination sites (ii) their charge state affects how the electric field is distributed within the *i* layer [30]. As an effective method for enhancing the light stability of a-Si:H-based solar cells, the stacked-cell idea has gained popularity.

Multi-bandgaps in multijunction cells allow response to a variety of wavelengths. Bandgaps in multijunction technologies can easily be tuned through different alloying, however a higher bandgap at the top junction than the bottom junction is necessary. In this way, a portion of the energy often lost in single-junction solar cells can be collected and transformed. On top of that multijunction technologies offer reduced light-induced deterioration, and it has also been indicated that greater performances may result by optically connecting two mechanically distinct α-Si cells in succession [31].

Because of the stronger electric field and lower absorption in each sub cell, a multi-junction device experiences less light-induced degradation than a single-junction solar cell with the same total absorber layer thickness. Combination of an a-Si:H top cell with additional sub-cells in the bottom in the order of decreasing bandgap

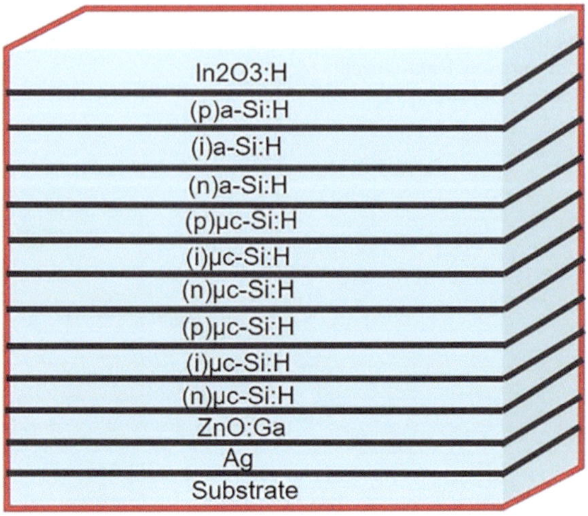

Fig. 5 The layout of a triple-junction α-Si:H/µc-Si:H/µc-Si:H solar cell

can result in even better efficiencies. A larger current density from the top cell than the bottom cell is required to maximize solar cell efficiency. Micromorph solar cell refers to the tandem connection of a a-Si:H on top of µc-Si:H cell. Lately, the a-Si:H/µc-Si:H/µc-Si:H triple-junction solar cell demonstrated the highest efficiency [33]. A schematic of the triple junction is shown in Fig. 5.

For tandem structures, honeycomb textured substrates are often employed. Usually, silver (Ag) and zinc oxide (ZnO:Ga) doped with gallium (Ga) coatings are applied on the substrate to minimize the reflective losses and to enhance the conductivity. Afterwards, hydrogenated-α-Si (α-Si:H) is normally coated via a diode or triode plasma-enhanced chemical vapour deposition (PECVD). Following this procedure, RF magnetron sputtering is used to deposit the transparent conducting oxide (TCO) layer. This layer which serves as the front window, is commonly made of ITO or hydrogenated indium oxide (In_2O_3:H). Ag can be used as the grid electrode, and to enhance cell performance, an antireflection coating can then be applied [15].

Initially, the α-Si solar cell was deposited in a p–i–n configuration; however, it can also be manufactured with an n–i–p formation sequence [15]. In both p–i–n and n–i–p configurations, light enters the device through the p layer since the p layer efficiently facilitates the hole collection. This is because holes are less mobile than electrons. For the p–i–n arrangement, the superstrate must withstand temperature changes during the deposition and should be transparent. As a result, glass superstrates were used to create p–i–n solar modules. The n–i–p arrangement works well with opaque substrates and is frequently used with flexible substrates like polymers and stainless steel [33].

3.2 *Fabrication*

Characterized by a short-range order of disordered Si atoms, the continuous random lattice structures result in higher grain boundary defects causing defect state energy levels in the bandgap. The so-formed trap levels influence electron transport, reducing electron mobility. The entire a-Si fabrication technique is classified into top-down and bottom-up approaches, where the former enables high-temperature deposition of thin film layers over a rigid substrate followed by their transfer onto flexible ones and the latter mostly favours low-temperature deposition [34].

The dangling and non-geometric bond angles of a-Si result in enhanced light scattering within the material. The resultant light scattering creates a light-trapping effect, which improves the light absorption coefficient of the material. A higher absorption coefficient reduces the need for higher material thickness, which favours lightweight, flexible device fabrication.

Hydrogenation of a-Si causes the bonding of hydrogen atoms with the dangling bonds of a-Si, forming hydrogenated a-Si. This results in a-Si passivation, reducing defect density with enhanced electron mobility. The short-range order results in a wider bandgap, which helps the material function as an excellent visible light absorber.

The two most efficient thin film deposition techniques are plasma-enhanced chemical vapour deposition (PECVD) and e-beam evaporation. The bottom-up approach generally involves Si atom assembly following the low-temperature deposition over suitable substrates, which is mainly achieved by PECVD techniques.

3.2.1 Plasma-Enhanced Chemical Vapour Deposition (PECVD)

PECVD is the primary approach for fabricating high-quality hydrogenated amorphous silicon (a-Si:H) absorber material for thin film solar cells. This technique has the benefit of fabricating large-area devices at lower temperatures. Intrinsic hydrogenated amorphous silicon (i-a-Si:H) films have better passivation quality for silicon heterojunction (SHJ) solar cells and create higher V_{oc}. Commercialized SHJ production mainly uses PECVD technique.

CVD generally involves incorporating reactant precursor gas or powder, which needs to be deposited, followed by chemical reaction and subsequent deposition upon suitable substrates.

The principal process associated with the PECVD is the formation of gas plasma and the subsequent deposition of the desired material. When subjected to high voltages, the gas within the reaction chamber ionizes and forms plasma. The ionized atoms chemically react and get deposited over substrates.

Silane gas diluted with any inert gases is used as the process gas for the PECVD technique. Argon and hydrogen are the most commonly used dilutants. The process involves decomposition of silane gas to ionized Si atoms and hydrogen. The Si atoms are deposited, and the unreacted H atoms are exhausted. The reaction is given in equation,

$$SiH_4 \rightarrow Si + 2H_2$$

The deposition parameters are crucial for film fabrication as it can affect the film properties [34].

The silane to Ar gas flow ratio in the gas mixture significantly affects the plasma growth process and the microstructure of fabricated thin films. The significant parameters affecting the film quality and crystallinity are the glow discharge power, silane gas flow rate and reactor gas pressure. A higher silane gas flow rate favours higher deposition and absorption coefficients with an increased optical bandgap. On the contrary, the higher absorption coefficient arises from the generation of higher defect density, with the increased silane gas flow rate. This results in a reduction of solar cell device efficiency [35].

Instead of using high-temperature treatments for material activation, the technique uses highly energetic plasma for deposition, enabling low-temperature thin film fabrication.

Substrate temperature also plays a crucial role as it determines the electronic qualities of thin films. The fabricated film remains amorphous when the substrate temperature is kept below 580 °C, above which it turns to μc-Si thin films. Optimum film qualities like higher structure order, low defect density, and high photosensitivity are achieved at nearly 200 °C [36]. Lowering the temperature allows their fabrication on flexible substrates but at the expense of reduced film quality. Such a situation can be eliminated in PECVD fabrication which allows the low-temperature fabrication of a-Si thin films where the extra energy needed for the lattice ordering is provided by the process gas flux. This allows film fabrication below 100 °C, suitable for flexible substrates.

The significant benefit of this technology is the easiness of doping. It involves mixing the optimum concentration of dopant gas into the process gas or partial doping, which can be achieved by interaction over deposited a-Si thin film layer. The type of doping and the performance enhancement with doping depends on the dopant gas incorporated. Studies demonstrated PECVD deposition of a-Si at 250 °C with doping by incorporating CO_2, PH_3, and B_2H_6 gases [37].

Flexible solar cell devices must sustain their flexibility while keeping the electrical properties unaltered. Subsequent thinning of a-Si absorber layers enables its deposition over non-planar geometries. The working parameters, especially the process temperature, are determined depending on whether film deposition is over a permanent substrate or on an intermediate substrate, which is then transferred onto the permanent substrate [36].

The performance of a-Si:H/c-Si heterojunction-based solar cells predominantly lies in their interface properties. So, the deposition technique is crucial in eliminating the probable recombinations and setting up of the individual layers for device fabrication. The works of Korte et al. emphasized the requirement of c-Si surface pretreatment and the following a-Si:H deposition. A combined c-Si pretreatment and surface passivation could create a device with (n) a-Si:H/(p)c-Si heterojunction with an efficiency of 18.5%.

Apart from being an absorber material, a-Si can function as a passivation layer and enables BSF. a-Si helps to achieve the necessary band bending associated with the p–n junction formation and also passivates the c-Si dangling bonds. The passivation lowers recombination velocity to as low as 3 cm/s. Combined with the PECVD technique, this large-area low-temperature processing method can reduce production costs while enhancing device performance.

3.2.2 Electron Beam Evaporation Technique

E-beam evaporation uses highly energetic electrons to incident on the source material, followed by source vaporization. The highly energetic electrons emitted from heated filaments are guided through a magnetic field and incident over the source material. This causes local heating of the source material and subsequent material emission. The vapours are further deposited on substrates. The schematic of the process is shown in Fig. 6.

It has a very high deposition rate of 1 μm/min. However, the fabrication of very thin films is limited because of the higher deposition rates. The thicker films have reduced surface texturing and are unfit for light trapping. The method mostly follows substrate texturing prior to film deposition for efficient light trapping. It further creates a problem of shadowing near the textured surfaces, as it is a line-of-sight process. Shadowing can make the film more porous while reducing the performance [38]. So, highly compatible substrate texturing, with reduced shadowing, is essential for better film deposition.

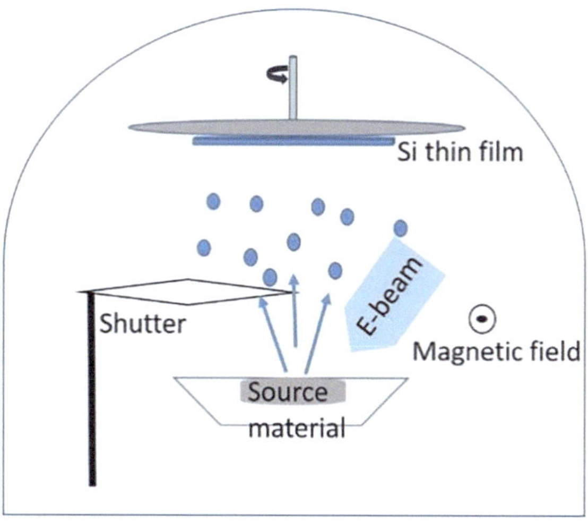

Fig. 6 Schematic of the electron beam deposition process

4 Gallium Arsenide (GaAs)

The most effective way to supply satellites with electricity is through the use of photovoltaic solar generators. The initial use was proven by US satellite Vanguard 1 in 1958. Following several years of gradual expansion for the space photovoltaic industry, the late 1990s saw a surge in large-scale applications, with the primary users being military, communications, and science space missions. Airmass and spectrum illumination are two important space environment conditions that must be included in the specifications of solar cells intended for space travel [39].

In 1960, the Russian Ioffe Physico Technical Institute (PTI) developed the first GaAs solar cells, which had a 3% efficiency. It soon became apparent that this material was more temperature stable and radiation resistant than Si. Regarding Si, the first significant use of GaAs as a photovoltaic device occurred in space, when GaAs solar cells were used to power the Russian spacecraft Venera-2 and Venera-3 when they arrived at Venus in 1965. Until the 1980s, when epitaxial deposition techniques like metal organic vapour phase epitaxy (MOVPE) and molecular beam epitaxy (MBE) started to be widely used to deposit thin films on GaAs or Ge substrates, III-V development was impeded by challenges with crystal growth technology (epitaxial GaAs was mainly grown from the liquid phase) and obtaining sharp junction (doping was realized by diffusion) [40].

Gallium and arsenic elements combine to form gallium arsenide. This substance has a zinc blende crystal structure and is a direct bandgap (1.42 eV) semiconductor derived from a mixture of groups III and V that finds extensive application in high-efficiency solar cells along with high-speed microelectronics monolithic microwave frequency integrated circuits (MMIC), infrared LED, lasers fibre optic drivers, and receivers [40, 41].

GaAs exhibits improved electronic characteristics compared to Si, including greater electron mobility and electron saturation velocity, which enable operation at frequencies higher than 290 GHz. Moreover, high frequency GaAs devices have lower noise levels than Si devices, and their greater breakdown voltage allows them to operate at higher power than comparable Si devices. These features allowed GaAs IC to be used in certain radar systems, satellite communication, mobile phones, and microwave point-to-point communications [40].

GaAs-based solar cells have a higher absorption coefficient than Si-based solar cells, making them the preferred absorber material. GaAs solar cells provide the following benefits over Si-based solar cells, which have led to their usefulness in the aerospace sector: excellent efficiency of photoelectric conversion, excellent radiation resistance, and broad spectrum coverage [42]. The direct bandgap of GaAs allows for the absorption of the entire solar spectrum in less than 3 mm. According to theoretical calculations, the ideal bandgap for a single-junction solar cell to achieve the best AM1.5—1 sun conversion efficiency is approximately 1.34 eV, which is closer to the GaAs gap than the Si one (1.12 eV), despite some limitations in the fabrication of microelectronic devices. III-V semiconductors have developed more slowly than Si semiconductors because they are considerably harder to produce and optimize [40]. Si still has a few significant benefits over GaAs, though, like being less expensive, having a bigger substrate diameter (up to 12 inches), being able to employ SiO_2 as a great gate oxide or as an insulator, and having greater hole mobility.

Despite their expensive production, GaAs/Ge cells demonstrated notable advancements in the 1990s, such as decreased weight and area, increased efficiency, and smaller stowage space per launch. Multijunction cells are much more expensive than Si cells because of the substrate and growth process costs. The cost has been reasonable for space uses, but for terrestrial and commercial applications, strategies to bring down the cost were needed. The most effective way to do this has been to employ concentration [39].

4.1 Typical Cell Structure

The typical structure of GaAs solar cell had transformed a lot from homojunction GaAs solar cells to heterojunction, then dual heterojunction and finally to multiple heterojunctions as shown in Fig. 7. The GaAs/Ge heterojunction was widely investigated configuration in GaAs/Ge technology. Reducing crystal flaws in GaAs epilayers produced on Ge substrates is necessary for continued advancements in GaAs/Ge technology. The minority carrier lifetime in the epilayers is significantly impacted by these defects, which lowers the performance of GaAs solar cells made on Ge substrates. Crystal defects in the epilayers are thought to result from mismatches of those characteristics between the heterocrystals, despite the fact that GaAs's lattice constant and thermal expansion coefficient are quite similar to those of Ge.

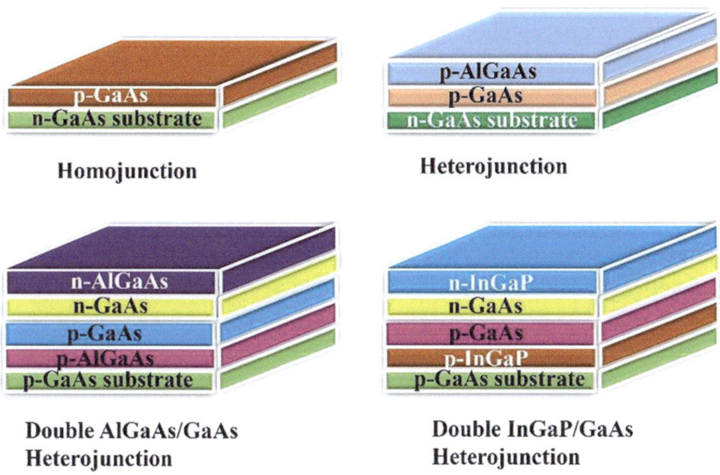

Fig. 7 Configuration improvement in GaAs solar cells

As a result, the buffer layer structure plays a crucial role in minimizing the hetero crystal incompatibilities. In order to achieve a high-quality GaAs layer on Ge substrate by metalorganic chemical vapour deposition (MOCVD), the scientists have investigated a novel buffer layer configuration. A ~ 1 μm thick GaAs layer and a preliminary ~ 0.05 μm thick $Al_xGa_{1-x}As$ (1.5 < x < 0.35) layer made up the buffer layer structure. GaAs solar cells manufactured on Ge substrate with the novel buffer layer structure showed notable gains in V_{oc} and conversion efficiency compared to those fabricated on Ge substrate with the traditional buffer layer. The new type cell's fill factor was restricted by its series resistance; nonetheless, its Vo~ and J_{sc} values were about equivalent to those of a 25.39% efficient GaAs solar cell that was constructed on a GaAs substrate [43].

By stacking, one may increase the performance of GaAs solar cells, creating multi-junction or cascade solar cells. In order to reduce reflectance and create a difference in refractive index between the window and absorber layers, the window layer modifies the electrical and optical characteristics. Every layer has a different bandgap and composition, which allows it to absorb light in certain spectral ranges. Typically, materials with a larger bandgap at the top absorb light in the visible spectrum, whereas materials with a smaller bandgap are found at the bottom. Indium Phosphide (InP), Indium Gallium Phosphide (InGaP), Gallium Indium Arsenide (GaInAs), and Aluminium Gallium Arsenide (AlGaAs) are common and efficient materials for thin window layers [42].

Another configuration having three-junction architecture was $Ga_xIn_{1-x}P$/GaAs/Ge solar cell with $Ga_xIn_{1-x}P$ as the top cell (bandgap: 1.8–1.9 eV) and GaAs as the bottom cell produced monolithically over a lattice-matched interconnecting tunnel junction. $Ga_xIn_{1-x}P$ has the same lattice constant as GaAs with a band for $x \approx 0.5$ between 1.8 and 1.9 eV for the gap energy [44].

On comparing the $Ga_{0.5}In_{0.5}P$ and GaAs junctions, the Ge junction absorbs around twice as many low energy photons. A 1.0 eV junction that is lattice matched with the other junctions should ideally take the place of the Ge bottom junction. At first, it seemed that the dilute nitride alloy $Ga_xIn_{1-x}N_yAs_{1-y}$ met these requirements, but it was later shown to be constrained by what appear to be intrinsic defects. Although $In_{0.3}Ga_{0.7}As$ has a lattice constant that is around 2% more than GaAs', it may still be utilized as the 1.0 eV junction. Gradient composition buffer layers sometimes known as "metamorphic" growth, can be used to lessen the dislocations caused by this large lattice mismatch. However, the residual threading dislocations would severely deteriorate any subsequently grown junctions with higher band-gaps. One way to prevent this deterioration of the top junctions is to develop in an inverted configuration.

With an inverted, monolithic triple-junction device that combines a lattice-matched 1.8 eV $Ga_{0.5}In_{0.5}P$ and 1.4 eV GaAs junction with a metamorphic 1.0 eV $In_{0.3}Ga_{0.7}As$ junction, exceptional solar cell performance was noticed. Organometallic vapour phase epitaxy (OMVPE) at atmospheric pressure was used to develop it on a (001) GaAs substrate that was miscut 2° toward (111). Because the top $Ga_{0.5}In_{0.5}P$ and middle GaAs lattice-matched junctions were generated before any lattice-mismatched layers, their performance was not compromised by threading dislocations that arise during mismatched growth. For the best possible performance of the solar cell, the top two junctions, which generate the majority of the electricity, were therefore produced with high crystal perfection. Three $Al_{0.3}Ga_{0.7}As$:C/GaAs:Se tunnel junctions were linked in sequence with the three junctions.

A graded $Ga_xIn_{1-x}P$ layer, which is transparent to the light meant for the bottom junction, was developed in order to reduce the dislocations in the junction. Using eight 0.25-μm-thick intermediate compositions of $Ga_xIn_{1-x}P$, the composition of the material was step graded from $Ga_{0.51}In_{0.49}P$, with a nominal lattice constant equal to that of GaAs, to $Ga_{0.22}In_{0.78}P$. The composition decreased back to $Ga_{0.25}In_{0.75}P$, which is lattice matched to the $In_{0.3}Ga_{0.7}As$ active junction, following the expansion of $Ga_{0.22}In_{0.78}P$ by 1.0 μm. The back surface field layers and passivating window were made of this $Ga_{0.25}In_{0.75}P$ composition [45].

With and without an antireflection layer, the graphene/GaAs solar cell has a conversion efficiency of 15.5% and 10.4%, respectively. The immense potential of graphene/GaAs heterostructure solar cells for practical applications has led to strong support for these studies. Certain features of graphene include unusual quantum effect, micro-scale ballistic transport, high carrier mobility, and 2.3% constant absorption of visible light. Furthermore, it has a high mechanical strength and excellent heat conductivity. The conversion efficiency can be increased by up to 15.5% using AntiReflecting Coating (ARC) [41].

4.2 *Fabrication*

Three industrial processes—Vertical Gradient Freeze (VGF), Bridgman-Stockbarger method, and Liquid Encapsulated Czochralski (LEC) growth—are used to prepare GaAs. The VGF approach grows the crystals uniformly to a certain diameter, slices them finely, rounds their edges, polishes them, and packs the wafers. By allowing vapours of gallium and arsenic to react with the furnace and deposit free molecules on the seed crystal at the end of the furnace, the Bridgman-Stockbarger process develops crystals in a horizontal zone furnace. Producing superior single crystals that may be avoided by using semi-insulating materials is the purpose of the LEC growth process [41].

Molecular Beam Epitaxy (MBE) is an advanced vacuum deposition method mainly utilized to create GaAs semiconductor and other complicated material single crystals. Since the initial reports on the use of MBE to fabricate semiconductor devices were published in the 1970s, the method has grown in importance within the semiconductor industry and become a potent instrument in the domains of materials science and condensed matter physics.

The main benefit of MBE is its capacity to precisely regulate the development and structure of the materials it creates, such as controlling the thickness of layers at the atomic-layer level and creating a clean interface between layers. This fine control, particularly the capacity to produce sharp layer transitions, has created a plethora of opportunities for low-dimensional system study.

To give a brief synopsis of the MBE process, beams of elemental or compound molecules are directed onto a substrate that is heated, where they condense and create a crystalline layer. Ultra-high vacuum (UHV) conditions are used throughout this procedure to guard against contamination and guarantee the purity of the crystal that is developing. Heat is applied to the source materials, which are frequently put into effusion cells, causing them to sublimate and emit a focused stream of atoms or molecules onto the substrate. A large-area, single-crystal film may be produced by precisely registering the atoms in the formed film by control of growth parameters, such as substrate temperatures.

MBE technology has become much more capable and applicable over time as a result of technology improvements. Enhancements in real-time monitoring methods like Reflection High-Energy Electron Diffraction (RHEED) and improved effusion cell design have allowed for real-time diagnostics of the growth process and significantly improved the quality of growth processes [46]. Numerous considerations, including its semiconductor qualities, strong thermal stability, and simplicity of handling in a vacuum environment, influenced this decision [46].

The Group V source's trihydrides and the Group II source's room temperature volatile, thermally stable precursors were the foundation for MOVPE's early development. The fundamental characteristics of both are that they are both chemically stable at room temperature and volatile in an appropriate carrier gas stream, often hydrogen [47]. Either the hot vapour above the surface or the hot surface itself can be the site of the precursors' reaction to produce the III-V compound on the substrate.

The following is the stoichiometric reaction for the development of GaAs:

$$(CH_3)_3Ga + AsH_3 = GaAs + 3CH_4$$

The concentration of the carbene (CH_2) reaction product decreases as AsH_3 increases. Gas-phase homolysis of TMGa ($Ga(CH_3)_3$) initiates the reaction pathway, producing methyl radicals ($.CH_3$) and dimethylgallium ($Ga(CH_3)_2$). Methyl radicals have two major functions that they may play, and this is crucial in all MOCVD procedures for III-V semiconductor deposition. Stable methane and a hydrogen radical can be produced when methyl radicals react with the surrounding hydrogen carrier gas. Arsine (AsH_3) and methyl radicals can react to produce stable methane and AsH_2 [7]. TMGa will undergo pyrolysis around 500 °C, whereas arsine breakdown needs a temperature of about 700 °C [47].

Both in the vapour and on the surface, ligands from TMGa and arsine may be gradually removed. The temperature of the vapour above the substrate as well as the surface temperature will determine the dominating route. As a general rule, when the vapour reaction rate is slower and surface adsorption is more effective, more of the reaction process will take place on the surface at lower temperatures. It's also critical to recognize that the surface has two distinct adsorption sites: the As sites, which will bind to cations, and the Ga sites, which will bind to anion species.

The thin film practicality, flexibility, and high efficiency of GaAs photovoltaic (PV) cells have made them the subject of intense research for flexible energy harvesting systems. GaAs-based thin film photovoltaic cells, however, are incompatible with fabric carriers, which limits their wearable platform applications. Thin film transfer methods utilizing a double-layered sacrificial layer, Au–Au bonding, a Cr/Au bilayer to provide metal stress, and an etchant for rapid epitaxial lift-off (ELO) are used to address this issue. Furthermore, a polyimide layer adhered to the underside of the fabric substrate speeds up the lateral etching process by spontaneous bending while simultaneously shielding the delicate epi layer. By using these methods, GaAs thin film PV epi structures may be successfully transferred onto fabric platforms without compromising crystal quality. The standard PV cell production method is used to create fabric-based GaAs PV cells [48].

Highly crystalline long GaAs nanowires (NWs) with minimal crystal defects are synthesized economically by chemical vapour deposition and configured into novel Schottky photovoltaic structures by simply using asymmetric Au–Al contacts, thereby enabling cost-effective GaAs solar cells for smart systems such as transparent and flexible photovoltaics. By normalizing to the projection area of the NW, the single NW Schottky device exhibits a record high apparent energy conversion efficiency of 16% under air mass 1.5 global illumination, without any doping profiles such as p–n junction and complex coaxial junction architectures. The exceptional versatility and bright future of these GaAs NW Schottky photovoltaics for the next generation of smart solar energy harvesting devices are clearly demonstrated by the ease with which these Schottky cells can be integrated upon glass and plastic substrates for transparent and flexible solar cells [49].

5 Cadmium Telluride (CdTe)

Thin film technology can reduce the material requirement for device fabrication as it utilizes materials having nano thickness. This can reduce the "cost per Watt" of the device. Highly uniform and repeatable absorbers can be developed with proper film deposition techniques, which favour large-scale device fabrication. Minimization of material precursors, precursor availability, ease of deposition, and post-deposition conditions are researched as they can affect the fabrication cost and improve efficiency. When properly designed, with reduced interface defects, tandem devices with multiple junctions of thin films can be developed. The stacked thin films with different bandgaps can utilize most of the solar spectrum, improving efficiency.

A II-VI chalcogenide family member, Cadmium telluride (CdTe) benefits from its low cost, optimum direct bandgap (1.43–1.74 eV), higher absorption coefficient (10^4–10^5 cm^{-1}), conductivity tuning, and higher theoretical conversion efficiency of 32%. The high absorption coefficient of CdTe enable the thin films having a thickness of 1 μm to absorb 90% of visible light in the range of 550–800 nm. The greatest benefit of II-VI semiconductors is the smaller influence of grain boundaries on the electronic properties of thin films. A relatively lower recombination centre density and the presence of boundary depletion layers could be the reason for the reduced influence [50].

Depending on the presence of Cd or Te vacancy, the material can be made in p or n-type. Higher vapour pressure of Cd can create Cd re-evaporation, resulting in Cd deficiency in thin films and is responsible for the p-type conductivity [51]. The higher mobility and significant crystallinity are responsible for the higher conductivity in CdTe thin films. The CdTe film resistivity ranges from 2–25 Ω-cm. It can be further decreased by proper doping. Efficiencies > 22% imply the importance of CdTe thin films in PV applications.

The thin film crystal structure influences the growth orientation of CdTe thin films. Films are mostly (111) oriented at low temperatures but are statistically oriented above 450 °C. Irrespective of the substrate's nature, the films grow parallel to the substrate when Cd and Te atoms are arranged hexagonally as alternative layers when viewed perpendicular to 111 direction. The high material stability arises from their strong binding energy and high melting temperature of 1041 °C. The heavy Cd metal can create environmental hazards, but their higher binding energy in CdTe compound permits Cd liberation only above 900 °C. This permits their application as a functional material [52].

The long wavelength absorption of visible light is limited by the bandgap of CdTe (~1.45 eV). This is relaxed either by doping or using different absorber layer structures. Se with a lower lattice mismatch with Te is a suitable choice of dopant, as it provides good solubility in CdTe. Se inclusion can enhance the short circuit current density of the device and also the graded $CdSe_xTe_{1-x}$ layer can reduce surface defect density [53].

The p-type conductivity of CdTe can be enhanced by an improvement in hole density made by copper incorporation. Optimum doping of copper can reduce CdTe film defects and strain and improve the average crystallite size. With doping concentration, crystallite size is enhanced from 25 to 36 nm. It can also create crystal lattice reorientation, causing a phase change in CdTe films. The band-gap of Cu-doped films lies in the range of 1.46–2.04 eV, enhancing visible absorption. Copper enhances carrier concentration in films with reduced recombination and results in better conductivity [50].

Bandgap tunability, lower resistivity, robustness, high chemical stability, and well-established numerous fabrication techniques of CdTe favour their applications in the PV industry. However, the formation of interfacial defects and vacancies needs to be tackled, which can further reduce the bulk and grain boundary recombinations and improve efficiency [54]. CdTe has significant applications as solar cells and substrates for superior infrared detectors based on HgCdTe, gamma and X-ray detectors.

Bulk carrier transport can be achieved by the larger grains in CdTe films, the formation of which depends on the synthesis technique. A p-CdTe/n-CdS with suitable passivation treatments creates a good heterojunction, which most device designs utilize. Physical and chemical methods of thin film fabrication have their own pros and cons. The former suffers from vacuum conditions, increasing the cost of production. Chemical fabrication creates difficulty in purity and composition control. Both methods can be used for large-area fabrication. However, different synthesis approaches alter the material features, such as conductivity type and carrier lifetime.

The two major types of CdTe fabrication include confined space sublimation (CSS) and pulsed laser deposition (PLD).

5.1 Typical Cell Structure

The quality of thin films and device efficiency dramatically depend on the fabrication approach followed.

Most commonly, CdTe-based solar cell designs use a CdTe/CdS heterojunction, following a conventional solar cell device design: glass/SnO_2/CdS/CdTe/back contact. A typical routine for improving the intrinsic, optical and electrical qualities of CdTe and CdS thin films is $CdCl_2$ treatment. The associated morphological enhancements include film density improvement and grain size modification. An increase in grain size can improve carrier mobility, thereby electrical conduction. $CdCl_2$ treatment can also cause recrystallization by improving atomic mobility near the annealing temperature. Apart from morphological and structural enhancements, the treatment provides p-type doping for films. CdTe resistivity is rapidly reduced (from 10^4–10^5 Ω-cm to 100 Ω-cm) by the inclusion of $(V_{Cd}\text{–}Cl)^-$ complexes, behaving like acceptor-like dopants. The lower formation energy of the complexes creates shallower energy levels than that of Cd vacancies and improves the conductivity. However, the treatment needs to be optimized for proper film adhesion [55, 56].

Figure 8 gives the schematic of the most common CdTe solar cell structure. Both substrate and superstrate structures are implemented, but the latter influences more on device efficiency. Post-deposition treatments on CdTe can be well executed in a superstrate structure, thus forming a compact CdTe layer with less interlayer diffusion. The substrate choice depends on application requirements and can be rigid, flexible, transparent, or opaque. Substrate design should be employed while using an opaque substrate. Hence, depending on the fabrication approaches, deposition and post-deposition conditions, and application requirements, the substrate can be chosen [57].

Highly conductive but transparent semiconductor oxides with a bandgap above 3.2 eV are preferred as front contacts, as they can ensure light transmittance along with charge collection. ITO, FTO, AZO, etc. are the common TCOs, but non-conventional TCOs like tin compounds, Cd_2SnO_4, and Zn_2SnO_4 are also developed. Deposition of an intrinsic oxide layer upon TCOs are appreciated for optimum band alignment and maximum carrier collection, reducing device shunting, lowering the buffer layer thickness, and reducing the diffusion of impurity elements either from TCOs or substrates into CdTe absorber [57]. A window layer is deposited upon the intrinsic oxide layer to absorb unutilized light by the absorber layer with a p–n junction formation. The most common approach requires a thicker CdS window layer. But the bandgap of the material ~ 2.4 eV causes non-negligible active wavelength absorption and reduces the light incidenting on the CdTe layer. This is enhanced by the SnO_2 TCO layer having only 80% transmittance and ~ 10–15 Ω/sq sheet resistivity. A lower window layer thickness is favoured, and the thickness is limited to

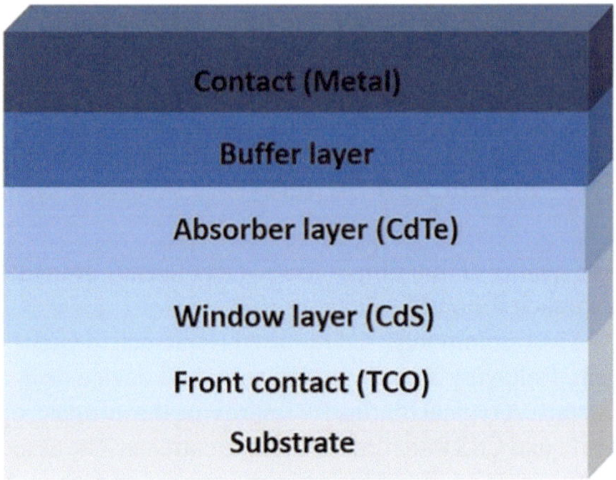

Fig. 8 Schematic of typical CdTe/CdS solar cells

less than 100 nm. CdTe is a highly stable, most explored absorber material with an optimum bandgap. Numerous approaches have been developed for the deposition, giving higher efficiency devices [51].

Back contact selection is complex for CdTe absorbers, as their higher electron affinity demands metals with higher work functions. Most designs use copper-incorporated back contacts for higher charge collection. But copper diffusion creates device instability together with a higher device efficiency. Thus, Cu-free back contacts like Sb_2Te_3/Ni, Ni:P, and Sb_2Te_3/Mo are developed at the expense of reduced efficiency. A heavily doped absorber surface or a buffer layer upon the absorber can reduce the probable formation of Schottky barriers [55]. A schematic of band alignment in CdTe absorbers is shown in Fig. 9.

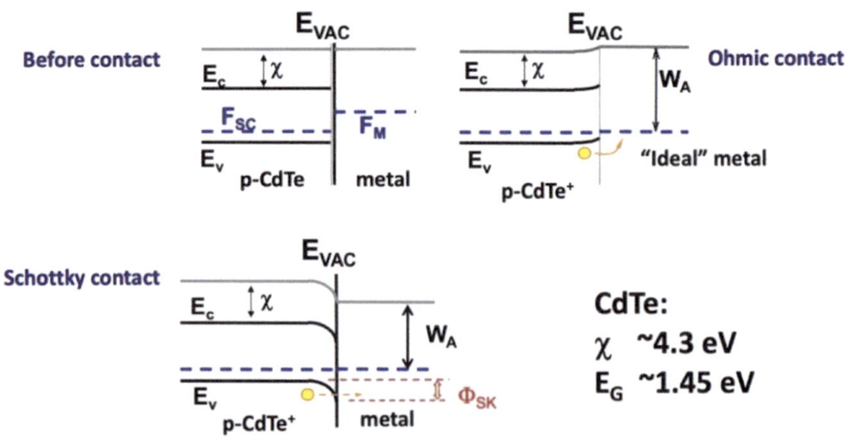

Fig. 9 Representation of band alignment in CdTe absorbers with ohmic and Schottky back contact. Reproduced from Romeo et al. [57]

5.2 *Fabrication*

5.2.1 Confined Space Sublimation (CSS)

As discussed earlier, the reduced light incidence on the active layer and the conventional wet fabrication processes can form additional contaminants and cause further interfacial defects. Altogether, the conventional device design and fabrication approaches need to be re-evaluated to get better film quality and efficiency. Meanwhile, PVD techniques use a vacuum environment for film fabrication and provide precise control of deposition parameters. Confined space sublimation (CSS) and pulsed laser deposition (PLD) are the two thin film deposition techniques that can produce high-quality stoichiometric CdTe thin films with enhanced efficiency.

Figure 10 shows the schematic of the CSS fabrication set-up. The method undergoes precursor sublimation followed by condensation over a substrate, where the target and substrate are closely arranged between 1 and 3 mm distance and the whole set-up is maintained in an inert ambient. High-temperature source heating can be done, which sublimes the source material and imparts high KE to the sublimed atoms. CdTe thin film deposition usually requires 500–700 °C source temperature. This technique can also achieve higher deposition rates. Since Cd has higher vapour pressure, the entire process is carried out at a higher Ar pressure, typically between 1 and 100 mbar, to avoid the possible material re-evaporation. Higher sample temperature treatments for large grain formation can be achieved by the CSS method. CdTe films fabricated by CSS have a unique feature of as deposited large grain size greater than 1 μm. When treated with $CdCl_2$, the grains get enhanced by smaller grain elimination and grain boundary re-growth, modifying film quality. The treatment successfully passivates the grain boundaries and also reduces CdS/CdTe interfacial lattice mismatch [55].

In CSS deposition of CdTe, the film growth rates depend on source temperature, chamber pressure source-substrate temperature difference and distance. 600 °C

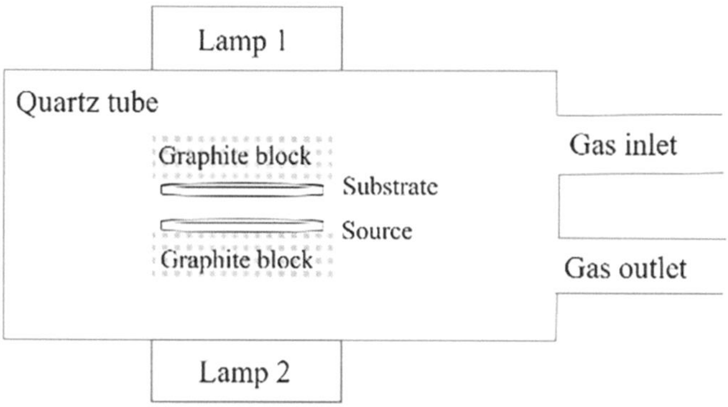

Fig. 10 Schematic of the CSS fabrication set-up

source temperature, 10^{-2} torr pressure and 2–3 mm separation distance are generally used. In most cases, the CdTe films were found to have 1–2 μm sized grains.

Research focussing on altering the back contact and TCO materials successfully achieved higher efficiency devices. Device having a structure: glass/ITO/SnO_2/CdS/CdTe/Sb_2Te_3/Mo is of particular interest because of the replacement of Cu-graphite back contact with a stable ohmic p-type Sb_2Te_3 semiconductor layer. The possibility of Cu diffusion into CdTe is eliminated here, which reduces device degradation. Such a configuration exhibits an efficiency of 14.6% with J_{sc} of 23 mA cm^{-2} and V_{oc} of 858 mV. The importance of the order of absorber layer deposition clearly suggests the necessity of CdTe deposition upon the $CdCl_2$-treated CdS layer. The highly compact CdTe layers have lower diffusion into the CdS layer, thus creating the necessary heterojunction properties [55].

Replacing SnO_2 with more transparent and lower resistive (~1.5×10^{-4} Ω-cm) Cd_2SnO_4 TCO enhances light absorption by the CdTe absorber layer. The lower resistivity further reduces TCO layer thickness, thereby enhancing device J_{sc}. The homogeneous single-phase spinel polycrystalline microstructure of Cd_2SnO_4 is responsible for the enhanced optical properties and electrical conductivity. A $ZnSnO_x$ buffer layer with a higher bandgap can enable ZTO/CdS interdiffusion, which improves device features. The most probable Te diffusion to the CdS layer and the corresponding generation of $CdS_{1-y}Te_y$ alloy, limiting heterojunction features and device efficiency, can be suppressed by oxygenation of the CdS window layer. Highly efficient CdTe solar cells can be developed with modified fabrication approaches and device designs [58].

5.2.2 Pulsed Laser Deposition (PLD)

When highly energetic laser pulses incident on target material, the target temperature rapidly rises to $> 10^{11}$ Ks^{-1}. This causes the emission of target elements with high KE and produces plasma, which condenses over substrates and recrystallizes, forming CdTe films. The high kinetic energy (1–100 eV) of the emitted atoms provides greater surface mobility and demands lower substrate heating for crystallization [59]. The stoichiometry, deposition rate, and thickness of films can be controlled easily by using pulse power. The highest advantage of PLD is the extensive area deposition and control of deposition parameters. Single crystalline, polycrystalline and epitaxial thin films of CdTe on different substrates are fabricated by PLD.

To analyse the impact of fabrication technique, Bhattacharya et al. compared CdTe films deposited by PLD with thermal evaporated (TE) films. PLD holds higher deposition rates (500 Å/30 ns) than thermal evaporation (100 Å/s). While keeping the thickness and deposition temperature constant, PLD fabricated CdTe films show better crystallinity with higher (111) preferred orientation. The high KE of the emitted atoms are responsible for better grain growth and enhanced crystallinity [51].

With a lower resistivity of $6–7 \times 10^4$ Ω-cm and 1.5 eV bandgap, PLD deposited films perform better than their TE counterpart, having resistivity of $3–4 \times 10^5$ Ω-cm and no specified absorption edge. Also, mass dependence of ejected atom energy is more pronounced in TE than in PLD. Thus, PLD fabrication maintains film stoichiometry more than in TE. As already mentioned, Cd vacancies create holes that are responsible for conductivity. Though TE generates more Cd vacancies than PLD, the more minor crystallization in TE limits conductivity [51].

Another aspect of PLD is the easier formation of the hexagonal CdTe phase, which is otherwise difficult to achieve. The metastable nature of the hexagonal phase limits its development. CdTe phase sensitivity towards pulse energy can be tuned easily for developing hexagonal phases. Compared to the cubic phase, the hexagonal phase has a lower bandgap, respectively, 1.6 eV and 1.54 eV [59].

Since PLD and CSS are the two major fabrication techniques for CdTe-based device fabrication, both methods are incorporated together, forming the device FTO/CdS/CdTe/ZnTe:Cu/(ZnTe|CdTe)n/ZnTe:Cu/Au with an efficiency of 12.80% and V_{oc} of 781 mV. Here, the ZnTe/CdTe layers are deposited using cyclical PLD, and the CdTe layer is deposited using the CSS method. The study suggests the importance of ZnTe/CdTe composite layers, as their absence reduces the efficiency to 8.08%. The composite layers form single offset superlattice structures with a bandgap lower than the two constituents, absorbing lower energy photons. The resultant sub-band formation enhances carrier conduction and enhances efficiency. A similar trend can be observed in QE, showing a maximum of 82% at 580 nm for the device with a composite structure [53].

CdTe thin films are well-explored materials and achieved device efficiencies > 22%. Further research focuses on achieving the theoretical efficiency of 32%, and one of the significant approaches towards that extent is the selection of suitable p-type back surface field materials for creating a highly ohmic junction. ZnTe:Cu with optimum copper content is one such choice providing device enhancements. The Cu incorporation reduces ZnTe resistivity because of the Zn replacement by Cu atoms. However, a higher Cu content increases its interaction with the CdTe layer, creating point defects [60].

Though new absorber materials are being developed for better efficiency, research focusing on analysing the underlying recombination mechanisms and band alignment at front and back surfaces, absorber-light interactions, dopant influence on defect densities, minority carrier lifetime, etc. needs to be carried out for extracting maximum yield from well-established device designs. Fabrication approaches that satisfy good film morphology and grain boundaries, stoichiometry, uniform doping, and electrode contact must be chosen for efficient device generation [52].

6 Conclusion

Thin film photovoltaics, which provide notable benefits over conventional crystalline silicon (c-Si) solar cells, have completely changed the field of solar cells. These technologies, which lower material prices and streamline production processes, make solar energy more accessible. Innovative applications including wearable electronics, portable solar devices, and building-integrated photovoltaics (BIPV) are made possible by their ability to be deposited on flexible substrates. This adaptability and versatility increase solar energy's potential applications across several industries. Moreover, improvements in thin film performance and efficiency, especially in CIGS and GaAs, show that they have the ability to compete with conventional c-Si cells. Their attractiveness is further enhanced by the environmental benefits of simplified production methods and lower material utilization.

For both terrestrial and space photovoltaics, thin film solar cells (TFSC) are a promising technology that provide a multitude of options for device design and manufacturing. Different layers (contact, buffer, absorber, reflector, etc.) can be deposited on a range of substrates (flexible or rigid, metal or insulator) using a number of processes (PVD, CVD, ECD, plasma-based, hybrid, etc.). Because of their adaptability, the layers may be engineered and tailored to enhance device performance. Large-area devices, which are necessary for practical applications, necessitate complicated thin film device manufacturing that calls for careful control throughout the whole production sequence. As has been shown commercially for various cells, a thorough understanding of thin film deposition methods may aid in the creation of high-efficiency devices over broad regions. Focused research and development efforts are necessary for novel materials and gadgets, as well as creative production techniques.

Second-generation solar cells including a-Si, CdTe, CIGS, and GaAs, provide an appealing blend of stability, efficiency, affordability, and future potential. While a-Si offers cost-effective options for large-area applications, CdTe stands out for its affordability and modest efficiency. The a-Si:H-based thin film solar cell has the potential to be built on a megawatt scale for certain applications. Despite limited production output, low efficiency, stability issues, and comparable cost to crystalline Si cells, this cell technology offers a straightforward and standardized manufacturing method for integrated modules on hard and flexible substrates. Even though there are improvements in efficiency and stability, the multijunction cell will still fall behind the more stable, efficient, and cost-effective c-Si cell. It also gets degrades when exposed to light. This problem halts its advancement in the PV market. Despite huge industrial endeavours to make CdTe and CIS-based cells on a megawatt scale, progress has been slow because to manufacturing reliability issues and high device costs. Simple production processes, such as screen printing, spray pyrolysis, electroplating, or chemical solution deposition on flexible substrates, with lesser efficiencies, might improve the economic viability of these cells. CIGS has improved efficiency and long-term stability under varying circumstances, making it a formidable competitor for advanced applications. Additionally, they utilize less material due to their high absorption coefficient in visible range. GaAs, while very efficient, is nevertheless confined to high-performance applications because to its high production costs.

Continuous improvements in manufacturing processes and materials research promise an even brighter future for these technologies, increasing their practicality and widespread use in a variety of solar energy applications. Thin film photovoltaics will be essential to the worldwide shift to affordable and sustainable solar energy as long as research and development keeps pushing technical progress, which will have a significant influence on the solar cell market.

Acknowledgements Arya Mohan acknowledge the Kerala State Council for Science, Technology, and Environment, Government of Kerala, India, for providing a research fellowship (Ref. 317). Heera S acknowledge DST, Government of India (IF200250) for providing fellowship through the Inspire scheme. Deepa KG thank KSCSTE, Government of Kerala, India for funding through SRS scheme (KSCSTE/268/2024/SRSPS).

References

1. M. Inès et al., Progress and prospects for ultrathin solar cells. Nat. Energy **5**, 959–972 (2020). https://doi.org/10.1038/s41560-020-00714-4
2. M.A. Green et al., Solar cell efficiency tables (Version 64). Prog. Photovoltaics Res. Appl. **32**, 425–441 (2024). https://doi.org/10.1002/pip.3831
3. T.D. Lee, A.U. Ebong, A review of thin film solar cell technologies and challenges. Renew. Sustain. Energy Rev. **70**, 1286–1297 (2017). https://doi.org/10.1016/j.rser.2016.12.028
4. T. Salameh et al., Review of solar photovoltaic cooling systems technologies with environmental and economical assessment. J. Clean. Prod. **326**, 129421 (2021). https://doi.org/10.1016/j.jclepro.2021.129421
5. A.E. Delahoy et al., New technologies for CIGS photovoltaics. Sol. Energy **77**(6), 785–793 (2004). https://doi.org/10.1016/j.solener.2004.08.012
6. J. Ramanujam et al., Flexible cigs, CdTe and a-Si:H based thin film solar cells: a review. Prog. Mater. Sci. **110**, 100619 (2020). https://doi.org/10.1016/j.pmatsci.2019.100619
7. K.L. Chopra, P.D. Paulson, V. Dutta, Thin-film solar cells: an overview. Prog. Photovoltaics Res. Appl. **12**(2–3), 69–92 (2004). https://doi.org/10.1002/pip.541
8. D.A. Barkhouse et al., Device characteristics of a 10.1% hydrazine-processed $Cu_2ZnSn(SE,S)_4$ solar cell. Prog. Photovoltaics Res. Appl. **20**(1), 6–11 (2011). https://doi.org/10.1002/pip.1160
9. N. Mufti et al., Review of CIGS-based solar cells manufacturing by structural engineering. Sol. Energy **207**, 1146–1157 (2020). https://doi.org/10.1016/j.solener.2020.07.065
10. Md.M. Moon et al., Comparative study of the second generation a-si:H, CdTe, and CIGS thin-film solar cells. Adv. Mater. Res. **1154**, 102–111 (2019). https://doi.org/10.4028/www.scientific.net/amr.1154.102
11. A. Bouloufa, K. Djessas, A. Zegadi, Numerical simulation of $CuIn_xGa_{1-x}Se_2$ solar cells by amps-1d. Thin Solid Films **515**(15), 6285–6287 (2007). https://doi.org/10.1016/j.tsf.2006.12.110
12. J. Ramanujam, U.P. Singh, Copper indium gallium selenide based solar cells—a review. Energy Environ. Sci. **10**(6), 1306–1319 (2017). https://doi.org/10.1039/c7ee00826k
13. Federal Laboratory for Materials Testing and Research. Flexible solar cells with record efficiency of 22.2% (n.d.). The Federal Council [Preprint]. Available at: http://www.empa.ch
14. K.H. Ong, R. Agileswari, B. Maniscalco, P. Arnou, C.C. Kumar, J.W. Bowers, M.B. Marsadek, Review on substrate and molybdenum back contact in CIGS thin film solar cell. Int. J. Photoenergy 2018, 1–14 (2018). https://doi.org/10.1155/2018/9106269
15. E.T. Efaz, A review of primary technologies of thin-film solar cells. Eng. Res. Express **3**, 032001 (2021). https://doi.org/10.1088/2631-8695/ac2353
16. D. Shin et al., Effects of the incorporation of alkali elements on $Cu(In,Ga)Se_2$ thin film solar cells. Sol. Energy Mater. Sol. Cells **157**, 695–702 (2016). https://doi.org/10.1016/j.solmat.2016.07.015
17. A. Chirilă et al., Potassium-induced surface modification of $Cu(In,Ga)Se_2$ thin films for high-efficiency solar cells. Nat. Mater. **12**(12), 1107–1111 (2013). https://doi.org/10.1038/nmat3789
18. M. Kaelin, D. Rudmann, A.N. Tiwari, Low cost processing of CIGS thin film solar cells. Sol. Energy **77**(6), 749–756 (2004). https://doi.org/10.1016/j.solener.2004.08.015

19. H. Karaagac et al., Characterization of $CuIn_{0.7}Ga_{0.3}Se_2$ thin films deposited by single stage thermal evaporation process. Phys. Status Solidi C **14**(12) (2017). https://doi.org/10.1002/pssc.201700145
20. K. Machkih, R. Oubaki, M. Makha, A review of CIGS thin film semiconductor deposition via sputtering and thermal evaporation for solar cell applications. Coatings **14**(9), 1088 (2024). https://doi.org/10.3390/coatings14091088
21. P.M.P. Salomé et al., A comparison between thin film solar cells made from co-evaporated $CuIn_{1-x}Ga_xSe_2$ using a one-stage process versus a three-stage process. Prog. Photovoltaics Res. Appl. **23**(4), 470–478 (2014). https://doi.org/10.1002/pip.2453
22. P. Sigmund, Theory of sputtering. I. Sputtering yield of amorphous and polycrystalline targets. Phys. Rev. **184**(2), 383–416 (1969). https://doi.org/10.1103/physrev.184.383
23. B. Kim, B.K. Min, Strategies toward highly efficient CIGSE thin-film solar cells fabricated by sequential process. Sustain. Energy Fuels **2**(8), 1671–1685 (2018). https://doi.org/10.1039/c8se00158h
24. W. Xin et al., Investigation on the diffusion mechanism of GA element and regulation of the V-shaped bandgap in CIGS thin films based on magnetron sputtering with a quaternary target. Sol. Energy **273**, 112510 (2024). https://doi.org/10.1016/j.solener.2024.112510
25. M. Syabriyana, Z. Ulma, One-Step Sputtering $CuInGaSe_2$ (CIGSe) Process for Thin Film Solar Cells: Progress and Challenges (Preprint, n.d.), p. 131. https://doi.org/10.32672/picmr.v6i1.1162
26. M. Acciarri et al., CIGS thin films grown by hybrid sputtering-evaporation method: properties and PV Performance. Sol. Energy **175**, 16–24 (2018). https://doi.org/10.1016/j.solener.2018.02.024
27. S. Guha, J. Yang, A. Banerjee, Amorphous silicon alloy photovoltaic research? Present and future. Prog. Photovoltaics Res. Appl. **8**(1), 141–150 (2000). https://doi.org/10.1002/(sici)1099-159x(200001/02)8:1%3c141::aid-pip305%3e3.0.co;2-i
28. K. Morigaki, C. Ogihara, Amorphous semiconductors: structure, optical, and electrical properties, in *Springer Handbook of Electronic and Photonic Materials* (2006), pp. 565–580. https://doi.org/10.1007/978-0-387-29185-7_25
29. H. Kang, Crystalline silicon vs. amorphous silicon: the significance of structural differences in photovoltaic applications. IOP Conf. Ser. Earth Environ. Sci. **726**(1), 012001 (2021). https://doi.org/10.1088/1755-1315/726/1/012001
30. B. Rech, H. Wagner, Potential of amorphous silicon for solar cells. Appl. Phys. A Mater. Sci. Process. **69**(2), 155–167 (1999). https://doi.org/10.1007/s003390050986
31. T.D. Lee, A. Ebong, Thin film solar technologies: a review, in *2015 12th International Conference on High-capacity Optical Networks and Enabling/Emerging Technologies (HONET)* (2015), pp. 1–10. https://doi.org/10.1109/honet.2015.7395441
32. S. Sivaraj, R. Rathanasamy, G.V. Kaliyannan, H. Panchal, A. Jawad Alrubaie, M. Musa Jaber, Z. Said, S. Memon, A comprehensive review on current performance, challenges and progress in thin-film solar cells. Energies 15(22), 8688 (2022). https://doi.org/10.3390/en15228688
33. M. Stuckelberger et al., Review: progress in solar cells from hydrogenated amorphous silicon. Renew. Sustain. Energy Rev. **76**, 1497–1523 (2017). https://doi.org/10.1016/j.rser.2016.11.190
34. H. Meng, X. Wu, F. Ma, Q. Zeng, L. Zhou, Structural optimization and growth of intrinsic hydrogenated amorphous silicon films by HWCVD. Sol. Energy Mater. Sol. Cells **271**, 112835 (2024). https://doi.org/10.1016/j.solmat.2024.112835
35. C.B. Singh, S. Bhattacharya, S. Prayogi, U.S. Patel, P.B. Bhargav, N. Ahmed, A new look at an explanation of band gap of PECVD grown a-Si:H thin films using absorption spectra, spectroscopic ellipsometry, Raman, and FTIR spectroscopy. Opt. Mater. **154**, 115809 (2024). https://doi.org/10.1016/j.optmat.2024.115809
36. J.K. Rath, M. Brinza, Y. Liu, A. Borreman, R.E.I. Schropp, Fabrication of thin film silicon solar cells on plastic substrate by very high frequency PECVD. Sol. Energy Mater. Sol. Cells **94**(9), 1534–1541 (2010). https://doi.org/10.1016/j.solmat.2010.01.013
37. J.Y. Lee, J.E. Ju, C. Lee, S.M. Won, K.J. Yu, Novel fabrication techniques for ultra-thin silicon based flexible electronics. Int. J. Extreme Manuf. **6**(4), 042005 (2024). https://doi.org/10.1088/2631-7990/ad492e

38. H. Cui, P.R. Campbell, M.A. Green, Compatibility of glass textures with e-beam evaporated polycrystalline silicon thin-film solar cells. Appl. Phys. A **111**, 935–942 (2013). https://doi.org/10.1007/s00339-012-7318-3
39. L. El Chaar, L.A. Lamont, N. El Zein, Review of photovoltaic technologies. Renew. Sustain. Energy Rev. **15**(5), 2165–2175 (2011). https://doi.org/10.1016/j.rser.2011.01.004
40. M. Bosi, C. Pelosi, The potential of III-V semiconductors as terrestrial photovoltaic devices. Prog. Photovoltaics Res. Appl. **15**(1), 51–68 (2006). https://doi.org/10.1002/pip.715
41. H. Victor, A detailed review on Si, GaAs, and CIGS/CdTe based solar cells and efficiency comparison. Przegląd Elektrotechniczny **1**(12), 11–20 (2020). https://doi.org/10.15199/48.2020.12.02
42. A. Adeyinka et al., a-review-of-current-trends-in-thin-film-solar-cell-technologies (2023), p. 17. https://www.researchgate.net/publication/368925684_a-review-of-current-trends-in-thin-film-solar-cell-technologies
43. K. Takahashi et al., Characteristics of GaAs solar cells on GE substrate with a preliminary grown thin layer of AlGaAs. Sol. Energy Mater. Sol. Cells **50**(1–4), 169–176 (1998). https://doi.org/10.1016/s0927-0248(97)00141-4
44. S.P. Philipps, F. Dimroth, A.W. Bett, High efficiency III–V multijunction solar cells. Solar Cells 353–381 (2013). https://doi.org/10.1016/b978-0-12-386964-7.00012-3
45. J.F. Geisz et al., High-efficiency GaInP/GaAs/InGaAs triple-junction solar cells grown inverted with a metamorphic bottom junction. Appl. Phys. Lett. **91**(2) (2007). https://doi.org/10.1063/1.2753729
46. T. Wang, H. Song, K. He, Structural design and molecular beam epitaxy growth of GaAs and in as heterostructures for high mobility two-dimensional electron gas. Quantum Front. **3**(1) (2024). https://doi.org/10.1007/s44214-024-00061-5
47. S.J.C. Irvine, P. Capper, Introduction to metalorganic vapor phase epitaxy, in *Metalorganic Vapor Phase Epitaxy (MOVPE)* (2019), pp. 1–18. https://doi.org/10.1002/9781119313021.ch1
48. Y. Yun et al., Flexible fabric-based GaAs thin-film solar cell for wearable energy harvesting applications. Sol. Energy Mater. Sol. Cells **246**, 111930 (2022). https://doi.org/10.1016/j.solmat.2022.111930
49. N. Han et al., High-performance GaAs nanowire solar cells for flexible and transparent photovoltaics. ACS Appl. Mater. Interfaces **7**(36), 20454–20459 (2015). https://doi.org/10.1021/acsami.5b06452
50. D. Adhikari, M.M. Junda, E. Bastola, P. Koirala, R.J. Ellingson, R.W. Collins, N.J. Podraza, Glancing angle deposited CdTe: nanostructured films and impact on solar cell performance. Surf. Coat. Technol. **381**, 125127 (2020). https://doi.org/10.1016/j.surfcoat.2019.125127
51. P. Bhattacharya, D.N. Bose, Pulsed laser deposition of CdTe thin films for heterojunctions on silicon. Semicond. Sci. Technol. **6**(5), 384 (1991). http://iopscience.iop.org/0268-1242/6/5/012
52. D. Bonnet, The CdTe thin film solar cell-an overview. Int. J. Sol. Energy **12**(1–4), 1–14 (1992). https://doi.org/10.1080/01425919208909746
53. S. He, H. Lu, B. Li, J. Zhang, G. Zeng, L. Wu, W. Li, W. Wang, L. Feng, Study of CdTe/ZnTe composite absorbing layer deposited by pulsed laser deposition for CdS/CdTe solar cell. Mater. Sci. Semicond. Process. **67**, 41–45 (2017). https://doi.org/10.1016/j.mssp.2017.05.009
54. A.F. Butt, M. Azhar, H. Yousaf, K.M. Batoo, D. Khan, M. Noman, M.U. Chaudhry, S. Naseem, S. Riaz, Chemically processed CdTe thin films for potential applications in solar cells–effect of Cu doping. Heliyon **10**(3) (2024). https://doi.org/10.1016/j.heliyon.2024.e24492
55. N. Romeo, A. Bosio, R. Tedeschi, V. Canevari, Growth of polycrystalline CdS and CdTe thin layers for high efficiency thin film solar cells. Mater. Chem. Phys. **66**(2–3), 201–206 (2000). S0254-0584(00)00316-3
56. T. Potlog, N. Spalatu, V. Fedorov, N. Maticiuc, C. Antoniuc, V. Botnariuc, J. Hiie, T. Raadik, V. Valdna, The performance of thin film solar cells employing photovoltaic ZnSe/CdTe, CdS/CdTe and ZnTe/CdTe heterojunctions, in *2011 37th IEEE Photovoltaic Specialists Conference* (IEEE, 2011), pp. 001365–001370. https://doi.org/10.1109/PVSC.2011.6186211

57. A. Romeo, E. Artegiani, CdTe-based thin film solar cells: past, present and future. Energies **14**(6), 1684 (2021). https://doi.org/10.3390/en14061684
58. X. Wu, High-efficiency polycrystalline CdTe thin-film solar cells. Sol. Energy **77**(6), 803–814 (2004). https://doi.org/10.1016/j.solener.2004.06.006
59. S.K. Pandey, U. Tiwari, R. Raman, C. Prakash, V. Krishna, V. Dutta, K. Zimik, Growth of cubic and hexagonal CdTe thin films by pulsed laser deposition. Thin Solid Films **473**(1), 54–57 (2005). https://doi.org/10.1016/j.tsf.2004.06.157
60. N.I. Ahmad, T.S. Kiong, C. Doroody, K.S. Rahman, M.N. Norizan, M.F. Ahmad, Y.B. Kar, M.N. Harif, N. Amin, Copper doping effect in the back surface field layer of CdTe thin film solar cells. Alex. Eng. J. **88**, 155–163 (2024). https://doi.org/10.1016/j.aej.2024.01.020

Conversion of Solar to Thermal Energy

Principles and Applications of Concentrated Solar Power Systems

Manikandan Krishnamurthy, Gururaj Govindaraja, and A. Sreekumar

Abstract The demand for energy supply has increased drastically while the conventional energy source dwindled to its core. Saving natural resources for our future has become indispensable although this path traces out to be nostalgic there are alternate ways to stock up the void created by the conventional resources, concentrated solar power systems (CSP) is one such with enticing potential. This chapter provides a rundown of the fundamental principles and applications of the CSP systems. It initiates with the applicable factors that diminishes achievable concentration level of the solar radiation, Optical techniques, its topologies, parameters and margins on the extent to which radiation can be concentrated, the state of thermal energy losses in the receiver. The power cycle variants and its principles capitalized for power generation. Accompanied by the applications of CSP system which includes power generation with distinctive modes such as parabolic trough collector (PTC), solar power tower (SPT), linear Fresnel reflector (LFS), parabolic dish systems (PDS), concentrated photovoltaic (CPV). Industrial applications like smelting, melting, sterilizing, water desalination, process heating, thermal energy storage and so forth are to be discussed.

Keywords Concentrated solar power system · Fundamental principles · Power generation · Industrial applications · Thermal energy storage

Abbreviations

CSP	Concentrated Solar Power
PTC	Parabolic Trough Collector
SPT	Solar Power Tower
LFC	Linear Fresnel Reflector
PDS	Parabolic Dish System
CPV	Concentrated Photovoltaic

M. Krishnamurthy · G. Govindaraja · A. Sreekumar (✉)
Department of Green Energy Technology, Madanjeet School of Green Energy Technology, Pondicherry University (A Central University), Puducherry, India
e-mail: sreekumar@pondiuni.ac.in

P. Ravindran et al. (eds.), *Photon to Power*, Progress in Optical Science and Photonics 32, https://doi.org/10.1007/978-981-96-5914-2_4

HTF	Heat Transfer Fluid
S.D	Standard Deviation
CSR	Concentrated Solar Receiver
CFD	Computational Fluid Dynamics
ORC	Organic Rankine Cycle
CST	Concentrated Solar Tower
AC	Alternating Current
LPG	Liquefied Petroleum Gas

Special Characters

η_{CSP}	Overall efficiency
η_{focusing}	Concentrator efficiency
η_{receiver}	Receiver efficiency
η_{transfer}	Transport efficiency
η_{storage}	Storage efficiency
$\eta_{\mathrm{generation}}$	Conversion efficiency
C_o	Optical concentration ratio
G_r	Irradiation at receiver surface
G	Incident solar irradiance
C_g	Geometrical concentration ratio
A_c	Aperture area
A_r	Receiver area
G_{cs}	Solar intensity (annulus + solar aureole)
G_s	Solar disc intensity
ε_o	Emissivity of sun
σ	Stefan Boltzmann constant
T_R	Absolute temperature at the receiver
T_s	Sun's surface temperature
μ_{in}	Refractive index at inlet

1 Introduction

The most prudentially attained sustainable energy resource is the solar energy; the annual energy capitalized for both industrial and domestic applications shall be easily tackled with the solar energy received by the earth in sixty minutes [1]. Concentrated solar power (CSP) is proven as one of the tangible technologies to convert solar energy into electricity through thermal route [2]. The typical CSP system comprises of five basic pathways: (1) Focusing/Accumulation. (2) Absorption. (3) Transfer. (4) Containerization/Storage. (5) Generation [3].

(1) Focusing/Accumulation: The irradiation from the sun that incidents on a huge accumulator or focusing unit is redirected onto a relatively smaller receiver.
(2) Absorption: The receiver which receives the sunlight, absorbs to produce heat.
(3) Transfer: The fluid used in heat exchanger carries away the heat from the receiver.
(4) Containerization: The heat gets sustained on to the system through thermal storage unit for non-solar hour usage.
(5) Generation: The heat gets utilized by the heat engine of application it preferred.

A CSP system shall be illustrated graphically by Fig. 1. As mentioned earlier, it starts with the focusing unit called concentrator which are available in different configuration like trough, Fresnel, tower and dish commonly categorized under point and line focusing system. The point focusing system scales up the radiation 500–1000 times whereas the line focusing system up to 50–100. Receiver intercepts the radiation converts it to thermal energy. Currently, CSP dominantly uses trough for focusing and a mono-steel tubes with annular evacuated glass covering for a convection loss free heat transfer. Yet another technique followed is arranging multiple tubes in formation of cavity structure for both line and point source. Also, there are cases where the receiver work is to directly allow the fluid to absorb the radiation which can be also called as volumetric receiver. This can be achieved by accommodating a provision of opening in the cavity with a matrix through which the HTF pass over.

The absorbed heat energy from receiver takes two paths: it shall be converted directly to desired form or it can be switched to some other location for terminal conversion. The receiver in CSP shall be operated individually or can be integrated with other devices in order to enhance its performance. The Stirling engine, semi-conductor devices like concentrating photovoltaic and thermoelectric converters are used as an integral component of a receiver.

The HTF carries away the receiver's heat if power conversion carried out separately. Heat transfer oil is predominantly used as HTF in trough plants for its superior

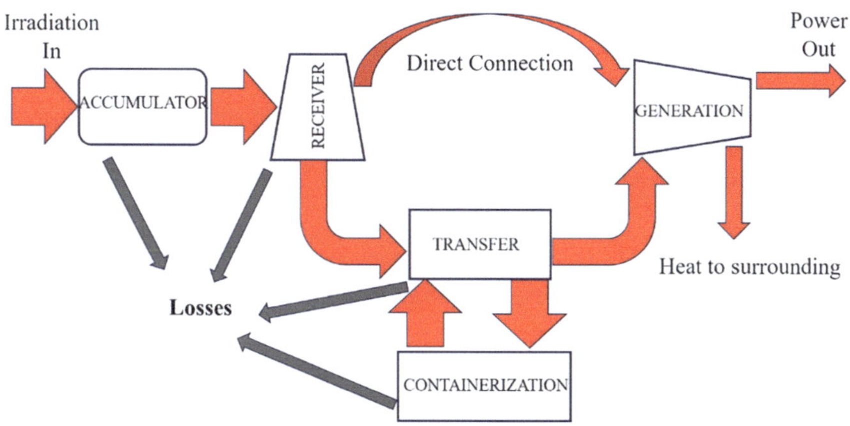

Fig. 1 Graphical representation of concentrated solar power plant system [4]

thermal stability and locomotive property. The need of a heat exchanger in CSP shall be avoided by direct stream generation system where the fluid that transfers the heat and the fluid that work in it are one and the same. In tower CSP system, molten salts are the most preferred material as they not only do the job of heat transfer but also sustains the thermal energy for long time. As the result, it is introduced in trough-based CSP system too. The heat transfer mechanism through chemical reaction, air even alloys are also considered as option which are under development stage. The transport path of power generation paves a way to an option of thermal energy storage system before final conversion of electric power generation. Insulated tank is used to maintain the thermal energy within absorbed molten salts is the current commercialized predominant technique used in CSP.

The rap-up stage in a concentrated solar power system is its power generation. The successfully proven methods of power generation are stream turbine integrated with Stirling engine, Rankine cycle, Photovoltaic and Brayton cycle. The efficiency of each sub unit in the CSP shall be evaluated by the ratio of obtained energy and the energy received. The efficiency of the total CSP system (η_{CSP}) is obtained through multiplying the efficiency of the individual sub units: accumulator/focusing unit (η_{focusing}), absorber/receiver unit (η_{receiver}), transport/transfer unit (η_{transfer}), containerization/storage unit (η_{storage}) and generation/conversion unit ($\eta_{\mathrm{generation}}$). The valve obtained can be at certain instant or it can be over different timescale say over a day or a month or a year.

The development of CSP needs the following driving principles:

Higher the temperature in the sub systems, more effective will be the final energy conversion rate.

Contradictory to this, higher the temperature more will be the losses in the receiver. This shall be tackled by decreasing the receiver size through accumulation of the irradiance.

Some tolerance in deciding the optimal operating ambience is needed based on the stock availability and cost factor.

This chapter discusses the fundamental principles and applications of the CSP systems. The applicable factors that diminishes the expected concentration level of the solar radiation, optical techniques, its topologies, parameters and margins on the extent to which radiation can be concentrated and the state of thermal energy losses in the receiver are discussed here. Optimization of performance of the CSP system for various practicable temperature and receiver perforation size are illustrated here accompanied by the applications of CSP system.

2 Topology in Concentrated Solar Power System

CSP is a technology that harnesses solar energy in aim of generating electricity [5]. The process initiates with development of heat by accumulating solar irradiation onto a small area with the help of a polished surface. The accumulation of radiation on to a receiver in a CSP system is distinguished by various configurations which mostly based on the arrangement and orientation of the mirrors [6]. The four major configurations are as follows:

Parabolic trough collector

Solar power tower

Linear Fresnel reflector

Parabolic dish system.

Parabolic Trough Collector

It is a system in which the sunlight is directed onto an absorber pipe, on result of reflection by a parabolic mirror as illustrated in Fig. 2 [7].

Solar Power Tower

Solar power tower is a single point concentration system, where hundreds and thousands of mirror array arrangement which collectively called heliostats are used for an effective concentration as illustrated in Fig. 3 [7].

Linear Fresnel Reflector

Linear Fresnel reflector system is a combination of both SPT and PTC systems. Unlike PTC, the absorber pipe is not placed within the reflectors but placed in a separate tower setup as illustrated in Fig. 4 [8]. The more improvised version of LFR is a design where a pair of receivers placed parallel to each reflector rows making it

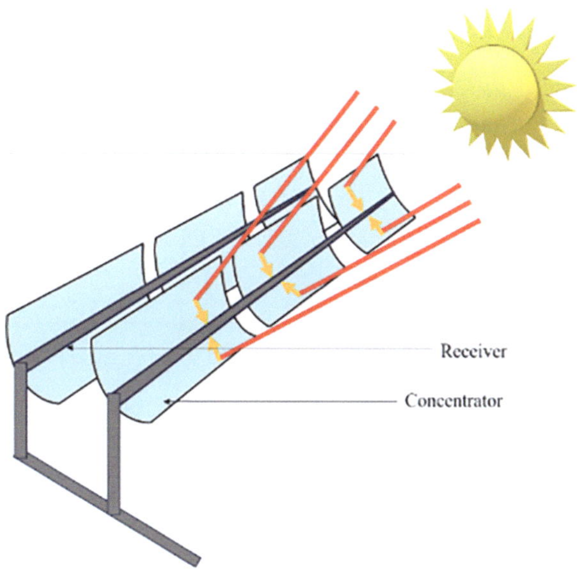

Fig. 2 Parabolic trough collector [7]

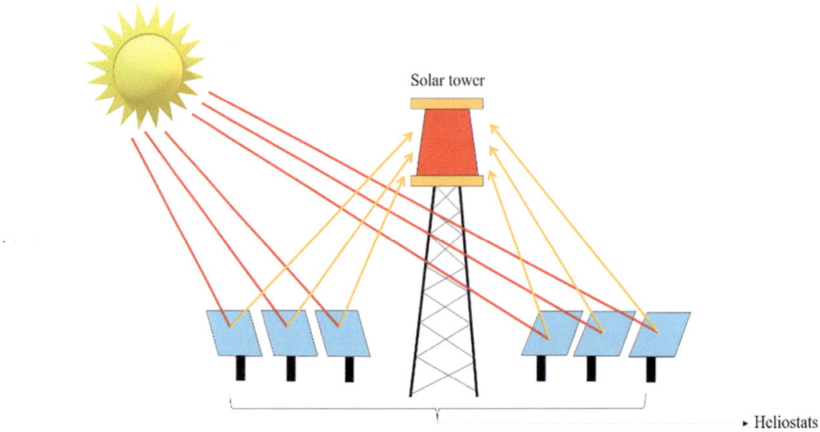

Fig. 3 Solar power tower [7]

compact when compared with PTC. This version is named as compact linear Fresnel reflector (CLFR) which are examined in working plants like Puerto Err ado in Spain [9].

Parabolic Dish System

A cluster of polished surfaces forming a parabolic structure that accumulates sunlight towards a single point as shown in Fig. 5. These types of systems are more flexible and efficient as it can be operated in two directions following the source [10].

Fig. 4 Linear Fresnel reflector [10]

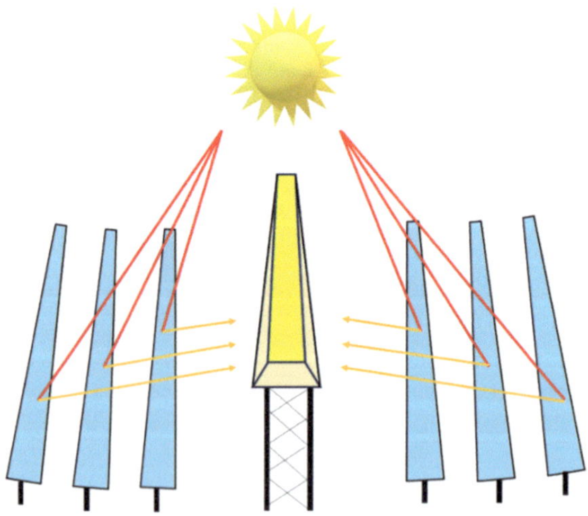

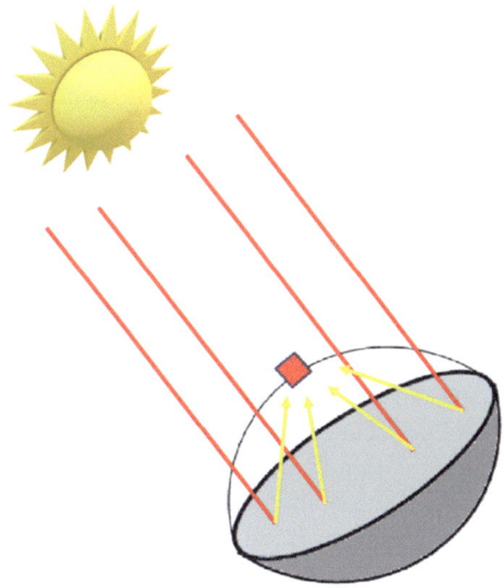

Fig. 5 Parabolic dish system [7]

3 Applicable Factors That Diminish Achievable Concentration Level of the Solar Radiation

The errors experienced in solar accumulators causes to diminishes the achievable level of irradiance concentration. The errors can be of micron level or even scaling up to centimetres. The errors that diminish the level of concentration shall be categorized as: (i) Specularity error (ii) Surface slope error (iii) Shape error (iv) Tracking error and (v) Groupism of errors.

(i) **Specularity Error**

Specularity error relies totally on mirror, like standard of reflection particularly the degree to which the rays from the mirror falls within the constraints of law of reflection. The reflection law states that the angle created while incident must be same as the angle created during refection. Contradictory to this is a diffuse reflector; the role of this component is to fade away the rays that arrives from the reflector by scattering it in different directions considering the effects it causes to the surrounding. The specularity error $\mathbf{S.D_{sp}}$ shall be stated as the standard deviation of the angle made by the rays distributed on the surface at particular incident angle. In reality, all surfaces hold some angularity value of specularity error, originates from the surface's microscopic properties.

(ii) **Surface Slope Error**

The probability of surface slope error predominantly occurs in large concentrating mirrors which are easily vulnerable to damages such as distortion and ripple at their surfaces, degree to which the aberration occurs are generally termed as surface slope error $\mathbf{S.D_{surf}}$. The performance with respect to optics of a concentrator is directly proportional to the surface slope error, which means more the slope error at the surface, lower will be its performance. Also, higher slope error leads to broadening of the focal spot thereby decreasing its concentration ratio to greater extent. The focus made by concentrator can be measured in terms of **mrad**. An optimized focal system shall measure a value of 0.4 mrad, although a tolerance of around 5 mrad shall be allowed based on the variant of concentrator and its applications.

(iii) **Shape Error**

The concentrators which are used in large-scale power plants are generally made of facets. The facets refer to a highly polished flat surface normally polygon in shape. There are chances where these large polished surfaces might undergo distortion or the orientation in which they are placed might be incorrect. The factor that causes distortion can be thermal expansion, intensity of the wind flow and expose of stresses. Photogrammetry is the preferred technique to measure shape error $\mathbf{S.D_{sh}}$. The shape error can be evaluated by the standard deviation of the normal angle made at the surface by the misaligned facets.

(iv) **Tracking Error**

The primary job of a tracking system is to orient the concentrator such a way that it must point towards the sun without any deviation. The tracking systems performance cannot be 100%, because it does not deliver the desired result of pointing the exact position of the sun always whenever it is used. This offset in the angle is often occurs with respect to time, notably in switch ON/OFF variant of tracking system. The tracking error $\mathbf{S.D}_{\mathrm{tr}}$ shall be explained in terms of S.D with respect to angular distribution.

(v) **Groupism of Errors**

In a normal CSP system, the probability of errors occurring as individual is rare rather it always arrives as clusters-specularity, slope, shape and tracking error together which shall be termed as total optical error, $\mathbf{S.D_{tot}}$ of a concentrator. If these randomly distributed error is optimized to closer accuracy, then the total optical error shall be evaluated by taking the root of sum of all errors.

$$\mathbf{S.D}_{\mathrm{overall}} = \sqrt{S.D_{sp}^2 + S.D_{sh}^2 + S.D_{surf}^2 + S.D_{tr}^2} \quad (1)$$

The errors that occur in surface and slope cannot be calibrated into normal distribution, so these constraints must be taken into consideration while handling Eq. (1) also, if the shape of the sun is considered as one of the distributions in formulation, then an additional parameter $\mathbf{S.D^2_{sun}}$ shall be added into Eq. 1.

4 Optical Techniques

4.1 Solar Radiation

The main source of energy from the sun shall be considered as a black body radiance which emits irradiation. This spectrum of energy will get absorbed by various matter in outer space at various wavelengths in course of travel towards earth, so a value of around 5200 K shall be closer to the optimized value of the black body.

The only irradiance which are directly radiated are captured by the concentrating systems avoiding diffused and scattered radiations. The density of the flux generated from an un-scattered energy source which is measured perpendicularly at a flat surface to the sun's irradiance is termed as direct normal irradiance shortly called as DNI. The rate at which the solar radiation flows per unit area in terms of energy can be measured by a S.I. unit-watts per sq metre (W/m^2) and shall be denoted by the symbol (**G**). This measurement can be called in various terms like insolation, irradiance, radiant flux and flux density which are valid technical terms in solar technology. The same irradiance measurement on a plane outside earth's atmosphere takes a value of around 1367 W/m^2, which is usually called the solar constant. Meanwhile the irradiance measured within earth's atmosphere might takes different values based on various factors like surrounding condition, weather and location but generally taken to be 1000 W/m^2 for approximation. The soul of a CSP lies in its capability to intensify the available solar irradiance to its maximum.

The concentration ratio is a vital metric while discussing about CSP. The more predominantly used ways to describe the concentration ratio takes dual paths. Optically, the concentration ratio shall be denoted by C_o is the irradiance which is received at the receiver surface which is denoted by G_r divided by the solar irradiance which is incident on the concentrator which shall be denoted by G.

$$C_o = \frac{G_r}{G} \tag{2}$$

The flux which is distributed to its peak at the point of the highest light intensity, which can be at any spot on a delivered flux distribution is normally the concentration ratio typically defined to be. Whereas the concentration ratio in terms of geometry shall be defined to be the collector's area of aperture A_c divided by the receiver area A_r. It is denoted by C_g.

$$C_g = \frac{A_c}{A_r} \tag{3}$$

The concentration ratio in terms of geometry C_g totally depends on the receiver's area of exposure to irradiance. The portion of the surface that expose to the maximum radiation or the design which captures the maximum radiation along the focal region are the major criteria taken into account while choosing the receivers aperture. The

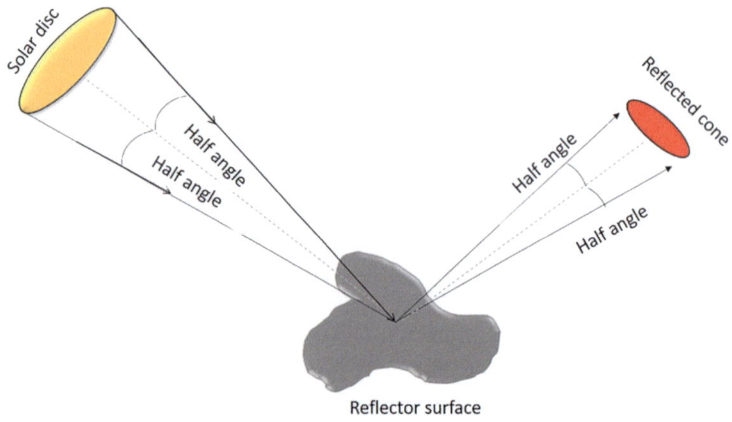

Fig. 6 Reflection of a direct solar irradiance upon a reflecting element [11]

calculated concentration ratio is validated by the count of suns. For example, a solar flux of around 1000 W/m^2 can be written as 1.2 MW/m^2 at surface of the receiver. The approximate angle subtends by the sun when viewed from the earth is around 4.65 milliradians which is called as the half angle as illustrated in Fig. 6.

The circumsolar ratio (CSR) is one of the key aspects in sun optics which is defined as the ratio of solar intensity obtained by integrating the annulus and the solar aureole which is denoted by 'G_{cs}' to the sum of 'G_{cs}' and the intensity obtained from the solar disc 'G_s'. The sun aureole is the glow around the sun usually taken to be 2.5° for easy calibration using lab equipment.

$$\mathrm{CSR} = \frac{G_{cs}}{G_{cs} + G_s} \tag{4}$$

The collectors with high concentration may experience deficit in its capture efficiency if the circumsolar ratio takes higher value. This occurs generally due to high fraction of spillage in flux. The circumsolar ratio varies with change in sun shape distribution. The atmospheric conditions greatly influence the sun shape distribution factor to greater extent particularly based on the amount of water content in the air.

The way in which distribution of the sun shape proposed takes different ways, the one that predominantly followed is given by Eq. 5:

$$\mathbf{I_r}(\theta) = \left\{ \begin{array}{l} \frac{\cos(0.326.\theta)}{\cos(0.308.\theta)}, 0 \le \theta \le \theta_s \\ \theta^{\gamma} \exp k,\ \theta > \theta_s \end{array} \right\} \tag{5}$$

Here,

$\gamma = 2.2.\,\mathrm{CSR}^{0.43}.\,\ln{(0.52.\mathrm{CSR})} - 0.1$

$\mathrm{k} = 0.9.\mathrm{CSR}^{-0.3}.\,\ln{(13.5.\mathrm{CSR})}.$

I_r is the intensity measured relatively at $\theta = 0^{\circ}$. There are also other modelling techniques to validate the sun shape distribution such as sun shape distribution by Kuiper and Rabl and Bendt.

The sun shape distribution is more likely similar in all concentrators especially if the concentrators modelled at higher ratio of concentration which shall be around > 10,000 suns that are seen in concentrators in solar furnace and dish type concentrators provided there is an accurate optical surface. Apart from these high-level sun distributions, there are also distribution for low level concentration collectors which is termed as pill box distribution is the intensity distributed flat-topped.

4.2 Concentration Limits

The receiver with small apertures experiences high thermal loss. A concentration with higher precision shall overcome this crisis. There lies a limitation thermodynamically that restricts the concentration level which can be achieved. There are also other limitations which are based on geometry and range of a concentrators. Any polished surface in any concentrator system can reflect a cone of radiation perfectly with similar angular spread because of a fixed angular width possessed by the sun as illustrated earlier in Fig. 6. The 2nd law of thermodynamics plays a vital role in deciding the energy transfer between the source and the receiver. This tells the fact that the maximum temperature that can attain by a solar receiver cannot surpass the temperature of the sun. Considering these facts the limits on concentration ratio shall be proposed.

In order to calibrate the distance at which the irradiance fall off, it initiates with a consideration of sun as a sphere of black body with a radius of 'r' at a distance of 'L' from a receiver or observer as illustrated in Fig. 7. There will be an evenly spread radiation across the observer sphere of surface area $4\pi L^2$ at range of 'L' from the sun is given by:

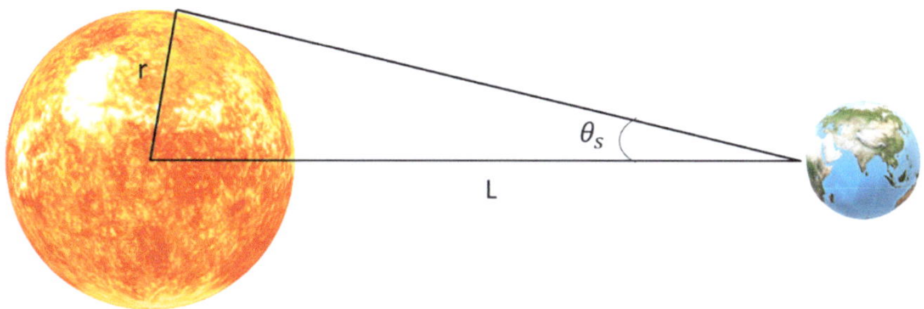

Fig. 7 Irradiance flux from a black body surface [12]

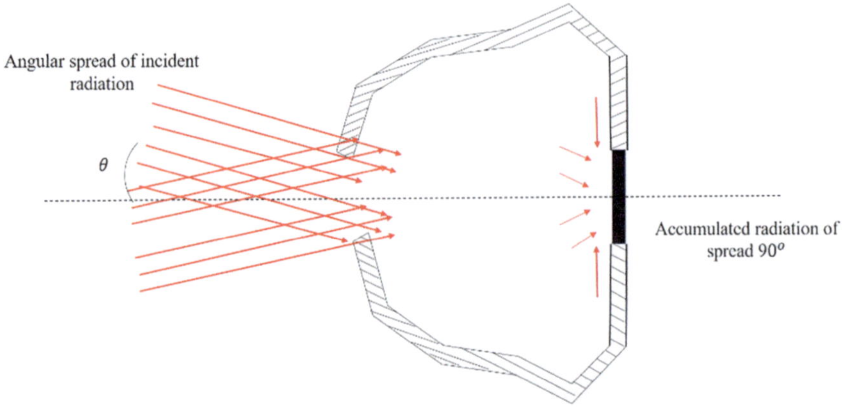

Fig. 8 Half angle angular spread of radiation to 90° angular spread

$$\mathrm{G} = \varepsilon_o\left(\frac{r^2}{L^2}\right) = \varepsilon_o\left(\frac{(L\sin\theta_S)^2}{L^2}\right) = \sigma T^4 \sin^2\theta_S, \tag{6}$$

Here, $\varepsilon_o = \sigma T^4$ is the emissivity of the sun (Black body), and Stefan Boltzmann constant ($5.670 \times 10^{-8}\,\mathrm{W/m^2K^4}$) for '$\sigma$'.

Consider, an ideal surface which concentrates the solar irradiance received at an angular spread of 'θ' at a particular area of collector aperture denoted by 'A_c', accumulating it onto a receiver of area 'A_R' in such a way that it produces a 'right-angled' (90°)-half angle at point of incidence as illustrated in Fig. 8.

The emitted radiation takes a path, back to the source direction after it heats up the receiver. Hypothetically, if no other losses take place in the receiver, the temperature will rise such a way that it equalizes the source temperature. Which means both will be in equilibrium, is mentioned by the relation given below:

$$\sigma T_s^4 A_c \sin^2\theta = \sigma T_R^4 A_R \tag{7}$$

Here, T_R—Absolute temperature at the receiver; T_s—sun's surface temperature.

The concentration ratio of an any solar concentrator, particularly point-focus types shall not be more than,

$$C_g = \frac{A_C}{A_R} = \frac{1}{\sin^2\theta} \tag{8}$$

There are also cases where the planar receiver (black body) is substituted with an opening at another end leading to further transformation of angular spread creating a new optical system and 3rd aperture area. For a newly formed optical system as the result of the above modification, the same procedure shall be carried out as that of single aperture system considering it as black body and relation is given by Eq. 9. The modified system is illustrated in Fig. 9.

$$\sigma T_{\mathrm{R}}^4 A_{\mathrm{R}} = \sigma T_{\mathrm{S}}^4 A_{\mathrm{out}} \sin^2 \phi \tag{9}$$

The above stated system shall be considered without any particular reasons as an ideal concentrator optically for which the relation shall be mentioned with,

$$\sigma T_{\mathrm{S}}^4 A_{\mathrm{in}} \sin^2 \theta = \sigma T_{\mathrm{S}}^4 A_{\mathrm{out}} \sin^2 \phi \tag{10}$$

Therefore, the concentration ratio for the above system shall be

$$C_{g3D} = \frac{A_{\mathrm{in}}}{A_{\mathrm{out}}} = \frac{\sin^2 \phi}{\sin^2 \theta} \tag{11}$$

From Eq. 11, for a concentrator, θ is angle of incident while, ϕ is the spread of the radiation which is concentrated onto the receiver that incident on it.

Thus, the 2nd law of thermodynamics paves the way to a conclusion that the accumulation of radiation shall be achieved only by enhancing angular spread. This intrinsically reduces the probability of concentration of radiations which are non-parallel. Detailed analysis includes the permutations of materials with various refractive indices μ_{in} and μ_{out} at inlet and outlet, respectively, and its relation is given by the result in Eq. 12.

$$C_{\mathrm{g3D}} = \left(\frac{\mu_{out} \sin \phi}{\mu_{\mathrm{in}} \sin \theta} \right)^2 \tag{12}$$

Air is the medium in which the concentration takes places in most of the concentrated solar power system, hence the refractive index of the materials both at inlet and outlet will remains the same, $\mu_{\mathrm{in}} = \mu_{\mathrm{out}}$.

The acceptance angle of CSP system decides the extent to which the irradiation can be concentrated rather than the angle of spread. A concentrator can be made with higher ratio of concentration, by narrowing down the acceptance angle which is still less than the spread angle of the source itself. But this configuration does not

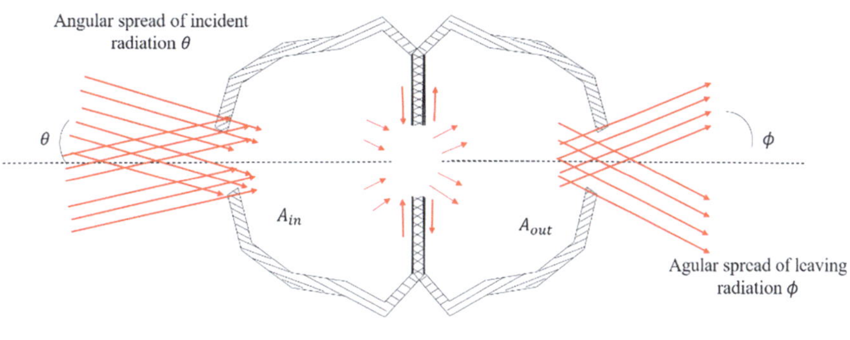

Fig. 9 Incident radiation of angle θ in area A_{in} and leaving angular spread of ϕ in area A_{out} in an arbitrary concentrator

aid much, as it collects only the fraction of source's radiation. On the other hand, the acceptance angle shall be chosen wide for a concentrator without tracking system by giving away the favour of higher concentration ratio. In general, the angle of acceptance is almost same as that of the angular size of the source.

The same protocol if repeated in a concentrator which are line focused, the limitations with respect to geometry during incidence is restricted as unidirectional. Also, the density of the flux drops to R^{-1} instead of R^{-2}, is stated by the relation in Eq. 13.

$$\mathrm{C_{g2D}} = \frac{\mu_{\mathrm{out}} \sin\phi}{\mu_{\mathrm{in}} \sin\theta} \tag{13}$$

The highest concentration ratio that can be attained in concentrators which are line focused is by considering the medium's refractive index which is air as 1 and the half angle in which the radiation exit $\phi = 90°$, shall bring to relation given in Eq. 14.

$$\mathrm{C_{g2D}}(\max) = \frac{1}{\sin\theta} \tag{14}$$

The acceptance angle is also a key factor when it comes to the linear concentrators type too. Still, it is constrained within the limits favoured by Eq. 14 since it has 90° position with respect to the plane of the source with a linear receiver.

Analysing the above restrict equations, it is found that the concentration limit thermodynamically for a line source is 215 and that of point focus is 46,250 for the source (Sun) with angular size of 4.65 mrad.

4.3 Optics in Paraboloids and Parabolas

The first and foremost role of a parabolas and its variants in CSP lies in its capability to focus the radiation that incidents parallel onto it, towards a point at distance '*d*' form its vertex. The major difference between a parabola and paraboloids is that the paraboloids is a surface which takes a parabolic shape and it's the surface structure given to a concentrator that resembles the dish shape. Whereas the parabola is a cross-sectional structure capitalized as a linear concentrator that resembles the trough shape. The parabola constrained to curve which are two-dimensional. The various relations which are related to the optics of parabolas and paraboloids were discussed under this sub heading.

The relation that defines the parabola functionally with its axis directed to y-axis is given by, Eq. 15 (Fig. 10).

$$\mathrm{y} = \frac{x^2}{4d} \tag{15}$$

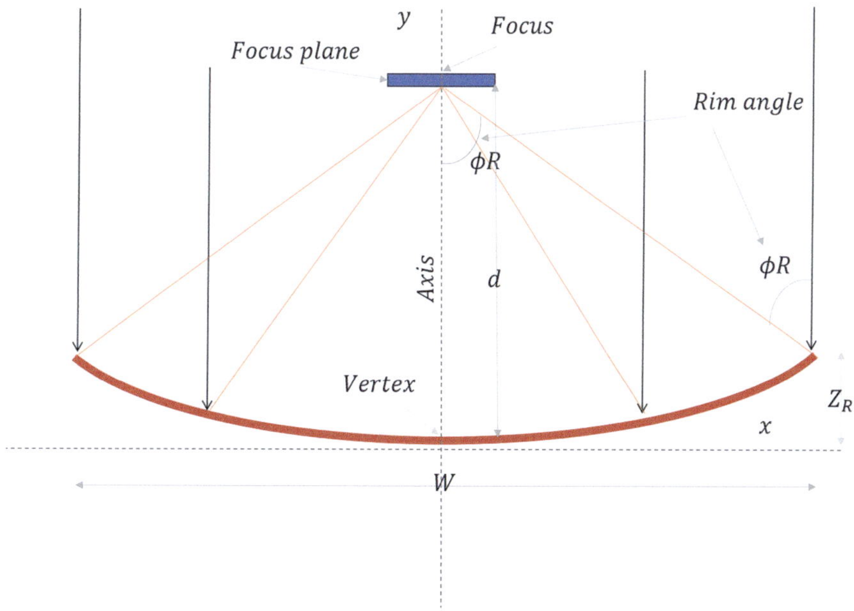

Fig. 10 Optics involved in parabolas (accumulating all radiations that incidents parallel, towards a single point focus) [4]

when it comes with paraboloids in which the axis coinciding with *z*-axis is given by, the Eq. 16.

$$z = \frac{r^2}{4d} = \frac{x^2 + y^2}{4d} \tag{16}$$

The angle that formed between the central axis and the line of focus at extreme edge of the concentrator is termed as a rim angle which is denoted by ϕ_R as mentioned in Fig. 10. The geometry of a concentrator cross-sectionally shall be defined by correlating the focal length denoted by '*d*' as mentioned in Fig. 10 and the rim angle 'ϕ_R'. Thus, the rim angle of a concentrators (parabola and paraboloids) is related by

$$\tan \phi_R = \frac{W/2}{d - Z_R} \text{(Parabolas)} \tag{17}$$

and

$$\tan \phi_R = \frac{4dW/2}{4d^2 - (W/2)^2} \text{(Paraboloids)} \tag{18}$$

Here, '*W*' is the concentrator width and 'Z_R' is the depth from the edge of the concentrator's rim.

There lies a fact which reduces the achievable optical concentration in a parabolas and paraboloids. This occurs due to wide range of angles in the irradiance that incident onto the concentrators on result of huge physical extent of source (Sun). Whereas the radiation which are perfectly parallel alone can aid the focus towards a single point, this is quite contradictory to the fact seen earlier. These issues are addressed in the content under the sub heading below.

4.4 Concentration Constraints with Flat Receivers

Every single point on the surface of a parabolic mirror will reflect the radiation which are angularly distributed in a form of cone of rays by the source with a half angle 'θ_s'. Consider a spot on a flat receiver placed in focal zone (plane) as a result of reflection of a cone of rays from the reflector mirror surface as illustrated in Fig. 11. The distance from a point of reflection on the rim and axis is represented by 'x' and the measure of width of the focal spot on the plane on result of reflection is mentioned as 'f'. Equations 19 and 20 states the relation involved in it.

$$x = 2r \sin \phi_R \tag{19}$$

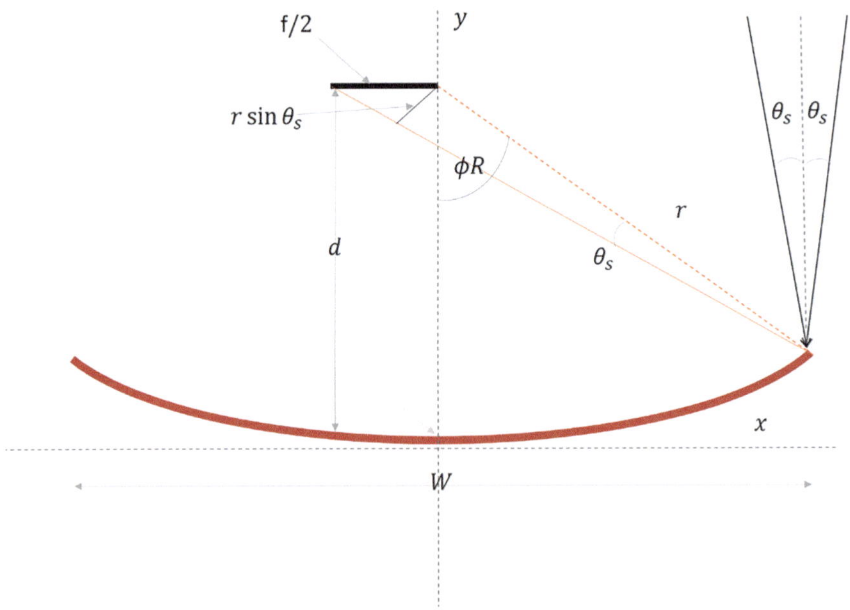

Fig. 11 Concentration of cone of rays from the source towards a flat receiver [4]

For both dish and trough type reflectors, the spot formed on result of reflecting the cone of rays from single point will resembles as elliptical shape on receiver with major axis of length 'f'.

$$f = \frac{2r \sin \theta_s}{\cos \phi_R} \tag{20}$$

The spot size which is created on the focal plane relies totally on the angle at which the radiation incidents ϕ, , hence the real image of the source (Sun) cannot be reproduced on the focal plane at the end of gathering of all such spots throughout the mirror. So, this is why the optics in relates with solar concentrators are mentioned to be non-imaging. Lower the rim angle higher will be possibility of 'r' becoming equal with the focal distance 'd' for entire mirror surface. This is exactly the condition of producing the real image of the source onto the focal plane. The resultant diameter of the spot formed is given by the relation as stated in Eq. 21.

$$\mathrm{D} = \frac{2d \sin \theta_S}{\cos \theta} = 2\mathrm{d} \sin \theta_S \tag{21}$$

When it comes to a receiver with flat surface of rim angle 90°, the elements on the mirror at extreme edge will produce an infinitely lengthened spot, widening their radiation spread over entire plane of focus.

If the receiver size is in such a way that if it can accommodate all the spots that incidents on to it by the mirror surface, then the receiver's diameter shall be called by the size of the spot which is reflected from the points located at the edge of the mirror in which 'x' equals the 'W' and $\phi = \phi_R$. For a flat receiver accompanied by a parabolic trough, the concentration ratio in terms of geometry is given by

$$C_g = \frac{A_C}{A_R} = \frac{LW}{Lf} = \frac{W}{f} \tag{22}$$

Substituting Eqs. 19 and 20 in Eq. 22 gives,

$$C_g = \frac{\sin 2\phi_R}{2 \sin \theta_s} \tag{23}$$

The derivative in terms of ϕ_R, gives an optimal rim angle in a parabolic trough.

$$\frac{dC_g}{d\phi_R} = \frac{d}{d\phi_R}\left(\frac{\sin 2\phi_R}{2 \sin \theta_S}\right) = \frac{2 \cos 2\phi_R}{2 \sin \theta_S} = 0 \tag{24}$$

(for various $\phi_R = 0, 45^\circ, 90^\circ$)

At an angle $\phi_R = 45°$, the concentration ratio reaches its maximum value for a trough with flat receiver and acceptance angle $\theta_s = 4.65$ mrad is given by the relation,

$$C_{g\,max}(\text{flat, trough}) = \frac{1}{2\sin\theta_s} = 108 \tag{25}$$

The procedure shall be repeated for a dish type concentrator too and the resulting concentration ratio is given by,

$$C_g = \left(\frac{\sin 2\phi}{2\sin\theta_s}\right)^2 \tag{26}$$

Differentiating Eq. 26 with respect to ϕ, gives the peak concentration ratio which is given by the relation,

$$\frac{dC_g}{d\phi} = \frac{d}{d\phi}\left(\frac{\sin^2 2\phi}{4\sin^2\theta_s}\right) = \frac{\sin 2\phi \cos 2\phi}{\sin^2\theta_s} = 0 \tag{27}$$

(for various $\phi_R = 0, 45^\circ, 90^\circ$)with the maximum at an angle 45°,

Hence, the relation that gives the maximum concentration ratio in paraboloidal dish is given by,

$$C_{g\,max}(\text{dish, flat}) = \frac{1}{4\sin^2\theta_s} = 11{,}600 \tag{28}$$

These values are constrained under the thermodynamic limits which were discussed earlier in this chapter. Mostly the rim angle is set close to 45°, in a system consisting of a cavity receiver for both dish and trough type concentrators.

4.5 Concentration Constraints with Round and Cylindrical Receivers

In case of a receiver holding a circular cross-section as illustrated in Fig. 12, the target's diameter shall be rewritten as,

$$D = 2r\sin\theta_s \tag{29}$$

For a concentrator with round receiver, the concentration ratio is given by the relation,

$$C_g = \frac{A_C}{A_R} = \frac{\sin\phi}{\pi\sin\theta_s} \tag{30}$$

At $\phi = 90^\circ$ the concentration ratio takes the maximum which is given by,

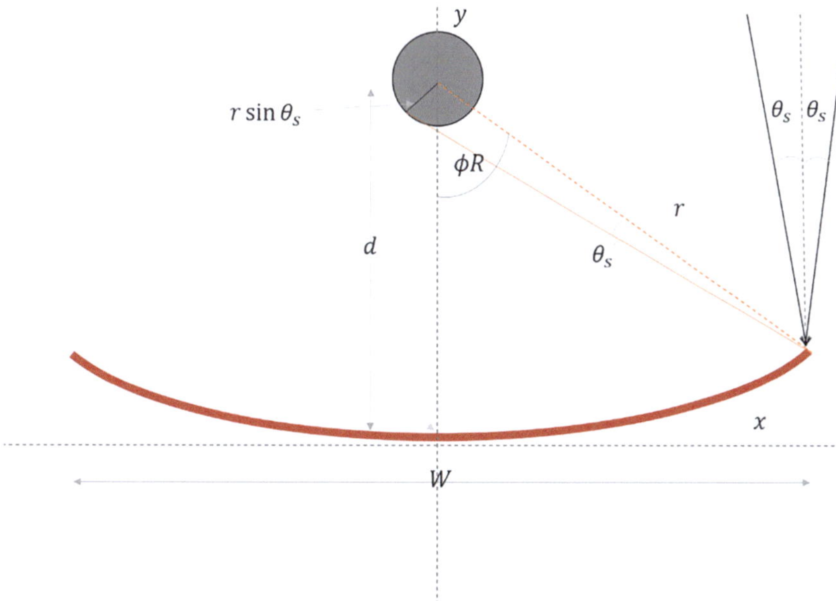

Fig. 12 Concentration of cone of rays from the source towards a round/cylindrical receiver [4, 13]

$$C_{g\,max}(\text{round, trough}) = \frac{1}{\pi \sin \theta_s} = 68.5 \tag{31}$$

Same for a dish type concentrator with round receiver, the concentration ratio is given by the relation,

$$C_g = \frac{A_C}{A_R} = \frac{\frac{\pi}{4} W^2}{4\pi r^2 \sin^2 \theta_s} = \frac{\sin^2 \phi}{4 \sin^2 \theta_s} \tag{32}$$

At $\phi = 90°$, the concentration ratio takes the maximum which is given by,

$$C_{g\,max}(\text{round, dish}) = \frac{1}{4 \sin^2 \theta_s} = 11{,}600 \tag{33}$$

the trough takes the value from the optical analysis, states that the ratio of thermodynamic limits to 'π' that is expressed as $\frac{1}{\pi \sin \theta_s}$ and for dish type it is one fourth of the thermodynamic limit which shall be expressed as $\frac{1}{4 \sin^2 \theta_s}$.

Thus, the concentration ratio based on both geometry and optical analysis delivers that, at the focal spot Centre the limits expressed by the optical analysis are higher than that of limits expressed by the geometric analysis. This is because the focal spot size shall be varied with varying reflection radius 'x'.

Furthermore, an extensive result that includes the mirror area required per aperture area for the four major configurations discussed earlier-trough with flat receiver;

Table 1 Optical analysis in trough and dish type concentrators with flat, round receivers

Concentrator	Rim angle ($\tan \phi_R$)	Concentration ratio (C_g) [Flat receiver]	Concentration ratio (C_g) [Round receiver]
Trough	$\frac{W/2}{d-Z_R}$	$\frac{1}{2\sin\theta_s}$	$\frac{1}{\pi\sin\theta_s}$
Dish	$\frac{4dW/2}{4d^2-(W/2)^2}$	$\frac{1}{4\sin^2\theta_s}$	$\frac{1}{4\sin^2\theta_s}$

dish with flat receiver and trough with round/cylindrical receiver; dish with round/cylindrical receiver were discussed in [8].

Details consolidating all relations is given in Table 1. Thus, for a trough and dish with optimal size the quantity of mirror required for an aperture is less in case of receiver with flat surface as compared to the receiver with cylindrical or round surface. Since, the glass reflection is closer to a normal angle.

5 Thermal Energy Losses in the Receiver

At steady state, the total energy flow in a receiver on effect of concentrated radiation is equalized with the energy carried away by the heat transfer fluid, loss in energy due to unnecessary reflections, loss due to radiative emission and other process like conduction and convection. Thus, the overall loss in a receiver is contributed by these four major losses.

$$\dot{Q}_{\text{loss}} = \dot{Q}_{\text{conduction}} + \dot{Q}_{\text{convection}} + \dot{Q}_{\text{radiation}} + \dot{Q}_{\text{reflection}} \tag{34}$$

The ratio of energy usefully absorbed to the total energy intercepted from the concentrator gives the efficiency of a receiver. It is given by Eq. 35.

$$\eta_{\text{receiver}} = \frac{\dot{Q}_{\text{capitalized}}}{\dot{Q}_{\text{intercpeted}}} = \frac{\dot{Q}_{\text{input}} - \dot{Q}_{\text{loss}}}{\dot{Q}_{\text{intercpeted}}} \tag{35}$$

where the energy that is intercepted by the receiver is given by the relation as mentioned in Eq. 36.

$$\dot{Q}_{\text{intercepted}} = \int_{\text{time}} \int_{\text{aperture}} G_{\text{input}}(t)\,\mathrm{d}A\,\mathrm{d}t \tag{36}$$

The stagnation temperature in receiver is referred to a condition where, the receiver gets heated until the total losses precisely equalizes the total intercepted radiation, considered only if no useful energy conversion takes places with zero receiver efficiency.

5.1 Radiative Losses

The loss in terms of radiation in a receiver includes the net energy emitted on result of its temperature attained by the quantum of radiation intercepted and reflection of some concentrated radiation that incidents on it. The energy which are emitted and absorbed are essentially independent process and the combination of these two processes gives the net energy transferred. There is always a situation where some fraction of the radiation that incident might get reflected and this shall happen from any surface of a receiver. This reflection by the receiver mainly depends on the geometry. The receiver that contains glass covering will also enhance the reflection loss, for which an anti-reflective layer is used as a mitigating agent. In case of cavity receiver, the net energy absorbed shall be estimated by undergoing a simultaneous solution of all the process taking place inside the receiver. The software like ray-tracing shall aid in tracking down the reflected radiations but it does not detect the emitted one.

The grey bodies are one which having a constant emissivity in almost all wavelengths that many surfaces were modelled based on the same. But in case of selective absorption surface used in solar thermal applications, it is customized to absorb wavelength associated with solar radiation and low emissivity related to infrared radiation from hot receiver.

The procedure for estimating the final distribution of solar radiation that incident on a diffusely reflecting surfaces are same as those in determining the emitted radiation leaving the hot receiver surfaces.

The radiations that are emitted from a surface will be partially intercepted by surfaces around the field, in proportion to radiation shape factor which in other words known as view factor. A typical presentation of heat transfer through radiation can be seen in [14].

$F_{ij} \equiv$ Fraction of radiation emitted from the surface 'i' and

$$\text{fraction of radiation reaching surface 'j'} \tag{37}$$

The energy balance in radiation for a diffusely reflecting surface is illustrated in Fig. 13. The diagram elaborates the integral energy entering and leaving the receiver which made of a diffusively emitting and reflecting surface.

Hence, the radiation loss from a receiver when aperture is considered as a single surface interacting only with the surrounding is given by the relation,

$$\dot{Q}_{\text{radiation}} = \sigma A \varepsilon F_{\text{RS}} \left(T_{\text{surface}}^4 - T_{\infty}^4 \right) \tag{38}$$

where F_{RS} is shape factor between receiver and the environment. A single effective total absorptivity shall be considered for a given receiver aperture area for calculating reflection loss and the resultant relation is given by,

$$\dot{Q}_{\text{reflection}} = (1 - \alpha) \mathrm{A} \dot{Q}_{\text{sol}} \tag{39}$$

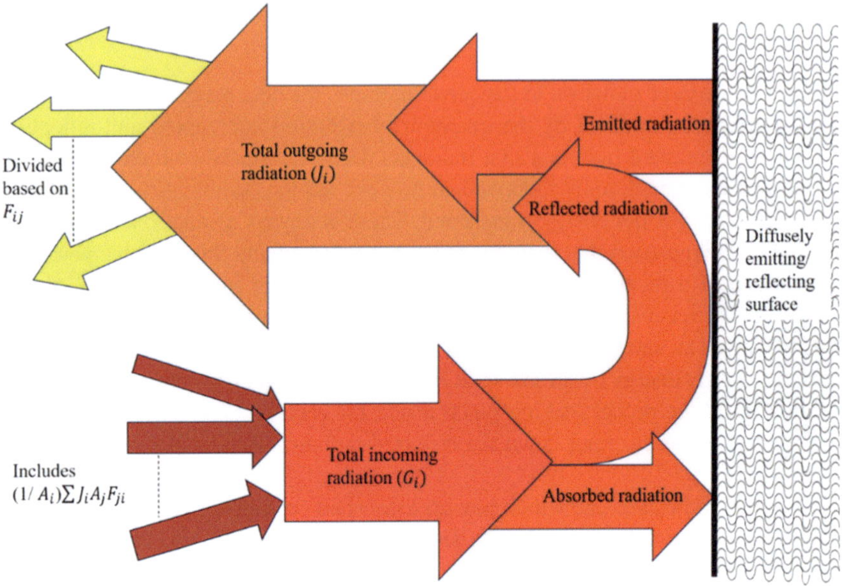

Fig. 13 Diffusely emitting and reflecting surface—energy balance (radiation) [15]

5.2 *Convective Losses*

The flow of air over a high temperature receiver surface takes away the heat paving way to a convection loss. This can be minimized with aid of an evacuated glass tube or a glass cover upon the surface of the receiver. There are dish and tower type systems that are equipped with quartz layer designed to endure extremely high level of solar flux. There lies a complexity in modelling and measuring both natural and forced convection losses. Where the natural convection losses are due to buoyancy effect and forced convection losses occurs due to wind. One of such challenge is that the loss due to convection and radiation are interdependent, as the magnitude of both is influenced by the surface temperature. This temperature, in turn depends on external heat losses like convection and radiation, also the internal heat transfer between the fluid and the receiver. To simplify the analysis, it is common to decouple these processes by assuming either a uniform receiver wall temperature or a uniform heat flux into the HTF. Direct measurement of convection losses is highly challenging, so they are often estimated by subtracting known losses from the total energy balance. Detailed convection heat transfer simulation can be conducted using commercial CFD tools like FLUENT and Star-CCM+. OpenFOAM (OpenFOAM Foundation, 2011) offers an open-source alternative. Semi-empirical formulas addressing the heat transfer coefficients for both convections shall be undertaken for a simplified analysis, though their accuracy is generally limited to ±20%.

Hence, the convective loss in a receiver shall be evaluated once the correlated value of heat transfer coefficient is determined.

$$\dot{Q}_{convection} = hA(T_{receiver} - T_{\infty}) \tag{40}$$

5.3 Conduction Losses

Thermal losses will occur through the insulating covers behind the receivers, as well as through any thermal pathways connecting the surface of the hot receiver to the surrounding environment. These losses are inversely proportional to the overall resistance (thermally) and directly proportional to difference in temperature, which is influenced by the conductivities and geometry of a material.

$$\dot{Q}_{conduction} = \frac{(T_{receiver} - T_{\infty})}{R_{thermal}} \tag{41}$$

In one dimensional scenario, where heat loss occurs through a homogeneous insulating layer with thickness '*L*' with thermal conductivity '*k*', is given by the relation:

$$\dot{Q}_{conduction} = kA\frac{(T_{receiver} - T_{\infty})}{L} \tag{42}$$

5.4 End Losses and Cosine Losses

A significant source of 'loss' in solar concentrators evolves from the inability to consistently align mirrors normal to the solar rays that incidents. Based on cosine of the incidence angle the apparent area, perceived from the sun as the result of mirror reflection off-axis will tends to decrease. Assuming the aperture area of the concentrator that equals the mirror area. This reduction in apparent area directly decreases the concentration ratio of the concentrator, which is thus termed cosine loss [4].

This cosine loss effect is present in all types of solar concentrators. Except for dishes, cosine losses in all concentrators vary with sun position. Dishes consistently face the sun, so cosine losses result solely from mirror's curvature. The off-axis arrangement of most central receiver systems aims to minimize average cosine losses.

End losses are specific to trough and linear Fresnel concentrators. End losses refer to the radiation which are reflected from the mirrors on account of sun not directly overhead, misses the receiver and concentrates beyond the receiver's edge. End losses arises mostly during forenoon and dawn periods or might persist throughout the year, which mainly decided by the orientation.

6 Power Generation Principles in CSP System

There are various solar-to-electric conversion systems that are employed across different concentrator designs.

6.1 Steam Turbine

A significant portion of global electricity is produced using stream turbines, primarily powered by heat from fossil or nuclear sources. An advantage of CSP is the simplicity of integrating new heat sources into the leading power generation technologies. As a result, most of the CSP system currently in operation utilize steam turbines. Almost all concentrator designs are utilized for generating steam in turbine energy conversion.

A Rankine cycle plant operating with steam turbine functions by:

Pressurizing feedwater to high levels, such as above 10 MPa.

Superheated stream in a boiler is result of heating via a heat exchanger using an alternative HTF.

Allowing the steam to expand at low pressure through a series of turbines that in turn drives the generator.

Upon completing the expansion process, the low-pressure steam is condensed using a cooling tower and subsequently recirculated within the system.

The Rankine cycle achieves greater efficiency in conversion with increased steam temperature and high pressure while entering the turbine. An important feature for enhancing efficiency is incorporation of multi staged steam extraction from the turbine, which are capitalized to heat the feedwater before it enters the boiler. The amount of liquid condensed within the turbine must be minimized to prevent blade erosion. This can be attained by ensuring the vapour is adequately superheated before expansion.

Systems achieve greater efficiency when designed as larger units and operating at full capacity. Larger systems incur lower costs per unit of capacity. Coal power stations typically use large-scale turbines with a capacity of around 500 MW. In CSP applications, larger turbine necessitates a larger field, leading to increased thermal line losses. Therefore, a trade-off exists regarding turbine size, with many experts suggesting 250 MW unit as it provides the most cost-effective energy solution. As of early 2012, no CSP systems of this size have been constructed, though several are in planning stages. Advanced steam turbine can operate efficiently at inlet temperature up to 700 °C. Linear Fresnel concentrators and trough are limited to approximately 400 °C with thermal oil as the heat transfer fluid, and can go up to 500 °C with alternative HTF like direct steam generation. Dish and tower systems can achieve the temperature required for the highest steam turbine inlet temperature and pressure; however, the challenge shifts to ensuring material durability in either the turbine or the solar receiver. Advanced steam turbines are currently being manufactured to operate under supercritical conditions, optimizing maximum conversion efficiency. Supercritical steam refers to steam at pressure and temperature exceeding the critical point (22 MPa, 374 °C), where phase transition occurs continuously instead of through nucleate boiling, at elevated temperature. These turbines, feasible only at extremely large scales, have not yet been implemented in CSP plants. At very small scales, reciprocating steam engines outperform turbine in efficiency. Though still produced commercially, they remains relatively unpopular because of their complexity and maintenance challenges.

6.2 Organic Rankine Cycles

An Organic Rankine Cycle (ORC) operates similarly to a steam Rankine cycle but utilizes an organic fluid with a lower boiling point to optimize performance with lower temperature heat sources. These systems can attain higher efficiencies than steam turbines in smaller applications. However, the capital, operating and maintenance costs per MW installed are higher than those of a water/steam system. ORC technology is actively sought for geothermal power applications because it is well-suited for low temperature sources. ORC systems are utilized in a number of modest-sized linear concentrator systems. The more potential use of ORC systems in concentrated solar power system is a 'bottoming cycle', where a high temperature cycle generates exhaust heat sufficient to drive an ORC system for supplementary power generation.

6.3 Stirling Engines

It is a type of engine which is externally heated featuring a piston that reciprocates and operating with a fixed and enclosed volume of working fluid usually air. An ideal Stirling cycle consists of a combination of adiabatic and constant temperature process. In the ideal limits, it achieves the highest possible thermodynamic conversion efficiencies between two temperature boundaries. To date, Stirling engines considered for CST applications are generally small in range of kilowatts. Dish mounted Stirling engine integrate the receiver, engine and generator into a single unit at the focus. Stirling engines are used in dish concentrators for an extended period. This extensive history and dominance lead many in the CSP field to refer to dish systems broadly as 'dish-Stirling' also their application in direct steam generation, photovoltaics and other methods. Stirling engines have yet to be significantly applied to other types of collectors. Stirling engines is of two types: piston-crankshaft engines similar to IC and free-piston engines, that uses an oscillating piston linked to a linear generator without physical linkage. Both types have been effectively implemented.

Dish-Stirling mechanism maintain the highest efficiency in terms of solar-to-electric conversion among technologies, achieving approximately 30% solar-to-AC electric efficiency. Stirling system are suitable for smaller applications compared to Rankine cycles; however, thermal storage for dish-Stirling configurations remains under development.

6.4 Brayton Cycles

The Brayton cycle forms the foundation for jet engine functionality and turbo generators employed in power stations that utilizes gas turbines. It is often misunderstood that gas turbines are named for burning gas; in reality, the term refers as the working fluid employed usually air. In fuel-fired operation, any hydrocarbon fuel, including aviation fuel, propane, diesel biogas or LPG can be combusted to produce the required heat. Alternatively, solar energy can be used to elevate the compressed air temperature prior to expansion. Temperature around 1000 °C is required before expansion for efficient operation, which is typically achievable only with tower or dish-based systems. A CSP systems utilizing solar heating within a Brayton cycle have been successfully operated. A combined cycle power plant utilizes a gas turbine, channelling its hot gases at the exhaust to a heat recovery steam generator that produces steam for the steam turbine cycle. The combined conversion efficiency can potentially exceed 50%, making it the most efficient thermal-to-electric conversion solution currently available on the market. A key advantage of using the Brayton cycle for CSP application is the potential to integrate combined cycle operation with ORC bottoming cycles, achieving high efficiency. In case of dish applications, the Brayton cycle offers the possibility of reduced operating and maintenance costs in comparison with Stirling engine systems.

6.5 Concentrating Photovoltaics

Concentrating photovoltaic typically utilizes costly, high-efficiency cells to maximize the benefits from the investment in concentrator system. Uni-crystal silicon cells of high quality, with efficiency around 20%, are used. Beyond this, multi-junction cells have seen a rapid increase in efficiency over the past ten years, rising from 30 to 43%.

A CPV system experiences parasitic losses due to the operation of tracking system, control mechanisms, wiring inefficiencies, inverter performance and the cooling system's operation. These parasitic losses diminish the overall system's useful AC output. A major challenge for PV employed in high concentration league is the heat generated by photons not directly converted into electricity. A triple junction cell would fail in seconds without an effective cooling to dissipate the heat probably under 500 suns.

6.6 Other Methods

Other methods for converting solar radiation may eventually prove competitive for example:

The Kalina cycle is kind of a modified Rankine cycle utilizing ammonia-water mixtures of varying concentrations demonstrates enhanced performance for temperature ranging from 200 to 300 °C. It is being actively pursued commercially especially for geothermal applications.

Thermoelectric converters generate electricity directly from thermal energy. Semiconductor-based systems function like photovoltaic cells, but electron excitation into conduction band is achieved through thermal energy instead from photon absorption.

Thermionic converters are the system that directly produces the electricity via heat through electron excitation from active surface through a vacuum to a collector.

Thermo-photovoltaics employ PV cells optimized for specific thermal radiation wavelength in order to convert radiation re-emitted from the surface which is been heated.

Mageto-hydrodynamics converters generate a potential difference by expanding heated ionized gas through a magnetic field.

7 Power Generation and Distinctive Modes

Concentrating solar thermal power is one of the antiquated technologies that is adapting to the current practices in the power generation sector and will soon be taking part in contributing large amount of energy production. CSP not only plays a crucial role in power utilities, it can also serve for water heating, steam generation, desalination, thermal energy storage and other industrial process heating like food processing, pasteurization, etc. CSP is categorized based on their nature of utilization, source and geometry.

7.1 Based on Source

1. **Point source**—The reflected radiation is focused on a single concentrated area in the centre of the system where the fluid gets heated.
2. **Line source**—The reflected radiation is focused on a radial line path with heating fluid in the middle of the concentrator.

7.2 Based on Geometry

1. **PTC, LFR and CPV**—Single axis sun tracking and two-dimensional radiation concentrating are done to extract heat or power.
2. **PDS and SPT**—Double axis sun tracking and three-dimensional radiation concentrating are done to extract heat or power [16].

8 Parabolic Trough Collector

The parabolic trough collectors can be accommodated for both the power plant and domestic power generation or water heating [17]. The parabolic troughs are affordable to produce and store energy in the form of heat when analogizing with the prevailing photovoltaic modules corresponding to the area requirement and material processing [18]. There won't be any devastation of the trough system in awful weather conditions. The power density and efficiency of the system can be attained high with proper arrangement. The production capacity can be augmented effortlessly in the local domain without any skilled labour. The stumbling blocks faced while installing parabolic troughs are the proper structural design for solar tracking with moving parts. The stipulation of land cover to achieve proper concentration without shading and the demand of frequent maintenance for relentless functioning. The molten salt used to conserve heat can cause corrosion formation and in higher temperatures between 120 and 220 °C, it might freeze which will deface the performance of the system [19]. The receiver is cooled down with the direct exposure of wind and due to receiver employment in the focal line shadow falls on the concentrator, this affects the efficacy of the heat retainment. The life span of PTC systems are around 30 years with proper maintenance and the thermal efficiency ranges from 15 to 20%.

8.1 Assorted Applications of PTC

Solar water heating parabolic trough concentrators are designed as a system which is adequate for one community or a place (hospitals, hotels, prisons, orphanages and barracks) instead of a stand-alone solar trough. Since they trap only direct radiation, to increase effectiveness these troughs are aligned in rows and interlinked together. This function is suitable for the precondition of a minimum of 500 gallons of hot water. Each row is assimilated with an evacuated tube encasing an absorption tube in the focal line of the concentrators to haul the heat transfer fluid and to confiscate the reflected radiation into the flowing medium. The absorption tube is selectively coated (black) to minimize the heat losses due to the environment and to intensify the solar radiation capture. The concentrators are mounted and fastened on structures with solar tracking arrangements utilizing the gravitational force for the reduction of operating energy. In this application, the heat transfer fluid from the heat exchanger is pumped through the blackened receiver tube where it gains the reflected heat (above 400 °C) and passes back to the heat exchanger making a closed loop. The utility tank water is straightaway allowed to flow through the tubes by the thermosyphon principle in the heat exchanger, where it absorbs the trough-directed heat from the HTF (thermic fluid) and distributes the hot water to the faucets of the public via the same tank [20]. Figure 14 represents the schematic diagram of PTC assisted water heater.

On the other hand, solar concentrated trough for electricity generation, two different fluids are acquainted to circulate the collected heat. The hot working fluid from the heat exchanger gets superheated in an auxiliary heater to change it completely to steam. This high-pressure steam rotates the turbine, i.e., producing electricity in a generator. Then the exhausted low-pressure steam is again condensed by the condenser and recirculated to the heat exchanger making a secondary cycle. The primary cycle is same as the water heating application, in special cases an extra intermediate HTF is used in heat exchangers. The troughs are towards the east and west direction for tracking the sun [14]. Figure 15 represents the working process of a PTC power plant.

The immediate necessity of pure water for daily usage in household activities and industrial processes initiates the avocation of desalination systems for converting brackish water into consumable water. The conventional solar still integrated with PTC is one of the ways to boost the evaporation rate of the water. The water from solar still is allowed to flow through the parabolic troughs to accrete heat and circulate back to still to gain heat from direct radiation. The evaporated water is condensed and collected in a tank [22, 23]. In arid regions for large-scale production, cogeneration plants are installed to generate both electricity and fresh water simultaneously from seawater. Multistage flash (MSF) and multi-effect desalination (MED) are the

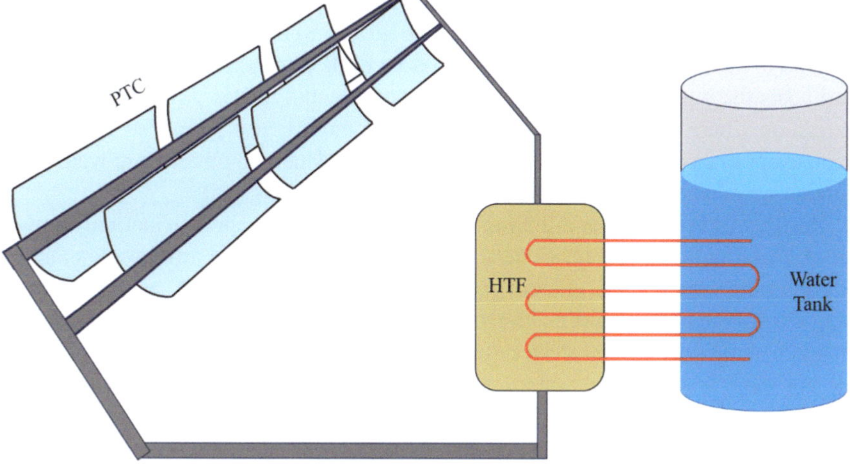

Fig. 14 Schematic diagram of PTC—water heater

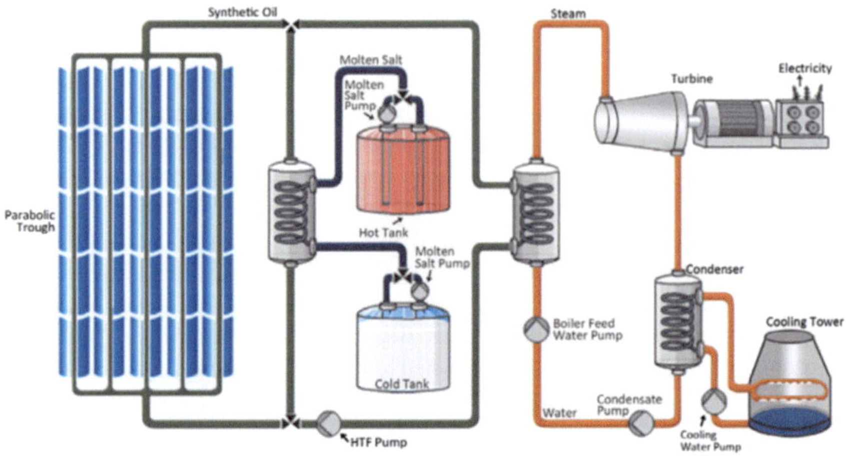

Fig. 15 Schematic diagram of the PTC power plant working process reproduced from [21]

eminent methods to produce fresh water from brine water. The Rankine cycle of the PTC power plant is advanced with an MSF/MED unit to trap the steam heat from the turbine. The condensed water is pumped to the heat exchanger for recurring heat from HTF and the vaporized brine water is condensed to fresh water. An additional reverse osmosis (RO) plant is employed to provide desalinated water in cold climates and TES tanks were included to operate MSF whenever there is a lack of solar radiation [24]. Figure 16 illustrates the desalination process combinedly with power generation from a PTC system. The research carried out by Jafari Mosleh et al. [25] presented in Fig. 17a, b has demonstrated that PTC independently can be used for desalination by constructing it with a heat pipe as an absorber. The absorber is encircled with aluminium foil and on top of that with twin glass evacuated tubes. The evacuated tube is coated black on the inner surface to attract radiation and ethanol is filled in the heat pipe. The heat absorbed on each layer transmits to another layer and heats the ethanol. The ethanol gets evaporated and moves upwards towards its extension where brackish water is enclosed around the heat pipe. The water gains heat, evaporates and starts flowing through the condenser to obtain pure water. The condensed ethanol again shifts down the heat pipe due to gravity.

Refrigeration and air conditioning for preserving food and human thermal comfort are engrossing for the countries near the equatorial region. The refrigeration and air conditioning with PTC can curtail global power consumption. PTC assisted adsorption, absorption or dehumidification cooling systems have been studied by numerous researchers. In adsorption refrigeration system, the PTC receiver heats the water and the hot water is stored in a water tank. This hot water is passed through steel tubes inside the adsorber and circulated back to the tank. This heats the adsorber bed and the vapour methanal (refrigerant) drifts through condenser tubes to convert it into a liquid state. The refrigerant is collected in a liquid receiver and then allowed to flow in the evaporator via flow control valves by gravity. The refrigerant evaporates

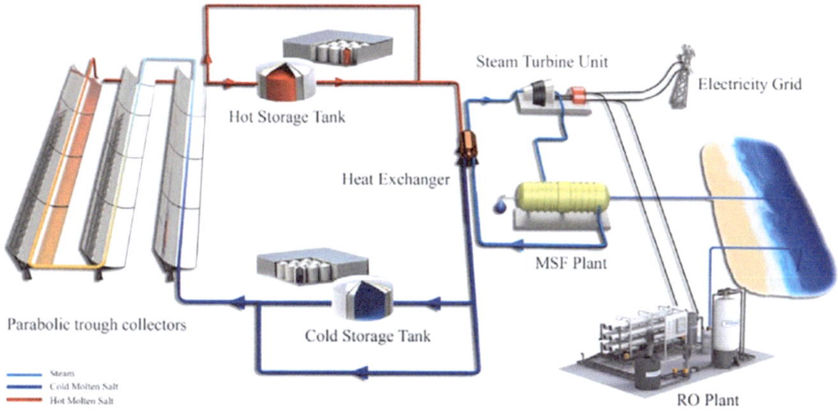

Fig. 16 Dual functional PTC system for power generation and seawater desalination adapted from [24]

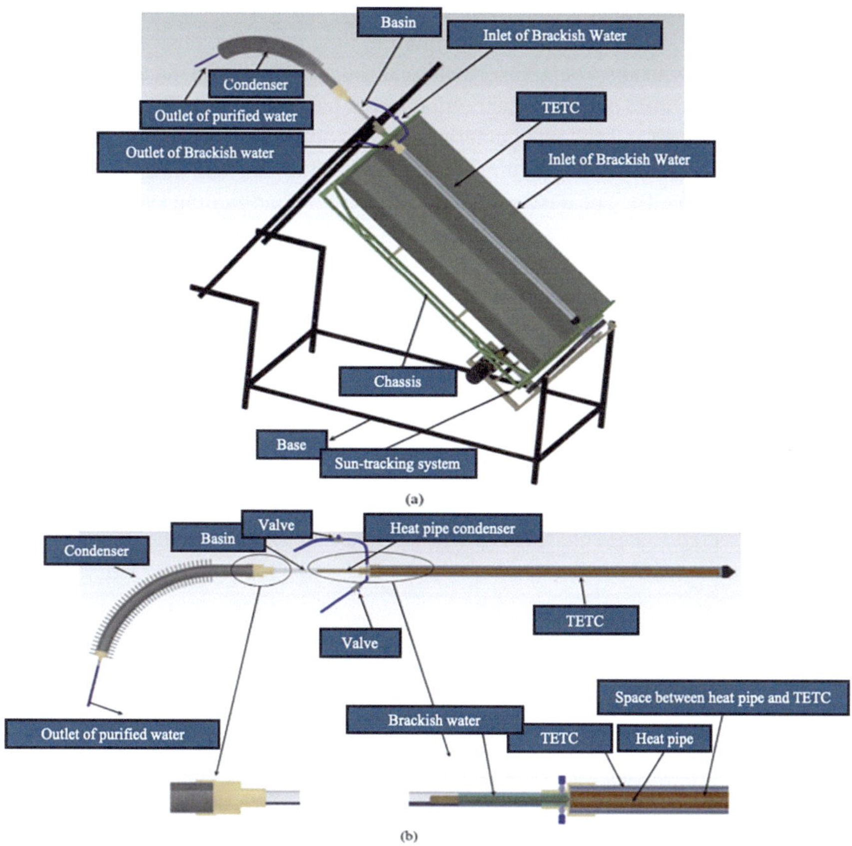

Fig. 17 **a** Stand-alone PTC desalination model; **b** Components adopted from [25]

by absorbing the heat from the cooling space and recirculated to an adsorber [26]. This system is driven without a hot water tank by employing heat pipes from the receiver directly to the adsorber [27]. PTC promoted ejector absorption chiller is operated with therminol (VP-1) working fluid. The cold thermal fluid is transferred to Eurotrough-II from the bottom of the storage tank and the hot fluid from PTC to the upper portion of the tank. The hot thermal oil is distributed to generator to exchange its warmth to refrigerant ($LiBr/H_2O$) and again recirculated to the storage tank. The steam from the generator and evaporator is received in an ejector to pass into a condenser in which the steam releases its heat to the surroundings. The condensed liquid is expanded to low-pressure steam mixture in an expansion valve and passed to evaporator where the high-pressure steam is partly sent to the absorber. Again, the heat is released to the atmosphere in absorber creating a weak solution. This is pumped to a heat exchanger which is directed to generator [28]. In a similar pattern, a solar air heater running with a double-effect absorption chiller is integrated with PTC. A three-phase energy storage unit is installed for backup at night hours [29]. Figure 18a, b represent the working mechanisms of PTC—refrigeration/air conditioning units. A desiccant air conditioning system is designed with PTC for building thermal comfort. The ambient air is permitted to flow through a desiccant revolving wheel employed with adsorbent/absorbent materials (silica gel) inside a dehumidifier to remove the moisture content. The outdoor air is passed through PTC receiver and the outlet hot dry air is used as regeneration air to absorb the moisture collected in the zeolite drum. The processed air is circulated in the building and the operation cycle is continued [30]. The desiccant dehumidifier air conditioning system powered by PTC is shown in Fig. 19.

District heating in cold climates and countries with freezing temperatures are constrained by PTC. The residential buildings are heated by hot water released from the PTC power plant after power generation or directly heated by hybrid systems with flat plat collector (FPC) and PTC. Underfloor or radiative heating with insulated

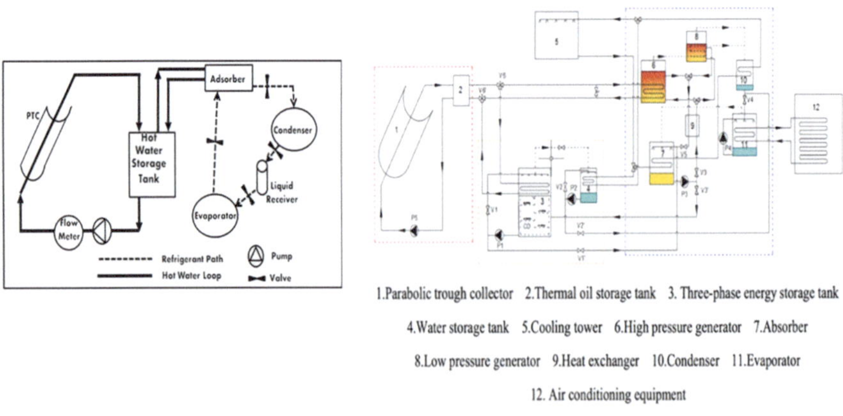

Fig. 18 Schematic diagram of PTC assisted, **a** adsorption refrigeration, **b** absorption air conditioning system reproduced from [26, 29]

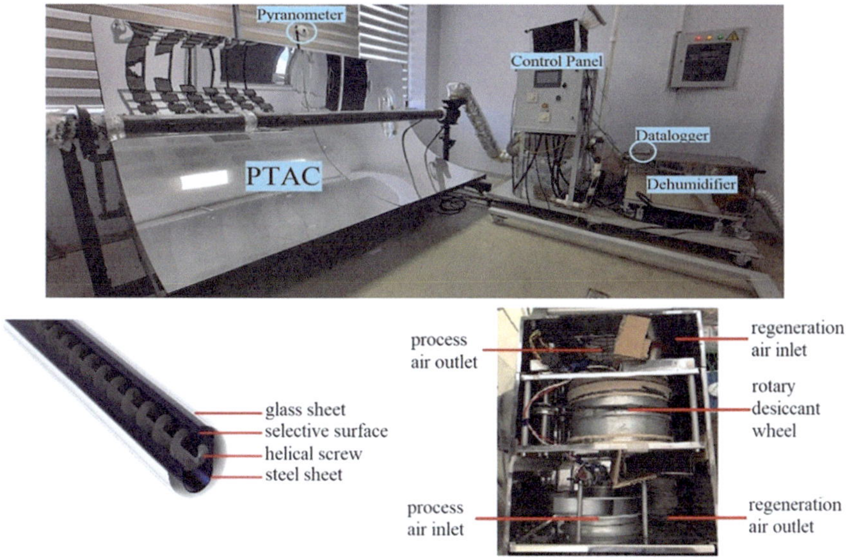

Fig. 19 PTC integrated desiccant dehumidifier for air conditioning adopted from [30]

pipe is the most common practice followed to manage the comfort temperature of the occupant [31]. The building underfloor heating system for thermal comfort is portrayed in Fig. 20. The temperature is modified according to each house requirement with control units. The water is passed back to the power plants for reprocessing. In a similar way, greenhouse systems are operated to raise the inside temperature and efficiency.

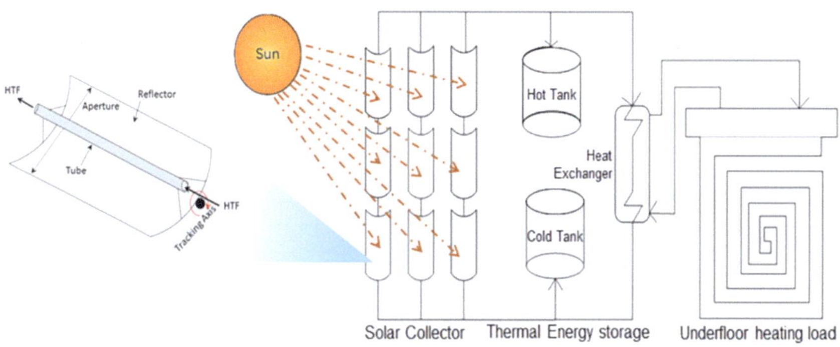

Fig. 20 PTC—underfloor heating in residential buildings for human thermal comfort reproduced from [31]

8.2 Case Studies of the Stated PTC Applications

Solar energy generating systems (SEGS) in Mojave Desert, USA, has nine separate power plants constructed with PTC concertedly producing 2700 kWh/m^2 every year to fulfil the energy demand of 2,32,500 households in southern California. The entire system was fabricated by Luz Industries in the term of 1984–1991 stage by stage with 936,384 mirrors having a cumulative capacity of 354 MW. In this, SEGS 1 was installed with a storage unit containing a high-pressure fluid (caloria) to accumulate heat around 240 °C by circulating the same fluid as HTF which attains 307 °C. Due to the cost of building high-pressure tank (42% of total plant cost), this architecture was abandoned in the further SEGS. The other plants work with two fluids one as HTF (synthetic Oil) and another as working fluid (water) reducing the usage and cost of caloria. The area occupied by SEGS is 6,400,000 m^2, which is around 369 km of parabolic troughs aligned in rows. The energy scarcity formed on the occasion of poor weather was compensated by operating the steam turbine of the Rankine cycle with natural gas in the power plant. This arises only 10% of the gross power generated from the system. The power plant was improvised after the year of 1995 by modifying the former systems in which one was using molten salt ($NaNO_3$: KNO_3—60:40) for storing energy. The power plant was in the superannuated stage after 30 years of its lifespan in 2021, so the power generating stations were decommissioned one after another and replaced with solar photovoltaic modules [32, 33]. The pictures from the aerial view and PTC row arrangement are presented in Fig. 21.

Genesis Solar Energy Project (GSEP), located in Blythe, California, USA, was constructed by NextEra Energy Resources in 2014. The power plant is generating electricity around 580 GWh/year with a plant capacity of 250 MW covering a land area of 7.9 km^2. The aperture area of the power station is 1928320 m^2 accommodating 600,000 RP3 parabolic mirrors. The power station is running under a Rankine cycle turbine with water as a secondary fluid to produce energy and therminol VP-1 (Biphenyl/Diphenyl oxide) was enforced as receiver HTF in the elementary cycle. The cost expended to install the GSEP station was 1215.50 million dollars [35].

Solnova Solar Power Station (SSPS) was established in Sanlúcar la Mayor municipality of Seville city, Spain in 2010 with five individual stations with a plant capacity

Fig. 21 Pictorial view of SEGS PTC power plant adopted from [6, 34]

of 250 MW. Currently, SSPS 1, 3 and 4 are alone in operating condition with each producing 50 MW of electrical energy, the remaining other sites are still in the construction phase. The power station solnova 1 was manufactured with EuroTrough (ET-150) by Abengoa Solar. The modules were arranged in 90 rows, employing 4 modules in each row. The dimensions of EuroTrough modules are 12.5 m in length and 5.76 m in width. Synthetic oil was used as receiver HTF to generate electrical energy of 114.6 GWh/year. The occupied land cover of the system is 1,200,000 m^2 and achieved an overall estimated efficiency of 19%. The turbine of this power plant can also run on natural gas after sunset. Similarly, SSPS 3 and 4 are constructed and working without storage units [33, 36].

Adams County detention facility located in Brighton, Colorado was designed with a PTC water heating system having a production capacity of 18,000 gallons of hot water. This facility supports around 800 prisoners to use in their daily routine for sanitation, bathing and cooking. The system is arranged with PTC in a land cover of 699 m^2 to preheat the water before sending it to six natural gas boilers. This saves 40% energy consumption of the boilers. A stainless steel non-pressurized storage tank was additionally installed to retain the heat. The system was equipped in 1987 with a budget of $ 21,840 with subsidy and the payback/month was $ 200. The PTC is operated with an ethylene glycol mixture as HTF with a temperature of 124 °C [37].

The demand for cold water in a research laboratory in Borj Cedria, Tunisia is executed by a 16 kW double-effect absorption chiller integrated with 39 m^2 PTC. The chiller consists of storage tanks, cooling tower, backup gas-fired boiler and fan coils linked to the building for conditioning. The PTC modules have an aperture area of 13.34 m^2 and a receiver area of 0.68 m^2 with a concentration ratio of 19.6. Figure 22a, b shows the installed system operation and real-time image of the aligned PTC [38].

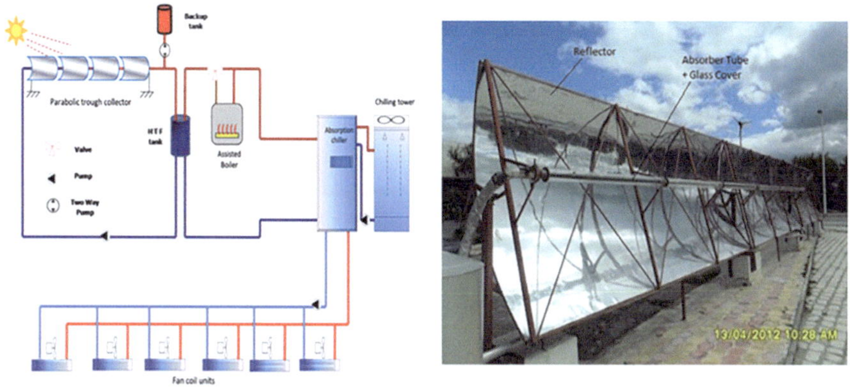

(a) (b)

Fig. 22 PTC—double-effect absorption chiller, **a** process flow diagram and **b** Image of implemented PTC system reproduced from [38]

Fig. 23 Taars PTC and FPC hybrid district heating system Courtesy from [40]

A hybrid system with PTC and FPC for district heating was constructed in Taars, Denmark, in August 2015. The PTC was placed in six rows having an aperture area of 4039 m^2. 12.6 m spacing was given between each row of PTC. The glycol/water mixture is 75 °C after absorbing the solar radiation and heat is transferred to the inlet water of PTC through a heat exchanger. The outlet water of PTC is around 95 °C which is passed in underlaid pipes of buildings. The Taars district heating plant is visualized in Fig. 23 [39, 40].

9 Solar Power Tower

Solar power tower have accomplished the second highest receiver temperature and immense heat storage capacity when contrasting with the other CSP systems. Simplicity and the possibility of hybridization of central power systems are another asset. Larger the SPT system with proper alignment of heliostats, more effectively the plant is operated with low cost. Remarkably, for this reason and to avoid the reflector shadow as well as blocking losses, large land cover area is consumed to run a SPT power plant [41]. The efficiency of the central receiver systems directly relies on the optical properties, size, cleanliness, solar tracking and arrangement of the focal mirrors, receiver design and materials used as receiver and HTF. The wind flowing between the reflector and receiver affects the radiation projected causing convective heat losses; this can be minimized by providing optimum reflector size and receiver air curtains [42, 43]. SPT power stations lifespan is around 25 years and thermal efficiency ranges from 20 to 30% [16].

9.1 Assorted Applications of SPT

The SPT by its name itself indicates that it is more apt for power generation. In SPT power plants, ceramic or metal receiver is implanted on an elevated structural tower in the central focal point of heliostats. The mirrors divert the collected radiation from its surface towards the receiver where the heat gets drawn by HTF (water, air or molten salt). This forms high temperature steam or hot fluid (250–1000 °C) which can either be directly utilized to operate a Rankine cycle or transported to a heat exchanger then generating power via turbines. The produced power is distributed to the utility grid. There are two types of heliostat arrangement according to the latitude of the located power plant. A surround field is used in the equator nearby regions which is circular in shape and while going far apart from the equator polar field heliostat arrangement is considered which is wedge in shape. This minimizes the space requirement of mirrors and tower height. The receivers are classified into cavity tubular, volumetric and particle types. The functioning diagrams of these SPT plants are displayed in Fig. 24. At present volumetric receiver containing air/gas as HTF is emerging since it has a high efficiency of up to 75% and lower reflective heat losses.

SPT for water heating application works on the same principle as a power generation system except the heating rate is low. The reflected radiation from the heliostats falling on the receiver heats the incoming pumped working fluid (water). This is reserved in a storage tank at 60 °C and recirculated to the receiver to maintain the required temperature depending on the consumers. Figure 25 manifests the water heating SPT system [45].

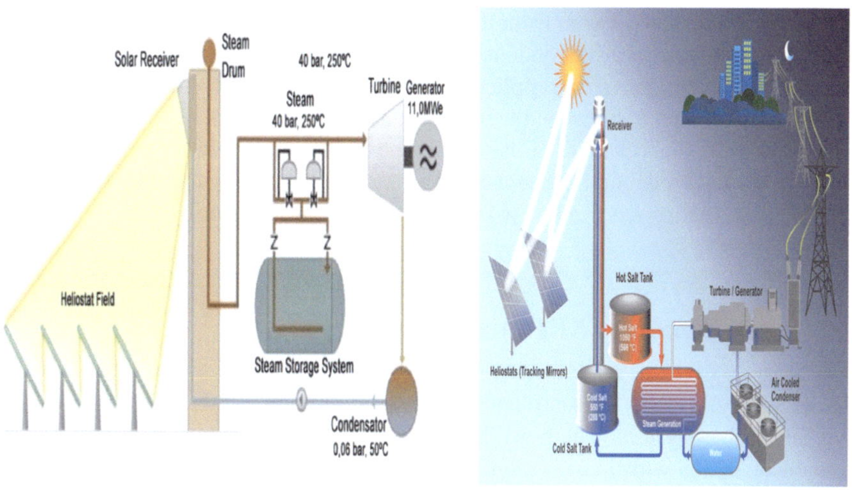

a b

Fig. 24 Schematic diagram of **a** Volumetric type SPT and **b** Cylindrical tubular type SPT working process adopted from [36, 44]

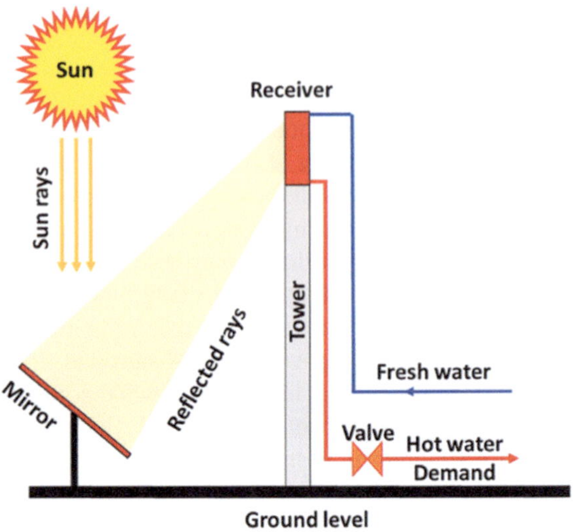

Fig. 25 SPT water heating system reproduced from [45]

RO or MED plant is coupled with SPT for purifying saline water in three different modes. In the RO process, the pump is used to force the water to percolate in a semi-permeable membrane leaving salt content behind is operated by the electrical energy produced from the generator of SPT. The MED plants need both electrical and heat energy, in one mode heat alone is derived from a small SPT power plant and electricity is appropriated from electrical grids. In another mode, heat and power are delivered from the receiver and Rankine cycle of the SPT power plant [46]. A SPT energized direct contact membrane distillation (DCMD) system is revealed by Soomro and Kim [47]. The exhaust steam from the power generating turbine of the SPT plant is condensed in an evaporative condenser by using seawater as coolant. The warm brine water from the condenser is supplied to one side of a DCMD unit. A hydrophobic membrane is placed in between the hot water and cold permeate. The incongruity of vapour pressure on both sides of the membrane creates the driving force to pass pure water to the cold side, and the remaining is drained to sea. Figure 26 expresses the (a) SPT desalination system and (b) various modes of energy utilization.

9.2 *Case Studies of SPT Power Plants*

Planta Solar 10 (PS-10) power plant was the breakthrough and dawn of the commercial SPT power system. The powerhouse was commissioned by Abengoa Solar in Andalusia, Spain, with 11 MW plant capacity in 2007. Similarly ensuing this, Planta Solar 20 (PS-20) was constructed nearby it with a capacity of 20 MW in 2009. In the PS-10 and PS-20 cavity type receivers were mounted on 115 and 165 m tall structures amalgamated with steam turbines to produce power with the assistance of 624 and 1255 heliostats. Both of the power stations use water as HTF with an outlet

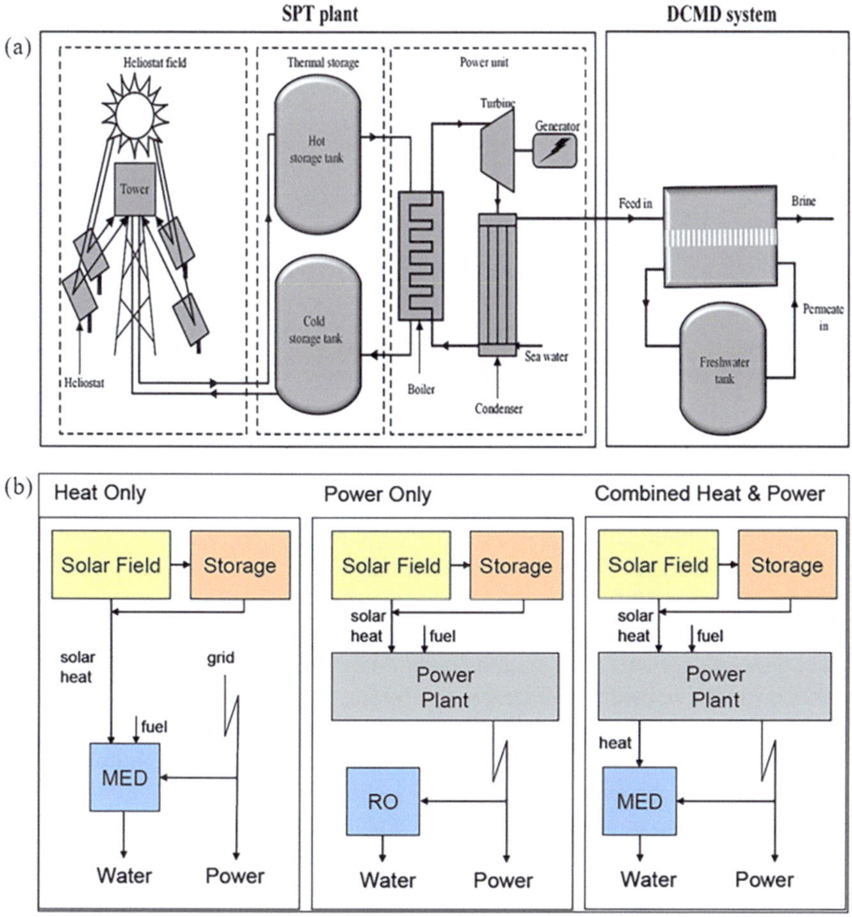

Fig. 26 **a** SPT-DCMD desalination system; **b** Process flow chart adopted from [46, 47]

temperature of 250–300 °C from the receiver. This steam can be stored for 1 h in a ceramic alumina bed and operated with 45 bar pressure in a power cycle. Each mirror placed in these systems has an aperture area of 120 m^2 and the plant occupy 135 hectares of land cover [33, 36]. The PS-10, PS-20 plants and its receiver is provided in Fig. 27a, b.

Jülich Solar Tower stationed at North-Rhine Westphalia, Germany in 2009 by Kraftanlagen München. It was the combined research outcome of the German Aerospace Centre and Solar Institut Jülich. The powerhouse is with a production capacity of 1.5 MW to electrify 350 houses. This is a pioneer SPT executing with ambient air HTF in a volumetric receiver embracing 680–900° C of temperature with 27 bar pressure. In the power cycle steam produced from the water carrying the heat (480 °C) from the hot air in a heat recovery steam generator is allowed to run the turbines. The power station includes 2150 heliostats each having an aperture area of

Fig. 27 **a** PS-10 and PS-20 power field; **b** Cavity receiver reproduced from [48]

8 m^2 covering around 18,000 m^2 land area and a 60 m high tower. The thermal energy storage unit is additionally employed to compensate for 1.5 h sof cloudy hours in the HTF cycle [46]. Figure 28 represents the (a) aerial view of the Jülich power plant and (b) Volumetric SPT.

Ivanpah SEGS is currently allocated in the first position among the largest SPT power generation plants in operating condition having a capacity of 392 MW. The system is located in the Mojave Desert, California, under the Clark Mountains in the USA and was operational from the year 2014. The power plant occupies 1417 hectares of land area with 1,73,500 heliostats and a field aperture area of 2,295,960 m^2. The receiver comes in the external type category with a rectangular shape, situated on a 140 m tower. The power station receiver directly transforms water into steam to revolve the turbines in the Rankine cycle for producing electricity. Therefore, the heat exchanger is neglected in this system and the formed steam is around 566 °C. The power plant has a thermal-to-electrical conversion efficiency of 17.30%. Figure 29 displays the (a) Ivanpah SEGS field and (b) Rectangular tubular receiver.

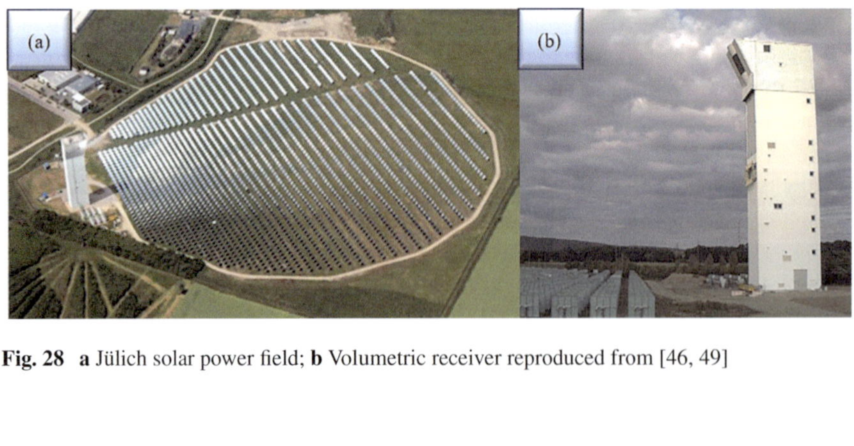

Fig. 28 **a** Jülich solar power field; **b** Volumetric receiver reproduced from [46, 49]

Fig. 29 **a** Ivanpah SEGS field; **b** Rectangular tubular receiver [48]

10 Linear Fresnel Reflector

Linear Fresnel Reflectors (LFR) are alluring in a broad variety of applications due to their upper edge in linear concentrator technologies. The cost associated with mirrors and tracking systems is merest while comparing with PTC since they use flat mirrors and immovable receivers [50]. The operation and maintenance of these systems are effortless in various prospects. When coming to tracking motor size is reduced, hoses are directly interlinked between the receivers without ball joints and cleaning is easy even a provision of automation can be done. Simple supporting structures are adequate to face the wind loads on the reflectors and receivers [28, 51]. Besides these LFR has several drawbacks to amend in which the shading on reflective surfaces caused by the adjoining mirrors and secondary reflector affects the optical performance. Blocking of transmitted radiation from mirrors to receiver by contiguous mirrors is another obstacle. This can be compensated by enlarging the gap of the reflectors or spacing between the reflectors and receivers. But the other factors such as land cover utilization, receiver heat losses, glitches in tracking systems are altered [52, 53]. Thermal efficiency of LFR systems are between 8 to 12% and even can cross PTC with appropriate design [54]. Lifespan of these system is 20–30 years and more with frequent maintenance and replacement of minor parts.

10.1 Assorted Applications of LFR

Power production in LFR coincides with PTC systems, apart from the steam generation section. LFR is constituted by an array of narrow flat or slightly curved reflectors which can be rotated for ensuing the sun path to focus the propagated solar radiation. The concentrated beam gets diverted to fixed receiver plates or tubes where the heat is drawn by HTF/water with a temperature range of 250–450 °C. In lieu of direct evacuated receiver tubes, a trapezoidal or semi-cylindrical cavity secondary reflector can be employed to enhance the concentration efficiency. The steam produced is supplied to the Rankine cycle for distributing electricity in the utility grids [55, 56]. Dabwan and Mokheimer [57] revealed a cogeneration power plant design and simulation with the combination of LFR and gas turbine systems and stated that cost can be reduced by 76–85% of the conventional plant. Reciprocating steam engine integrated LFR system to generate power was demonstrated in a study conducted by Dellicompagni et al. [58]. The layout of a LFR power plant is presented in Fig. 30.

LFR water heater can be either operated as a rooftop system with low concentration ratio reflectors or as a larger system for delivering hot water to residential or industrial areas. Figure 31 displays the portrait of the LFR water heater. The mirrors arranged with solar tracking focuses the radiation on a secondary reflector and then reflected to a black chromed surface where absorber tubes are inserted. The water pumped through the tube carries the heat (60–90 °C) and stored in a utility tank for future use. Several studies were conducted on this water heater with copper absorber tubes incorporated aluminium trapezoidal cavity [60, 61]. A LFR water heater model with 41.2% thermal efficiency was fabricated with varied width mirrors by Babu [62].

Desalination utilizing heat from LFR concentrators can be drifted by some approaches allied with MED, RO, DCMD and so on [47, 63, 64]. A double-effect eccentric absorber tube coupled with LFR reflectors was proposed by Gang et al. [65] as a direct desalination system. The absorber tube consists of two crescent shaped sleeves one on top of another for evaporation and condensation of brine water. The distillation process inside the absorber tube takes place in two stages in which the internal bottom layer of the first chamber is heated by the reflected radiation from mirrors in stage-1. This evaporates the water supplied from the distribution pipe and forms a condensate film on the top of the chamber. Further, the formed film gets condensed to pure water and collected in a storage tank by transmitting its heat to the second chamber's bottom layer. In stage-2, the remaining seawater from the first chamber is recirculated to the upper chamber and evaporated. This gets condensed

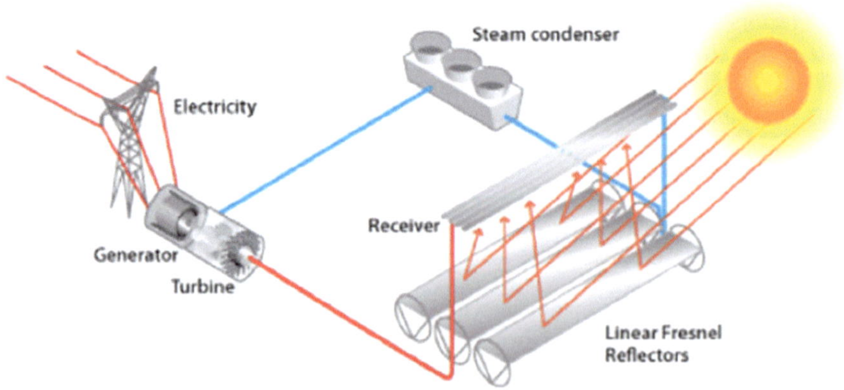

Fig. 30 Layout of a LFR power plant [59]

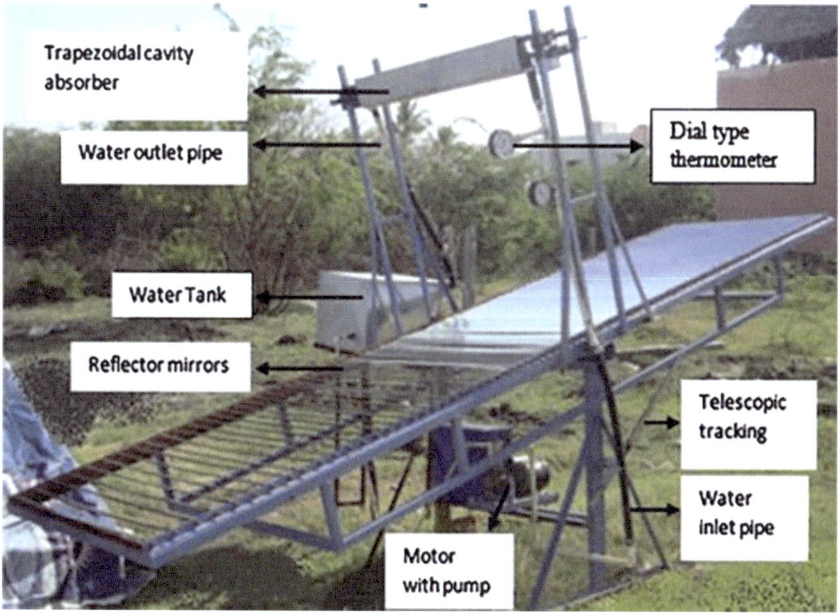

Fig. 31 LFR water heater adopted from [61]

again by releasing its gained heat to the surrounding environment and collected in the tank. The seawater and distillate separation is carried out by the layers formed with the difference in temperature and water vapour density. Figure 32 unveils a LFR seawater desalination apparatus.

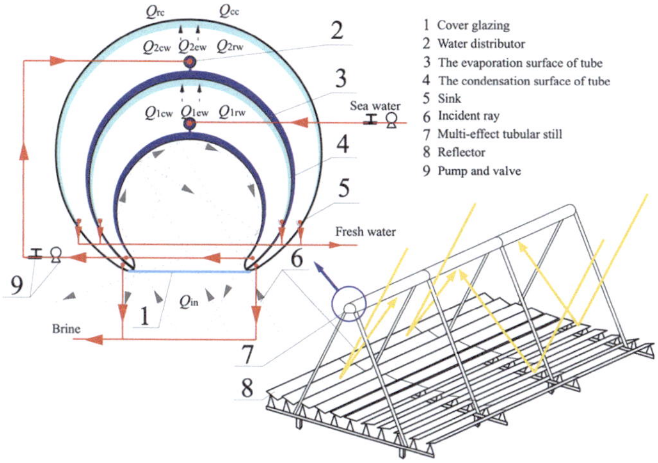

Fig. 32 Eccentric tube LFR desalination apparatus [65]

LFR concentrator driven absorption refrigeration and air conditioning systems are pertinent for serving the cooling demand in hot seasons. The heat appropriate for vaporizing refrigerant in the generator of an absorption chiller can be obtained from a LFR system. A LFR assisted LiBr-H_2O single-double-effect hybrid (SDHE) chiller system was demonstrated by Zhou et al. [31]. The system setup consists of 24 tubular absorber LFR concentrators each separately comprising 10 curved reflectors to power the chiller and supply energy to charge the molten salt in a shell and tube heat exchanger. The heat from this is carried by HTF (Therminol 55) and transferred to the circulated hot water from the chiller in a heat exchanger. This delivers the high-pressure refrigerant to operate the chiller in a single effect when the water reaches above 85 °C, similarly in double-effect mode above 145 °C with 0.7 and up to 1.2 COP, respectively. Figure 33 illustrates a layout of SDHE chiller propelled with LFR setup. In the same manner, an absorption chiller operated with roof integrated LFR mathematical model for air conditioning was compared with a vapour compression air conditioning unit running with photovoltaic for deserts [66].

Solar cooker hooked with absorber tubes of LFR collectors are suitable for indoor cooking in food processing industries and residential buildings. A cooker attached to the end of a LFR evacuated absorber tube performance analysis was carried out by Farooqui [67]. The water converted to steam through the heat gained from the LFR collector was used to heat the cooking pot employed inside an evacuated and insulated chamber. The varying HTF loads (water) of 3–7 L was attempted in a 7 L capacity cooking container. It was concluded that for this capacity cooker 6 L of water load is optimum and the system was having maximum energy and exergy efficiency in the range of 20–25% and 2.3–3.8%, respectively. Figure 34. showcases a LFR facilitated cooker and its pot configuration. This setup is preferred for utilizing to cover the cooking needs of an individual house with an ultimate operating temperature of 250 °C.

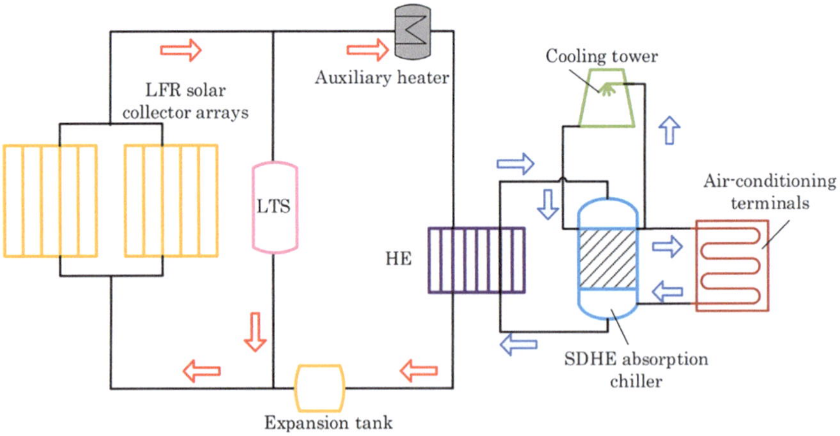

Fig. 33 Structural diagram of a LFR–SDHE absorption chiller [31]

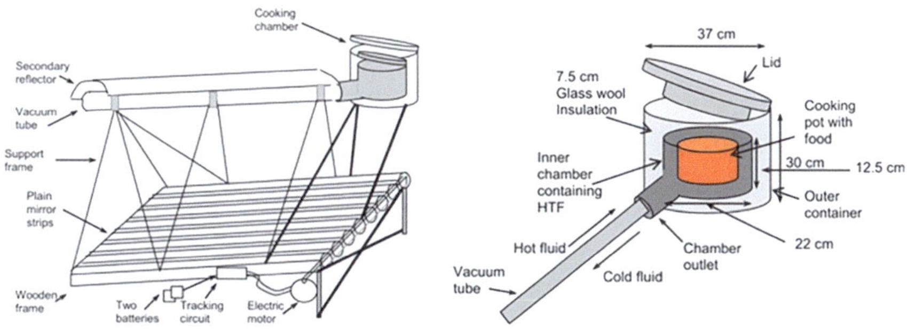

Fig. 34 Design of a LFR cooker [67]

10.2 Case Studies of the LFR Power and Air Conditioning Applications

Dhursar LFR power station located in Rajasthan, India, was installed by Rajasthan Sun Technique Energy Pvt, Ltd with the collaboration of Areva company, France in 2014. The plant has a nominal capacity of 125 MW and occupies a land area of 3.4 km^2. It was constituted with 35 steam generating units each having 33 mirrors with the dimension of 540 m length and 2.25 m width. The receivers are mounted with nine absorber tubes at a height of 30 m above the reflectors. These eight tubes placed on both sides of the central tube are known as economizer cum evaporators in which water is converted to steam. This again flows through the middle absorber tube for changing it to superheated steam hence called a superheater tube. The outlet superheated steam is around 390 °C in temperature and 90 bar in pressure. To operate the plant at full load capacity to generate power via turbines requires 568 tonnes/h of steam production [68].

LFR driven double-effect absorption chiller with a cooling capacity of 174 kW was constructed in the Engineering School of Seville, Spain, in 2008. The LFR had an aperture area of 352 m^2 and generated around 75% of the heat required to operate the generator of the chiller. The remaining energy was supplied from a natural gas burner in addition to this system. The refrigerant and absorbent used in the chiller are water + NH_3 and LiBr + water mixtures. A provision of 4 m gap was given between the 64 m absorber tube and reflectors. The LFR collectors are capable of producing steam with 180 °C temperature and 13 bar pressure. Each reflective mirrors are designed with a dimension of 2 m length and 0.5 m width, arranged altogether in eleven rows. Figure 35 presents a LFR image that enacts an absorption chiller. The daily average LFR collator efficiency and absorption chiller COP were found to be 0.35 and 1.1–1.25.

Fig. 35 Pictorial view of a LFR setup for double-effect absorption chiller [69]

11 Parabolic Dish Systems

Parabolic dish concentrators (PDC) stands with the highest concentration ratio and receiver temperature among the other CSP systems as a single concentrating unit. The hybrid mode operation of PDC can be availed for the residential consumers for utilizing in multiple applications concurrently. The PDC systems are with high optical, thermal conversion efficiency and power density. These are superior in some aspects such as modularity, lifespan, moisture and wind load resistance. The low initial capital and labour cost is another advantage of PDC plants. It has a minimal amount of cosine losses due to smaller acceptance angle, but if reflector mirrors are imperfectly aligned, there is a chance of reflective loss occurrence. The structural complexity allied in dual-axis tracking and moving parts engaged for converting thermal energy to power inflates the maintenance cost. The space required to install PDC power plants is more when comparing with PTC due to their shape and shading factors [70, 71]. Thermal efficiency and lifespan of the PDC power stations ranges from 25 to 30% and 20 to 30 years [54].

11.1 Assorted Applications of PDC

Power generation from PDC systems is executed in various power conversion approaches such as micro gas turbines (MGT), Stirling engines (SE), thermoelectric generators (TEG) and Organic Rankine Cycle systems (ORCS). Among these Stirling engines are attractive for producing electricity due to their high efficiency, adoptability and capability to operate in elevated temperatures. The concentrated radiation diverted from PDC is acquired by a cavity receiver containing HTF, like sodium. The heat from the HTF is transferred to helium or hydrogen (working fluid) in the SE after reaching its required temperature to vaporize. The working fluid creates the mechanical energy by compression and expansion motions of pistons in the engine. This reciprocating force is converted to electricity with the aid of a generator coupled to an engine [72]. The power conversion unit comprising a MGT utilizes heat from a PDC receiver (cavity, tubular or volumetric type) to produce electricity. Aichmayer [73] developed a MGT-PDC system in which exhaust from the turbine is directed to a recuperator for transferring its heat (540 °C) to incoming compressed ambient air (3 bar) from a compressor. Further, this air draws heat from the HTF in the receiver (800 °C) and then circulates to expand in a mini turbine to drive a generator for electricity production. The thermoelectric generator works on the Seebeck effect in which the temperature gradient maintained in two junctions of dissimilar materials produces a voltage by inducing charge carriers. The TEG is directly employed in the cavity receiver where one side is heated by an absorber plate and the other side is cooled by heat transfer fluid to retain the temperature variation of the junctions in the module to produce electricity. Sahu et al. [74] demonstrated a TEG-PDC system with a dual trumpet reflector for heating an aluminium absorber plate (Max: 363 °C) clamped with one side of the module. The other side is fastened to the bottom surface of the receiver where HTF (water) flow provision is arranged to cool down below 50 °C. This allows to extract of 24–194 W electricity from a Bi_2Te_3 TEG module with the formed temperature gradient. Figure 36 represents the a), c) and e) components of different power conversion units and b), d) and f) images of PDC setups for power generation.

The PDC water heaters are designed with cylindrical, spherical or cone tube wound cavity shaped receivers situated in the central point of parabolic concentrators. Selective coating is applied on the receiver tubes to enhance the absorbance and capped with insulation materials on the external surface of the tubes to prevent heat losses by surroundings. The reflectors are fabricated with polished aluminium sheets with reflective coatings as the operating temperature of this system is in the range of 50–60 °C. The photographs of a prototype PDC water heater are provided in Fig. 37. In special cases for obtaining higher efficiency, small pieces of mirrors are clubbed together to form the dish shaped concentrator [76]. The cold water from the storage tank is either distributed to the receiver by thermosyphon principle or circulating pump for acquiring the heat absorbed on the tubes. This gets transferred to the utility tank to cover consumer demands. Copper is preferred as receiver tubes due to its high thermal conductivity. Numerous research were carried out with this application in PDC prototypes in different shaped receivers and heat transfer enhancements [77, 78]

Single slope solar still positioned in the focal point of a PDC is pertinent for seawater desalination. Figure 38 illustrates the prototype designed by Bahrami et al. [79]. In this black chromed solar still made of stainless steel sheets is employed as an absorber. The reflected radiation from the collector heats the seawater in the receiver which is pumped from sea or salt water tank and converted to vapour. The

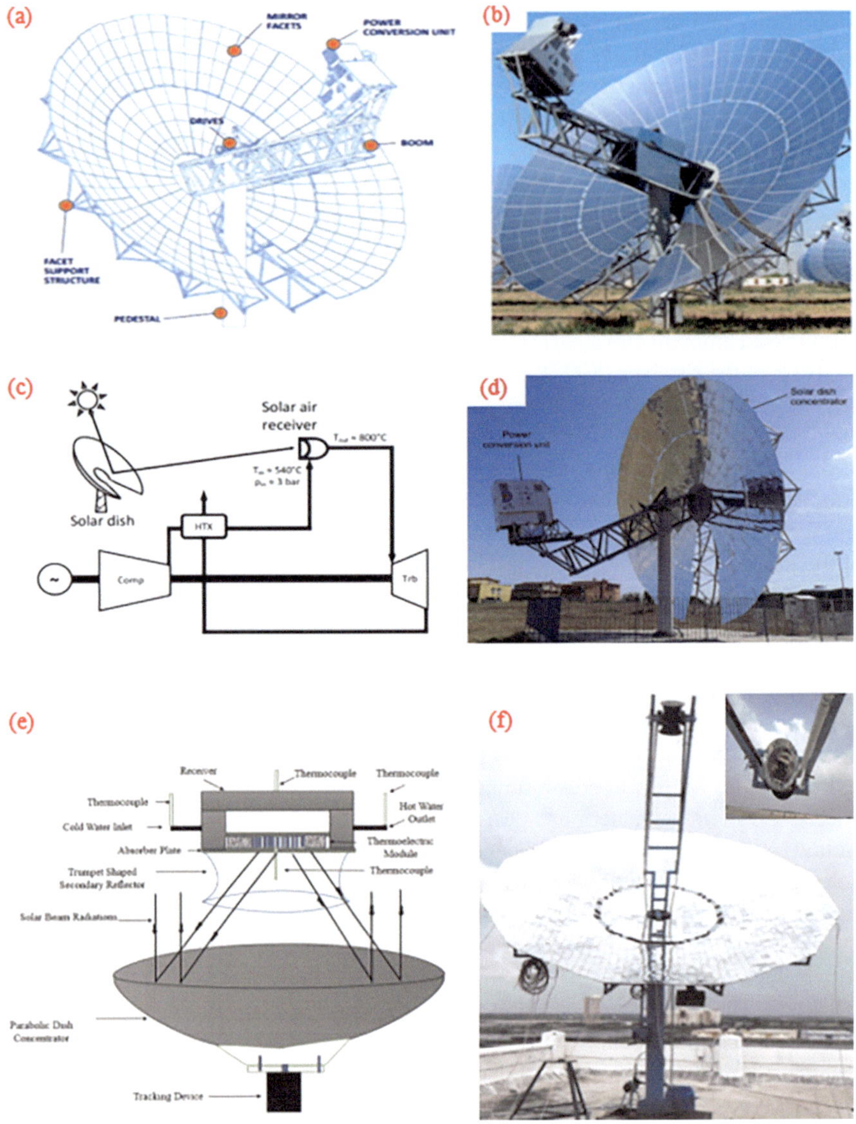

Fig. 36 Schematic and pictorial view of **a**, **b** SE-PDC courtesy from [75]; **c**, **d** MGT-PDC courtesy from [73]; **e**, **f** TEG-PDC courtesy from [74]

makeup inlet seawater passing through a flat plate condenser placed on top of the solar still chills the vapour to condensate and collected in a storage tank with the help of gravity. The preheated brine water is delivered to the evaporator for steam production. In a similar pattern a hemispherical absorber was attached to the bottom of the solar still containing paraffin-filled copper balls were constructed and demonstrated

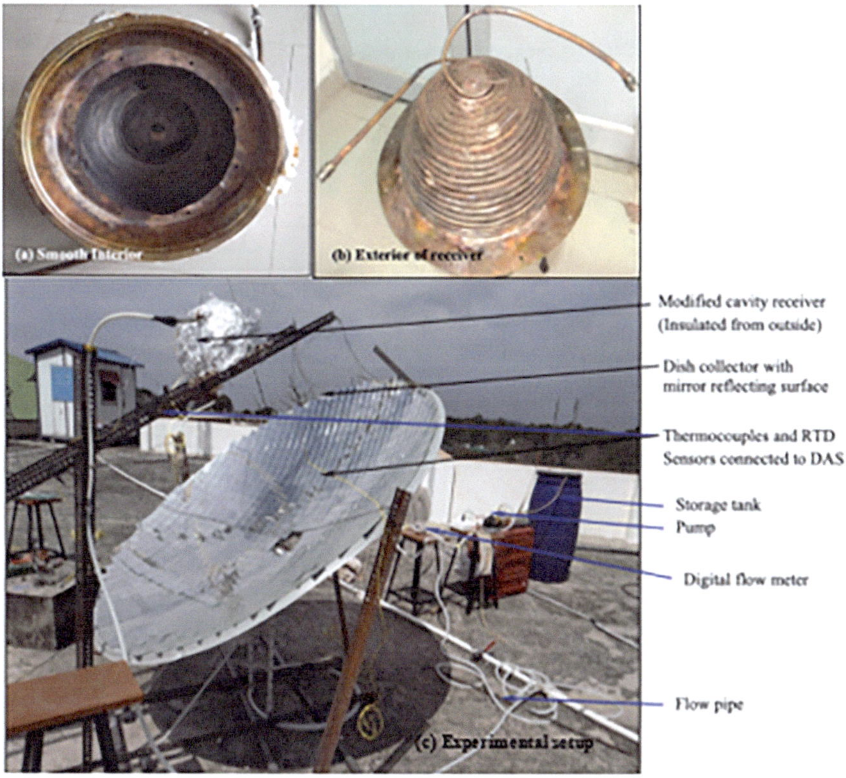

Fig. 37 Photographs of **a** receiver internal view, **b** receiver external view and **c** prototype [76]

for desalinating saline water [80]. Esfanjani et al. [81] developed a PDC propelled dehumidifier-humidifier desalination system. The saline water from a tank is pumped to copper tubes arranged as a coil in the dehumidifier where it gets preheated by moist hot air from the humidifier. The humid air relinquishes its heat energy and converted into a product of fresh water droplets and dry air. The clean water is collected from the dehumidifier with the aid of cotton strings and dry air is supplied to the bottom of the humidifier. The partially warmed brine water is further heated in a PDC and sprayed inside the humidifier by the dint of a nozzle. The released droplets penetrating from the top of the polypropylene porous bed hand over their moisture and heat to the incoming counterflow dry air. The cooled strong brine water is drained from the humidifier and moist air is recirculated to the dehumidifier after removing droplets with salt content in the drift eliminator. Figure 39 revealed the functioning PDC dehumidifier-humidifier desalination.

A Desiccant air conditioning system amalgamated with PDC designed by [82] is portrayed in Fig. 40. The Parabolic dish collector reflects the solar radiation in a cylindrical cavity receiver comprising with eight turns of helical tubes inside an annular gap around the absorber. A HTF oil (SeraSence SF 350) was passed into the

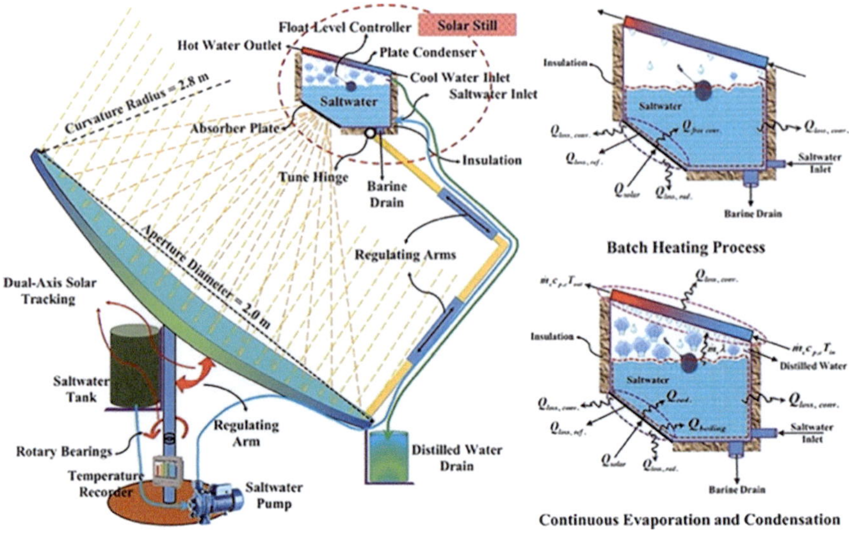

Fig. 38 Schematic diagram of PDC energized solar still desalination model adopted from [79]

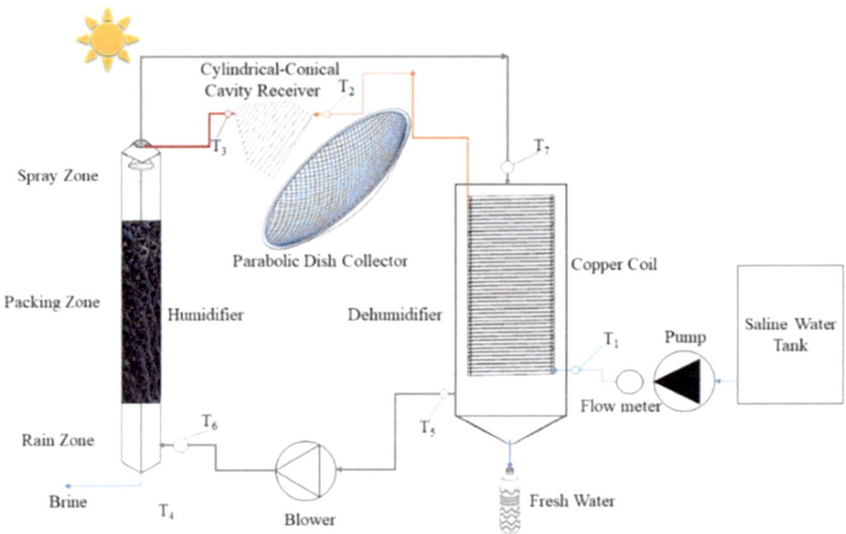

Fig. 39 Structural diagram of a PDC driven dehumidifier-humidifier desalination adopted from [81]

helical tube to gain the heat captivated in the absorber. The hot oil is then transferred into two fined tube heat exchangers arranged one after the other inside a rectangular channel. The surrounding air is forced on the surface of rectangular heat exchangers to form dry air by obtaining the heat from HTF. The oil is supplied back to the storage tank and hot air progressing to the outlet of the channel is diverted to the desorption column. The distributed dry air is used as regeneration for the silica gel desiccant. The moisture from the adsorbent is carried by the regeneration air and comes out as cooled air.

PDC cookers are designed for both indoor and outdoor cooking according to the consumer's convenience and budget. The cooker utensil is directly placed in the focal point of PDC acting as a receiver for stewing water with ingredients in outdoor cookers whereas steam is generated from the receiver and supplied to cooking pots for boiling the food products in indoor cooking. There are two types of receiver arrangements in one the absorber is mounted on the structure which is fastened to the dish focal region. Here parabolic polished aluminium reflectors or Fresnel mirrors are used for concentrating the solar radiation. In the other type, the absorber is connected to steam supply pipes and the radiation is projected from a separate reflector like Scheffler collectors [77]. Singh [83] has constructed a tube panel heater where high temperature steam is passed from the PDC receiver for heating cooking utensils like pans by conduction mode in a kitchen. The heat lost steam is again

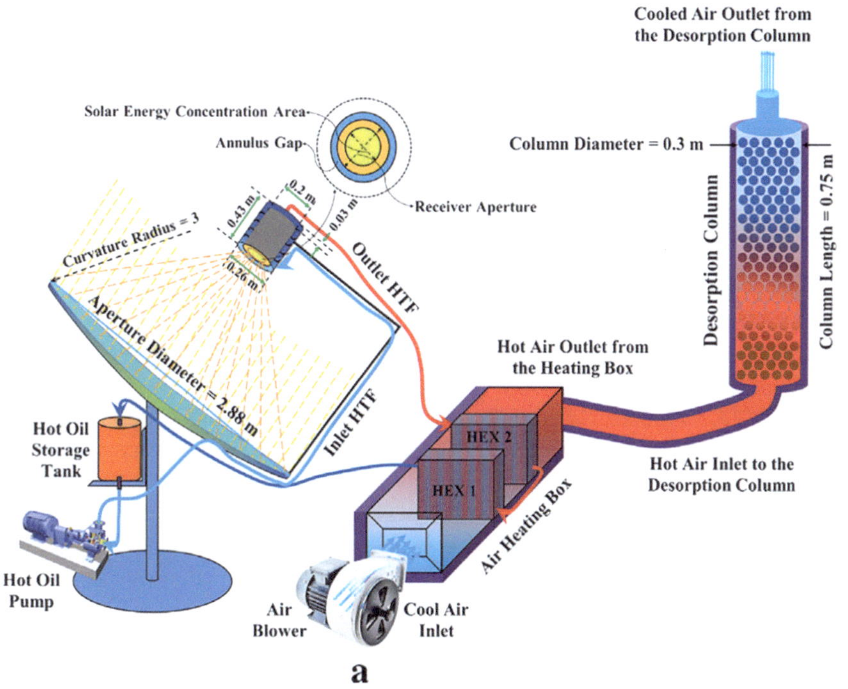

Fig. 40 Design of PDC—desiccant air conditioning unit reproduced from [82]

sent back to a collector for regeneration. Banitia et al. [90] simulated a beam-down collector with dual reflectors in which the concentrated radiation from the primary reflector hits the secondary reflector. This diverts the solar beam back to a tube where HTF absorbs the heat. Further, the working fluid circulated inside a U-shaped thin tube devours the heat from the surrounded HTF and permitted to flow to the cooker bottom by thermosyphon principle for indoor cooking. Figure 41a–c presents an outlook of indoor and outdoor PDC cooking systems through layouts, pictures and simulated images.

11.2 Case Studies of the PDC Power and Cooking Applications

Solar cooker with Scheffler collectors was constructed in Om Shanti retreat centre located in Gurugram city of Haryana, India in 2001 for cooking 6000 meals/day with 10,000 USD. The total system efficiency was about 25% considering all losses. Each collector and receiver covers an area of 10 and 0.07 m^2 individually. The absorbers are inter-connected with a 200 L capacity steam storage tank where partially heated water (60 °C) is supplied from a boiler. This hot water is converted to steam when allowed to flow through absorbers. Each absorber is heated to 150–170° C by two Scheffler concentrators placed facing both the sides of receiver. The operating pressure for this system is 3 kg/cm^2 and safety valves are provided to release the pressure if it exceeds 12 kg/cm^2. The steam from the tank is supplied to the bottom of five 50L cooking utensils with the aid of flow control valves. Figure 42 shows the pictures of Sheffler arrangement and cooking vessels in the Om Shanti retreat centre. The sun tracking of concentrators is on a single axis and the payback of the system is within 2 years [85].

Stirling engine PDC power plant in Maricopa City of Arizona, USA was installed by Tessera Solar in 2013. The power plant had a capacity of 1500 kW electricity production with sixty SE-PDC. The reflectors are built with silver glass coated mirrors having a reflectance of 94%. The land cover area consumed by the plant was 60,000 m^2 and a yearly efficiency of 26% [86]. The aerial view of the Maricopa PDC power station is unveiled in Fig. 43.

A SE-PDC was constructed at Tianjin University, China, with a 12.5 m diameter reflector and 0.25 m cavity receiver. The cavity was formed with 12 turns of 25.4 m length and 3.26 cm inner diameter tubes. This cavity was linked to a 1800 rpm SE with helium working gas. The SE was operated in 20 bar and 646 °C average temperature. The tracking was carried out in dual axis and a triple-phase asynchronous generator was used to produce electricity. The system had a gross efficiency of 23.05% and maximum power capacity of 23.11 kW [75]. The image of SE-PDC of Tianjin University is displayed in Fig. 44.

In 1980, a solar bowl was installed in Crosbyton, Texas Tech University, USA. The whole bowl structure was fixed on round cylindrical beams and comprised of

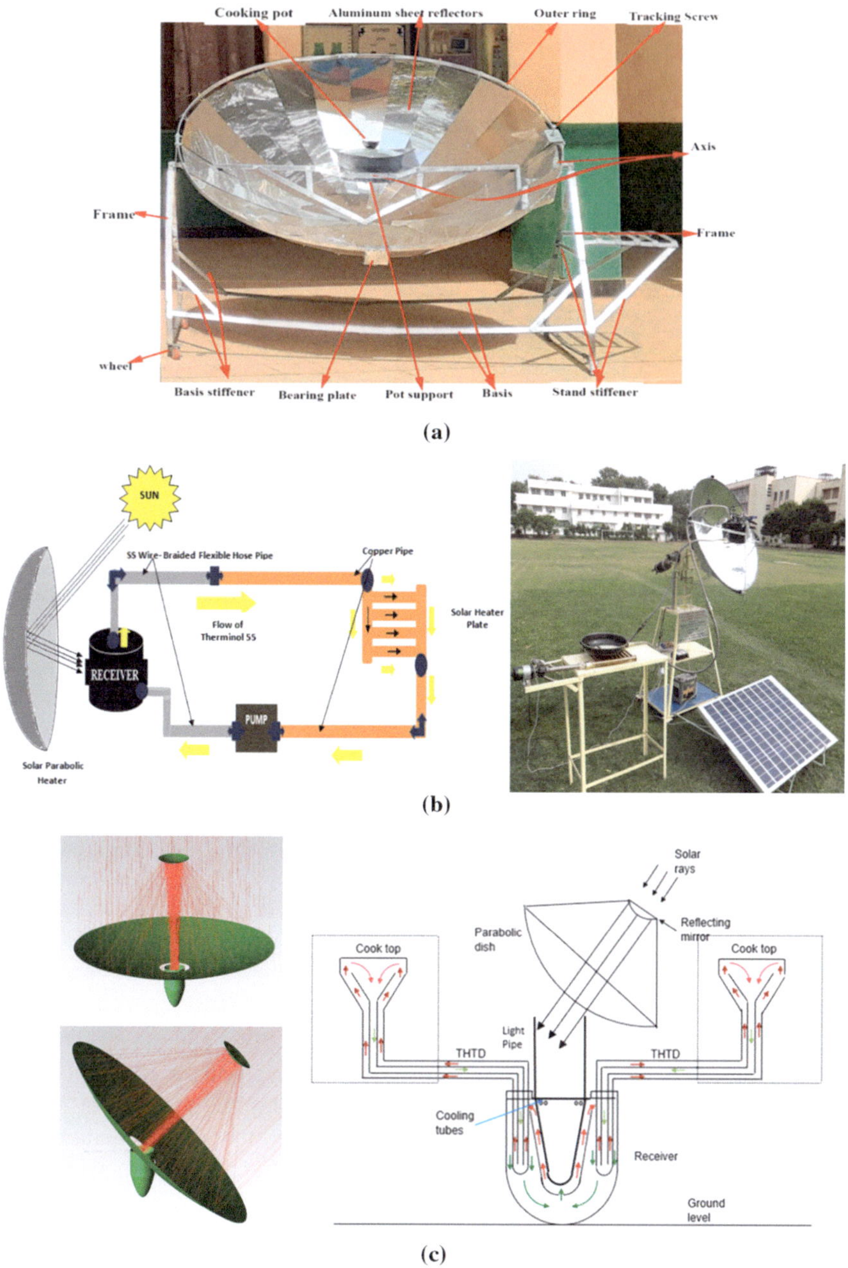

Fig. 41 **a** Image of an outdoor PDC cooker, **b** schematic diagram and image of indoor PDC cooker and **c** schematic diagram and simulated image of indoor PDC cooker. Reproduced from [77, 83, 84]

Fig. 42 Om shanti retreat centre Scheffler cooker [85]

Fig. 43 Maricopa SE-PDC power plant [86]

430 glass mirrors. The system had an aperture area of 20 m and the receiver was constructed with a cylindrical coil arrangement for tracking in dual axis. The collector uses water as HTF in the receiver which is directly converted to steam in 540 °C temperature and 6.8 MPa Pressure. The system was used for power generation and for educating students at Texas University [86]. The picture of the solar bowl at Taxis Tech University is provided in Fig. 45.

Fig. 44 Photograph of Tianjin University SE-PDC reproduced from [75]

Fig. 45 Texas Tech University solar bowl [86]

12 Concentrated Photovoltaics Application

Concentrated photovoltaics (CPV) utilizes Fresnel lenses or parabolic mirrors to boost the efficiency of solar modules. These systems comes under the classification of power generated from concentrated solar energy. However, the concentrator focuses light rather than heat to produce electricity. On the other hand, concentrated photovoltaic/thermal (CPV/T) an advanced version of this captures both light and heat for hybrid applications. Fresnel lenses have several advantages over conventional lenses such as slimmer in thickness, lighter in weight, capable of focusing with a higher degree of precision, mass production with various sizes and are cost-effective [87]. The drawbacks of the Fresnel lens associated with PV applications are shorter lifespan, requirement of maintenance and hot spots on PV active region caused by chromatic abrasion or defects of the Fresnel lens. The complexity of sun tracking in CPV is created by additional load and volume due to a lengthy focal gap. Discolouration as a consequence of material degradation, dust accumulation and so on leads to efficiency reduction [88, 89].

The Fresnel lenses are either aligned on PV modules to concentrate incident rays in point focus on a single cell or line focus on a series of cells. Figure 46 represents a pictorial view of point focused SPV. The multi-junction solar cells are preferred for amalgamating Fresnel lenses, since they have low thermal deterioration and demand higher luminance focus for raising their energy conversion efficiency with less space requirement. The antireflecting coating on cells, chromatic filter coating on the lens or introducing a homogenizer prism between the concentrator and cells are provided to avoid the non-uniform collimated rays from the Fresnel lenses. In common, Fresnel lenses are made with glass, polymer or acrylic materials with concentric rings. Yamaguchi et al. enhanced the efficiency of tandem cells (InGaP/InGaAs/Ge) with the combination of dome shaped Fresnel lens and homogenizers. The examined $400 \times 7000\ cm^2$ CPV module containing 36 lens-paired cells obtained an outdoor efficiency of 27%. Wang et al. proposed and simulated a band focused Fresnel lens for directing incident rays on a series of cells in CPV modules [90, 91].

The parabolic mirrors used in CPV modules have some advantages over the Fresnel lenses such as a higher concentration ratio, negligible amount of optical abrasion, high durability and easier maintenance. The drawbacks faced by these mirrors are high manufacturing cost, mismatch in alignment and tracking reduces efficiency,

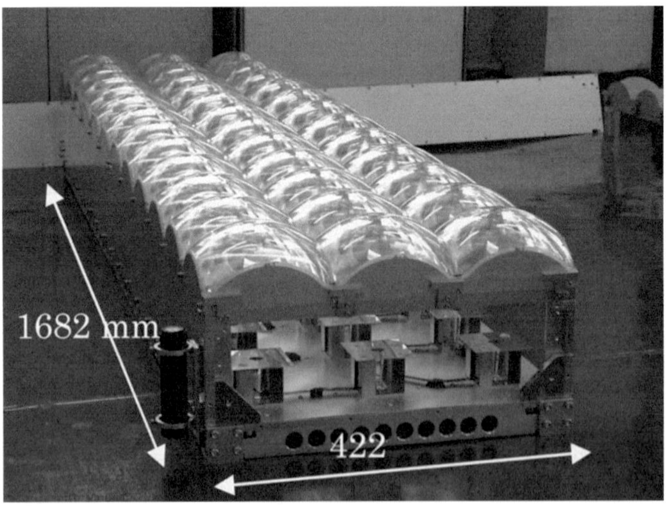

Fig. 46 Point -focus SPV system adopted from [90]

Fig. 47 Portrait of PDC steered photovoltaic adopted from [93]

stronger structural support and larger area requirements due to their size and weight. Parabolic mirrors are arranged in an array to create a CPV module similar to the Fresnel lens. Each mirrors are inserted with multi-junction solar cells in the midpoint. The irradiation falling on the parabolic mirrors reflects to a secondary mirror situated in the focal region, which diverts the light rays on the surface of employed solar cells. The high-efficiency cells convert the light into electricity instantly and supply it to utility grids [92]. Figure 47 depicts a parabolic reflector directed PV module.

The function of CPV/T is to produce heat and power individually or combinedly with the integration of solar cells instead of receiver to the conventional CSP reflectors (parabolic trough, parabolic dish, compound parabolic concentrator, heliostat or Fresnel type). In addition, a heat exchanger is directly coupled with the solar modules to extricate the absorbed heat for wielding different thermal systems. An array of CPV/T constructed with a compound parabolic concentrator was developed by Wang et al. in Xi'an City (China). The CPV/T modules in this system are composed of twenty (156 mm $\times$ 78 mm) polycrystalline solar cells which are embedded in the top portion of a rectangular heat exchanger made up of aluminium. Water supplied from a cold tank is used as HTF in the intended heat exchanger to carry the concentrated heat to store in a hot storage tank. Only a small variation in the performance of the modules was noticed while comparing with two-axis and south-north tracking arrangements. From this evidence, the system was scaled up with south-north tracking covering an aperture area of 810 m^2. Figure 48 displays the pictures of a CPV/T array that uses PTC for concentrating the light intensity and solar radiation. The maximum power generation capacity of the system was observed as 91.8 kW with an electrical efficiency of 12.2–13.2% range and thermal efficiency of 55.3%. The entire system comprises 810 CPV/T modules in 18 different arrays connected in series [94].

Fig. 48 Photographs of a CPV/T array reproduced from [94]

12.1 Case Study of CPV

A high concentration photovoltaic (HCPV) power plant with a 30 MW capacity was commissioned in Alamosa, Colorado state of USA in 2012. The station is constituted of more than 500 units of HCPV Amonix 7700 tracker in 225 acres of land cover area. These units are exclusively comprised of 7560 Fresnel lenses for delivering 500 times more daylight on gallium arsenide multi-junction solar cells. The modules were implanted on a 70 ft wide and 50 ft elevated structure with dual-axis tracking. The generated power is distributed to Colorado public service by the Xcel energy branch [95, 96]. The HCPV Arrays with their structures are shown in Fig. 49.

Fig. 49 Alamosa HCPV power station adopted from [95]

13 Industrial Applications

13.1 *Pasteurization and Sterilization*

Detrimental microorganisms from the food beverages are eliminated in the pasteurization process. Especially in milk to destroy the microorganisms and products such as mycobacterium tuberculosis, escherichia coli, phosphatase and peroxidase behoves a temperature of 50–75 °C. Subsequently, the pasteurized beverages expeditiously cooled down to prevent the multiplying of unannihilated microbes. Franco et al. [97] fabricated a focal length reduced parabolic dish concentrator by arranging Fresnel ring mirror strips for 10 L milk pasteurization. Unlike the usual parabolic dish, in this the concentration is done in line focus. The steam from the absorber situated in the focal point of the PDC is supplied to bain-marie water where the milk is constantly heated. Here, the milk is maintained at the required pasteurization temperature (63 °C) and duration (1 h). A milk pasteurizing tank operated with the heat source from a 10 m^2 Scheffler concentrator is designed and compared with an evacuate tube collector integrated pasteurizer by [98]. The steam produced in a fixed receiver by absorbing concentrated radiation is allowed to flow inside a tank surrounding another tank (pasteurizer) containing milk. The heat from the steam/hot water is transferred to milk to retain the 63 °C temperature for 30 min. The pasteurizing tank consists of a stirrer for even distribution of heat. An image of PDC for milk pasteurization is shown in Fig. 50.

Sterilizing is a process in which the microbes, viruses, fungi, bacteria and spores are extirpated from medical or food instruments to prevent life-threatening diseases. In common practice, autoclaves are used for sterilizing by introducing high pressure and temperature steam (120–250 °C) for a certain duration (5–15 min). A review of different solar thermal collectors coupled with autoclaves was reported by [99]. In this, they have classified sterilizing based on the collectors used and based on heat transfer to the autoclave. The PDC, PTC and LFR are CSPs that have been integrated with autoclaves for sterilizing. There are three types of autoclave heat utilization: (a) The heat derived from concentrators is directly focused on the autoclave containing heating fluid. (b) The concentrated heat absorbed by the receiver HTF is supplied to the autoclave where heat gets transmitted to working fluid. c) A heat exchanger is employed between the absorber and autoclave and is operated with two working fluids. The sterilizing devices commanded with various CSPs are provided in Fig. 51.

13.2 *Smelting and Melting*

The raw materials from mines are segregated into pure metals or minerals after crossing several stages of processing such as preparation, concentration, reduction, smelting and refining. In the smelting process, impurity-free metals are extracted from the ore material by heating at higher temperatures. Reducing agents (coke,

Fig. 50 PDC—Pasteurization system reproduced [97]

hydrogen, carbon monoxide etc.,) or flux (limestone, borax, etc.,) are used in high temperature smelting for forming chemical reactions to remove impurities as different products. Eglinton et al. [100] differentiated CSP utilization in various phases of smelting. In the thermal process, furnaces or kilns are employed to separate minerals. A study was conducted to extract lime from limestone in a solar kiln assisted by heliostats. Another application is supplying preheated air from CSP systems as regenerative heating in Cowper stoves for iron extraction in blast furnaces. Syngas production through SPT in the CISRO energy centre is an example of a chemical process. This system heats the combination of methane and steam together in the presence of catalyst at 850 °C to obtain syngas (CO_2 and H_2) with other products. Figure 52 represents the CSIRO SPT plant for the production of syngas. The process in which the conversion of metal oxides to metals is known as carbothermic reduction. Metal oxides like ZnO are heated in the presence of natural gas where it gets converted into metals (Zn) and syngas.

Galvanizing, casting, recycling and thermal storage are a few material manufacturing processes that craves high temperatures for reforming it to molten state. Bezuidenhout et al. [101] developed a PDC with six elliptical shaped reflectors individually concentrating radiation towards a cylindrical cavity receiver. This cavity is encompassed by another zinc containing cavity, where the zinc metal melts by enthralling the heat from the absorber. The metal rehabilitated from feedstock in this

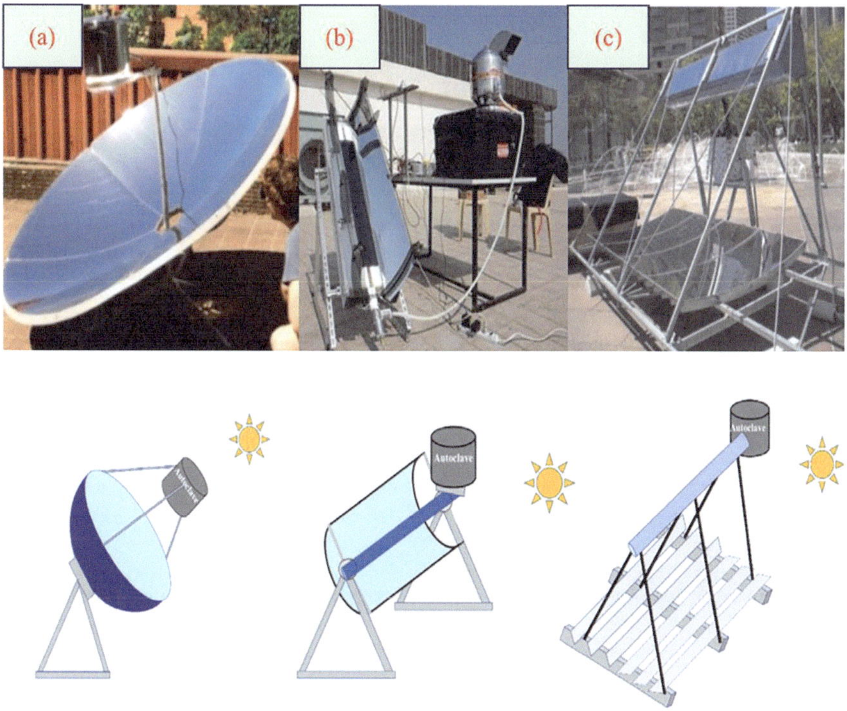

Fig. 51 Structural design and images of **a** PDC, **b** PTC, **c** LFR sterilizing unit. Reproduced from [99]

setup is converted into components that are necessary to construct usable products. Two different parabolic troughs are fabricated by [102] with stainless steel sheets as reflecting surfaces and copper absorbers incorporated with non-evacuated glass/quartz tubes as receivers. These troughs were employed to melt the plastic wastes and to recycle them into useful commodities. In the study conducted by Romero et al. [103], 1.5 m PDCs positioned on the vertical axis with a shutter radiation controller are used to convert the raw material into glasses inside a tubular crucible made up of alumina. This system was compared with an electric furnace and the same alumina crucible for the production of glass. Apart from industrial reforming processes in cold regions ice is melted to water for daily consumption. Agrawal et al. [104]demonstrated ice melting with the aid of a 1 m^2 Scheffler concentrator and three distinct arrangement receivers. Figure 53 displays the images of (a) zinc melting, (b) glass melting, (c) plastic melting devices.

Fig. 52 CSIRO SPT plant for syngas production adopted from [100]

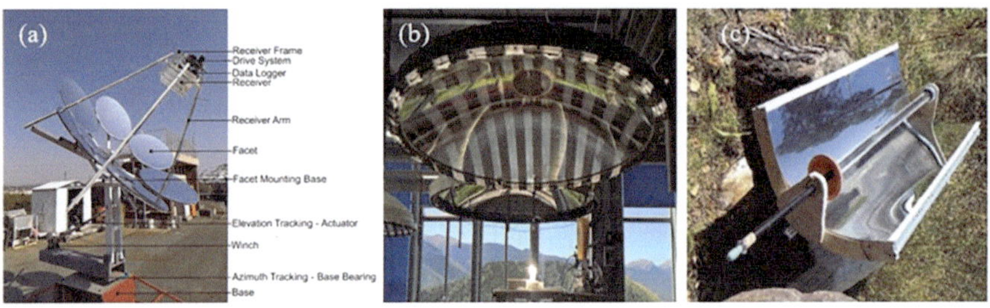

Fig. 53 **a** Zinc melting, **b** glass melting and **c** plastic melting devices. Adopted from [101–103]

13.3 Pyrolysis

Pyrolysis is a method for the production of fuels from organic materials such as solid wastes, biomass and coal by heating it in an oxygen-free environment. The operating temperature of this process lies between 300 to 900 °C, which is apt to be attained with CSP concentrators. A pyrolysis bed type circular reactor integrated with a 16 m^2 Scheffler collector was constructed by Chintala et al. [68] to convert jatropha seeds into biofuel. The reactor was heated up to 250–320 °C on average to produce bio-oil securing a yield of 20%. The non-condensable gas (syngas) and remaining bio-char

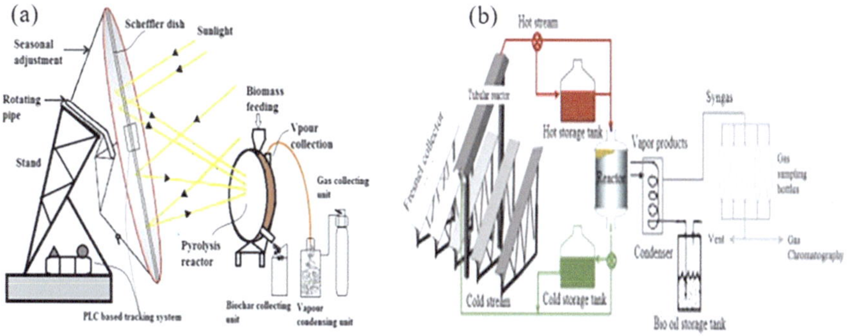

Fig. 54 Schematic diagram of **a** Scheffler collector pyrolysis **b** LFR pyrolysis. Reproduced from [68, 105]

are also collected separately after thermal decomposition. In a similar way, a LFR system was modelled with an evacuated tube and secondary reflector for heating HTF. The hot fluid is collected in a tank as thermal energy storage and distributed to the reactor for heating biomass like lignins, hemicellulose or cellulose with different decomposition temperatures. The heat transfer converts the feedstocks into vapour and bio-char. The vapour is further condensed to bio-oil and the rest is collected as syngas and the HTF is recirculated to LFR through a storage tank [105]. The schematic diagrams of the Scheffler collector and LFR pyrolysis are given in Fig. 54.

13.4 Welding

Two similar or dissimilar metals are joined together along with or without filler materials as a single unit in the welding process. Romero et al. [106] determined the effectiveness of different welding joints acquired by concentrated radiation beams on H13 steel and stainless steel. The system was designed with a heliostat that deflects solar radiation on a 1.5 m^2 PDC which further focused onto hollow hemispherical quartz. The base and filler (E6013 electrodes) materials are placed in centre of a quartz chamber enclosed with an argon atmosphere to absorb the heat obtained from reflectors. A cooling arrangement and shutter control was provided in the chamber to regulate the heat. The welding speed was coordinated by a triple-axis tacker. A similar system was developed by Cambronero et al. [107] for joining two aluminium foam plates by using aluminium alloy (AlSi) as filler material. The concentrated radiation was allowed to fall on a consumable electrode placed in between the parent materials. This foams the filler alloy and merges the base metals. Ceramic plates are used as insulators on top of base metal to avoid thermal decomposition and to prevent oxidation on foams gas shielding was enforced. A line diagram and prototype image of a PDC operated welding are provided in Fig. 55.

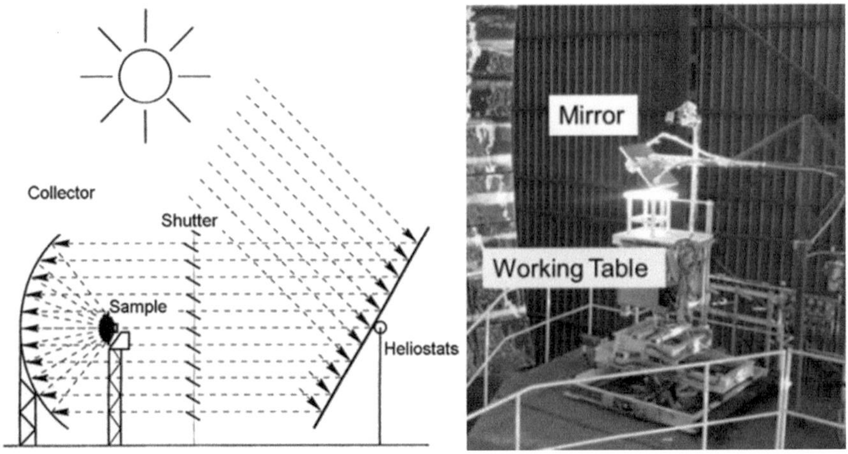

Fig. 55 PDC-welding line diagram and prototype image adopted from [107]

13.5 Distillation

In crude oil refineries, the distillation process is one of the methods that need an enormous amount of heat energy to vaporize it for separating various petroleum fuels, valuable products and asphalt according to their density. Prestigiacomo et al. [108] proposed and simulated a 100 m^2 PTC system with the usage of molten salt as both HTF and thermal energy storage for integrating in the topping section of two existing distillation columns. The modelled PTC system heats the molten salts to 550 °C and stored in the hot storage tank. This heat is transferred to crude oil in the heat exchanger. The partially heated oil is converted to vapour in furnaces and sent to distillation columns for separating the fuels and products. A lab-scale prototype was fabricated for steam distillation of extract oils from agricultural crops. The operating temperature reached 295 °C with the integration of a 6 m PTC system. Therminol 66 was used as HTF in this system to transmit heat to the oil for vaporizing. This vapour was condensed in a distillation column to obtain the fuel. The yields of lemon, orange and mandarin were noticed as 0.53, 0.67 and 1.09% in the conducted experiment [109]. Figures 56 and 57 show an oil extraction prototype unit and line diagram of a crude oil distillation plant.

13.6 Food Processing

Baking, boiling, fermentation, emulsification, caramelization and drying are some of the industrial food processes that require a restrained amount of heat. The heat and electricity generated from CSP plants are capable of supporting these processes and supplying power for mixing, grinding, chopping, etc. machines. Ebadi et al. [110]

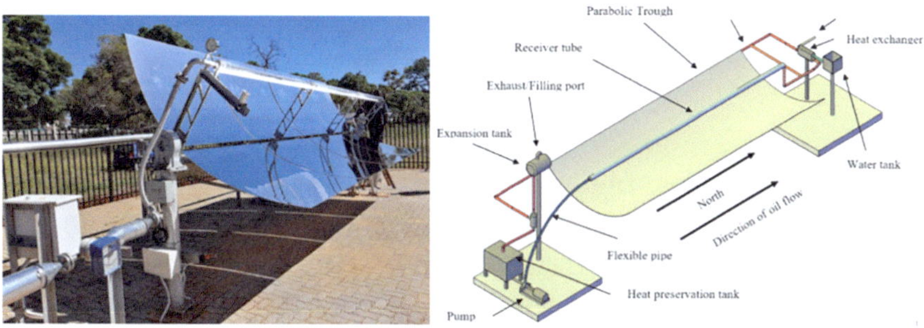

Fig. 56 Image and Schematic diagram of a prototype distiller reproduced from [109]

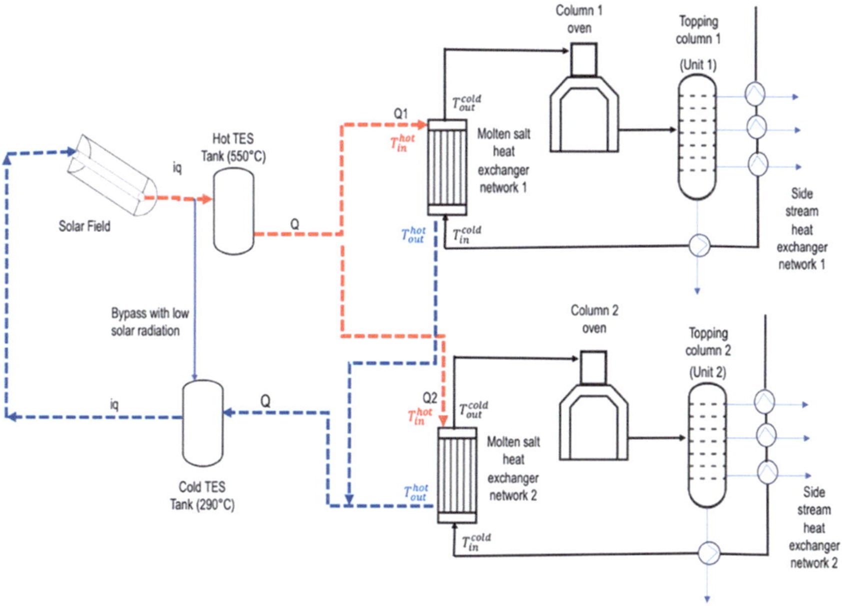

Fig. 57 Layout of a crude oil distillation process. Adopted from [108]

developed a prototype for drying tomatoes with five compound parabolic concentrators interlinked with a drying chamber. The heat collected from the compound parabolic collectors by ambient air is sucked by a fan arrangement and channelled to the dryer for dehydrating the tomato for preservation. An auxiliary heater is employed between the concentrator chamber for heating in non-sunshine hours. Gunes et al. [111] proposed a PTC plant with TES for producing peach puree. The heat from the system is utilized in preheating, aseptic packing and evaporating of peach extract. Camci et al. [112] evaluated a PTC coupled spray dryer for milk powder production. The milk is sprayed into a drying chamber where hot air (200 °C) supplied by the concentrators absorbs the moisture content to form milk powder instantly. Initially, the milk was preheated in a vacuumed evaporator and the energy required to run the compressor, pump are provided by a separate PV system.

13.7 Textile Processing

In textile industries, heat is essential for washing, dyeing and drying of fibber/yarn used for sewing cloths. Youssef et al. [113] simulated a CPV/T system for textile dyeing and electricity production. The CPV/T consists of a PTC employed with an evacuated tube containing a PV encircled absorber tube. The solar beam reflected to the receiver generates electricity from the PV and absorbs heat by coolants for colouring the fibre and drying. El Golli et al. [114] investigated the performance of a PTC system for purifying water/dye mixture or degrading antibiotic medicines. The solar radiation absorbed by a quartz tube inserted in the focal point of the trough decomposes RBB dye/Ciprofloxacin from the textile industry effluents.

13.8 Water Irrigation

CSP system assisted water pumps are used for agricultural water irrigation. The small power generating units like the PDC Stirling engine explained in 11.1 are accommodated to drive field water pumps for distributing water from underground bore wells or waterbodies to pipes and sprinklers. The recycled wastewater produced by utilizing heat from the CSPs for agriculture and gardening is another approach.

13.9 Gasification and Incineration

CSP gasification is a process of heating waste products or coal at a high rate of temperature with a controlled amount of steam or oxygen supply for producing syngas. A multi-functional system operated with a SPT power plant was designed by Ozturk and Dincer [115]. The heat collected by HTF from a SPT plant is utilized to produce electricity in a ranking cycle, hydrogen in a PEM electrolyser, syngas in a coal gasifier, heating and cooling effect in an absorption chiller, steam from a heat exchanger and electricity in Brayton cycle. The steam produced is supplied with compressed air for the gasification of coal in a chamber. The syngas after eliminating H_2S and Tar are stored for further usage. Solid waste management through incineration is a way to compress the volume of waste products. Burning waste materials with a CSP system produces enormous heat and other by-products (ash and flue gas) that can be induced for steam generation to produce electricity. Wang et al. [116] proposed a waste hybrid system with Fresnel lens collectors and incinerators for power generation, district and industrial process heating. The heat collected from the Fresnel lenses is directed

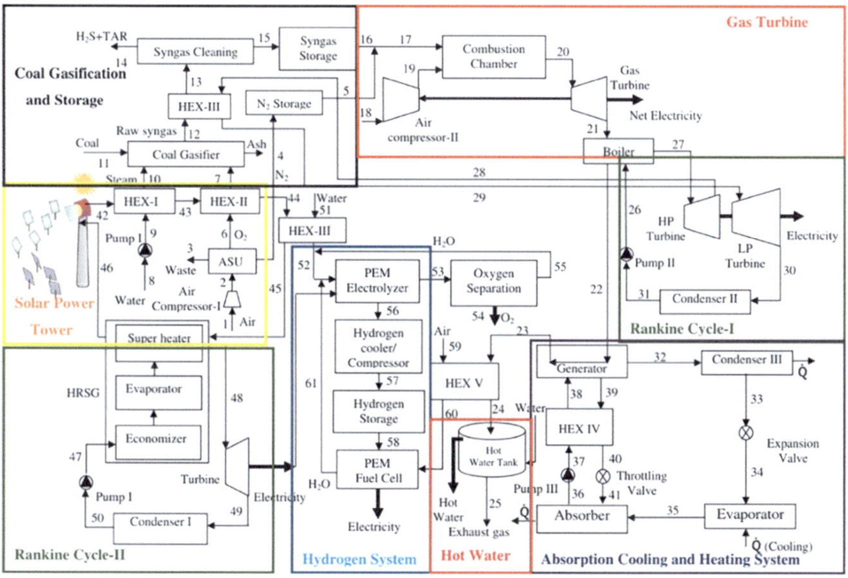

Fig. 58 Multi-purpose SPT coal gasification plant reproduced from [115]

to the incinerator chamber via HTF and then evaporator to provide steam for various applications. The incinerator is used as an auxiliary heater in off-sunshine hours. The layout of multi-purpose SPT coal gasification power plant and hybrid power generation plant with Fresnel concentrator and incinerator are presented in Figs. 58 and 59.

14 Thermal Energy Storage

Thermal energy storage (TES) materials are employed to store and retrieve energy from CSP collectors to extend the duration of heat retainment in off sunshine and night hours. Thermal energy storage is classified into sensible, latent and thermochemical energy storage in which all are applicable to accommodate in CSP power plants. Water, rocks, sand, synthetic oils (HTF) and concrete are some of the sensible heat storage materials that are directly used in CSP storage units, receiver beds etc. Molten salts, salt hydrates, paraffin and metalloids are the most preferred latent heat storage that is capable of escalating the plant's thermal efficiency due to their high heat storage capacity. Sodium sulphate decahydrate and ammonium nitrates are materials that are used for creating exothermic and endothermic reactions for retracting energy from CSP. The working process of the system operated with TES are same as applications that have been stated in the previous part of this chapter [117, 118]. Schematic diagrams and existing CSP plants integrated with TES are provided in Fig. 60a–c

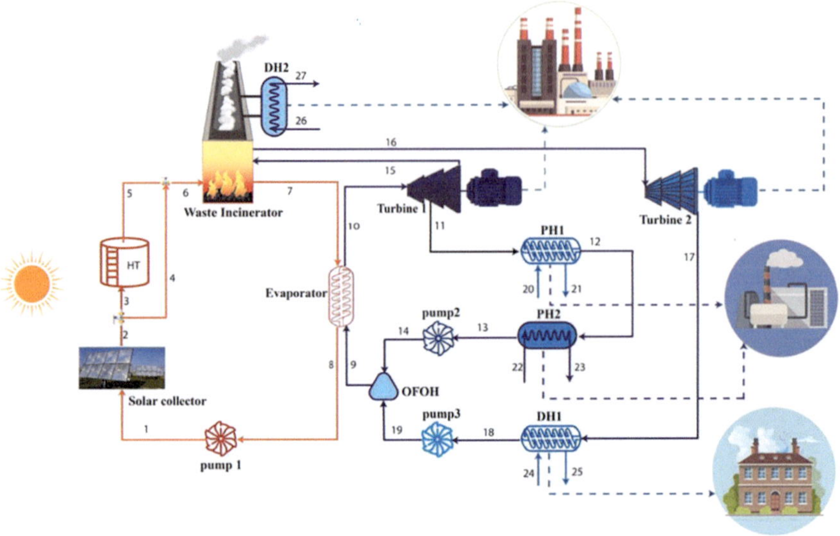

Fig. 59 Hybrid Fresnel collector and incineration multi-generation plant. Adopted from [116]

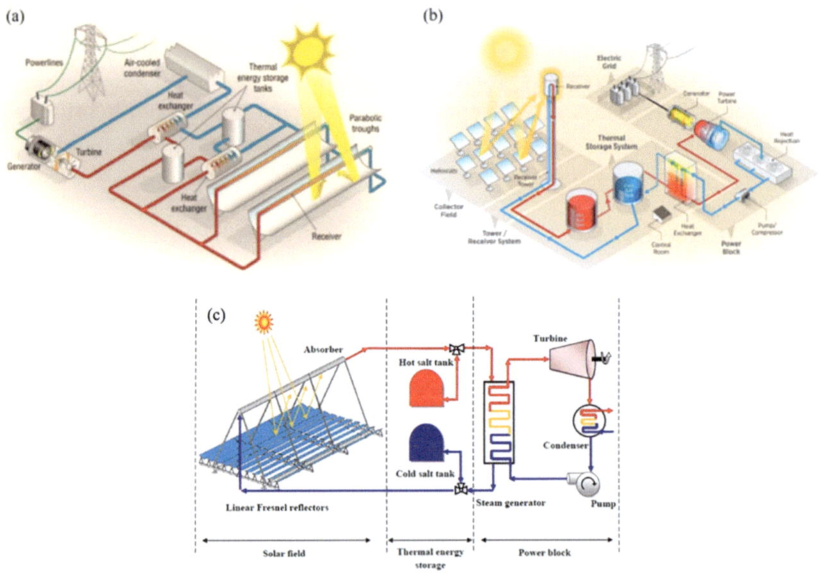

Fig. 60 **a–c** Schematic diagrams of CSP plants with TES adopted from [55, 119]

and 61a–c. Thermal energy storage utilized in these plants are provided in Table 2 with their storage capacities.

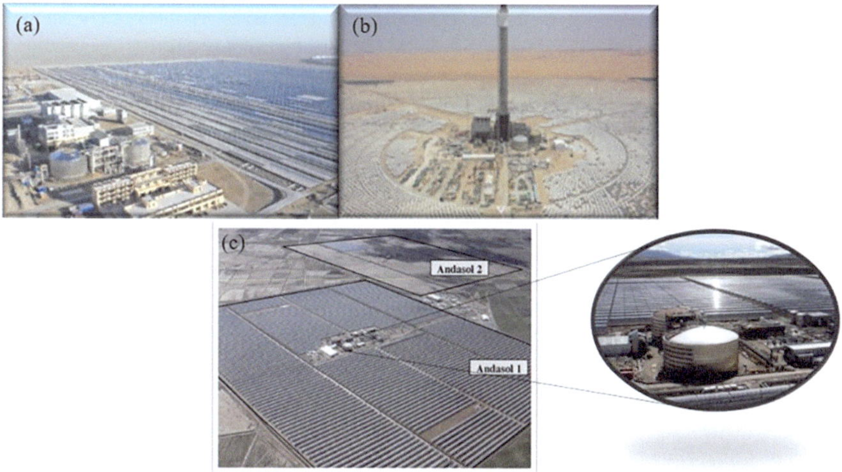

Fig. 61 Real-time view of CSP plans with TES. **a** LFR power plant (Dacheng dunhuang), **b** SPT power plant (Noor-Ouarzazate solar power station) and **c** PTC power plant (Andasol 1 and 2). Reproduced from [117, 120]

Table. 2 Thermal energy storage materials used in the stated CSP power stations [117, 118]

Power plant	Developer	Location	Power capacity/ plant (kW)	TES materials	Working temperature (°C)	Storage capacity (hours)
Dacheng Dunhuang (LFR)	Lanzhou Dacheng Technology Co., LTd	Gansu, China	50	Sodium sulphate decahydrate	290–550	15
Noor solar power station (SPT)	ACWA Power and Abengoa	Ouarzazate, Morocco	150	Potassium nitrate and sodium nitrate mixture	293–393	7
Andasol 1 and 2 (PTC)	Solar Millennium AG	Andalusia, Spain	50	Potassium nitrate and sodium nitrate mixture	293–393	7.5

References

1. G.W. Crabtree, N.S. Lewis, Solar energy conversion. Phys. Today. **60**(3), 37–42 (2007). https://doi.org/10.1063/1.2718755
2. M.T. Islam, N. Huda, A.B. Abdullah, R. Saidur, A comprehensive review of state-of-the-art concentrating solar power (CSP) technologies: current status and research trends. Renew. Sustain. Energy Rev. **91**, 987–1018 (2018). https://doi.org/10.1016/j.rser.2018.04.097

3. L.A. Weinstein, J. Loomis, B. Bhatia, D.M. Bierman, E.N. Wang, G. Chen, Concentrating solar power. Chem. Rev. **115**(23), 12797–12838 (2015). https://doi.org/10.1021/acs.chemrev.5b00397
4. K. Lovegrove, J. Pye, Fundamental principles of concentrating solar power (CSP) systems, in *Concentrating Solar Power Technology* (Elsevier, 2012), pp. 16–67. https://doi.org/10.1533/9780857096173.1.16
5. A. Peinado Gonzalo, A. Pliego Marugán, F.P. García Márquez, A review of the application performances of concentrated solar power systems. Appl. Energy **255**, 113893 (2019). https://doi.org/10.1016/j.apenergy.2019.113893
6. R. Pitz-Paal, Concentrating solar power systems. EPJ. Web Conf. **148**, 00008 (2017). https://doi.org/10.1051/epjconf/201714800008
7. H.L. Zhang, J. Baeyens, J. Degrève, G. Cacères, Concentrated solar power plants: review and design methodology. Renew. Sustain. Energy Rev. **22**, 466–481 (2013) https://doi.org/10.1016/j.rser.2013.01.032
8. R. Abbas, M.J. Montes, M. Piera, J. M. Martínez-Val, Solar radiation concentration features in linear Fresnel reflector arrays. Energy Convers. Manag. **54**(1), 133–144 (2012). https://doi.org/10.1016/j.enconman.2011.10.010
9. D.R. Mills, G.L. Morrison, Compact Linear Fresnel Reflector solar thermal powerplants. Solar Energy **68**(3), 263–283 (2000). https://doi.org/10.1016/S0038-092X(99)00068-7
10. K.H. Kumar, A.M. Daabo, M.K. Karmakar, H. Hirani, Solar parabolic dish collector for concentrated solar thermal systems: a review and recommendations. Environ. Sci. Pollut. Res. **29**(22), 32335–32367 (2022). https://doi.org/10.1007/s11356-022-18586-4
11. W.E. Reifsnyder, Radiation geometry in the measurement and interpretation of radiation balance. Agric. Meteorol. **4**(4), 255–265 (1967). https://doi.org/10.1016/0002-1571(67)90026-X
12. M. Ballico, A simple technique for measuring the infrared emissivity of black-body radiators. Metrologia **37**(4), 295–300 (2000). https://doi.org/10.1088/0026-1394/37/4/5
13. N. Hernández et al., Conical receiver for a paraboloidal concentrator with large rim angle. Sol. Energy **86**(4), 1053–1062 (2012). https://doi.org/10.1016/j.solener.2011.09.008
14. A.J. Abdulhamed, N.M. Adam, M.Z.A. Ab-Kadir, A.A. Hairuddin, Review of solar parabolic-trough collector geometrical and thermal analyses, performance, and applications. Renew. Sustain. Energy Rev. **91**, 822–831 (2018). https://doi.org/10.1016/j.rser.2018.04.085
15. M. Wild, D. Folini, C. Schär, N. Loeb, E.G. Dutton, G. König-Langlo, The global energy balance from a surface perspective. Clim. Dyn. **40**(11–12), 3107–3134 (2013). https://doi.org/10.1007/s00382-012-1569-8
16. R.P. Merchán, M.J. Santos, A. Medina, A. Calvo Hernández, High temperature central tower plants for concentrated solar power: 2021 overview. Renew. Sustain. Energy Rev. **155**, 111828 (2022). https://doi.org/10.1016/j.rser.2021.111828
17. A. Ahmad, O. Prakash, R. Kausher, G. Kumar, S. Pandey, S.M.M. Hasnain, Parabolic trough solar collectors: a sustainable and efficient energy source. Mater. Sci. Energy Technol. **7**, 99–106 (2024). https://doi.org/10.1016/j.mset.2023.08.002
18. E. Bellos, C. Tzivanidis, Alternative designs of parabolic trough solar collectors. Prog. Energy Combust. Sci. **71**, 81–117 (2019). https://doi.org/10.1016/j.pecs.2018.11.001
19. W. Fuqiang, C. Ziming, T. Jianyu, Y. Yuan, S. Yong, L. Linhua, Progress in concentrated solar power technology with parabolic trough collector system: a comprehensive review. Renew. Sustain. Energy Rev. **79**, 1314–1328 (2017). https://doi.org/10.1016/j.rser.2017.05.174
20. B. Zou, J. Dong, Y. Yao, Y. Jiang, An experimental investigation on a small-sized parabolic trough solar collector for water heating in cold areas. Appl. Energy **163**, 396–407 (2016). https://doi.org/10.1016/j.apenergy.2015.10.186
21. T. eddine Boukelia, M.-S. Mecibah, Parabolic trough solar thermal power plant: potential, and projects development in Algeria. Renew. Sustain. Energy Rev. **21**, 288–297 (2013). https://doi.org/10.1016/j.rser.2012.11.074
22. A. Kumar, S. Vyas, D. Nchelatebe Nkwetta, Experimental study of single slope solar still coupled with parabolic trough collector. Mater. Sci. Energy Technol. **3**, 700–708 (2020). https://doi.org/10.1016/j.mset.2020.07.005

23. A.M. Aqlan, M. Aklan, A.E. Momin, Solar-powered desalination, a novel solar still directly connected to solar parabolic trough. Energy Rep. **7**, 2245–2254 (2021). https://doi.org/10.1016/j.egyr.2021.04.041
24. N.A. Moharram, S. Bayoumi, A.A. Hanafy, W.M. El-Maghlany, Hybrid desalination and power generation plant utilizing multi-stage flash and reverse osmosis driven by parabolic trough collectors. Case Stud. Therm. Eng. **23**, 100807 (2021). https://doi.org/10.1016/j.csite.2020.100807
25. H. Jafari Mosleh, S. Jahangiri Mamouri, M.B. Shafii, A. Hakim Sima, A new desalination system using a combination of heat pipe, evacuated tube and parabolic trough collector. Energy Convers. Manag. **99**, 141–150 (2015). https://doi.org/10.1016/j.enconman.2015.04.028
26. N.H. Abu-Hamdeh, K.A. Alnefaie, K.H. Almitani, Design and performance characteristics of solar adsorption refrigeration system using parabolic trough collector: experimental and statistical optimization technique. Energy Convers. Manag. **74**, 162–170 (2013). https://doi.org/10.1016/j.enconman.2013.04.043
27. A. El Fadar, A. Mimet, M. Pérez-García, Study of an adsorption refrigeration system powered by parabolic trough collector and coupled with a heat pipe. Renew. Energy **34**(10), 2271–2279 (2009). https://doi.org/10.1016/j.renene.2009.03.009
28. E. Bellos, C. Tzivanidis, Parametric analysis and optimization of a cooling system with ejector-absorption chiller powered by solar parabolic trough collectors. Energy Convers. Manag. **168**, 329–342 (2018). https://doi.org/10.1016/j.enconman.2018.05.024
29. Y. Bi, L. Qin, J. Guo, H. Li, G. Zang, Performance analysis of solar air conditioning system based on the independent-developed solar parabolic trough collector. Energy **196**, 117075 (2020). https://doi.org/10.1016/j.energy.2020.117075
30. M.S. Büker, H. Parlamış, M. Alwetaishi, O. Benjeddou, Experimental investigation on the dehumidification performance of a parabolic trough solar air collector assisted rotary desiccant system. Case Stud. Therm. Eng. **34**, 102077 (2022). https://doi.org/10.1016/j.csite.2022.102077
31. L. Zhou, X. Li, Y. Zhao, Y. Dai, Performance assessment of a single/double hybrid effect absorption cooling system driven by linear Fresnel solar collectors with latent thermal storage. Sol. Energy **151**, 82–94 (2017). https://doi.org/10.1016/j.solener.2017.05.031
32. M. Medrano, A. Gil, I. Martorell, X. Potau, L.F. Cabeza, State of the art on high-temperature thermal energy storage for power generation. Part 2—Case studies. Renew. Sustain. Energy Rev. **14**(1), 56–72 (2010). https://doi.org/10.1016/j.rser.2009.07.036
33. A. Poullikkas, Economic analysis of power generation from parabolic trough solar thermal plants for the Mediterranean region—a case study for the island of Cyprus. Renew. Sustain. Energy Rev. **13**(9), 2474–2484 (2009). https://doi.org/10.1016/j.rser.2009.03.014
34. E.Z. Moya, Parabolic-trough concentrating solar power systems, in *Concentrating Solar Power Technology* (Elsevier, 2021), pp. 219–266. https://doi.org/10.1016/B978-0-12-819970-1.00009-8
35. Genesis Solar, L. L. C., *Genesis Solar Energy Project* (2009)
36. T.M. Pavlović, I.S. Radonjić, D.D. Milosavljević, L.S. Pantić, A review of concentrating solar power plants in the world and their potential use in Serbia. Renew. Sustain. Energy Rev. **16**(6), 3891–3902 (2012). https://doi.org/10.1016/j.rser.2012.03.042
37. W. Gretz, Federal Technology Alert A publication series designed to speed the adoption of energy-efficient and renewable technologies in the Federal sector Parabolic-Trough Solar Water Heating. Renewable technology for reducing water-heating costs
38. M. Balghouthi, A.B.H. Ali, S.E. Trabelsi, A. Guizani, Optical and thermal evaluations of a medium temperature parabolic trough solar collector used in a cooling installation. Energy Convers. Manag. **86**, 1134–1146 (2014). https://doi.org/10.1016/j.enconman.2014.06.095
39. Z. Tian, B. Perers, S. Furbo, J. Fan, Annual measured and simulated thermal performance analysis of a hybrid solar district heating plant with flat plate collectors and parabolic trough collectors in series. Appl. Energy **205**, 417–427 (2017). https://doi.org/10.1016/j.apenergy.2017.07.139

40. Z. Tian, B. Perers, S. Furbo, J. Fan, Analysis and validation of a quasi-dynamic model for a solar collector field with flat plate collectors and parabolic trough collectors in series for district heating. Energy **142**, 130–138 (2018). https://doi.org/10.1016/j.energy.2017.09.135
41. S. Alexopoulos, B. Hoffschmidt, Advances in solar tower technology. WIREs Energy Environ. **6**(1) (2017). https://doi.org/10.1002/wene.217
42. U. Pelay, L. Luo, Y. Fan, D. Stitou, M. Rood, Thermal energy storage systems for concentrated solar power plants. Renew. Sustain. Energy Rev. **79**, 82–100 (2017). https://doi.org/10.1016/j.rser.2017.03.139
43. L. L. Vant-Hull, Central tower concentrating solar power systems, in *Concentrating Solar Power Technology* (Elsevier, 2021), pp. 267–310. https://doi.org/10.1016/B978-0-12-819970-1.00019-0
44. M.J. Wagner, W.T. Hamilton, A. Newman, J. Dent, C. Diep, R. Braun, Optimizing dispatch for a concentrated solar power tower. Sol. Energy **174**, 1198–1211 (2018). https://doi.org/10.1016/j.solener.2018.06.093
45. M. Qenawy, H.S. El-Mesery, J. Wang, S. Abdelhady, Design and thermal performance analysis of concentrating solar power tower for water heating systems. Case Stud. Therm. Eng. **48**, 103141 (2023). https://doi.org/10.1016/j.csite.2023.103141
46. S. Alexopoulos, B. Hoffschmidt, Solar tower power plant in Germany and future perspectives of the development of the technology in Greece and Cyprus. Renew. Energy **35**(7), 1352–1356 (2010). https://doi.org/10.1016/j.renene.2009.11.003
47. M.I. Soomro, W.-S. Kim, Performance and economic investigations of solar power tower plant integrated with direct contact membrane distillation system. Energy Convers. Manag. **174**, 626–638 (2018). https://doi.org/10.1016/j.enconman.2018.08.056
48. T. Conroy, M.N. Collins, R. Grimes, A review of steady-state thermal and mechanical modelling on tubular solar receivers. Renew. Sustain. Energy Rev. **119**, 109591 (2020). https://doi.org/10.1016/j.rser.2019.109591
49. R. Buck, P. Schwarzbözl, 4.17 solar tower systems, in *Comprehensive Energy Systems* (Elsevier, 2018), pp. 692–732. https://doi.org/10.1016/B978-0-12-809597-3.00428-4
50. D. Pulido-Iparraguirre, L. Valenzuela, J.-J. Serrano-Aguilera, A. Fernández-García, Optimized design of a Linear Fresnel reflector for solar process heat applications. Renew. Energy **131**, 1089–1106 (2019). https://doi.org/10.1016/j.renene.2018.08.018
51. H. Beltagy, The effect of glass on the receiver and the use of two absorber tubes on optical performance of linear Fresnel solar concentrators. Energy **224**, 120111 (2021). https://doi.org/10.1016/j.energy.2021.120111
52. J. Zhu, H. Huang, Design and thermal performances of Semi-Parabolic Linear Fresnel Reflector solar concentration collector. Energy Convers. Manag. **77**, 733–737 (2014). https://doi.org/10.1016/j.enconman.2013.10.015
53. Z.-D. Cheng, X.-R. Zhao, Y.-L. He, Y. Qiu, A novel optical optimization model for linear Fresnel reflector concentrators. Renew. Energy **129**, 486–499 (2018). https://doi.org/10.1016/j.renene.2018.06.019
54. A. Kumar, O. Prakash, A. Dube, A review on progress of concentrated solar power in India. Renew. Sustain. Energy Rev. **79**, 304–307 (2017). https://doi.org/10.1016/j.rser.2017.05.086
55. B. Belgasim, Y. Aldali, M.J.R. Abdunnabi, G. Hashem, K. Hossin, The potential of concentrating solar power (CSP) for electricity generation in Libya. Renew. Sustain. Energy Rev. **90**, 1–15 (2018). https://doi.org/10.1016/j.rser.2018.03.045
56. M. Ghodbane, B. Boumeddane, Z. Said, E. Bellos, A numerical simulation of a linear Fresnel solar reflector directed to produce steam for the power plant. J. Clean. Prod. **231**, 494–508 (2019). https://doi.org/10.1016/j.jclepro.2019.05.201
57. Y.N. Dabwan, E.M.A. Mokheimer, Optimal integration of linear Fresnel reflector with gas turbine cogeneration power plant. Energy Convers. Manag. **148**, 830–843 (2017). https://doi.org/10.1016/j.enconman.2017.06.057
58. P. Dellicompagni, L. Saravia, M. Altamirano, J. Franco, Simulation and testing of a solar reciprocating steam engine. Energy **151**, 662–674 (2018). https://doi.org/10.1016/j.energy.2018.03.110

59. M.S. Guney, Solar power and application methods. Renew. Sustain. Energy Rev. **57**, 776–785 (2016). https://doi.org/10.1016/j.rser.2015.12.055
60. G. Mokhtar, B. Boussad, S. Noureddine, A linear Fresnel reflector as a solar system for heating water: theoretical and experimental study. Case Stud. Therm. Eng. **8**, 176–186 (2016). https://doi.org/10.1016/j.csite.2016.06.006
61. R. Manikumar, A. Valan Arasu, Heat loss characteristics study of a trapezoidal cavity absorber with and without plate for a linear Fresnel reflector solar concentrator system. Renew. Energy **63**, 98–108 (2014). https://doi.org/10.1016/j.renene.2013.09.005
62. M. Babu, S.S. Raj, A. Valan Arasu, Experimental analysis on Linear Fresnel reflector solar concentrating hot water system with varying width reflectors. Case Stud. Therm. Eng. **14**, 100444 (2019). https://doi.org/10.1016/j.csite.2019.100444
63. I.B. Askari, M. Ameri, Techno economic feasibility analysis of Linear Fresnel solar field as thermal source of the MED/TVC desalination system. Desalination **394**, 1–17 (2016). https://doi.org/10.1016/j.desal.2016.04.022
64. I.B. Askari, M. Ameri, F. Calise, Energy, exergy and exergo-economic analysis of different water desalination technologies powered by Linear Fresnel solar field. Desalination **425**, 37–67 (2018). https://doi.org/10.1016/j.desal.2017.10.008
65. W. Gang, Y. Qichang, Z. Hongfei, Z. Yi, F. Hui, J. Rihui, Direct utilization of solar linear Fresnel reflector on multi-effect eccentric horizontal tubular still with falling film. Energy **170**, 170–184 (2019). https://doi.org/10.1016/j.energy.2018.12.150
66. M.A. Serag-Eldin, Thermal design of a roof-mounted CLFR collection system for a desert absorption chiller. Int. J. Sustain. Energ. **33**(3), 506–524 (2014). https://doi.org/10.1080/14786451.2012.761998
67. S.Z. Farooqui, Impact of load variation on the energy and exergy efficiencies of a single vacuum tube based solar cooker. Renew. Energy **77**, 152–158 (2015). https://doi.org/10.1016/j.renene.2014.12.021
68. V. Chintala, S. Kumar, J.K. Pandey, A.K. Sharma, S. Kumar, Solar thermal pyrolysis of non-edible seeds to biofuels and their feasibility assessment. Energy Convers. Manag. **153**, 482–492 (2017). https://doi.org/10.1016/j.enconman.2017.10.029
69. P. Bermejo, F.J. Pino, F. Rosa, Solar absorption cooling plant in Seville. Sol. Energy **84**(8), 1503–1512 (2010). https://doi.org/10.1016/j.solener.2010.05.012
70. M.Z. Malik et al., A review on design parameters and specifications of parabolic solar dish Stirling systems and their applications. Energy Rep. **8**, 4128–4154 (2022). https://doi.org/10.1016/j.egyr.2022.03.031
71. A.Z. Hafez, A. Soliman, K.A. El-Metwally, I.M. Ismail, Design analysis factors and specifications of solar dish technologies for different systems and applications. Renew. Sustain. Energy Rev. **67**, 1019–1036 (2017). https://doi.org/10.1016/j.rser.2016.09.077
72. M.E. Zayed, J. Zhao, A.H. Elsheikh, W. Li, S. Sadek, M.M. Aboelmaaref, A comprehensive review on Dish/Stirling concentrated solar power systems: design, optical and geometrical analyses, thermal performance assessment, and applications. J. Clean. Prod. **283**, 124664 (2021). https://doi.org/10.1016/j.jclepro.2020.124664
73. L. Aichmayer, J. Garrido, B. Laumert, Thermo-mechanical solar receiver design and validation for a micro gas-turbine based solar dish system. Energy **196**, 116929 (2020). https://doi.org/10.1016/j.energy.2020.116929
74. S.K. Sahu, A.S.K, S.K. Natarajan, Electricity generation using solar parabolic dish system with thermoelectric generator—an experimental investigation. Heat Transf. **50**(8), 7784–7797 (2021). https://doi.org/10.1002/htj.22253
75. M.E. Zayed, J. Zhao, A.H. Elsheikh, Z. Zhao, S. Zhong, A.E. Kabeel, Comprehensive parametric analysis, design and performance assessment of a solar dish/Stirling system. Process. Saf. Environ. Prot. **146**, 276–291 (2021). https://doi.org/10.1016/j.psep.2020.09.007
76. S. Bopche, K. Rana, V. Kumar, Performance improvement of a modified cavity receiver for parabolic dish concentrator at medium and high heat concentration. Sol. Energy **209**, 57–78 (2020). https://doi.org/10.1016/j.solener.2020.08.089

77. B.A. Mekonnen, K.W. Liyew, M.T. Tigabu, Solar cooking in Ethiopia: Experimental testing and performance evaluation of SK14 solar cooker. Case Stud. Therm. Eng. **22**, 100766 (2020). https://doi.org/10.1016/j.csite.2020.100766
78. A. Mawire, S.H. Taole, Experimental energy and exergy performance of a solar receiver for a domestic parabolic dish concentrator for teaching purposes. Energy Sustain. Dev. **19**, 162–169 (2014). https://doi.org/10.1016/j.esd.2014.01.004
79. M. Bahrami, V. Madadi Avargani, M. Bonyadi, Comprehensive experimental and theoretical study of a novel still coupled to a solar dish concentrator. Appl. Therm. Eng. **151**, 77–89 (2019). https://doi.org/10.1016/j.applthermaleng.2019.01.103
80. T. Arunkumar, D. Denkenberger, R. Velraj, R. Sathyamurthy, H. Tanaka, K. Vinothkumar, Experimental study on a parabolic concentrator assisted solar desalting system. Energy Convers. Manag. **105**, 665–674 (2015). https://doi.org/10.1016/j.enconman.2015.08.021
81. P. Esfanjani, A. Mahmoudi, S. Rashidi, M.S. Valipour, Experimental investigation of a novel design of cavity receiver for a parabolic dish collector humidification-dehumidification desalination system. Energy Convers. Manag. **299**, 117845 (2024). https://doi.org/10.1016/j.enconman.2023.117845
82. V. Madadi Avargani, R. Karimi, T. Tavakoli Gheinani, Mathematical modeling of an integrated system for regeneration of solid desiccants using a solar parabolic dish concentrator. Int. J. Heat Mass. Transf. **142**, 118479 (2019). https://doi.org/10.1016/j.ijheatmasstransfer.2019.118479
83. O.K. Singh, Development of a solar cooking system suitable for indoor cooking and its exergy and enviroeconomic analyses. Sol. Energy **217**, 223–234 (2021). https://doi.org/10.1016/j.solener.2021.02.007
84. D. Banitia, S. Ramachandran, S.S. Bhogilla, P.K. Vijayan, Optical analysis and design of a novel solar beam down concentrator for indoor cooking. Solar Compass **12**, 100083 (2024). https://doi.org/10.1016/j.solcom.2024.100083
85. T. Kanyowa, G. Victor Nyakujara, E. Ndala, S. Das, Performance analysis of Scheffler dish type solar thermal cooking system cooking 6000 meals per day. Solar Energy **218**, 563–570 (2021). https://doi.org/10.1016/j.solener.2021.03.019
86. J. Coventry, C. Andraka, Dish systems for CSP. Sol. Energy **152**, 140–170 (2017). https://doi.org/10.1016/j.solener.2017.02.056
87. W.T. Xie, Y.J. Dai, R.Z. Wang, K. Sumathy, Concentrated solar energy applications using Fresnel lenses: a review. Renew. Sustain. Energy Rev. **15**(6), 2588–2606 (2011). https://doi.org/10.1016/j.rser.2011.03.031
88. L. Jing et al., Design of novel compound Fresnel lens for high-performance photovoltaic concentrator. Int. J. Photoenergy **2012**, 1–7 (2012). https://doi.org/10.1155/2012/630692
89. T.T. Pham, N.H. Vu, S. Shin, Design of curved Fresnel lens with high performance creating competitive price concentrator photovoltaic. Energy Procedia **144**, 16–32 (2018). https://doi.org/10.1016/j.egypro.2018.06.004
90. M. Yamaguchi, T. Takamoto, K. Araki, N. Ekins-Daukes, Multi-junction III–V solar cells: current status and future potential. Sol. Energy **79**(1), 78–85 (2005). https://doi.org/10.1016/j.solener.2004.09.018
91. G. Wang, Z. Chen, P. Hu, X. Cheng, Design and optical analysis of the band-focus Fresnel lens solar concentrator. Appl. Therm. Eng. **102**, 695–700 (2016). https://doi.org/10.1016/j.applthermaleng.2016.04.030
92. A. Baiju, M. Yarema, Status and challenges of multi-junction solar cell technology. Front Energy Res. **10** (2022). https://doi.org/10.3389/fenrg.2022.971918
93. M. Martínez et al., *8 Years of CPV: ISFOC CPV Plants, Long-Term Performance Analysis and Results* (2017), p. 020007. https://doi.org/10.1063/1.5001406
94. Z. Wang, J. Wei, G. Zhang, H. Xie, M. Khalid, Design and performance study on a large-scale hybrid CPV/T system based on unsteady-state thermal model. Sol. Energy **177**, 427–439 (2019). https://doi.org/10.1016/j.solener.2018.11.043
95. K. Mohammadi, S. Khanmohammadi, H. Khorasanizadeh, K. Powell, Development of high concentration photovoltaics (HCPV) power plants in the US Southwest: economic assessment

and sensitivity analysis. Sustain. Energy Technol. Assess. **42**, 100873 (2020). https://doi.org/10.1016/j.seta.2020.100873
96. R.Y. Lin, *Advanced Solar Power Technology-Multiple Junction Photovoltaics* (2020), pp. 505–527. https://doi.org/10.1007/978-981-13-9012-8_22
97. J. Franco, L. Saravia, V. Javi, R. Caso, C. Fernandez, Pasteurization of goat milk using a low cost solar concentrator. Sol. Energy **82**(11), 1088–1094 (2008). https://doi.org/10.1016/j.solener.2007.10.011
98. K.S. Khan, Y. Latif, A. Munir, O. Hensel, Comparative thermal analyses of solar milk pasteurizers integrated with solar concentrator and evacuated tube collector. Energy Rep. **8**, 7917–7930 (2022). https://doi.org/10.1016/j.egyr.2022.06.001
99. J.-F. Ituna-Yudonago, Y.-R. Galindo-Luna, O. García-Valladares, R.B.y Brown, R. Shankar, J. Ibarra-Bahena, Review of solar-thermal collectors powered autoclave for the sterilization of medical equipment. Alexandria Eng. J. **60**(6), 5401–5417 (2021). https://doi.org/10.1016/j.aej.2021.04.007
100. T. Eglinton, J. Hinkley, A. Beath, M. Dell'Amico, Potential applications of concentrated solar thermal technologies in the Australian minerals processing and extractive metallurgical industry. JOM **65**(12), 1710–1720 (2013). https://doi.org/10.1007/s11837-013-0707-z
101. P.J.A. Bezuidenhout, W.G. le Roux, Solar cavity receiver for melting zinc metal. Appl. Therm. Eng. **247**, 122984 (2024). https://doi.org/10.1016/j.applthermaleng.2024.122984
102. D. Mewes, L. Sujessy, *The Application of a Solar Parabolic Trough Collector for the Melting of Plastic Waste* (2019), pp. 169–178. https://doi.org/10.1007/978-3-030-10856-4_17
103. M. Romero, J.I. Robla, I. Padilla, J. García-Hierro, A. López-Delgado, Eco-efficient melting of glass frits by concentrated solar energy. Sol. Energy **174**, 321–327 (2018). https://doi.org/10.1016/j.solener.2018.08.077
104. H. Agrawal, A. Yadav, Experimental investigation of ice-chamber for melting of ice based on Scheffler solar concentrator for high altitude regions. Heat Transfer **49**(5), 2472–2493 (2020). https://doi.org/10.1002/htj.21731
105. M.A. Rahman, A.M. Parvej, M.A. Aziz, Concentrating technologies with reactor integration and effect of process variables on solar assisted pyrolysis: a critical review. Therm. Sci. Eng. Progr. **25**, 100957 (2021). https://doi.org/10.1016/j.tsep.2021.100957
106. A. Romero, I. García, M.A. Arenas, V. López, A. Vázquez, High melting point metals welding by concentrated solar energy. Sol. Energy **95**, 131–143 (2013). https://doi.org/10.1016/j.solener.2013.05.019
107. L.E.G. Cambronero, I. Cañadas, J.M. Ruiz-Román, M. Cisneros, F.A. Corpas Iglesias, Weld structure of joined aluminium foams with concentrated solar energy. J. Mater. Process. Technol. **214**(11), 2637–2643 (2014). https://doi.org/10.1016/j.jmatprotec.2014.05.032
108. C. Prestigiacomo et al., Concentrated solar heat for the decarbonization of industrial chemical processes: a case study on crude oil distillation. Energy **293**, 130718 (2024). https://doi.org/10.1016/j.energy.2024.130718
109. K.E. Manyako, I. Chiyanzu, J. Mulopo, J. Abdulsalam, Pilot-scale evaluation of concentrating solar thermal technology for essential oil extraction and comparison with conventional heating sources for use in agro-based industrial applications. ACS Omega **7**(24), 20477–20485 (2022). https://doi.org/10.1021/acsomega.1c06879
110. H. Ebadi, D. Zare, M. Ahmadi, G. Chen, Performance of a hybrid compound parabolic concentrator solar dryer for tomato slices drying. Sol. Energy **215**, 44–63 (2021). https://doi.org/10.1016/j.solener.2020.12.026
111. N. Colak Gunes, E. Gungor, M. Gunes, Solar process heat for sustainable production of concentrated peach puree. J. Food Process Eng. **44**(7) (2021). https://doi.org/10.1111/jfpe.13744
112. M. Camci, Thermodynamic analysis of a novel integration of a spray dryer and solar collectors: a case study of a milk powder drying system. Drying Technol. **38**(3), 350–360 (2020). https://doi.org/10.1080/07373937.2019.1570935
113. W. Ben Youssef, T. Maatallah, C. Menezo, S. Ben Nasrallah, Assessment viability of a concentrating photovoltaic/thermal-energy cogeneration system (CPV/T) with storage for a textile

industry application. Solar Energy **159**, 841–851 (2018). https://doi.org/10.1016/j.solener.2017.11.058
114. A. El Golli et al., Parabolic trough concentrator design, characterization, and application: solar wastewater purification targeting textile industry dyes and pharmaceuticals—techno-economic study. EuroMediterr J. Environ. Integr. (2024). https://doi.org/10.1007/s41207-024-00531-1
115. M. Ozturk, I. Dincer, Thermodynamic assessment of an integrated solar power tower and coal gasification system for multi-generation purposes. Energy Convers. Manag. **76**, 1061–1072 (2013). https://doi.org/10.1016/j.enconman.2013.08.061
116. L. Wang, M.P.R. Teles, H. Yu, B.V.F. Silva, A. Arabkoohsar, Waste incineration and heat recovery hybridized with low-focus Fresnel lens solar collectors for sustainable multi-generation; A thorough techno-economic-environmental analysis and optimization. Chemosphere **346**, 140409 (2024). https://doi.org/10.1016/j.chemosphere.2023.140409
117. L. Schnatbaum, Solar thermal power plants. Eur. Phys. J. Spec. Top. **176**(1), 127–140 (2009). https://doi.org/10.1140/epjst/e2009-01153-0
118. M.I. Khan, F. Asfand, S.G. Al-Ghamdi, Progress in research and technological advancements of thermal energy storage systems for concentrated solar power. J. Energy Storage **55**, 105860 (2022). https://doi.org/10.1016/j.est.2022.105860
119. S. Bayoumi, N.A. Moharram, A.I. Shehata, M.M. Imam, W.M. El-Maghlany, A multi-criteria performance assessment of concentrated solar power plants for site and technology selection in Egypt. Int. J. Environ. Sci. Technol. **21**(3), 2989–3004 (2024). https://doi.org/10.1007/s13762-023-05114-1
120. M. Imran Khan, F. Asfand, S.G. Al-Ghamdi, Progress in research and technological advancements of commercial concentrated solar thermal power plants. Solar Energy **249**, 183–226 (2023). https://doi.org/10.1016/j.solener.2022.10.041

Design and Optimization of Solar Thermal Collectors

G. Raam Dheep

Abstract The design and optimization of solar thermal collectors play a crucial role in harnessing renewable energy from the sun, a resource that is both abundant and sustainable. Solar thermal collectors capture solar radiation and convert it into thermal energy. This thermal energy is used for heating water, air, or other fluids in residential, commercial, and industrial applications. As the global demand for clean energy continues to grow, the efficiency and effectiveness of solar thermal collectors have become increasingly important. Artificial intelligence-based machine learning methods (AI-ML) to design and optimize solar thermal collectors involves a multidisciplinary approach that integrates principles of thermodynamics, fluid dynamics, materials science, and energy engineering. This chapter provides the insights, key considerations, and strategies to design and optimize solar thermal collectors.

Keywords Solar thermal collectors · Artificial intelligence · Machine learning · Geometric optimization · Thermal performance

1 Introduction

Solar thermal energy is a technology that harnesses sunlight to generate heat, which is used for applications like electricity generation, water and space heating. Photovoltaic systems converts sunlight directly into electricity, whereas solar thermal systems focus on capturing and utilizing the thermal energy from the sun [1–3]. Solar thermal technology is a proven and effective way to harness the sun's energy for heating applications, with the potential to contribute significantly to reducing greenhouse gas emissions and dependency on fossil fuels.

Solar thermal systems can convert sunlight into heat with relatively high efficiency. After the initial investment, solar thermal systems can significantly reduce energy bills, particularly for heating and hot water. Solar thermal energy is renewable and

G. Raam Dheep (✉)
Department of Robotics and Automation, School of Electronics and Electrical Engineering, Lovely Professional University, Punjab, India
e-mail: raam.25227@lpu.co.in

P. Ravindran et al. (eds.), *Photon to Power*, Progress in Optical Science and Photonics 32, https://doi.org/10.1007/978-981-96-5914-2_5

produces no direct emissions, making it an environmentally friendly option. However, the following disadvantages makes the system less reliable. Some solar thermal systems, especially those for large-scale applications, require significant space for collectors [4–8]. The efficiency of solar thermal systems are intermittent in nature as it depends on sunlight, weather conditions, and geographic location. Solar thermal technology is a proven and effective way to harness the sun's energy for heating applications, with the potential to reduce greenhouse gas emissions from fossil fuels.

Solar thermal energy systems are used for wide range of applications in which thermal energy is the source of input [9–15]. Few applications includes as follows,

- **Domestic Hot Water**: One of the most common uses, solar thermal systems can provide a significant portion of a household's hot water needs.
- **Space Heating**: Solar thermal systems can be integrated with a building's heating system to provide space heating.
- **Industrial Process Heat**: In industries that require large amounts of heat (e.g., food processing, chemical manufacturing), solar thermal energy can be used to reduce reliance on fossil fuels.
- **Solar Thermal Power Plants**: Also known as Concentrated Solar Power (CSP), these plants use large arrays of mirrors or lenses to focus sunlight onto a small area, generating high temperatures that drive turbines to produce electricity.

The components of solar thermal system includes: **Solar Collectors to** absorb sunlight and convert it into heat. **Heat Transfer Fluid** that circulates through the solar collector, absorbing heat and transporting it to where it's needed [16–21]. **Heat Exchanger to** transfers the heat from the transfer fluid to the water or air used for heating. **Storage Tanks to** store the heat for later use, allowing the system to provide energy even when the sun isn't shining. And control system that manage the flow of the heat transfer fluid, ensuring optimal operation and efficiency.

Designing and optimizing solar thermal collectors is a critical aspect of advancing solar thermal technology, enabling more efficient and cost-effective harnessing of the sun's energy. By focusing on the materials, geometries, and configurations of collectors, engineers can maximize energy absorption and minimize heat losses [22–28]. Ultimately, the continuous refinement of solar thermal collectors not only contributes to greater energy efficiency and sustainability but also positions solar thermal technology as a viable and competitive alternative to traditional energy sources. As the global demand for clean energy grows, the role of optimized solar thermal collectors will be increasingly pivotal in meeting energy needs while mitigating environmental impacts.

2 Classification of Solar Thermal Collectors

The classification of solar thermal collectors is essential for understanding the diversity and applicability of this technology across various contexts. Solar thermal collectors can be classified based on various criteria such as their design, temperature range,

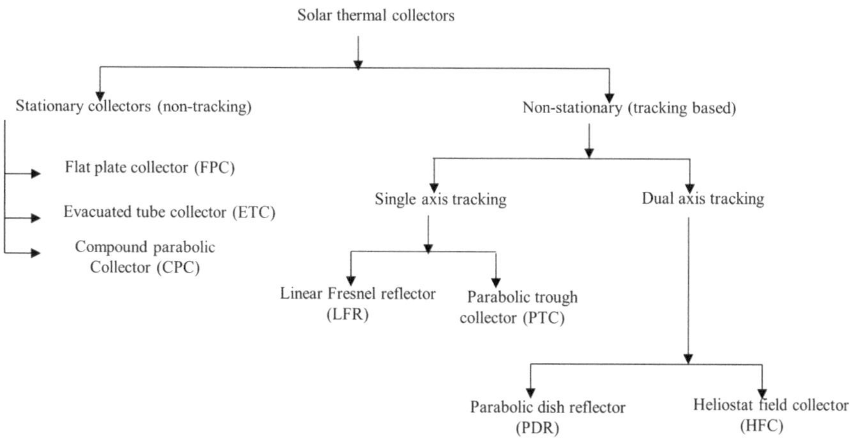

Fig. 1 Classification of solar thermal collectors

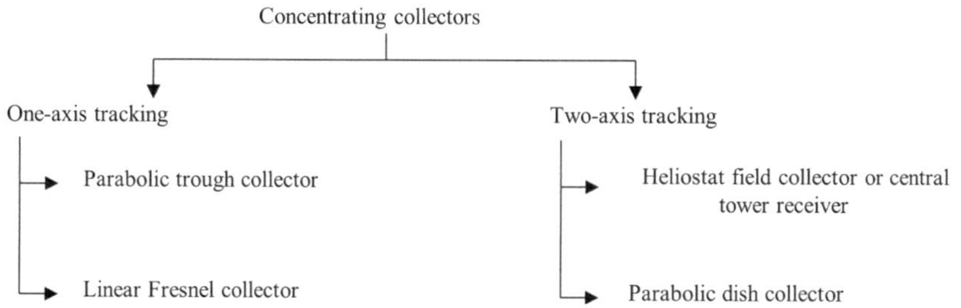

Fig. 2 Classification of concentrating solar thermal collectors

and the type of heat transfer fluid used [29–39]. These classifications help in selecting the appropriate solar thermal collector based on specific application needs, environmental conditions, and desired efficiency levels. Classification of solar thermal collectors is shown in Figs. 1 and 2. Below are the main classifications.

2.1 *Based on Design and Configuration*

Flat-Plate Collectors: These are the most common type of solar thermal collectors, consisting of a flat, dark-colored absorber plate encased in an insulated box with a transparent cover. **Applications**: Domestic hot water, space heating, and pool heating.

Evacuated Tube Collectors: Composed of multiple glass tubes, each containing an absorber tube within a vacuum. The vacuum minimizes heat loss, making these collectors more efficient, especially in colder climates. **Applications**: Residential and commercial water heating, space heating, and industrial processes.

Concentrating Collectors: These collectors use mirrors or lenses to focus sunlight onto a small area, generating higher temperatures. **Applications**: Industrial process heat, electricity generation (Concentrated Solar Power or CSP). They can be further classified into:

- **Parabolic Trough Collectors**: Curved mirrors focus sunlight onto a receiver tube.
- **Fresnel Collectors**: Use flat mirrors to focus light onto a fixed receiver.
- **Parabolic Dish Collectors**: Parabolic mirrors concentrate sunlight onto a single point, typically used for high-temperature applications.
- **Solar Power Towers**: Use an array of mirrors (heliostats) to focus sunlight onto a central receiver on a tower.

2.2 *Based on Temperature Range*

- **Low-Temperature Collectors**: Operate at temperatures below 100 °C. Typically used for applications like swimming pool heating and residential hot water. **Examples**: Unglazed and glazed flat-plate collectors.
- **Medium-Temperature Collectors**: Operate at temperatures between 100 °C and 300 °C. Used for applications requiring higher temperatures such as space heating and industrial processes. **Examples**: Evacuated tube collectors glazed flat-plate collectors.
- **High-Temperature Collectors**: Operate at temperatures above 300 °C. Used for electricity generation and industrial applications requiring very high temperatures. **Examples**: Parabolic trough collectors, Fresnel collectors, parabolic dish collectors, and solar power towers. The operating characteristic features of various solar thermal collectors are listed in Tables 1 and 2.

Table 1 Operating characteristics of stationary solar thermal collectors

Tracking system	Type of collector	Type of absorber	Concentration ratio	Operating temperature
Stationary collectors	Flat-plate collector	Flat	1	40–90
	Evacuated tube collector	Flat	1	60–120
	Compound parabolic collector	Tubular	1–5	80–150

Table 2 Operating characteristics of concentrating solar thermal collectors

Tracking system	Type of collector	Focus	Concentration ratio	Operating temperature
Single axis	Linear Fresnel Reflector (LFR)	Line focus	10–40	100–250
	Parabolic Trough Collector (PTC)	Line focus	15–45	150–300
Dual axis	Parabolic Dish Reflector (PDR)	Point focus	100–1000	500–1000
	Heliostat Field Collector (HFC)	Point focus	300–1500	1000–2000

2.3 *Based on the Type of Heat Transfer Fluid*

- **Liquid-Based Collectors**: Use a liquid (usually water, antifreeze, or oil) as the heat transfer medium. These are the most common types of collectors. **Examples**: Flat-plate collectors, evacuated tube, and concentrating collectors.
- **Air-Based Collectors**: Use air as the heat transfer medium. These are less common but are used for space heating and drying applications. **Examples**: Unglazed air collectors, glazed air collectors.

2.4 *Based on the Method of Heat Transfer*

- **Active Systems**: Utilize pumps or fans to circulate the heat transfer fluid. **Examples**: Most liquid-based and some air-based collectors.
- **Passive Systems**: Rely on natural convection or gravity to circulate the heat transfer fluid. **Examples**: Thermosiphon systems, batch water heaters.

2.5 *Based on the Mode of Operation*

- **Direct Systems**: The fluid that is directly used (e.g., water for domestic use) is heated in the collector. **Examples**: Direct water heating systems.
- **Indirect Systems**: The collector heats a separate fluid, which then transfers the heat to the end-use fluid via a heat exchanger. **Examples**: Systems using antifreeze solutions for cold climates.

By categorizing collectors based on design, temperature range, heat transfer fluid, heat transfer method, and mode of operation, we gain insights into their strengths and limitations. This systematic approach allows for informed decisions in selecting the most suitable collector for specific applications, whether for residential water heating, industrial process heat, or large-scale power generation [40–42]. As solar

Table 3 Comparison on solar thermal collectors

Feature	Non-concentrating collectors	Concentrating collectors
Absorber area	Same as the collector area	Smaller than the collector area
Temperature	Lower temperatures (up to 100 °C)	Higher temperatures (up to 1000 °C)
Efficiency	Lower efficiency	Higher efficiency
Cost	Lower initial cost	Higher initial cost
Applications	Domestic hot water heating, space heating	Industrial process heat, electricity generation
Types	Flat-plate collectors, evacuated tube collectors	Parabolic trough, parabolic dish, linear Fresnel reflector, central receiver

thermal technology continues to evolve, these classifications will remain vital in optimizing energy efficiency and expanding the use of renewable energy solutions.

3 Comparison Between Concentrating and Non-concentrating Thermal Collectors

As we have discussed various types of solar thermal collectors, Table 3 gives a comparison on classification of solar thermal collectors and its applications [43–45].

4 Design Parameters of Solar Thermal Flat-Plate Collector

The European Standard EN 12975 (2001) is commonly used to evaluate the thermal performance of flat plate and concentrating solar collectors. It provides information on testing solar collectors using single-phase fluids and no significant internal storage. The thermal performance is determined by obtaining instantaneous efficiency values under steady state or quasi-steady state conditions, requiring measurement of incident solar radiation and energy addition to the transfer fluid [46–57]. Transient thermal response characteristics and angle dependence of the collector are also evaluated. The collector efficiency is calculated by including correction factors for collector efficiency and heat removal. The incidence angle modifier is used to account for the effect of the angle of incidence on the cover system and absorber plate interaction. The collector time constant is determined to evaluate its transient behavior, and the stagnation temperature is obtained from the conditions at zero efficiency. Useful heat from collector is given by,

$$Q_u = A_c\left[I(\tau\alpha) - U_L\left(T_p - T_a\right)\right]$$

The heat removal factor (F_R) of solar flat-plate collector is defined as the ratio of actual useful energy collected to the useful energy collected if the absorber is at inlet fluid temperature (T_i) is given by,

$$F_R = \frac{\text{actual useful energy collected}}{\text{useful energy collected if absorber is at inlet fluid temperature}}$$

$$\frac{Q_u}{A_c} = F_R[I(\tau\alpha) - U_L(T_i - T_a)]$$

The instantaneous efficiency (η) of solar flat-plate collector is defined as,

$$\eta = \frac{\text{useful energy collected}}{\text{solar energy incident on collector}}$$

$$\eta = \frac{Q_u}{IA_c}$$

$$\eta = F_R(\tau\alpha) - F_R U_L \frac{(T_i - T_a)}{I}$$

$$F_R = \frac{\dot{m}C_p}{U_L A_c}\left(1 - \exp\left[-\frac{U_L A_c F'}{\dot{m}C_p}\right]\right)$$

4.1 Collector Efficiency Factor (F′)

The collector efficiency factor (F') is defined as the ratio of useful energy collected to the useful energy collected when the temperature of absorber is at local fluid temperature (T_m). It is expressed as,

$$F' = \frac{\text{useful energy gain}}{\text{useful energy gain if absorber is at local fluid temperature}}$$

$$\frac{Q_u}{A_c} = F'[I(\tau\alpha) - U_L(T_m - T_a)]$$

From the values of F', F_R and U_L, the efficiency of solar flat-plate collector is calculated using the formula,

$$\eta = F_R(\tau\alpha) - F_R U_L \frac{(T_i - T_a)}{I}$$

The instantaneous efficiency (η) of flat-plate collector is defined as the ratio of useful heat energy gained to the solar radiation incident per unit aperture area.

$$\eta = \frac{q_u}{IA_c}$$

The useful heat gain (q_u) is measured using fluid flow rate, specific heat of fluid inlet and outlet fluid temperature. It is given by,

$$q_u = \dot{m}C_p(T_o - T_i)$$

Therefore,

$$\eta = \frac{\dot{m}C_p(T_o - T_i)}{IA_c}$$

5 Design Parameters of Solar Thermal Concentrating Collectors

Concentration Ratio (C): Concentration ratio is defined as the ratio of aperture area to absorber area. Depending upon the type of solar thermal collector concentration ratio varies from unity to thousands. Concentration ratio is given by,

$$C = A_a/A_r.$$

where A_a is aperture area and A_r is absorber area. For FPC with no concentrators, the concentration ratio $C = 1$. The maximum concentration ratio for solar concentrators with single axis tracking is given by,

$$C_{\max} = 1/\sin(\theta_m)$$

and for two axes tracking collector,

$$C_{\max} = 1/\sin^2(\theta_m)$$

where θ_m is the half accept angle. For concentrating collectors $C_{\max}$ depends on the size of the sun's disk which has a width of 0.53° or 32'. Therefore $C_{\max}$,

For single axis tracking collectors $C_{\max} = 1/\sin(\theta_m) = 1/\sin(16') = 216.$

For two axes tracking collectors $C_{\max} = 1/\sin(\theta_m) = 1/\sin^2(16') = 46747.$

Intercept Factor (γ): Intercept factor is the ratio of solar radiation intercepted by the absorber to the total solar radiation focused by the reflector. It is given by,

$$\gamma = \frac{\int_{-\omega/2}^{\omega/2} I(x\mathrm{d}x)}{\int_{-\infty}^{\infty} I(x\mathrm{d}x)}$$

Table 4 Angle of incidence for various tracking method

Tracking method	Angle of incidence
Two axis tracking	$\cos(\theta) = 1$
Collector axis in N–S axis polar E–W tracking	$\cos(\theta) = \cos(\delta)$
Collector axis in N–S axis horizontal E–W tracking	$\cos(\theta) = \sqrt{\sin^2(\delta) + \cos^2(\delta)\sin^2(h)}$ or $\cos(\theta) = \cos(\phi)\cos(h) + \cos(\delta)\sin^2(h)$
Collector axis in E–W axis horizontal N–S tracking	$\cos(\theta) = \sqrt{\sin^2(\delta) + \cos^2(\delta)\cos^2(h)}$ or $\cos(\theta) = \sqrt{1 - \cos^2(\delta)\sin^2(h)}$

Intercept factor depends on the size of the absorber. Intercept factor has a value greater than 0.9. The intercept factor is 1 if the solar radiation is normal to the aperture.

5.1 *Optical Efficiency* (η_O)

Optical efficiency of solar concentrating collectors is defined as the ratio of energy absorbed by the absorber to the energy incident on the collector aperture area. The optical efficiency of solar concentrating collectors depends on geometry of collector, angle of incidence, optical properties of materials and type of tracking. Angle of incidence of solar radiation depends upon the type of tracking method used to track the sun. Angle of incidence for different tracking method is given in Table 4. Optical efficiency of solar concentrating collectors is given by,

$$\eta_o = \rho\tau\alpha\gamma\left[\left(1 - A_f\tan(\theta)\cos(\theta)\right)\right]$$

where A_f is the geometric factor.

$$A_f = \frac{2}{3}A_a h_p + fA_a\left[1 + \frac{A_a^2}{48f^2}\right]$$

5.2 *Thermal Efficiency* (η)

Thermal performance of solar concentrating collectors is determined based on the thermal efficiency of the collector. Thermal efficiency is defined as the ratio of useful energy delivered to the energy incident on the aperture of concentrating collector.

$$\eta = \frac{q_u}{IA_a}$$

Useful energy delivered by the concentrating collector is given by

$$q_u = I\eta_o A_a - A_r U_L (T_r - T_a)$$

If q_u is the useful energy then useful energy gain per unit length $q_u\prime$ of the collector at receiver temperature (T_r),

$$q_u\prime = \frac{q_u}{L}$$

$$q_u\prime = \frac{I\eta_o A_a}{L} - \frac{A_r U_L}{L}(T_r - T_a)$$

Amount of energy transferred to the heat transfer fluid from absorber at fluid temperature (T_f) is given by,

$$q_u\prime = \frac{\left(\frac{A_r}{L}\right)(T_r - T_f)}{\frac{D_o}{h_{fi} D_i} + \left(\frac{D_o}{2K}\ln\frac{D_o}{D_i}\right)}$$

From (1) and (2), useful energy gain per unit length $q_u\prime$ is given by,

$$q_u\prime = F' \frac{A_q}{L}\left[\eta_o I - \frac{U_L}{C}(T_f - T_a)\right]$$

where F' is collector efficiency factor and concentration ratio $C = A_a/A_{r.}$

$$F' = \frac{\frac{1}{U_L}}{\frac{1}{U_L} + \frac{D_o}{h_{fi} D_i} + \left(\frac{D_o}{2K} + \ln\frac{D_o}{D_i}\right)}$$

$$q_u = F_R[I\eta_o A_a - A_r U_L (T_i - T_a)]$$

Thermal efficiency of solar concentrating collectors is given by,

$$\eta = \frac{q_u}{IA_a}$$

$$\eta = F_R\left[{}_o - U_L\left(\frac{T_i - T_a}{I.C}\right)\right]$$

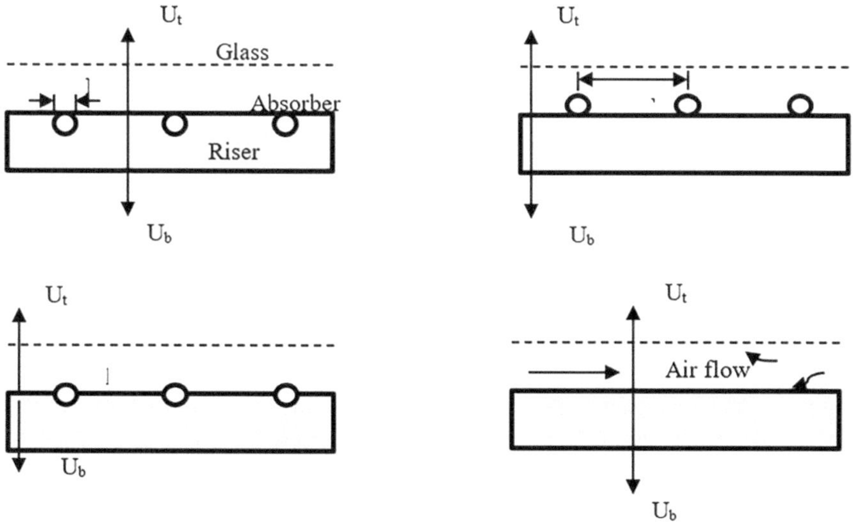

Fig. 3 Absorber configuration in solar water heater and space heater

6 Various Configurations of Solar Flat-Plate Collector

A **solar flat-plate collector** is a common device used to absorb and transfer thermal energy for heating purposes, especially for water or air. It is widely used due to its simplicity, durability, and cost-effectiveness [58–66]. There are various absorber configurations of flat-plate collectors which are being used for space and water heating applications is shown Fig. 3 as follows,

7 Machine Learning Approaches to Design and Optimize Solar Thermal Collectors

The application of artificial intelligence techniques in mechanical engineering has been steadily growing during the past 20 years. This is primarily due to the fact that artificial intelligence modeling systems are now much more effective in the engineering field. Performance forecasting is crucial for many solar and air conditioning applications. It is advised that thermal system performance be predicted for engineering applications using artificial intelligence techniques [67–71]. Accurate performance characteristic projections are crucial for studies, designers, and end users. The accuracy of thermal device performance estimation can be improved through testing. Even with the constraints of physical tests, precise results can be obtained by using numerical computer models. Improved heat transfer can result from actions like boundary layer disruption and airflow turbulence. Many research investigations and computations have been carried out in an effort to improve the absorber plate roughness's heat transfer rate.

Machine learning (ML) outperforms conventional approaches to problem-solving in terms of the speed and quality of the solutions it produces. It is very helpful in handling difficult problems where other soft computing techniques are insufficient. Machine learning is used to estimate solar energy systems' performance in heat transfer. In computer intelligence, neural networks—a non-linear analog technique—are employed. Artificial neural networks (ANNs) enable the identification of patterns in data across various dimensions [72–77]. ANNs are a popular machine learning technology that leverages the ability of the human brain to comprehend intricate non-linear correlations within particular situations. Computational fluid dynamics has been used extensively to study a SAH's maximum performance (CFD). While computational fluid dynamics (CFD) has become a widely used method for studying solar air heater design, its use is still complicated, expensive, and time-consuming due to the demand for high-resolution grids and the ongoing need for CFD validation and verification. Researchers are investigating a novel approach based on artificial neural networks (ANNs) to get around the shortcomings of CFD.

Previous studies on solar air heaters (SAHs) have explored numerous design modifications to enhance their thermal and hydraulic performance. Certain parameters, such as solar radiation, mass flow rate, and air temperature, have a direct and linear impact on performance [76, 78]. Other factors, like relative humidity and wind speed, exert a more indirect, non-linear influence, affecting efficiency in a less straightforward manner. Optimizing both direct and indirect parameters is crucial for improving overall system performance.

SACs with fixed-length collectors and variable-width absorbers, finding that a trapezoidal absorber plate outperformed a concave design in terms of efficiency. Double-pass parallel and counterflow SACs through analytical analysis, observing that performance remained consistent at higher flow rates for both configurations. Perforated glazed solar collector with mass flow rates ranging from 0.017 to 0.036 kg/s, finding that black and white absorbers achieved maximum and minimum efficiencies of 85% and 54%, respectively. Computational and empirical studies on the thermal efficiency of a flat-plate SAC, which ranged from 22 to 40% and the thermal inertia of solar air heaters varied between 0.4 and 0.7. Mathematical model for SACs, adjusting solar radiation levels between 100 and 1000 W/m^2 and mass flow rates between 0.02 and 0.05 kg/s, resulting in thermo-hydraulic efficiencies between 35 and 55% [79–81]. Semi-circular plastic SAC achieved a thermal efficiency of 80%, significantly higher than triangular (25%) and flat (35%) cover designs.

Numerous researchers have explored surface modification techniques to enhance the performance of solar air collectors (SACs), with key goals being to reduce energy losses, accelerate airflow redistribution, and disrupt boundary layer development. The effect of surface roughness on pressure drop and heat transfer in SACs, finding that heat transfer improved by up to 2.4 times, though the friction coefficient increased by 4.25 times. Compared double-pass SACs with flat and V-corrugated absorbers, noting that the V-corrugated collector achieved 11–14% greater efficiency than the flat-plate design. Theoretically studied an upward-type SAC with a finned absorber and external air recycling, discovering that external recycling improved heat convection and thermal efficiency [49, 82–87]. A double-pass SAC with fins and baffles was analyzed, revealing that baffles had a more significant impact on performance than fins. Five different absorber surfaces, concluding that a ribbed receiver provided 55% efficiency at a flow rate of 0.05 kg/s, outperforming other designs like corrugated, trapezoidal, and flat-plate collectors.

An experimental study on double-pass SACs with flat and V-corrugated absorbers, finding that the Nusselt number for V-corrugated absorbers increased by 4.5 times compared to flat plates, with a 3.1-fold increase in friction factor. Kabeel et al. [91] investigated SACs with flat, finned, and baffled absorbers, reporting a maximum efficiency of 83.8% for a baffled fin absorber at a flow rate of 0.04 kg/s. Sevik and SACs with flexible aluminum foil absorbers, with and without glass covers, finding that the aluminum foil absorber achieved 81.3% efficiency at a flow velocity of 0.044 kg/s. Minimizing losses through the glass cover resulted in a thermal efficiency of approximately 75%. Attaching aluminum cans to a flat absorber increased collector efficiency by up to 45% during peak sunlight. Double-pass SACs with fins and wire mesh, showing that the double-pass collector improved efficiency by up to 19% over a simple absorber [88–91]. Conical spring attached to a flat-plate SAC achieved a maximum efficiency of about 68% and ribbed SAC provided up to 9% higher overall thermal efficiency than a flat absorber. The thermal performance of double pass, V-corrugated solar air heater improved collector efficiency at higher mass flow rates compared to a plate-based absorber.

The usage of neural networks to predict the performance parameters of solar thermal flat-plate collectors has gained interest in recent years, owing to their capability to model the complex non-linear relationships characteristic of solar energy systems [92–95]. This literature review compiles and analyzes existing research in this field, emphasizing the methodologies used, key outcomes, and gaps in the current knowledge base.

ANN algorithms are being used to analyze solar collector thermal performance without the requirement of extensive testing. According to the European Standard EN 12975, which defines testing procedures for solar collectors, data from these tests are utilized to create ANN-based predictive models. Algorithms of neural network are particularly effective for solving complex, poorly defined problems and have been applied successfully across various fields. This study explores the suitability of ANNs for estimating solar collector performance parameters, such as coefficients from standard tests and incidence angle modifiers, using data of different solar thermal collectors studied at SPF laboratories, Switzerland. Figure 4 shows the architecture of artificial neural network for input, hidden, output layers and activation functions. Algorithms used in predictive tools could significantly reduce the time required by design engineers, enabling them to estimate collector performance without conducting physical experiments. The data used in the study comes from the LTS database, which contains information on 130 thermal solar collectors, categorized by collector type, absorber material, absorber type, coating, and insulation type. The study focused on performance data, primarily from flat-plate collectors, as they make up the bulk of the dataset and are the most commonly used collector type. Six types of ANNs were developed to predict various collector performance parameters, including dimensions, constructional characteristics, and performance coefficients. Key data for performance prediction includes collector dimensions, absorber type, insulation type, and performance coefficients [96–102]. The dataset was used for training, testing, and validating the ANN models. However, missing data and variations in testing conditions introduced different patterns into the neural network models, affecting prediction consistency.

The neural network model's performance was evaluated by comparing its predictions with the actual performance coefficients of eight unknown test datasets, including collector efficiency, time constant of collector and angle of incidence modifier under the conditions of various wind flow. The differences between the predicted and actual results were minimal, with the maximum deviations within acceptable limits. The model accurately predicted collector efficiency values close to the actual ones for two different temperature difference levels, and the variations in incidence angle modifier and collector time constant were also within acceptable ranges. However, the absence of key data, such as fluid content, absorber mass, and stagnation temperature, limited the accuracy of the predictions. Despite this, the model required only a few minutes for training and less than a second for validation, offering a fast and accurate solution. Several studies have explored the use of ANNs for predicting various performance parameters associated with flat-plate collectors.

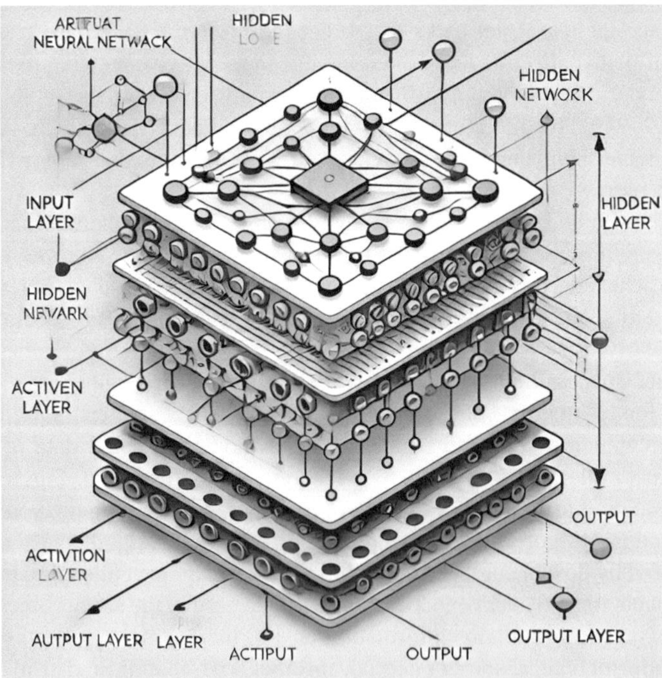

Fig. 4 Architecture for artificial neural network

For instance, dendritic neuron model integrated with effective learning algorithms that showcased robust capabilities for classification, approximation, and prediction tasks. Findings suggest that such models could be adapted to forecast the thermal performance of flat-plate collectors, emphasizing the potential for enhanced accuracy and efficiency in predictions.

In a comparative analysis, it was noted that traditional methods like multivariate linear regression often fall short in accurately predicting complex phenomena such as thermal performance. Both ANNs and linear regression for water quality parameters, demonstrating that ANNs significantly outperformed conventional techniques in terms of prediction accuracy. This suggests that similar approaches could be beneficial in the realm of solar collector performance, where multifaceted environmental and operational variables typically influence outcomes. The effectiveness of Long Short-Term Memory (LSTM) networks in an Internet of Things (IoT) environment for performance prediction [103–107]. While their focus was on water quality, the methodology employed could be translated to solar energy applications, providing a framework for real-time data analysis and prediction. The adaptability of LSTM networks in addressing temporal dependencies aligns well with the operational dynamics of flat-plate collectors, where performance can vary significantly throughout the day and across seasons. Furthermore, the work of utilizing machine learning models for prediction sheds light on the importance of feature selection and model tuning. Their comparative analysis identified key parameters influencing, a practice that could be mirrored in flat-plate collector studies to isolate critical factors impacting thermal efficiency.

The artificial neural network (ANN) can efficiently predict collector performance parameters, offering benefits such as speed and reducing the need for extensive physical testing. Its accuracy could be further enhanced by incorporating a wider range of cases and diverse design parameters, potentially leading to the development of a versatile design tool [108–111]. This tool could enable engineers to confidently predict the performance of new collector designs without the need for initial testing, knowing the results are close to actual values. However, final validation through testing remains essential. As the network continues to learn, its predictive capabilities are expected to improve over time. Despite promising results from current studies, there are few research gaps in the literature. One among them is the lack of comprehensive datasets specifically tailored to flat-plate collectors, which limits the development of robust predictive models. Most research relies on generic datasets that may not fully account for the unique operational characteristics of solar collectors under varying environmental conditions.

Additionally, while various neural network architectures have been proposed, there is insufficient exploration of hybrid models which compares the performance of different algorithms, such as combining ANNs with other machine learning techniques like Support Vector Machines (SVMs) or Adaptive Neuro-Fuzzy Inference Systems (ANFIS). The performance of these hybrid approaches in predicting flat-plate collector parameters remains largely unexplored [112–114]. Another gap lies in the real-world applicability of these models. Many existing studies focus on theoretical developments without extensive field validation. The translation of laboratory results to practical scenarios, where factors such as installation orientation, local climate conditions, and maintenance routines come into play, is essential for ensuring the reliability of ANN predictions.

8 Optimizing Solar Thermal Collectors with AI and Machine Learning

Artificial intelligence (AI) and machine learning (ML) offer powerful tools to optimize the performance of solar thermal collectors. By analyzing vast amounts of data and identifying patterns, these techniques can significantly enhance the efficiency and effectiveness of these systems [115–117]. Some of the techniques and tools used in optimizing the thermal performance includes,

Machine Learning Algorithms: Supervised Learning: Train models on historical data to predict future performance. Unsupervised Learning: Discover hidden patterns in data without explicit labels. Reinforcement Learning: Learn optimal control strategies through trial and error. Deep Learning: Use neural networks to analyze complex data patterns. Internet of Things (IoT): Collect real-time data from sensors to feed into AI models. Cloud Computing: Process and analyze large amounts of data efficiently. By leveraging AI and ML, solar thermal collector systems can become more intelligent, efficient, and reliable, contributing to a sustainable energy future. Here are some specific applications:

1. Predictive Maintenance:
   - Sensor Data Analysis: AI algorithms can analyze data from sensors monitoring the system's components (e.g., pumps, valves, heat exchangers) to predict potential failures.
   - Preventive Maintenance Scheduling: By identifying early warning signs, predictive maintenance can be scheduled to minimize downtime and optimize system performance.
2. Real-time Optimization:
   - Dynamic Tilt Angle Adjustment: AI can analyze real-time weather data (solar irradiance, temperature, wind speed) to dynamically adjust the tilt angle of the collector to maximize solar energy absorption.
   - Flow Rate Optimization: AI can optimize the flow rate of the heat transfer fluid to balance heat extraction and thermal losses, improving system efficiency.
3. Energy Yield Forecasting:
   - Weather Data Analysis: AI models can analyze historical weather data and current weather forecasts to predict future solar energy yields.
   - Optimized Energy Management: This information can be used to optimize energy storage and distribution, ensuring efficient utilization of solar energy.

4. Fault Detection and Diagnosis:
   - Anomaly Detection: AI algorithms can identify anomalies in system behavior, such as unusual temperature fluctuations or pressure drops, indicating potential faults.
   - Root Cause Analysis: By analyzing the data, AI can help pinpoint the root cause of the problem, enabling faster and more accurate troubleshooting.
5. Design Optimization:
   - Simulation and Modeling: AI can be used to simulate the performance of different collector designs under various conditions, helping to identify the optimal configuration.
   - Material Selection: AI can analyze the properties of different materials to select the most suitable ones for the collector components.

9 Conclusion

Solar energy represents a cornerstone of sustainable energy solutions, offering vast potential for addressing global energy demands while minimizing environmental impacts. Within this domain, solar thermal energy plays a pivotal role by efficiently converting sunlight into heat for various applications, from domestic water heating to industrial processes. The development and deployment of solar thermal collectors are critical for harnessing this energy. These devices, which capture and transfer solar radiation, form the backbone of solar thermal systems. Their efficiency and effectiveness depend significantly on their design and optimization.

Design and optimization of solar thermal collectors involve integrating principles of thermodynamics, material science, and engineering to enhance performance, reduce costs, and ensure long-term reliability. Advanced simulation tools and methodologies enable the creation of innovative collector designs that maximize energy capture even under diverse climatic conditions. Optimization methods are indispensable in the design and enhancement of thermal performance in solar thermal collectors, enabling these systems to operate at peak efficiency while reducing costs and resource use. By systematically refining critical parameters such as geometry, materials, fluid dynamics, and operational conditions, optimization ensures that collectors achieve superior energy capture and minimal thermal losses. Advanced techniques, including multi-objective optimization, computational modeling, and AI-driven algorithms, allow for innovative solutions that balance performance, cost, and sustainability. These approaches have proven vital in maximizing absorption efficiency, improving heat transfer, and minimizing energy losses, thereby elevating the overall effectiveness of solar thermal systems.

As global energy demands grow and the shift toward renewable energy accelerates, optimization methods will continue to play a central role in the evolution of solar thermal collectors. By improving design and thermal performance, these methods contribute to making solar thermal energy more accessible, efficient, and impactful in meeting the needs of a sustainable future. As we advance into a future increasingly shaped by renewable energy, refining the technologies underpinning solar thermal systems remains a pressing priority. The ongoing evolution in the design and optimization of solar thermal collectors will undoubtedly solidify their role as essential components in achieving energy security and sustainability worldwide. By leveraging cutting-edge research and technology, we can unlock the full potential of solar thermal energy as a transformative force for a greener, more sustainable future.

References

1. N. Kannan, D. Vakeesan, Solar energy for future world:-a review. Renew. Sustain. Energy Rev. **62**, 1092–1105 (2016)
2. E. Kabir, P. Kumar, S. Kumar, A.A. Adelodun, K.-H. Kim, Solar energy: potential and future prospects. Renew. Sustain. Energy Rev. **82**, 894–900 (2018)
3. Z. Şen, Solar energy in progress and future research trends. Prog. Energy Combust. Sci. **30**(4), 367–416 (2004)
4. C. Philibert, *The Present and Future Use of Solar Thermal Energy as a Primary Source of Energy* (International Energy Agency, 2005), pp. 1–16
5. T.M. Tritt, H. Böttner, L. Chen, Thermoelectrics: direct solar thermal energy conversion. MRS Bull. **33**(4), 366–368 (2008)
6. Y. Tian, C.-Y. Zhao, A review of solar collectors and thermal energy storage in solar thermal applications. Appl. Energy **104**, 538–553 (2013)
7. H.P. Garg, S.C. Mullick, V.K. Bhargava, *Solar Thermal Energy Storage* (Springer Science & Business Media, 2012)
8. K. Ravi Kumar, N.V.V. Krishna Chaitanya, N.S. Kumar, Solar thermal energy technologies and its applications for process heating and power generation–a review. J. Clean. Prod. **282**, 125296 (2021)
9. S.A. Kalogirou, Solar thermal collectors and applications. Prog. Energy Combust. Sci. **30**(3), 231–295 (2004)
10. W. Lipiński, E. Abbasi-Shavazi, J. Chen, J. Coventry, M. Hangi, S. Iyer, A. Kumar et al., Progress in heat transfer research for high-temperature solar thermal applications. Appl. Therm. Eng. **184**, 116137 (2021)
11. M. Karagiorgas, A. Botzios, Th. Tsoutsos, Industrial solar thermal applications in Greece: economic evaluation, quality requirements and case studies. Renew. Sustain. Energy Rev. **5**(2), 157–173 (2001)
12. M. Thirugnanasambandam, S. Iniyan, R. Goic, A review of solar thermal technologies. Renew. Sustain. Energy Rev. **14**(1), 312–322 (2010)
13. S. Mesgari, R.A. Taylor, N.E. Hjerrild, F. Crisostomo, Q. Li, J. Scott, An investigation of thermal stability of carbon nanofluids for solar thermal applications. Sol. Energy Mater. Sol. Cells **157**, 652–659 (2016)
14. S.S.S. Baljit, H.-Y. Chan, K. Sopian, Review of building integrated applications of photovoltaic and solar thermal systems. J. Clean. Prod. **137**, 677–689 (2016)
15. K.Y. Leong, H.C. Ong, N.H. Amer, M.J. Norazrina, M.S. Risby, K.Z. Ku Ahmad, An overview on current application of nanofluids in solar thermal collector and its challenges. Renew. Sustain. Energy Rev. **53**, 1092–1105 (2016)
16. M.S. Bretado-de los Rios, C.I. Rivera-Solorio, K.D.P. Nigam, An overview of sustainability of heat exchangers and solar thermal applications with nanofluids: a review. Renew. Sustain. Energy Rev. **142**, 110855 (2021)
17. Q. Li, G. Flamant, X. Yuan, P. Neveu, L. Luo, Compact heat exchangers: a review and future applications for a new generation of high temperature solar receivers. Renew. Sustain. Energy Rev. **15**(9), 4855–4875 (2011)

18. A.H. Elsheikh, H.N. Panchal, S. Sengottain, N.A. Alsaleh, M. Ahmadein, Applications of heat exchanger in solar desalination: current issues and future challenges. Water **14**(6), 852 (2022)
19. A. Patel, Enhancing heat transfer efficiency in solar thermal systems using advanced heat exchangers. Multidiscip. Int. J. Res. Dev. (MIJRD) **2**(06), 31–51 (2023)
20. E. Ebrahimnia-Bajestan, M.C. Moghadam, H. Niazmand, W. Daungthongsuk, S. Wongwises, Experimental and numerical investigation of nanofluids heat transfer characteristics for application in solar heat exchangers. Int. J. Heat Mass Transfer **92**, 1041–1052 (2016)
21. V. Kapsalis, D. Karamanis, Solar thermal energy storage and heat pumps with phase change materials. Appl. Therm. Eng. **99**, 1212–1224 (2016)
22. G.N. Kulkarni, S.B. Kedare, S. Bandyopadhyay, Design of solar thermal systems utilizing pressurized hot water storage for industrial applications. Sol. Energy **82**(8), 686–699 (2008)
23. Z. Wang, *Design of Solar Thermal Power Plants* (Academic Press, 2019)
24. Q. Li, C. Zheng, A. Shirazi, O.B. Mousa, F. Moscia, J.A. Scott, R.A. Taylor, Design and analysis of a medium-temperature, concentrated solar thermal collector for air-conditioning applications. Appl. Energy **190**, 1159–1173 (2017)
25. A. Allouhi, Y. Agrouaz, M.B. Amine, S. Rehman, M.S. Buker, T. Kousksou, A. Jamil, A. Benbassou, Design optimization of a multi-temperature solar thermal heating system for an industrial process. Appl. Energy **206**, 382–392 (2017)
26. M.C.M. Probst, C. Roecker, *Architectural Integration and Design of Solar Thermal Systems* (EPFL Press, 2011)
27. M. Picón-Núñez, G. Martínez-Rodríguez, A.L. Fuentes-Silva, Design of solar collector networks for industrial applications. Appl. Therm. Eng. **70**(2), 1238–1245 (2014)
28. R. Palumbo, M. Keunecke, S. Möller, A. Steinfeld, Reflections on the design of solar thermal chemical reactors: thoughts in transformation. Energy **29**(5–6), 727–744 (2004)
29. A. Herez, H. El Hage, T. Lemenand, M. Ramadan, M. Khaled, Review on photovoltaic/thermal hybrid solar collectors: classifications, applications and new systems. Sol. Energy **207**, 1321–1347 (2020)
30. G. Barone, A. Buonomano, C. Forzano, A. Palombo, Solar thermal collectors, in *Solar Hydrogen Production* (Academic Press, 2019), pp. 151–178
31. Ž. Jesko, Classification of solar collectors. rN **1**(21) 21 (2008)
32. L. Evangelisti, R. De LietoVollaro, F. Asdrubali, Latest advances on solar thermal collectors: a comprehensive review. Renew. Sustain. Energy Rev. **114**, 109318 (2019)
33. V. Pranesh, R. Velraj, S. Christopher, V. Kumaresan, A 50 year review of basic and applied research in compound parabolic concentrating solar thermal collector for domestic and industrial applications. Sol. Energy **187**, 293–340 (2019)
34. S.A. Kalogirou, S. Karellas, K. Braimakis, C. Stanciu, V. Badescu, Exergy analysis of solar thermal collectors and processes. Prog. Energy Combust. Sci. **56**, 106–137 (2016)
35. P.K. Nagarajan, J. Subramani, S. Suyambazhahan, R. Sathyamurthy, Nanofluids for solar collector applications: a review. Energy Procedia **61**, 2416–2434 (2014)
36. S. Suman, Mohd.K. Khan, M. Pathak, Performance enhancement of solar collectors—a review. Renew. Sustain. Energy Rev. **49**, 192–210 (2015)
37. F. Hussain, Md. Hasanuzzaman, Solar thermal collector, in *Technologies for Solar Thermal Energy* (Academic Press, 2022), pp. 93–122
38. M. d'Antoni, O. Saro, Massive solar-thermal collectors: a critical literature review. Renew. Sustain. Energy Rev. **16**(6), 3666–3679 (2012)
39. M. Saini, A. Sharma, V.P. Singh, G. Dwivedi, S. Jain, Solar thermal receivers—a review, in *Advancement in Materials, Manufacturing and Energy Engineering, Vol. II: Select Proceedings of ICAMME 2021* (2022), pp. 311–325
40. J. Freeman, K. Hellgardt, C.N. Markides, An assessment of solar–thermal collector designs for small-scale combined heating and power applications in the United Kingdom. Heat Transfer Eng. **36**(14–15), 1332–1347 (2015)
41. I. Iparraguirre, A. Huidobro, A. Fernández-García, L. Valenzuela, P. Horta, F. Sallaberry, T. Osório, A. Sanz, Solar thermal collectors for medium temperature applications: a comprehensive review and updated database. Energy Procedia **91**, 64–71 (2016)

42. V. Siva Reddy, S.C. Kaushik, K.R. Ranjan, S.K. Tyagi, State-of-the-art of solar thermal power plants—a review. Renew. Sustain. Energy Rev. **27**, 258–273 (2013)
43. A. Giostri, M. Binotti, P. Silva, E. Macchi, G. Manzolini, Comparison of two linear collectors in solar thermal plants: parabolic trough versus Fresnel. J. Sol.Energy Eng. **135**(1), 011001 (2013)
44. K.K. Matrawy, I. Farkas, Comparison study for three types of solar collectors for water heating. Energy Convers. Manage. **38**(9), 861–869 (1997)
45. L. Garcia-Rodriguez, A.I. Palmero-Marrero, C. Gómez-Camacho, Comparison of solar thermal technologies for applications in seawater desalination. Desalination **142**(2), 135–142 (2002)
46. P. Obstawski, T. Bakoń, D. Czekalski, Comparison of solar collector testing methods—theory and practice. Processes **8**(11), 1340 (2020)
47. C. Grass, W. Schoelkopf, L. Staudacher, Z. Hacker, Comparison of the optics of non-tracking and novel types of tracking solar thermal collectors for process heat applications up to 300 °C. Sol. Energy **76**(1–3), 207–215 (2004)
48. J.E. Hill, *Development of Proposed Standards for Testing Solar Collectors and Thermal Storage Devices*, vol. 899 (US Department of Commerce, National Bureau of Standards, 1976)
49. D. Rojas, J. Beermann, S.A. Klein, D.T. Reindl, Thermal performance testing of flat-plate collectors. Sol. Energy **82**(8), 746–757 (2008)
50. H.P. Garg, S.N. Garg, Measurement of solar radiation–I. Radiation instruments. Renewable Energy **3**(4/5), 321–333 (1993)
51. C.K. Pandey, A.K. Katiyar, Solar radiation: models and measurement techniques. J. Energy (2013)
52. G.A. Zerlaut, Solar radiometry instrumentation, calibration techniques, and standards. Solar Cells **18**, 189–203 (1986)
53. G.N. Tiwari, *Solar Energy: Fundamentals, Design, Modelling and Applications* (Narosa Publishing House, 2005)
54. J.H. Lienhard IV, J.H. Lienhard IV, *A Heat Transfer Textbook*, 4th edn (Phlogiston Press, Cambridge, Massachusetts, USA, 2016)
55. T.L. Bergman, A.S. Lavine, F.P. Incropera, D.P. Dewitt, *Fundamentals of Heat and Mass Transfer*, 7th edn (Wiley, USA)
56. Y.A. Cengel, *Heat Transfer: A Practical Approach*, 2nd edn (Tata McGraw-Hill Company Limited, 2002)
57. J.P. Holman, *Heat Transfer*, 10th edn. (Tata McGraw-Hill Company, Higher Education, NY, USA, 2010)
58. C.P. Kothandaraman, *Fundamentals of Heat and Mass Transfer*, Revised 3rd edn (New Age International (P) Limited, India, 2006)
59. S.E. Frid, N.V. Lisitskaya, State-of-the-art solar collectors: typical parameters and trends. Appl. Solar Energy **54**, 279–286 (2018)
60. S. Mehnert, K. Kramer, P. Kovacs, P. Dias, S. Fischer, E. Mateu, P. Navarro, F. Helminger, *Towards a New Global Standard for Solar Thermal Collectors* (ISES, 2011)
61. T. Osório, M.J. Carvalho, Testing of solar thermal collectors under transient conditions. Sol. Energy **104**, 71–81 (2014)
62. R.W. Bertram, Testing and standards for thermal solar collectors, in *Solar Energy Conversion II* (Pergamon, 1981), pp. 73–91
63. I. Visa, A. Duta, M. Comsit, M. Moldovan, D. Ciobanu, R. Saulescu, B. Burduhos, Design and experimental optimisation of a novel flat plate solar thermal collector with trapezoidal shape for facades integration. Appl. Therm. Eng. **90**, 432–443 (2015)
64. A. Sreekumar, G. Raam Dheep, *Solar Thermal Energy: Including Laboratory Experiments* (New Academic Science, 2020). ISBN 1523140704, 9781523140701
65. R.W. Bliss Jr, The derivations of several "plate-efficiency factors" useful in the design of flat-plate solar heat collectors. Sol. Energy **3**(4), 55–64 (1959)

66. D.E. Prapas, B. Norton, S.D. Probert, *Thermal Design of Compound Parabolic Concentrating Solar-Energy Collectors* (1987), pp. 161–168
67. Z. Yao, Y. Lum, A. Johnston, L.M. Mejia-Mendoza, X. Zhou, Y. Wen, A. Aspuru-Guzik, E.H. Sargent, Z.W. Seh, Machine learning for a sustainable energy future. Nat. Rev. Mater. **8**(3), 202–215 (2023)
68. C. Correa-Jullian, J.M. Cardemil, E.L. Droguett, M. Behzad, Assessment of deep learning techniques for prognosis of solar thermal systems. Renewable Energy **145**, 2178–2191 (2020)
69. S. Shamshirband, T. Rabczuk, K.-W. Chau, A survey of deep learning techniques: application in wind and solar energy resources. IEEE Access **7**, 164650–164666 (2019)
70. D. Rangel-Martinez, K.D.P. Nigam, L.A. Ricardez-Sandoval, Machine learning on sustainable energy: a review and outlook on renewable energy systems, catalysis, smart grid and energy storage. Chem. Eng. Res. Des. **174**, 414–441 (2021)
71. A. Mosavi, M. Salimi, S.F. Ardabili, T. Rabczuk, S. Shamshirband, A.R. Varkonyi-Koczy, State of the art of machine learning models in energy systems, a systematic review. Energies **12**(7), 1301 (2019)
72. S.A. Kalogirou, Optimization of solar systems using artificial neural-networks and genetic algorithms. Appl. Energy **77**(4), 383–405 (2004)
73. A.H. Elsheikh, S.W. Sharshir, M.A. Elaziz, A.E. Kabeel, W. Guilan, Z. Haiou, Modeling of solar energy systems using artificial neural network: a comprehensive review. Sol. Energy **180**, 622–639 (2019)
74. T.E. Boukelia, O. Arslan, M.S. Mecibah, ANN-based optimization of a parabolic trough solar thermal power plant. Appl. Therm. Eng. **107**, 1210–1218 (2016)
75. S. Fischer, P. Frey, H. Drück, A comparison between state-of-the-art and neural network modelling of solar collectors. Sol. Energy **86**(11), 3268–3277 (2012)
76. W. Yaïci, E. Entchev, Performance prediction of a solar thermal energy system using artificial neural networks. Appl. Therm. Eng. **73**(1), 1348–1359 (2014)
77. M. Vakili, S.A. Salehi, A review of recent developments in the application of machine learning in solar thermal collector modelling. Environ. Sci. Pollut. Res. **30**(2), 2406–2439 (2023)
78. S. Kalogirou, S. Lalot, G. Florides, B. Desmet, Development of a neural network-based fault diagnostic system for solar thermal applications. Sol. Energy **82**(2), 164–172 (2008)
79. K.M. Pandey, R. Chaurasiya, A review on analysis and development of solar flat plate collector. Renew. Sustain. Energy Rev. **67**, 641–650 (2017)
80. F. Cruz-Peragon, J.M. Palomar, P.J. Casanova, M.P. Dorado, F. Manzano-Agugliaro, Characterization of solar flat plate collectors. Renew. Sustain. Energy Rev. **16**(3), 1709–1720 (2012)
81. F. Struckmann, Analysis of a flat-plate solar collector. Heat and Mass Transport, Project Report, 2008MVK160 (2008)
82. H.K. Elminir, A.E. Ghitas, F. El-Hussainy, R. Hamid, M.M. Beheary, K.M. Abdel-Moneim, Optimum solar flat-plate collector slope: case study for Helwan, Egypt. Energy Convers. Manag. **47**(5), 624–637 (2006)
83. A. Fudholi, K. Sopian, A review of solar air flat plate collector for drying application. Renew. Sustain. Energy Rev. **102**, 333–345 (2019)
84. H.C. Hottel, B.B. Woertz, The performance of flat-plate solar-heat collectors. Trans. Am. Soc. Mech. Eng. **64**(2), 91–103 (1942)
85. S. Aggarwal, R. Kumar, S. Kumar, M. Bhatnagar, P. Kumar, Computational fluid dynamics based analysis for optimization of various thermal enhancement techniques used in evacuated tubes solar collectors: a review. Mater. Today Proc. **46**, 8700–8707 (2021)
86. A.A. Badran, I.A. Al-Hallaq, I.A. Eyal Salman, M.Z. Odat, A solar still augmented with a flat-plate collector. Desalination **172**(3), 227–234 (2005)
87. Z. Chen, S. Furbo, B. Perers, J. Fan, E. Andersen, Efficiencies of flat plate solar collectors at different flow rates. Energy Procedia **30**, 65–72 (2012)
88. R.P. Garcia, S. del Rio Oliveira, V.L. Scalon, Thermal efficiency experimental evaluation of solar flat plate collectors when introducing convective barriers. Sol. Energy **182**, 278–285 (2019)

89. S.K. Saha, D.K. Mahanta, Thermodynamic optimization of solar flat-plate collector. Renewable Energy **23**(2), 181–193 (2001)
90. C. Cristofari, G. Notton, P. Poggi, A. Louche, Influence of the flow rate and the tank stratification degree on the performances of a solar flat-plate collector. Int. J. Therm. Sci. **42**(5), 455–469 (2003)
91. J.K. Nayak, E.H. Amer, Experimental and theoretical evaluation of dynamic test procedures for solar flat-plate collectors. Sol. Energy **69**(5), 377–401 (2000)
92. J.A. Duffie, W.A. Beckman, *Solar Engineering of Thermal Processes* (Wiley, New York, 1991). European Standard EN 12975-2, *Thermal Solar Systems and Components—Solar Collectors—Part 2: Test Methods* (2001)
93. S. Haykin, *Neural Networks: A Comprehensive Foundation* (Macmillan, New York, 1994)
94. S. Kalogirou, C. Neocleous, C. Schizas, Artificial neural networks for modelling the starting-up of a solar steam generator. Appl. Energy **60**(2), 89–100 (1998)
95. S. Kalogirou, S. Panteliou, A. Dentsoras, Artificial neural networks used for the performance prediction of a thermosyphon solar water heater. Renewable Energy **18**(1), 87–99 (1999)
96. S.A. Kalogirou, A.D. Panteliou, A. Dentsoras, Modeling of solar domestic water heating systems using artificial neural networks. Sol. Energy **65**(6), 335–342 (1999)
97. S. Kalogirou, Long-term performance prediction of forced circulation solar domestic water heating systems using artificial neural networks. Appl. Energy **66**(1), 63–74 (2000)
98. S. Kalogirou, M. Bojic, Artificial neural networks for the prediction of the energy consumption of a passive solar building. Energy Int. J. **25**(5), 479–491 (2000)
99. S. Kalogirou, S. Panteliou, Thermosyphon solar domestic water heating systems long-term prediction using artificial neural networks. Sol. Energy **69**(2), 163–174 (2000)
100. S. Kalogirou, Artificial neural networks in renewable energy systems: a review. Renew. Sustain. Energy Rev. **5**(4), 373–401 (2001)
101. J.F. Kreider, X.A. Wang, Artificial neural network demonstration for automated generation of energy use predictors for commercial buildings, in *The Use of Artificial Intelligence in Building Systems*, ed. by J.S. Haberl, R.M. Nelson, and C.C. Culp (ASHRAE, 1995), pp. 193–198
102. C.W. Anderson, D.C. Hittle, A.D. Katz, R.M. Kretchman, Reinforcement learning neural networks and PI control applied to a heating coil, in *Proceedings of the International Conference EANN* (London, UK), pp. 135–142
103. S.A. Kalogirou, C.C. Neocleous, C.N. Schizas, A comparative study of methods for estimating intercept factor of parabolic trough collectors, in *Proceedings of the International Conference EANN* (London, UK), pp. 5–8
104. S. Kalogirou, Artificial neural networks for predicting the local concentration ratio of parabolic trough collectors, in *Proceedings of the International Conference EuroSun* (Freibrug, Germany), pp. 470–475
105. S.A. Kalogirou, C.C. Neocleous, C.N. Schizas, Artificial neural networks in modeling the heat!up response of a solar steam generation plant, in *Proceedings of the International Conference EANN* (London, UK), pp. 1–4
106. S.A. Kalogirou, C.C. Neocleous, C.N. Schizas, Artificial neural networks for the estimation of the performance of a parabolic trough collector steam generation system, in *Proceedings of the International Conference EANN* (Stockholm, Sweden), pp. 227–232
107. P. Liu, J. Wang, A.K. Sangaiah, Y. Xie, X. Yin, Analysis and prediction of water quality using LSTM deep neural networks in IoT environment. Sustainability **11**(7), 2058 (2019)
108. C. Chen, Y. Zuo, W. Ye, X. Li, Z. Deng, S.P. Ong, A critical review of machine learning of energy materials. Adv. Energy Mater. **10**(8), 1903242 (2020)
109. M. Sridharan, B. Prakash, Predicting the performance of solar photovoltaic thermal water collectors using hybrid fuzzy logic expert system. Int. J. Ambient Energy **43**(1), 8098–8112 (2022)
110. M. Sridharan, Short review on various applications of fuzzy logic-based expert systems in the field of solar energy. Int. J. Ambient Energy **43**(1), 5112–5128 (2022)
111. B. Das, Jagadish, Investigation of thermal performance of solar collector variables using fuzzy logic-based expert system, in *Evolutionary Methods Based Modeling and Analysis of Solar*

Thermal Systems: A Case Studies Approach (Springer International Publishing, Cham, 2023), pp. 63–76
112. Y. Varol, A. Koca, H.F. Oztop, E. Avci, Forecasting of thermal energy storage performance of phase change material in a solar collector using soft computing techniques. Expert Syst. Appl. **37**(4), 2724–2732 (2010)
113. S.M. Alirahmi, A. Khoshnevisan, P. Shirazi, P. Ahmadi, D. Kari, Soft computing based optimization of a novel solar heliostat integrated energy system using artificial neural networks. Sustain. Energy Technol. Assess. **50**, 101850 (2022)
114. J.-L. Casteleiro-Roca, P. Chamoso, E. Jove, A. González-Briones, H. Quintián, M.-I. Fernández-Ibáñez, R.A. Vega Vega et al. Solar thermal collector output temperature prediction by hybrid intelligent model for smartgrid and smartbuildings applications and optimization. Appl. Sci. **10**(13), 4644 (2020)
115. Z.Q. Fard, Z.S. Zomorodian, S.S. Korsavi, Application of machine learning in thermal comfort studies: a review of methods, performance and challenges. Energy Build. **256**, 111771 (2022)
116. Z. He, W. Guo, P. Zhang, Performance prediction, optimal design and operational control of thermal energy storage using artificial intelligence methods. Renew. Sustain. Energy Rev. **156**, 111977 (2022)
117. E. Assareh, A. Riaz, M. Ahmadinejad, S. Hoseinzadeh, M.Z. Abdehvand, S. Keykhah, T. Jafarinejad, R. Moltames, M. Lee, Enhancing solar thermal collector systems for hot water production through machine learning-driven multi-objective optimization with phase change material (PCM). J. Energy Stor. 73, 108990 (2023)

Thermochemical Processes in Solar Fuel Production

Jaimy Gebbeken, Zhen Cao, Tianchao Xie, and Abhishek K. Singh

Abstract Solar energy provides abundant renewable energy resources that can be utilized for power, high-temperature heat, and fuel generation. To tackle the intermittency issue of solar energy, fuel generation provides a high energy density storage solution that can also act as a feedstock for chemical production. Solar fuels can be generated directly using thermal energy (e.g. solar thermochemical processes), photo pathways (e.g., photoelectrochemistry), or a hybrid of both. The solar thermochemical processes utilize concentrated solar thermal energy as the high-temperature heat input. These processes could achieve high conversion efficiency due to the potential of full solar spectrum utilization and produce primary fuels such as hydrogen, and synthetic gas (with CO and H_2 as main components) or derived fuels such as methanol, and synthetic liquid fuels from H_2O and CO_2. This chapter describes various thermochemical pathways that can be used for fuel generation. Among these, a metal oxide-based redox looping thermochemical cycle for H_2O and CO_2 splitting is the most viable option due to its direct use of solar energy, process simplicity, and scalability. It is a two-step process in which a metal oxide is first reduced at high-temperature and low pressure, releasing oxygen and forming a lower valance metal oxide. In the second step, the reduced metal oxide reacts with steam/CO_2 and generates H_2/CO. The metal oxide then re-oxidizes back to its original state, making it a closed-loop process. The current Technology Readiness Level of this process remains low due to several technical challenges at the material, reactor, and system levels. From the material perspective, a cheaper material with faster reaction kinetics and good cycle stability is needed. Challenges at the reactor level include reactor material limitation and sealing issues due to high operating temperature, energy penalty for low-pressure creation, and heat transfer in the porous material. Since the metal oxide reduction and water-splitting steps occur at different temperatures, effective heat recuperation between these two steps is crucial to achieve higher process efficiencies. However, this is extremely difficult due to severe operating conditions.

J. Gebbeken · Z. Cao · T. Xie · A. K. Singh (✉)
Department of Thermal and Fluid Engineering, University of Twente, Enschede, The Netherlands
e-mail: a.k.singh@utwente.nl

Z. Cao
e-mail: caozhen0520@gmail.com

P. Ravindran et al. (eds.), *Photon to Power*, Progress in Optical Science and Photonics 32, https://doi.org/10.1007/978-981-96-5914-2_6

The chapter explains the thermodynamics of the thermochemical cycles, reactive material requirements, reactor development, and system-level analysis. The opportunities and challenges of these processes are described, and a future outlook of the technology is provided.

1 Introduction

Global warming represents one of the most urgent challenges of the twenty-first century, requiring immediate and innovative solutions. Initiatives like 'A Clean Planet for All', proposed by the European Union [1], emphasize the critical need to address climate change through sustainable practices. Solar energy, due to its abundant and renewable characteristics, is considered to play a vital role in these efforts [2]. Nevertheless, solar energy systems encounter inherent limitations, particularly intermittency, compromising their reliability as a stand-alone energy source [3]. Solar fuels present a compelling solution to address this issue. These fuels offer energy storage capabilities through high-energy-density chemical bonds, ensuring availability even when direct sunlight is inaccessible [4].

Solar fuels include a variety of compounds, such as hydrogen, hydrocarbons, ammonia, and hydrazine, each of which provides distinct advantages in energy storage and utilization [4–6]. Solar fuels can be produced through various processes, including photochemical, thermochemical, electrochemical, and photobiological pathways. These methods convert sunlight into chemical energy stored in bonds, facilitating the sustainable production of fuels that can serve as alternatives to fossil fuel sources [7].

The production of solar fuels can be facilitated by the thermal decomposition of a reactant using solar energy that will result in a product or fuel of interest. Consider, for example, the thermal decomposition of water to produce hydrogen, a process known as water splitting or direct thermolysis when performed through a singular reaction [8]. The governing endothermic reaction for direct thermolysis is denoted by [9],

$$H_2O \rightleftharpoons H_2 + \tfrac{1}{2}O_2 \tag{1.1}$$

Although direct thermolysis is a feasible strategy for producing hydrogen, it is not a viable solution due to the high reaction temperature requirement ($> 2200\,^\circ C$). In addition, this temperature must be maintained to ensure the safe separation of the product gases to prevent an explosive mixture through recombination [8, 9]. An alternative approach is to perform the thermal decomposition of reactants in a series of consecutive reactions to segregate the release of specific products in space or time [2]. This is known as a multistep cycle, in which a specific reaction is considered at each step including endothermic steps to utilize solar energy and the production of the fuel in the final step. The most commonly considered is the two-step cycle, as the inclusion of more steps is associated with a significant decrease in reaction temperatures, an increase in energy losses, and lower system stability [2, 10, 11].

Thermochemical cycling has recently gained significant attention for solar fuel production due to its reaction stability, reduced exergy loss, and compatibility with existing energy supply systems [11, 12]. These cycles typically involve thermochemical materials that enable the separation of specific products from reactants during endothermic reduction reactions. A key application of this process is the use of metal-oxide materials, which facilitate oxygen reduction in water- or carbon-splitting reactions. This approach is used to produce hydrogen from water or facilitate the generation of synthesis gas (syngas) or hydrocarbons from the mixture of carbon dioxide and water [5, 9]. When solar energy is used to drive the endothermic metal oxide reduction step in a two-step thermochemical cycle for hydrogen production, it is commonly called the solar thermochemical hydrogen (STCH) process.

The STCH process can be categorized into two classes (1) stoichiometric and (2) non-stoichiometric, based on how the metal oxide enables the water splitting [8]. In the stoichiometric reactions, one mole of metal oxide produces one mole of hydrogen and ½ mole of oxygen for water splitting and thermal reduction processes, respectively. They can produce a much higher yield of hydrogen because the majority of oxygen is released (reduction process) and incorporated (oxidation process) due to phase change during the cycling process. On the contrary, the non-stoichiometric chemistries do not encompass the phase change because these materials perform the redox reactions through a vacancy (or non-stoichiometric) facilitated mechanism [10]. In these reactions, the thermal reduction of the metal oxide is not complete, thus avoiding the phase change. Although the hydrogen yield is generally lower than the stoichiometric reactions, the redox process is usually much faster due to better kinetics. Therefore, these reactions can improve overall hydrogen production. Due to these advantages, most of the current research is focused on the development of a non-stoichiometric-based STCH process.

The arbitrary governing high-temperature endothermic reduction reaction for a two-step thermochemical cycle using a non-stoichiometric metal oxide material is denoted by [13],

$$\tfrac{1}{\delta}\mathrm{MO}_x(s) \rightleftharpoons \tfrac{1}{\delta}\mathrm{MO}_{\mathrm{x}-\delta}(s) + \tfrac{1}{2}\mathrm{O}_2(g)\,, \tag{1.2}$$

where δ denotes the degree of non-stoichiometry in the metal oxide and is generally much smaller than unity, that is, $\delta \ll 1$ [14]. This reduction reaction ensures the metal oxide material has a sufficient oxygen affinity to facilitate the water- or carbon-splitting reaction safely via exothermic oxidation of the metal oxide [9], respectively, denoted by [9, 13],

$$\frac{1}{\delta}\mathrm{MO}_{\mathrm{x}-\delta}(s) + \mathrm{H}_2O(g) \rightleftharpoons \frac{1}{\delta}\mathrm{MO}_x(s) + \mathrm{H}_2(g), \tag{1.3}$$

$$\frac{1}{\delta}\mathrm{MO}_{\mathrm{x}-\delta}(s) + \mathrm{CO}_2(\mathrm{g}) \rightleftharpoons \frac{1}{\delta}\mathrm{MO}_x(s) + \mathrm{CO}(g). \tag{1.4}$$

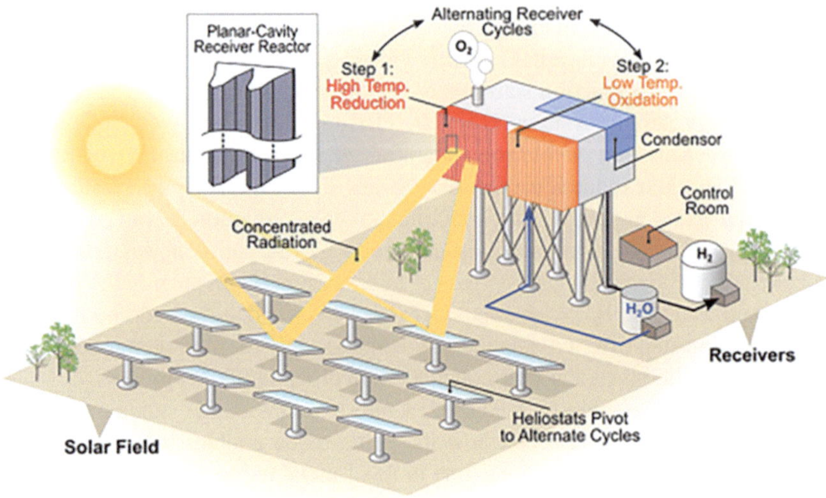

Fig. 1 Schematic of a conceptual STCH platform of NREL. Images reprinted from [15] with permission from Elsevier

Due to its high-temperature requirement, the STCH process is integrated with concentrated solar thermal (CST) systems having high concentration ratios such as parabolic dish collectors or solar tower systems. Figure 1 shows a schematic of a planar cavity-receiver concept from NREL that uses a two-receiver system for alternative water splitting (oxidation reaction) and metal-oxide reduction processes. The concentrated solar flux is directed onto the reactor performing the metal oxide reduction reaction by moving the heliostat field's focal point, accordingly. This concept avoids having any high-temperature reactive material transport between the reactors thereby easing the operation and reducing the heat losses. Each reactor should be optimized for both reduction and oxidation reactions, which constrains the reactor design space.

2 Reaction Feasibility and Thermodynamic Limits

The design process for a new two-step thermochemical reactor for solar fuel production starts with establishing the feasibility of the reaction. The feasibility is largely governed by the reaction kinetics of the metal oxide materials considered in the reactor and provides insight into the allowable operating conditions. In the past, many studies have been devoted to evaluating the feasibility of a predetermined material, consider [16–18], but this is a laborious task when no specific material is considered and recent advances aim to consider a generalized approach [2, 9, 13].

This section introduces the basic considerations of this generalized approach, which is usually referred to as thermodynamic analysis.

2.1 The Gibbs Free Energy of a Reaction

The feasibility analysis of a reaction is typically considered through the change in the Gibbs free energy during the reduction and oxidation reactions of a metal-oxide material. Gibbs free energy, G, represents the maximum reversible work that can be obtained from a system at constant temperature and pressure, which are representative conditions assumed to hold during most reactions. The Gibbs function in extensive form is defined from elementary thermodynamic quantities, as described by [19],

$$G := H - TS \equiv U + pV - TS \tag{2.1}$$

where H denotes the enthalpy of the system, T the temperature, S the entropy, U the internal energy, p the pressure and V the volume.

2.1.1 Multi-component System Analysis

To analyze the thermodynamic behavior of a multi-component system, it is essential to consider how the Gibbs free energy changes with variations in system conditions and the composition of the reacting species. The change in Gibbs free energy during a reaction process is denoted by [2, 20],

$$dG = Vdp - SdT + \sum_i \mu_i dn_i \leq 0 \tag{2.2}$$

where μ_i and n_i denote the chemical potential and the number of moles of the pure component i, and $dG \leq 0$ ensures the spontaneity of the reaction. A reaction is considered spontaneous, or feasible, when it is allowed to occur, provided that the activation energy barrier is overcome. The system pressure and temperature are typically considered to be (nearly) constant, i.e., dp, $dT \approx 0$, throughout the reaction processes such that the change in Gibbs for multi-component systems is denoted by,

$$(dG)_{p,T} = \sum_i \mu_i dn_i \leq 0 \tag{2.3}$$

The Gibbs free energy of the reaction then only depends on the reacting species and their chemical potential. The chemical potential represents the chemical work potential of each component in the system and can be denoted in terms of the system parameters by [20],

$$\mu_i = \overline{g}_i^\circ + \mathrm{RT}\ln(a_i)., \tag{2.4}$$

where the superscript o denotes the consideration of a quantity at a reference state, $\overline{g}_i^\circ := \overline{g}_i(T, p = p^\circ)$ the Gibbs per mole at a reference state, R the universal gas constant, and a_i the activity of a component i in the mixture. In consideration of component i being a gaseous species that abides the ideal gas law, the activity is denoted by [20],

$$a_i := \frac{y_i p}{p^\circ} \equiv \frac{p_i}{p^\circ}., \tag{2.5}$$

where y_i and p_i denote the molar fraction and partial pressure of the pure component i.

2.1.2 Reaction Equilibrium

The change in Gibbs free energy for a reaction of interest is determined by having a closer look at the stoichiometric representation, which may for an arbitrary reaction be denoted by [20],

$$\sum_i \nu_i^R R_i \rightleftharpoons \sum_j \nu_j^P P_j \tag{2.6}$$

where non-reacting components are excluded because they do not exert an influence on the reaction and ν denotes the stoichiometric coefficient for the specific reactant, R, or product, P. The stoichiometric coefficients do not inherently dictate the amounts of components present, but do influence the relation between each component, as can be denoted by [20],

$$\underbrace{\frac{-\mathrm{d}n_i^R}{\nu_i^R}}_{\text{Reactants}} = \underbrace{\frac{\mathrm{d}n_j^P}{\nu_j^P}}_{\text{Products}} = \mathrm{d}\varepsilon \tag{2.7}$$

where n denotes the number of moles of each component and ε the extent of reaction, which is a quantity that measures the extent to which the reaction has proceeded. The contributions of reactants are negative, reflecting their consumption, whereas products contribute positively due to their formation. Non-reacting components can be excluded from the analysis, as their change in the number of moles is identically zero. The individual components and their dependence on the extent of reaction can now be substituted into the expression for the change in Gibbs free energy at a specified pressure and temperature, as given in Eq. (2.3), to derive the relevant relation for the system under consideration, denoted by [20],

$$(dG)_{p,T} = \left[\sum_i -\nu_i^R \mu_i^R + \sum_j \nu_j^P \mu_j^P\right] d\varepsilon \leq 0 \tag{2.8}$$

A useful and often initially considered assumption is to regard the reaction at equilibrium. The equilibrium condition is denoted by $(dG)_{p,T} = 0$, such that the equation of reaction equilibrium can be denoted by [20],

$$\sum_i -\nu_i^R \mu_i^R + \sum_j \nu_j^P \mu_j^P = 0 \tag{2.9}$$

2.2 *Solar Fuel Production Systems*

The behavior of the solar fuel production systems is for an initial analysis assumed to be governed by the equation of reaction equilibrium for both the reduction and oxidation reaction. This section introduces the derivation and simplification, using common assumptions, of these relations and how the parameters relate to the system processes.

2.2.1 Reduction Reaction

The equation of reaction equilibrium for the reduction reaction, denoted in Eq. (1.2), can now be derived using the arguments introduced above and is denoted by,

$$-\frac{1}{\delta}\left(\overline{g}_{ox}^{\circ} + RT_{red}\ln(a_{ox})\right) + \frac{1}{\delta}\left(\overline{g}_{red}^{\circ} + RT_{red}\ln(a_{red})\right) + \frac{1}{2}\left(\overline{g}_{O_2}^{\circ} + RT_{red}\ln\left(\frac{p_{O_2}}{p_{red}^{\circ}}\right)\right) = 0$$

where the activities of the oxidized and reduced states are typically considered equivalent. Furthermore, the contributions denoted by Gibbs per mole in a reference state, $\overline{g}_i^{\circ}$, are combined into a singular quantity that describes the Gibbs free energy of the formation of oxygen vacancies in a reference state [2], $d\overline{g}_{\mathrm{MO}}^{\circ} := -\frac{1}{\delta}\overline{g}_{\mathrm{ox}}^{\circ} + \frac{1}{\delta}\overline{g}_{\mathrm{red}}^{\circ} + \frac{1}{2}\overline{g}_{\mathrm{O_2}}^{\circ}$. The equation of reaction equilibrium for the reduction reaction is now denoted by,

$$d\overline{g}_{MO}^{\circ} + \tfrac{1}{2}RT_{\mathrm{red}}\ln\left(\frac{p_{O_2}}{p_{\mathrm{red}}^{\circ}}\right) = 0 \tag{2.11}$$

The Gibbs free energy of the oxygen vacancy formation, $d\overline{g}_{\mathrm{MO}}^{\circ}$, can be expressed in terms of enthalpy and entropy, $d\overline{h}_{\mathrm{MO}}^{\circ}$ and $d\overline{s}_{\mathrm{MO}}^{\circ}$, for a constant temperature. The final equation of reaction equilibrium for the reduction reaction now only depends on familiar metal-oxide material parameters and a singular operational parameter, and is denoted by [2, 13],

$$d\overline{h}^{\circ}_{\mathrm{MO}} - T_{\mathrm{red}} d\overline{s}^{\circ}_{\mathrm{MO}} = -\tfrac{1}{2}\mathrm{RT}_{\mathrm{red}} \ln\left(\frac{p_{\mathrm{O}_2}}{p^{\circ}_{\mathrm{red}}}\right) \tag{2.12}$$

2.2.2 Oxidation Reaction

Consider the water-splitting reaction denoted in Eq. (1.3) to illustrate the derivation of the reaction equilibrium equation for the oxidation reaction. The derivation starts, similarly to the derivation of the reduction reaction, by simple substitution of previously introduced arguments and is denoted by,

$$\begin{aligned} &-\tfrac{1}{\delta}\left(\overline{g}^{\circ}_{\mathrm{red}} + \mathrm{RT}_{\mathrm{ox}} \ln(a_{\mathrm{red}})\right) - \left(\overline{g}^{\circ}_{\mathrm{H_2O}} + \mathrm{RT}_{\mathrm{ox}} \ln\left(\frac{p_{\mathrm{H_2O}}}{p^{\circ}_{\mathrm{ox}}}\right)\right) \\ &+\tfrac{1}{\delta}\left(\overline{g}^{\circ}_{\mathrm{ox}} + \mathrm{RT}_{\mathrm{ox}}\ln(a_{\mathrm{ox}})\right) + \left(\overline{g}^{\circ}_{\mathrm{H_2}} + \mathrm{RT}_{\mathrm{ox}}\ln\left(\frac{p_{\mathrm{H_2}}}{p^{\circ}_{\mathrm{ox}}}\right)\right) = 0 \end{aligned} \tag{2.13}$$

where both water vapor and hydrogen are treated as ideal gases for simplicity. Analogously to the derivation of the reduction reaction, the activities of the reduced and oxidized states are assumed to be equivalent. Two simplified contributions are defined through the addition and subtraction of the Gibbs free energy per mole at a reference state for oxygen, $\frac{1}{2}\overline{g}^{\circ}_{\mathrm{O_2}}$. The first considers again the Gibbs free energy associated with the formation of oxygen vacancies at a reference state [2], $d\overline{g}^{\circ}_{\mathrm{MO}}$, whereas the second considers the Gibbs free energy pertinent to the water-splitting process, $d\overline{g}^{\circ}_{WS} := -\overline{g}^{\circ}_{\mathrm{H_2O}} + \overline{g}^{\circ}_{\mathrm{H_2}} + \overline{g}^{\circ}_{\mathrm{O_2}}$. Finally, considering the gas constants for water vapor and hydrogen to be equivalent, the two logarithmic contributions can be combined to derive the simplified equation that characterizes the equilibrium of the oxidation or water-splitting reaction denoted by,

$$-d\overline{g}^{\circ}_{MO} + d\overline{g}^{\circ}_{WS} + \mathrm{RT}_{\mathrm{ox}}\ln\left(\frac{p_{\mathrm{H_2}}}{p_{\mathrm{H_2O}}}\right) = 0 \tag{2.14}$$

The Gibbs free energy of the water-splitting reaction can alternatively be rewritten in terms of the equilibrium constant of the reaction, K_{WS}, as denoted by [2],

$$d\overline{g}^{\circ}_{WS} := -RT_{\mathrm{ox}}\ln(K_{WS}). \tag{2.15}$$

where the equilibrium constant is a function of the temperature of the oxidation reaction and is commonly considered available through tabulated data [20]. The ratio of partial pressures can be related to the feedstock conversion extent as denoted by [9],

$$\frac{p_{\mathrm{H_2}}}{p_{\mathrm{H_2O}}} \equiv \frac{X_{\mathrm{H_2O}}}{1-X_{\mathrm{H_2O}}} \tag{2.16}$$

where X_{H_2O} denotes the feedstock conversion extent, which can be used to measure the effectiveness of the chemical conversion process. The final equation of reaction equilibrium for the oxidation or water-splitting reaction is now denoted by,

$$d\overline{h}^{\circ}_{MO} - T_{ox} d\overline{s}^{\circ}_{MO} = -\tfrac{1}{2} R T_{ox} \ln\left(\tfrac{1-X_{H_2O}}{X_{H_2O}} K_{WS}\right) \quad (2.17)$$

2.3 Operational Window

The operational window for an arbitrary solar fuel production system can be explored by rewriting the reduction and oxidation equations of reaction equilibrium in terms of the condition for spontaneity, $dG \leq 0$, and assuming a strictly positive change in extent of reaction, $d\varepsilon > 0$. This enables illustration of the span of allowable enthalpy and entropy by rewriting Eqs. (2.12) and (2.17) [13],

$$d\overline{h}^{\circ}_{MO} \leq \underbrace{T_{red}}_{m_{red}} d\overline{s}^{\circ}_{MO} \underbrace{- \frac{1}{2} R T_{red} \ln\left(\frac{p_{O_2}}{p^{\circ}_{red}}\right)}_{b_{red}}, \; dG_{red} \leq 0 \quad (2.18)$$

$$d\overline{h}^{\circ}_{MO} \geq \underbrace{T_{ox}}_{m_{ox}} d\overline{s}^{\circ}_{MO} \underbrace{- \frac{1}{2} R T_{ox} \ln\left(\frac{1 - X_{H_2O}}{X_{H_2O}} K_{WS}\right)}_{b_{ox}}, \; dG_{ox} \leq 0 \quad (2.19)$$

where the reduction and oxidation temperatures denote the limit line slopes, m_{red} and m_{ox}, and the y-intercepts, b_{red} and b_{ox}, are dependent on the operation conditions. The operational window for an arbitrary solar fuel production system is illustrated in green in Fig. 2.

Within the window, normal operation conditions for both the reduction and oxidation reactions are expected and thus present a desirable preliminary consideration of the design of a solar fuel production system. The size of this window strongly depends on the design of the fuel production cycles, the physical limiting values for the included parameters, and potential system safety factors.

2.3.1 Cycle Temperature Design

The temperatures for the reduction and oxidation reactions need to be selected with relative care in the design of the solar fuel production system to ensure efficiency and longevity. A first and foremost consideration is the behavior of the thermochemical material:

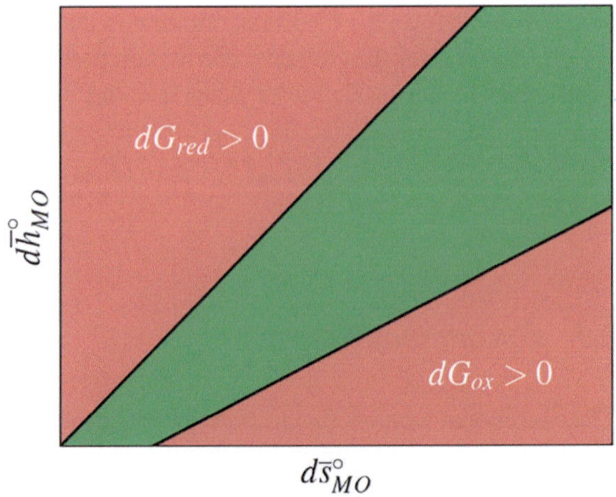

Fig. 2 Typical operational window for an arbitrary solar fuel production system in a swing cycle. The thermochemical material is required to possess a reaction enthalpy and entropy receding in the operational window, depicted by green, to promote reaction feasibility

Low temperatures will cause the reactions to progress at insufficient rates or even induce unfavorable thermodynamic conditions when the activation temperatures are not reached.

High temperatures could induce rapid degradation of metal-oxide materials or introduce undesirable structural transformations, usually leading to lower system efficiency [21–23].

The difference in temperature between the two reactions is another important consideration in the design of solar fuel production systems.

Temperature swing cycles capitalize on the temperature difference between the reduction and oxidation reactions. They offer a wider range of favorable thermodynamic properties, improving material selection and increasing oxygen storage capabilities per cycle, leading to better feedstock conversion and overall system efficiency [2, 9, 13, 24]. Furthermore, temperature changes create stronger driving forces for reduction and oxidation reactions, further enhancing performance [9]. However, these advantages are counterbalanced by challenges such as increased heating times during reduction, slower oxidation kinetics, and material degradation due to thermal stresses [13, 24].

In contrast, isothermal cycles maintain consistent temperatures throughout the process, offering greater exergy efficiency, improved solar-to-fuel efficiency, and reduced system complexity [9, 11, 21, 24]. They also benefit from higher power density and reduced pre-heating requirements, making them simpler and more robust in operation [9, 24]. However, this simplicity comes at the expense of reduced feedstock conversion and efficiency, as well as narrower operating windows and higher reaction temperatures [2, 11, 13].

In essence, temperature swing cycles are better suited for maximizing conversion and efficiency but demand robust materials to withstand extreme conditions. In contrast, isothermal cycles emphasize operational simplicity and robustness at the cost of reduced flexibility and efficiency. The optimal choice depends on the specific priorities and constraints of the production system.

2.3.2 Physical Limitations

In addition to the restrictions and requirements introduced by the cycle design, physical limitations should be considered. These limitations are present in both the operational parameters and in the material coefficients.

The *feedstock conversion extent* at the output of the reactor, X_{H_2O}, is a parameter that in an ideal case approaches zero, signifying a full conversion of the supplied steam to hydrogen. Although this is often set as a consideration or target value in material screening studies [9], achieving such a low value in practice is highly unlikely. Realistic conversion extents, even in well-optimized processes, often fall short due to the inherent balance required between thermodynamic feasibility, reaction kinetics, and practical scalability.

The *entropy* of the oxygen vacancy formation, $d\bar{s}^{\circ}_{MO}$, should, in theory, be maximized to promote reactor efficiency, just as the enthalpy should [13]. However, specifically increasing the entropy is not a trivial task and could even result in a decrease in the reduction reaction temperature, potentially complicating reactor design and optimization of performance [9].

3 Reactive Material

As discussed earlier, the STCH process is usually classified into stoichiometric and non-stoichiometric cycles. The reactive materials for stoichiometric cycles can be further classified into volatile and non-volatile metal oxides. Due to the high temperatures of the reduction reaction, the "volatile" redox pairs typically exhibit a solid-to-gas transition of the reduced material that evolves with the oxygen in the process. The reduced material could be a metal or a lower valance metal oxide. Although this phase change is thermodynamically beneficial because of high entropy gain, the effective gas–gas separation (oxygen and metal/metal oxide gas) is challenging at higher temperatures. Without the effective separation, a recombination of these gases occurs which readily forms the original reactant (metal oxide) thereby reducing the yield of the reaction. The most studied volatile STCH cycles include: ZnO/Zn, CdO/Cd, SnO_2/SnO, and GeO_2/GeO [25–31].

ZnO/Zn pair is well studied by the research group from ETH/PSI Zurich [25]. They characterized the material and developed a prototype scale reactor for assessing this cycle. The reaction temperature of ZnO is 2300 K at atmospheric pressure, whereas Zn melts at 692 K and vaporizes at 1180 K. The theoretical solar-to-chemical energy efficiency of 39% could be achieved by using this cycle [26]. The primary challenge of this cycle relates to the recombination of the product mixture (Zn and oxygen). To avoid this recombination, the produced gas mixture needs to be quenched rapidly which is sensitive to the temperature and the ratio of inert gas and Zn (g). Due to this limitation, the product usually contains a significant amount of ZnO thereby reducing the overall hydrogen generation from this cycle. Another volatile cycle that was investigated by several groups utilized the CdO/Cd pair. First, it was investigated by PROMES France and later General Atomics and the University of Nevada, Las Vegas [28]. A fluidized bed reactor was developed to demonstrate the thermal decomposition of the metal oxide [32, 33]. This cycle also has the same challenge of rapid quenching which hinders the practical implementation of this process. SnO_2/SnO, and GeO_2/GeO pairs were also investigated by researchers from PROMES France, and Korea [28, 34]. These cycles also have the same challenge of product recombination.

In the volatile cycles, the metal oxides stay in the solid state during the complete cycling process. Several reactive materials have been studied for these cycles. First, Nakamura proposed the use of Fe_3O_4/FeO pair for water splitting aided by solar energy in 1977 [28] This pair was first tested in a prototype reactor in the 1980s at the solar furnace, PROMES, France [35]. The same research group tested Mn_3O_4/ MnO in the same research facility [36]. Although these demonstrations showed a proof-of-concept of the STCH process, the reactive materials reached their melting points, forming the liquid phase of Fe_3O_4 and FeO. It drastically decreased the available surface area of the reactive materials, making them deactivated. This redox pair was further researched by a research group at the University of Florida. They determined the thermodynamic limit of the process and recommended measures to optimize performance, including minimizing particle size to increase the surface-to-volume ratio, extending the gas-phase residence time in the reactor, and partially reducing or oxidizing the iron-based reactants to prevent operation in the diffusion-limited regime [37].

Further research showed that the substitution of iron cations in the lattice of Fe_3O_4 by a transition or an alkaline earth metal forms mixed metal iron oxides or ferrites $(Fe_{1-x}M_x)_3O_4)$ which can be reduced at lower temperatures in comparison to pure Fe_3O_4 thus avoiding the phase transition [38]. Ferrites belong to the family of spinel materials which, together with perovskites, are two of the most known families of oxides containing multiple metal cations that can exhibit various valence states. The primary challenge with the non-volatile materials includes the sintering and slag formations that reduce the active surface area thereby decreasing the hydrogen generation after certain numbers of cycles.

Although the reactive materials for non-stoichiometric cycles usually do not suffer from these issues, generally they have low hydrogen generation capacity due to the lower extent of metal oxide reduction which consequently reduces the hydrogen production in the oxidation reaction. Ceria (CeO_2) has long been a benchmark material in thermochemical solar fuel production due to its stability and efficient reaction kinetics compared to other metal oxides [11, 13]. Its ability to undergo reversible redox cycles makes it suitable for high-temperature processes aimed at producing solar fuels.

In 2010, a Caltech (USA) research group proposed a ceria-based cycle utilizing the non-stoichiometric reaction scheme $((CeO_2)/CeO_{2-\delta})$ [38]. Afterward, several other groups have extensively investigated this cycle. It is observed that this cycle provides higher yield at a faster rate only when the reduction temperature is high (Tred = 1500 °C) and outperforms other redox materials (Fig. 3) [8]. At lower reduction temperatures, the reduction extent is reduced which decreases the hydrogen output.

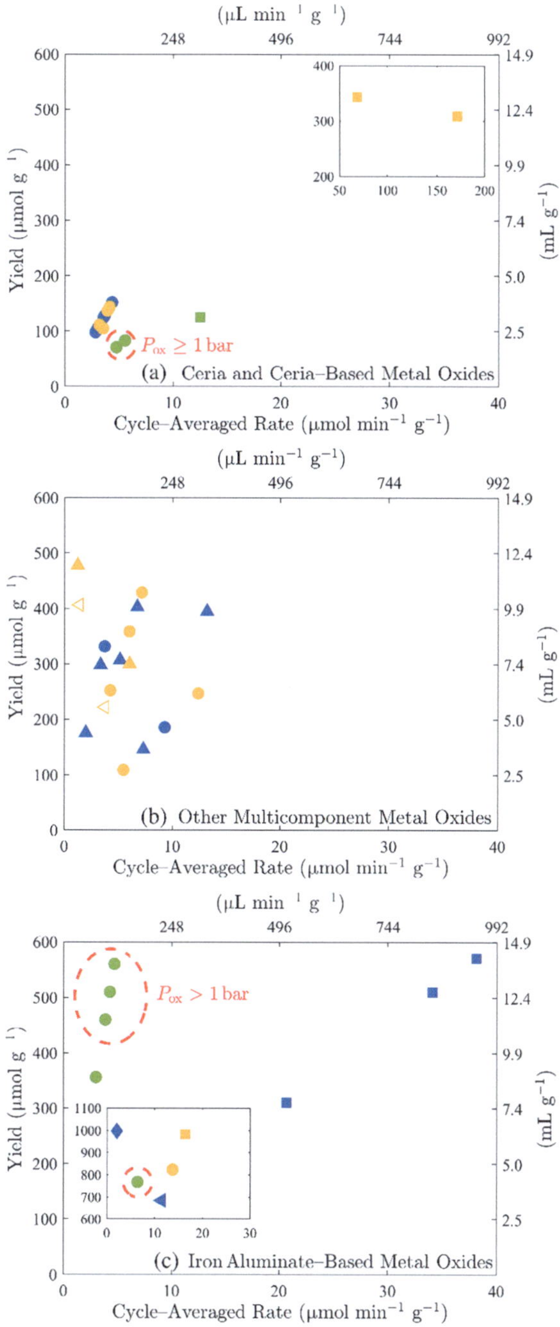

Fig. 3 Thermochemical yield of ceria and ceria-based metal oxides. Images reprinted from [8] with permission from WIREs

Isothermal cyclic operation showed an overall cycle-average increase even though the hydrogen production is reduced. The use of pressurized steam during the water-splitting step shows a slight improvement in the yield and rate (Fig. 3, circle region).

In two-step thermochemical solar fuel production systems, several material properties are critical to the efficiency and long-term stability of the process. Among the most important are the enthalpy and entropy of oxygen vacancy formation, denoted as $d\overline{h}^{\circ}_{MO}$ and $d\overline{s}^{\circ}_{MO}$, respectively, as well as the degree of non-stoichiometry, δ, of the metal oxide. These properties directly impact the material's ability to undergo efficient redox cycles, which is central to the production of solar fuels.

However, Ceria also has limitations, particularly its relatively high reduction temperature, which can impact efficiency. To address this, Ceria is often doped with elements such as zirconium (Zr) to improve its performance [2]. Despite these advancements, newer alternatives, such as perovskites, are emerging as more promising materials that offer potentially better efficiency and lower operational temperatures.

Perovskites are a class of materials that consider the chemical structure ABO_3, where A and B are cations. The A-site cation is typically larger, while the B-site cation is smaller. This unique crystal structure imparts perovskites with excellent thermal, chemical, and mechanical properties, making them attractive candidates for energy conversion and storage applications, including solar thermochemical processes [14]. These materials can be tuned to obtain certain properties that could improve the performance of the STCH process. Several material properties are critical to the efficiency and long-term stability of the process. Among the most important are the enthalpy and entropy of oxygen vacancy formation, denoted as $d\overline{h}^{\circ}_{MO}$ and $d\overline{s}^{\circ}_{MO}$, respectively, as well as the degree of non-stoichiometry, δ, of the metal oxide. These properties directly impact the material's ability to undergo efficient redox cycles, which is central to the production of solar fuels.

Examples of material properties for nonstoichiometric Ceria ($CeO_{2-\delta}$), doped alternatives, and perovskites considered in the literature are presented in Table 1, where it is apparent how slight changes in the degree of non-stoichiometry (δ) can significantly influence the material properties and equilibrium temperatures, and how doping Ceria or considering perovskites could lower reduction temperatures.

In many cases, external limitations predetermine the material under consideration and the desired operating conditions. In these instances, it is important to iteratively optimize the material based on the degree of non-stoichiometry to achieve the appropriate enthalpy and entropy, thereby ensuring that the reduction and oxidation temperatures are suitable.

4 Thermochemical Reactors for Fuel Production

Solar reactors are critical components to fulfill the solar-fuel conversion. The reactor is a cavity geometry to harness concentrated solar energy to drive high-temperature chemical reactions. These reactors facilitate the splitting of water or carbon dioxide

Table 1 Thermodynamic properties for materials found in the literature at a specific degree of non-stoichiometry with the equilibrium temperatures determined at a pressure ratio of 10^(-5) and feedstock conversion extent of 0.1, adapted from [2]

Material	$d\bar{h}^{\circ}_{MO}$[kJ/mol]	$d\bar{s}^{\circ}_{MO}$[J/(mol K)]	$\delta[-]$	T_{red} [K]	T_{ox} [K]
$CeO_{1.9186}$	410	176	0.0814	1833	1590
$CeO_{1.9427}$	420	191	0.0573	1762	1475
$(Ce_{0.95}Zr_{0.05})O_{1.9583}$	397	188	0.0417	1683	1304
$(Ce_{0.80}Zr_{0.20})O_{1.9116}$	357	158	0.0884	1734	1294
$(Ce_{0.90}Nb_{0.10})O_{1.9852}$	414	212	0.0148	1593	1199
$(Ce_{0.90}Y_{0.10})O_{1.9899}$	444	230	0.0101	1598	1254
$(Ce_{0.90}La_{0.10})O_{1.9859}$	453	235	0.0141	1602	1271
$(La_{0.90}Sr_{0.10})MnO_{2.9502}$	327	134	0.0498	1798	1311
$(La_{0.60}Sr_{0.40})MnO_{2.93}$	270	128	0.07	1603	656
$(La_{0.50}Sr_{0.50})MnO_{2.93}$	242	110	0.07	1533	303
$(La_{0.6}Sr_{0.4})(Mn_{0.6}Al_{0.4})O_{2.9511}$	282	150	0.0489	1425	1000

into hydrogen or carbon monoxide, respectively, which can then be used to produce synthetic fuels. There are many ways to categorize the reactors. For example, depending on the interaction between the reactive material and the solar heating, the reactor can be classified into two types: directly heating reactors and indirectly heating reactors. In direct heating reactors, the reactive material or the reaction chamber directly absorbs concentrated solar radiation. In contrast, in indirect heating reactors, solar energy heats an intermediate medium, e.g., gas, or molten salt, which transfers heat to the reactive material. The reactors can also be categorized according to thermochemical cycles (volatile and non-volatile cycles), and the form and state of reactive materials (particles, foams, and honeycombs or moving bed, fluidized bed, packed bed, etc.). Table 2 shows the classification of various solar reactors for water/carbon dioxide splitting.

As discussed by Agrafiotis et al. [38], there are some problems or downsides when using indirectly irradiated reactors because solar energy is transferred to materials through tubes and heat transfer fluids. This limits the real heat flux to the reactive material, resulting in low efficiency. In addition, indirectly irradiated reactors require heat exchangers or other mechanisms to transfer heat, increasing system complexity and cost. Thus, directly irradiated reactors are mostly used for thermochemical fuel production, using particles or structures directly exposed to concentrated solar radiation, while only very few indirectly irradiated reactors have been proposed.

When designing a solar reactor, many aspects must be considered. For example, different thermochemical processes lead to different reactor designs, and directly/indirectly irradiated reactors also have different designs [39]. However, the critical process that affects reactor performance is heat and mass transfer, and thus, its role is the most critical consideration in designing reactors [40]. As aforementioned,

Table 2 The reactors of two-step thermochemical processes [38]

Reactor concept	Thermochemical cycles							
	Volatile cycles				Non-volatile cycles			
	Non-structured reactors		Structured reactors		Non-structured reactors		Structured reactors	
	Non-moving particles	Moving particles	Non-moving structures	Moving structures	Non-moving particles	Moving particles	Non-moving structures	Moving structures
Directly irradiated		Rotating cavities		Moving fronts	Packed beds	Spouted beds	Honeycombs	Rotary cylinders
		Entrained beds				Moving beds	Foams	Rotating fins
Indirectly irradiated		Aerosols	Sintered plates		Packed beds			

reactors can be classified based on different criteria. To make the storyline consistent and clear, the reactor will be introduced based on the type of thermochemical cycle (material), i.e., volatile cycle and non-volatile cycle of metal oxides.

In the two-step thermochemical process, the first step is a thermal reduction process, and the second step is an H_2O/CO_2 splitting process, which is favored by various conditions [41]. The thermal reduction process is thermodynamically favored by a high temperature (e.g., > 2000 K) and a low oxygen partial pressure, while the splitting process requires a relatively low temperature (e.g., 1000–1300 K) and a high H_2O/CO_2 partial pressure. As a result, some materials might experience solid-to-gas phase changes in the thermal reduction process, while some materials could keep the condensed state. The former materials-based cycle is the volatile cycle, and the latter one is the non-volatile cycle.

4.1 *Volatile-Cycle-Based Reactors*

The volatile cycles include the ZnO/Zn cycle, CdO/Cd cycle, SnO_2/SnO cycle, and GeO_2/GeO cycle. The two-step volatile cycle is explained with ZnO/Zn as an example. The first step is the solar-driven reduction of ZnO: ZnO(s) à Zn(g) + $0.5O_2$, (T > 2000 K). The second step is H_2O/CO_2 splitting process: Zn + H_2O/CO_2àZnO(s) + H_2/CO, (T < 1300 K). It is seen that the reduced gaseous metals or low-valence metal oxides and oxygen occur together at the first step. They have the risk of being re-oxidized if not well managed. Therefore, the second step cannot directly proceed after the first step. Normally, the volatile-cycle system includes two reactors. One is for thermal reduction, and the gaseous metal or low-valence metal oxide condenses for the next step. the other reactor is for splitting, which is not necessarily solar-driven. Thus, solar-driven reactors are designed only for the thermal reduction process in volatile cycles.

4.1.1 The Rotating Kiln (Cavity) Reactor

The first-generation rotating kiln reactor was proposed and designed by Haueter et al. [42], called ROCA in Fig. 4. Solar directly irradiates the reactant ZnO powders through a quartz window, an aperture, and a compound parabolic concentrator. ZnO powders are continuously fed by a feeder. After the reduction, the gaseous Zn(g) and oxygen are purged and cooled t by inert gas from the purge inlet. The gaseous Zn is then isolated and condensed for subsequent reactions. Following this, PSI developed an improved rotating kiln reactor, ZIRRUS, which utilized an inner cavity wall as insulation [43]. In 2011, ZIRRUS was scaled up from 10 to 100 kW [44]. After improving thermal insulation design and addressing the protection of the quartz window and occasional clogging of the reactor outlet, the 100 kW ZIRRUS was successfully demonstrated in 2014 [45].

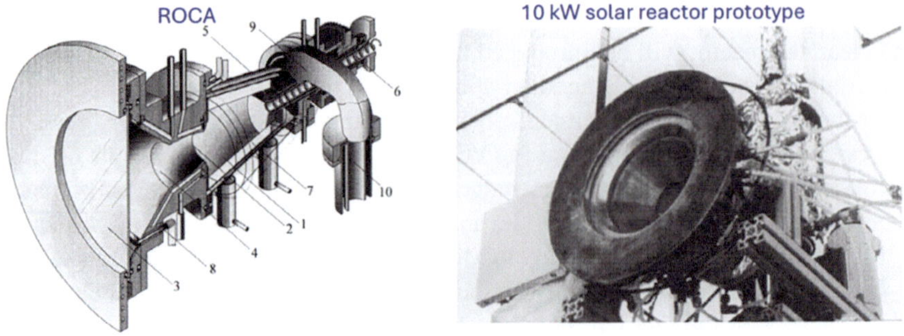

Fig. 4 ROCA schematics and 10 kW solar reactor prototype 1-rotating cavity-receiver, 2-aperture, 3-quartz window, 4-compound parabolic concentrator, 5-shell, 6-reactant feeder, 7-ZnO layer, 8-purge gas inlet, 9-product outlet port, 10-quench device. Images reprint from [42] with permission from Elsevier

4.1.2 Gravity-Fed Entrained Bed Reactor (GRAFSTR)

The rotating kiln reactor could achieve high and uniform temperatures in the cavity, and the reactant could reside in the cavity to sustain enough reaction time. However, the rotating configuration requires external power and might need efforts to maintain its reliability, which brings out engineering challenges. In contrast, Koepf et al. [46]developed a stationary cavity reactor with gravity-driven reactive flows. This reactor is supposed to have high chamber temperatures and uses powders as solar absorbers and reactants.

Figure 5 shows the gravity-fed entrained bed reactor, which was designed and investigated by the ETH/PSI group in collaboration with the University of Delaware. Solar energy is concentrated and directed to the cavity through a window and an aperture. In the stationary cavity, ZnO powders are fed through a series of 15 vibrating hoppers around the reactor's top perimeter. The powders fall into the cavity along tilted inner surfaces (Al_2O_4 tiles) of the cavity, generating a thin sheet under gravity to absorb the concentrated solar and be thermally reduced. After the reduction, the gaseous production (Zn and Oxygen) is drained from a tubular exit. Numerical modeling was implemented to optimize and refine the design. Experiments were carried out with the solar simulator of PSI, only reaching a temperature of 1173 K to check the adhesion and pre-sintering of powders to the reaction surface.

4.1.3 The Moving Front Reactor

As discussed by Chambon et al. [47], rotary reactors have several downsides. For example, cavity rotation makes it difficult to measure local temperatures by thermocouples and to monitor generated oxygen due to the risk of air leakage at the sealing of the cavity. The EHT/PSI group [48]designed a reactor in which pre-sintered ZnO blocks replaced ZnO powders, as shown in Fig. 6a. The front surface of the ZnO

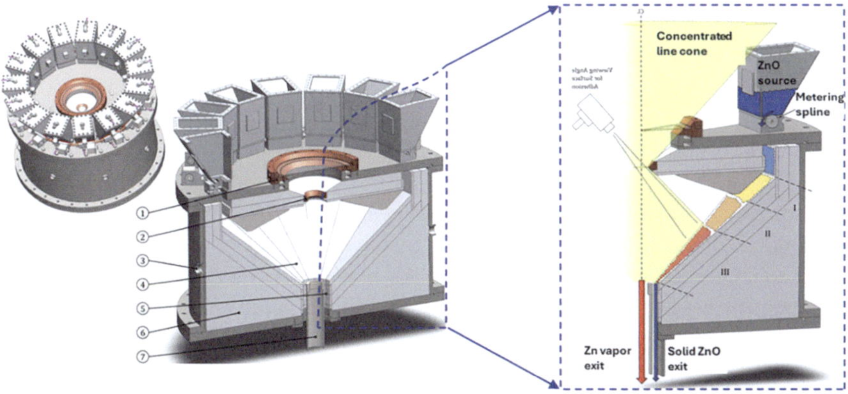

Fig. 5 The gravity-fed entrained bed reactor 1-window mount and vortex-flow generator, 2-cavity aperture, 3-data-acquisition port, 4-alumina-tile reaction surface, 5-solid ZnO exit, 6-insulation and support, 7-vapor product and gas exit. Images reprint from [46] with permission from Elsevier

block is subjected to concentrated solar and is thermally reduced, while the ZnO block moves forward gradually to keep its front surface always aligned with the focal point of the concentrated solar. They also developed a one-dimensional model to simulate this process. In addition, making the window clear is essentially important for solar irritation to the block. Inert gas flow paths were carefully designed to prevent Zn vapor from condensing on the window and keep the wind clear.

Chambon et al. [47] reported a moving-front solar thermochemical reactor for thermal dissociation of ZnO and SnO_2 at about 1900 K, as shown in Fig. 6b. The reactive material was prepared as pellets. The pellets were put into a vertical ceramic cavity and could be pushed upward by a rotary piston. The front surface of the pile of pellets is always at the focal point of the concentrated solar energy. Through this

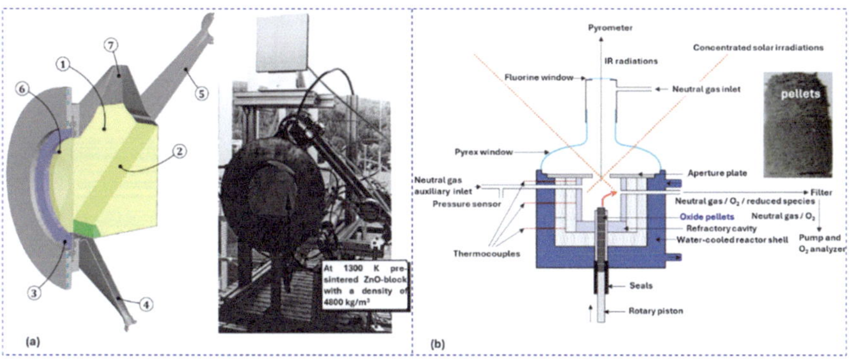

Fig. 6 Moving front solar reactors: (a) ETH/PSI reactor, image reprint from [48] with permission from Elsevier. (b) CNRS-PROMES reactor, image reprint from [47] with permission from AIChE

reactor, temperature and oxygen concentration could be collected for ZnO and SnO_2 reduction to reveal the thermochemical performances and the reduction kinetics.

4.1.4 Indirectly-Irradiated Aerosol and Structured Reactors

Aerosol reactors use very fine particles suspended in a gas flow (like fluidized) as the solid reactant. In such a reactor, the reactive material is not directly exposed to the concentrated solar energy. Instead, the material is heated by an intermediate which receives solar irradiation. Perkins et al. [49] from the University of Colorado designed and tested an aerosol reactor in the lab for ZnO reduction at temperatures between 1873 and 2023 K. Figure 7a shows the schematics of the reactor. A graphite wall is electrically heated, and a muffle tube is placed between the hot graphite wall and the aerosol flow. Then, the particles are heated up rapidly upon being exposed to the hot muffle tube walls. The electric furnace mimics concentrated solar energy, and the particles are heated through the hot graphite wall and muffle tube, i.e., indirectly irradiated.

In the moving front reactors, structured pellets were used to replace powders. Similarly, structured reactors were also developed with indirectly-irradiated heating. Melchior et al. [49]designed an indirectly irradiated reactor for ZnO reduction by placing pre-sintering plates into a circular absorber, as shown in Fig. 7b. The absorber is irradiated by concentrated solar power, and then the reactive plates are heated. Batch-mode experiments were carried out, and it was found that the reduction did

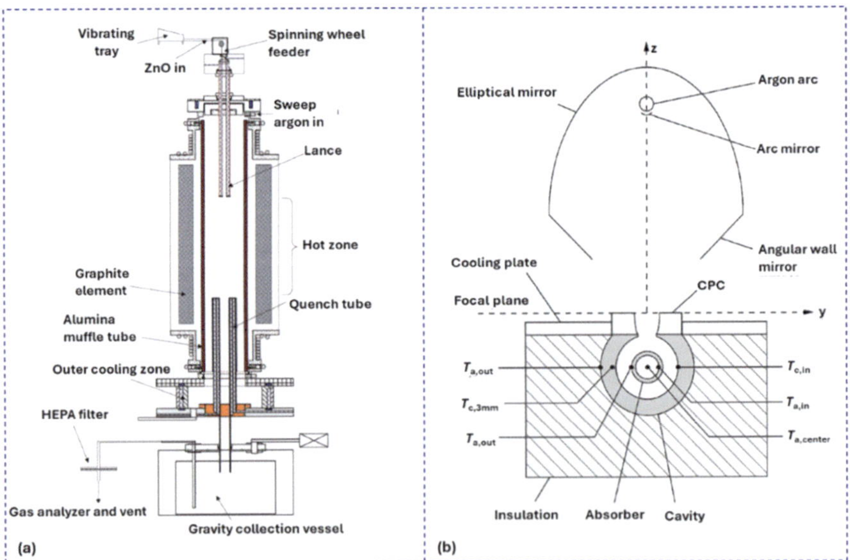

Fig. 7 Indirectly-irradiated reactors: **a** aerosol reactor [49] **b** structured reactor [50]. Images reprint from [49, 50] with permission from Elsevier

not appear at temperatures below 1750 K. Numerical models were also developed to simulate a continuous thermochemical process, with validation by comparing the measured and numerical temperatures and reaction rates in the 1780–1975 K range.

4.2 Non-volatile-Cycle-Based Reactors

The reduced metal oxide remains solid or liquid in non-volatile cycles without experiencing the solid-to-vapor phase change. Thus, the production is less prone to be re-oxidized. The reduction reaction in the non-volatile cycles could be either stoichiometric or nonstoichiometric, while current research interests lie in the nonstoichiometric oxides. The non-volatile materials were compared in detail by the Germany Aerospace Centre (DLR) [51], and Sandia National Laboratories (SNL) [52]. The non-volatile cycles include Fe_3O_4/FeO cycle, CeO_2/$CeO_{2-\delta}$ cycle, and perovskite cycle. There are two types of reactor concepts and designs to attain continuous or batch-wise fuel production. One idea is to move reactive materials from a thermal reduction reactor/zone to a splitting reactor/zone, achieving continuous production. The other idea is to make the thermal reduction and splitting in one reactor, avoiding any solid transportation. Through this design, batch-wise fuel production could be achieved by switching gas flow, and meanwhile, solar energy density needs to be adjusted to satisfy the temperature, and heat demands of the two processes (reduction and splitting).

4.2.1 Packed Bed Reactor

The first reactor driven by concentrated solar is a packed bed reactor with ferrite powders, as reported by the Tokyo Institute of Technology and Paul Scherrer Institute [53]. A quartz tube is located in the focus of a solar furnace which consists of a sun-tracking heliostat and a stationary paraboloidal dish. The quartz tube is packed with reactants of $Ni_{0.5}Mn_{0.5}Fe_2O_4$ powder mixed with Al_2O_3 grains. A secondary concentrator composed of a CPC and an involute is to provide uniform irradiation on the reactor tube. Figure 8a shows the schematics of the experimental system and the reactor. During the thermal reduction, Ar gas passed through the packed bed, while during the splitting, a mixture of Ar gas and H_2O steam was used to stimulate the water splitting.

The powders in the packed bed above is stationary. In contrast, a moving packed bed reactor was proposed by Ermanoski from SNL, Siegel from Bucknell University and Stechel from Arizona State University [54], in which a powder flow is presents. Figure 8b shows the schematics of the moving packed particle bed reactor which is composed of three main parts: a thermal reduced chamber, a recuperator and a fuel production chamber. CeO_2 powders in the reduction chamber are directly heated by concentrated solar power through a window-covered aperture. Then, the reduced powders move downward to the reaction chamber through the recuperator. In the

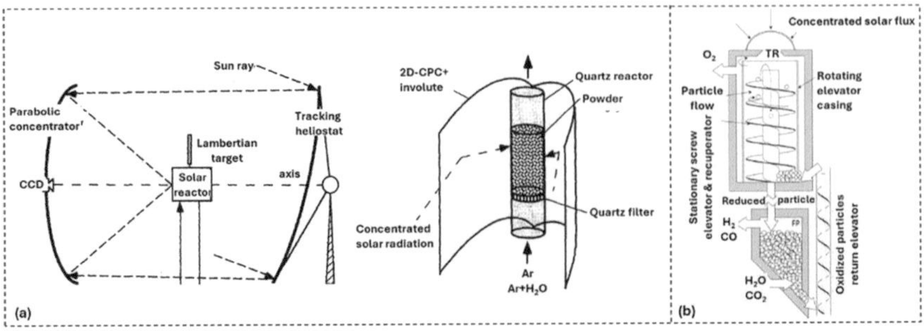

Fig. 8 Packed bed reactor: **a** stationary packed bed [53], **b** moving packed bed [54]. Images reprint from [53] and [54] with permission from Elsevier

reaction chamber, splitting takes place by exposing the reduced powders to H_2O steam or CO_2. After the splitting, the product is removed from the reaction chamber, and the reoxided powders return to the reduction chamber through a return elevator.

This combined system was developed by Sandia National Lab in collaboration with the Germany Aerospace Centre, Cologne, Germany. They successfully achieved a high temperature during the thermal reduction and the motion of particles at higher temperatures [55]. Currently, it is further developed by this consortium.

4.2.2 Fluidized Bed Reactor

An internal circulating fluidized bed reactor is developed by researchers from Niigata University [56]. Figure 9a demonstrates the conceptual diagram of a fluidized bed reactor combined with a solar concentrator. Inside the reactor is a draft tube, and then an internal circulating flow of particles and gas is generated in the reactor between the internal tube and the space between the tube and the reactor wall. Concentrated solar radiation directly heats the reactive particles that circulate in the bed. An experimental setup is designed to investigate two-step water splitting using $NiFe_2O_4$/m-ZrO_2 particles, as shown in Fig. 9b. A high-power solar simulator is used to heat the internally circulating reactive particles in the reactor. The thermal reduction and water splitting take place in a single reactor by switching the carried gas between nitrogen and steam. It is found that the thermal reduction occurs at a temperature range of 1773–1873 K inside the tube, and 1373–1523 K in the annulus region, while the water splitting occurs at 1373 K to produce hydrogen. Finally, a prototype of the internally circulating fluidized bed is established, as shown in Fig. 9c.

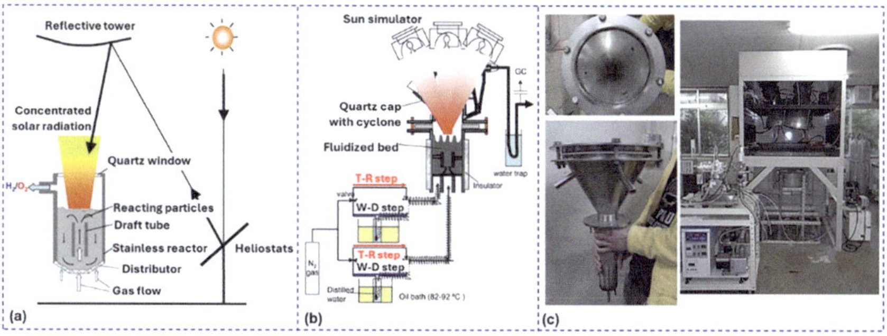

Fig. 9 Fluidized bed reactor: **a** conceptual diagram, **b** experimental setup of two-step water splitting, **c** prototype fluidized bed reactor. Images reprint from [56] with permission from Elsevier

4.2.3 Structured Honeycomb Reactor

The packed bed reactor uses reactive material in the form of powders. In addition to the powder-packed bed reactor, the honeycomb reactor is also proposed and investigated. In the honeycomb reactor, one idea is to deposit the reactive material on a ceramic multi-channel monolithic honeycomb. In comparison to the packed bed, the honeycomb could provide better heat transfer between gas and solid and reduce the flow resistance. Agrafiotis et al. [57] synthesized iron-oxide-based redox materials and coated ceramic honeycombs with the materials. They designed and constructed a honeycomb reactor for solar water splitting, as shown in Fig. 10a. The honeycomb reactor was able to achieve uniform temperatures in the order of 1573 K. However, a chamber reactor of honeycomb cannot achieve continuous production because the thermal reduction and water splitting have to take place sequentially. To address this issue, a dual-chamber reactor was developed by Roeb et al. from DLR [58]. Ferrite-coated monolithic honeycombs were prepared. One chamber is to split water while the other is for thermal reduction, switching the role for the next cycle, as shown in Fig. 10b. Later, a 100 kW-pilot scale plant for solar water splitting was achieved by Roeb et al. [59], as shown in Fig. 10c. The pilot-scale reactor was installed on a solar tower in the Plataforma Solar de Almeria Spain.

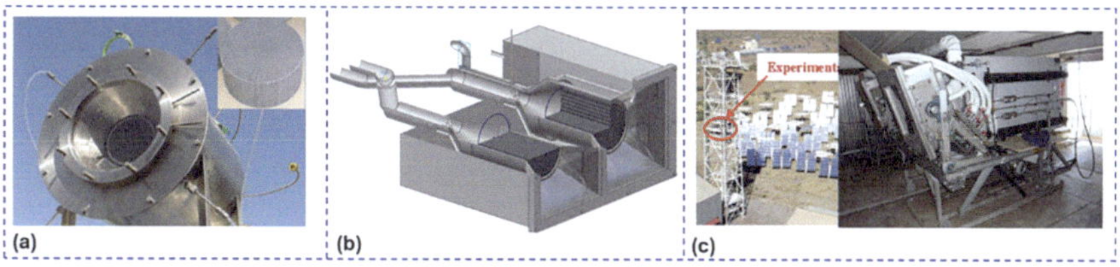

Fig. 10 Honeycomb reactor: **a** single-chamber reactor [57], **b** dual-chamber reactor [58], **c** 100 kW-pilot scale [59]. Images reprint from [57–59] with permission from Elsevier

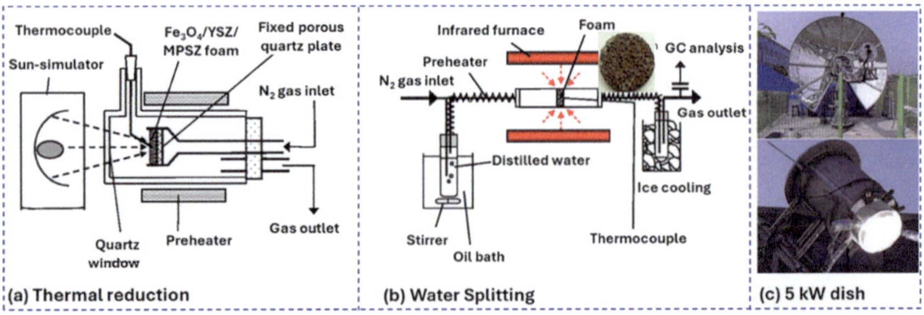

Fig. 11 Foam reactor: **a** the thermal reduction reactor rig [60], **b** the water-splitting reactor rig [60]. **c** 5 kW solar dish concentrator at Inha University [61]. Images reprint from [60, 61] with permission from Elsevier

4.2.4 Structured Foam Reactor

Another type of structure reactor is to use foam as reactive material supports. Gokon et al. [60] prepared Fe_3O_4/YSZ/MPSZ foam devices by coating the Magnesia-Partially-Stabilized Zirconia (MPSZ) foam with YSZ and Fe_3O_4 particles. Then, they developed a lab-scale experimental setup for thermal reduction and water splitting, respectively, as shown in Fig. 11a, b. This means that the two steps were performed in different test rigs. In the reduction step, the foam was heated by a sun-simulator which concentrated Xe-beams, reaching temperatures of 1673–1723 K, while an infrared furnace was used to heat the reduced foam to 1373 K for water splitting. Later, they developed ferrite/zirconia-coated foam reactors [61], and installed the reactor on a 5 kW solar dish concentrator, as shown in Fig. 11c.

Reticulated ceramic foams directly manufactured entirely using reactive materials, CeO_2, was achieved by the EHT/PSI group in collaboration with researchers from Caltech [62]. Figure 12 shows the prepared CeO_2 foam and the reactor. In the test, the thermal reduction was performed at 1800 k, and the splitting of H_2O and CO_2 mixture was performed at 1100 K, producing syngas with H_2/CO molar ratios from 0.25 to 2.34.

4.2.5 Rotary-Type Reactor

Kaneko et al. from Tokyo Institute of Technology developed a rotary-type reactor [63]. They coated cylindrical rotors with CeO_2 and Ni, Mn-ferrite, separately. The reactor consists of a rotor and two reaction cells, as shown in Fig. 13a. The rotor was rotated by an electric motor during the reaction. Simultaneous and continuous production of O_2 and H_2 can be achieved in the reduction cell and the water splitting cell. Regarding CeO_2, successive evolution of hydrogen gas was confirmed at 1623 K and 1273 K in the reduction cell and splitting cell, respectively. The reactor also successfully ran using the Ni, Mn-ferrite at 1473 K and 1173 K in the reduction cell and splitting cell, respectively. Another concept of rotatory reactor was proposed by

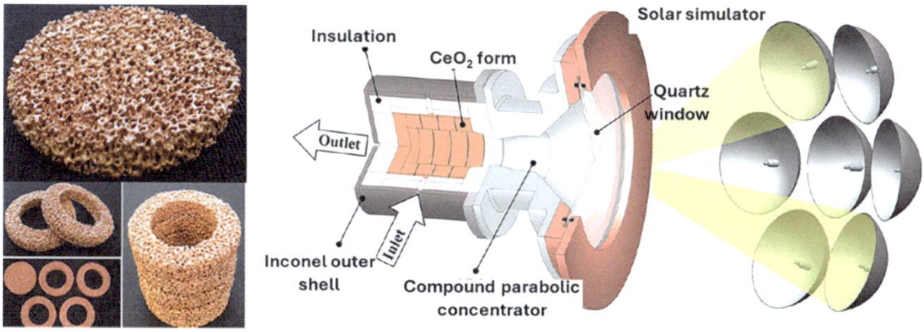

Fig. 12 Ceria-based monolithic foam reactor. Images reprint from [62] with permission from ACS Publications

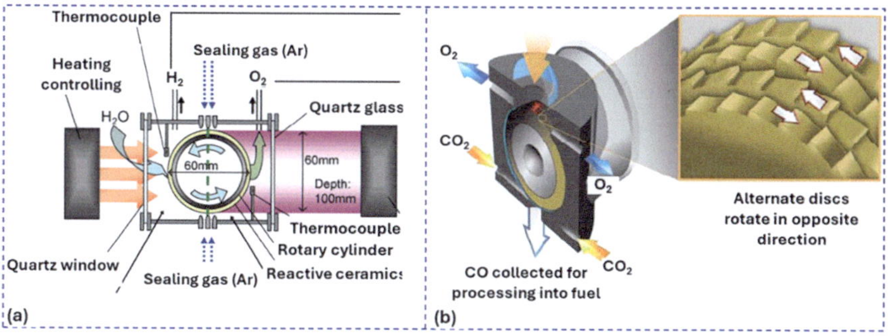

Fig. 13 Rotatory-type reactors: **a** solar reactor of Tokyo Institute of Technology [63], **b** CR5 reactor [64]. Images reprint from [63] with permission from ACS Publications, and permission to reprint images from [64] from the report authors

SNL, named counter-rotating-ring receiver-reactor-recuperator (CR5) [64], as shown in Fig. 13b. CR5 consists of rings and disks rotating reversely, and these rings and disks are equipped with reactive material-made fins around the circumference.

5 Technology Readiness Levels

The development of solar fuel technology holds significant potential to accelerate progress in decarbonization. However, comprehensive economic assessments and Technology Readiness Level (TRL) evaluations are crucial for facilitating its commercial application and broader adoption. These evaluations not only provide a robust foundation for informed decision-making but also help identify potential technical bottlenecks and economic risks, paving the way for the successful deployment of the technology.

5.1 Economic Assessment System

Various methods and systems are available for the economic assessment of thermochemical hydrogen production. The annuity method distributes the total cost of a project's life cycle into fixed annual payments through annualization, providing the average annual cost of the project [65]. This approach is widely applied in cost analysis for both projects and energy systems. The Levelized Cost of Hydrogen (LCOH) offers a comprehensive evaluation of the unit cost of hydrogen production by considering capital expenditure, operating costs, fuel costs, and total hydrogen output over the project's lifetime [66]. It is particularly effective for comparing the unit costs of different technologies and scenarios. The Hydrogen Economic Evaluation Program (HEEP), developed by the International Atomic Energy Agency (IAEA), serves as a specialized tool for the economic analysis of hydrogen production, enabling cost comparisons across various energy-driven hydrogen production technologies.

Based on these analytical frameworks, researchers have conducted some economic assessments of thermochemical hydrogen production systems. A summary of key findings is presented in Table 3.

To comprehensively evaluate the cost of hydrogen production, facilitate comparisons across different technologies, and ensure clarity and ease of understanding, the Levelized Cost of Hydrogen (LCOH) is chosen as the primary method for the economic assessment of two-step thermochemical cycle hydrogen production.

The LCOH is defined as the ratio of the total comprehensive cost of a hydrogen production system to the total hydrogen produced every year over its entire lifetime [68],

Table 3 Key findings of previous research

Year	Researchers	Heat sources	Cost calculation method	Thermochemical system	Cost
2002	A. Steinfeld [67]	Solar energy	-	Zn/ZnO Two-step cycle	5 $/kg
2008	D. Graf et al. [65]	Solar energy	Annualized Cost Method	Metal oxides Two-step cycle	6.7 €/kg
				Hybrid-sulfur cycle	5.4 €/kg
2019	M. Moser et al. [68]	Solar energy	LCOH	CeO2 Two-step cycle	6.7–13.1 €/kg
2020	R.S. El-Emam et al. [69]	Nuclear energy	HEEP	S-I Three-step cycle	2.5–5.7 $/kg
				Hybrid-sulfur cycle	2.3–6.3 $/kg
2023	S. Lee et al. [66]	Solar electricity and microwave heating	LCOH	CeO2 Two-step cycle	12.9 $/kg

$$\mathrm{LCOH} = \frac{\text{total annual cost}}{m_{\mathrm{Hydrogen}}}. \tag{5.1}$$

In calculating the total annual cost, all relevant expenses, including construction costs, capital costs, operation and maintenance costs, and other associated expenditures, are considered. Large-scale initial construction costs are assumed to be fully financed through loans and are converted into annual instalments, according to [68],

$$\mathrm{Inst}_{\mathrm{annual}} = \mathrm{Cost}_{\mathrm{const}} \frac{i(1+i)^n}{(1+i)^n - 1}. \tag{5.2}$$

The basic assumption in economic assessment and the fundamental parameters of the system are given in Table 4 [68].

Table 4 Basic assumptions and fundamental parameters in the economic model [68]

Parameter	Unit	Value
Construction time	y	2
Debt period	y	20
Plant lifetime	y	30
Interest rate	%/y	0.08
Discount rate	%/y	0.05
Absorber life	y	1
Absorber specific cost	€/kg	10
Water use for mirror cleaning	m^3/m^2	0.02
Water cost	€/m^3	3
N2 pumping consumption	kWh/mol	0.001356
Pressure ratio	–	5
Adiabatic compressor efficiency	–	0.7
Cost of electricity	€/kWh	0.2
O&M cost	%inv/y	2
Insurance factor	%inv/y	0.5
Solar field specific investment	€/m^2	150
Thermal storage specific investment	€/kWh_{th}	30
Hydrogen storage specific investment	€/kWh_{th}	30
Hydrogen storage efficiency	%	90
Reactor shell cost	€/piece	5000
Specific cost of insulation	€/m	10,000
Specific cost of absorber	€/kg	10
Absorber mass	kg	150–250
Reactor total mass	kg	700
Reactor efficiency	%	13.4

5.2 *Economic Assessment Result*

Taking the given assumption and parameters as the basic case, the resulting LCOH is 13.06 €/kg, and the total investment for the ceria solar thermochemical water splitting system is 32.5 Mio. €, which accounts for more than 60% of the LCOH [68]. The cost shares from different parts in LCOH for the basic case are given in Fig. 14. The construction cost and the loan interest occupy the largest proportion which numbers 60%. This means that the most efficient methods to reduce the LCOH are reducing construction costs, reducing loan interest rates, and improving production efficiency to spread construction costs. At the same time, continuously reducing the unit cost of reaction material and extending its life with technological advancements are the main ways to lower the reaction material cost which constitutes 21% of the LCOH.

Based on the economic assessment model, the sensitivity analysis of LCOH is carried out and the results are illustrated in Fig. 15. Similar to the previous analysis, the increase in hydrogen production (derived from efficiency increase) has the most significant effect on reducing LCOH. When hydrogen production doubles, the LCOH goes to 6.53 €/kg. And some researchers pointed out that the established cost of CSP will decrease by 46% till 2030 [70], in this situation, the LCOH will be reduced by 39.7%. At the same time, when the life of reaction material is improved from 1 to 10 years, the LCOH is reduced by 18.5%. However, the reduction effect of it is not significant after the life is 5 years, at which time the LCOH is reduced by 16.4%. Meanwhile, when the interest rate and material unit cost are reduced by 50%, the LCOH decreases by 17.1% and 10.3% respectively. Additionally, in the best case,

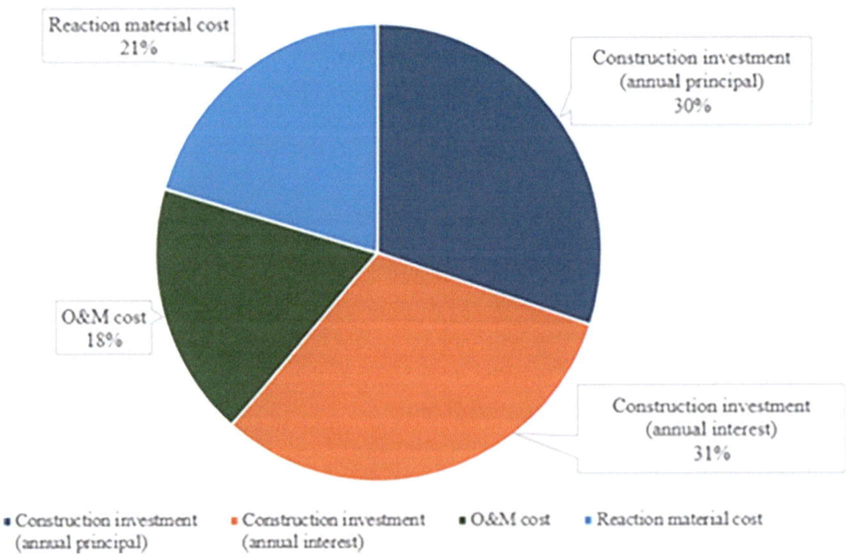

Fig. 14 The cost shares from the different parts in the LCOH for the basic case

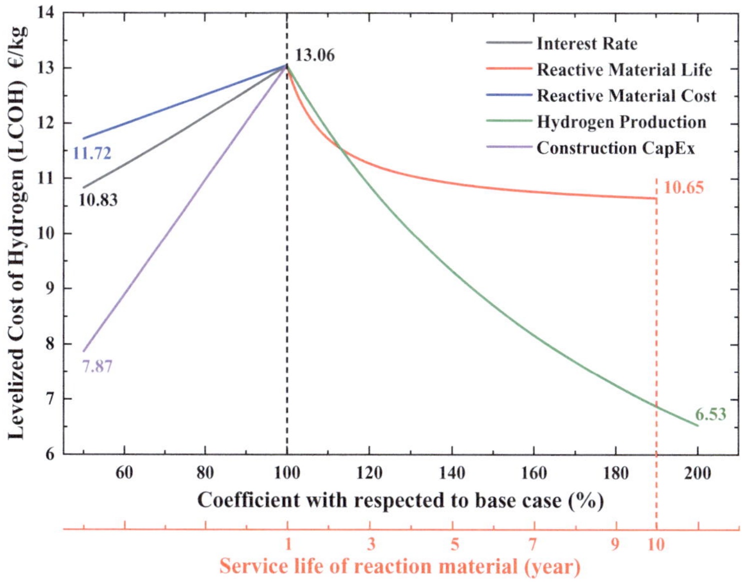

Fig. 15 Sensitivity analysis of LCOH

in which all the parameters including construction costs, interest rate, material life, material unit cost, and production efficiency are assumed to be the best, the LCOH will reach 2.11 €/kg. In contrast, the US Department of Energy (DOE) targets the centralized hydrogen cost from electrolysis to be 2 $/kg by 2020 [69].

5.3 *Technology Readiness Level (TRL)*

Technology Readiness Level (TRL) is a standardized method proposed by NASA for evaluating the development stage of a technology. There has been some research that summarized the TRLs of various hydrogen production technologies from different perspectives.

Shown in Table 5, Alberto Boretti compared the TRL of direct and indirect solar fuel production method [71]. The TRL of the direct solar fuel production method using ceria is still poor, only in the phase of TRL 2–4. The best demonstration of this technology which is established by ETH Zurich still cannot provide sufficient hydrogen production under long operating time. While the indirect method (using the S-I cycle) has a better TRL, which is 5–7. There have been many demonstrations of plants from different organisms like the Japan Atomic Energy Agency (JAEA), the French Alternative Energies, Atomic Energy Commission (CEA), General Atomics, and Sandia National Laboratories.

Table 5 The TRL of direct and indirect solar fuel production method [71]

Method	Reaction System	TRL	Example
Direct	CeO_2 Tow-step cycle	2–4	Ceria solar fuel production demonstration plant at ETH Zurich
Indirect	S-I Three-step cycle	5–7	Many demonstration plants have been built and successfully operated

Table 6 TRL comparison of different hydrogen production technologies

Hydrogen color	Process	TRL
Gray	Steam reforming	9 [72]
Blue	Steam reforming + CCS	8–9 [72]
Green	Electrolysis	8–9 [73]
Green	Solar thermochemical water splitting (two-step cycle)	2–4 [71]
Aquamarine	Solar thermal methane pyrolysis	3–4 [70]

Alberto Boretti also compared the TRL of different hydrogen production technologies from a broader perspective, which is shown in Table 6.

6 Key Challenges and Future Outlook

The STCH process can potentially provide a scalable water-splitting solution compared to the modular-based solution of the electrolyser technology. It could have a much higher hydrogen generation efficiency than electrolysis because of the utilization of the whole solar spectrum, which is impossible with electrolysers. In the last few decades, several redox pairs suitable for the STCH process have been investigated, worldwide. Several bench-scale and pilot-scale reactors were demonstrated. All these efforts enhanced the research community's understanding of these cycles, and the learning could be utilized in other high-temperature thermochemical processes. Despite all these efforts, the STCH process is not yet commercialized due to three key challenges that require further efforts before it can be commercially exploited. In the rest of this section, these challenges are described.

Redox pair: The development of a cheap, and cycle-stable redox pair that can be reduced at a lower temperature with a faster reaction rate is a big challenge for this technology. Currently, the research community has a consensus that the non-stoichiometric cycles have the most potential, and the majority of the research efforts are in this direction. The non-stoichiometric ceria is a good candidate because it has good chemical kinetics and has shown cycle stability, but its thermal reduction temperature is much higher (T_{red} > 1500 °C) which is needed to attain a higher reduction extent due to its much higher reduction enthalpy [8]. For any practical viability of this material, its reduction temperature must be decreased which can potentially be achieved by doping the material. Besides ceria, another new class of materials perovskites ($ABO_{3-\delta}$) are gaining traction nowadays [60]. These materials can consist of several different elements, they are amenable to doping, require moderate reduction temperature, and show good stability at a much larger extent of reduction. This new class of material could provide a future solution for STCH reactive material.

Solid heat recuperation: Another major challenge is the efficient heat recuperation from the solid reactants at high temperatures. In the non-isothermal cycles, reduction and oxidation reactions occur at different temperatures. If the sensible heat due to this temperature difference can be extracted and utilized within the cycle completely then we need to provide heat to perform the reduction reaction only. It will drastically improve the solar-to-fuel efficiency of the process. Several reactor designs have been proposed to tackle this issue without much practical success. For the most optimal heat recuperation, moving particle/bed configuration is conducive but high operational temperatures hinder any successful implementation of such a concept. Another possibility is to use an inert gas for heat extraction from a stationary bed. Although it is theoretically less efficient but a practical solution. In such a configuration, the inert gas would be switched between the reactors for the heat extraction/addition processes. It requires the development of high-temperature valves that can handle the expected temperatures.

Vacuum pump work: Since the thermal reduction of metal oxide occurs at a high temperature and low pressures (~ 10^{-3} bar in case of non-stoichiometric ceria), the work penalty associated with the generation of such a low pressure is significant and must be reduced for increasing the overall efficiency of the process. This pressure reduction could be achieved either by purging the reactor with an inert gas or using a vacuum pump. In case of an inert gas purge, a gas separation is required to re-use the inert gas. Due to lower pressures, the amount of inert gas needed is extremely large, therefore, the energy penalty associated with this method is also significant [76]. Alternatively, the vacuum pump could be a viable solution provided they have higher electrical-to-pump work efficiency. The current commercially available vacuum pumps are less efficient and there is enough scope for their efficiency improvement. Another possibility is the use of a thermochemical pump which also works on similar redox reaction-based cycles for the generation of low-pressures [77]. In these systems, the available waste heat is the primary driving force.

The successful solutions to these issues could pave the path for the commercial implementation of STCH technology which in turn could support the generation of green hydrogen using solar energy.

References

1. https://eur-lex.europa.eu/legal-content/

2. A. Bayon, A. De La Calle, E.B. Stechel, C. Muhich, Operational limits of redox metal oxides performing thermochemical water splitting. Energy Tech **10**(1), 2100222 (2022). https://doi.org/10.1002/ente.202100222
3. T.-Z. Ang, M. Salem, M. Kamarol, H.S. Das, M.A. Nazari, N. Prabaharan, A comprehensive study of renewable energy sources: classifications, challenges and suggestions. Energ. Strat. Rev. **43**, 100939 (2022). https://doi.org/10.1016/j.esr.2022.100939
4. N.S. Lewis, Research opportunities to advance solar energy utilization. Science **351**(6271), aad1920 (2016). https://doi.org/10.1126/science.aad1920
5. S. Haussener, Solar fuel processing: comparative mini-review on research, technology development, and scaling. Sol. Energy **246**, 294–300 (2022). https://doi.org/10.1016/j.solener.2022.09.019
6. S. Fukuzumi, Production of liquid solar fuels and their use in fuel cells. Joule **1**(4), 689–738 (2017). https://doi.org/10.1016/j.joule.2017.07.007
7. S. Licht, 'STEP: efficient carbon capture and solar thermal electrochemical production of ammonia, fuels, cement, carbon nanotubes, metals and bleach (2019). arXiv. https://doi.org/10.48550/ARXIV.1901.07134
8. J.T. Tran, K.J. Warren, S.A. Wilson, C.L. Muhich, C.B. Musgrave, A.W. Weimer, An updated review and perspective on efficient hydrogen generation via solar thermal water splitting. WIREs Energy Environ. **13**(4), e528 (2024). https://doi.org/10.1002/wene.528
9. A. Lidor, B. Bulfin, A critical perspective and analysis of two-step thermochemical fuel production cycles. Solar Compass **11**, 100077 (2024). https://doi.org/10.1016/j.solcom.2024.100077
10. S. Abanades, P. Charvin, G. Flamant, P. Neveu, Screening of water-splitting thermochemical cycles potentially attractive for hydrogen production by concentrated solar energy. Energy **31**(14), 2805–2822 (2006). https://doi.org/10.1016/j.energy.2005.11.002
11. Y. Guo et al., A review of solar thermochemical cycles for fuel production. Appl. Energy **357**, 122499 (2024). https://doi.org/10.1016/j.apenergy.2023.122499
12. T. Liu, Q. Liu, J. Lei, J. Sui, H. Jin, Solar-clean fuel distributed energy system with solar thermochemistry and chemical recuperation. Appl. Energy **225**, 380–391 (2018). https://doi.org/10.1016/j.apenergy.2018.04.133
13. S. Li et al., Thermodynamic guiding principles for designing nonstoichiometric redox materials for solar thermochemical fuel production: ceria Perovskites, and Beyond. Energy Tech **10**(1), 2000925 (2022). https://doi.org/10.1002/ente.202000925
14. T. Liu, H. Zheng, H. Wang, J. Wang, Z. Wang, H. Kong, A review of methane-driven two-step thermochemical cycle hydrogen production. J. Clean. Prod. **466**, 142668 (2024). https://doi.org/10.1016/j.jclepro.2024.142668
15. Z. Ma, P. Davenport, G. Saur, System and technoeconomic analysis of solar thermochemical hydrogen production. Renew. Energy **190**, 294–308 (2022). https://doi.org/10.1016/j.renene.2022.03.108
16. I. Ermanoski, J.E. Miller, M.D. Allendorf, Efficiency maximization in solar-thermochemical fuel production: challenging the concept of isothermal water splitting. Phys. Chem. Chem. Phys. **16**(18), 8418–8427 (2014). https://doi.org/10.1039/C4CP00978A
17. B.D. Ehrhart, C.L. Muhich, I. Al-Shankiti, A.W. Weimer, System efficiency for two-step metal oxide solar thermochemical hydrogen production—Part 2: impact of gas heat recuperation and separation temperatures. Int. J. Hydrogen Energy **41**(44), 19894–19903 (2016). https://doi.org/10.1016/j.ijhydene.2016.07.110
18. A. De La Calle, A. Bayon, Annual performance of a thermochemical solar syngas production plant based on non-stoichiometric CeO_2. Int. J. Hydrogen Energy **44**(3), 1409–1424 (2019). https://doi.org/10.1016/j.ijhydene.2018.11.076
19. D.V. Schroeder, *An Introduction to Thermal Physics*, 1st edn (Oxford University Press, 2021). https://doi.org/10.1093/oso/9780192895547.001.0001
20. M.J. Moran, H.N. Shapiro, D.D. Boettner, M.B. Bailey, *Fundamentals of Engineering Thermodynamics*, 8th edn. (Wiley, Hoboken, NJ, 2014)

21. A.J. Carrillo, J. González-Aguilar, M. Romero, J.M. Coronado, Solar energy on demand: a review on high temperature thermochemical heat storage systems and materials. Chem. Rev. **119**(7), 4777–4816 (2019). https://doi.org/10.1021/acs.chemrev.8b00315
22. A. Vidal, D. Martinez, An ageing protocol for testing high temperature solar materials for thermochemical applications. Sol. Energy Mater. Sol. Cells **212**, 110572 (2020). https://doi.org/10.1016/j.solmat.2020.110572
23. A. Haeussler, S. Chuayboon, and S. Abanades, Solar redox cycling of ceria in a monolithic reactor for two-step H_2O/CO_2 splitting: isothermal methane-induced reduction versus temperature-swing cycle. *Presented at the SOLARPACES 2019: International Conference on Concentrating Solar Power and Chemical Energy Systems, Daegu, South Korea* (2020), p. 170009. https://doi.org/10.1063/5.0028582
24. S. Abanades, A review of oxygen carrier materials and related thermochemical redox processes for concentrating solar thermal applications. Materials **16**(9), 3582 (2023). https://doi.org/10.3390/ma16093582
25. A. Steinfeld, Solar-processed metals as clean energy carriers and water-splitters. Int. J. Hydrogen Energy **23**(9), 767–774 (1998). https://doi.org/10.1016/S0360-3199(97)00135-3
26. T. Rager, J. Golczewski, Solar-thermal zinc oxide reduction assisted by a second redox pair. Z. Phys. Chem. **219**(2), 235–246 (2005). https://doi.org/10.1524/zpch.219.2.235.57308
27. R. Palumbo et al., The production of Zn from ZnO in a high-temperature solar decomposition quench process—I. The scientific framework for the process. Chem. Eng. Sci. **53**(14), 2503–2517 (1998). https://doi.org/10.1016/S0009-2509(98)00063-3
28. E. Bilgen, C. Bilgen, Solar hydrogen production using two-step thermochemical cycles. Int. J. Hydrogen Energy **7**(8), 637–644 (1982). https://doi.org/10.1016/0360-3199(82)90188-4
29. P. Charvin, S. Abanades, F. Lemont, G. Flamant, Experimental study of SnO_2 /SnO/Sn thermochemical systems for solar production of hydrogen. AIChE J. **54**(10), 2759–2767 (2008). https://doi.org/10.1002/aic.11584
30. S. Abanades, P. Charvin, F. Lemont, G. Flamant, Novel two-step SnO_2/SnO water-splitting cycle for solar thermochemical production of hydrogen. Int. J. Hydrogen Energy **33**(21), 6021–6030 (2008). https://doi.org/10.1016/j.ijhydene.2008.05.042
31. M. Chambon, S. Abanades, G. Flamant, Solar thermal reduction of ZnO and SnO_2: Characterization of the recombination reaction with O2. Chem. Eng. Sci. **65**(11), 3671–3680 (2010). https://doi.org/10.1016/j.ces.2010.03.005
32. Ave, H.: Support for Cost Analyses on (2011)
33. J. Nie, Y. Chen, B. Wong, L.C. Brown, 'Numerical modeling of vapor condensation during cadmium quenching process in a solar receiver. In: ASME 2009 3rd International Conference on Energy Sustainability, Volume 1, San Francisco, California, USA: ASMEDC, (2009), pp. 943–950. https://doi.org/10.1115/ES2009-90382
34. J. Ambriz, M. Ducarroir, F. Sibieude, Preparation of cadmium by thermal dissociation of cadmium oxide using solar energy. Int. J. Hydrogen Energy **7**(2), 143–153 (1982). https://doi.org/10.1016/0360-3199(82)90141-0
35. A. Tofighi, F. Sibieude, Dissociation of magnetite in a solar furnace for hydrogen production. Tentative production evaluation of a 1000 kW concentrator from small scale (2 kW) experimental results. Int. J. Hydrogen Energy **9**(4), 293–296 (1984). https://doi.org/10.1016/0360-3199(84)90079-X
36. F. Sibieude, M. Ducarroir, A. Tofighi, J. Ambriz, High temperature experiments with a solar furnace: the decomposition of Fe304, Mn304, CdO'.
37. A. Singh, F. Al-Raqom, J. Klausner, J. Petrasch, Production of hydrogen via an Iron/Iron oxide looping cycle: thermodynamic modeling and experimental validation. Int. J. Hydrogen Energy **37**(9), 7442–7450 (2012). https://doi.org/10.1016/j.ijhydene.2012.01.074
38. C. Agrafiotis, M. Roeb, C. Sattler, A review on solar thermal syngas production via redox pair-based water/carbon dioxide splitting thermochemical cycles. Renew. Sustain. Energy Rev. **42**, 254–285 (2015). https://doi.org/10.1016/j.rser.2014.09.039
39. E. Alonso, M. Romero, Review of experimental investigation on directly irradiated particles solar reactors. Renew. Sustain. Energy Rev. **41**, 53–67 (2015). https://doi.org/10.1016/j.rser.2014.08.027

40. E. Koepf, I. Alxneit, C. Wieckert, A. Meier, A review of high temperature solar driven reactor technology: 25years of experience in research and development at the Paul Scherrer Institute. Appl. Energy **188**, 620–651 (2017). https://doi.org/10.1016/j.apenergy.2016.11.088
41. M. Roeb, C. Sattler, Isothermal water splitting. Science **341**(6145), 470–471 (2013). https://doi.org/10.1126/science.1241311
42. P. Haueter, S. Moeller, R. Palumbo, A. Steinfeld, The production of zinc by thermal dissociation of zinc oxide—solar chemical reactor design. Sol. Energy **67**(1–3), 161–167 (1999). https://doi.org/10.1016/S0038-092X(00)00037-2
43. W. Villasmil, A. Meier, A. Steinfeld, Dynamic modeling of a solar reactor for zinc oxide thermal dissociation and experimental validation using IR thermography'.
44. W. Villasmil, M. Brkic, D. Wuillemin, A. Meier, A. Steinfeld, Pilot scale demonstration of a 100-kWth solar thermochemical plant for the thermal dissociation of ZnO. J. Sol.Energy Eng. **136**(1), 011016 (2014). https://doi.org/10.1115/1.4025512
45. E. Koepf, W. Villasmil, A. Meier, Pilot-scale solar reactor operation and characterization for fuel production via the Zn/ZnO thermochemical cycle. Appl. Energy **165**, 1004–1023 (2016). https://doi.org/10.1016/j.apenergy.2015.12.106
46. E. Koepf, S.G. Advani, A. Steinfeld, A.K. Prasad, A novel beam-down, gravity-fed, solar thermochemical receiver/reactor for direct solid particle decomposition: Design, modeling, and experimentation. Int. J. Hydrogen Energy **37**(22), 16871–16887 (2012). https://doi.org/10.1016/j.ijhydene.2012.08.086
47. M. Chambon, S. Abanades, G. Flamant, Thermal dissociation of compressed ZnO and SnO_2 powders in a moving-front solar thermochemical reactor. AIChE J. **57**(8), 2264–2273 (2011). https://doi.org/10.1002/aic.12432
48. S. Mo¨ller, R. Palumbo, The development of a solar chemical reactor for the direct thermal dissociation of zinc oxide. J. Solar Energy Eng. **123**(2), 83–90 (2001) https://doi.org/10.1115/1.1349717
49. C. Perkins, Thermal ZnO dissociation in a rapid aerosol reactor as part of a solar hydrogen production cycle. Int. J. Hydrogen Energy **33**(2), 499–510 (2008). https://doi.org/10.1016/j.ijhydene.2007.10.021
50. T. Melchior, C. Perkins, A.W. Weimer, A. Steinfeld, A cavity-receiver containing a tubular absorber for high-temperature thermochemical processing using concentrated solar energy. Int. J. Therm. Sci. **47**(11), 1496–1503 (2008). https://doi.org/10.1016/j.ijthermalsci.2007.12.003
51. M. Roeb et al., Materials-related aspects of thermochemical water and carbon dioxide splitting: a review. Materials **5**(11), 2015–2054 (2012). https://doi.org/10.3390/ma5112015
52. J.E. Miller, A.H. McDaniel, M.D. Allendorf, Considerations in the design of materials for solar-driven fuel production using metal-oxide thermochemical cycles. Adv. Energy Mater. **4**(2), 1300469 (2014). https://doi.org/10.1002/aenm.201300469
53. Y. Tamaura, Production of solar hydrogen by a novel, 2-STEP, water-splitting thermochemical cycle'
54. I. Ermanoski, N.P. Siegel, E.B. Stechel, A new reactor concept for efficient solar-thermochemical fuel production. J. Sol. Energy Eng. **135**(3), 031002 (2013). https://doi.org/10.1115/1.4023356
55. A. Singh et al., Design of a pilot scale directly irradiated, high temperature, and low pressure moving particle cavity chamber for metal oxide reduction. Sol. Energy **157**, 365–376 (2017). https://doi.org/10.1016/j.solener.2017.08.040
56. N. Gokon, T. Mataga, T. Kondo, T. Kodama, Thermochemical two-step water splitting by internally circulating fluidized bed of $NiFe_2O_4$ particles: successive reaction of thermal-reduction and water-decomposition steps. Int. J. Hydrogen Energy **36**(8), 4757–4767 (2011). https://doi.org/10.1016/j.ijhydene.2011.01.076
57. C. Agrafiotis et al., Solar water splitting for hydrogen production with monolithic reactors. Sol. Energy **79**(4), 409–421 (2005). https://doi.org/10.1016/j.solener.2005.02.026
58. M. Roeb et al., Operational strategy of a two-step thermochemical process for solar hydrogen production. Int. J. Hydrogen Energy **34**(10), 4537–4545 (2009). https://doi.org/10.1016/j.ijhydene.2008.08.049

59. M. Roeb et al., Test operation of a 100kW pilot plant for solar hydrogen production from water on a solar tower. Sol. Energy **85**(4), 634–644 (2011). https://doi.org/10.1016/j.solener.2010.04.014
60. N. Gokon, T. Hasegawa, S. Takahashi, T. Kodama, Thermochemical two-step water-splitting for hydrogen production using Fe-YSZ particles and a ceramic foam device. Energy **33**(9), 1407–1416 (2008). https://doi.org/10.1016/j.energy.2008.04.011sss
61. N. Gokon, T. Kodama, N. Imaizumi, J. Umeda, T. Seo, Ferrite/zirconia-coated foam device prepared by spin coating for solar demonstration of thermochemical water-splitting. Int. J. Hydrogen Energy **36**(3), 2014–2028 (2011). https://doi.org/10.1016/j.ijhydene.2010.11.034
62. P. Furler, J. Scheffe, M. Gorbar, L. Moes, U. Vogt, A. Steinfeld, Solar thermochemical CO_2 splitting utilizing a reticulated porous ceria redox system. Energy Fuels **26**(11), 7051–7059 (2012). https://doi.org/10.1021/ef3013757
63. H. Kaneko et al., Rotary-type solar reactor for solar hydrogen production with two-step water splitting process. Energy Fuels **21**(4), 2287–2293 (2007). https://doi.org/10.1021/ef060581z
64. Johnson, T., et al.: Reimagining liquid transportation fuels: sunshine to petrol.', SAND2012-0307, 1035344 (2012). https://doi.org/10.2172/1035344
65. D. Graf, N. Monnerie, M. Roeb, M. Schmitz, C. Sattler, Economic comparison of solar hydrogen generation by means of thermochemical cycles and electrolysis. Int. J. Hydrogen Energy **33**(17), 4511–4519 (2008). https://doi.org/10.1016/j.ijhydene.2008.05.086
66. S. Lee, U.J. Na, H. Jo, Techno-economic assessment of green hydrogen production via two-step thermochemical water splitting using microwave. Int. J. Hydrogen Energy **48**(29), 10706–10723 (2023). https://doi.org/10.1016/j.ijhydene.2022.12.119
67. A. Steinfeld, Solar hydrogen production via a two-step water-splitting thermochemical cycle based on Zn/ZnO redox reactions. Int. J. Hydrogen Energy **27**(6), 611–619 (2002). https://doi.org/10.1016/S0360-3199(01)00177-X
68. M. Moser, M. Pecchi, T. Fend, Techno-economic assessment of solar hydrogen production by means of thermo-chemical cycles. Energies **12**(3), 352 (2019). https://doi.org/10.3390/en12030352
69. R.S. El-Emam, H. Ozcan, C. Zamfirescu, Updates on promising thermochemical cycles for clean hydrogen production using nuclear energy. J. Clean. Prod. **262**, 121424 (2020). https://doi.org/10.1016/j.jclepro.2020.121424
70. A. Boretti, There are hydrogen production pathways with better than green hydrogen economic and environmental costs. Int. J. Hydrogen Energy **46**(46), 23988–23995 (2021). https://doi.org/10.1016/j.ijhydene.2021.04.182
71. Boretti, A.: Technology readiness level of solar thermochemical splitting cycles. ACS Energy Lett., 1170–1174 (2021). https://doi.org/10.1021/acsenergylett.1c00181
72. M. Hermesmann, T.E. Müller, Green, turquoise, blue, or grey? environmentally friendly hydrogen production in transforming energy systems. Prog. Energy Combust. Sci. **90**, 100996 (2022). https://doi.org/10.1016/j.pecs.2022.100996
73. J. Rey, F. Segura, J. Andújar, Green hydrogen: resources consumption, technological maturity, and regulatory framework. Energies **16**(17), 6222 (2023). https://doi.org/10.3390/en16176222
74. S. Li, V.M. Wheeler, P.B. Kreider, R. Bader, W. Lipiński, Thermodynamic analyses of fuel production via solar-driven non-stoichiometric metal oxide redox cycling. Part 2. impact of solid-gas flow configurations and active material composition on system-level efficiency. Energy Fuels **32**(10), 10848–10863 (2018). https://doi.org/10.1021/acs.energyfuels.8b02082
75. B. Dou et al., Hydrogen production from the thermochemical conversion of biomass: issues and challenges. Sustainable Energy Fuels **3**(2), 314–342 (2019). https://doi.org/10.1039/C8SE00535D
76. A.K. Singh, et al., Thermal reduction of iron oxide under reduced pressure and implications on thermal conversion efficiency for solar thermochemical fuel production.Ind. Eng. Chem. Res. **54**(26), 6793–6803. https://doi.org/10.1021/ie504402x
77. S. Brendelberger, et al., Vacuum pumping options for application in solar thermochemical redox cycles–Assessment of mechanical-, jet-and thermochemical pumping systems.Solar Energy **141**, 91–102 (2017). https://doi.org/10.1016/j.solener.2016.11.023

Development of Advanced Heat Transfer Fluids for High-Temperature Solar Thermal Systems

Veerakumar Chinnasamy and Gopi Kannan Kalimuthu

Abstract High-temperature solar thermal systems are a key component of using solar energy for many industrial uses, such as process heat, desalination, and power production. However, the performance of these systems significantly relies on the properties of the heat transfer fluid (HTF) employed. Conventional HTFs, including molten salts and synthetic oils, have drawbacks that prevent them from being widely used, like low thermal stability, corrosion problems, and high viscosity at high temperatures. To overcome these challenges, scientists focus on developing novel HTFs with enhanced thermal stability, improved heat transfer properties, and reduced environmental impact. One of the promising strategies involves using nanofluids, which are colloidal suspensions of nanoparticles in a base fluid. Compared to traditional HTFs, nanofluids exhibit better thermal conductivity, making them attractive options for high-temperature solar applications. Liquid metals are another attractive avenue for the exploration of advanced HTFs. Efforts are being made in combinatorial material synthesis and high-throughput characterization techniques to identify optimal compositions of liquid metals. Moreover, research is being done to investigate alternative base fluids that provide better thermal stability and compatibility with solar collector materials, such as eutectic mixtures and organic compounds. Computational modeling and experimental studies are being conducted to comprehend the basic mechanisms governing heat transfer and fluid behavior in high-temperature solar systems using advanced HTFs. The development of advanced HTFs holds great potential for improving high-temperature solar thermal systems' efficiency, reliability, and cost-effectiveness, facilitating the transition toward a sustainable energy future. Continued research and innovation in this area are crucial to unlocking the full potential of solar energy as a clean and ample renewable energy source for meeting global energy demands. Comprehensive information on the development of advanced HTFs for high-temperature solar thermal systems is presented in this chapter.

V. Chinnasamy (✉)
Centre for Interdisciplinary Research (Mechanical), SRM University-AP, Andhra Pradesh, India
e-mail: veerakumar.c@hotmail.com

G. K. Kalimuthu
Department of Mechanical Engineering, Indian Institute of Technology (Indian School of Mines), Dhanbad, India

P. Ravindran et al. (eds.), *Photon to Power*, Progress in Optical Science and Photonics 32, https://doi.org/10.1007/978-981-96-5914-2_7

Keywords Solar thermal · Concentrated solar power · Heat transfer fluid · Nanofluid · Molten salt · Liquid metals · Ionic liquid

1 Introduction

Given the increasing demand for clean and sustainable power generation, cost-effective renewable energy sources are becoming increasingly attractive. Solar energy is a vital source that is widely studied due to its abundance [1]. Solar energy conversion systems are of two types: (i) solar thermal and (ii) solar photovoltaic systems. Solar thermal systems capture solar energy and convert it to heat. Solar thermal systems directly convert solar energy into heat for a variety of uses, in contrast to solar photovoltaic systems, which turn sunlight into electricity. The global status of solar heat capacity and the solar heat market is depicted in Fig. 1a, b.

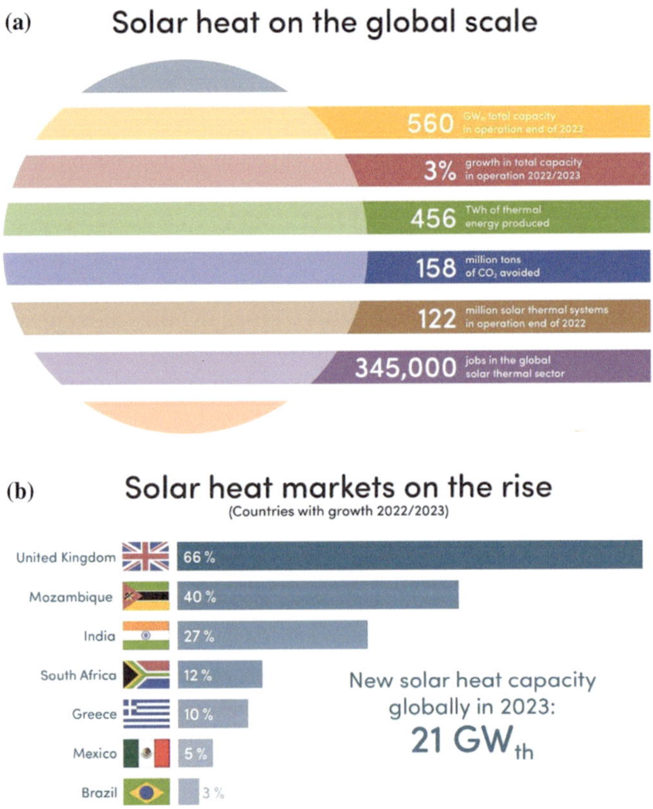

Fig. 1 **a** Solar heat on the global scale [2] (Licensed under Creative Commons). **b** Solar heat markets on the rise [2] (Licensed under Creative Commons)

Solar thermal systems can be categorized as low, medium, and high-temperature systems according to their operating temperature range [3]. Low-temperature solar thermal systems have operating temperatures below 100 °C. Low-temperature applications involve solar water heating and space heating. Medium-temperature solar thermal systems generally operate between 100 and 400 °C. Some of their applications are solar cooking and industrial process heating. High-temperature solar thermal systems are designed to capture and utilize solar energy to produce heat at extremely high temperatures, typically ranging from 400 °C to over 1000 °C. These systems are primarily used for electricity generation, industrial process heat, and advanced applications like hydrogen production. The most common technology that utilizes high-temperature solar thermal energy is Concentrating Solar Power (CSP) [4]. In high-temperature systems, the sun's energy is concentrated using mirrors or lenses, creating intense heat that can be stored or directly converted into mechanical energy and electricity. The ability of these systems to generate and store heat makes them a critical component of renewable energy strategies, offering a reliable, dispatchable energy source even when the sun is not shining. Concentrating solar collectors focus sunlight onto a small receiver, significantly increasing the temperature of the working fluid inside the system. There are several types of concentrating collectors, each suitable for different scales and applications. Parabolic Trough Collectors: This type of solar collector uses curved parabolic mirrors to focus sunlight onto a linear receiver pipe at the focal line. A heat transfer fluid (HTF) flows through the pipe and absorbs the concentrated heat. Solar Towers: Hundreds or thousands of flat mirrors (heliostats) track the sun and reflect sunlight onto a receiver at the top of a central tower. This system can achieve extremely high temperatures (up to 1000 °C) and is often paired with molten salt storage systems. Fresnel Reflectors: These use flat, linear mirrors to concentrate sunlight onto a receiver placed slightly above the mirrors. Fresnel systems are less expensive but typically achieve lower temperatures than parabolic troughs. Solar Dishes: Parabolic dish collectors focus sunlight onto a small receiver, often connected to a Stirling engine, which directly converts heat into mechanical work and then electricity. These systems achieve high efficiencies but are more suitable for small-scale applications. Figure 2 demonstrates the schematics of different concentrated solar thermal systems.

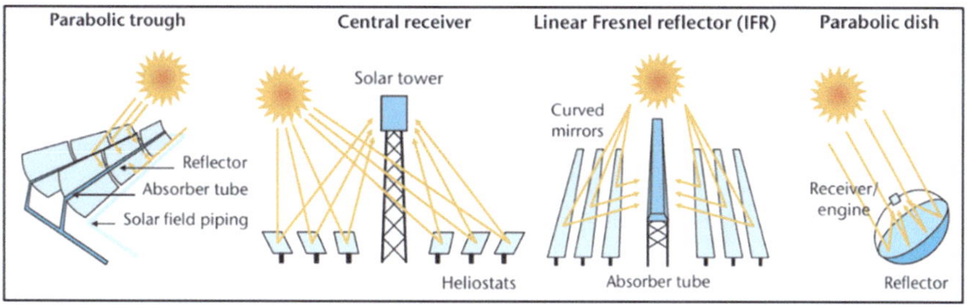

Fig. 2 Schematic of different concentrated solar thermal systems [5] (Permission obtained)

In solar thermal applications, HTFs are used to absorb, transport, and store thermal energy collected by solar collectors. The choice of HTF is critical for high-temperature applications, especially in CSP plants. High-temperature HTFs are crucial in industries where processes require elevated temperatures, such as chemical processing, power generation, and metalworking. These fluids must possess exceptional thermal properties, chemical stability, and the ability to withstand high temperatures without degrading. The most common traditional high-temperature HTFs are as follows: Mineral oils are derived from petroleum, which offers good thermal stability and relatively low viscosity at elevated temperatures. However, they have limitations regarding maximum operating temperature and can degrade over time. Synthetic oils are HTFs engineered to provide superior performance compared to mineral oils. They offer higher thermal stability and lower viscosity and can withstand higher temperatures. Molten salts are inorganic compounds that melt at relatively low temperatures and have excellent thermal properties. They are used in applications requiring extremely high temperatures, such as solar thermal power plants and nuclear reactors. Common molten salts include sodium nitrate, potassium nitrate, and sodium nitrite. Water is a common HTF, but its limitations at high temperatures (boiling point and pressure) necessitate other options in many industrial applications. Gases like air, steam, and nitrogen can be used as HTFs, but they often require higher flow rates and specialized equipment than liquids.

When selecting a high-temperature HTF (HTF) for solar thermal applications, several key considerations must be considered to ensure efficient, safe, and cost-effective operation. The choice of HTF directly influences the solar thermal system's performance, reliability, and economics. The key factors to be considered are as follows:

Operating Temperature Range: The fluid must withstand the system's maximum operating temperatures without degrading or losing its thermal properties. High-efficiency power cycles often require HTFs that can operate above 400 °C. The HTF should be thermally stable at high temperatures, resisting decomposition or chemical changes over time. This ensures long-term performance without needing frequent replacement.

Freezing Point: Some HTFs, such as molten salts, have high freezing points (e.g., around 120–220 °C). The system must be designed to prevent the HTF from solidifying, which can lead to operational challenges, especially in cold climates or during nighttime operations. A lower freezing point is desirable for easy startup and maintenance, avoiding additional heating systems to keep the fluid in liquid form.

Thermal Conductivity (TC) and Heat Capacity (HC): High TC is essential for efficient heat transfer between the HTF and the system components (e.g., solar collectors and heat exchangers). It also leads to faster heat absorption and transfer. The HTF's HC influences the amount of heat it can store and transfer. A higher HC means the fluid can transport more thermal energy, and improving system efficiency.

Viscosity: The HTF's viscosity affects the pumping power required to circulate the fluid through the system. Low viscosity at operating temperatures is preferred to minimize energy consumption and reduce wear on pumps and pipes. Some fluids become highly viscous at lower temperatures, which can complicate startup and reduce efficiency during cold conditions.

Pressure Requirements: The HTF's vapor pressure should be low at operating temperatures to avoid needing high-pressure containment. Fluids with low vapor pressures are preferable because they reduce system complexity, material stress, and safety risks. Fluids that require high-pressure operation (e.g., steam) lead to higher containment and operational costs due to the need for strong, pressure-resistant materials.

Safety Considerations: Many organic HTFs (e.g., mineral and synthetic oils) are flammable at high temperatures, which poses a fire risk. Non-flammable alternatives like molten salts or water/steam are safer. The HTF should be non-toxic and environmentally benign to minimize health risks to workers and environmental damage in case of leaks. The HTF should be chemically stable and non-reactive with system components to prevent corrosion, deposits, or equipment degradation over time.

Corrosion Potential: The selected HTF must be compatible with the materials used in the system (e.g., pipes, heat exchangers). Some fluids, such as molten salts, can be highly corrosive, requiring expensive alloys or special coatings to protect the equipment.

Cost and Availability: The cost of purchasing the HTF can vary significantly. For example, synthetic oils are more expensive than mineral oils, while molten salts are relatively cost-effective but require more complex handling systems. Operating costs include maintaining fluid quality (e.g., replacing degraded fluid), the energy required to pump the fluid, and additional infrastructure for freezing protection or high-pressure containment. The longer the HTF remains stable and functional, the less frequently it needs to be replaced, reducing long-term operating costs.

Environmental Impact: HTF should ideally have a low environmental impact. Some organic fluids can degrade into toxic or harmful by-products, while others, like molten salts, are environmentally benign. Non-toxic, non-hazardous fluids are preferred to minimize environmental contamination and health risks in case of a leak or accidental release.

Energy Storage Compatibility: In some systems, the HTF may also serve as the thermal storage medium (as is common with molten salts in CSP plants). In such cases, the HTF should have high HC and thermal stability to store energy effectively over time. Some advanced systems may integrate HTFs that can undergo phase changes (e.g., from liquid to solid or gas), allowing higher energy storage densities.

Ease of Handling and Maintenance: Fluids that require special handling (e.g., due to toxicity, freezing, or flammability) increase the complexity and cost of operating the system. The ease with which the fluid can be monitored, maintained, and replaced is critical for long-term operational efficiency. Some HTFs require regular analysis and purification to maintain optimal performance.

2 HTFs for High-temperature Applications

2.1 *Liquid HTFs*

2.1.1 Molten Salts

Molten salts are high-temperature liquid salts that play a crucial role in industries that require efficient heat transfer and storage, particularly in the solar energy sector. They are solid salts heated to their melting point and can be used as heat transfer and storage mediums in liquid form. Molten salts are mainly used in CSP systems, industrial heat transfer, and thermal energy storage. The key characteristics of molten salts are their high melting point, high HC, and low vapor pressure. Based on their composition, molten salts generally become liquid above 120 °C. They have high thermal stability and can absorb and store large quantities of heat. However, due to chlorides and carbonates, molten salts are corrosive at high temperatures, making them difficult to use. Another disadvantage is that they require a heating system to maintain them in the liquid phase, as they have a high freezing point.

Common types of molten salts are chlorides, carbonates, and nitrates. Sodium chloride and potassium chloride are examples of chlorides and potassium carbonates; sodium and lithium carbonates fall under carbonates and have high operating temperatures but are corrosive. In a mixed form called solar salt, potassium nitrate and sodium nitrate are used as a binary mixture in solar thermal applications. Ternary mixtures of the salts are also used as molten salts with controlled melting points. With the addition of nitrates, the melting point can be brought down. Due to the corrosive nature of molten salts, specialized coatings are required for the piping and storage structures. Chloride salts have significant potential as high-temperature thermal energy storage (TES) mediums for CSP systems. In a study by Zhao et al. [6], LiCl, KCl, and $CaCl_2$ were chosen to increase the energy storage density and broaden the working temperature range. With mass ratios of 37.85%–53.38%–8.77%, 30.90%–13.82%–55.28%, and 1.78%–18.61%–79.61%, respectively, new ternary chloride salts, LiCl–KCl–$CaCl_2$, were created. These corresponded to melting points of 340.93, 433.57, and 626.85 °C. Only the first two salts were tested because of the third salt's high melting point. DSC tests showed only a 0.46% and 1.64% difference between the actual and predicted melting points. According to TG findings, the maximum working temperature in air was 650 °C. After short-term cycling, the salts retained excellent thermal properties, indicating strong potential for high-temperature thermal energy storage (TES) applications. The eutectic salts NaCl–KCl–$ZnCl_2$ and NaCl–KCl–$MgCl_2$ are exciting candidates for high-temperature HTF and TES materials in CSP systems with an operational temperature range of 230–750 °C. Zhang et al. [7] constructed a circulation system equipped with flow and heat transfer equipment to test the heat transfer coefficient (HTF) of molten salts at high temperatures. Convective HTFs for NaCl-KCl-$ZnCl_2$ (molar fraction: 13.8–41.9–44.3%) were experimentally measured in tubes to find appropriate heat transfer correlations. It has been determined by Wang et al. [8] that a suitable high-temperature HTF for heat transfer and thermal storage in third-generation CSP systems is the eutectic chloride molten salt $MgCl_2$-KCl-NaCl (45.98–38.91–15.11 wt%). Important thermophysical characteristics, including density, vapor pressure, heat of fusion, viscosity, HC, and TC, have all been experimentally measured for the first time throughout a 450–700 °C temperature range. Convective heat transfer in CSP systems, which include solar receivers, heat exchangers, and thermal storage units, must be calculated using these principles. Molten salt is used as an HTF in solar fields using linear Fresnel and parabolic trough collector systems in direct molten salt technology. Phase diagrams, composition, melting ranges, melting temperature, density, HC, TC, viscosity, and handling of five molten salt systems—Solar Salt, HitecXL (CaKNa//NO3), LiNaK-Nitrate, Hitec (NaK//NO23), and CaLiNaK//NO23—as well as their minimum and maximum operating temperatures are evaluated by Bonk et al. [5], with the most suggested values being emphasized and discussed.

2.1.2 Organic HTFs

Organic HTFs comprise carbon-based compounds widely used in various thermal systems, including CSP, chemical processing, and other industries requiring efficient heat transfer at high temperatures. In solar thermal systems, high-temperature organic fluids are essential for systems that operate at moderate to high temperatures, typically ranging from 200 to 400 °C. These fluids offer unique advantages, such as low viscosity, chemical stability, and operational flexibility. High-temperature organic HTFs are typically formulated using synthetic organic compounds, such as aromatic hydrocarbons, silicone oils, and diphenyl-based mixtures. These fluids are designed to operate efficiently in systems where the temperatures are too high for water or glycol-based fluids but not as extreme as molten salts or liquid metals.

The essential properties of organic HTFs are:

- Good stability: They are stable up to 400 °C and suitable for medium- and high-temperature applications.
- Low viscosity: They have relatively low viscosity and require low pumping power.
- Low freezing point: They generally have low freezing points, making them easier to handle. No additional heating is required when they are not in operation.
- Non-corrosive: They are generally non-corrosive and compatible with metals, requiring reduced maintenance.
- High specific heat: They can store and transport large amounts of heat with high specific heat.
- Low vapor pressure: With low vapor pressure at high temperatures, they do not vaporize easily.

Despite their many advantages, organic HTFs are not without their drawbacks, mainly when used in high-temperature systems like CSP plants. At temperatures approaching or exceeding their upper limit (around 400 °C), organic HTFs can degrade, breaking down into smaller molecules and losing their heat transfer efficiency. Degradation can also lead to the formation of solids or sludge, which can clog pipes and reduce the overall system performance. Regular fluid replacement or maintenance is required to mitigate these issues. Organic fluids are typically flammable, posing a potential fire risk in systems that operate at high temperatures. Proper safety protocols and monitoring systems must be in place to minimize the risk of accidents, particularly in large-scale industrial systems. Compared to other high-temperature fluids like molten salts, liquid metals, or supercritical CO_2, organic HTFs have lower maximum operating temperatures. This limits their use in systems that require temperatures above 400 °C, reducing their applicability in next-generation solar thermal plants aiming for higher efficiency. Some organic HTFs may pose environmental or health risks if disposed of improperly. The breakdown of certain organic fluids can produce toxic by-products, making disposal and environmental regulations an important consideration.

Developing new organic HTFs with improved thermal stability, higher temperature limits, and lower environmental impacts is ongoing. Advances in chemical formulation and nanotechnology may allow for creating organic HTFs that can operate at higher temperatures while maintaining their beneficial properties. Furthermore, integrating nanoparticles into organic fluids can enhance heat transfer capabilities, increase efficiency, and reduce degradation. High-temperature organic HTFs are vital in parabolic trough solar thermal systems and many industrial processes. They ensure thermal stability, low viscosity, and easy operation, offering reliability for medium-temperature systems. However, their temperature constraints and risk of thermal degradation are challenges as CSP plants aim for higher operating temperatures and efficiencies. Advances in fluid chemistry and materials science may soon enhance the potential of organic HTFs, making them more effective for high-temperature applications. Mineral and synthetic oils are frequently employed as HTFs in operations that require temperatures higher than 200 °C. Vegetable oils are becoming increasingly popular because they are eco-friendly and renewable. Gomma et al. [9] have suggested refined Jatropha curcas oil as a possible thermal oil for high-temperature applications such as CSP. The oil displayed less than 5% volatility, a consistent HC, little density growth, and a similar viscosity after being thermally aged at 210 °C for up to 2160 h. After aging, its flash point held steady at 175 °C, down from 220 °C when it was first discovered. These findings imply that jatropha oil has great promise as a thermal storage and HTF material.

2.1.3 Ionic Liquids

Ionic liquids (ILs) have gained significant attention as potential high-temperature HTFs for solar thermal applications, particularly in CSP systems. Traditional HTFs like molten salts, water, and thermal oils have limitations like high melting points, decomposition at lower temperatures, or poor thermal stability. Ionic liquids, with their unique properties, offer a promising alternative. Commonly used thermal oils degrade at temperatures above 400 °C and pose a fire risk. ILs surpass these limitations with better thermal stability. While molten salts are widely used in CSP systems, their high melting points (typically 220–240 °C) require additional energy for maintaining liquid states, which can lower system efficiency. Ionic liquids have much lower melting points and do not suffer this drawback.

As discussed in this section, there are several advantages of ILs as high-temperature HTFs. Many ILs are thermally stable at temperatures exceeding 400 °C and can go as high as 600 °C in some cases. This makes them suitable for high-temperature solar applications, requiring efficient heat storage and transfer at elevated temperatures. ILs have near-zero vapor pressure, even at high temperatures. Unlike traditional HTFs, this makes them safer to use at elevated temperatures as there is no risk of vaporizing or building excessive pressure. The structure of ionic liquids can be tuned by changing the cation–anion combination, allowing for the adjustment of properties like viscosity, TC, and HC to optimize their performance for specific solar thermal applications. Many ILs are non-flammable, a critical feature for safety in high-temperature operations, particularly in CSP systems that may encounter extreme conditions. The potential to use ILs as HTFs and thermal storage mediums offers a dual-purpose advantage, which could simplify system design and reduce costs.

ILs have emerged as a potential alternative for high-temperature HTFs in CSP systems and other thermal applications. However, several significant challenges are associated with using ILs as high-temperature HTFs. ILs are generally more expensive than traditional HTFs, though their performance and extended lifetime at high temperatures could offset initial costs in the long run. Some ILs exhibit high viscosities, which can hinder efficient heat transfer and flow in CSP systems and require further optimization. Although ILs are stable at high temperatures, some still decompose above certain thresholds or in the presence of moisture, necessitating the development of even more thermally stable ILs. Corrosion can be a concern depending on the system materials and the IL composition. Research is ongoing to ensure that ionic liquids are compatible with the structural materials of CSP systems. Ongoing research focuses on designing ILs that can operate efficiently above 500 °C, remain cost-effective, and be environmentally friendly. If these challenges can be overcome, ionic liquids may become a key technology in improving the efficiency and sustainability of solar thermal energy systems. In preparation for its possible use in CSP, Paul et al. [10] examined the thermophysical characteristics and high-temperature thermal performance of N-butyl-N,N,N-trimetylammoniumbis(trifluormethylsulfonyl)imide ([N4111] [NTf2]) IL. The experiment's findings showed that TC varied from 0.124 to 0.121 W/m K for temperature changes between 283 and 343 K, slightly decreasing as the temperature increased. It was discovered that temperature significantly impacted the viscosity of IL and that the two variables had an exponential connection. HC rose linearly from 298 to 618 K at the same time as temperature. Despite having a lower HTF than de-ionized water, IL may still be a good option for solar collector use due to its other desirable qualities, including thermal stability. Wadekar [11] assessed the feasibility of using 1-butyl-3-methylimidazolium bis[(trifluoromethyl) sulfonyl]imide, ([bmim][Tf2N]) IL as HTF. It was concluded that ILs are good candidates for HTFs due to their low volatility, low flammability, and excellent thermal stability; nevertheless, using heat transfer improvement to enhance their thermal performance may require careful attention.

2.1.4 Liquid Metals

Liquid metals have emerged as promising HTFs for solar thermal applications due to their superior thermal properties compared to conventional HTFs such as water and oils. This overview will cover the critical aspects of liquid metals as HTFs, including their properties, advantages, challenges, and potential applications in solar thermal systems. Liquid metals, such as mercury, sodium, and gallium, exhibit exceptionally high TC. This property allows for rapid heat transfer, which is crucial for enhancing the efficiency of solar thermal systems. Liquid metals generally have lower specific heat capacities than water, so they can quickly absorb and release heat. This can lead to improved response times in solar thermal systems. Many liquid metals have high boiling points, allowing them to operate at elevated temperatures without vaporizing. For example, sodium has a boiling point of 883 °C, making it suitable for high-temperature solar thermal applications. Liquid metals typically have lower viscosities than traditional HTFs, which enhances their flow characteristics and reduces pumping power requirements. Some liquid metals, particularly gallium, are less corrosive than conventional HTFs, which can degrade over time. This property can extend the lifespan of the solar thermal system components. The thermal properties of several potential liquid metals are presented in Table 1.

The key advantages of liquid metals as high-temperature HTFs are discussed here. The high TC of liquid metals allows for efficient heat transfer between the solar collector and the heat storage or power generation system. Liquid metals can operate effectively at much higher temperatures than water or organic fluids. This capability is critical for CSP systems, where higher temperatures can lead to greater efficiency in energy conversion. The low thermal expansion of liquid metals can reduce heat losses in storage and transport, improving overall system efficiency. Liquid metals are well-suited for integration with advanced energy technologies, such as thermochemical and thermal energy storage systems. Some liquid metals can directly participate in thermochemical reactions, allowing novel energy conversion methods.

Table 1 Properties of potential liquid metals [12, 13]

Liquid metals	Melting point (°C)	Boiling point (°C)	Specific HC ($Jkg^{-1} K^{-1}$)
Sodium	98	883	1.26
Tin	232	2687	0.24
Gallium	29.8	2403	3.75
Lithium	180	1347	4.16
Lead–bismuth (44.5–55.5)	125	1638	0.146

While liquid metals have several advantages as HTFs, this section discusses their application challenges. The cost of liquid metals can be prohibitive, particularly for materials like mercury and gallium. The economic feasibility of using these fluids in large-scale solar thermal systems is still a subject of research. Liquid metals like mercury are toxic, posing health and environmental risks. Strict handling and containment measures are necessary, complicating system design and operation. Not all construction materials are compatible with liquid metals, particularly at elevated temperatures. Ensuring that system components can withstand prolonged exposure to liquid metals without degradation is crucial. The unique properties of liquid metals require specialized pumping systems and designs to effectively manage flow and heat transfer. Even though this field has a lot of promise, research, and development are still in their infancy. More field research and experiments are required to completely comprehend the performance of liquid metals in solar thermal applications.

Liu et al. [14] conducted a numerical simulation utilizing liquid sodium, a viable option for next-generation solar power tower systems, to study heat transport in a solar receiver tube. Three media are compared: liquid sodium, solar salt, and Hitec in uniform and nonuniform heat flux environments. The average heat transfer and friction coefficients and their effects on the maximum heat flux, inlet temperature, and mass flow rate are examined. The average heat transfer coefficient of liquid sodium is found to be 2.5–5 times higher than that of the other two salts when the maximum heat flux is 0.1 MW/m^2, the mass flow rate is 1.0 kg/s, and the input temperature is between 550 and 800 K. Quantitative thermodynamic analyses of how liquid metals can improve the efficiency of concentrating solar power receiver systems were conducted by Pacio et al. [15]. It measures the temperature ranges and HTFs of two liquid metals, sodium and lead–bismuth eutectic. These media can raise HTFs by three to twenty times, although the highest temperatures at which they can operate are approximately 900 °C for sodium and 1600 °C for LBE. There is also a discussion of constraints like material availability and safety standards. The more significant HTFs and operating temperatures can significantly increase receiver efficiency and fluid outlet temperatures to mitigate efficiency losses from ambient heat dissipation. For example, sodium can nearly achieve the theoretical top limit of heat transmission efficiency. For ordinary fluids, a standard two-equation turbulence model with a constant turbulent Prandtl number is sufficient for accurate CFD simulations. However, this model fails to predict heat transfer accurately for liquid metals with low Prandtl numbers. Manservisi and Menghini [16] proposed a four-parameter turbulence model to improve predictions for liquid metal flows that will enhance CSP receiver efficiency. Liquid metals present a promising avenue for enhancing the efficiency and effectiveness of solar thermal systems. Their unique properties, including high TC, high boiling points, and reduced heat losses, make them suitable for high-temperature applications like CSP and thermal energy storage. However, cost, safety concerns, and material compatibility must be addressed to facilitate broader adoption. Continued research and development in this field could unlock new possibilities for renewable energy technologies and improve the sustainability of solar thermal energy systems.

2.1.5 Nanofluids

Nanofluids are an emerging type of HTF comprising a base fluid (such as water, oil, molten salt, or synthetic oils) dispersed with nanoparticles. These nanoparticles, typically composed of metals, metal oxides, or carbon-based materials, range from 1 to 100 nm. Incorporating these particles enhances the thermal properties of the base fluid, significantly improving its heat transfer capability. In high-temperature applications, particularly in CSP systems and industrial processes, nanofluids are a promising solution because of their enhanced TC, HC, and heat transfer efficiency. This section will explore the properties, advantages, challenges, and potential applications of nanofluids as high-temperature HTFs.

The advantages of nanofluids as high-temperature HTFs are presented here. The addition of nanoparticles increases the effective TC of the base fluid. This improvement is crucial for maximizing heat transfer efficiency, especially at high temperatures, which is critical in solar thermal systems where efficient heat collection and transfer are needed. Nanofluids can exhibit higher specific HC than their base fluids, allowing them to store more thermal energy. This can lead to better TES systems in solar power plants. Nanoparticles can significantly enhance the convective HTF, resulting in more efficient heat transfer in systems such as solar receivers or heat exchangers. Due to the improved TC, the system requires less pumping power to achieve the same heat transfer rates, which can lower operational costs and improve energy efficiency. By changing the type, size, and concentration of nanoparticles, the properties of nanofluids can be tailored for specific high-temperature applications. This tunability allows optimization of TC, viscosity, and specific heat based on the needs of the solar thermal system. The advantages of nanofluids over traditional HTFs are as follows: Due to their enhanced thermal properties, nanofluids can extract more energy from solar radiation than conventional HTFs like water, oils, or molten salts. For some nanofluids, especially when used with molten salts, the melting point of the base fluid can be lowered by adding nanoparticles, reducing the energy losses required to maintain the fluid in a liquid state. Nanofluids can operate effectively over a broad temperature range, allowing them to be used in various parts of a solar thermal plant (e.g., heat collectors, storage systems, and heat exchangers).

The challenges in using nanofluids for solar thermal applications are presented in detail in this part. Nanoparticles tend to agglomerate over time, reducing the long-term stability of the nanofluid and compromising its performance. This issue needs to be addressed using surfactants or other stabilization techniques. Adding nanoparticles often increases the viscosity of the base fluid, which can negatively impact flow characteristics and increase pumping power requirements. Optimizing the concentration of nanoparticles is critical to balancing viscosity and thermal performance. Depending on the type of nanoparticles and the base fluid, corrosion of system components could be a concern, particularly at high temperatures. Research is focused on minimizing the corrosive effects while retaining the enhanced thermal properties of the nanofluids. Nanoparticles, particularly those made of metals or advanced materials like graphene, can be expensive, raising the cost of using nanofluids on a large scale in solar thermal plants. However, the improved efficiency and potential long-term cost savings could offset this initial investment. The potential environmental and health impacts of nanoparticles are not fully understood, and their production, handling, and disposal require careful consideration.

Merino et al. [17] described the fabrication and characterization of nanofluids based on 2D-Tungsten disulfide (WS2) within a typical HTF for CSP plants. The nanosheets in the nanofluids generated via exfoliation were stable and clearly defined. The characterization results showed that including WS2 nanosheets boosted TC by up to 30% while not affecting surface tension, viscosity, or isobaric-specific heat. The nanofluid showed potential for CSP applications, as seen by a 22% improvement in the Ur factor, which measures thermal efficiency in solar collectors. Furthermore, the Reynolds number and friction factor remained mostly unchanged with the addition of the nanomaterial, which is advantageous for real-world CSP applications. Numerical simulations of three nanofluids (MgO–VP1, Al_2O_3–VP1, Fe_2O_3–VP1) in CSP plants by Begdouri and Fadar [18] showed that while they improve heat transfer, they also increase pumping energy consumption. This leads to lower net energy production and higher GHG emissions. While nanofluids offer benefits, their drawbacks outweigh them in CSP applications. Therminol 66, a high-temperature oil used in the oil and gas industry for heat exchangers, was used as the base fluid for the fabrication of nanofluids by Safaei et al. [19]. At 0.1 weight percent, SiO_2 and Al_2O_3 nanoparticles were introduced. Density, viscosity, and specific HC were measured thermophysical characteristics. A high-temperature apparatus assessed the performance of nanofluids at 280–320 °C. The specific HC increased, although neither density nor viscosity changed significantly, according to the data. According to the heat exchanger test, the efficiency of the nanoparticles was up to 88% higher than that of the base fluid.

Chen et al. [19] thoroughly analyzed the developments in carbon nanofluids based on molten salt for high-temperature applications. This study investigates the effects of concentration, size, and type of carbon nanoparticles on the heat transmission and storage capabilities of molten salt. Additionally, the role of nanoparticle shape on heat transfer is analyzed. Additionally, more efficient ways of preparing carbon-based molten salt nanofluids are suggested. Future directions for molten salt-based carbon nanofluids at high temperatures are also briefly addressed. Wang et al. [20] evaluated the synthesis, thermal characteristics, and summary of nanofluids based on molten salt and ionic liquid. The Krieger-Dougherty model has been shown in earlier research to predict viscosity accurately; however, no generic model exists for predicting TC or specific heat. Using binary carbonate as the base and 20 nm SiO_2 and MgO nanoparticles as additions, Yaxuan et al. [21] synthesized carbonate salt nanofluids using the aqueous solution approach. Important properties were examined using differential scanning calorimetry (DSC), thermogravimetric analysis (TGA), laser flash analysis (LFA), and micromorphology techniques. The findings demonstrated that whereas 20 nm MgO enhanced sensible heat, TC, and stability, 20 nm SiO_2 substantially impacted latent heat. In SiO_2- and MgO-based nanofluids, distinct nanostructures were noted, and cloud nuclei generation mechanisms were suggested to account for the increases. Yu et al. [22] reported the formulation, production, and characterization of a novel molten salt nanocomposite for high-temperature heat storage, aiming to improve specific heat and TC using the microwave technique. Figure 3 shows the process for making molten salt nanocomposites. Based on eutectic quaternary salt, the nanocomposite has different concentrations of SiO_2 and TiO_2 nanoparticles added. It was discovered that adding nanoparticles lowered the melting point and latent heat using simultaneous thermal analysis (STA) and LFA. The specific heat rose 6.3% and 9.8% over SiO_2 and TiO_2 alone and 28.1% over pure salt. 53.7% more TC was achieved, and the improved characteristics were mainly attributed to a high-density nanostructure. There is potential for high-temperature thermal storage systems with this nanocomposite. Sang and Liu [23] used a two-step solution approach to develop ternary carbonate nanofluids (K_2CO_3-Li_2CO_3-Na_2CO_3, 4:4:2) containing several nanoparticles (SiO_2, CuO, TiO_2, Al_2O_3) and SiO2 of varied sizes (5, 20, 30, 60 nm). For particles sized 5–30 nm, SiO_2 was the most effective in increasing specific HC, increasing it by 78.0–116.8% between 500 and 540 °C. In that order, specific heat was increased by 50.6–73.9%, 31.1–56.5%, and 50.6–66.5% in CuO, TiO_2, and Al_2O_3. The amplification relied on needle-like nanostructures and the dispersion of the nanoparticles; there were no chemical interactions between the carbonates and the nanoparticles.

In a study by Cavieres et al. [24], two equimolar IL mixes were utilized to form ionanofluids (INFs): [Omim][PF6] + [Odmim][PF6] and [Omim][PF6] + [Odmim][BF4]. The base fluids were supplemented with MWCNTs at a weight percentage of 0.04–0.1. The thermophysical properties were evaluated and compared with materials frequently used to store solar energy. The HC rose to 3.7 times for thermonuclear VP-1 and 3.2 times for nitrated salts, while the thermal storage

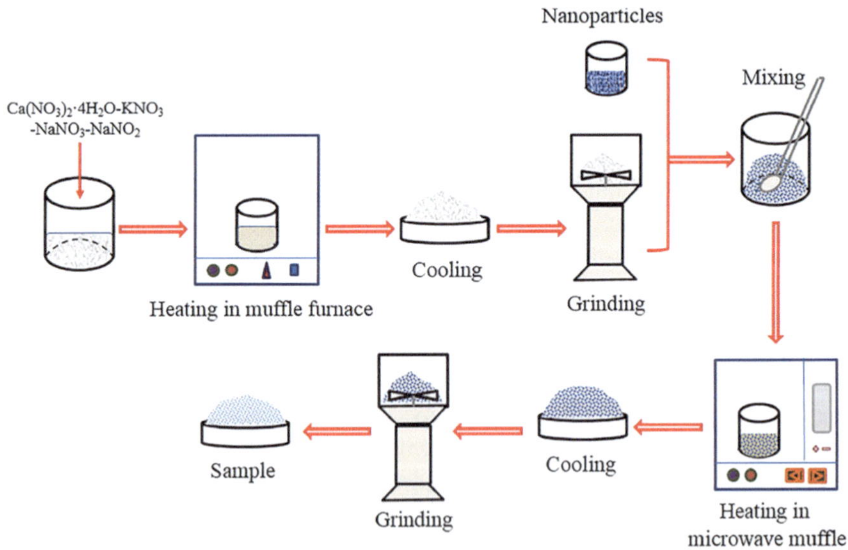

Fig. 3 Schematic diagram for the preparation of molten salt nanocomposites [22] (Permission obtained)

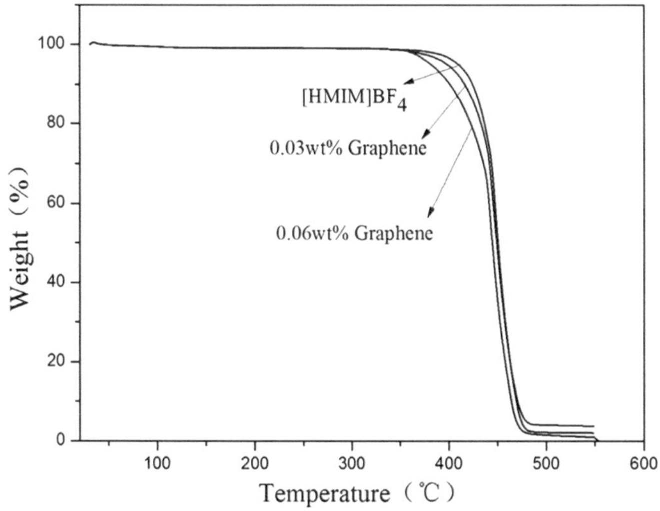

Fig. 4 TC of [HMIM]BF4 and graphene-dispersed nanofluids as a function of temperature [25] (Permission obtained)

density increased by 9.4 and 284%, respectively. Their decreased thermal diffusivity supports their potential for energy storage applications. Liu et al. [25] devised an experimental technique to estimate nanofluid's TC over 100 °C. Utilizing the ionic liquid [HMIM]BF4, a thorough investigation of the thermodynamic characteristics of graphene-dispersed nanofluids, including TC, viscosity, specific heat, and density, was carried out at temperatures as high as 200 °C. TC increased 15.2%–22.9% in the nanofluid containing 0.06 weight percent graphene between 25 and 200 °C. The TC of [HMIM]BF4 and graphene-dispersed nanofluids as a function of temperature is presented in Fig. 4. At 210 °C, viscosity dramatically dropped, making it appropriate for medium-to-high temperature applications. With a breakdown temperature of about 440.6 °C, the graphene nanofluid showed good thermal stability and exceeded Therminol VP-1 in thermodynamic properties.

2.1.6 Gaseous Fluids

Gaseous fluids are less commonly used than liquids and can effectively transfer heat media in high-temperature applications. There are several advantages to using gases as HTFs. Gases can transfer large amounts of heat due to their high specific HC. The low viscosity of gases reduces pressure drop in the system, leading to lower pumping costs. Gases can operate at extremely high temperatures without vaporizing or decomposing, making them suitable for metallurgical and chemical reactions. Gases do not wet surfaces, preventing fouling and corrosion issues. Some factors to be considered when using gases are discussed here. Gaseous fluids often require much higher volumetric flow rates, leading to larger piping and equipment

Table 2 Common gaseous fluids used in high-temperature applications

Gaseous fluid	Operating temperature (°C)
Air	500
Carbon dioxide	> 550
Nitrogen	1000
Helium	> 550

to achieve the same heat transfer rate as a liquid. Compressing gases to increase their density and improve heat transfer can be energy-intensive. Expansion can also lead to significant cooling effects. Gases can be more prone to leakage than liquids, requiring careful containment measures. High-velocity gas flow can generate noise and vibration, necessitating appropriate noise control measures. Common gaseous fluids used in high-temperature applications are listed in Table 2.

Air is a commonly used gaseous fluid that is widely available and inexpensive. It is also chemically inert, non-toxic, and non-corrosive. However, the disadvantages of air as HTF are Low heat transfer efficiency due to low specific HC and density compared to other fluids. It requires high pressures or large volumes for efficient energy transfer. Carbon dioxide is non-toxic, non-flammable, and relatively inexpensive. Supercritical CO_2 can achieve high efficiencies, making it a promising fluid for next-generation solar thermal systems. However, High pressures are needed to maintain CO_2 in its supercritical state, which may increase material and safety requirements. It causes high-temperature corrosion, requiring advanced piping and containment materials. Nitrogen is commonly used in closed-loop systems for heat transfer at moderate to high temperatures. It is inexpensive, non-flammable, and chemically stable. However, Similar to air, it has low TC and HC, making it less efficient than some liquid alternatives. Helium is an inert gas with high TC, making it a good conductor of heat. It can withstand extremely high temperatures (up to 1000 °C or more). But it is expensive and not as readily available as air or CO_2. While gaseous fluids offer unique advantages for high-temperature heat transfer, their suitability depends on specific application requirements. Factors such as temperature range, heat transfer rate, pressure drop, and equipment limitations must be carefully considered. In many cases, a combination of liquid and gaseous fluids may be optimal to achieve desired performance.

Using air as HTF, Famiglietti and Lecuona [26] conducted an experimental investigation of a linear Fresnel solar collector. The exit air temperature reached up to 500 °C. A commercial turbocharger lowers the amount of blower power needed by pressurizing the air. In the spring and summer, a carefully designed test campaign was carried out in Madrid, Spain, to offer experimental validation for a comprehensive 1D opto-thermal numerical model. The receiver tube's static and specially created dynamic models underwent analysis and testing. Good et al. [27] described an ingenious 9.7 m aperture parabolic trough mirror. It was constructed from inflated, metalized polymer films that were affixed to a stiff concrete framework. The receiver uses air as the HTF at ambient pressure and combines an array of absorber cavities with secondary concentrating optics at temperatures exceeding 600 °C. The air heated by the sun is directed toward a closely packed bed of rocks in order to retain thermal energy and provide a steady, high-temperature heat source for the power block. Air was used as the HTF in a computational analysis of heat storage in a honeycomb ceramic TES system for a solar thermal power plant by Li et al. [28]. The system's thermal performance was assessed using a one-dimensional TES model created in Modelica. Under different charging and discharging settings, the simulation results closely match experimental data for two honeycomb ceramic materials. The ceramic length and cross-sectional area have the most effects on beginning costs, according to the model's analysis of the effects of honeycomb geometric parameters on storage performance and material costs.

Detailed reviews on technology advancement and research on supercritical CO_2 as HTF are presented by several researchers [29, 30]. Bovard et al. [31] conducted a numerical study of heat transfer in supercritical CO_2 and water turbulent flow inside a circular tube using RSM turbulence models for a two-dimensional grid. Analysis was done on how the mass flow rate and pressure affected the buoyancy, flow acceleration, and HTF. The findings indicate that increased pressure also lessens the impacts of buoyancy and flow acceleration, whereas increasing mass flux increases the former. A supercritical CO_2–water test loop was developed and studied by Ma et al. [32] to evaluate the heat transfer efficiency of a twin pipe heat exchanger. They examined the impacts of supercritical CO_2-side pressure, mass flow, buoyancy force, and water-side mass flux. The results show that the total and supercritical CO_2-side HTFs decrease with increasing supercritical CO_2-side pressure and peak above the pseudo-critical threshold. However, buoyancy at low supercritical CO_2 mass flux significantly impacts heat transfer, but this effect decreases with increasing mass flux. Due to high-pressure requirements, high expense, limited availability, and difficult system design, little experimental research has been conducted on helium and nitrogen.

3 Impacts of Advanced HTFs

Solar thermal technologies, particularly CSP systems, rely on efficient HTFs to capture, transport, and store solar energy for conversion into electricity. As the demand for renewable energy rises, selecting advanced high-temperature HTFs becomes crucial due to their potential to improve system efficiency and reduce costs. However, these choices also have significant environmental and economic impacts.

3.1 Environmental Impact

Due to their chemical composition, advanced HTFs, such as molten salts, synthetic oils, or liquid metals, may introduce environmental hazards. For instance, some synthetic oils may contain toxic or non-biodegradable components. These substances could contaminate soil and water if released into the environment through spills or leaks. Molten salt mixtures like sodium nitrate ($NaNO_3$) and potassium nitrate (KNO_3) are considered environmentally benign compared to synthetic oils. However, manufacturing these salts requires significant energy and resources, contributing to their environmental footprint. HTFs are exposed to high temperatures for extended periods, leading to thermal degradation. This degradation may result in the formation of harmful by-products, such as volatile organic compounds (VOCs) in synthetic oils. These VOCs can contribute to air pollution and greenhouse gas emissions. Decomposed HTFs may also require special disposal methods, increasing the environmental burden. Energy-intensive production processes such as refining synthetic oils or preparing molten salts contribute to their environmental impact. For example, extracting raw materials for molten salts can generate a significant carbon footprint. Reducing the energy demand for HTF production or choosing naturally abundant and easily processed materials could mitigate environmental harm. Proper disposal or recycling of HTFs is crucial to minimizing environmental impact. Some advanced HTFs, like molten salts, can be recycled or reused with minimal environmental harm, while others, such as synthetic oils, require careful treatment to avoid toxic emissions or contamination. High-temperature HTFs present risks of accidents, including leaks, spills, or fires, particularly when handling combustible materials like synthetic oils. These incidents could have serious environmental and health consequences if not appropriately managed.

3.2 Economic Consideration

The cost of advanced HTFs varies significantly depending on the material and the complexity of production. For example, synthetic oils tend to be more expensive than molten salts due to refining costs. Meanwhile, liquid metals, promising for their TC, can be prohibitively expensive due to their rare and costly raw materials (e.g., liquid sodium or gallium). Advances in manufacturing processes or sourcing cheaper, abundant materials could help reduce HTF costs and make solar thermal power more economically viable. High-temperature HTFs are essential for improving the overall efficiency of CSP systems. Fluids capable of operating at higher temperatures allow for more efficient power cycles, such as supercritical steam turbines, which can lower the levelized cost of energy (LCOE). While advanced HTFs can improve system performance, the associated high costs of advanced materials must be balanced against potential efficiency gains. In some cases, long-term operational savings may offset the additional capital investment in expensive HTFs. Advanced HTFs can influence the operational costs of CSP plants. Fluids with higher stability and lower degradation rates require less frequent replacement, reducing maintenance costs. For example, molten salts, which are thermally stable at high temperatures, can lower the frequency of fluid replacement and decrease overall maintenance expenses. On the other hand, some HTFs, particularly synthetic oils, may degrade faster under high-temperature conditions, necessitating more frequent replacements and increased operational costs. The long-term economic feasibility of advanced HTFs depends on their lifecycle costs, including production, operation, maintenance, and disposal. While some fluids may have higher upfront costs, their durability and performance may result in lower costs over the lifecycle of the CSP plant. Economic considerations also include the cost of environmental compliance, especially if the HTF poses contamination risks or requires specialized disposal methods at the end of its life.

The choice of advanced high-temperature HTFs for solar thermal applications has far-reaching environmental and economic implications. Molten salts, synthetic oils, and liquid metals offer distinct advantages and drawbacks. While some HTFs improve the efficiency and reliability of CSP systems, they may introduce environmental hazards or high costs. Ultimately, the best choice of HTF will depend on balancing performance with economic and environmental considerations. Continued research and development in this field could lead to more sustainable and cost-effective solutions, driving the growth of CSP technology as a key contributor to renewable energy generation.

4 Challenges and Future Directions

Advanced high-temperature HTFs have the potential to improve efficiency, reduce costs, and increase the deployment of CSP plants. However, significant challenges remain. Future directions focus on addressing these challenges to create more efficient, cost-effective, and environmentally sustainable HTFs.

HTFs must maintain stability at high temperatures over long periods. However, many currently available fluids, particularly synthetic oils, degrade at elevated temperatures. Thermal degradation reduces the fluid's efficiency and necessitates regular replacement, which increases operational costs and decreases the reliability of CSP plants. Many advanced HTFs, such as synthetic oils and liquid metals, are expensive due to complex manufacturing processes or the scarcity of raw materials. Liquid metals, such as liquid sodium, offer excellent thermal properties but are costly and potentially hazardous. Advanced HTFs must be compatible with the materials used in CSP system components, such as pipes, pumps, and heat exchangers. Some high-temperature HTFs, particularly liquid metals, can be corrosive to specific materials, while molten salts may solidify at lower temperatures, leading to blockages in the system. HTFs operating at high-temperatures pose safety risks, including leaks, spills, fires, and explosions. Synthetic oils are flammable, while molten salts and liquid metals can present operational hazards if not carefully managed. The environmental footprint of HTFs, including their production, use, and disposal, must be minimized. Some fluids, particularly synthetic oils, have environmental drawbacks due to their non-biodegradable nature and potential to release toxic by-products during degradation. Liquid metals, while promising, often involve complex and energy-intensive production processes. Many advanced HTFs, especially molten salts, have high melting points, leading to freezing risks during low sunlight or plant downtime.

This necessitates additional heating to keep the fluid in a liquid state, which can consume energy and increase operating costs.

Ongoing research is focused on developing novel HTFs that can operate at ultra-high temperatures (above 600 °C) with improved thermal stability and low environmental impact. Ionic liquids, nanofluids, and other advanced composites are being explored as potential alternatives to traditional fluids. Combining HTFs with dual functionalities for heat transfer and storage is gaining attention. Research into hybrid systems, where the HTF also functions as the energy storage medium, such as with molten salts, aims to simplify and optimize CSP systems by integrating heat transfer and storage into a single process. Researchers are investigating ways to lower the melting points of molten salt mixtures through innovative chemical formulations or additives. For example, mixing chlorides, carbonates, or fluorides with nitrates can reduce the freezing point without compromising thermal properties. Supercritical CO_2 and other high-temperature gases are being investigated as HTFs because of their high thermal efficiency at extreme temperatures. Supercritical fluids can improve power cycle efficiency and reduce the physical size of the heat exchange components. Improving the corrosion resistance of CSP system materials is a priority for enabling the use of more advanced HTFs, such as molten salts and liquid metals. Research into advanced alloys and coatings that can withstand the high temperatures and corrosive environments created by HTFs is underway.

5 Conclusion

The development of advanced HTFs is pivotal to the evolution of high-temperature solar thermal systems, especially for CSP technologies. As the demand for efficient, cost-effective, and sustainable renewable energy sources grows, HTFs capable of operating at elevated temperatures play a critical role in enhancing system performance, improving thermal energy storage integration, and reducing operational costs. This chapter has examined the key challenges in the field, including thermal stability, material compatibility, cost considerations, environmental impact, and operational safety. While current HTFs such as molten salts, synthetic oils, and liquid metals offer various benefits, they also present limitations that impede the full optimization of CSP systems. Issues like thermal degradation, corrosion, and high melting points require innovative solutions to ensure the reliability and efficiency of solar thermal applications. Future research directions, including exploring novel HTFs such as ionic liquids, nanofluids, and supercritical gases, hold significant promise for overcoming these barriers. Advances in lowering the freezing points of molten salts, improving corrosion resistance, and integrating HTFs with thermal energy storage systems will be key drivers for the next generation of CSP plants. Furthermore, the push toward more environmentally friendly, recyclable, and cost-efficient HTF will support the broader adoption of solar thermal technologies in the global energy market. In conclusion, the development of advanced HTFs is essential for the future of high-temperature solar thermal systems. Continued innovation, alongside integrating hybrid renewable energy systems, will be instrumental in positioning CSP as a leading solution in the transition toward a cleaner and more sustainable energy landscape.

References

1. P. Kosmopoulos, Chapter 1 - Solar irradiance and exploitation of the sun's power, in *Planning and Management of Solar Power from Space,* ed. by P. Kosmopoulos (Academic Press, 2024), pp. 1–20. https://doi.org/10.1016/B978-0-12-823390-0.00006-5
2. Solar Heat Worldwide, International Energy Agency (IEA) (2024). https://www.iea-shc.org/solar-heat-worldwide (Accessed 9 Sept 2024)
3. S.A. Kalogirou, Chapter 10 - Solar thermal power systems, in *Solar Energy Engineering* 3rd edn. ed. by S.A. Kalogirou (Academic Press, 2024), pp. 559–599. https://doi.org/10.1016/B978-0-323-99350-0.00002-8
4. M. Imran Khan, F. Asfand, S.G. Al-Ghamdi, Progress in research and technological advancements of commercial concentrated solar thermal power plants. Solar Energy **249**, 183–226 (2023). https://doi.org/10.1016/j.solener.2022.10.041
5. S. Bonk, N. Sau, M. Uranga, T. Hernaiz, Bauer, advanced heat transfer fluids for direct molten salt line-focusing CSP plants. Prog. Energy Combust. Sci.. Energy Combust. Sci. **67**, 69–87 (2018). https://doi.org/10.1016/j.pecs.2018.02.002
6. L. Zhao, J. Wang, L. Cui, B. Li, X. Du, H. Wu, Performance design of high-temperature chloride salts as thermal energy storage material. J. Therm. Sci. **33**, 479–490 (2024). https://doi.org/10.1007/s11630-024-1921-4
7. Y. Zhang, X. Wang, Q. Hu, P. Li, Q. Liu, B. Xu, Experimental study of eutectic molten salts NaCl/KCl/ZnCl2 heat transfer inside a smooth tube for high-temperature application. J. Sol. Energy Eng. **144** (2022). https://doi.org/10.1115/1.4053974
8. X. Wang, J. Del Rincon, P. Li, Y. Zhao, J. Vidal, Thermophysical properties experimentally tested for NaCl-KCl-MgCl2 eutectic molten salt as a next-generation high-temperature heat transfer fluids in concentrated solar power systems. J. Sol. Energy Eng. **143** (2021). https://doi.org/10.1115/1.4049253
9. K.E. Gomna, N.L. N'Tsoukpoe, Y. Pierrès, Coulibaly, Thermal stability of a vegetable oil-based thermal fluid at high temperature. Afr. J. Sci. Technol. Innov. Dev.Innov. Dev. **12**, 317–326 (2020). https://doi.org/10.1080/20421338.2020.1732080
10. T.C. Paul, A.K.M.M. Morshed, E.B. Fox, A.E. Visser, N.J. Bridges, J.A. Khan, Thermal performance of ionic liquids for solar thermal applications. Exp. Therm. Fluid. Sci. **59**, 88–95 (2014). https://doi.org/10.1016/j.expthermflusci.2014.08.002
11. V.V. Wadekar, Ionic liquids as heat transfer fluids—an assessment using industrial exchanger geometries. Appl. Therm. Eng. **111**, 1581–1587 (2017). https://doi.org/10.1016/j.applthermaleng.2016.04.156
12. N. Lorenzin, A. Abánades, A review on the application of liquid metals as heat transfer fluid in concentrated solar power technologies. Int. J. Hydrogen Energy **41**, 6990–6995 (2016). https://doi.org/10.1016/j.ijhydene.2016.01.030
13. J. Pacio, T. Wetzel, Assessment of liquid metal technology status and research paths for their use as efficient heat transfer fluids in solar central receiver systems. Solar Energy **93,** 11–22 (2013). https://doi.org/10.1016/j.solener.2013.03.025
14. J. Liu, Y. He, X. Lei, Heat-transfer characteristics of liquid sodium in a solar receiver tube with a nonuniform heat flux. Energies (Basel) **12** (2019). https://doi.org/10.3390/en12081432
15. J. Pacio, Cs. Singer, Th. Wetzel, R. Uhlig, Thermodynamic evaluation of liquid metals as heat transfer fluids in concentrated solar power plants. Appl. Therm. Eng. **60**, 295–302 (2013). https://doi.org/10.1016/j.applthermaleng.2013.07.010

16. S. Manservisi, F. Menghini, A CFD four parameter heat transfer turbulence model for engineering applications in heavy liquid metals. Int. J. Heat. Mass. Transf. **69**, 312–326 (2014). https://doi.org/10.1016/j.ijheatmasstransfer.2013.10.017
17. P. Martínez-Merino, S.D. Midgley, E.I. Martín, P. Estellé, R. Alcántara, A. Sánchez-Coronilla, R. Grau-Crespo, J. Navas, Novel WS2-based nanofluids for concentrating solar power: performance characterization and molecular-level insights. ACS Appl. Mater. Interfaces **12**, 5793–5804 (2020). https://doi.org/10.1021/acsami.9b18868
18. O. Achkari Begdouri, A. El Fadar, Effectiveness of the use of nanofluids in concentrated solar power plants—Electrical and environmental assessment. Renew. Energy Focus **45**, 10–20 (2023). https://doi.org/10.1016/j.ref.2023.02.003
19. A.H.N. Safaei, A. Rashidi, High temperature nanofluids based on therminol 66 for improving the heat exchangers power in gas refineries. Appl. Therm. Eng. **170**, 114991 (2020). https://doi.org/10.1016/j.applthermaleng.2020.114991
20. W. Wang, Z. Wu, B. Li, B. Sundén, A review on molten-salt-based and ionic-liquid-based nanofluids for medium-to-high temperature heat transfer. J. Therm. Anal. Calorim.Calorim. **136**, 1037–1051 (2019). https://doi.org/10.1007/s10973-018-7765-y
21. X. Yaxuan, W. Huixiang, W. Zhenyu, W. Yuting, X. Qian, W. Gang, L. Chuan, D. Yulong, M. Chongfang, Insights into the enhancement mechanisms of molten salt nanofluids. Int. J. PhotoenergyPhotoenergy **2022**, 4912922 (2022). https://doi.org/10.1155/2022/4912922
22. Q. Yu, Y. Lu, X. Zhang, Y. Yang, C. Zhang, Y. Wu, Comprehensive thermal properties of molten salt nanocomposite materials base on mixed nitrate salts with SiO_2/TiO_2 nanoparticles for thermal energy storage. Solar Energy Mater. Solar Cells **230**, 111215 (2021). https://doi.org/10.1016/j.solmat.2021.111215
23. L. Sang, T. Liu, The enhanced specific heat capacity of ternary carbonates nanofluids with different nanoparticles. Sol. Energy Mater. Sol. Cells **169**, 297–303 (2017). https://doi.org/10.1016/j.solmat.2017.05.032
24. J. Cavieres, M. José Inestrosa-Izurieta, D.A. Vasco, J.I. Urzúa, Ionanofluids based on ionic liquid mixtures, a new approach as an alternative material for solar energy storage. J. Mol. Liq. **351**, 118677 (2022). https://doi.org/10.1016/j.molliq.2022.118677
25. J. Liu, F. Wang, L. Zhang, X. Fang, Z. Zhang, Thermodynamic properties and thermal stability of ionic liquid-based nanofluids containing graphene as advanced heat transfer fluids for medium-to-high-temperature applications. Renew. Energy **63**, 519–523 (2014). https://doi.org/10.1016/j.renene.2013.10.002
26. A.L. Famiglietti, Small-scale linear Fresnel collector using air as heat transfer fluid: Experimental characterization. Renew. Energy **176**, 459–474 (2021). https://doi.org/10.1016/j.renene.2021.05.048
27. P. Good, G. Zanganeh, G. Ambrosetti, M.C. Barbato, A. Pedretti, A. Steinfeld, Towards a commercial parabolic trough CSP system using air as heat transfer fluid. Energy Proc. **49**, 381–385 (2014). https://doi.org/10.1016/j.egypro.2014.03.041
28. Q. Li, F. Bai, B. Yang, Y. Wang, L. Xu, Z. Chang, Z. Wang, B. El Hefni, Z. Yang, S. Kubo, H. Kiriki, M. Han, Dynamic simulations of a honeycomb ceramic thermal energy storage in a solar thermal power plant using air as the heat transfer fluid. Appl. Therm. Eng. **129**, 636–645 (2018). https://doi.org/10.1016/j.applthermaleng.2017.10.063
29. L. Cheng, G. Xia, Progress and prospects for research and technology development of supercritical CO_2 thermal conversion systems for power. Energy Storage, and Waste Heat Rec. Heat Transfer Eng. (n.d.) 1–18. https://doi.org/10.1080/01457632.2023.2282765
30. L.F. Cabeza, A. de Gracia, A.I. Fernández, M.M. Farid, Supercritical CO_2 as heat transfer fluid: a review. Appl. Therm. Eng. **125**, 799–810 (2017). https://doi.org/10.1016/j.applthermaleng.2017.07.049

31. S. Bovard, M. Abdi, M.R.K. Nikou, A. Daryasafar, Numerical investigation of heat transfer in supercritical CO_2 and water turbulent flow in circular tubes. J. Supercrit Fluids **119**, 88–103 (2017). https://doi.org/10.1016/j.supflu.2016.09.010
32. T. Ma, W. Chu, X. Xu, Y. Chen, Q. Wang, An experimental study on heat transfer between supercritical carbon dioxide and water near the pseudo-critical temperature in a double pipe heat exchanger. Int. J. Heat Mass Transf. **93,** 379–387 (2016) https://doi.org/10.1016/j.ijheatmasstransfer.2015.10.017

Conversion of Solar Energy to Electrochemical Energy

Materials and Interfaces in Photoelectrochemical Cells

Md. Sultan Mahmud, Al Jumlat Ahmed, Md. Shafiul Alam, Md. Shamim Sarker, and M. S. Hossain Lipu

Abstract The increasing world population and expanding economy demand a steady and reliable power supply. Most of the power supply comes from fossil fuels, raising concerns regarding environmental issues, natural disasters, and diseases. Therefore, the search for alternative fuel is highly crucial, and timely essential to save our planet. The United Nations (UN) declared the SDG goal 2030, whereas goal 7 emphasized the use and accessibility of clean and sustainable energy technologies. For instance, hydrogen fuel could be considered a blessing to achieve this goal due to its lightweight atom, requiring less energy to dissociate from naturally abundant water sources, high energy density comparable to fossil fuel, and zero emission. At present, hydrogen fuel is mainly produced through non-renewable routes such as steam methane reforming, nuclear energy, coal gasification, and other hydrocarbon. In contrast, all methods are neither eco-friendly nor sustainable. Hence, the available suitable options are electrolysis or photolysis of water to produce eco-friendly hydrogen fuel. The electrolysis of water requires a semiconductor, where electrical power serves as the input energy to split water and produce hydrogen. It should be noted that the semiconductor used in this process is called an electrocatalyst. Although the production of hydrogen fuel through electrocatalyst is clean and

---

Md. S. Mahmud (✉)
Department of Electrical and Electronic Engineering, University of Asia Pacific, Dhaka 1205, Bangladesh
e-mail: sultan_mahmud@uap-bd.edu

A. J. Ahmed
CSIRO Manufacturing, Victoria 3169, Australia
e-mail: aljumlat.ahmed@csiro.au

Md. S. Alam
Department of Electrical Engineering, College of Engineering, King Faisal University, Al-Ahsa 31982, Saudi Arabia

Md. S. Sarker
Department of Electrical Engineering and Information Systems, Department of Bioengineering, Graduate School of Engineering, The University of Tokyo, Tokyo 113-8656, Japan

M. S. H. Lipu
School of Electrical, Computer and Telecommunications Engineering, University of Wollongong, NSW 2522, Australia

P. Ravindran et al. (eds.), *Photon to Power*, Progress in Optical Science and Photonics 32, https://doi.org/10.1007/978-981-96-5914-2_8

environmentally safe, the direct consumption of electrical energy is not considered a viable option. On the other hand, solar-driven hydrogen production, also known as solar water splitting, photolysis of water, or photoelectrochemical (PEC) water splitting, is regarded as a promising approach, attracting researchers and scientists worldwide. PEC water splitting requires only a semiconductor (serve as a photocatalyst), a counter electrode, and water. When sunlight strikes the (either n-type or p-type) semiconductor surface, it generates electron–hole pairs, which are efficiently separated. The charge carriers flow through an external circuit to the counter electrode, leading to the production of hydrogen fuel. Here, the most important work is to develop an efficient semiconductor that can absorb a large portion of the light, split the water, and achieve the benchmark solar-to-hydrogen (STH) conversion efficiency. Since the invention of PEC hydrogen production, numerous semiconductor materials extensively explored to reach the benchmark efficiency but no semiconductor achieved the goal till now because of various limitations such as large bandgap, unavailability, slower oxygen evolution reaction kinetics, and less stability. In this chapter, we discussed the fundamental principles of PEC solar water splitting, focusing on both three-electrode and two-electrode system mechanisms. We also explored the various semiconductor material, synthesis methods for various semiconductor nanostructures, along with their modification techniques and inherent limitations. Additionally, an overview of solar-to-hydrogen (STH) evaluation techniques also provided, offering valuable insights for beginners and researchers into PEC materials.

Keywords PEC hydrogen production · PEC materials · Modification of materials · STH efficiency

1 Introduction

Energy is a fundamental pre-requisite for the overall progress of a society and a country, particularly essential for fostering industrial growth and driving economic development. Therefore, humans needed energy for survival, which drove them to discover fire as an initial source of energy in ancient periods. With changing times, humans gradually started to seek sustainable energy sources. In this continuous progress, wood was initially used, followed by coal, later replaced by oil and gas as primary energy sources [1]. Throughout history, human energy sources have transferred from wood to coal with Watt's steam engine in 1769, followed by electricity generation in France in 1875, and later, Daimler's 1886 invention of the internal combustion engine increased the demand for oil and gas as efficient energy options [1]. Over time, these energy sources have been harnessed to drive societal progress, foster industrial and economic revolutions, and improve the quality of life. While fossil fuels have accelerated the overall progress of humanity, the toxic gases emitted from these sources, such as CO_2, CO, nitrous oxides, and sulfur dioxides, significantly contribute to environmental pollution. For example fossil fuels produce about

75% of the world's greenhouse gases and 90% of its carbon dioxide [2]. This pollution is responsible for the greenhouse effect, natural disasters, and various diseases. On the other hand, energy demand is predicted to grow by 56% by 2040, and if we keep depending on fossil fuels, this will lead to higher greenhouse gas emissions [2]. In addition, approximately 80% of the global population relies on fossil fuels, which are often imported from foreign countries, leading to energy crises caused by geopolitical factors that affect around 6 billion people [2]. Consequently, the widespread adoption of renewable energy is timely essential. Historically, renewable energy dates back more than 5,500 years, with early examples including wind energy in Mediterranean sailing ships, and evidence of windmill used in India over 2,500 years ago [3]. Apart from this, solar energy was used in the eighteenth century with solar furnaces for melting metals, and its utilization expanded in 1839 with the development of photovoltaic effect by Becquerel for the using of solar energy, and 1883 with the development of Ericsson's hot air engine, improving solar thermal energy conversion [3]. Presently, in the twenty-first century, solar and wind energy are the widely used renewable energy sources. Solar energy captured by photovoltaic cell and wind energy is captured by wind turbines. Apart from this, other renewable sources such as hydro-energy, bio-energy, ocean energy, and geothermal energy contribute to eco-friendly power generation which helps to reduce the carbon emissions and addressing the global demand for clean energy. The currently used renewable energies are weather-dependent which imposes the challenge to ensure the continuous and reliable power supply due to the fluctuation of energy generation. For instance, climate change significantly affects wind speed and their distribution based on place and time, as wind energy output is directly proportional to the cube of wind speed, with extreme weather causing fluctuations in wind speeds and thereby disrupting power production, the increasing temperature and cloud coverage impact the efficiency and performance of solar PV modules, due to the changes in rainfall, runoff, and extreme weather, like (e.g., heavy rain or droughts) significantly affect the hydroelectric power generation [4]. Therefore, considering the drawbacks of conventionally used renewable energy, environmentally friendly hydrogen fuel generation and storage systems could not only help decarbonize the energy sector but also ensure continuous power production. Hydrogen fuel is seen as a promising solution to achieve the goal because its lightweight atom requires less energy to extract from abundant water sources, has a high energy density similar to fossil fuels, and produces zero emissions. Currently, around 80% of hydrogen fuel is produced from various fossil fuels, such as natural gas (steam methane reforming), coal (coal gasification), and nuclear sources (high-temperature water radiolysis), whereas most of the method releases greenhouse gases and toxic element [5]. In addition, some renewable sources are also currently being used to produce the hydrogen fuel such as biomass gasification, geothermal energy, algea, and wind energy, which suffered from numerous disadvantages like greenhouse emission, impurities H_2S, toxic element, complex structure, and low efficiencies [5]. Therefore, the only available options for producing sustainable and eco-friendly hydrogen are electrolysis and photolysis of water. In brief, electrochemical water splitting, or electrolysis, uses a semiconductor as an electrocatalyst, along with a reference electrode and a counter electrode, with electrical energy applied as the input to split water. Since electrolysis of water splitting consumes electrical energy, it cannot be considered an economically or environmentally viable technique. Therefore, the photolysis of water, widely known as solar water splitting or photoelectrochemical (PEC) hydrogen fuel production, is considered environmentally friendly due to its zero emissions and sustainability. Around 57 years ago, in 1967, Akira Fujishima, a doctoral student under the supervision of Professor Kenichi Honda at Tokyo University, made a groundbreaking discovery in the field of water photolysis. He observed that when titanium dioxide (TiO_2) was exposed to intense light which facilitated the decomposition of water. Later, the discovery became known as the Fujishima-Honda effect, attracting researchers worldwide. Since the discovery, researchers have attempted to develop efficient single semiconductors as photocatalysts for practical applications. However, these efforts have faced many challenges, such as large band gaps, limited availability of materials, slow oxygen evolution reaction kinetics, and low stability in alkaline solutions. It should be mentioned that the semiconductor need to achieve a benchmark solar-to-hydrogen (STH) conversion efficiency of 10% [6]. In this chapter, we focused on and described the basic principles of solar water splitting, various semiconductor materials used as photoelectrode so far, their limitations and fabrication of nanostructuring, further developments, and finally, the techniques for evaluating their efficiency. This will provide overall valuable insights for beginners and researchers in this field.

2 Photoelectrochemical (PEC) Hydrogen Production and Measurement

Photoelectrochemical (PEC) hydrogen production, commonly referred to as solar water-splitting, is a clean, emission free, and sustainable way to generate hydrogen fuel using sunlight. PEC hydrogen production is a thermodynamically uphill reaction, splitting water (H_2O) molecule into hydrogen (H_2), and oxygen ($1/2\ O_2$) requires 2 electron–hole pairs [7]. These means producing 1 mol of hydrogen (H_2) and oxygen (O_2) requires 4 electron–hole pairs to complete the overall reaction, which is equivalent to the Gibbs free energy of 237.2 kJ/mole. In the PEC mechanism, when sunlight strikes the photoelectrode (considering an n-type semiconductor) immersed in water, electron–hole pairs are generated. These pairs separate efficiently, with the electrons flowing through the external circuit to the counter electrode, whereas electrons reduce the water into hydrogen. Meanwhile, the holes in the semiconductor oxidize water into produce the oxygen. It is important to note that if the semiconductor is p-type (photocathode), the reduction reaction occurs on the semiconductor surface to produce hydrogen, while the oxidation reaction takes place at the counter electrode. Conversely, if the semiconductor is n-type (photoanode), the oxidation reaction occurs on the semiconductor surface, and the reduction reaction takes place at the counter electrode. On a laboratory scale, PEC performance uses three-electrode and two-electrode setups which connected with a potentiostat. In a three-electrode setup, a semiconductor is used as the photoelectrode, a platinum (Pt) electrode serves as the dark electrode (commonly known as the counter electrode), a reference electrode, electrolyte solution (acidic or base), and Xenon lamp or solar simulator as a light source (ideally 100 mW/cm^2). As a reference electrode, the most common type used is silver/silver chloride (Ag/AgCl) with saturated KCl, depending on the solution type. For example, in alkaline solutions like sodium hydroxide (NaOH), silver/silver chloride (Ag/AgCl) (saturated potassium chloride KCl) is often used to ensure the stability and accuracy of the measurement. In addition, the mercury/mercury oxide (Hg/HgO) electrode is also used as a reference electrode in alkaline solutions. In a two-electrode setup, only the semiconductor and the dark electrode are used. The difference between the three-electrode and two-electrode setups is the reference electrode, and biasing potential. The reference electrode used to ensure the stable potential of photoelectrode without interference from the counter electrode or electrochemical reaction in the three-electrode system. A two-electrode setup is used when no external bias potential is required (known as unassisted PEC) to split water, meaning that light alone is sufficient to generate the necessary potential. In this setup, there is no precise control over the potential of the photoelectrode, which is called as real world configuration. Here, the biasing potential refers to the minimum potential required to split water into hydrogen and oxygen. For instance, theoretically, the minimum biasing potential required for water splitting is 1.23 V, but in practice, it can range from 1.5 to 1.9 V due to energy losses, such as from heat dissipation. It should be mentioned that the biasing potential is only applicable for the three-electrode system in laboratory scale hydrogen production. In an ideal case, or for practical PEC applications, when photon energy hits the semiconductor surface, it produces electron–hole pairs and separate if the photon energy is greater than the band gap energy, whereas, band gap energy must be at least 1.23 eV to straddle the water. This means that, in practical applications, both the photon energy and band gap energy must be satisfied, and the system should operate in a two-electrode setup without any external bias potential. However, the three-electrode experimental setup in the laboratory consists of a working electrode (either a p-type or n-type semiconductor), a reference electrode, and a counter electrode, all connected to a potentiostat as previously mentioned, as shown in Fig. 1.

The primary focus is the photoelectrochemical (PEC) performance analysis of the working electrode, which is the semiconductor. Here, the performance of the working electrode is typically evaluated by measuring the photocurrent density as a function of the applied potential, using the linear sweep voltammetry (LSV) method selected on the potentiostat. A higher photocurrent density generally indicates increased hydrogen fuel production. In addition, the stability of the photoelectrode is measured by the photocurrent density versus time, using the amperometry method. However, the actual hydrogen production by the photoelectrode should be measured using gas chromatography.

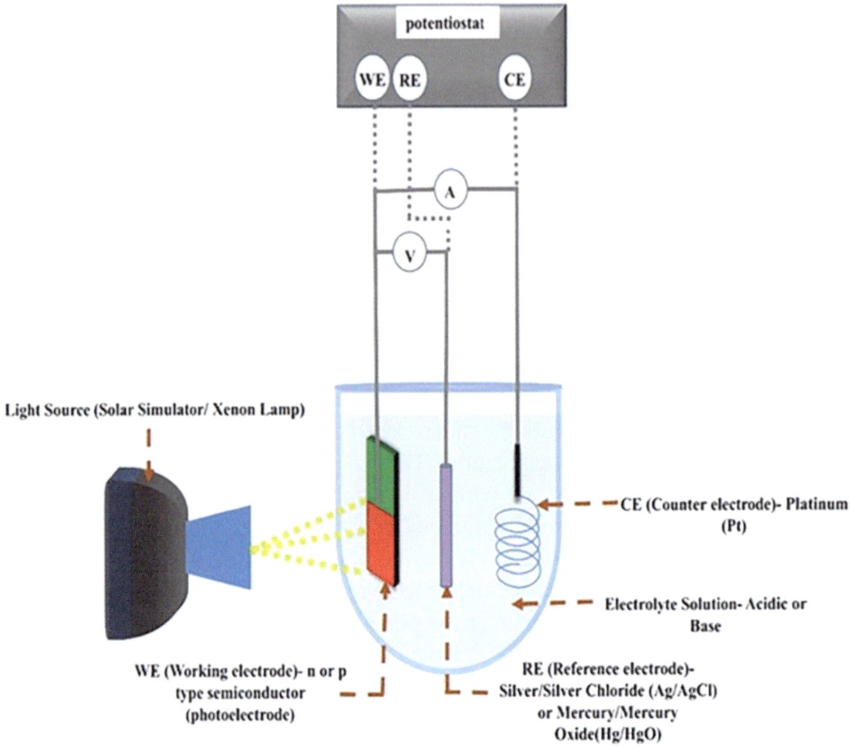

Fig. 1 PEC (three-electrode laboratory setup) hydrogen production

3 Semiconductor Materials as Photoelectrodes

The development of highly efficient semiconductor materials is a key challenge for the successful application of PEC hydrogen production. Since the invention of PEC technology, numerous semiconductor materials have been investigated to achieve practical efficiency and application. However, due to various limitations, no semiconductor material has achieved the desired efficiency. P-type and n-type semiconductors are used as photoelectrodes in PEC systems. Specifically, if the material is p-type, it functions as a photocathode, while an n-type material serves as a photoanode. In this section, we have discussed different semiconductor materials, their advantages, and limitations. This will aid in identifying the most efficient materials for further development. Based on the widely used, we have classified the photocatalyst semiconductors into several sub-sections below.

3.1 *Oxides Based Semiconductor*

P-type and n-type oxide semiconductors are widely used in the field of photoelectrochemical (PEC) performance. The commonly used semiconductor materials for this purpose include Cu_2O, CuO, NiO, TiO_2, ZnO, WO_3, and α-Fe_2O_3. Another type of oxide based semiconductor is bimetallic semiconductors, composed of two different metal elements, are promising materials for solar water splitting because of their unique electronic and catalytic properties such as delafossite $CuFeO_2$, lanthanum ferrite oxide $LaFeO_3$, calcium iron oxide $CaFe_2O_4$, Bismuth vanadate $BiVO_4$, and Copper tungstate $CuWO_4$.

3.1.1 Cuprous Oxide and Cupric Oxide

When categorizing photoelectrochemical (PEC) materials, Cu_2O (cuprous oxide) would be positioned at the forefront due to its numerous advantageous properties which includes suitable direct bandgap of 2 eV, available, non-toxic, high solar radiation absorption (300–620 nm) which covers 50% of visible photons, higher theorctical photocurrent density of -14.7 mA/cm^2, and higher theoretical solar-to-hydrogen conversion efficiency of 18% which exceed the benchmark efficiency of 10% for the practical application [8]. Here, most importantly the band positions significantly enhance photocurrent density and efficiency of Cu_2O. The conduction band is more negative than the hydrogen evolution reaction (-0.7 eV vs. RHE), while the valence band is more positive than the oxygen evolution reaction (1.3 eV vs. RHE), making it ideal for PEC applications [9]. Apart from this, CuO (cupric oxide), a p-type semiconductor from the same family possesses a lower bandgap of 1.2 eV– 1.7 eV, excellent band alignment, and higher light absorption, thereby, higher theoretical photocurrent density of 35 mA/cm^2 compared to Cu_2O [10]. Although CuO demonstrates a higher photocurrent density due to its narrow band gap, it has several disadvantages: its conduction band and valence band level is less negative and positive than those of Cu_2O, and it is also less stable than Cu_2O in aqueous solution, therefore, Cu_2O is considered to be superior to CuO. In summary, the main drawbacks associated with copper oxide based photocathode is poor minority carrier diffusion length (20 nm -100 nm), higher light absorption depth (10 μm), poor stability, and photo corrosion (photo generated electron –hole reduce the photocathode rather than production of H_2) [10, 11].

3.1.2 Nickel Oxide

NiO (Nickel Oxide) is less studied comparatively than other p-type semiconductor material for the solar water splitting as a pristine single photoelectrode. NiO is an attractive candidate due to its wide bandgap (3.6–4.0 eV), low cost, stability, high mobility of charge carrier, and resistance to photo corrosion [12–15]. On the other hand, NiO is widely used as a co-catalyst in photoelectro-chemical (PEC) applications instead of as a single photocathode because it reduces the overpotential, increases the photocurrent density of the primary semiconductor, and passivates unstable semiconductors [16]. These properties make NiO suitable for efficient and durable PEC applications when combined with various semiconductor materials.

3.1.3 Titanium Dioxide

Titanium dioxide (TiO_2) was the first semiconductor used in photoelectrochemical (PEC) applications. Since its introduction, extensive efforts have been made to enhance its photocatalytic activity for more efficient PEC performance. TiO_2 is an n-type semiconductor with numerous advantages such as being photoactive, stable under extreme conditions (resist chemical and photo corrosion), cheap, and environmentally friendly [17]. Titanium dioxide (TiO_2) has properties that make it a suitable candidate for PEC applications, but it has also some drawbacks such as wide bandgap of 3 eV that allows it only to absorb the UV light, resulting in high recombination rate of electron–hole and lower the theoretical solar-to-hydrogen (STH) conversion efficiency of just 2.2%, which is much lower than what is needed for practical applications [17].

3.1.4 Zinc Oxide

Zinc Oxide (ZnO) is a third-generation n-type semiconductor material with excellent photocatalytic properties. ZnO offers several advantages, including a direct wide bandgap of 3.37 eV, similar to that of TiO_2, but with electron mobility that is 100 times higher than TiO_2 [18]. It has a strong capacity to maintain an electric field, along with a suitable band structure: the bottom of the conduction band is more negative than the hydrogen reduction potential, while the top of the valence band is more positive than the oxidation potential [19]. Additionally, ZnO exhibits low reflectivity, high light absorption, and an ability to adsorb O_2 molecules due to its surface states, making it a strong candidate for photoelectrochemical (PEC) applications [19]. However, ZnO has some limitations, such as its wide bandgap, which restricts light absorption primarily to the UV range, covering only around 4% of the solar spectrum [19]. Apart from this, it exhibits a high rate of electron–hole recombination at the interface between the surface and the electrolyte, along with limited surface reactivity [19].

3.1.5 Tungsten Oxide

Tungsten oxide (WO_3) is a widely used n-type semiconductor known for its abundance, adaptable composition, good chemical stability in various pH environments, and excellent electrical conductivity [20]. WO_3 can form several crystal structures, such as tetragonal (α-WO_3), orthorhombic (β-WO_3), monoclinic I (γ-WO_3), triclinic (δ-WO_3), and monoclinic II (ε-WO_3) [21]. Among all other phases, the monoclinic phase is particularly suitable for PEC applications because it has a balanced bandgap ($E_g = 2.5–2.8$ eV), a moderate hole diffusion length of about 150 nm (which is higher than α-Fe_2O_3 at 2–4 nm), and better electron transport properties than TiO_2 [20, 21]. Furthermore, WO_3 has a theoretical solar-to-hydrogen conversion efficiency of 4.8%, making it a strong candidate for use in solar-driven hydrogen production [20]. Nevertheless, WO_3 exhibits limitations, including the slow kinetics of holes, rapid recombination of charge carriers, and inefficient charge transfer at the semiconductor/electrolyte interface [22].

3.1.6 Hematite

The n-type α-Fe_2O_3 (hematite) is widely recognized as a promising material for photoelectrochemical (PEC) applications due to its high theoretical solar-to-hydrogen conversion efficiency. Fe_2O_3 exists in five crystalline forms: α- Fe_2O_3 (hematite), β- Fe_2O_3 (magnetite-like or bixbyite structure), γ- Fe_2O_3 (spinel structure), ε- Fe_2O_3 (orthorhombic structure), and ζ- Fe_2O_3 (monoclinic structure) [23, 24]. Among these, only α- Fe_2O_3 and β- Fe_2O_3 demonstrate the properties needed for effective solar water splitting, primarily because of their suitable bandgap energy and high theoretical solar-to-hydrogen (STH) conversion efficiency. Although β- Fe_2O_3 has a favorable 1.9 eV bandgap and a theoretical STH efficiency of 20.9%, its instability at high temperatures, where it often transforms into hematite, and its rarity in nature limit its practical application [23, 25]. Consequently, α-Fe_2O_3 is the preferred phase for PEC applications due to its abundance, cost-effectiveness, stability in alkaline environments, optimal 2.2 eV bandgap for absorbing visible light, high thermal stability, non-toxicity, and high STH potential of 15%, making it especially suited for efficient solar water splitting [26–28].

3.1.7 Delafossite

Several copper-based delafossite materials, such as $CuMO_2$ (where M = Cr, Al, Fe, Ga, Rh), have shown significant potential for photoelectrochemical (PEC) applications [29]. Among these, only $CuFeO_2$, a p-type semiconductor, is considered a highly suitable photocathode material for PEC [29]. It provides numerous advantages, including a suitable bandgap of 1.5 eV, an optimal conduction band position of –0.4 V vs. the reversible hydrogen electrode (RHE) that accelerates the efficient reduction reactions, and additional benefits like abundance, excellent stability, and high absorptivity [30]. Although $CuFeO_2$ possesses an ideal bandgap, its limited visible light absorption reduces charge separation efficiency, negatively impacting its solar-to-hydrogen conversion efficiency [31].

3.1.8 Lanthanum Ferrite Oxide

The lanthanum ferrite oxide ($LaFeO_3$) is a recently developed p-type semiconductor with promising applications in photoelectrochemical (PEC) systems. It provides key advantages, such as an ideal band gap of 2.1–2.6 eV, which facilitates absorption across a broad light spectrum, particularly in the visible range [32]. Additionally, $LaFeO_3$ demonstrates notable chemical stability, photo stability, and structural flexibility [32, 33]. However, it also faces challenges that limit its efficiency in PEC applications, such as high charge carrier recombination, restrict the charge transfer rate, and lower visible light response [33].

3.1.9 Calcium Iron Oxide

Another bimetallic p-type semiconductor, calcium iron oxide ($CaFe_2O_4$), is a suitable candidate for photoelectrochemical (PEC) applications. To the best of knowledge, Omae et al. first discovered that $CaFe_2O_4$ act as a photocathode after observing the narrow band gap and visible light absorption [34]. $CaFe_2O_4$ has a band gap of 1.9 eV, with a conduction band position of –0.6 V vs. the normal hydrogen electrode (NHE) and a valence band position of 1.3 V vs. NHE and low cost of material indicates the suitable candidate [35] but the diffusion length of charge carrier causes limit the performance [36].

3.1.10 Bismuth Vanadate

Bismuth vanadate ($BiVO_4$) has recently attracted significant attention from researchers due to its theoretical solar-to-hydrogen conversion efficiency of 9.1%, which is close to the benchmark efficiency needed for practical applications [37]. $BiVO_4$, an n-type semiconductor, presents several key advantages for solar applications. Its direct bandgap of 2.4 eV allows it to absorb approximately 11% of the sunlight reaching Earth [37]. Furthermore, with a conduction band positioned at 0.02 V vs. RHE, it requires minimal bias potential to initiate the reaction, making it an efficient candidate for solar-driven processes [38]. Although $BiVO_4$ has many advantages, it still suffers from low electron conductivity and slower kinetics of oxygen evolution [38].

3.1.11 Copper Tungstate

Copper tungstate ($CuWO_4$) is an n-type semiconductor with a bandgap around 2.3 eV that allows light absorption with wavelengths less than 540 nm, possesses a valence band position at 2.60 V versus NHE (normal hydrogen electrode) that is beneficial for water oxidation, and corresponding theoretical STH of 13% [39, 40]. In addition, $CuWO_4$ is environmentally friendly with high Faradaic efficiency, and maintains stability across a wide pH range (up to 9.5), making it reliable for long-term photoelectrochemical application [39]. Nevertheless, the photoelectrochemical (PEC) performance of $CuWO_4$ significantly restricted because of its relatively low light absorption coefficient, and higher internal resistance to charge transfer, which further constrains its ability to achieve the theoretical photocurrent densities anticipated for this material [40].

3.2 Chalcogenides-Based Semiconductor

Chalcogenide semiconductors are materials formed when chalcogen elements, such as sulfur (S), selenium (Se), or tellurium (Te), combine with transition metals to create compounds known as chalcogenide semiconductors. Various p-type and n-type chalcogenide semiconductors are available for photoelectrochemical (PEC) applications such as Cadmium sulfide CdS, Zinc sulfide ZnS, and Copper gallium disulfide $CuGaS_2$.

3.2.1 Cadmium Sulfide

CdS (Cadmium sulfide) is one of the most widely investigated chalcogenide semiconductor materials. The CdS hold the narrow bandgap of 2.4 eV as well, the position of conduction band level is – 0.5 V versus normal hydrogen electrode (NHE) and valence band position of 1.7 V versus NHE [41] that facilities the absorption of large portion of solar spectrum, consequently, accelerate the reaction mechanism. However, the limitation such as photo corrosion and rapid recombination charge carrier of CdS prevent from the maximum PEC performance [41]. On the other hand, due to the drawbacks of CdS, it is not used alone as a single photoelectrode [42]. Instead, it is widely used as a light-absorbing sensitizer to enhance the PEC performance of other wide band gap materials, such as TiO_2, providing insightful and effective strategies for the forming of hetero-structures [42].

3.2.2 Zinc Sulfide

Zinc sulfide (ZnS) is an attractive semiconductor material for photoelectrochemical (PEC) systems due to its numerous advantages. The benefits of ZnS include rapid charge carrier generation, a relatively high conduction band position, and abundant active sites that enhance the hydrogen generation reaction [43]. On the other hand, ZnS has a wide band gap of 3.6 eV, limiting its light absorption to the UV range and preventing it from absorbing visible light [44]. This limitation negatively impacts overall performance by leading to higher charge carrier recombination, lower photocurrent density, and reduced solar-to-hydrogen (STH) conversion efficiency

[43, 44]. Furthermore, compared to other wide band gap materials, ZnS has a relatively larger band gap, making it crucial to find ways to reduce it. One approach to address this is through nanostructuring, which may help to lower the band gap and enhance its PEC performance.

3.2.3 Copper Gallium Disulfide

Copper gallium disulfide ($CuGaS_2$), a member of the chalcopyrite family of I-III-VI compound semiconductors, provides significant advantages. These semiconductors are characterized by a relatively small bandgap, high light absorption coefficient, and excellent stability under illumination [45]. Additionally, they demonstrate robust chemical stability in electrolyte solutions, making them highly suitable for PEC applications [45]. Among chalcopyrite materials, $CuGaS_2$ is particularly attractive for photoelectrochemical (PEC) research due to its direct bandgap of 2.43 eV, which enables effective visible light absorption [45]. This material can be easily doped into either n-type or p-type, is non-toxic, and exhibits a high absorption coefficient, making it an excellent candidate for PEC applications [45].

3.3 Group III-V Based-Semiconductors

III-V semiconductors are the part of main group family, such as Gallium Phosphide (GaP), Indium Phosphide (InP), and Silicon (Si), which are considered effective materials for photoelectrochemical (PEC) systems. The advantages of III-V semiconductors include excellent optical properties, favorable band positions that facilitate redox reactions, and high theoretical solar-to-hydrogen (STH) conversion efficiency [46]. However, the main problem associated with III-V semiconductors are highly corrosive in the strong acidic and base media which result in photoelectrodes degradation [46].

3.3.1 Gallium Phosphide and Indium Phosphide

Gallium Phosphide (GaP) is used in a range of applications, from light-emitting diodes (LEDs) to photoelectrochemical (PEC) application. In PEC systems, GaP is a proven semiconducting material with a bandgap of 2.3 eV, which allows for a potential theoretical solar-to-hydrogen (STH) conversion efficiency of up to 12% [47]. Gallium Phosphide (GaP) offers several advantages for application in PEC systems, including a high conduction band position suitable for hydrogen evolution and ability to absorb a significant portion of the light irradiation [46, 47]. On the other hand, the low light absorption of GaP presents several limitations, such as a slower oxidation reaction due to the reduced kinetics of hole transfer, therefore, it requires over potential [46, 48].

In addition, Indium Phosphide (InP), a semiconductor from the same family, offers high solar radiation absorption due to its narrow bandgap of 1.3 eV [49]. Its conduction band position is also higher than the water reduction potential, making it suitable for charge transfer to facilitate water reduction [49].

3.3.2 Silicon

Silicon (Si), one of the most abundant and cost-effective semiconducting materials, is the backbone of photovoltaic devices, with silicon-based photovoltaic cells already achieving an efficiency of 25% [50]. The small bandgap of silicon (1.12 eV) enables it to absorb a large portion of the solar spectrum, making it an excellent candidate for PEC applications [50]. Therefore, commercial silicon photovoltaic (Si PV) systems have already been employed in solar water splitting for hydrogen production, achieving theoretical solar-to-hydrogen (STH) conversion efficiency [50]. However, these systems are associated with complexity and high costs. For instance, it is necessary to apply a single Si photoelectrode alone to avoid the limitation, but the single Si photoelectrode is unstable without any protection layer, high recombination rate, and slower reaction kinetics at the surface [50, 51].

3.4 Nitride–Oxynitride-Based Semiconductor

Tantalum oxynitride is a semiconducting material from the nitride-oxynitride family. Tantalum oxynitride-based photoelectrode showed the PEC performance due to its bandgap energy between 1.9 and 2.5 eV and the suitable band position, but the solar-to-hydrogen (STH) conversion efficiency is poor which is less than 1% [52]. However, tantalum oxynitride is classified into three forms: tantalum oxynitride (TaON), tantalum nitride (Ta_3N_5), and perovskite-related tantalum oxynitrides [52], whereas TaON, and Ta_3N_5 are the potential candidate and extensively investigated material for the solar-driven hydrogen production.

3.4.1 Tantalum Oxynitride

Tantalum oxynitride (TaON) is an n-type semiconductor material, exhibits as an effective photoanode material for PEC system because of its valence and conduction band edge positions which are the favorable for oxidation and reduction reaction, and narrow bandgap of 2.4 eV enables visible light absorption [53, 54]. However, TaON PEC efficiency still far from the benchmark efficiency which results from the self-deactivation under irradiation, converted into Ta_2O_5, high exciton binding energy, and rapid electron–hole recombination [53]. In addition, its needed external bias to operate the reaction [53].

3.4.2 Tantalum Nitride

The PEC properties of the tantalum nitride (Ta_3N_5) was the first discovered by Japanese researchers in 2004 [55]. Since the discovery of PEC properties of Ta_3N_5, it has been regarded as an excellent photoanode. Ta_3N_5 is an n-type semiconductor with a bandgap of 2.1 eV, and its band positions are suitable for splitting water into hydrogen and oxygen [56]. The most exciting thing about these materials is the higher theoretical photocurrent density of 12.9 m A/cm^2, which is the corresponding theoretical STH efficiency of 15.9% [56] comparable with single metallic oxides semiconductors (e.g., α- Fe_2O_3 and Cu_2O). In addition, its abundance, low supply risk, visible light absorption, and zero bias potential make this material suitable for PEC applications [56]. Despite recognized as an excellent semiconductor, the unstable during PEC operation, and poor efficiency of surface charge accumulation restrict the overall performance of Ta_3N_5 [57].

4 Nanostructure Surface of Semiconductors and Modification Strategy

Photoelectrochemical (PEC) semiconductors face numerous limitations; therefore, it is essential to fabricate them in nanostructured forms to address these challenges effectively. The nanostructuring of semiconductors can potentially alter their band structure, enhancing light absorption up to the maximum visible range, and increasing the surface area available for chemical reactions. The fabrication of nanostructures can increase the surface area, reduce recombination, improve charge separation, and increase transport efficiency, consequently, these advantages can lead to higher photocurrent density and solar-to-hydrogen (STH) conversion efficiency. This section will provide a step-by-step outline of key focus areas. Initially, emphasis should be placed on fabricating pristine semiconductor nanostructures to reduce costs while striving to achieve the maximum theoretical solar-to-hydrogen (STH) conversion efficiency for the respective materials. Although no pristine single semiconductor has achieved the desired solar-to-hydrogen (STH) conversion efficiency to date, it is crucial to fabricate and investigate various nanostructures, such as nanowires, films, nanotubes, nanoflowers, and nanorods, which have the potential to enhance overall performance. Here, we will focus on widely studied materials, such as n-type α-Fe_2O_3 (hematite) and p-type Cu_2O (cuprous oxide), as model semiconductors. These materials will be used to demonstrate various nanostructure fabrication strategies and their subsequent enhancements through different modification approaches, providing valuable insights for other semiconductors as well. For instance, hematite nanostructures can be fabricated using various methods. High-density pristine nanowires, for example, can be produced through stress-induced atomic diffusion or thermal oxidation. In the both methods, an iron substrate is simply heated in an atmospheric environment after pre-treatment of the iron substrate

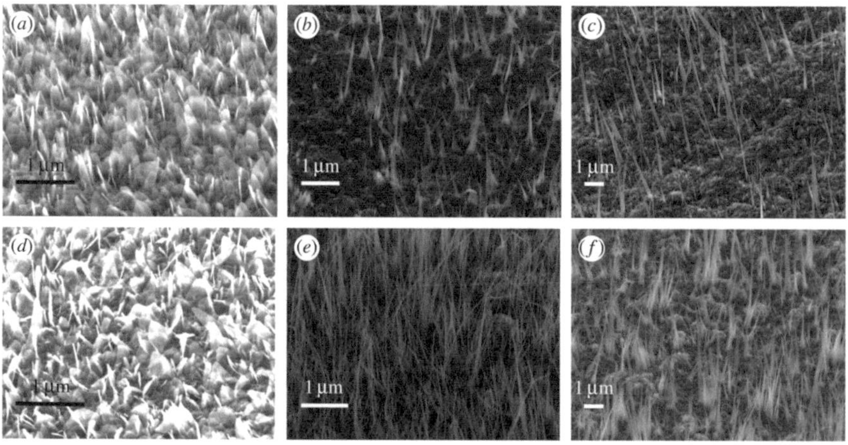

Fig. 2 FESEM images of fabricated nanowires by varying the heating conditions for PEC hydrogen production [58]

to generate high-quality, dense 1D nanowires without using any chemicals, as shown in Fig. 2 [58–60].

Hence, these methods could be considered as eco-friendly and cost-effective method. In addition, hematite films can be fabricated using the electrodeposition technique, where metallic iron is first deposited onto an FTO (fluorine-doped tin oxide) coated glass substrate, subsequently, annealing treatment converts the deposited iron into hematite films [61]. The FTO layer also serves as a doping agent, further enhancing the PEC performance of the hematite film. Apart from this, the preparation of hematite nanorod structures using the hydrothermal method can potentially enhance PEC performance. Initially, a solution of ferric chloride and sodium nitrate is prepared by adjusting the pH level [62]. The cleaned FTO substrate is then sonicated with ethanol, acetone, and deionized water before being placed into a Teflon liner with the solution [62]. The mixture is heated, resulting in the formation of an iron oxyhydroxide (FeOOH) film. Finally, the prepared film is transferred to a muffle furnace and heated at the desired temperature to convert it into hematite nanorods [62]. The hematite nanorod can be also prepared using RF (Radio frequency) sputtering, where an iron target is sputtered onto the FTO substrate [63]. The resulting film is subsequently annealed at a high temperature in a muffle furnace, forming vertically aligned hematite nanorods [63], as shown in Fig. 3.

The two-step heating method is another strategy for preparing nanowires with a higher density. In the first step, nanowires are prepared using the stress-induced atomic diffusion method at a lower temperature, whereas, in the second step, the sample is heated at a higher temperature in a furnace to convert the nanowires into single-crystalline, high-density 3D hematite nanowires, which are potentially applicable for photoelectrochemical (PEC) performance [64]. The nanotube structure is also a potential candidate for PEC applications. Therefore, the crystalline nanotube

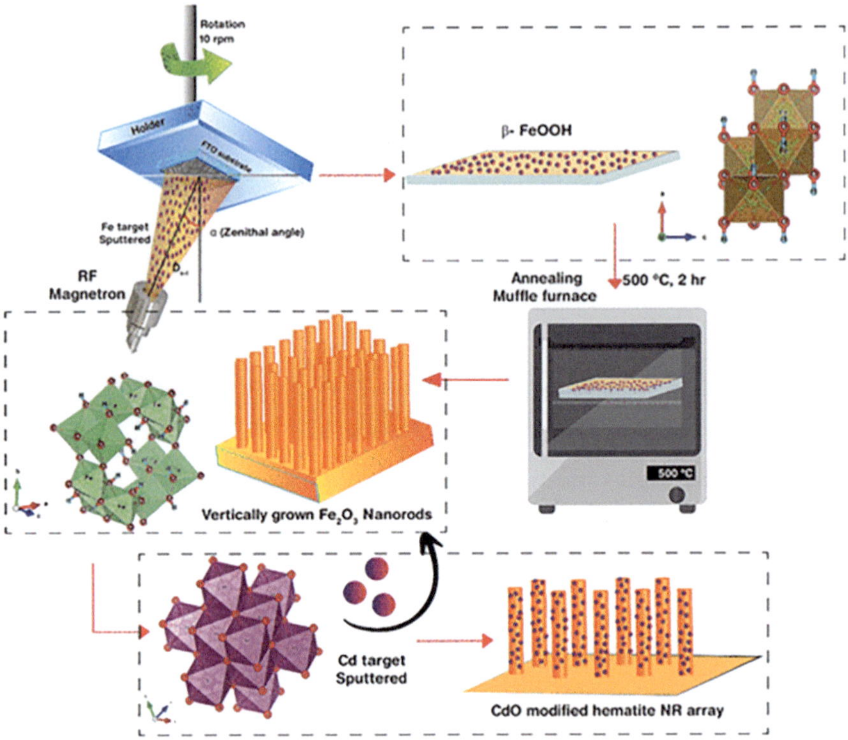

Fig. 3 RF sputtering and annealing treatment fabrication of hematite nanorod structure for PEC application [63]

structure can be achieved using hydrothermal techniques for potential PEC application [65]. Additionally, the nanoflower structure is a promising nanostructure for PEC applications. Nanoflowers can be synthesized using a cost-effective template route method. For example, cupric oxide (CuO)/copper hydroxide (Cu $(OH)_{2)}$ flower-like structure can be formed on copper foil by preparing a chemical mixture of sodium hydroxide (NaOH) and ammonium persulfate ($(NH_4)_2S_2O_{8)}$, thereafter, the resulting structure is then heated at a higher temperature in a furnace under a nitrogen atmosphere, converting it into Cu_2O nanoflowers [66], as shown Fig. 4.

These methods could be further improved using chemical vapor deposition (CVD) techniques instead of conventional furnaces after the initial chemical preparation. By adjusting the temperature and annealing the material in a noble gas environment, 3D Cu_2O nanoflowers could be produced, potentially generating a higher photocurrent density. Besides the nanoflower, Cu_2O film can be produced using magnetron sputtering techniques [67]. The most important aspect here is that achieving nanostructures such as 1D, 2D, or 3D semiconductors on metal substrates, Fluorine-doped Tin Oxide (FTO) or Indium Tin Oxide (ITO) can improve the overall PEC performance compared to bulk structures. However, this alone is not sufficient. While FTO or

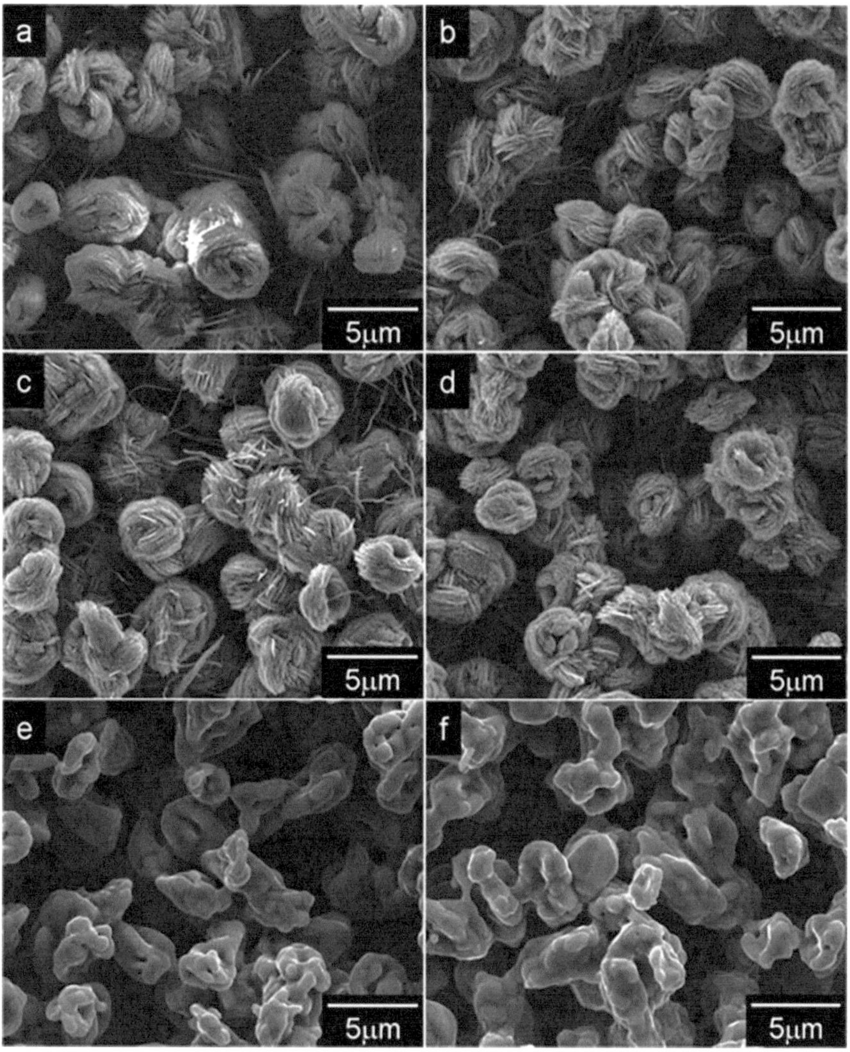

Fig. 4 FESEM images **a** nanoflower (CuO) on the nanowires $Cu(OH)_2$, **b–f** Cu_2O nanoflowers by varying the temperature conditions, Reproduced with permission from Elsevier, Ref. [66]

ITO substrates can assist in doping the pristine semiconductor, further improvements can be achieved through doping, co-catalyst loading, surface coating, or forming heterojunctions. The doping strategy, particularly the homogeneous doping approach, has already been proven. However, the newly developed gradient doping method is considered superior for further improvements. For instance, the enhanced photocurrent density of hematite compared to the pristine hematite nanorod structure can be achieved through tantalum (Ta) gradient doping [68]. The reason gradient

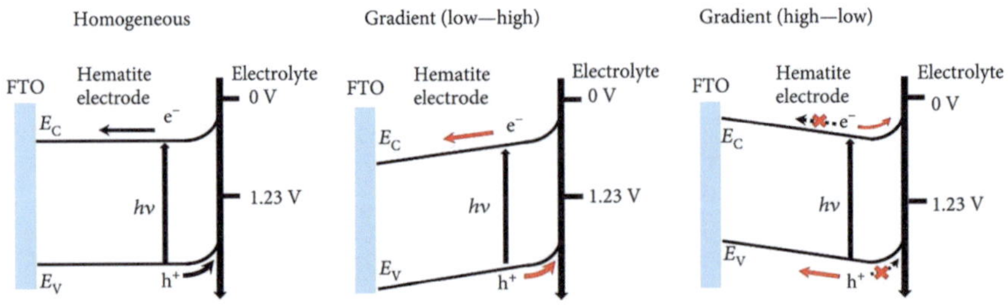

Fig. 5 Band banding effect due to the gradient doping which accelerates the charge separation efficiency [70].

doping is preferred over homogeneous doping is that it generates a built-in electric field, induces a band bending effect, increases electron concentration, enhances conductivity, and facilitates faster charge separation, as shown in Fig. 5 [69].

On the other hand, loading a co-catalyst onto the semiconductor can enhance charge separation by reducing surface recombination, reduce the turn on voltage, and improve the charge separation efficiency, in this regard, the iridium oxide/ruthenium oxide (IrO_2/RuO_2) co-catalyst can increase photocurrent density during PEC operation when coated on the hematite film [71]. Apart from this, the combination of different metal oxide semiconductors through the alignment of their band structures is referred to as forming a heterostructure which is another strategy to improve the PEC performance. In this setup, the first semiconductor, known as the pristine semiconductor, serves as the actual photoelectrode, while the second semiconductor is utilized as a heterojunction layer. Careful consideration of the band alignment between the first and second semiconductors is crucial because imperfect alignment could be significantly reduced the performance (i.e., Type I, Type II, Type III). By forming a heterostructure, light absorption, charge separation, and charge collection at the surface can be enhanced, while reaction kinetics are improved and the recombination rate is reduced, consequently, creating a ferric titanate/iron oxyhydroxide ($Fe_2TiO_5/FeOOH$) heterostructure with hematite nanorods and CuO-Cu_2O thin film heterostructure can significantly boost overall performance [72, 73]. To summarize, improving the performance of photoelectrodes in photoelectrochemical (PEC) systems involves the fabrication of nanostructuring of metal oxide semiconductors. While this approach improves performance, it has not yet reached practical STH efficiency levels. Many efforts have been made to enhance photoelectrode performance further, but significant challenges remain. A key goal should be to increase the photocurrent density of the pristine semiconductor to at least 50–60% of its theoretical STH value. Afterward, modifying the material with foreign elements or other semiconductors could help achieve the target photocurrent density and solar-to-hydrogen (STH) conversion efficiency, while also reducing the use of chemicals and costs.

5 Photocurrent Density and Solar–to-Hydrogen (STH) Conversion Efficiency

To evaluate the performance of a single photoelectrode semiconductor, it is essential to determine its efficiency. Therefore, the solar-to-hydrogen (STH) conversion efficiency of semiconductor relies on three key steps such as capturing the sunlight to generate charge carrier, efficient separation of charges, and accumulation of charges on the surface (i.e., holes to oxidize the water) to split water [71]. These procedures ensure sunlight is converted into energy with minimal waste, making the system work effectively. The experimental solar-to-hydrogen (STH) conversion efficiency of a single photoelectrode can be evaluated using two key metrics: Solar-to-Hydrogen (STH) Conversion Efficiency and Applied Bias Photon-to-Current Efficiency (ABPE). These metrics are derived from the photocurrent density generated by the photoelectrode, as shown in Fig. 6.

Essentially, the higher the photocurrent density, the greater the hydrogen production, as hydrogen generation is directly proportional to the photocurrent density. Here, although both formulas are quite suitable for measuring the performance of a single photoelectrode, they should only be applied to a two-electrode system without an external bias [75–77]. Therefore, it is preferable to use a gas chromatograph for directly evaluating the actual performance. However, the first formula is described as follows [77].

$$\mathrm{STH} = \frac{J_{sc}\left(\mathrm{mA/cm^2}\right)\times(1.23\,\mathrm{V})\times\eta_F}{P\left(\mathrm{mW/cm^2}\right)} \times 100\% \tag{1}$$

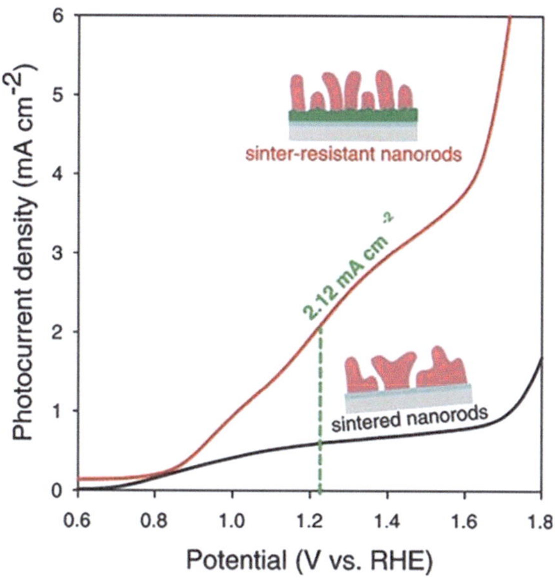

Fig. 6 Photocurrent density (J-V curve) at 1.23 V vs. RHE for unmodified hematite nanorods [75].

whereas J_{sc} denoted as photocurrent density generated at 1.23 V under zero bias potential, η_F refers to the hydrogen evolution efficiency, which is considered to be 100%, and P corresponds to the light intensity, standardized (100 mW/cm^2).

Alternatively, another equation, commonly referred to as the applied bias photon-to-current conversion efficiency (ABPE), has been derived from the thermodynamic perspective of the photoelectrode, as shown below [76].

$$\mathrm{ABPE} = \frac{J_{sc}\left(\mathrm{mA/cm^2}\right)\times(1.23\ V - V)\times\eta_F}{P\left(\mathrm{mW/cm^2}\right)} \times 100\% \tag{2}$$

From the Eq. 1, all parameters remain the same, with the only change being the added applied bias potential from the J-V curve in Eq. 2. A key drawback of Eq. 2 is that it may overestimate the efficiency value.

6 Future Outlook

Although researchers have made significant improvements in the development of nanostructured semiconductor materials and their modification strategies to achieve benchmark efficiency, but it is crucial to enhance the photocurrent density and STH efficiency of pristine semiconductors, which have been very poor so far. Although most of the researchers put effort into making the nanostructure but most of the cases nanostructure is not homogenous. Therefore, future research could be further improved by enhancing the homogeneity of pristine nanostructures, for example, vertically grown homogeneous nanowires without attached to another nanowire with the same size and shape could provide a large portion of light access and a more uniform path for electron transport, therefore reduced the charge carrier recombination and facilitate more efficient electron collection, improving the photocurrent. In addition, compared to irregular nanostructures, uniform nanowires could have fewer defects or surface traps as well as enhance light trapping.

7 Conclusion

Photoelectrochemical (PEC) hydrogen production represents a promising solution to the environmental issues caused by fossil fuel-based energy production. The dependence on fossil fuels has led to numerous adverse effects on both living organisms, and environment. While PEC technology shows a pathway toward sustainable fuel production, its practical implementation remains challenging due to the limitations of current semiconductor materials. The poor efficiency and stability of these semiconductors hinder their widespread application. In the twenty-first century, ensuring eco-friendly and sustainable fuel production is essential to protect humanity, biodiversity, and the planet. To address this, critical efforts must focus on developing efficient semiconductors that can overcome the current limitations of PEC systems. Such advancements could pave the way for entirely sustainable and environmentally friendly energy solutions.

This chapter provides a comprehensive discussion on PEC hydrogen production, including its fundamental mechanisms, major materials used, and their limitations. We explore various fabrication strategies and highlight advanced modification techniques to enhance the performance of PEC systems. In addition, methods to evaluate semiconductor efficiency are detailed to guide researchers in this field. This organized discussion aims to inspire innovation and progress in PEC technology for sustainable energy production.

References

1. C. Zou, Q. Zhao, G. Zhang, B. Xiong, Energy revolution: from a fossil energy era to a new energy era. Nat. Gas. Ind. B **3**, 1–11 (2016). https://doi.org/10.1016/j.ngib.2016.02.001
2. A.I. Osman, L. Chen, M. Yang et al., Cost, environmental impact, and resilience of renewable energy under a changing climate: a review. Environ. Chem. Lett. **21**, 741–764 (2023). https://doi.org/10.1007/s10311-022-01532-8
3. B. Sørensen, A history of renewable energy technology. Energy Pol. **19**, 8–12 (1991). https://doi.org/10.1016/0301-4215(91)90072-V
4. M.A. Russo, D. Carvalho, N. Martins, A. Monteiro, Forecasting the inevitable: a review on the impacts of climate change on renewable energy resources. Sustain Energy Technol. Assess **52**, 102283 (2022). https://doi.org/10.1016/j.seta.2022.102283
5. M. Amin, H.H. Shah, A.G. Fareed et al., Hydrogen production through renewable and non-renewable energy processes and their impact on climate change. Int. J. Hydrog. Energy **47**, 33112–33134 (2022). https://doi.org/10.1016/j.ijhydene.2022.07.172
6. J. Xiao, C. Li, X. Jia et al., Enabling high low-bias performance of Fe_2O_3 photoanode for photoelectrochemical water splitting. J. Colloid Interface Sci. **633**, 555–565 (2023). https://doi.org/10.1016/j.jcis.2022.11.134
7. M.G. Walter, E.L. Warren, J.R. McKone et al., Solar water splitting cells. Chem. Rev. **110**, 6446–6473 (2010). https://doi.org/10.1021/cr1002326
8. I.V. Bagal, N.R. Chodankar, M.A. Hassan et al., Cu_2O as an emerging photocathode for solar water splitting—a status review. Int. J. Hydrog. Energy **44**, 21351–21378 (2019). https://doi.org/10.1016/j.ijhydene.2019.06.184
9. I.R. Hamdani, A.N. Bhaskarwar, Cu_2O nanowires based p-n homojunction photocathode for improved current density and hydrogen generation through solar-water splitting. Int. J. Hydrog. Energy **46**, 28064–28077 (2021). https://doi.org/10.1016/j.ijhydene.2021.06.067
10. R. Siavash Moakhar, S.M. Hosseini-Hosseinabad, S. Masudy-Panah et al., Photoelectrochemical water-splitting using CuO-based electrodes for hydrogen production: a review. Adv. Mater. **33**, 2007285 (2021). https://doi.org/10.1002/adma.202007285
11. A. Paracchino, V. Laporte, K. Sivula et al., Highly active oxide photocathode for photoelectrochemical water reduction. Nat. Mater. **10**, 456–461 (2011). https://doi.org/10.1038/nmat3017
12. J. Mou, Y. Ren, J. Wang et al., Nickel oxide nanoparticle synthesis and photocatalytic applications: evolution from conventional methods to novel microfluidic approaches. Microfluid Nanofluidics **26**, 25 (2022). https://doi.org/10.1007/s10404-022-02534-2
13. C.G. Morales-Guio, M.T. Mayer, A. Yella et al., An optically transparent iron nickel oxide catalyst for solar water splitting. J. Am. Chem. Soc. **137**, 9927–9936 (2015). https://doi.org/10.1021/jacs.5b05544

14. A. Abdelmoneim, M.S. Abdel-wahab, M.A.K. Elfayoumi et al., Enhancing the solar hydrogen generation performance of nickel-oxide nanostructured thin films doped with molybdenum. Phys. Scr. **99**, 045906 (2024). https://doi.org/10.1088/1402-4896/ad2c4a
15. H.L.S. Santos, P.G. Corradini, M.A.S. Andrade, L.H. Mascaro, CuO/NiOx thin film–based photocathodes for photoelectrochemical water splitting. J. Solid State Electrochem. **24**, 1899–1908 (2020). https://doi.org/10.1007/s10008-020-04513-5
16. C. Jiang, A. Moniz SJ, A. Wang, et al.: Photoelectrochemical devices for solar water splitting—materials and challenges. Chem. Soc. Rev. **46**, 4645–4660 (2017). https://doi.org/10.1039/C6CS00306K
17. K. Arifin, R.M. Yunus, L.J. Minggu, M.B. Kassim, Improvement of TiO_2 nanotubes for photoelectrochemical water splitting: review. Int. J. Hydrog. Energy **46**, 4998–5024 (2021). https://doi.org/10.1016/j.ijhydene.2020.11.063
18. X. Long, L. Gao, F. Li et al., Bamboo shoots shaped $FeVO_4$ passivated ZnO nanorods photoanode for improved charge separation/transfer process towards efficient solar water splitting. Appl. Catal. B Environ. **257**, 117813 (2019). https://doi.org/10.1016/j.apcatb.2019.117813
19. M. Ma, Y. Huang, J. Liu et al., Engineering the photoelectrochemical behaviors of ZnO for efficient solar water splitting. J. Semicond. **41**, 091702 (2020). https://doi.org/10.1088/1674-4926/41/9/091702
20. S.S. Kalanur, I.-H. Yoo, K. Eom, H. Seo, Enhancement of photoelectrochemical water splitting response of WO_3 by Means of Bi doping. J. Catal. **357**, 127–137 (2018). https://doi.org/10.1016/j.jcat.2017.11.012
21. S.S. Kalanur, L.T. Duy, H. Seo, Recent progress in photoelectrochemical water splitting activity of WO_3 photoanodes. Top. Catal. **61**, 1043–1076 (2018). https://doi.org/10.1007/s11244-018-0950-1
22. X. Feng, Y. Chen, Z. Qin et al., Facile fabrication of sandwich structured WO_3 nanoplate arrays for efficient photoelectrochemical water splitting. ACS Appl. Mater. Interfaces **8**, 18089–18096 (2016). https://doi.org/10.1021/acsami.6b04887
23. E.A. Campos, D.V.B.S. Pinto, J.I.S. de Oliveira et al., Synthesis, characterization and applications of iron oxide nanoparticles—a short review. J. Aerosp. Technol. Manag. **7**, 267–276 (2015). https://doi.org/10.5028/jatm.v7i3.471
24. J. Tuček, L. Machala, S. Ono et al., Zeta-Fe_2O_3—a new stable polymorph in iron(III) oxide family. Sci. Rep. **5**, 15091 (2015). https://doi.org/10.1038/srep15091
25. N. Zhang, Y. Guo, X. Wang et al., A beta-Fe_2O_3 nanoparticle-assembled film for photoelectrochemical water splitting. Dalton Trans. **46**, 10673–10677 (2017). https://doi.org/10.1039/C7DT00900C
26. J. Li, S.K. Cushing, P. Zheng et al., Plasmon-induced photonic and energy-transfer enhancement of solar water splitting by a hematite nanorod array. Nat. Commun. **4**, 2651 (2013). https://doi.org/10.1038/ncomms3651
27. J.-W. Jang, C. Du, Y. Ye et al., Enabling unassisted solar water splitting by iron oxide and silicon. Nat. Commun. **6**, 7447 (2015). https://doi.org/10.1038/ncomms8447
28. K.-Y. Yoon, J. Park, M. Jung et al., NiFeOx decorated Ge-hematite/perovskite for an efficient water splitting system. Nat. Commun. **12**, 4309 (2021). https://doi.org/10.1038/s41467-021-24428-7
29. Y.J. Jang, Y.B. Park, H.E. Kim et al., Oxygen-intercalated $CuFeO_2$ photocathode fabricated by hybrid microwave annealing for efficient solar hydrogen production. Chem. Mater. **28**, 6054–6061 (2016). https://doi.org/10.1021/acs.chemmater.6b00460
30. M.S. Prévot, N. Guijarro, K. Sivula, Enhancing the performance of a robust sol–gel-processed p-type delafossite $CuFeO_2$ photocathode for solar water reduction. Chemsuschem **8**, 1359–1367 (2015). https://doi.org/10.1002/cssc.201403146
31. Y. Oh, W. Yang, J. Tan et al., Boosting visible light harvesting in p-type ternary oxides for solar-to-hydrogen conversion using inverse opal structure. Adv. Funct. Mater. **29**, 1900194 (2019). https://doi.org/10.1002/adfm.201900194

32. A. Aadenan, N.A. Arzaee, M.F.M. Noh et al., Photoelectrochemical water splitting performance of $LaFeO_3$ thin film photocathode prepared by spray gun deposition. Thin Solid Films **777**, 139892 (2023). https://doi.org/10.1016/j.tsf.2023.139892
33. J. Fan, X. Ran, H. Fu et al., The roles of oxygen vacancies in $LaFeO_3$ photocathodes in photoelectrochemical water splitting. Appl. Surf. Sci. **649**, 159011 (2024). https://doi.org/10.1016/j.apsusc.2023.159011
34. Y. Matsumoto, M. Omae, K. Sugiyama, E. Sato, New photocathode materials for hydrogen evolution: calcium iron oxide ($CaFe_2O_4$) and strontium iron oxide ($Sr_7Fe_{10}O_{22}$). In: ACS Publ. https://pubs.acs.org/doi/pdf/10.1021/j100287a018. Accessed 6 Nov 2024 (2002)
35. S. Ida, K. Yamada, T. Matsunaga et al., Preparation of p-type $CaFe_2O_4$ photocathodes for producing hydrogen from water. J. Am. Chem. Soc. **132**, 17343–17345 (2010). https://doi.org/10.1021/ja106930f
36. K. Kirchberg, R. Marschall, Sol–gel synthesis of mesoporous $CaFe_2O_4$ photocathodes with hierarchical pore morphology. Sustain Energy Fuels **3**, 1150–1153 (2019). https://doi.org/10.1039/C8SE00627J
37. W. Matsumoto, T. Fukushima, S. Heguri et al., Boosting charge transport in the $BiVO_4$ photoanode interface modified with an aluminum hydroxide layer for solar water oxidation. Sustain Energy Fuels **8**, 1626–1635 (2024). https://doi.org/10.1039/D3SE01656K
38. M. Tayebi, B.-K. Lee, Recent advances in $BiVO_4$ semiconductor materials for hydrogen production using photoelectrochemical water splitting. Renew. Sustain. Energy Rev. **111**, 332–343 (2019). https://doi.org/10.1016/j.rser.2019.05.030
39. Li, C., Guo, B., Peng, B., et al.: Copper tungstate ($CuWO_4$) Nanoflakes coupled with cobalt phosphate (Co-Pi) for effective photoelectrochemical water splitting. Int. J. Electrochem. Sci. **14**, 9017–9029 (2019). https://doi.org/10.20964/2019.09.74
40. M. Valenti, D. Dolat, G. Biskos et al., Enhancement of the photoelectrochemical performance of $CuWO_4$ thin films for solar water splitting by plasmonic nanoparticle functionalization. J. Phys. Chem. C **119**, 2096–2104 (2015). https://doi.org/10.1021/jp506349t
41. D. Pelayo, E. Pérez-Peña, M.J. Rivero, I. Ortiz, Shedding light on the photocatalytic hydrogen generation from seawater using CdS. Catal. Today **433**, 114672 (2024). https://doi.org/10.1016/j.cattod.2024.114672
42. Y. Xie, Y. Sun, R. Jiang et al., Controllable synthesis of CdS nanospheres photoelectrode for photoelectrochemical water splitting. Resour. Chem. Mater **3**, 38–45 (2024). https://doi.org/10.1016/j.recm.2023.09.001
43. F. Kurnia, Y.H. Ng, R. Amal et al., Defect engineering of ZnS thin films for photoelectrochemical water-splitting under visible light. Sol. Energy Mater. Sol. Cells **153**, 179–185 (2016). https://doi.org/10.1016/j.solmat.2016.04.021
44. G. Wang, B. Huang, Z. Li et al., Synthesis and characterization of ZnS with controlled amount of S vacancies for photocatalytic H_2 production under visible light. Sci. Rep. **5**, 8544 (2015). https://doi.org/10.1038/srep08544
45. K. Sangaré, H. Cherfouh, B. Marsan, Synthesis and characterization of N-type $CuGaS_2$ nanoparticles and films for purpose of photoelectrocatalytic water splitting. J. Electrochem. Soc. **168**, 086506 (2021). https://doi.org/10.1149/1945-7111/ac1cc6
46. I.V. Bagal, M. Arunachalam, A. Waseem et al., Gallium phosphide photoanodes coated with nickel oxyhydroxide cocatalyst for stable photoelectrochemical water splitting reactions. Appl. Surf. Sci. **558**, 149873 (2021). https://doi.org/10.1016/j.apsusc.2021.149873
47. K. Becker, C. Xiao, S. Assavachin et al., 14.8 % Quantum efficient gallium phosphide photocatalyst for hydrogen evolution. J. Am. Chem. Soc. **146**, 7723–7733 (2024). https://doi.org/10.1021/jacs.3c14545
48. Amorphous TiO_2 coatings stabilize Si, GaAs, and GaP photoanodes for efficient water oxidation | Science. https://www.science.org/doi/10.1126/science.1251428. Accessed 11 Nov 2024
49. M.H. Lee, K. Takei, J. Zhang et al., P-Type InP nanopillar photocathodes for efficient solar-driven hydrogen production. Angew. Chem. Int. Ed. **51**, 10760–10764 (2012). https://doi.org/10.1002/anie.201203174

50. C. Cheng, W. Zhang, X. Chen et al., Strategies for improving photoelectrochemical water splitting performance of Si-based electrodes. Energy Sci. Eng. **10**, 1526–1543 (2022). https://doi.org/10.1002/ese3.1087
51. J. Zheng, H. Zhou, Y. Zou et al., Efficiency and stability of narrow-gap semiconductor-based photoelectrodes. Energy Environ. Sci. **12**, 2345–2374 (2019). https://doi.org/10.1039/C9EE00524B
52. C. Zhen, R. Chen, L. Wang et al., Tantalum (oxy)nitride based photoanodes for solar-driven water oxidation. J. Mater. Chem. A **4**, 2783–2800 (2016). https://doi.org/10.1039/C5TA07057K
53. H. Ullah, A.A. Tahir, S. Bibi et al., Electronic properties of β-TaON and its surfaces for solar water splitting. Appl. Catal. B Environ. **229**, 24–31 (2018). https://doi.org/10.1016/j.apcatb.2018.02.001
54. R. Abe, M. Higashi, K. Domen, Facile fabrication of an efficient oxynitride TaON photoanode for overall water splitting into H2 and O2 under visible light irradiation. J. Am. Chem. Soc. **132**, 11828–11829 (2010). https://doi.org/10.1021/ja1016552
55. A. Ishikawa, T. Takata, J.N. Kondo et al., Electrochemical behavior of Thin Ta_3N_5 semiconductor film. J. Phys. Chem. B **108**, 11049–11053 (2004). https://doi.org/10.1021/jp048802u
56. B. Zhang, Z. Fan, Y. Li, Engineering tantalum nitride for efficient photoelectrochemical water splitting. Sci. China Chem. **67**, 2171–2180 (2024). https://doi.org/10.1007/s11426-024-2058-9
57. H. Hajibabaei, D.J. Little, A. Pandey et al., Direct deposition of crystalline Ta_3N_5 thin films on FTO for PEC water splitting. ACS Appl. Mater. Interfaces **11**, 15457–15466 (2019). https://doi.org/10.1021/acsami.8b21194
58. Y. Xie, Y. Ju, Y. Toku, Y. Morita, Synthesis of a single-crystal Fe_2O_3 nanowire array based on stress-induced atomic diffusion used for solar water splitting. R Soc Open Sci **5**, 172126 (2018). https://doi.org/10.1098/rsos.172126
59. L. Wang, H. Hu, N.T. Nguyen et al., Plasmon-induced hole-depletion layer on hematite nanoflake photoanodes for highly efficient solar water splitting. Nano Energy **35**, 171–178 (2017). https://doi.org/10.1016/j.nanoen.2017.03.035
60. R.-T. Gao, J. Zhang, T. Nakajima et al., Single-atomic-site platinum steers photogenerated charge carrier lifetime of hematite nanoflakes for photoelectrochemical water splitting. Nat. Commun. **14**, 2640 (2023). https://doi.org/10.1038/s41467-023-38343-6
61. A.B. Silva, C.D.F. da Silva, F.L. Souza et al., All-electrochemically synthesized tin and nickel oxide-modified hematite as photo-electrocatalyst anodes for solar-driven water splitting. J. Catal. **391**, 273–281 (2020). https://doi.org/10.1016/j.jcat.2020.08.014
62. Y. Xu, H. Zhang, D. Gong et al., Solar water splitting with nanostructured hematite: the role of oxygen vacancy. J. Mater. Sci. **57**, 19716–19729 (2022). https://doi.org/10.1007/s10853-022-07885-3
63. M. Alhabradi, S. Nundy, A. Ghosh, A.A. Tahir, Vertically aligned CdO-decked α-Fe_2O_3 nanorod arrays by a radio frequency sputtering method for enhanced photocatalytic applications. ACS Omega **7**, 28396–28407 (2022). https://doi.org/10.1021/acsomega.2c02996
64. M.S. Mahmud, S. Arumugam, Y. Kimura et al., Enhancement of efficiency of solar water splitting based on highly dense hematite single-crystal nanowires. J. Mater. Sci. **58**, 14426–14439 (2023). https://doi.org/10.1007/s10853-023-08721-y
65. C. Feng, S. Fu, W. Wang et al., High-crystalline and high-aspect-ratio hematite nanotube photoanode for efficient solar water splitting. Appl. Catal. B Environ. **257**, 117900 (2019). https://doi.org/10.1016/j.apcatb.2019.117900
66. Y.-K. Hsu, C.-H. Yu, Y.-C. Chen, Y.-G. Lin, Synthesis of novel Cu_2O micro/nanostructural photocathode for solar water splitting. Electrochim. Acta **105**, 62–68 (2013). https://doi.org/10.1016/j.electacta.2013.05.003
67. C. Qin, X. Chen, N. Jiang et al., Surface densification strategy assisted efficient Cu_2O heterojunction photocathode for solar water splitting. Mater Today Nano. **21**, 100294 (2023). https://doi.org/10.1016/j.mtnano.2022.100294

68. H. Zhang, D. Li, W.J. Byun et al., Gradient tantalum-doped hematite homojunction photoanode improves both photocurrents and turn-on voltage for solar water splitting. Nat. Commun. **11**, 4622 (2020). https://doi.org/10.1038/s41467-020-18484-8
69. F. Feng, C. Li, J. Jian et al., Gradient Ti-doping in hematite photoanodes for enhanced photoelectrochemical performance. J. Power. Sources **449**, 227473 (2020). https://doi.org/10.1016/j.jpowsour.2019.227473
70. P. Liu, C. Wang, L. Wang, et al. Ultrathin hematite photoanode with gradient Ti doping. Research (2020). https://doi.org/10.34133/2020/5473217
71. P. Dias, L. Andrade, A. Mendes, Hematite-based photoelectrode for solar water splitting with very high photovoltage. Nano Energy **38**, 218–231 (2017). https://doi.org/10.1016/j.nanoen.2017.05.051
72. S. Farhoosh, B. Eftekharinia, M. Tayebi et al., Newly designed ternary hematite-based heterojunction for PEC water splitting. Appl. Surf. Sci. **550**, 149374 (2021). https://doi.org/10.1016/j.apsusc.2021.149374
73. A.M. Alotaibi, E. Muayqil, N. Al Abass et al., Surface engineering of CuO-Cu_2O heterojunction thin films for improved photoelectrochemical water splitting. Renew. Energy **235**, 121326 (2024). https://doi.org/10.1016/j.renene.2024.121326
74. D. Wang, Y. Zhang, C. Peng et al., Crystallinity engineering of hematite nanorods for high-efficiency photoelectrochemical water splitting. Adv. Sci. **2**, 1500005 (2015). https://doi.org/10.1002/advs.201500005
75. C.S.Y. Gurudayal, M.H. Kumar et al., Improving the efficiency of hematite nanorods for photoelectrochemical water splitting by doping with manganese. ACS Appl. Mater. Interfaces **6**, 5852–5859 (2014). https://doi.org/10.1021/am500643y
76. H. Dotan, N. Mathews, T. Hisatomi et al., On the solar to hydrogen conversion efficiency of photoelectrodes for water splitting. J. Phys. Chem. Lett. **5**, 3330–3334 (2014). https://doi.org/10.1021/jz501716g
77. P. Biswas, A. Ainabayev, A. Zhussupbekova et al., Tuning of oxygen vacancy-induced electrical conductivity in Ti-doped hematite films and its impact on photoelectrochemical water splitting. Sci. Rep. **10**, 7463 (2020). https://doi.org/10.1038/s41598-020-64231-w

Reaction Mechanisms and Catalysis in Solar Water Splitting

Esdras J. Canto-Aguilar, Elena Puentes-Prado, Alice Kuzhikandathil, and Eduardo Gracia-Espino

Abstract Solar water splitting, through electrocatalysis, photoelectrocatalysis, and photocatalysis, presents a promising pathway for converting abundant solar energy into clean hydrogen fuel. However, to meet the increasing demand for green hydrogen, catalysts for the water splitting reaction must not only achieve high efficiency but also be primarily composed of abundant materials with minimal environmental impact. Understanding the interplay between surface properties, optical properties, and chemical reactivity is crucial to identify suitable alternative materials capable of achieving high solar-to-hydrogen conversion. In this chapter, we discuss the fundamentals of electrochemical, photoelectrochemical, and photocatalytic water splitting. We explore how material's properties and surface characteristics influence the hydrogen evolution reaction and oxygen evolution reaction. We revise their working principles and key concepts, from light absorption and charge separation to surface characteristics and adsorption energies. Additionally, we discuss key physical, chemical, and optical properties of electrodes and photoelectrodes that are crucial for an efficient water splitting, while presenting strategies for achieving high performance.

Keywords Water electrolysis · Electrocatalysis · Photocatalysis · Hydrogen evolution · Oxygen evolution

1 Introduction

The increasing global energy demand and the urgent need to mitigate climate change requires a significant shift toward sustainable energy sources such as solar energy. While solar energy offers great potential, the collection and storage of solar energy remain a significant challenge. Here, electrocatalytic, photoelectrochemical, and photocatalytic water splitting offer promising ways to convert renewable solar energy and store it as chemical energy, primarily in the form of hydrogen gas. Hydrogen,

E. J. Canto-Aguilar · E. Puentes-Prado · A. Kuzhikandathil · E. Gracia-Espino (✉)
Department of Physics, Umeå University, Umeå, Sweden
e-mail: eduardo.gracia@umu.se

P. Ravindran et al. (eds.), *Photon to Power*, Progress in Optical Science and Photonics 32, https://doi.org/10.1007/978-981-96-5914-2_9

with its high energy density and zero-emission combustion, emerges as a versatile energy carrier. It can efficiently store excess electricity generated from solar power and balance seasonal fluctuations in energy output. This stored energy can then be utilized in various applications, such as hydrogen fuel cell vehicles, which produce zero CO_2 emissions.

The efficiency and hydrogen yield of water splitting is highly dependent on the process of interest and the choice of the catalysts/electrocatalysts. These catalysts play a crucial role in enhancing the efficiency of water splitting by lowering the energy barriers for the two half-reactions involved in water splitting, namely hydrogen evolution reaction (HER) and oxygen evolution reaction (OER). Electrocatalytic water splitting involves the production of hydrogen and oxygen gases from water molecules using electricity. However, water electrolysis is an energy demanding process due to the high overpotential required to drive the overall reaction. While noble metals offer the most efficient electrocatalysts for HER and OER, their limited availability and high cost hinder their widespread adoption. Therefore, the development of earth-abundant and cost-effective electrocatalysts with comparable or better performance to noble metals is essential. Photoelectrocatalytic (PEC) water splitting can harvest the energy of the sun to perform the oxidation or reduction of water molecules with the photogenerated charge carriers in the absorbing material. In PEC water splitting the generation of oxygen and hydrogen gases at the surface of light-absorbing materials takes place through catalytic charge transfer reactions between the photogenerated charge carriers and the reaction precursors adsorbed at the surface of the material. Thus, the performance of a photoelectrocatalytic system is set by the kinetic equilibrium and the efficiency of several phenomena taking place at different time scales. Photocatalysis is a process that uses the energy of light to accelerate a photoreaction in the presence of a catalyst, typically a semiconductor. Light is used by the photocatalyst to generate electron–hole pairs that ultimately participate in redox reactions driving the desired chemical transformations.

This chapter explores the fundamental principles and recent advancements in solar water splitting driven by electrocatalysis, photoelectrocatalysis, and photocatalysis. We explore key factors influencing the performance of these systems, such as material properties, surface characteristics, and light absorption efficiencies. Furthermore, we discuss the design strategies for developing advanced catalysts and photocatalysts that can enhance the overall efficiency and stability of solar water splitting devices.

2 Electrocatalytic Water Splitting

Electrocatalytic water splitting involves the production of hydrogen and oxygen gases from water molecules using electricity. Hydrogen gas is produced via the hydrogen evolution reaction which takes place at the electron-rich electrode, namely the cathode. Meanwhile, oxygen gas is produced at the electron-deficient electrode, the anode, where the oxygen evolution reaction occurs. Water splitting is an energy demanding process due to the high overpotential required to drive the overall reaction.

The use of electrocatalysts for both HER and OER are required to improve the energy conversion efficiency and minimize the overpotential [1–3]. Currently, the most efficient electrocatalysts for HER and OER are based on noble metals [4]. However, their limited availability and high cost have hindered their extensive use in current technologies such as alkaline water electrolysis, anion exchange membrane water electrolysis [5–7] and proton exchange membrane water electrolysis.

Several key parameters for evaluating the activity include overpotential, Tafel slope, turnover frequency (TOF), faradic efficiency, and operational stability. Overpotential addresses the additional potential needed to overcome kinetic barriers and internal resistances. The Tafel slope provides insights into the kinetics of electron transfer, with smaller slopes indicating faster kinetics. TOF measures the number of desired products generated per catalytic site per second, reflecting intrinsic activity. Faradic efficiency indicates electron utilization efficiency, typically measured by the ratio of experimentally produced gas to theoretical gas. And operational stability assesses the durability of electrocatalysts under operational conditions.

2.1 Reaction Mechanism and Kinetics

2.1.1 Hydrogen Evolution Reaction

The hydrogen evolution reaction comprises a two-electron transfer process that converts protons into hydrogen molecules when subjected to an external potential. This process is highly influenced by environmental conditions [8]. HER can occur in either acidic or alkaline electrolytes. The acid HER mechanism is illustrated in Fig. 1a, while the alkaline HER mechanism is depicted in Fig. 1b. In an acid medium, the catalyst surface undergoes two consecutive steps [9]. Initially, a hydrogen atom binds to the active sites on the catalyst, forming an adsorbed proton (H^*), known as the Volmer step or discharge reaction (Eq. 2.1). Subsequently, H^* reacts with another proton-electron pair to produce H_2, a process referred to as Heyrovsky step or electrochemical desorption step (Eq. 2.2). Alternatively, two H^* intermediates on the catalyst surface can directly produce H_2, this step is known as Tafel step or combination reaction (Eq. 2.3). The overall HER is represented by Eq. 2.4.

$$H^+ + e^- \rightarrow H^* \tag{2.1}$$

$$H^* + H^+ + e^- \rightarrow H_{2(g)} \tag{2.2}$$

$$H^* + H^* \rightarrow H_{2(g)} \tag{2.3}$$

$$2H^+ + 2e^- \rightarrow H_{2(g)} \tag{2.4}$$

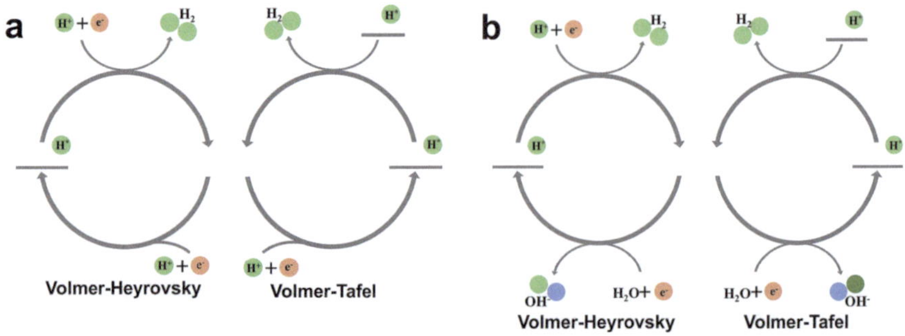

Fig. 1 Schematic diagram of the HER mechanism in **a** acid electrolyte, and **b** alkaline electrolyte

In an alkaline electrolyte, the HER starts with the dissociation of water molecules to supply protons, as there are no H^+ ions present [10]. This process includes the Volmer step (Eq. 2.5), the Heyrovsky step (Eq. 2.6), and the Tafel step (Eq. 2.7), similar to the mechanism in acid solution. The overall reaction is shown in Eq. 2.8.

$$H_2O + e^- \rightarrow H^* + OH^- \tag{2.5}$$

$$H^* + H_2O + e^- \rightarrow OH^- + H_2 \tag{2.6}$$

$$H^* + H^* \rightarrow H_2 \tag{2.7}$$

$$2H_2O + 2e^- \rightarrow 2OH^- + H_2 \tag{2.8}$$

The HER activity is generally lower in alkaline media compared to acid media because the slow water dissociation step (Eq. 2.5) [11]. It has been reported that locally generated hydronium ions (H_3O^+) form an acid-like environment at the surface of the catalyst, even under high concentration of OH^-. This phenomenon is responsible for the acid-like HER activity observed in electrocatalysts in alkaline electrolytes. Further studies have shown that these H_3O^+ ions are produced by a high-rate water dissociation process, aiding the desorption of hydrogen atoms (H^*) from the catalyst surface [12, 13].

The HER activity is closely related to strength of the hydrogen adsorption (H_{ad}), with the adsorption free energy of hydrogen (ΔG_H) being a key parameter [14–16]. The efficiency of HER catalysts is often depicted using volcano plots (Fig. 2) that relate ΔG_H to the exchange current density (j_0). The volcano plot indicates that metals with nearly zero ΔG_H leads to an optimal HER performance. Pt is currently the best HER catalyst due to its nearly ideal H_{ad} that results in a large exchange current density, explaining its superior HER activity [17].

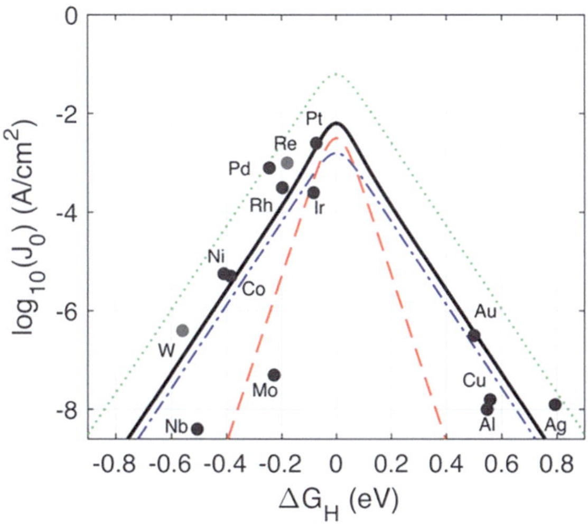

Fig. 2 HER volcano plot for polycrystalline metals. The discharge reaction (green), the electrochemical desorption (blue), the combination reaction (red), and the total reaction (black) are shown. Reproduced with permission from ref. [15], under creative commons CC-BY license

2.1.2 Oxygen Evolution Reaction

The oxygen evolution reaction, the counterpart to HER in water splitting, involves the transfer of four electrons, breaking the O–H bond, and forming the O–O bond. This makes OER a thermodynamically critical process that requires a high potential to overcome its kinetic energy barrier [18]. Currently, OER remains the bottleneck in water splitting. The reaction mechanism of OER varies with pH, differing between acid and alkaline environments [19]. The mechanisms of OER in acid and alkaline are shown in Fig. 3. In an acid electrolyte, the oxidation of two water molecules results in the production of O_2 gas through four proton-electron transfer steps [20]. Initially, water is adsorbed onto an active site (*), where it loses a proton and an electron to form adsorbed hydroxyl species (OH^*, Eq. 2.9). A subsequent proton-electron transfer yields adsorbed oxygen (O^*, Eq. 2.10). $O_{2(g)}$ can later be generated through the direct combination of two O^* species, or through the formation of a hydroperoxo intermediate (OOH^*, Eqs. 2.11 and 2.12) [21]. The OOH^* intermediate subsequently loses a proton-electron pair realizing O_2 (Eq. 2.13).

$$H_2O_{(l)} + * \rightarrow OH^* + H^+ + e^- \tag{2.9}$$

$$OH^* \rightarrow O^* + H^+ + e^- \tag{2.10}$$

$$2O^* \rightarrow 2^* + O_{2(g)} \tag{2.11}$$

$$O^* + H_2O_{(l)} \rightarrow OOH^* + H^+ + e^- \tag{2.12}$$

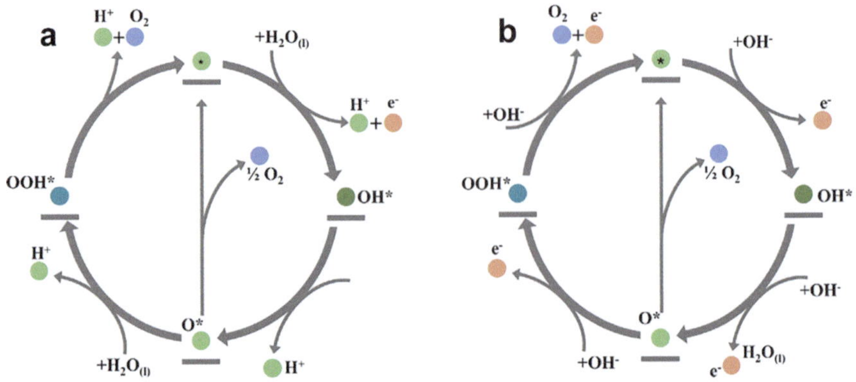

Fig. 3 Schematic diagram of OER reaction mechanisms in **a** acid, and **b** alkaline environments

$$OOH^* \rightarrow * + O_{2(g)} + H^+ + e^- \tag{2.13}$$

$$2H_2O_{(l)} \rightarrow O_{2(g)} + 4H^+ + 4e^- \tag{2.14}$$

In alkaline electrolytes, oxygen molecules are formed by converting OH^- ions through a series of four electron transfer steps [22–25].

$$* + OH^- \rightarrow OH^* + e^- \tag{2.15}$$

$$OH^* + OH^- \rightarrow O^* + H_2O_{(l)} + e^- \tag{2.16}$$

$$O^* + OH^- \rightarrow OOH^* + e^- \tag{2.17}$$

$$OOH^* + OH^- \rightarrow O_{2(g)} + H_2O_{(l)} + e^- \tag{2.18}$$

$$4OH^- \rightarrow O_{2(g)} + 2H_2O_{(l)} + 4e^- \tag{2.19}$$

In both acid and alkaline conditions, each fundamental step has a distinct free energy associated with its corresponding intermediate combination. The step with the highest free energy difference is identified as the rate-limiting step, which determines the minimum required reaction potential [26]. Theoretical calculations are used to provide insight to understand the thermodynamics of each reaction step [27–29]. The efficiency of OER catalysts is evaluated using the Gibbs free adsorption energies of these intermediates, with optimal catalysts showing appropriate bonding strengths for oxygen intermediates, often visualized through a volcano plot [30].

2.2 *Electrocatalysts for Water Electrolysis*

2.2.1 Electrocatalysts for HER

While Pt- and Ir-based electrocatalysts excel in performing the HER and OER, respectively, their high cost and scarcity hinder their widespread use. Alloying noble metals with transition metals not only reduces the quantity of noble metals, but also enhances electrochemical performance at the atomic level, making it a more cost-effective strategy. The HER activity of noble metal-based catalysts has been enhanced by increasing the intrinsic activity of the catalytic active site. Examples include PdIr alloy nanosheets with large electrochemically active surface area, and unique surface electronic structure [31]. IrAg alloys with low HER overpotential and excellent durability due to their large surface area and abundant oxidized Ir species [32]. IrNi alloy shows superior HER activity due to a more optimal adsorption energy of hydrogen [33]. Defect-rich RhCu nanotubes [34] are highly active electrocatalysts for overall water splitting under all pH conditions. And ultra-small hollow PtNiCu nanoparticles [35] and RuNi nanostructures [36] outperform commercial Pt/C and Ru/C electrocatalysts. Non-precious metal alloys such as nickel-based materials are increasingly replacing Pt-based catalysts due to their lower cost and excellent catalytic performance for HER. For example, hollow nanorods of NiMo alloy on Ti mesh showed low onset potential and good stability in alkaline solutions [37]. Other nickel alloys, such as NiW, NiMo, NiZn, CuNi, and NiFeMo also demonstrated improved HER activity [38–43]. Cobalt-based materials such as CoMo and CoMoCu alloys have shown corrosion resistance and enhanced catalytic activity [44].

Another strategy to produce inexpensive HER electrocatalysts is the use of heterojunction. Heterojunction catalysts involve the use of two or more different semiconducting materials forming a junction. This configuration creates a built-in electric field at the interface and helps separate the charge carriers (electrons and holes), reducing recombination losses and enhancing the overall catalytic activity. The inherent electric field formed at the interface of the two materials facilitates the movement of electrons to the active sites responsible for the reduction reaction. Protons from the electrolyte adsorb to the surface of the catalyst, and then electrons are transferred from the catalyst to the adsorbed protons, leading to H_2 evolution [45]. The combination of materials allows for a better alignment of the bands for maximum efficiency in transfer of the electrons. Also, heterojunction structures tend to have a larger surface area thus providing more active sites for the HER process. The synergistic effect of different materials might also enhances the stability and durability of the catalyst under operational conditions [46].

Heterojunction catalysts such as 2D/2D Schottky heterojunctions comprised of transition metal phosphates and MXene have demonstrated outstanding HER activity in freshwater and seawater. Phase-controlled heterojunctions operated by iron phosphide were reported to be active toward HER, OER, and the oxygen reduction reaction

[45]. Furthermore, the addition of nanowires and nanosheets have been used in addition to the existing heterojunction catalysts to improve the surface area and electron transport, resulting in enhanced catalytic activity [47].

2.2.2 Electrocatalysts for OER

2.3 Metal Oxides Catalysts

RuO_2 and IrO_2 are key electrocatalysts for OER in acid and alkaline solutions, and although both are prompt to corrosion, RuO_2 is less stable than IrO_2 under high anodic potential [48]. To enhance the stability of RuO_2, researchers have developed doped bimetallic oxides ($Ru_xIr_{1-x}O_2$), core–shell structures ($IrO_2@RuO_2$), and even quaternary mixed oxides (Sn-SbMo-W) which improved both stability and performance [49, 50]. Perovskites (ABO_3), which consist of alkali- and/or rare-earth metals (A) and transition metals (B), are also materials with great potential for band engineering. The oxidation state of B can vary, allowing for tailored properties through doping [51]. For example, increasing the amount of Sr and Co in $La_{1-x}Sr_xFe_{1-y}Co_yO_3$ enhances OER activity due to changes in the *d*-band, [52] where the strength of B–OH bonding, influenced by electron populations in the σ^* band, also affects the OER performance [53].

Spinels ($A'B'_2O_4$) are composed of group 2, group 13, and first-row transition metals, featuring both octahedral (O_h) and tetrahedral (T_d) sites. There are two types of spinels: normal spinels with divalent cations (A^{2+}) in T_d sites, and inverse spinels with divalent cations in O_h sites [51]. Spinels are excellent electric conductors and stable in alkaline solutions, making them suitable electrocatalysts for OER [54]. Iron-based and cobalt-based spinels, often doped with metals like Mn, Ni, Cu, Zn, and Li, exhibit varying OER activities. There are studies reporting higher activity of $CoFe_2O_4$ than $CuFe_2O_4$, $NiFe_2O_4$, and $MnFe_2O_4$ [55]. Bimetallic spinels generally perform better than monometallic ones, as different metal ions play distinct roles during OER. For example, Co^{2+} ions dominate the OER activity in Co_3O_4, and replacing Co^{2+} with Zn^{2+} or Co^{3+} with Al^{3+} affects the performance [56, 57]. A strong coupling between electrocatalysts and their substrates, have also an effect on the OER performance. Electrodepositing Co_3O_4 on Au substrates or dispersing spinels on carbon nanotubes, nanosheets, or graphene improves catalytic efficiency [58].

Layered oxides, including metal hydroxides and metal oxyhydroxides, exhibit high electrocatalytic activity for OER due to their layered-stacking conformation with protons between layers. These oxides, particularly cobalt- and nickel-based compounds, are known for their excellent OER activity. Phase transformations between different forms of $Ni(OH)_2$ and NiOOH during charge/discharge cycles also influence their OER activity [59]. Incorporating foreign elements such as Fe, Mn, and Cr into these compounds can further improve their performance [60]. Additionally, hybrid structures, such as those combining metal oxides with graphene oxide, have demonstrated superior OER activity and stability [61]. Despite the challenges of poor electronic conductivity and stability in some materials, innovative approaches like hybridizing FeOOH with CeO_2 have shown promise in enhancing long-term durability and reducing the OER overpotential [62].

Nonmetal Oxides Catalysts

Metal chalcogenides, particularly those based on cobalt and nickel, are emerging as highly effective materials for water-splitting due to their superior electroconductivity and resistance to corrosion in both acid and alkaline environments [63, 64]. Initially utilized for HER, these materials are now being explored for OER. They exhibit OER activities comparable to metal oxides with overpotentials typically ranging from 200 to 300 mV at 10 mA cm^{-2} [65, 66]. Notably, materials such as NiS and CoSe undergo surface transformations into oxides during OER, which enhances their catalytic activity. The incorporation of dopants, or generation of vacancies has been shown to further improve OER performance, as demonstrated by $FeNiS_2$, (Ni, Co)Se, and $CoSe_2$ which exhibit lower overpotentials, increased conductivity, and long-term stability [67–69]. Overall, the robust properties, bi-functionality, and stability of metal chalcogenides make them excellent candidates for efficient and sustainable water-splitting systems.

Metal pnictides, such as nitrides and phosphides of cobalt and nickel, are promising for HER due to their excellent conductivity and stability in harsh electrolytes [70]. They exhibit remarkable HER activities with overpotentials below 200 mV in alkaline environments, making them potential candidates for overall water-splitting [71]. The OER mechanism in metal pnictides involves the formation of a thin metal oxide layer on the surface, which facilitates electron transfer [72]. Doping with elements like iron also boosts OER performance, though long-term stability can be an issue due to surface oxidation [73].

Organometallic compounds, particularly those containing transition metals such as Mn, Fe, Co, Ru, and Ir, are being developed to mimic the oxygen-evolving complex in natural photosynthesis [74]. These compounds, such as the well-known ruthenium blue dimer, were initially thought to require di-atomic or multi-atomic structures for effective OER [75]. However, recent studies have shown that mononuclear organometallics can also exhibit excellent OER activities. Despite their poor stability and low electroconductivity, advancements in composite materials, including mononuclear, multinuclear, and cluster organometallic catalysts, have shown promising OER activity [76]. For instance, manganese corrole complexes and bimetallic porphyrin systems incorporating Fe and Co have demonstrated enhanced stability and performance [77]. Additionally, recent developments in organometallics, such as stable dinuclear Ir catalysts and pentanuclear iron catalysts, suggest that both oxyl radical coupling and water nucleophilic attraction are viable pathways for oxygen evolution [78].

Developing non-metal materials for OER is a promising approach to reduce costs associated with metal-based electrocatalysts. Carbon-based materials, including carbon black, carbon nanotubes, and graphene, have demonstrated good electroconductivity and chemical stability. Nitrogen-doped carbon nanomaterials, for example, show significant OER activity but suffer from long-term stability issues [79, 80]. Enhancements in OER performance have been achieved by oxidizing carbon nanotubes and creating composite materials like nitrogen and oxygen dual-doped graphene mixed with carbon nanotubes. Graphitic carbon nitride (g-C_3N_4) combined with graphene or carbon nanotubes also shows promise due to its high surface area and active sites [81]. Additionally, black phosphorus, with its adjustable electroconductivity, has emerged as a potential non-metal OER catalyst, exhibiting performance comparable to RuO_2 [82]. These advancements highlight the potential of non-metal compounds in developing cost-effective and efficient OER catalysts.

2.4 *Strategies for Improving Catalytic Activity*

Doping and Vacancy Engineering

Doping engineering is a prominent research area in the field of electrocatalytic water splitting. Element doping is primarily utilized to adjust the electronic structure, enhance electronic conductivity, and increase the quantity of active sites. It can effectively tune the electronic structure of adjacent regions, which further optimizes the adsorption of reactants or intermediates [83–85]. Anion doping can modify the surface electronic structure to optimize intermediate adsorption/desorption [86]. For example, N-doped CoS_2 shows improved HER and OER due to better adsorption energies [87]. Elements with higher electronegativity, like fluorine, can effectively modulate electronic structures, enhancing electrocatalytic performance [88]. Cation doping can tune the electronic structure, optimizing adsorption energies and enhancing intrinsic activity. Multi-metal catalysts often show better performance than single-metal oxides [40, 89, 90]. Transforming semiconductive materials into metallic ones through cation doping, such as V-doped NiS_2, can significantly improve electrocatalytic performance [91]. Combining anion and cation doping can achieve superior electrocatalytic performance through synergistic effects. For instance, Fe and O co-doped Co_2P shows excellent HER and OER performance [73, 90, 92].

Vacancy engineering in electrocatalysts has become a significant research focus due to its ability to modulate intrinsic activities. Experimental and theoretical studies have shown that vacancy engineering can increase the density of active sites, ensure fast charge transfer, and optimize adsorption/desorption energy of reaction intermediates. [16, 93, 94] Anion vacancies in transition-metal catalysts have been produced through plasma treatments. Both Ar- and H_2-plasma treatments have been used to create O-vacancies in Co_3O_4, [95] or sulfur-vacancies in MoS_2, [16, 96, 97] significantly improving the catalytic activity. Cation vacancies have been shown to enhance electrocatalytic properties by promoting electron transfer and optimizing electronic configurations [98]. These strategies demonstrate the potential of vacancy engineering to tune the number of active sites, stability, and conductivity.

Phase and Interface Engineering

The phase modulation can change the oxidation state and affect the adsorption energy of reaction intermediates for spinel-type electrocatalysts [99]. Key areas of focus include designing and constructing pure amorphous phases, converting amorphous phases into crystalline phases, and designing and regulating allotropes. Pure amorphous phases often outperform crystalline counterparts in OER catalysis due to better adsorption of OH^* and formation of OOH^* [100, 101]. But also, transforming amorphous phases into crystalline phases can induce lattice stress, improving catalytic performance [102]. Surface reconstruction of amorphous phases can also lead to superior OER catalysis [103]. Despite these advancements, there is still a need for better understanding of phase evolution during catalytic processes. Techniques like in-situ XRD and Raman spectroscopy are paving the way for tracking the evolution of active species.

Interface engineering has been used to tune the kinetic behavior of intermediate reaction species. Heterointerfaces play a vital role in electrocatalysis by adapting the adsorption free energy of intermediate species and accelerating electron transport [104]. For instance, Co-Ni_3N nanowires with enhanced confinement effects at the interface, promotes fast electron transfer and catalytic HER [105]. Similarly, combining NiFe layered double hydroxide nanosheets with defective graphene resulted in low overpotential for OER due to electron redistribution at the interface [106]. A strong electronic interactions at interfaces are key to interface engineering, though further research is needed on the evolution of interfaces for overall water splitting reactions.

3 Photoelectrocatalytic Water Splitting

Production routes able to convert the energy coming from renewable sources in electricity and the direct or indirect storage of this electrical energy as chemical energy are attractive alternatives to overcome the issues associated with conventional hydrogen synthesis processes, where the photoelectrocatalytic (PEC) water splitting is a promising technology able to harvest the energy of the sun to perform the oxidation or reduction of water with the photogenerated charge carriers in the absorbing material. In PEC water splitting, the generation of O_2 or H_2 at the surface of light-absorbing materials takes place through catalytic charge transfer reactions between the photogenerated charge carriers and the reaction precursors adsorbed at the materials surface. Thus, the performance of a PEC system is set by the kinetic equilibrium and the efficiency of several phenomena taking place at different time scales.

In this section we will discuss some of the most important processes involved in the functioning and degradation of photoelectrochemical systems able to perform water splitting. As in this type of systems at least one of the electrodes consist of a light-absorbing semiconductor material in contact with an electrolyte solution, we start by describing the characteristics of the semiconductor-liquid junction in the dark at equilibrium conditions or under an applied bias. We continue with the effect of light-perturbation on electron–hole generation, and the redox reactions at the surface by the majority/minority charge carriers that reach the interface. We will also review some of the high performance photoanodes and photocathodes currently studied, the reaction mechanisms in these materials, their efficiency limiting factors, as well as the most important strategies to improve their optical properties, electrochemical activity and particularly, their long-term stability during the water splitting half reactions.

3.1 Working Principles of PEC Water Splitting

3.1.1 The Semiconductor-Liquid Junction

A semiconducting material shows electrical properties between a metal and an insulator, being classified in elemental semiconductors (as Si and Ge) and compound semiconductors (TiO_2, ZnO, GaAs, etc.). Their distinctive electrical and optical properties are related to their complex electronic structure, leading to energy levels that will control the transport of charge carrier toward the interface and the redox reactions. Figure 4 depicts the energy diagram for a semiconductor without band-bending at the surface. Here we find the ionization energy (IE), the work function (WF), and the electron affinity (EA) as energetic parameters referred against the vacuum level (E_{VAC}). E_{VAC} is the energy level where an electron is at rest with zero kinetic energy, implying that the electron is outside the surface and no longer attached to the solid.

The charge transport levels in a solid arise from the overlapping of several molecular orbitals forming a continuous of electronic states, known as bands. The molecular orbitals in the valence band (VB) are full of electrons which are located at the chemical bonds in the material. Molecular orbitals in the conduction band (CB) are completely empty. The difference in energy between the CB and the VB is known as the energy bandgap (E_g). For a semiconductor, the jump of an electron from the VB to the CB and the creation of a positive hole (h^+) can be achieved by applying a perturbation (electromagnetic or thermal) with energy equal or larger than E_g. This process increases the density of mobile electrons and holes in the material. Another important energetic parameter is the Fermi level (E_F), which corresponds to the energy level where the electronic occupancy probability is 50%. The Fermi level sets the electrochemical potential of the electrons in the material, and it is the driving force to reach equilibrium at the interface of two different materials. This equilibrium is dependent on the density of states, temperature, dopants, among other things [107, 108]. The position of E_F in the bandgap provides information about the conductivity and type of majority charge carriers. For an intrinsic semiconductor (Fig. 4b), E_F lies at the center of the gap. For an *n*-type semiconductor, it lies close to the conduction band minimum (Fig. 4c). And for a *p*-type semiconductor it lies close to the valence band maximum (Fig. 4d). Thus, usually the conductivity of an intrinsic semiconductor at room temperature is low due to the small amount of mobile charge carriers in the

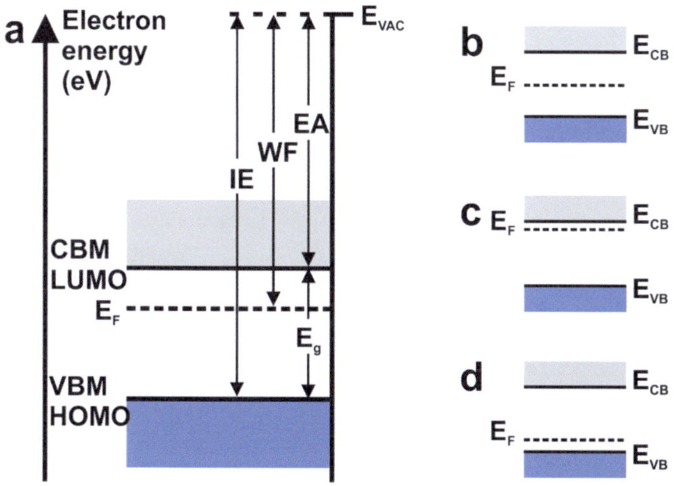

Fig. 4 **a** Energy diagram for a hypothetical semiconductor without band bending at the surface. The most important energetic parameters and electronic levels are indicated. Energy diagrams of an **b** intrinsic, **c** n-type, and **d** p-type semiconductor

material. The opposite is observed for *n*-type (*p*-type) semiconductors where a large density of free electrons (holes) in the conduction (valence) band results from the thermal excitation of charge carriers from (to) donor (acceptor) levels close to the bands [109].

When a semiconductor electrode is placed in contact with an electrolyte solution containing a redox couple, the difference between the E_F and the formal potential of the redox couple (E^0_{redox}) leads to a spontaneous charge transfer process until thermodynamic equilibrium is achieved. This process is depicted in Fig. 5 for an *n*-type semiconductor in contact with a redox couple with a E^0_{redox} more positive than that of the E_F of the material (Fig. 5a). At equilibrium conditions, and ignoring surface states, both E^0_{redox} and E_F are at the same position. As a result, electrons move from the surface of the semiconductor leaving behind a positively charged layer which extends a distance *d* inside the material (Fig. 5b). The formation of this layer, space charge region (SC), creates an electric field represented as an upward bending of the bands since their energy at the interface is fixed. At the electrolyte side, there is also a potential drop giving place to the electrical double layer, named the Helmholtz layer and the Gouy-Chapman region. However, in a concentrated electrolyte solution the capacitance of the Helmholtz layer is greater than the space charge region capacitance, and due to its higher concentration of charge carriers, the potential drop occurs almost exclusively within the semiconductor. The band bending at the semiconductor surface decreases the concentration of majority charge carriers in the conduction band, setting a barrier that hinders spontaneous reactions between free electrons and the accepting species in the redox couple.

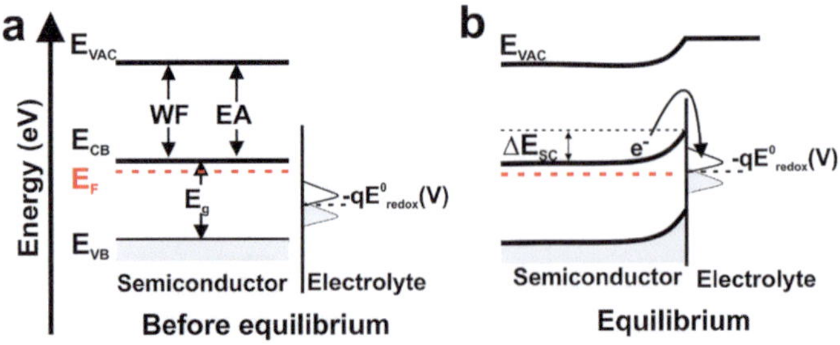

Fig. 5 Energy diagram of an n-type semiconductor. **a** Before, and **b** after contacting an electrolyte solution with a redox couple having a Fermi level or formal potential (E^0_{redox}) more positive than the materials Fermi level (E_{F})

Figure 6 depicts other possible situations once the semiconductor and electrolyte are in equilibrium. At *flat band* conditions (Fig. 6a), there is no band bending since the Fermi level of the semiconductor and the electrolyte are at the same position, thus the concentration of free electrons in the conduction band and near the surface is high. At *accumulation* conditions (Fig. 6c), a downward bending of the bands is observed leading to a negatively charged layer inside the semiconductor. This is caused because the formal potential of the redox couple is more negative than the material E_F, which promotes electron transfer from the electrolyte to the semiconductor. If the E^0_{redox} is close to the semiconductor valence band (Fig. 6d), a massive upward bending of the bands is observed, known as *inversion* [109]. The latter results in a thick SC region and a large energy barrier for the transfer of free electrons from the bulk to the surface. A similar situation occurs in *p*-type semiconductor, but the charge in the SC region and the direction of the band bending are opposite as for those described in *n*-type semiconductors.

3.1.2 Redox Reactions at Semiconducting Electrodes in the Dark

In 1960, Heinz Gerischer developed a model to explain the differences observed in the reaction kinetics between metallic and semiconducting electrodes in contact with an electrolyte. According to the Gerischer model, electron transfer takes place through electronic levels in the solid and energy levels in the solution, an isoenergetic process. Due to the interactions of the solvent molecules with the oxidized/reduced species of the redox couple the description of the energy levels in solution is more complex than in a solid, leading to a distribution of empty (Ox) or occupied (Red) energy levels as shown in Fig. 7.

From Fig. 7a-b, it is clear that for metallic electrodes, an increase in the anodic or cathodic current associated to the oxidation of the occupied levels or the reduction of the empty levels can be easily achieved by polarizing the electrode by some energy around the formal potential of the redox couple, $E^0{}_{redox,}$ due to the wide-ranging overlap of states between both the metal and the electrolyte. For the case

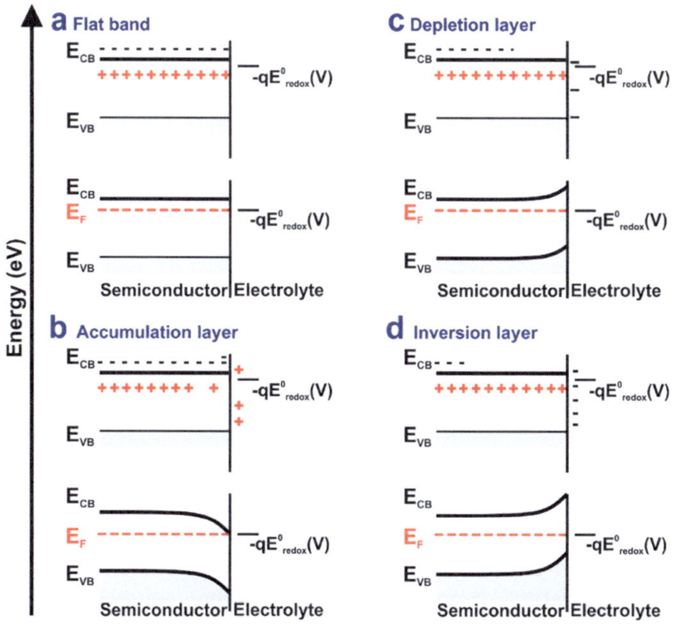

Fig. 6 Energy diagram of an n-type semiconductor in contact with an electrolyte solution containing a redox couple (E^0_{redox}) with a Fermi level **a** at the same position as the material (Flat band), **b** more positive than that of the material (depletion), **c** more negative than that of the material (accumulation), and **d** close to the materials valence band (inversion)

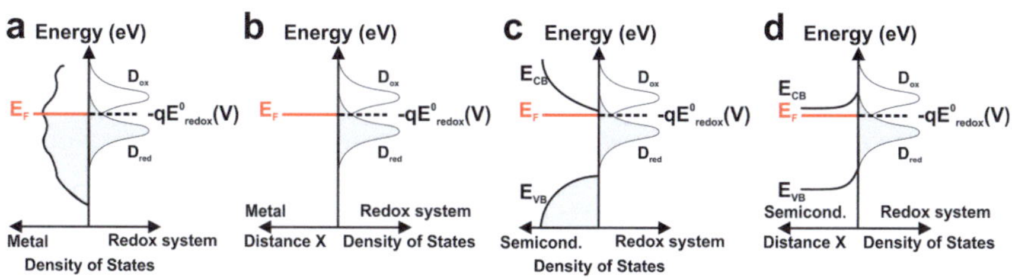

Fig. 7 Density of states and energy diagrams for a **a**–**b** metal-electrolyte, and **c**–**d** semiconductor-electrolyte interface at equilibrium

of a semiconductor electrode (Fig. 7c-d), charge transfer reactions can only occur through bands. In other words, the distribution of occupied or empty states in the redox couple must overlap with the valence or the conduction band. Figure 7d presents the energy diagram of an *n*-type semiconductor in contact with a redox system, with a strong overlap between the conduction band edge at the surface and the distribution of empty states in the redox couple. As mentioned before, the band bending due to the formation of the SC region at equilibrium conditions sets a barrier for the spontaneous transfer of electrons in the conduction band to the empty states in the redox couple. However, by changing the applied bias to the semiconductor, its Fermi level changes as well as the bending of the bands at the surface as described in Fig. 6. An increase in the concentration of free electrons in the conduction band (Fig. 6a-c) leads to reduction reactions. A similar reasoning applies for oxidation reactions on a *p*-type semiconductor. It is worth noting that in agreement with this model, electron transfer between surface states and empty/occupied states in the redox couple is also possible, as widely reported in the literature [110–116].

3.1.3 Effect of Illumination

As reviewed in the last section for an *n*-type semiconductor at equilibrium and in the dark, the SC region sets an electric field that hinders the transfer of free electrons in the conduction band from the bulk to the surface (Fig. 8a). However, upon light absorption, this electric field helps to separate the electron–hole ($e^- - h^+$) pairs photogenerated in the material, where the minority charge carriers (holes) are responsible for performing the electrochemical reactions at the interface (Fig. 8b). The magnitude of this electric field (or band bending) determinates the efficiency of the charge separation process (usually limited by recombination) and direction of the photocurrent achieved. Figure 8c depicts this situation for an *n*-type semiconductor in the dark and under illumination [109, 117]. It can be observed that when the potential applied is more negative than the flat band potential (Fig. 6c) a cathodic current is observed in both dark and light conditions due to the accumulation of the majority charge carriers at the surface, going down to zero as the potential approaches to the flat band (Fig. 6a) due the amount of free electrons in the CB decrease (dark) or by the high recombination rate of the photogenerated charge carriers (light). No current is observed at potentials more positive than the flat band potential for the electrode in the dark owed to the lack of electron vacancies in the VB. The opposite is observed for the electrode under light conditions, since the growing upward bending of the bands efficiently separates the photogenerated charge carriers (dash red line) leading to a fast increase in the anodic current to an almost constant value dependent on the light intensity and the kinetics of the reaction performed. Figure 8c also shows the effect of charge carrier recombination on the current–potential curve (solid red line). Unlike the case without recombination, the band bending at the depletion regime (Fig. 6b) is not enough to separate the charge carriers and there is a wide potential window in where recombination is the dominant process and no current is observed, shifting the photocurrent onset potential to the inversion regime (Fig. 6d) where the density of majority charge carriers in the SC region and surface is negligible and the recombination is suppressed. It is clear that in a real PEC systems a certain degree of band bending is needed to develop a current related to the separation and transfer of charge carriers to electroactive species at the interface, however the presence of structural defects and a high concentration of surface states in the absorbing material leads to a situation called Fermi-level pining where the applied potential has not a big

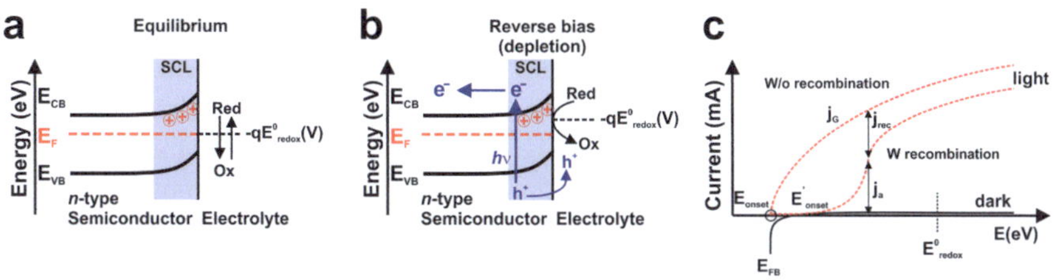

Fig. 8 Energy diagram of an n-type semiconductor in contact with an electrolyte solution **a** at equilibrium and in the dark, and **b** under illumination and applying a positive potential to the semiconducting electrode. **c** Shows the current–potential response of an n-type semiconductor in the dark and under illumination performing an electrochemical reaction

effect on the position of the Fermi level with the band edges at the surface in consequence decreasing the band bending experienced, something that also contributes to the photocurrent onset potential.

High chemical stability under light irradiation is an important requirement for any absorbing material used to perform photoelectrochemical reactions. The accumulation of charge carriers (h^+ or e^-) at the interface due to kinetic or mass transport limitations leads to the material photocorrosion in the form of dissolution or passivation of the interface. The overall effect of the photocorrosion is the gradual loss of the materials activity, something that nowadays limits the practical application of many PEC technologies based on different *n*- or *p*-type semiconducting materials, despite their interesting optical, electronic and electrocatalytic properties.

3.2 *PEC Oxidation and Reduction of Water*

3.2.1 Photoanodes and Photocathodes

The evolution of O_2 or H_2 gases at the surface of photoactive materials involves several steps whose efficiency determinates the overall performance of the system. To perform the photo oxidation/reduction of water, a material must absorb as much as possible of the solar spectrum. The incident photons must create charge carriers able to reach the material-electrolyte interface and transfer them to the reaction precursors adsorbed at the surface.

Both broadband light absorption and transfer of charge carriers completely depend on the electronic properties of the semiconductor. Ideally, a material with a bandgap of 1.48 eV (calculated from the dissociation enthalpy of water) might be able to carry out the water splitting reaction; however, owing to energy losses seen as reaction overpotentials and series resistances in the cell, it has been suggested a E_g value between 1.6 and 2.4 eV (775–516 nm) is required. On the other hand, as the solar irradiance dramatically decreases at wavelengths below 380 nm, where UV-light is ~ 5% of the total, a material with a $E_g \leq 3.2$ eV is desired [118, 119]. Upon light absorption, the minority charge carriers in the material participate in the corresponding water splitting reactions as illustrated in Fig. 9a. The h^+ in a n-type semiconductor reach the interface due to the upward band bending at the surface, promoting the OER (photoanode); while the downward band bending in a p-type material move the e^- to the surface leading to the HER (photocathode), all this if the position of the VB and the CB are more positive or more negative to the onset potential of that reactions, respectively. Figure 9b shows the band position of some of the most used n- and p-type semiconductors in the PEC water splitting. Metal oxides such as the TiO_2, WO_3, and α-Fe_2O_3 have been intensively investigated as photoanodes for this application, due to their convenient position in energy of their valence band, abundance of elements, and easy synthesis methods. However, the performance displayed by these materials during the OER is far from desirable, exhibiting poor light absorption, fast recombination of charge carrier (as in the case of TiO_2), [119, 120] low chemical stability, high photocorrosion (observed for the WO_3), [121, 122] or low light absorption coefficients, and short hole diffusion length (as seen for α-Fe_2O_3) [123, 124]. On the other hand, materials such as Si, InP, GaAs, and GaP have lead the research on photocathodes for the HER due to their narrow bandgap. Nevertheless, the high cost and scarcity of some of these materials, as well as their low stability under illumination (as in case of the p-Si) limit their use on commercial processes. Finding the ideal semiconducting material that meets all the requirements is a difficult task. However, extensive work carried out on this field by different groups has led to the development of novel and promising systems such as dye-sensitized semiconducting electrodes, [125, 126] and plasmonic photocatalytic structures [127, 128].

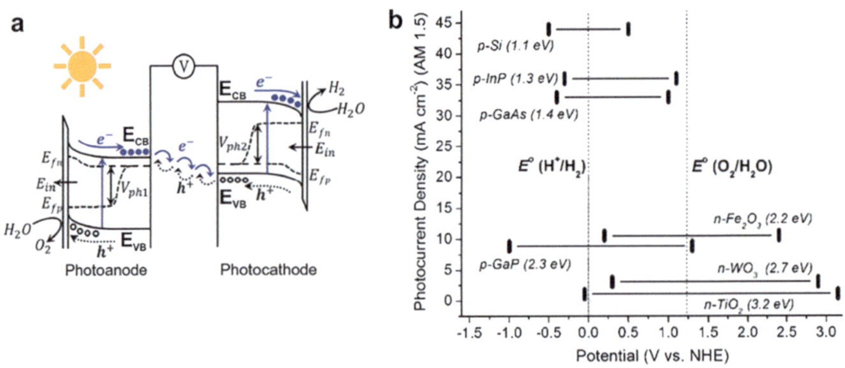

Fig. 9 **a** Energy diagram of a hypothetical photoelectrochemical cell, with both photoanode and photocathode performing the OER/HER reactions, respectively. **b** conduction band (left) and valence band (right) position of different n-type and p-type materials used in the PEC water splitting (x-axis) as well as the maximum theoretical photocurrent achieved at 1.5 AMG illumination conditions (y-axis). Images reproduced with permission from [124, 129]

3.2.2 Bandgap Engineering

Wide bandgap semiconductors usually meet the condition of having the VB/CB, or both, at potentials more positive or more negative than that of the OER/HER onset, something required to observe high reaction rates. However, the side effect on working with wide bandgap semiconducting electrodes is the low number of photons harvested from the solar spectra (Fig. 9b). The addition of new electronic states inside the bandgap of the material can be achieved through doping, generation of defects and oxygen vacancies, among others. Theoretical calculations on TiO_2-anatase phase have shown that the selection of adequate pairs of metal/non-metals as dopants can modify the size of the bandgap and position of the CBM and VBM without compromising the mobility of charge carriers. It has also been reported that the co-doping of TiO_2 with *n*-type/*p*-type pairs of transition metals increases the OER electrocatalytic activity by tuning the adsorption/desorption energies of reaction intermediates and products [130, 131]. Oxygen vacancies in the crystal lattice of TiO_2 and WO_3 photoanodes have shown to increase their photo response toward the visible range due to light absorption by the resulting surface states located in the middle of the band gap leading to the OER when there is a suitable energy coupling [132, 133].

3.2.3 Surface Modifications

Modification of the surface of light absorbing semiconductors with the addition of co-catalysts or the deposition of thin films may significantly boost the performance and stability of these systems. The photoexcitation and transfer of charge carriers from the bands and trap states are processes that occur in the scale of few to hundreds of fs, while the recombination of charge carriers with in the solid or at the surface is in μs, on the other hand the reactions involved in the splitting of the water molecule take place in the range of ms to s. Hence, the kinetics of the surface reactions involved is the rate-limiting step of the whole process. The deposition of co-catalyst has shown to decrease the energy barriers associated with the charge transfer reactions between the photoelectrode to the reaction precursors adsorbed on the material, in this way increasing the charge extraction due to a faster reaction rate at the surface. This effect is depicted in Fig. 10a for a photocathode before and after co-catalyst functionalization of the surface. Different co-catalyst materials based on noble metals, metal sulfides, metal phosphides, metal hydroxides, among others, have been tested in PEC water splitting systems, showing a positive impact on the OER/HER reaction rates and in consequence on the solar to hydrogen conversion efficiency; being the particle size, morphology, crystal facets, and loading of these materials the determining factors on their overall effect [134–136].

The deposition of thin and compact films of another semiconductor (usually oxides) on top of the absorbing material has been widely applied in PEC water splitting. The chemical and electronic properties of the thin film dictates if the coating will

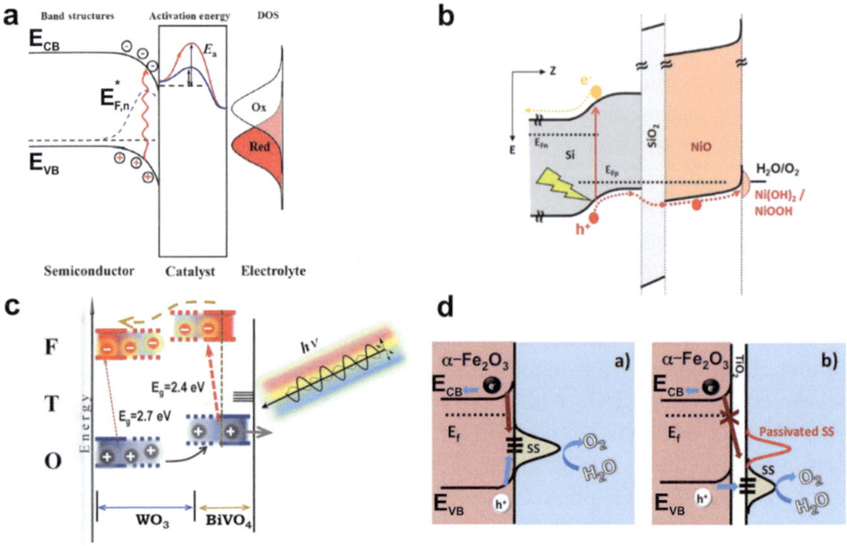

Fig. 10 **a** Energy diagram of a photocathode/catalyst/electrolyte junction where the effect of the catalyst on the energy barriers is depicted. Reproduced with permission from [109]. **b** Energy diagram of an n-p Si photoanode modified with a thin layer of NiO as protective layer. Reproduced with permission from [137]. **c** Energy diagram of a WO_3/$BiVO_4$ heterojunction under illumination. Reproduced with permission from [138]. **d** Energy diagram of a α-Fe_3O_4 photoanode without (left) and with (right) a thin film of TiO_2 and its effect on the recombination via surface states and h^+ transport to the surface. Reproduced with permission from [139]

behave as a corrosion protective layer, or as a charge extraction layer (heterojunction). NiO is a good example of a high conductive, chemically stable and OER active metal oxide in alkaline solutions. It has been reported that a ~ 50 nm NiO film on top of a *n-p* Si photoanode lead to an increase in the electrode lifetime and activity over 1000 h of operation under standard illumination conditions. The h^+ generated in the *n-p* Si photoelectrode reaches the NiO valence band after tunneling through the SiO_2 interlayer and transferred to the OER precursors via $Ni(OH)_2$ and NiOOH formed on the surface of the NiO film during operation (Fig. 10b) [137]. Another interesting and well-studied system is the WO_3/$BiVO_4$ heterojunction, where the suitable position of both VB and CB of these materials, the excellent electron conductivity in the WO_3, and the good light absorption of $BiVO_4$ results in enhanced current densities under illumination compared to their individual components (Fig. 10c) [138]. In materials such as α-Fe_2O_3, the passivation of recombination centers at the surface was achieved by depositing thin layers of TiO_2 and MoO_x. As a side effect of the presence of these passivating overlayers was a shift of the photocurrent onset. This was understood to be a new hole transport process to the interface through the surface states in these passivating overlayers, being necessary to functionalize the surface with a co-catalyst to improve OER reaction kinetics (Fig. 10d) [139]. Many more examples of materials and electrode architecture can be found in the current literature; and the cases discussed here give a general overview of some of the most important strategies directed to overcome common issues observed in photoanodes and photocathodes used in water splitting.

4 Photocatalytic Water Splitting

4.1 Basic Principles

Photocatalysis is a process that harnesses light energy to accelerate a photoreaction in the presence of a catalyst, typically a semiconductor. Light, typically the ultraviolet or visible part of the electromagnetic spectrum, is absorbed by the photocatalyst, generating electron–hole pairs. These charge carriers then participate in redox reactions on the surface of the catalyst, driving the desired chemical transformations [140, 141]. The process of water splitting through photocatalysis unfolds in several essential photophysical and chemical stages, encompassing four main steps depicted in Fig. 11: (1) light absorption, (2) charge excitation, (3) charge separation and transfer, and (4) surface redox reactions on the photocatalyst.

Light absorption begins when the photocatalyst absorbs photons with energy equal to or greater than its bandgap. Upon light absorption, an electron in the VB is promoted to the CB, generating an electron–hole pair ($e^- - h^+$). Electrons in the CB now have enough energy to drive reduction reactions, while the holes in the VB serve as oxidizing agents. Effective charge separation and transfer are crucial for successful photocatalytic activity. Finally, electrons migrate to the surface to reduce protons in water, while holes move toward the surface to oxidize water. Electrons at the CB need sufficient energy for H^+ reduction. This condition is satisfied when

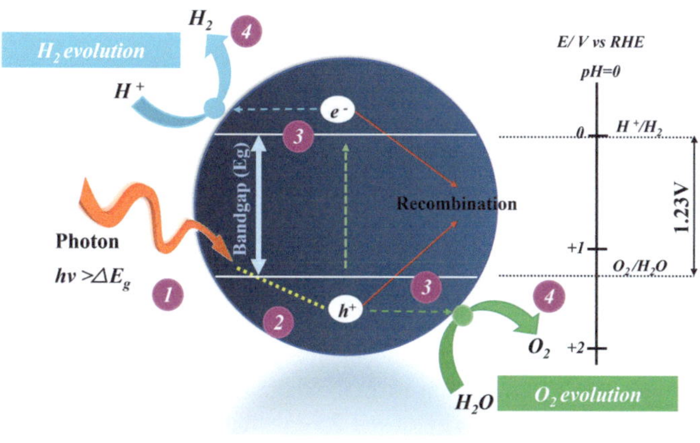

Fig. 11 Process of photocatalytic water splitting on a heterogeneous photocatalyst

the CB potential is more negative than the H^+/H_2 reduction potential, which is 0 V relative to the reversible hydrogen electrode (RHE). For water oxidation, holes in the VB need sufficiently positive potential to facilitate the oxidation process. This requires the VB to be at least 1.23 V (vs. RHE).

The nature of the bandgap, whether direct or indirect, significantly affects efficiency, as it determines how electrons are excited within the semiconductor. Direct bandgap materials experience rapid electronic transitions with high absorption efficiency. Electron excitation in indirect bandgap semiconductors requires both photon absorption and phonon interaction, as the minimum of the CB and maximum of the VB are not aligned. This process reduces the probability of electronic transitions, but it can be compensated by increasing the thickness of the photocatalysts. However, this might increase material synthesis costs and impact charge carrier extraction efficiency. Still, it offers some benefits, photogenerated $e^- - h^+$ pairs tend to have longer lifetimes, reducing recombination rates.

The careful selection and design of photocatalysts that maximize light absorption while minimizing recombination is crucial to achieve optimal redox reactions and efficient hydrogen and oxygen production. These photophysical and electronic properties establish a foundation upon which the thermodynamic aspects of the water-splitting process can be further explored [142, 143].

4.2 Thermodynamic Principles

Thermodynamics is fundamental in assessing whether a chemical reaction will proceed spontaneously or require additional energy input. The Gibbs free energy change (ΔG^o) associated with the reaction provides such information. When ΔG° is negative, the reaction is thermodynamically favorable and proceeds in the forward direction. Conversely, a positive ΔG^o indicates that the reverse reaction is favored. The value of ΔG^o reflects both the enthalpy (ΔH^o) and entropy (ΔS^o) contributions according to the Eq. 4.1:

$$\Delta G^0 = \Delta H^0 - T\Delta S^0 \tag{4.1}$$

Thermodynamically, water splitting is endothermic and does not occur spontaneously, as the ΔG^o equals + 237 kJ/mol under standard temperature and pressure conditions (25 °C and 1 atm). In this case, both enthalpy ($\Delta H^o > 0$) and entropy ($\Delta S^o > 0$) changes are positive, and even if the -$T\Delta S^o$ term is negative, its contribution alone is insufficient to make the process spontaneous, emphasizing the need for an additional energy source [144].

For the water splitting process to occur, the photocatalyst must absorb photons with a minimum energy equivalent to ΔG° of the reaction. This minimum energy corresponds to a photon an energy of ~ 1.23 eV (matching the oxidation potential of water under standard conditions). This value sets the minimum bandgap of the photocatalyst required to generate $e^- - h^+$ pairs with sufficient energy to participate in the water-splitting redox reactions. However, due to factors such as energy losses from $e^- - h^+$ recombination and absorption efficiency, it is required that the bandgap of the semiconductor is slightly larger, typically between 1.6 and 2.2 eV. The efficiency of this conversion is critical, and the quantum efficiency (fraction of absorbed photons that generate $e^- - h^+$ pairs capable of participating in redox reactions), and solar-to-hydrogen efficiency (solar energy converted into chemical energy) are two important metrics to take into consideration.

4.3 *Types of Photocatalysts*

The effectiveness of a photocatalyst depends on intrinsic properties such as electronic band structure, light absorption capacity and morphology. TiO_2 is one of the most widely researched photocatalysts due to its low cost, chemical stability, and ability to oxidize organic pollutants. TiO_2 has a band gap of ~ 3.2 eV in its anatase phase, which restricts its light absorption mainly to the UV range. Additionally, TiO_2 suffers from rapid $e^- - h^+$ pair recombination. Techniques such as doping with metal and non-metal elements and surface modification are commonly employed to overcome these limitations [145, 146]. Zinc oxide is another semiconductor widely used in photocatalytic applications due to its low cost and easy production. ZnO has a bandgap of ~ 3.3 eV, also limiting its light absorption to the UV range. ZnO faces challenges related to stability and photodegradation, especially under visible light conditions. Surface modification and doping with Ag or Cu are commonly used to improve its performance and stability [145, 147]. Unlike TiO_2 and ZnO, CdS has a narrower bandgap of ~ 2.4 eV, allowing it to absorb light in the visible spectrum. However, CdS is prone to photodegradation, especially in the presence of oxygen, limiting its stability and durability. Strategies to address this include encapsulating CdS in protective materials or combining it with other semiconductors in heterojunctions to facilitate charge separation and reduce $e^- - h^+$ recombination [148, 149].

Each of these materials has advantages and disadvantages in terms of photocatalytic efficiency, stability, and visible light response. Current research focuses on band structure modifications or creating heterojunctions to improve charge separation and, ultimately, photocatalytic efficiency.

4.3.1 Bandgap Engineering: Surface Modifications and Doping

A primary challenge in photocatalysis is the rapid recombination of $e^- - h^+$ pairs and their large bandgap, which reduces quantum efficiency and restricts light absorption primarily to the UV of the solar spectrum. This issue has driven advancements in band structure engineering via surface modifications, or by creating hybrid structures between semiconductors with complementary energy levels.

Surface Modification

Incorporating metals in the form of metal ions, metal compounds, or metallic nanoparticles, introduces electron and hole trapping sites. This affects the recombination rate of photoinduced charges by acting as electron (Eq. 4.2) or hole traps (Eq. 4.3).

$$M^{n+} + e_{cb} \rightarrow M^{(n-1)+} \tag{4.2}$$

$$M^{n+} + h_{vb} \rightarrow M^{(n+1)+} \tag{4.3}$$

where the energy level for $M^{n+}/M^{(n-1)+}$ lies below the CB edge (e_{cb}) and the energy level for $M^{n+}/M^{(n+1)+}$ lies above the VB edge (e_{vb}). Therefore, modification with metal ions can improve $e^{-} - h^{+}$ separation via selective trapping [150, 151] reduce carrier lifespan due to their role as recombination centers, and enhance optical absorption in the visible light range [152, 153].

Studies by Macyk and Kisch have demonstrated that metal compounds, like $PtCl_4$, induce photocatalytic activity under visible light. In this system, the excited platinum complex transfers electrons to the conduction band of TiO_2 [154, 155]. Iron oxides in TiO_2 systems have also been shown to enhance visible light absorption through the formation of FeO_x clusters on the photocatalyst surface [156]. Noble metals have been extensively studied due to their ability to absorb visible light via localized surface plasmon resonance. This phenomenon, utilized in systems such as Au/TiO_2 and Au/ZnO, facilitates the injection of Au electrons (hot electrons) into the conduction band of TiO_2 or ZnO. The involvement of hot holes has also been suggested in the photocurrent generation mechanism for Au nanoparticles deposited onto porous TiO_2. Other studies with Au/TiO_2 have reported alternative electron transfer mechanisms where the participation of the semiconductor is either minimal or negligible. Upon irradiation, hot electrons with energy between the vacuum level and the gold's work function can dissociate H_2 under visible-light irradiation [157].

Non-metallic modifications have also been applied to functionalize semiconductors, including the adsorption of anions, inorganic compounds, and organic compounds. One example of this is the modification of $NaTaO_3$ with glucose, which enhances visible light absorption due to the formation of carbonaceous species at the $NaTaO_3$ surface. These species act as sensitizers under visible light irradiation, facilitating the transfer of photogenerated electrons to the CB of $NaTaO_3$, where they react with adsorbed oxygen to form superoxide ions [158]. Another surface modification approach involves conductive polymers, such as poly(3-hexylthiophene) (P3HT), polyaniline, and polypyrrole. For instance, Shao et al. found that P3HT facilitates interfacial exciton dissociation when incorporated into black phosphorus. P3HT induces the accumulation of photogenerated electrons on the black phosphorus surface. One of the most promising options is using conjugated polymers made up of chains with π-conjugated bonds that facilitate efficient electron transport and light absorption in the visible spectrum. Examples include polymers based on triazine, phenyl, and thiophene units. Modifying these materials with electron-attracting functional groups such as halogens, cyano, and thiadiazol is crucial because it adjusts their LUMO and HOMO energy levels, improving photon capture and solar energy utilization. Furthermore, these modifications extend the π-conjugated network, reducing the bandgap and enhancing charge migration [159].

Elemental Doping

Unlike surface modification, doping involves the controlled incorporation of atoms into a crystalline structure of the material. This process can be classified into two main categories: cationic and anionic doping. Both strategies create additional energy levels in the material, either below the conduction band or above the valence band. In materials like TiO_2 or CdS, cationic doping typically reduces the bandgap energy, enhancing visible light absorption and charge separation. Conversely, anionic doping adjusts the position of the valence band, corrects structural defects, and improves the chemical stability of the material [160–162]. Doping can also occur in interstitial or substitutional forms, modifying the local electronic density distribution within the crystalline structure. For instance, in the anatase phase of TiO_2, the incorporation of metallic cations or non-metallic anions alters the chemical bonds in the TiO_6 octahedra, fine-tuning the electronic structure and reducing the bandgap through band engineering.

Specific criteria must be met for doped photocatalysts to be effective in water splitting: the conduction band minimum must be positioned above the H_2/H_2O potential to promote reduction, while the valence band maximum must be below the O_2/H_2O potential to facilitate oxidation. Additionally, impurity states generated in the bandgap should be shallow or interact with the main bands to ensure efficient charge transfer to reactive sites during their lifetime [163]. For example, doping ZnO with nitrogen narrows the bandgap by replacing oxygen atoms in the crystal lattice. As shown in Fig. 12, this substitution creates shallow acceptor states near the VB, raising its position and forming a new valence band. This change facilitates visible light absorption. Similarly, nickel-doped ZnO exhibit donor electrons that occupy energy levels close to the CB. This forms a new conduction band at a higher energy level, contributing to bandgap narrowing. Additionally, nickel acts as an active center, establishing electron traps that reduce electron–hole recombination. The combined doping of nitrogen and nickel results in an increase (decrease) in energy of the VB (CB), further narrowing the bandgap. This simultaneous doping enhances visible light absorption and optimizes charge separation and transfer [164].

4.3.2 Bandgap Engineering: Heterojunctions

Coupling two or more semiconductors has emerged as a key strategy for inhibiting charge carrier recombination and enhancing photocatalytic efficiency. Numerous mixed semiconductor colloid systems, such as CdS-TiO_2, CdS-ZnO, Cd_3P_2-TiO_2, Cd_3P_2-ZnO, and AgI-Ag_2S, have been investigated and shown to improve charge transfer processes [165]. However, other semiconductors, such as WO_3, might face

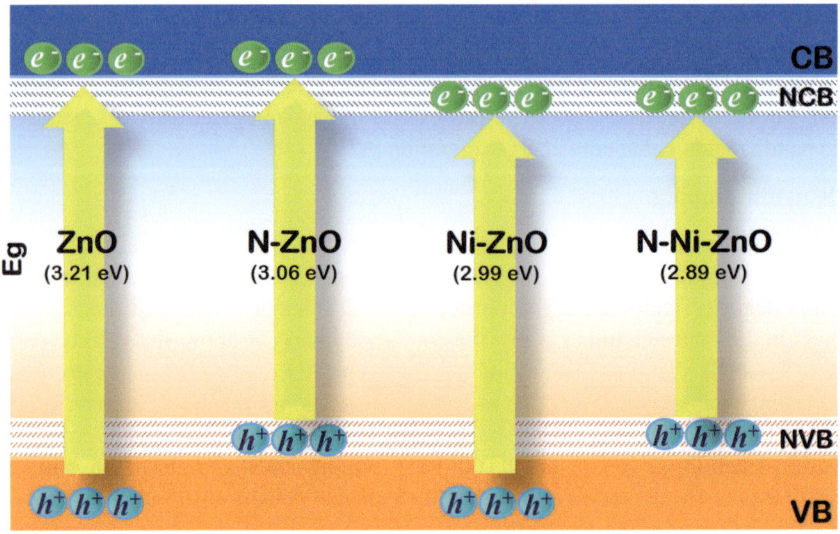

Fig. 12 Schematic of the bandgap narrowing mechanisms in ZnO, nitrogen-doped ZnO (N-ZnO), nickel-doped ZnO (Ni-ZnO), and N-Ni co-doped ZnO. Reproduced with permission from [164]

inherent limitations due to incorrect position of the CB requiring additional strategies such as the incorporation of co-catalysts to improve functionality under visible light irradiation [166].

Heterojunction type-II systems represent a key advancement in photocatalysis, particularly for water splitting, by combining two semiconductors with staggered band alignments. This configuration allows electrons and holes to separate into distinct spatial regions, prolonging the lifespan of photogenerated charge carriers and mitigating recombination issues, Fig. 13a. This separation allows the HER and OER to occur on different photocatalysts, enhancing the overall efficiency of the system. However, a limitation inherent to type-II heterojunctions is the reduction in redox potential due to the migration of charge carriers to less reactive CB and VB edges. Nevertheless, type-II systems remain a valuable strategy for enhancing charge separation, and their relatively straightforward design offers significant potential for integrating diverse semiconductor materials [167].

The Z-scheme photocatalytic system provides a superior alternative for water splitting by mimicking the natural photosynthetic process. Unlike type-II heterojunctions, Z-scheme systems employ two photocatalysts connected through a mediator, facilitating two-step photoexcitation. This configuration enables each photocatalyst to retain its intrinsic redox potential, allowing for highly efficient redox reactions (Fig. 13b). One photocatalyst (PS I) serves as the reduction site (HER), while the other (PS II) is the oxidation site (OER). A distinguishing feature of the Z-scheme system is its unique electron transfer mechanism. After photoexcitation, electrons in the CB of PS II recombine with holes in the VB of PS I via the mediator, preserving charge separation and sustaining high redox capabilities [167, 168]. Over

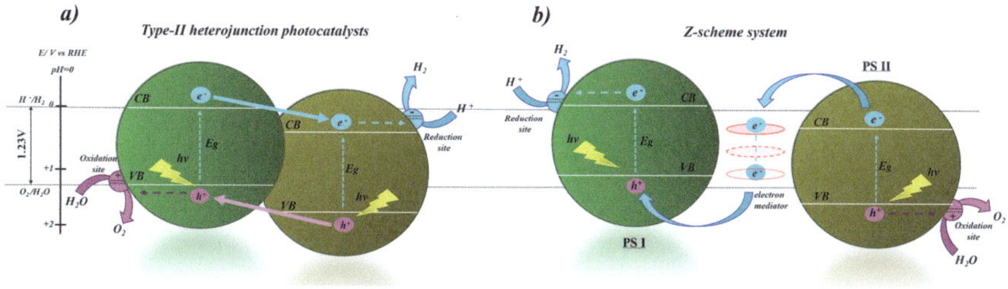

Fig. 13 Schematic illustration of **a** type-II heterojunction photocatalysts, and **b** Z-scheme photocatalytic system. Reproduced with permission from [167]

time, advancements in Z-scheme systems have included replacing dissolved mediators with interparticle electron transfer systems, which often cause undesirable side reactions. For example, Ru/$SrTiO_3$: Rh (for H_2 evolution) paired with $BiVO_4$ (for O_2 evolution) demonstrates the versatility and improved stability of mediator-free Z-scheme designs. Additionally, diverse semiconductors such as $SrTiO_3$, WO_3, and $BiVO_4$, combined with redox mediators like $NaIO_3$, $NaIO_3$/NaI, and Fe^{3+}/Fe^{2+}, have been optimized to achieve greater efficiency and durability.

4.4 *Intermediates and Facilitators*

4.4.1 Reactive Oxygen Species

Reactive oxygen species (ROS) play a critical role in photocatalytic water splitting, acting as intermediates in the oxidation reactions that generate oxygen. Key ROS include superoxide (O_2^-), hydrogen peroxide, and atomic oxygen, which are produced following the photonic excitation of photocatalysts. These species are essential for the OER by facilitating the cleavage of water molecules. In a photocatalytic system, superoxide (O_2^-) are formed when an electron combines with molecular oxygen and transforms either into H_2O_2 or atomic oxygen. In recent studies conducted with theraphthal, it has been observed that ROS generated under sonophotocatalytic conditions increase oxygen production, particularly when combined with ultrasonic treatment [169, 170]. Precisely controlling these reactive species is essential for developing more stable and efficient photocatalytic systems.

4.4.2 Sacrificial Agents

Sacrificial agents are compounds used in photocatalytic systems to improve the separation of charge carriers generated during the photocatalytic reaction. These agents act as electron donors or hole scavengers, effectively preventing recombination. Sacrificial agents can be classified into three main types based on their chemical nature and the reaction pathways: Organic electron donors, inorganic electron donors, and electron acceptors.

Organic compounds can be used as sacrificial agents in photocatalytic hydrogen production. For example, methanol is one of the most frequently used sacrificial agents. Its oxidation mechanism involves reacting with photogenerated holes, forming formaldehyde and hydrogen gas. The process can proceed through two pathways; direct oxidation by holes or indirect oxidation via hydroxyl radicals (·OH) as stated in Eq. 4.4.

$$CH_3OH + h^+ \rightarrow^{TiO2,hv} HCHO + H_2 \quad (4.4)$$

Other alcohols follow similar oxidation mechanisms. The essential advantage of using organic sacrificial agents is that they act as hole scavengers and generate valuable chemicals as byproducts, such as formic acid or formaldehyde, during the photocatalytic process (Fig. 14).

Inorganic sacrificial agents, such as sulfide (S^{2-}) and sulfite ($SO_3{}^{2-}$), also serve as efficient hole scavengers as stated in Eqs. 4.5–4.6. These agents are particularly useful in systems involving metal sulfide photocatalysts where they facilitate the separation of charge carriers and prevent photo-corrosion. In the case of sulfide, the oxidation process can result in the formation of sulfur species, which are further oxidized to sulfate or sulfur dioxide.

$$3S^{2-} + 4h^+ + 3H_2O \rightarrow 2S + SO_3^{2-} + 6H^+ \quad (4.5)$$

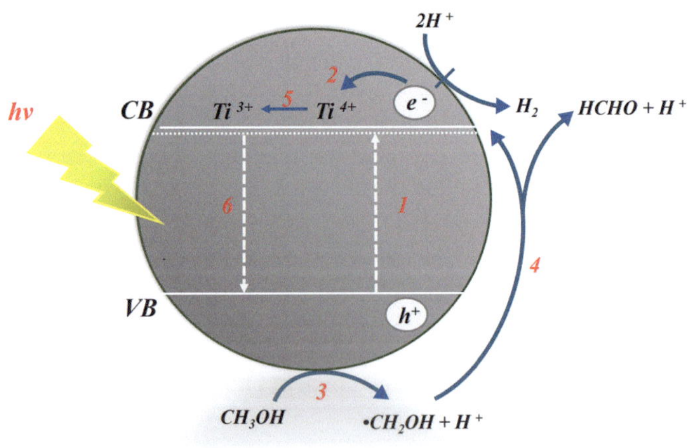

Fig. 14 Processes involved in the photocatalytic hydrogen evolution from aqueous methanol solution on **a** bare TiO_2. (1) Photogeneration of charge carriers. (2) Trapping of e^-. (3) First oxidation step of CH_3OH. (4) Formation of HCHO through e^- injection into the conduction band of TiO_2. (5) Formation of Ti^{3+} or reduction of H^+. (6) Recombination channel. For simplicity, the formation of $\cdot CH_2OH$ radicals by trapped holes ($\equiv Ti^{IV}OH\cdot^+$) or by ·OH radicals are represented by the hole oxidation step. Reproduced with permission from [171]

$$SO_3^{2-} + h^+ + H_2O \rightarrow SO_4^{2-} + 2H^+ \tag{4.6}$$

The presence of sulfide ions also helps regenerate the photocatalyst, like in the case of CdS where Cd^{2+} ions react with sulfide to rebuild the CdS structure. However, sulfide-based systems still face challenges, such as disulfide ions ($S_2{}^{2-}$) forming, which can reduce the photocatalytic efficiency by competing with protons for reduction [171, 172].

Electron acceptors, such as silver ions, are used in photocatalytic systems to facilitate water oxidation. These acceptors capture the photogenerated electrons by reducing Ag^+ to Ag^0, as depicted in Eqs. 4.7–4.9, while simultaneously promoting the OER.

$$4e^- + 4Ag^+ \rightarrow^{hv} 4Ag^0 \tag{4.7}$$

$$4h^+ + 2H_2O \rightarrow^{hv} O_2 + 4H^+_{aq} \tag{4.8}$$

$$nAg^0 \rightarrow^{hv} Ag_n^0 \tag{4.9}$$

Recent studies have demonstrated the feasibility of selectively oxidizing sacrificial agents to simultaneously produce value-added chemicals and H_2. For example, methanol can react with photogenerated holes to produce formalin (aqueous solution of formaldehyde) and formic acid, products with applications in the chemical industry. In addition to typical alcohols and sulfides, various readily available biomass and plastic waste materials, such as cellulose, furfural alcohols, and polylactic acid have been adopted as sacrificial agents. Diverse photocatalysts can selectively oxidize these compounds into products like ethylene glycol or fuel precursors, enabling the generation of value-added chemicals, making it extremely attractive from an economic and environmental standpoint [171, 173].

Acknowledgements The authors acknowledge support from the Kempe Foundation (JCSMK 23-0095), the Carl Tryggers Foundation (CTS 21-1581), the Olle Engkvist Foundation (219-0116) and the Swedish Foundation for Strategic Research (SSF-Agenda 2030–PUSH, and ID22-0062).

References

1. B. Fei, Z. Chen, J. Liu, H. Xu, X. Yan, H. Qing, M. Chen, R. Wu, Ultrathinning nickel sulfide with modulated electron density for efficient water splitting. Adv. Energy Mater. **10** (2020). https://doi.org/10.1002/aenm.202001963
2. D. Zhou, B. Jiang, R. Yang, X. Hou, C. Zheng, One-step synthesis of monodispersed pt nanoparticles anchored on 3d graphene foams and its application for electrocatalytic hydrogen evolution. Chin. Chem. Lett. **31**, 1540 (2020). https://doi.org/10.1016/j.cclet.2019.11.014

3. H. Zhou, F. Yu, Q. Zhu, J. Sun, F. Qin, L. Yu, J. Bao, Y. Yu, S. Chen, Z. Ren, Water splitting by electrolysis at high current densities under 1.6 volts. Energy Environ. Sci. **11**, 2858 (2018). https://doi.org/10.1039/C8EE00927A
4. J. Durst, A. Siebel, C. Simon, F. Hasché, J. Herranz, H.A. Gasteiger, New insights into the electrochemical hydrogen oxidation and evolution reaction mechanism. Energy Environ. Sci. **7**, 2255 (2014). https://doi.org/10.1039/C4EE00440J
5. Y. Zheng, Y. Jiao, A. Vasileff, S.-Z. Qiao, The hydrogen evolution reaction in alkaline solution: from theory, single crystal models, to practical electrocatalysts. Angew. Chem. Int. Ed. **57**, 7568 (2018). https://doi.org/10.1002/anie.201710556
6. S. Marini, P. Salvi, P. Nelli, R. Pesenti, M. Villa, M. Berrettoni, G. Zangari, Y. Kiros, Advanced alkaline water electrolysis. Electrochim. Acta **82**, 384 (2012). https://doi.org/10.1016/j.electacta.2012.05.011
7. I. Vincent, D. Bessarabov, Low cost hydrogen production by anion exchange membrane electrolysis: a review. Renew. Sustain. Energy Rev. **81**, 1690 (2018). https://doi.org/10.1016/j.rser.2017.05.258
8. Y. Li, W. Gao, L. Ci, C. Wang, P.M. Ajayan, Catalytic performance of pt nanoparticles on reduced graphene oxide for methanol electro-oxidation. Carbon **48**, 1124 (2010). https://doi.org/10.1016/j.carbon.2009.11.034
9. B.E. Conway, B.V. Tilak, Interfacial processes involving electrocatalytic evolution and oxidation of h2, and the role of chemisorbed h. Electrochim. Acta **47**, 3571 (2002). https://doi.org/10.1016/S0013-4686(02)00329-8
10. N. Danilovic, R. Subbaraman, D. Strmcnik, V.R. Stamenkovic, N. M. Markovic, Electrocatalysis of the her in acid and alkaline media. J. Serbian Chem. Soc. **78**, 2007 (2013). https://doi.org/10.2298/JSC131118136D
11. C. Niether, M.S. Rau, C. Cremers, D.J. Jones, K. Pinkwart, J. Tübke, Development of a novel experimental dems set-up for electrocatalyst characterization under working conditions of high temperature polymer electrolyte fuel cells. J. Electroanal. Chem. **747**, 97 (2015). https://doi.org/10.1016/j.jelechem.2015.04.002
12. X. Wang, C. Xu, M. Jaroniec, Y. Zheng, S.Z. Qiao, Anomalous hydrogen evolution behavior in high-ph environment induced by locally generated hydronium ions. Nat. Commun. **10** (2019). https://doi.org/10.1038/s41467-019-12773-7
13. H. Jin, X. Wang, C. Tang, A. Vasileff, L. Li, A. Slattery, S. Z. Qiao, (2021) Stable and highly efficient hydrogen evolution from seawater enabled by an unsaturated nickel surface nitride. Adv. Mater. **33**. https://doi.org/10.1002/adma.202007508
14. P. Su, W. Pei, X. Wang, Y. Ma, Q. Jiang, J. Liang, S. Zhou, J. Zhao, J. Liu, G.Q. Lu, Exceptional electrochemical her performance with enhanced electron transfer between ru nanoparticles and single atoms dispersed on a carbon substrate. Angewandte Chemie – Int. Edn. **60**, 16044 (2021). https://doi.org/10.1002/anie.202103557
15. J. Ekspong, E. Gracia-Espino, T. Wågberg, Hydrogen evolution reaction activity of heterogeneous materials: a theoretical model. J. Phys. Chem. C **124**, 20911 (2020). https://doi.org/10.1021/acs.jpcc.0c05243
16. J. Ekspong, E. Gracia-Espino, Theoretical analysis of surface active sites in defective 2h and 1t′ mos2 polymorphs for hydrogen evolution reaction: quantifying the total activity of point defects. Adv. Theo. Simulat. **3**, 1900213 (2020). https://doi.org/10.1002/adts.201900213
17. K. Ojha, S. Saha, P. Dagar, A.K. Ganguli, Nanocatalysts for hydrogen evolution reactions. Phys. Chem. Chem. Phys. **20**, 6777 (2018). https://doi.org/10.1039/C7CP06316D
18. Y. Sun, R. Li, X. Chen, J. Wu, Y. Xie, X. Wang, K. Ma, L. Wang, Z. Zhang, Q. Liao, Z. Kang, Y. Zhang, A-site management prompts the dynamic reconstructed active phase of perovskite oxide oer catalysts. Adv. Energy Mater. **11** (2021). https://doi.org/10.1002/aenm.202003755
19. J. Joo, T. Kim, J. Lee, S. I. Choi, K. Lee, Morphology-controlled metal sulfides and phosphides for electrochemical water splitting. Adv. Mater. **31** (2019). https://doi.org/10.1002/adma.201806682
20. C. Yang, Z.-D. Yang, H. Dong, N. Sun, Y. Lu, F.-M. Zhang, G. Zhang, Theory-driven design and targeting synthesis of a highly-conjugated basal-plane 2d covalent organic framework for

metal-free electrocatalytic oer. ACS Energy Lett. **4**, 2251 (2019). https://doi.org/10.1021/acsenergylett.9b01691
21. J.S. Kim, B. Kim, H. Kim, K. Kang, Recent progress on multimetal oxide catalysts for the oxygen evolution reaction. Adv. Energy Mater. **8**, 1702774 (2018). https://doi.org/10.1002/aenm.201702774
22. Q. Liang, G. Brocks, A. Bieberle-Hütter, Oxygen evolution reaction (oer) mechanism under alkaline and acidic conditions. J. Phys: Energy **3**, 026001 (2021). https://doi.org/10.1088/2515-7655/abdc85
23. S. Zhou, N. Liu, Z. Wang, J. Zhao, Nitrogen-doped graphene on transition metal substrates as efficient bifunctional catalysts for oxygen reduction and oxygen evolution reactions. ACS Appl. Mater. Interfaces **9**, 22578 (2017). https://doi.org/10.1021/acsami.7b05755
24. N. Snir, N. Yatom, M. Caspary Toroker, Progress in understanding hematite electrochemistry through computational modeling. Comput. Mater. Sci. **160**, 411 (2019). https://doi.org/10.1016/j.commatsci.2019.01.001
25. Y. Zuo, Y. Liu, J. Li, R. Du, X. Han, T. Zhang, J. Arbiol, N.J. Divins, J. Llorca, N. Guijarro, K. Sivula, A. Cabot, In situ electrochemical oxidation of cu2s into cuo nanowires as a durable and efficient electrocatalyst for oxygen evolution reaction. Chem. Mater. **31**, 7732 (2019). https://doi.org/10.1021/acs.chemmater.9b02790
26. S. Riyajuddin, K. Azmi, M. Pahuja, S. Kumar, T. Maruyama, C. Bera, K. Ghosh, Super-hydrophilic hierarchical ni-foam-graphene-carbon nanotubes-ni2p–cup2 nano-architecture as efficient electrocatalyst for overall water splitting. ACS Nano **15**, 5586 (2021). https://doi.org/10.1021/acsnano.1c00647
27. J. Rossmeisl, A. Logadottir, J.K. Nørskov, Electrolysis of water on (oxidized) metal surfaces. Chem. Phys. **319**, 178 (2005). https://doi.org/10.1016/j.chemphys.2005.05.038
28. J. Rossmeisl, Z.W. Qu, H. Zhu, G.J. Kroes, J.K. Nørskov, Electrolysis of water on oxide surfaces. J. Electroanal. Chem. **607**, 83 (2007). https://doi.org/10.1016/j.jelechem.2006.11.008
29. I.C. Man, H.Y. Su, F. Calle-Vallejo, H.A. Hansen, J.I. Martínez, N.G. Inoglu, J. Kitchin, T.F. Jaramillo, J.K. Nørskov, J. Rossmeisl, Universality in oxygen evolution electrocatalysis on oxide surfaces. ChemCatChem **3**, 1159 (2011). https://doi.org/10.1002/cctc.201000397
30. O. Diaz-Morales, I. Ledezma-Yanez, M.T.M. Koper, F. Calle-Vallejo, Guidelines for the rational design of ni-based double hydroxide electrocatalysts for the oxygen evolution reaction. ACS Catal. **5**, 5380 (2015). https://doi.org/10.1021/acscatal.5b01638
31. F. Lv, B. Huang, J. Feng, W. Zhang, K. Wang, N. Li, J. Zhou, P. Zhou, W. Yang, Y. Du, D. Su, S. Guo, A highly efficient atomically thin curved pdir bimetallene electrocatalyst. Nat. Sci. Rev. **8** (2021). https://doi.org/10.1093/nsr/nwab019
32. M. Zhu, Q. Shao, Y. Qian, X. Huang, Superior overall water splitting electrocatalysis in acidic conditions enabled by bimetallic ir-ag nanotubes. Nano Energy **56**, 330 (2019). https://doi.org/10.1016/j.nanoen.2018.11.023
33. F. Lv, W. Zhang, W. Yang, J. Feng, K. Wang, J. Zhou, P. Zhou, S. Guo, Ir-based alloy nanoflowers with optimized hydrogen binding energy as bifunctional electrocatalysts for overall water splitting. Small Methods **4** (2020). https://doi.org/10.1002/smtd.201900129
34. D. Cao, H. Xu, D. Cheng, Construction of defect-rich rhcu nanotubes with highly active rh3cu1 alloy phase for overall water splitting in all ph values, Adv. Energy Mater. **10** (2020). https://doi.org/10.1002/aenm.201903038
35. Z. Li, C. Yu, Y. Kang, X. Zhang, Y. Wen, Z. K. Wang, C. Ma, C. Wang, K. Wang, X. Qu, M. He, Y.W. Zhang, W. Song, Ultra-small hollow ternary alloy nanoparticles for efficient hydrogen evolution reaction. Nat. Sci. Rev. **8** (2021). https://doi.org/10.1093/nsr/nwaa204
36. G. Liu, W. Zhou, B. Chen, Q. Zhang, X. Cui, B. Li, Z. Lai, Y. Chen, Z. Zhang, L. Gu, H. Zhang, Synthesis of runi alloy nanostructures composed of multilayered nanosheets for highly efficient electrocatalytic hydrogen evolution. Nano Energy **66** (2019). https://doi.org/10.1016/j.nanoen.2019.104173
37. J. Tian, N. Cheng, Q. Liu, X. Sun, Y. He, A.M. Asiri, Self-supported nimo hollow nanorod array: an efficient 3d bifunctional catalytic electrode for overall water splitting. J. Mater. Chem. A **3**, 20056 (2015). https://doi.org/10.1039/c5ta04723d

38. S.H. Hong, S.H. Ahn, J. Choi, J.Y. Kim, H.Y. Kim, H.J. Kim, J.H. Jang, H. Kim, S.K. Kim, High-activity electrodeposited niw catalysts for hydrogen evolution in alkaline water electrolysis. Appl. Surf. Sci. **349**, 629 (2015). https://doi.org/10.1016/j.apsusc.2015.05.040
39. J. Zhang, Y. Zhou, S. Zhang, S. Li, Q. Hu, L. Wang, L. Wang, F. Ma, (2018) Electrochemical preparation and post-treatment of composite porous foam nizn alloy electrodes with high activity for hydrogen evolution. Sci. Rep. **8**. https://doi.org/10.1038/s41598-018-33205-4
40. M. Rafei, X. Wu, A. Piñeiro Garcia, V. Miranda la Hera, T. Wågberg, E. Gracia-Espino, Non-stoichiometric nifemo solid solutions; tuning the hydrogen adsorption energy via molybdenum incorporation. Adv. Mater. Interfaces **9**, 2201214 (2022). https://doi.org/10.1002/admi.202201214
41. X. Wu, A. Piñeiro-García, M. Rafei, N. Boulanger, E.J. Canto-Aguilar, E. Gracia-Espino, Scalable production of foam-like nickel–molybdenum coatings via plasma spraying as bifunctional electrocatalysts for water splitting. Phys. Chem. Chem. Phys. **25**, 20794 (2023). https://doi.org/10.1039/D3CP01444D
42. X. Wu, A. Piñeiro-García, M. Rafei, A. Kuzhikandathil, E. J. Canto-Aguilar, E. Gracia-Espino, Highly active and durable nanostructured nickel-molybdenum coatings as hydrogen electrocatalysts via solution precursor plasma spraying. ChemistryOpen **n/a**, e202400069. https://doi.org/10.1002/open.202400069
43. S. Saha, S. Vaidya, K.V. Ramanujachary, S.E. Lofland, A.K. Ganguli, Ternary alloy nanocatalysts for hydrogen evolution reaction. Bull. Mater. Sci. **39**, 433 (2016). https://doi.org/10.1007/s12034-016-1182-2
44. H.L.S. Santos, P.G. Corradini, M. Medina, L.H. Mascaro, Effect of copper addition on cobalt-molybdenum electrodeposited coatings for the hydrogen evolution reaction in alkaline medium. Int. J. Hydrogen Energy **45**, 33586 (2020). https://doi.org/10.1016/j.ijhydene.2020.09.128
45. Y. Li, G. Tang, Y. Wang, Y. Chai, C. Liu, Interfacial engineering of a phase-controlled heterojunction for high-efficiency her, oer, and orr trifunctional electrocatalysis. ACS Omega **7**, 13687 (2022). https://doi.org/10.1021/acsomega.1c07251
46. Z. Li, M. Hu, P. Wang, J. Liu, J. Yao, C. Li, Heterojunction catalyst in electrocatalytic water splitting. Coord. Chem. Rev. **439**, 213953 (2021). https://doi.org/10.1016/j.ccr.2021.213953
47. Y. Chen, Z. Xu, G.Z. Chen, Nano-scale engineering of heterojunction for alkaline water electrolysis. Materials **17**, 199 (2024). https://doi.org/10.3390/ma17010199
48. S. Cherevko, S. Geiger, O. Kasian, N. Kulyk, J.-P. Grote, A. Savan, B.R. Shrestha, S. Merzlikin, B. Breitbach, A. Ludwig, K.J.J. Mayrhofer, Oxygen and hydrogen evolution reactions on ru, ruo2, ir, and iro2 thin film electrodes in acidic and alkaline electrolytes: A comparative study on activity and stability. Catal. Today **262**, 170 (2016). https://doi.org/10.1016/j.cattod.2015.08.014
49. A.T. Marshall, R.G. Haverkamp, Electrocatalytic activity of iro2–ruo2 supported on sb-doped sno2 nanoparticles. Electrochimica Acta **55**, 1978 (2010). https://doi.org/10.1016/j.electacta.2009.11.018
50. A. Piñeiro-García, X. Wu, M. Rafei, P.J. Mörk, E. Gracia-Espino, A quaternary mixed oxide protective scaffold for ruthenium during oxygen evolution reaction in acidic media. Commun. Eng. **2**, 28 (2023). https://doi.org/10.1038/s44172-023-00080-5
51. N.-T. Suen, S.-F. Hung, Q. Quan, N. Zhang, Y.-J. Xu, H.M. Chen, Electrocatalysis for the oxygen evolution reaction: recent development and future perspectives. Chem. Soc. Rev. **46**, 337 (2017). https://doi.org/10.1039/C6CS00328A
52. Y. Matsumoto, S. Yamada, T. Nishida, E. Sato, Oxygen evolution on la1 − x srx fe1 − y coy o 3 series oxides. J. Electrochem. Soc. **127**, 2360 (1980). https://doi.org/10.1149/1.2129415
53. J. Suntivich, K.J. May, H.A. Gasteiger, J.B. Goodenough, Y. Shao-Horn, A perovskite oxide optimized for oxygen evolution catalysis from molecular orbital principles. Science **334**, 1383 (2011). https://doi.org/10.1126/science.1212858
54. I. Nikolov, R. Darkaoui, E. Zhecheva, R. Stoyanova, N. Dimitrov, T. Vitanov, Electrocatalytic activity of spinel related cobalties mxco3−xo4 (m = li, ni, cu) in the oxygen evolution reaction. J. Electroanal. Chem. **429**, 157 (1997). https://doi.org/10.1016/S0022-0728(96)05013-9

55. M. Li, Y. Xiong, X. Liu, X. Bo, Y. Zhang, C. Han, L. Guo, Facile synthesis of electrospun mfe2o4 (m = co, ni, cu, mn) spinel nanofibers with excellent electrocatalytic properties for oxygen evolution and hydrogen peroxide reduction. Nanoscale **7**, 8920 (2015). https://doi.org/10.1039/C4NR07243J
56. H.-Y. Wang, S.-F. Hung, H.-Y. Chen, T.-S. Chan, H.M. Chen, B. Liu, In operando identification of geometrical-site-dependent water oxidation activity of spinel co3o4. J. Am. Chem. Soc. **138**, 36 (2016). https://doi.org/10.1021/jacs.5b10525
57. T. Sharifi, E. Gracia-Espino, X. Jia, R. Sandström, T. Wågberg, Comprehensive study of an earth-abundant bifunctional 3d electrode for efficient water electrolysis in alkaline medium. ACS Appl. Mater. Interfaces **7**, 28148 (2015). https://doi.org/10.1021/acsami.5b10118
58. J.A. Koza, Z. He, A.S. Miller, J.A. Switzer, Electrodeposition of crystalline co3o4—a catalyst for the oxygen evolution reaction. Chem. Mater. **24**, 3567 (2012). https://doi.org/10.1021/cm3012205
59. H. Bode, K. Dehmelt, J. Witte, Zur kenntnis der nickelhydroxidelektrode—i.Über das nickel (ii)-hydroxidhydrat. Electrochimica Acta **11**, 1079 (1966) https://doi.org/10.1016/0013-4686(66)80045-2
60. L. Trotochaud, S.L. Young, J.K. Ranney, S.W. Boettcher, Nickel–iron oxyhydroxide oxygen-evolution electrocatalysts: the role of intentional and incidental iron incorporation. J. Am. Chem. Soc. **136**, 6744 (2014). https://doi.org/10.1021/ja502379c
61. X. Long, J. Li, S. Xiao, K. Yan, Z. Wang, H. Chen, S. Yang, A strongly coupled graphene and feni double hydroxide hybrid as an excellent electrocatalyst for the oxygen evolution reaction. Angew. Chem. Int. Ed. **53**, 7584 (2014). https://doi.org/10.1002/anie.201402822
62. J.-X. Feng, S.-H. Ye, H. Xu, Y.-X. Tong, G.-R. Li, Design and synthesis of feooh/ceo2 heterolayered nanotube electrocatalysts for the oxygen evolution reaction. Adv. Mater. **28**, 4698 (2016). https://doi.org/10.1002/adma.201600054
63. W. Zhu, X. Yue, W. Zhang, S. Yu, Y. Zhang, J. Wang, J. Wang, Nickel sulfide microsphere film on ni foam as an efficient bifunctional electrocatalyst for overall water splitting. Chem. Commun. **52**, 1486 (2016). https://doi.org/10.1039/C5CC08064A
64. I.G. McKendry, A.C. Thenuwara, J. Sun, H. Peng, J.P. Perdew, D.R. Strongin, M.J. Zdilla, Water oxidation catalyzed by cobalt oxide supported on the mattagamite phase of cote2. ACS Catal. **6**, 7393 (2016). https://doi.org/10.1021/acscatal.6b01878
65. J.S. Chen, J. Ren, M. Shalom, T. Fellinger, M. Antonietti, Stainless steel mesh-supported nis nanosheet array as highly efficient catalyst for oxygen evolution reaction. ACS Appl. Mater. Interfaces **8**, 5509 (2016). https://doi.org/10.1021/acsami.5b10099
66. P. Ganesan, M. Prabu, J. Sanetuntikul, S. Shanmugam, Cobalt sulfide nanoparticles grown on nitrogen and sulfur codoped graphene oxide: an efficient electrocatalyst for oxygen reduction and evolution reactions. ACS Catal. **5**, 3625 (2015). https://doi.org/10.1021/acscatal.5b00154
67. J.O.M. Bockris, T. Otagawa, The electrocatalysis of oxygen evolution on perovskites. J. Electrochem. Soc. **131**, 290 (1984). https://doi.org/10.1149/1.2115565
68. J. Jiang, S. Lu, H. Gao, X. Zhang, H.-Q. Yu, Ternary fenis2 ultrathin nanosheets as an electrocatalyst for both oxygen evolution and reduction reactions. Nano Energy **27**, 526 (2016). https://doi.org/10.1016/j.nanoen.2016.07.032
69. M.-R. Gao, X. Cao, Q. Gao, Y.-F. Xu, Y.-R. Zheng, J. Jiang, S.-H. Yu, Nitrogen-doped graphene supported cose2 nanobelt composite catalyst for efficient water oxidation. ACS Nano **8**, 3970 (2014). https://doi.org/10.1021/nn500880v
70. F. Safizadeh, E. Ghali, G. Houlachi, Electrocatalysis developments for hydrogen evolution reaction in alkaline solutions – a review. Int. J. Hydrogen Energy **40**, 256 (2015). https://doi.org/10.1016/j.ijhydene.2014.10.109
71. J. Chang, Y. Xiao, M. Xiao, J. Ge, C. Liu, W. Xing, Surface oxidized cobalt-phosphide nanorods as an advanced oxygen evolution catalyst in alkaline solution. ACS Catal. **5**, 6874 (2015). https://doi.org/10.1021/acscatal.5b02076
72. J. Ryu, N. Jung, J.H. Jang, H.-J. Kim, S.J. Yoo, In situ transformation of hydrogen-evolving cop nanoparticles: Toward efficient oxygen evolution catalysts bearing dispersed morphologies with co-oxo/hydroxo molecular units. ACS Catal. **5**, 4066 (2015). https://doi.org/10.1021/acscatal.5b00349

73. J. Duan, S. Chen, A. Vasileff, S.Z. Qiao, Anion and cation modulation in metal compounds for bifunctional overall water splitting. ACS Nano **10**, 8738 (2016). https://doi.org/10.1021/acsnano.6b04252
74. J.D. Blakemore, R.H. Crabtree, G.W. Brudvig, Molecular catalysts for water oxidation. Chem. Rev. **115**, 12974 (2015). https://doi.org/10.1021/acs.chemrev.5b00122
75. S.W. Gersten, G.J. Samuels, T.J. Meyer, Catalytic oxidation of water by an oxo-bridged ruthenium dimer. J. Am. Chem. Soc. **104**, 4029 (1982). https://doi.org/10.1021/ja00378a053
76. M. Okamura, M. Kondo, R. Kuga, Y. Kurashige, T. Yanai, S. Hayami, V.K.K. Praneeth, M. Yoshida, K. Yoneda, S. Kawata, S. Masaoka, A pentanuclear iron catalyst designed for water oxidation. Nature **530**, 465 (2016). https://doi.org/10.1038/nature16529
77. W. Schöfberger, F. Faschinger, S. Chattopadhyay, S. Bhakta, B. Mondal, J.A.A.W. Elemans, S. Müllegger, S. Tebi, R. Koch, F. Klappenberger, M. Paszkiewicz, J.V. Barth, E. Rauls, H. Aldahhak, W.G. Schmidt, A. Dey, A bifunctional electrocatalyst for oxygen evolution and oxygen reduction reactions in water. Angew. Chem. Int. Ed. **55**, 2350 (2016). https://doi.org/10.1002/anie.201508404
78. B. Wurster, D. Grumelli, D. Hötger, R. Gutzler, K. Kern, Driving the oxygen evolution reaction by nonlinear cooperativity in bimetallic coordination catalysts. J. Am. Chem. Soc. **138**, 3623 (2016). https://doi.org/10.1021/jacs.5b10484
79. K. Qu, Y. Zheng, S. Dai, S.Z. Qiao, Graphene oxide-polydopamine derived n, s-codoped carbon nanosheets as superior bifunctional electrocatalysts for oxygen reduction and evolution. Nano Energy **19**, 373 (2016). https://doi.org/10.1016/j.nanoen.2015.11.027
80. Y. Zhao, R. Nakamura, K. Kamiya, S. Nakanishi, K. Hashimoto, Nitrogen-doped carbon nanomaterials as non-metal electrocatalysts for water oxidation. Nat. Commun. **4**, 2390 (2013). https://doi.org/10.1038/ncomms3390
81. J. Tian, Q. Liu, A.M. Asiri, K.A. Alamry, X. Sun, Ultrathin graphitic c3n4 nanosheets/graphene composites: efficient organic electrocatalyst for oxygen evolution reaction. Chemsuschem **7**, 2125 (2014). https://doi.org/10.1002/cssc.201402118
82. Q. Jiang, L. Xu, N. Chen, H. Zhang, L. Dai, S. Wang, Facile synthesis of black phosphorus: an efficient electrocatalyst for the oxygen evolving reaction. Angew. Chem. Int. Ed. **55**, 13849 (2016). https://doi.org/10.1002/anie.201607393
83. J. Ekspong, N. Boulanger, E. Gracia-Espino, Surface activation of graphene nanoribbons for oxygen reduction reaction by nitrogen doping and defect engineering: an ab initio study. Carbon **137**, 349 (2018). https://doi.org/10.1016/j.carbon.2018.05.050
84. D.K. Perivoliotis, J. Ekspong, X. Zhao, G. Hu, T. Wågberg, E. Gracia-Espino, Recent progress on defect-rich electrocatalysts for hydrogen and oxygen evolution reactions. Nano Today **50**, 101883 (2023). https://doi.org/10.1016/j.nantod.2023.101883
85. A. Piñeiro García, D. Perivoliotis, X. Wu, E. Gracia-Espino, Benchmarking molybdenum-based materials as cathode electrocatalysts for proton exchange membrane water electrolysis: Can these compete with pt? ACS Sustainable Chem. Eng. **11**, 7641 (2023). https://doi.org/10.1021/acssuschemeng.2c07201
86. B. Wang, C. Tang, H.F. Wang, B.Q. Li, X. Cui, Q. Zhang, Anion-regulated hydroxysulfide monoliths as oer/orr/her electrocatalysts and their applications in self-powered electrochemical water splitting. Small Methods **2** (2018). https://doi.org/10.1002/smtd.201800055
87. N. Yao, P. Li, Z. Zhou, R. Meng, G. Cheng, W. Luo, Nitrogen engineering on 3d dandelion-flower-like cos2 for high-performance overall water splitting. Small **15**, 1901993 (2019). https://doi.org/10.1002/smll.201901993
88. W. He, L. Han, Q. Hao, X. Zheng, Y. Li, J. Zhang, C. Liu, H. Liu, H.L. Xin, Fluorine-anion-modulated electron structure of nickel sulfide nanosheet arrays for alkaline hydrogen evolution. ACS Energy Lett. **4**, 2905 (2019). https://doi.org/10.1021/acsenergylett.9b02316
89. D.-Y. Wang, M. Gong, H.-L. Chou, C.-J. Pan, H.-A. Chen, Y. Wu, M.-C. Lin, M. Guan, J. Yang, C.-W. Chen, Y.-L. Wang, B.-J. Hwang, C.-C. Chen, H. Dai, Highly active and stable hybrid catalyst of cobalt-doped fes2 nanosheets–carbon nanotubes for hydrogen evolution reaction. J. Am. Chem. Soc. **137**, 1587 (2015). https://doi.org/10.1021/ja511572q

90. A. Piñeiro-García, X. Wu, E.J. Canto-Aguilar, A. Kuzhikandathil, M. Rafei, E. Gracia-Espino, Quaternary mixed oxides of non-noble metals with enhanced stability during the oxygen evolution reaction. ACS Appl. Mater. Interfaces (2024). https://doi.org/10.1021/acsami.4c10234
91. H. Liu, Q. He, H. Jiang, Y. Lin, Y. Zhang, M. Habib, S. Chen, L. Song, Electronic structure reconfiguration toward pyrite nis2 via engineered heteroatom defect boosting overall water splitting. ACS Nano **11**, 11574 (2017). https://doi.org/10.1021/acsnano.7b06501
92. M. Rafei, A. Piñeiro-García, X. Wu, D.K. Perivoliotis, T. Wågberg, E. Gracia-Espino, Hydrogen evolution mediated by sulfur vacancies and substitutional mn in few-layered molybdenum disulfide. Mater. Today Energy **41**, 101524 (2024). https://doi.org/10.1016/j.mtener.2024.101524
93. G. Huang, Z. Xiao, R. Chen, S. Wang, *Defect Engineering of Cobalt-Based Materials for Electrocatalytic Water Splitting*. ACS Publications (2018)
94. H.-F. Wang, C. Tang, B.-Q. Li, Q. Zhang, A review of anion-regulated multi-anion transition metal compounds for oxygen evolution electrocatalysis. Inorganic Chem. Front. **5**, 521 (2018). https://doi.org/10.1039/C7QI00780A
95. L. Xu, Q. Jiang, Z. Xiao, X. Li, J. Huo, S. Wang, L. Dai, Plasma-engraved co3o4 nanosheets with oxygen vacancies and high surface area for the oxygen evolution reaction. Angew. Chem. **128**, 5363 (2016). https://doi.org/10.1002/anie.201600687
96. Y. Yin, J. Han, Y. Zhang, X. Zhang, P. Xu, Q. Yuan, L. Samad, X. Wang, Y. Wang, Z. Zhang, Contributions of phase, sulfur vacancies, and edges to the hydrogen evolution reaction catalytic activity of porous molybdenum disulfide nanosheets. J. Am. Chem. Soc. **138**, 7965 (2016). https://doi.org/10.1021/jacs.6b03714
97. A.-Y. Lu, X. Yang, C.-C. Tseng, S. Min, S.-H. Lin, C.-L. Hsu, H. Li, H. Idriss, J.-L. Kuo, K.-W. Huang, High-sulfur-vacancy amorphous molybdenum sulfide as a high current electrocatalyst in hydrogen evolution. Small **12**, 5530 (2016). https://doi.org/10.1002/smll.201602107
98. Z. Yuan, S.-M. Bak, P. Li, Y. Jia, L. Zheng, Y. Zhou, L. Bai, E. Hu, X.-Q. Yang, Z. Cai, Y. Sun, X. Sun, Activating layered double hydroxide with multivacancies by memory effect for energy-efficient hydrogen production at neutral ph. ACS Energy Lett. **4**, 1412 (2019). https://doi.org/10.1021/acsenergylett.9b00867
99. X.-M. Liu, X. Cui, K. Dastafkan, H.-F. Wang, C. Tang, C. Zhao, A. Chen, C. He, M. Han, Q. Zhang, Recent advances in spinel-type electrocatalysts for bifunctional oxygen reduction and oxygen evolution reactions. J. Energy Chem. **53**, 290 (2021). https://doi.org/10.1016/j.jechem.2020.04.012
100. C. Lv, C. Yan, G. Chen, Y. Ding, J. Sun, Y. Zhou, G. Yu, An amorphous noble-metal-free electrocatalyst that enables nitrogen fixation under ambient conditions. Angew. Chem. Int. Ed. **57**, 6073 (2018). https://doi.org/10.1002/anie.201801538
101. L. Li, Q. Shao, X. Huang, Amorphous oxide nanostructures for advanced electrocatalysis. Chem. A Eur. J. **26**, 3943 (2020). https://doi.org/10.1002/chem.201903206
102. J. Masa, P. Weide, D. Peeters, I. Sinev, W. Xia, Z. Sun, C. Somsen, M. Muhler, W. Schuhmann, Amorphous cobalt boride (co2b) as a highly efficient nonprecious catalyst for electrochemical water splitting: oxygen and hydrogen evolution. Adv. Energy Mater. **6**, 1502313 (2016). https://doi.org/10.1002/aenm.201502313
103. Y. Duan, Z.-Y. Yu, S.-J. Hu, X.-S. Zheng, C.-T. Zhang, H.-H. Ding, B.-C. Hu, Q.-Q. Fu, Z.-L. Yu, X. Zheng, J.-F. Zhu, M.-R. Gao, S.-H. Yu, Scaled-up synthesis of amorphous nifemo oxides and their rapid surface reconstruction for superior oxygen evolution catalysis. Angew. Chem. Int. Ed. **58**, 15772 (2019). https://doi.org/10.1002/anie.201909939
104. Y.P. Zhu, C. Guo, Y. Zheng, S.-Z. Qiao, Surface and interface engineering of noble-metal-free electrocatalysts for efficient energy conversion processes. Acc. Chem. Res. **50**, 915 (2017). https://doi.org/10.1021/acs.accounts.6b00635
105. C. Zhu, A.L. Wang, W. Xiao, D. Chao, X. Zhang, N.H. Tiep, S. Chen, J. Kang, X. Wang, J. Ding, In situ grown epitaxial heterojunction exhibits high-performance electrocatalytic water splitting. Adv. Mater. **30**, 1705516 (2018). https://doi.org/10.1002/adma.201705516

106. Y. Jia, L. Zhang, G. Gao, H. Chen, B. Wang, J. Zhou, M.T. Soo, M. Hong, X. Yan, G. Qian, A heterostructure coupling of exfoliated ni–fe hydroxide nanosheet and defective graphene as a bifunctional electrocatalyst for overall water splitting. Adv. Mater. **29**, 1700017 (2017). https://doi.org/10.1002/adma.201700017
107. A. Kahn, Fermi level, work function and vacuum level. Mater. Horiz. **3**, 7 (2016). https://doi.org/10.1039/C5MH00160A
108. C.P.C.V. Bernardo, R.A.M. Lameirinhas, J.P. de Melo Cunha, J.P.N. Torres, A revision of the semiconductor theory from history to applications. Discover Appl. Sci. **6**, 316 (2024). https://doi.org/10.1007/s42452-024-06001-1
109. X. Yang, D. Wang, Chapter two—photophysics and photochemistry at the semiconductor/electrolyte interface for solar water splitting. In: Semiconductors and Semimetals ed. by Z. Mi, , et al., vol. 97, Elsevier, pp. 47 (2017)
110. J. Bisquert, A. Zaban, M. Greenshtein, I. Mora-Seró, Determination of rate constants for charge transfer and the distribution of semiconductor and electrolyte electronic energy levels in dye-sensitized solar cells by open-circuit photovoltage decay method. J. Am. Chem. Soc. **126**, 13550 (2004). https://doi.org/10.1021/ja047311k
111. F. Fabregat-Santiago, G. Garcia-Belmonte, I. Mora-Seró, J. Bisquert, Characterization of nanostructured hybrid and organic solar cells by impedance spectroscopy. Phys. Chem. Chem. Phys. **13**, 9083 (2011). https://doi.org/10.1039/C0CP02249G
112. J. Bisquert, V.S. Vikhrenko, Interpretation of the time constants measured by kinetic techniques in nanostructured semiconductor electrodes and dye-sensitized solar cells. J. Phys. Chem. B **108**, 2313 (2004). https://doi.org/10.1021/jp035395y
113. J. Bisquert, A. Zaban, The trap-limited diffusivity of electrons in nanoporous semiconductor networks permeated with a conductive phase. Appl. Phys. A **77**, 507 (2003). https://doi.org/10.1007/s00339-002-1479-4
114. K. Strubbe, P.M. Vereecken, W.P. Gomes, Electroreduction of co2 + and ni2 + at iii–v semiconductors and properties of the semiconductor/metal interfaces formed. J. Electrochem. Soc. **146**, 1412 (1999). https://doi.org/10.1149/1.1391779
115. G. Oskam, P.C. Searson, M.W. Cole, Fabrication of n-type 4h–sic/ni junctions using electrochemical deposition. Appl. Phys. Lett. **76**, 1300 (2000). https://doi.org/10.1063/1.126015
116. Oskam, G., Long, J.G., Natarajan, A., Searson, P.C.: Electrochemical deposition of metals onto silicon. J Phys D: Appl. Phys. **31** (1927). https://doi.org/10.1088/0022-3727/31/16/001
117. Y.B. Vogel, J.J. Gooding, S. Ciampi, Light-addressable electrochemistry at semiconductor electrodes: redox imaging, mask-free lithography and spatially resolved chemical and biological sensing. Chem. Soc. Rev. **48**, 3723 (2019). https://doi.org/10.1039/C8CS00762D
118. M.G. Walter, E.L. Warren, J.R. McKone, S.W. Boettcher, Q. Mi, E.A. Santori, N.S. Lewis, Solar water splitting cells. Chem. Rev. **110**, 6446 (2010). https://doi.org/10.1021/cr1002326
119. C. Jiang, S.J.A. Moniz, A. Wang, T. Zhang, J. Tang, Photoelectrochemical devices for solar water splitting—materials and challenges. Chem. Soc. Rev. **46**, 4645 (2017). https://doi.org/10.1039/C6CS00306K
120. G.G. Bessegato, T.T. Guaraldo, J.F. de Brito, M.F. Brugnera, M.V.B. Zanoni, Achievements and trends in photoelectrocatalysis: from environmental to energy applications. Electrocatalysis **6**, 415 (2015). https://doi.org/10.1007/s12678-015-0259-9
121. S. Speldrich, M. Wark, G. Wittstock, Metal oxide protection layers for enhanced stability and activity of wo3 photoanodes in alkaline media. ACS Appl Energy Mater **6**, 9602 (2023). https://doi.org/10.1021/acsaem.3c01642
122. J. Knöppel, A. Kormányos, B. Mayerhöfer, A. Hofer, M. Bierling, J. Bachmann, S. Thiele, S. Cherevko, Photocorrosion of wo3 photoanodes in different electrolytes. ACS Phys. Chem. Au. **1**, 6 (2021). https://doi.org/10.1021/acsphyschemau.1c00004
123. K. Sivula, R. van de Krol, Semiconducting materials for photoelectrochemical energy conversion. Nat. Rev. Mater. **1**, 15010 (2016). https://doi.org/10.1038/natrevmats.2015.10
124. R. Mo, Q. Liu, H. Li, S. Yang, J. Zhong, Photoelectrochemical water oxidation in α-fe2o3 thin films enhanced by a controllable wet-chemical ti-doping strategy and co–pi co-catalyst

modification. J. Mater. Sci. Mater. Electron. **30**, 21444 (2019). https://doi.org/10.1007/s10854-019-02525-0
125. P. Xu, N.S. McCool, T.E. Mallouk, Water splitting dye-sensitized solar cells. Nano Today **14**, 42 (2017). https://doi.org/10.1016/j.nantod.2017.04.009
126. W.J. Youngblood, S.-H. A. Lee, K. Maeda, T.E. Mallouk, Visible light water splitting using dye-sensitized oxide semiconductors. Accounts Chem. Res. **42**, 1966 (2009). https://doi.org/10.1021/ar9002398
127. A. Dey, A. Mendalz, A. Wach, R.B. Vadell, V.R. Silveira, P.M. Leidinger, T. Huthwelker, V. Shtender, Z. Novotny, L. Artiglia, J. Sá, Hydrogen evolution with hot electrons on a plasmonic-molecular catalyst hybrid system. Nat. Commun. **15**, 445 (2024). https://doi.org/10.1038/s41467-024-44752-y
128. N. Wu, Plasmonic metal–semiconductor photocatalysts and photoelectrochemical cells: a review. Nanoscale **10**, 2679 (2018). https://doi.org/10.1039/C7NR08487K
129. T. Yao, X. An, H. Han, J.Q. Chen, C. Li, Photoelectrocatalytic materials for solar water splitting. Adv. Energy Mater. **8**, 1800210 (2018). https://doi.org/10.1002/aenm.201800210
130. T. Kutlusoy, S. Divanis, R. Pittkowski, R. Marina, A.M. Frandsen, K. Minhova-Macounova, R. Nebel, D. Zhao, S.F.L. Mertens, H. Hoster, P. Krtil, J. Rossmeisl, Synergistic effect of p-type and n-type dopants in semiconductors for efficient electrocatalytic water splitting. Chem. Sci. **13**, 13879 (2022). https://doi.org/10.1039/D2SC04585K
131. Y. Wang, R. Zhang, J. Li, L. Li, S. Lin, First-principles study on transition metal-doped anatase tio2. Nanoscale Res. Lett. **9**, 46 (2014). https://doi.org/10.1186/1556-276X-9-46
132. M. Liu, N. de Leon Snapp, H. Park, Water photolysis with a cross-linked titanium dioxide nanowire anode. Chem. Sci. **2**, 80 (2011). https://doi.org/10.1039/C0SC00321B
133. S.S. Kalanur, I.-H. Yoo, I.-S. Cho, H. Seo, Effect of oxygen vacancies on the band edge properties of wo3 producing enhanced photocurrents. Electrochim. Acta **296**, 517 (2019). https://doi.org/10.1016/j.electacta.2018.11.061
134. N.N. Rosman, R.M. Yunus, N.R.A.M. Shah, R.M. Shah, K. Arifin, L.J. Minggu, N.A. Ludin, An overview of co-catalysts on metal oxides for photocatalytic water splitting. Int. J. Energy Res. **46**, 11596 (2022). https://doi.org/10.1002/er.8001
135. M. Kumar, B. Meena, P. Subramanyam, D. Suryakala, C. Subrahmanyam, Recent trends in photoelectrochemical water splitting: the role of cocatalysts. NPG Asia Mater. **14**, 88 (2022). https://doi.org/10.1038/s41427-022-00436-x
136. C. Ros, T. Andreu, J.R. Morante, Photoelectrochemical water splitting: a road from stable metal oxides to protected thin film solar cells. J. Mater. Chem. A **8**, 10625 (2020). https://doi.org/10.1039/D0TA02755C
137. C. Ros, T. Andreu, J. David, J. Arbiol, J.R. Morante, Degradation and regeneration mechanisms of nio protective layers deposited by ald on photoanodes. J. Mater. Chem. A **7**, 21892 (2019). https://doi.org/10.1039/C9TA08638B
138. I. Rodríguez-Gutiérrez, E. Djatoubai, J. Su, A. Vega-Poot, G. Rodríguez-Gattorno, F.L. Souza, G. Oskam, An intensity-modulated photocurrent spectroscopy study of the charge carrier dynamics of wo3/bivo4 heterojunction systems. Sol. Energy Mater. Sol. Cells **208**, 110378 (2020). https://doi.org/10.1016/j.solmat.2019.110378
139. A. Vega-Poot, M. Rodríguez-Pérez, J. Becerril-González, I. Rodríguez-Gutiérrez, J. Su, G. Rodríguez-Gattorno, W.Y. Teoh, G. Oskam, Charge dynamics at surface-modified, nanostructured hematite photoelectrodes for solar water splitting. J. Electrochem. Soc. **169**, 056519 (2022). https://doi.org/10.1149/1945-7111/ac700b
140. A. Fujishima, K. Honda, Electrochemical photolysis of water at a semiconductor electrode. Nature **238**, 37 (1972). https://doi.org/10.1038/238037a0
141. X. Chen, S.S. Mao, Titanium dioxide nanomaterials: Synthesis, properties, modifications, and applications. Chem. Rev. **107**, 2891 (2007). https://doi.org/10.1021/cr0500535
142. K. Villa, J.R. Galán-Mascarós, N. López, E. Palomares, Photocatalytic water splitting: advantages and challenges. Sustainable Energy Fuels **5**, 4560 (2021). https://doi.org/10.1039/D1SE00808K

143. C. Bie, L. Wang, J. Yu, Challenges for photocatalytic overall water splitting. Chem **8**, 1567 (2022). https://doi.org/10.1016/j.chempr.2022.04.013
144. B. Ohtani, Revisiting the fundamental physical chemistry in heterogeneous photocatalysis: its thermodynamics and kinetics. Phys. Chem. Chem. Phys. **16**, 1788 (2014). https://doi.org/10.1039/C3CP53653J
145. R. Molinari, P. Argurio, M. Bellardita, L. Palmisano, 3.5 photocatalytic processes in membrane reactors. In: *Comprehensive Membrane Science and Engineering* (2nd edn), ed. by Drioli, E.,et al., (Elsevier, Oxford, 2017), pp 101
146. J.-P. Jeon, D.H. Kweon, B.J. Jang, M.J. Ju, J.-B. Baek, Enhancing the photocatalytic activity of tio2 catalysts. Adv. Sustainable Syst. **4**, 2000197 (2020). https://doi.org/10.1002/adsu.202000197
147. M.A. Hassaan, M.A. El-Nemr, M.R. Elkatory, S. Ragab, V.-C. Niculescu, A. El Nemr, Principles of photocatalysts and their different applications: a review. Top. Curr. Chem. **381**, 31 (2023). https://doi.org/10.1007/s41061-023-00444-7
148. X. Meng, S. Wang, C. Zhang, C. Dong, R. Li, B. Li, Q. Wang, Y. Ding, Boosting hydrogen evolution performance of a cds-based photocatalyst: In situ transition from type i to type ii heterojunction during photocatalysis. ACS Catal. **12**, 10115 (2022). https://doi.org/10.1021/acscatal.2c01877
149. S. Ma, Y. Deng, J. Xie, K. He, W. Liu, X. Chen, X. Li, Noble-metal-free ni3c cocatalysts decorated cds nanosheets for high-efficiency visible-light-driven photocatalytic h2 evolution. Appl. Catal. B **227**, 218 (2018). https://doi.org/10.1016/j.apcatb.2018.01.031
150. M. Graetzel, R.F. Howe, Electron paramagnetic resonance studies of doped titanium dioxide colloids. J. Phys. Chem. **94**, 2566 (1990). https://doi.org/10.1021/j100369a064
151. H. Zhang, G. Chen, D.W. Bahnemann, Photoelectrocatalytic materials for environmental applications. J. Mater. Chem. **19**, 5089 (2009). https://doi.org/10.1039/B821991E
152. D. Dvoranová, V. Brezová, M. Mazúr, M.A. Malati, Investigations of metal-doped titanium dioxide photocatalysts. Appl. Catal. B **37**, 91 (2002). https://doi.org/10.1016/S0926-3373(01)00335-6
153. T.H. Jun, K.S. Lee, Cr-doped tio2 thin films deposited by rf-sputtering. Mater. Lett. **64**, 2287 (2010). https://doi.org/10.1016/j.matlet.2010.07.069
154. W. Macyk, H. Kisch, Photosensitization of crystalline and amorphous titanium dioxide by platinum(iv) chloride surface complexes. Chem. A Eur. J. **7**, 1862(2001) https://doi.org/10.1002/1521-3765(20010504)7:9<1845::AID-CHEM1845>3.0.CO;2-2
155. W. Macyk, G. Burgeth, H. Kisch, Photoelectrochemical properties of platinum(iv) chloride surface modified tio2. Photochem. Photobiol. Sci. **2**, 322 (2003). https://doi.org/10.1039/b211583b
156. J.A. Libera, J.W. Elam, N.F. Sather, T. Rajh, N.M. Dimitrijevic, Iron(iii)-oxo centers on tio2 for visible-light photocatalysis. Chem. Mater. **22**, 409 (2010). https://doi.org/10.1021/cm902825c
157. S. Mukherjee, F. Libisch, N. Large, O. Neumann, L.V. Brown, J. Cheng, J.B. Lassiter, E.A. Carter, P. Nordlander, N.J. Halas, Hot electrons do the impossible: plasmon-induced dissociation of h2 on au. Nano Lett. **13**, 240 (2013). https://doi.org/10.1021/nl303940z
158. E. Kowalska, Z. Wei, M. Janczarek, Band-gap engineering of photocatalysts: surface modification versus doping. In: *Visible Light-active Photocatalysis*, pp. 447 (2018)
159. C. Zhao, Z. Chen, R. Shi, X. Yang, T. Zhang, Recent advances in conjugated polymers for visible-light-driven water splitting. Adv. Mater. **32**, 1907296 (2020). https://doi.org/10.1002/adma.201907296
160. Y. Yalçın, M. Kılıç, Z. Çınar, The role of non-metal doping in tio2 photocatalysis. J. Adv. Oxidation Technol. **13**, 281 (2010). https://doi.org/10.1515/jaots-2010-0306
161. A. Annamalai, H.H. Lee, S.H. Choi, S.Y. Lee, E. Gracia-Espino, A. Subramanian, J. Park, K.-J. Kong, J.S. Jang, Sn/be sequentially co-doped hematite photoanodes for enhanced photoelectrochemical water oxidation: Effect of be2+ as co-dopant. Sci. Rep. **6**, 23183 (2016). https://doi.org/10.1038/srep23183

162. A. Subramanian, E. Gracia-Espino, A. Annamalai, H.H. Lee, S.Y. Lee, S.H. Choi, J.S. Jang, Effect of tetravalent dopants on hematite nanostructure for enhanced photoelectrochemical water splitting. Appl. Surf. Sci. **427**, 1203 (2018). https://doi.org/10.1016/j.apsusc.2017.09.042
163. R. Asahi, T. Morikawa, H. Irie, T. Ohwaki, Nitrogen-doped titanium dioxide as visible-light-sensitive photocatalyst: designs, developments, and prospects. Chem. Rev. **114**, 9824 (2014). https://doi.org/10.1021/cr5000738
164. F. Maulana, Y. Engge, M. Nurhuda, Istiroyah, L. Hakim, A.M. Juwono, Boosting hydrogen production through water splitting: N, ni, and n-ni doped zno photocatalysts. J. Electrochem. Soc. 171, 046505 (2024). https://doi.org/10.1149/1945-7111/ad3d0f
165. P.V. Kamat, B. Patrick, Photophysics and photochemistry of quantized zinc oxide colloids. J. Phys. Chem. **96**, 6829 (1992). https://doi.org/10.1021/j100195a055
166. R. Abe, H. Takami, N. Murakami, B. Ohtani, Pristine simple oxides as visible light driven photocatalysts: highly efficient decomposition of organic compounds over platinum-loaded tungsten oxide. J. Am. Chem. Soc. **130**, 7780 (2008). https://doi.org/10.1021/ja800835q
167. B.-J. Ng, L.K. Putri, X.Y. Kong, Y.W. Teh, P. Pasbakhsh, S.-P. Chai, Z-scheme photocatalytic systems for solar water splitting. Adv. Sci. **7**, 1903171 (2020). https://doi.org/10.1002/advs.201903171
168. P. Chauhan, A. Kumar, Photocatalytic water splitting and charge carrier dynamics of janus ptsse/ζ-phosphorene heterostructure. Sci. Rep. **14**, 21618 (2024). https://doi.org/10.1038/s41598-024-72757-6
169. O.V. Kharissova, L.M. Torres Martínez, E. Luevano Hipólito, L.F. Garay-Rodríguez, M.R. Alfaro Cruz, B.I. Kharissov, High oxygen-yield homogeneous sonophotocatalysis for water-splitting using theraphthal. J. Photochem. Photobiol., A **437**, 114463 (2023). https://doi.org/10.1016/j.jphotochem.2022.114463
170. I.D. Rettig, T.M. McCormick, Enrolling reactive oxygen species in photon-to-chemical energy conversion: fundamentals, technological advances, and applications. Adv. Phys.: X **6**, 1950049 (2021). https://doi.org/10.1080/23746149.2021.1950049
171. J. Schneider, D.W. Bahnemann, Undesired role of sacrificial reagents in photocatalysis. J. Phys. Chem. Lett. **4**, 3479 (2013). https://doi.org/10.1021/jz4018199
172. E. Puentes-Prado, C.R. Garcia, J. Oliva, R. Galindo, J.J. Bernal-Alvarado, L.A. Diaz- Torres, C. Gomez-Solis, Enhancing the solar photocatalytic hydrogen generation of zns films by uv radiation treatment. Int. J. Hydrogen Energy **45**, 12308 (2020). https://doi.org/10.1016/j.ijhydene.2020.02.180
173. M. Sohail, S. Rauf, M. Irfan, A. Hayat, M.M. Alghamdi, A.A. El-Zahhar, D. Ghernaout, Y. Al-Hadeethi, W. Lv, Recent developments, advances and strategies in heterogeneous photocatalysts for water splitting. Nanoscale Adv. **6**, 1286 (2024). https://doi.org/10.1039/D3NA00442B

Progress in Organic Photovoltaics for Flexible and Low-Cost Devices

Duddu S. Sharada, Muruganandam Premkumar, Jayagopal Jayapratha, and Balu Balambiga

Abstract Organic photovoltaics (OPVs) have emerged as potential and sustainable alternative photovoltaics (PV) technologies to silicon-based and p–n junction solar cells. OPV materials due to their unique features and advantages such as low cost, light weight, flexibility, semitransparent, cost-effective large area printability, high absorption coefficient, tunable optical band gap, and energy levels have attracted great attention as a promising green energy technology from academic and industrial community. Due to the intrinsic property of OPV materials, flexibility feature of organic solar cells (OSCs) makes these materials highly attractive for building integrated photovoltaics systems, automotive OPV, wearable electronics, organic light-emitting diode (OLED), organic field-effect transistor (OFET), electrochromic devices for buildings, automotive windows, and glazing applications. In this chapter an overview of the history, concepts and various applications of flexible OPV device from the material perspective and fabrication technology is presented. The recent developments in OPV materials, structure–property relation, design aspects of materials and solar cell architecture are summarized. The critical role of flexible electrodes in enhancing the functionality, adaptability, conductivity, transparency, sheet resistance, flexibility, durability, and chemical stability to make FOPV pave the way for their integration into diverse applications. Stability being the main bottleneck for commercializing OPVs various limiting factors for stability and the recent developments in strategies to overcome these have been presented. From device fabrication perspective recent progress made in design aspects, material choice, processing/printing technologies for efficient and inexpensive OPVs is discussed. Finally, future perspectives, existing challenges with potential strategies for viable commercialization of OPVs is highlighted.

D. S. Sharada (✉) · M. Premkumar · J. Jayapratha
Molecular Engineering and Catalysis Laboratory, Department of Green Energy Technology, Madanjeet School of Green Energy Technologies, Pondicherry University, Puducherry 605014, India
e-mail: drdssharada@pondiuni.ac.in

B. Balambiga
PG and Research Deparment of Chemistry, Pachaiyappa's College, University of Madras, Chennai 600030, India

P. Ravindran et al. (eds.), *Photon to Power*, Progress in Optical Science and Photonics 32, https://doi.org/10.1007/978-981-96-5914-2_10

Keywords Flexible transparent electrode · Flexible organic photovoltaics · Conducting polymers · Metal nanowire · Ultrathin metal · Carbon-based electrode

1 Introduction

The growing need for renewable energy sources and escalating environmental challenges hassled the development of green and sustainable energy harvesting technologies. Among the various renewable energy sources, sunlight is one of the most abundant, powerful high energy density carbon-free, and greener sources [1]. Solar energy is a massive source. A total of 173,000 terawatts (trillion watts) of solar energy strikes the earth continuously every second, which is more than 10,000 times the Global energy consumption per year [2]. Among various generations of solar PV Technologies, second and third-generation solar cells have evened as thin film solar cells to reduce the cost and size of the well-developed and commercially used silicon-based first generation solar cell. However, to address the limitations of second generation solar cells, i.e., rare/have less accessibility and degradation of materials, the third generation solar cells have evolved with capabilities such as high power conversion efficiency and low production cost [3]. These film technologies have attracted great attention and are seen as game changers in solar PV technology due to low production tunable band gap, improved efficiency, and device lifetimes. Among the various third-generation solar cell technologies, there has been a growing interest in OPVs during the past one decade from both academia and industry [4, 5]. Despite lower Power Conversation Efficiency (PCE) compared to silicon-based solar cells, their considerably lower manufacturing cost makes them highly attractive as an accessible and affordable green energy technology. OSCs have attracted remarkable attention as potential sustainable energy technologies due to their characteristics and special merits, such as their light weight, low cost, non-toxic raw material intrinsic flexibility, and high throughout large area printing. [6] Intrinsic flexibility of the active layer is the most prominent characteristic of OSC, making it the most suitable for flexible electronic device applications [7]. Accordingly, all polymer OPV materials have provided an attractive opportunity for the fabrication of FOPV with good mechanical durability compatibility, excellent flexibility, and stretchability, resulting in application prospects for variable wearable electronics [8] self-powered electronics skins [9], sensors [10], portable devices for the Internet of things (IOT) [11] soft robots [12] and various medical applications. [13, 14]. Additionally, very thin photo-active layer of OSCs, with partial light transparency allowing visible light transmission makes semi-transparent organic photovoltaic cells (ST-OPVs) highly promising for their potential applications such as photovoltaics windows [15], skylights building integrated photovoltaics systems [16]. Indoor energy harvester technology for powering low-power electronics [17], agriphotovoltaics [18] and vehicle integrated photovoltaics [19].

A typical OSC consists of a photo-active layer surrounded by interface layers and sandwiched between the flexible bottom electrode and top metal electrode, after the early reports of OSC in 1958 by Kearns and Calvin of "Schottky-type solar cell" with poor exciton on separation [20], Tang developed the first OSCs in 1986, with donor-accepted bilayer-heterojunction to effectively separate excitons [21]. To further address this challenge in 1995, Yu et al. first proposed the concept of bulk-heterojunction OSCs, where the large interfaces generated in the donor–acceptor mixture resulted in effective diffusion and separation of excitons [22]. Till date, bulk-heterojunction (BHJ) structure remains the most widely explored active layer strategy in the fabrication of solution process OSCs devices [23].

During the last decade, impressive progress has been made in the field of OSCs, with a record performance of nearly 20% PCE (19.48% and NREL certified) in single-junction organic solar cells based on a ternary system [24]. Flexible novel non-fullerene acceptor (PM6: BTP-eC9 system)-based OSC based on a dual strategy involving the development of a novel acceptor and ternary blending achieved champion PCE 20.2% and stabilized efficiency attained 18.42% [25–27]. A DCP-based (OPV) module manufactured employing the blade coating approach has a maximum output voltage of 4.00 V and a remarkable PCE of 30.6% at 1000 lx, indicating its potential for efficient wireless power transfer applications [28].

This chapter provides a comprehensive overview of recent developments in OPV materials revealing important insights into structure–property relationships and design aspects that enhance OPV architecture. Stability remains a primary challenge for commercialization, prompting an exploration of various limiting factors and strategies to improve durability, which the discussion will then transition to the critical role of flexible electrodes in enhancing the functionality, adaptability, conductivity, transparency, sheet resistance, flexibility, durability, and chemical stability to make FOPV pave the way for their integration into diverse applications is being elaborated. From a fabrication perspective, advancements in design, material selection, and processing technologies are essential for OPVs' efficient and cost-effective production are discussed.

2 Device Structure

The typical device architecture was constructed with flexible transparent electrodes, an electron transport layer (ETL), a hole transport layer (HTL), a flexible substrate, and a passivation layer. The conversion of light energy (photon) to electrons usually described, (1) **Photon Absorption**: The photo-active material absorbs photons from incident light, which excites electrons from the valence band to the conduction band. (2) **Electron–Hole Pair Generation**: This excitation results in the formation of electron–hole pairs (excitons). (3) **Charge Separation**: An electric field, generated by the difference in energy levels between the donor and acceptor materials, facilitates the separation of these electron–hole pairs at the interface. (4) **Exciton Dissociation**: The excitons dissociate into free electrons and holes at the interface of the donor and acceptor layers. (4) **Charge Collection**: The electrodes collect the separated electrons and holes, allowing for the generation of an electric current [29]. The device architecture is classified into conventional and inverted structures [30]. The

a

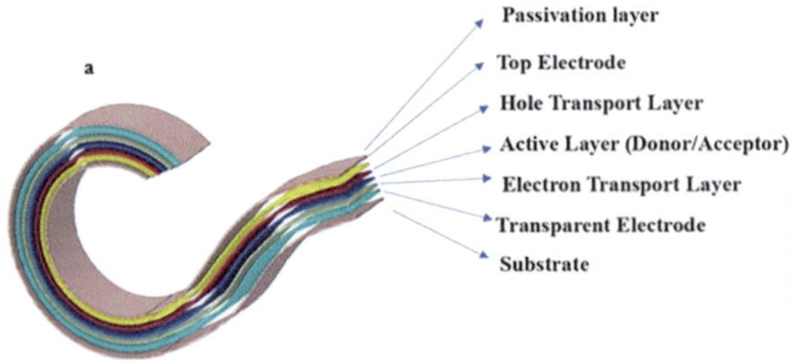

b

Thickness	
Flexible	Ultrathin
10 – 100 µm	1-10 µm
100 nm	Both
5 – 20 nm	Flexible & Ultrathin
100 - 300 nm	< 1 µm
5 - 20 nm	
100 nm	
10 - 100 µm	1-10 µm

Fig. 1 The development of flexible organic solar cells. **a** Typical device structure of FOSCs. **b** Typical thickness of each layer

device should be smaller than one micrometer except for the passivation layer and substrate for flexible devices. The passivation layer and substrate should be 10–100 µm large area should ($\geq$ 1 cm^{-1}) [31, 32] (Fig. 1).

2.1 Active Layer

The photovoltaic effect was first observed by Edmond Becquerel with wet cells in 1839. The electrically conducting conjugated polymers was first discovered by Alan J. Heeger, Alan G. Macdiamind, and Hideki Shirakawa in 1977 (Nobel prize in Chemistry awarded in 2000) [33]. This discovery paved the way for the development of organic electronics. The active layer of the flexible OPV should have good mechanical properties, effective absorption, proper morphology, diffusion length, and suitable band gap. Recent reports indicate tremendous research growth in this photo-active layer [34, 35].

2.1.1 Architecture

Active layer architecture can be made by using homo and heterojunction, which is composed of single or multiple components. The first generation of OSC was introduced with a single active layer with different work functions and poor PCE below 0.1% due to the poor dissociation of exciton (electron–hole) efficiency and several recombinations of electrons and holes [36]. The bilayer-heterojunction was introduced by Tang et al. in 1986 [20], a breakthrough report in the field of OSC. The bilayer-heterojunction consists of a donor and acceptor (D/A system). The D/A interface area has worked to improve the efficient exciton diffusion and charge separation. After a decade, the bulk-heterojunction (BHJ) was introduced by Yu et al. in 1995 to overcome the limitation of bilayer-heterojunction. The BHJ consists of a blended donors and acceptors as a single active layer, and the donors and acceptors were classified into polymers, small molecules, and fullerene derivatives. The discovery of BHJ is a notable development in the field of OSC, as it shows benchmark reports in PCE. Donor and acceptor systems are crucial to the active layer's efficiency and stability. The device performance depends on the layer's thickness, donor and acceptor ratio, and energy level alignment [36, 37].

2.1.2 Acceptors

In OSC, the acceptor should have a potential electron-accepting nature. To mitigate the energy level at the donor–acceptor interface, a BHJ with an electron donor and a lower bandgap acceptor must have an offset (ΔIE) of at least 0.45 eV between the ionization energies of the materials [38]. The evaluation of the acceptor materials has undergone two phases of research: in the early days, fullerene-based acceptors dominated in OSC and these were discovered by Robert F. Curl, Jr., Richard E. Smalley, and Sir Harold W. Kroto at Rice University in 1995 [22]. These fullerene materials boosted efficient electron mobility and isotropic charge transport character [36, 39]. The fullerene derivatives of [40] and PC71BM [41], were dominated as the acceptors in OSC. Following decades of fullerene-based acceptor dominance and increased interest in BHJ-based OSC, the limitations and disadvantages of fullerene acceptors spurred an investigation into non-fullerene acceptors. Over the past ten years, non-fullerene-based acceptors have demonstrated remarkable promising and unanticipated reports to surpass the limitations.

Non-Fullerene Acceptor (NFAs)

The fullerene acceptor has drawbacks, such as restricted functionalization, weak absorption in the visible spectral region, and poor band gap engineering. NFA has unique advantages, including simple synthesis procedures, chemically tunable shape, tunable band gap, modifiable absorption range, adjustable energy levels, NIR area absorption, and improved thermal and photo-stability [36].

NFA-Small Molecule (SM-NFAs)

The small molecular NFAs are classified into three types of structures: A-D-A type, A-D-A-D-A non-fused ring type, and A-DA D-A type, where the electron "push–pull" structure was applied. NFAs small molecules can be easily customized and tuned through chemical modification. The well-known acceptors for high-performing OSCs are ITIC, IDT-OB, IDIC, α-6 T, and BTP–4Cl [32, 42–45]. Yuan and coworkers reported revolutionary Y6 innovation in the field of NIF-SM; the Y6 is a versatile NFA-material with a low band gap and improved electron affinity. It has a highly conjugated core and enhanced optical absorption; researchers are trying to improve the efficiency of Y6 by designing the structure with different moieties [36, 39]

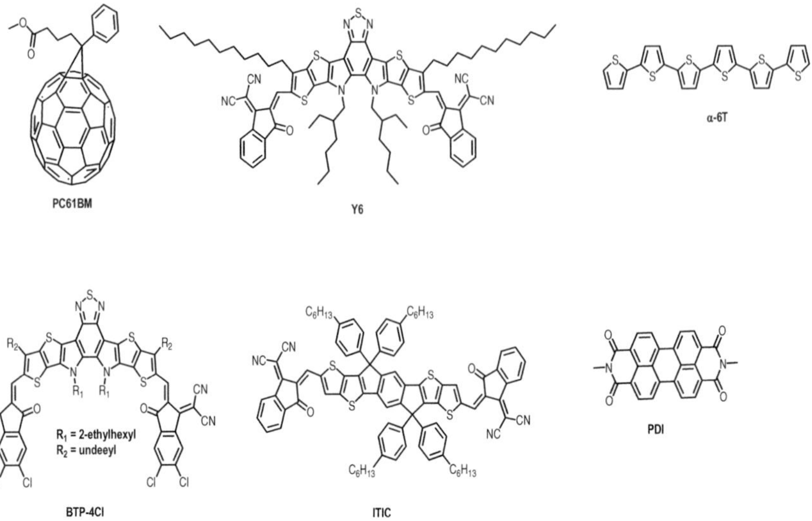

Fig. 2 Chemical structures of the acceptors

Polymer NFAs

Device stability and efficiency are crucial for both lab-scale and industrial applications. The limitations of SM-NFAs made the research community to be transmitted to by Polymer NFAs, which show good structural flexibility, excellent mechanical properties, and morphological strength [46]. In 2007, Zhan et al. [47] reported that the Perylene diimide unit is an acceptor in OSC devices. In 2013, PDI dimers were developed, which were linked with thiophene [17] having broad and planar structure, good photochemical stability, and tunable synthetic approach, followed by PDI Naphthalene diimide (NDI), thiophene imide (BTI) which are acting as the efficient polymer acceptors (Fig. 2).

2.1.3 Donors

Donor molecules are typically designed to have specific electronic properties, such as a suitable energy level alignment that facilitates efficient charge transfer. They usually possess a high absorption coefficient to capture a significant amount of the solar spectrum. The donor should have a more positive ionization value than acceptors. Before the NFA discovery, the donor was designed to match fullerene acceptors, e.g., NDMO-PPV, P3HT, etc. [48, 49]. When choosing a donor for a non-fullerene acceptor, it is essential to consider suitable energy level offsets, optical band gap, and miscibility. P3HT, PM6, PTB7, PBDB-T-2F, and PNTT are the used as the vital donors in the field of OSC (Fig. 3).

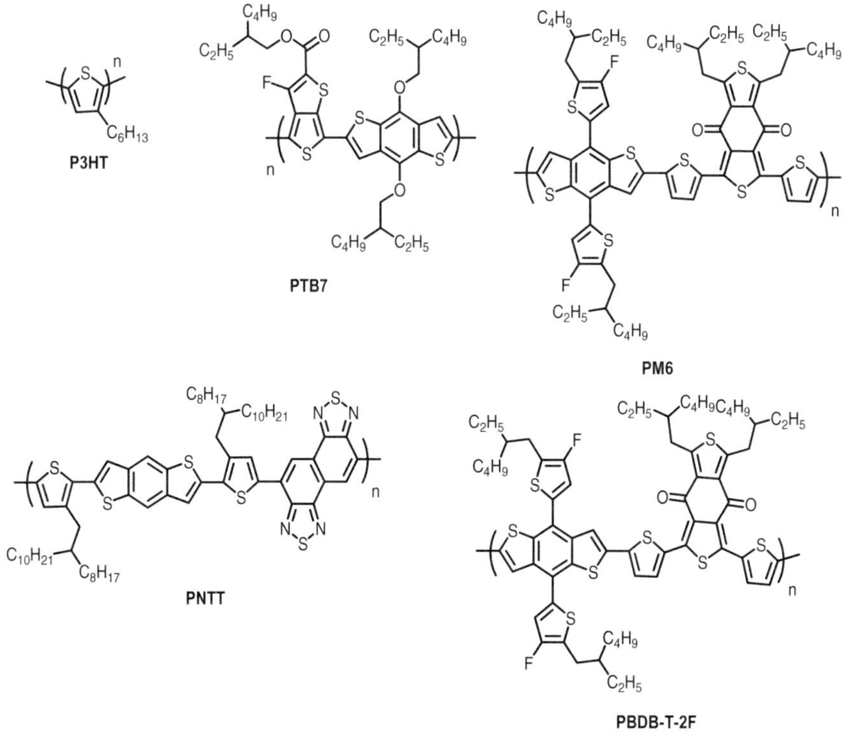

Fig. 3 Chemical structures of the donors

2.2 *Buffer Layers*

These electrodes' WFs are insufficient to match the majority of the most effective donor and acceptor materials, which results in an inadequate built-in electrical field and significant voltage loss in the devices. Large energy barriers are induced at the interfaces by the non-Ohmic connections between the active layer and electrodes, obstructing the charge transfer from the active layer to the electrodes. Much work has been done in interfacial engineering to solve these issues. To provide a favorable cascade energy alignment and significantly improve charge extraction efficiency and photovoltaic performance, the WF can be fine-tuned, and the interfacial contact can be improved [50, 51].

2.2.1 Electron Transport Layer (ETL)

In thin film FOPVs, the ETL plays several roles in conventional device structures like the work function (WF), PCE, and chemical stability. An inverted device structure introduces the n-type 'MO' metal oxide. (Substrate/electrode/ETL/Active layer/

HTL/electrode) [51, 52]. The main function of ETL is collecting and transporting the electron carriers as well as functioning as a blocking layer for the product. In the hole, during the recombination of OSC, the ETL must be thin enough (5–20 nm) to give good ohmic contact for the charge transfer. The thin ETL will increase the sheet resistance of the FOPVs. Therefore, the coating technique is also important for the ETLs. Chemical bath coating, spray pyrolysis, spin coating, and sol-gel are some of the significant techniques for getting a thin ETL for FOPVs [32, 44, 52]. Many researchers have reported that metal oxide-based ETL has excellent electron transport properties and suitable work function. Early ZnO-based ETLs show optimized morphological surface, good contact with the active layer, decreased charge recombination efficiency, and long-term stability of the device [51]. Tan et al. in 2019 first reported SnO_2-based ETL in semitransparent OSC with a high PCE of 12.88%, AVT of 25.6%, and excellent photo stability and chemically precipitated SnO_2 colloidal particles as an ETL in the device. The SnO_2 layer can be more efficient than light distribution because of the excellent transparency and high reflective index [53]. Yu et al. introduced MXenes, a novel intriguing family of two-dimensional (2D) transition metal carbide and nitride. MXene with ZnO as the ETL for OSCs was used, where Mxene is used as an additive for ZnO to fabricate novel ZnO/Mxene ($Ti_3C_2T_x$) composites film. The hybrid film creates a new electron transport pathway between the ZnO crystals which passivates the surface of ZnO by forming the Zn–O–Ti bond on the surface. This result shows that the PM6:Y6-based IPSCs achieved a champion PCE of 16.51% from 14.99% for the reference device, suggesting the good applicability of the ZnO/$Ti_3C_2T_x$ ETL [54].

2.2.2 Hole Transport Layer (HTL)

HTL plays a crucial role in the stability and PCE of the cell. The main function of HTL is to collect and transport the generated holes to the electrode, allowing them to recombine after passing through the load return. The HTL has excellent optical properties, water stability, good wettability, suitable work function (WF), high mechanical stability, and the HTL also has a moisture barrier in OSC [51, 52]. Generally, HTL should have good chemical and mechanical properties and excellent hole transport mobility. HTL should not affect the electrode and active layer. Both organic and inorganic HTL materials are used in OSC; inorganic materials, although having the potential to PCE, are not recombined for OSC and PSC due to their high-temperature need and high-cost deposition process [51, 52]. Additional materials such as MoO_x, Nio, Nb_2O_5, FeO_x, and CoO_x are some examples of inorganic HTLs [51]. Organic HTLs are essential components in organic solar cells, contributing to their efficiency, stability, and flexibility. PEDOT: PSS remains the most popular option for organic HTLs in traditional OSC due to its various benefits, including excellent mechanical properties and suitable work function (WF) [55]. At the same time, the acidic and hydrophilic nature of the PEDOT: PSS affects the electrode, so there are some chemical disadvantages there is pristine PEDOT: PSS HTL. So, the researcher focuses on hybrid PEDOT: PSS HTL [51]. Metal oxide-based hybrid HTLs having good device stability, higher efficient carrier mobility with longer charger carrier lifetime in the cell, suitable work function, and roughness reduction are reported [56, 57]. Modified material PETDOT: PSS, called doped PET: PSS, with the help of Polyethyleneimine derivatives like PA and TA, DA doped with PET: PSS as HTLs give different performance variations with different active layers due to the high number of hydrolysis delivers greater work functions and good electrical connectivity [58].

2.3 Substrate

Generally flexible substrates with a thickness ranging from 10 to 100 μm are utilized in flexible organic solar cells various flexible substrates such as polyimide (PI), polyethylene terephthalate (PET), polyethylene naphthalate (PEN), and polydimethylsiloxane (PDMS) [59] have been employed to fabricate FOPV. These conducting substrates have essential properties such as sheet resistance, optical transparency, flexibility, thermal properties, transparency, environmental stability, and oxygen and vapor parameters [43].

2.3.1 Polyimide

Polyimide (PI), due to its remarkable thermal and mechanical durability it has been recognized as a high-performance polymer. PI has been regarded as an alternative substrate for glass in the OSC when compared to other polymer substrates. It has good favorable mechanical properties such as lightweight, lower conducting transparent electrode, higher thermal properties, better dielectric qualities, and high resistance to solvents and chemicals [60]. Le et al. reported AgNWs with hybrid material of transparent conductive is encapsulated by a ZnO by KOH treatment and sowed low resistance 24 Ω sq-1and high optical transparency of 81% [61]. Lim et al. have developed the substrates of glass with AgNW electrode, it showed sheet resistance of 21.76 Ω sq^{-1} and an optical transparency of ~ 85% [62].

2.3.2 Polyethylene Terephthalate (PET)

PET an aromatic polymer is synthesized by reaction of ethylene glycol with either terephthalate acid or dimethyl terephthalate and it is also called aromatic polymers. PET has excellent properties such as dimensional stability at different temperatures and resistance to heat, chemicals and mechanical stress, opaque, translucent, transparent, and outstanding optical clarity. This set of characteristics makes it excellent for use as a substrate for FOSCs. Lee and their group developed an AgNW/SWCNT hybrid film with plasmon-coated devices resulting in sheet resistance of 26 sq^{-1}, > 90% optical transmittance and the RS drastically increased from 26 to 29 sq^{-1} at 5 mm bending radii [63].

2.3.3 Polyethylene Naphthalate (PEN)

The most widely used transparent polyester substrate is PEN, one of the aromatic polyesters. PEN has a similar structure to PET containing a naphthalene ring and because of this structural difference in the chain PEN has good mechanical properties. In 1996, Brolly et al. [64] conducted sorption and diffusion studies using PET and PEN substrates. According to his research, PEN has better barrier qualities than PET, and both materials have enhanced permeability when used with orientated polymer substrates. PEN absorbs at longer wavelengths than PET and has ~ 85% for PEN of transmittance. Rougier et al. have demonstrated PEN and AgNW-Gr as a substrate with Si-doped zinc oxide, and Hall effect measurements indicate resistivity ~ 8×10^{-4} Ω cm for films with a transmittance of 80% [65].

2.3.4 Polydimethylsiloxane (PDMS)

Polydimethylsiloxane generally referred to as polysiloxane, PDMS is typically used in OPV as an alternative to glass substrate because of its strong adhesion, high elasticity, and optical clarity [66, 67]. PDMS has greater stability to heat, easy and cheap manufacturing process, and it can exhibit linear extensions upto 30%. Despite these, PDMS also have some disadvantages like insufficient management of their surface characteristics and because of their extreme hydrophobicity, conductive materials will stick poorly to the PDMS substrate. Akter et al. has reported PDMS with AgNWs as substrate by spray deposition method, which showed the sheet resistance and transmittance of ~ 35 Ω^{-1} and 80%, respectively, [68]. Yang et al. have demonstrated PDMS with CuNws as substrate by electroplating method, as a result of 15.6% Ω^{-1} and transmittance of 82% was observed [69].

3 Flexible Transparent Electrode (FTE)

Indium tin oxide (ITO) is commonly used as a transparent conductive oxide electrode (TCOE) because of its enhanced properties such as high optical transparency of 80%–90%, low sheet resistance (Rsh) of 10–20 Ω sq^{-1} (on glass substrate) with suitable work function, band gap is 3.8 eV [82], and large mobile free carriers. The mechanical properties and efficiency are questionable if ITO electrodes move to the flexible OPVs. With the ITO coated on the plastic substrate, the electrical conductivity is gradually decreased, Rsh is increased, and finally, poor mechanical stress is achieved. These things make them unsuitable for applying flexible OPVs [83] and ITO coating techniques are also more expensive [31] ITO's replacement by FTEs has garnered significant attention over the past decade, exhibiting properties such as optical transparency, wettability, stability, sheet resistance, electrical characteristics, suitable WFs, and high mechanical robustness [82]. A Flexible Transparent electrode (FTE) is a major component in determining the performance (efficiency, flexibility, and stability). Most importantly, many promising and effective FTE materials have been developed, including conducting polymers, carbon materials, ultrathin metals, metal nanowires, graphene, Mxene, and their hybrid electrodes (Fig. 4) [84, 85]. Even though, conducting materials have limitations such as limited flexibility, weak adhesion, moderate conductivity, insufficient transparency, high surface roughness, poor wettability, and stability problems that hinder the fabrication of ITO-free flexible OSCs. However, this chapter systematically focuses on the doping treatments modifying the features and properties of the electrode (e.g., σ, Rsh, T%, WF, flexibility, wettability, stability, etc. and the design strategies of these FTEs involving PEDOT: PSS, graphene, Ag NW/mesh, metal oxide/metal/metal oxide, Mxene, and their hybrid composites.

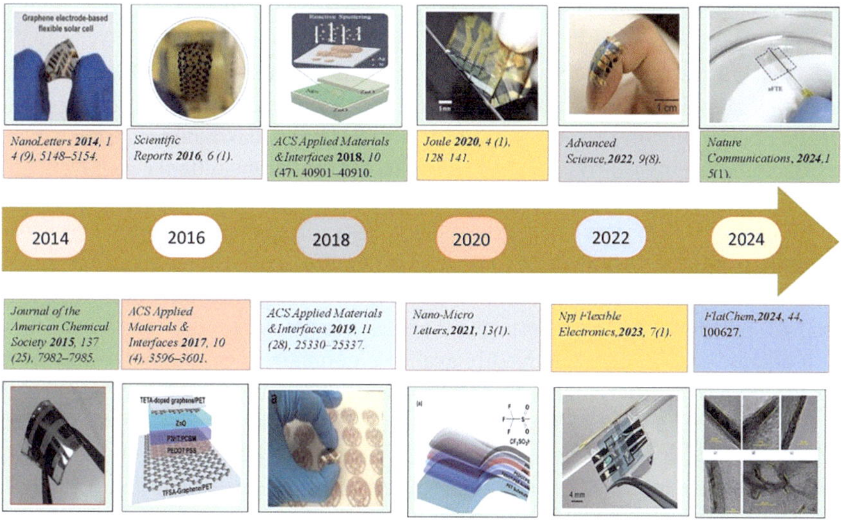

Fig. 4 Timeline illustrating the evolution of flexible transparent electrode in OPV over the years. Reproduced with permission [70] Copyright 2014, American Chemical Society. Reproduced with permission [71] Copyright 2015, American Chemical Society. Reproduced with permission [72] Copyright 2016, Springer Nature. Reproduce with permission [73] Copyright 2017, American Chemical Society. Reproduce with permission [74] Copyright 2018, American Chemical Society. Reproduced with permission [75] Copyright 2019, American Chemical Society. Reproduced with permission [76] Copyright 2020, Elsevier. Reproduced with permission [77] Copyright 2021, Springer Nature. Reproduced with permission [78] Copyright 2022, American Chemical Society. Reproduced with permission [79] Copyright 2023, Springer Nature. Reproduced with permission [80] Copyright 2024, Springer Nature. Reproduced with permission [81] Copyright 2024, Elsevier B.V.

3.1 Conductive Polymer-Based FTEs

In construct to traditional metal-to-metal oxide, band electrodes which have limited mechanical flexibility and poor elasticity, conductive polymers have gained tremendous attraction for integration into flexible electronic devices. Owing to flexibility, lightweight, cost-effectiveness, and environmentally friendly or sustainable [86–90].

Conductive polymers are organic materials with extended Pi conjugation that exhibit charge mobility, viscosity in metal-like optical properties, and electrical conductivity [91]. After the discovery of Polyacetylene (PA) in 1977 [92]. Conductive polymers have received great attention for various applications [89]. Among the conductive polymers, after the invention of Poly [3, 4-(ethylenedioxy) thiophene] PEDOT in 1988 by Bayer AG, it has been the most popular Conductive polymer beacuse of its high conductivity, optical transmittance and air stability [90]. To hydrophobic PEDOT conductive polymer upon addition of hydrophilic surfactant additives, i.e., PSS poly (styrene sulfonate) results in PEDOT: PSS which is another popular candidature studied for wide application, dye to its improved aqueous predominantly as thin films [93]. PEDOT: PSS have gained popularity as flexible transparent electrode due to its excellent optical transmittance (> 90% at 550 nm), high electrical conductivity, tunable work function (4.8–5.4 eV), good thermal stability, light weight, low surface roughness, high stiffness and strength. Moreover, due to their superior intrinsic flexibility, stretchablilty, cost-effectiveness, commercial availability, easy solution and film processing have gained attention over other electrode materials as flexible transparent electrode [94, 95].

In organic solar cells, aims readily available commercial aqueous dispersions of PEDOT: PSS with different ration of PEDOT and PSS are used, such as (Clevios PVP AI 4083, and PEDOT to PSS with 1: 6 is used as HTL and Clevios PH500, PH510, and PH1000, with a 1:2.5) is used as electrode [96]. In 2002, Zang et al. have for the first time demonstrated the modified (doped with glycerol and-sorbitol) PEDOT: PSS electrode as a potential replacement of ITO to fabricate the rigid organic solar cell [97]. Subsequently Aeronauts et al.in 2004, have for the first time deposited PEDOT: PSS as anode on polyethylene terephthalate (PET) substrate for flexible organic solar cell device [98]. A metallic Ag grid was introduced between PEDOT: PSS and substrate to enhance conductivity. To enhance the conductivity of pristine PEDOT: PSS (1 S cm-1) different preparation methods, pretreatment (doping) and post treatment many efforts have been made on.

Pretreatment

Several doping strategies have been studied to improve electrical conductivity of PEDOT: PSS using polar solvents Dimethyl sulfoxide (DMSO) [99], Ethylene glycol (EG) [100] (Polyethylene glycol (PEG) [101], and Ionic liquids [102], through inducing superior morphology and enhance carrier mobility. In 2020, Ge et al., have demonstrated micro doping with Polyhydroxy glycols [103] and D-Sorbitol [104] resulting in enhanced conductivity, transmittance and stability of PEDOT: PSS films, due to have strong interface bonding to substrate via Hydrogen bonding and Vander Waals interactions. Boa's et al. [105], Hu et al. [106] have demonstrated enhances conductivity and mechanical properties of PEDOT: PSS using sulfonate/ sulfonimide anions as ionic additives and fluorosurfactant dopant respectively, due to improved morphology. In 2017, Water soluble Polypyrrole (PPy) Nanoparticles have been employed as dopant materials to improve stability and PCE by 150%. Compared to pure PEDOT: PSS due to change in pH, conductivity and morphological changes [107]. The pretreatment with polar solvents and surfactants geometries large-domain aggregates, hence posttreatment and exposed to avoid formation of large-domain aggregates.

Post -reatment

In post-treatment using acids, sulfuric acid (H_2SO_4) treatment was found to be best way to fabricate PEDOT: PSS electrode with refined in improved conductivity [108, 109]. However, strong acid treatment led to destruction of flexible plastic substrate, to avoid this, Ouyang et al. [108] found, mild and weak acid and found methanesulfonic acid (CH_3SO_3H) to improve conductivity. Subsequently, many groups have examined mild acid post treatment (CH_3SO_3H, H_3PO_4) resulting in improved conductivity, superior optical transparency and smooth surface and excellent mechanical flexibility [110–112]. Recently, 0.8 M with tri-fluoromethanesulfonic acid (CF_3SO_3H) was doped with PH1000 electrode by Fan et al., resulting in a minimum sheet resistance of 32 Ω sq^1, increased work function of around 5.0 eV, improved transmittance (above 87%), efficiency of 16.6% (with PM6:Y6 as the active layer), 94.4% flexibility even at the 1000th cycle bending test of 1.rmm radius [77]. Further, Fan et al., have reported room-temperature CF_3SO_3H-doped PH1000 anode for annealing free solution-processed FOSC, with PCE 14.66% [113]. They have recently reported CF_3SO_3H doped PH1000 by a cost-effective CF3SO3H fumigation method for FOSC with 13.62% PCE.

Recently, Ge et al., have developed, FOSC with the efficiency of 14.17% by ecofriendly citric acid processed foldable PEDOT: PSS electrodes to mitigate corrosion damage of plastic substrate device retaining 90% of its efficiency after 1000th bending or folding cycles as shown in Fig. 5 [104]. To avoid damage of plastic substrates with acid, Kim et al., have developed H_2SO_4-assisted transfer printing strategy, resulting in high conductivity 42,000 Ω sq^1, of PEDOT: PSS electrode [114]. Recently, Fan et al., have demonstrated a unique perchloric acid ($HClO_4$) spray-doping treatment of PEDOT: PSS films resulting in 29.4 Ω sq^{-1} sheet resistance, high work function of 5.39 eV, 90.4% transmittance, better hydrophobicity and good wettability. This treatment method has a potential to make large area PEDOT: PSS electrode-based FOSC [115].

Recently Bihar et al., have developed ultrathin metal free OPV device with PEDOT: PSS electrodes as both bottom and top electrodes by inkjet printing, resulting in improved device performance, good stability in moist conditions [116]. In 2019, Wang et al., have demonstrated OPV cell with 1.6 cm^{-1} and with structure glass/ PEDOT: PSS PH1000/MoOx/BQR: PC71BM/Ca/Al with PH1000 electrode on ultrathin flexible glass substrate exhibiting transmittance of $\approx$ 77% at 550 nm, RS of $\approx$ 30 Ω sq^{-1} and PCE of 5.2% [117]. Further Meng et al. [118] developed large area

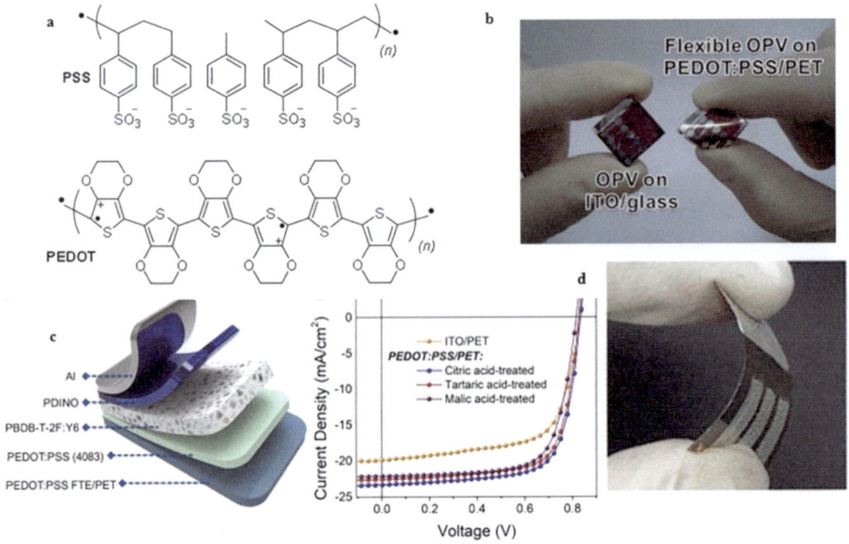

Fig. 5 Schematic representation of conducting polymer for the development of flexible OPV. **a** Chemical structures of polymer (PEDOT: PSS) electrode. **b** Photography of comparison between OSCs fabricated on rigid ITO/glass and flexible. Reproduced with permission [119] Copyright 2011, Elsevier B.V. **c** Device structure of the flexible OSC and J–V characteristics of flexible OSCs fabricated on ITO/PET and PEDOT:PSS/PET. Reproduced with permission [104] Copyright 2020, Elsevier. **d** Photography of the flexible all-plastic solar cell. Reproduced with permission [111] Copyright 2015, American Chemical Society

flexible OPV modules consisting of 6 cells with 15 cm^{-1} area, with substrate PET/hc-PEDOT:PSS/ZnO/PAL/MoO_3/Ag, and blend of active layer components PTB7-Th:PC71BM (7.58%), PBDB-T:ITIC (8.90%) resulting in high efficiency till date for PEDOT: PSS-based electrodes for OPVs.

3.2 Metal-Based Electrode

Metals offer the highest electrical conductivity compared to all other materials. Thin metal surfaces have mechanical flexibility and optical transparency. The combination of high conductivity, adjustable transmittance, and customizable flexibility makes materials made from metals highly capable for flexible transparent electrodes (FTEs). There are four main types of flexible electrodes produced using metals: ultrathin metal nanowire networks, metal films, metal meshes, foils, and grids. Each type has its own unique properties, advantages, and limitations when used in devices, which are discussed in detail below.

3.2.1 Metal Nanowires (MNWs)

Metallic nanowires are commonly used as electrodes in various electronic and optoelectronic devices because of their unique properties, such as high electrical conductivity, flexibility, and transparency. While silver (Ag), gold (Au), and copper (Cu) are the most commonly employed materials for these nanowires. It having low sheet resistance, solution processability, low cost, and affluence of scaling. The current research revolving around to improve the disadvantages [120], such as poor adhesion, air oxidation, and short circuiting involved in MNWs.

Silver Nanowires

Silver is a precious metal having work function of 4.3 eV which have high conductivity. The minimum energy essential for an electron to discharge from the surface of a material is known as work function. Ag has been commonly used as an electrode in various electronic devices. In 2008, Peumans' group introduced Ag NW electrodes for OSCs, fabricated through a fully solution-based process with low-temperature annealing, achieving 19% higher photocurrent than typical ITO-based devices [121]. Subsequent efforts focused on protective layers for Ag NWs, including PI, NOA63, polymethacrylate, polyvinylalcohol, graphene, and metal oxides. Seo et al. developed a cold isostatic pressing (CIP) method to create well-ordered silver NW junctions arranged PET substrates, offering low roughness, high conductivity, and excellent flexibility [122]. Lately, Li et al. [123] achieved an ultraflexible OSC with a semi-embedded Ag NW composite electrode, reaching a PCE of 13.55% (average 13.42%). An ionic liquid-assisted chemically welded FTE based on AgNWs prevents oxidation, achieving record power conversion efficiencies of 17.52% (0.062 cm^2) and 15.82% (1.0 cm^2) in single-junction flexible OSCs, thanks to in-situ twin-crystal growth at AgNW junctions [124] (Fig. 6(iv)) A fully flexible hybrid electrode was created by sandwiching a silver nanowire network between monolayer graphene and a polymer film. This design outperforms PET- or PEN-based electrodes, achieving 8.06 Ω/□ resistance with 88.3% transparency (Fig. 6(ii)) [125]. In 2024, Choy et al. [80] conducted an operation and discovered liquid-like behavior in ZnO nanoparticles (ZnONPs) on silver nanowires (AgNWs) using an in-situ solution process. This discovery resolved interfacial issues between ZnONPs and AgNWs, significantly improving multi-loading performance (Fig. 6(i), (iii)).

Cu Nanowires

Copper is 6% low conductive than silver, it is inexpensive and more abundant. Therefore, the Cu NWs will greatly decrease cost than Ag NWs [126]. In the year of 2013, Leo's group [127] established high-aspect-ratio Cu nanowires (NWs) using a hydrazine-free hydrothermal method. These Cu NWs were spray-coated onto glass and PET foils, constructing electrodes with exceptional mechanical flexibility, as their resistance remained stable after 1000 bending cycles at a 120° angle. A flexible device utilizing P3HT:PC61BM as the semiconducting layer achieved a PCE of 1.4% [128]. Thus, copper NWs are performing well for flexible OPVs.

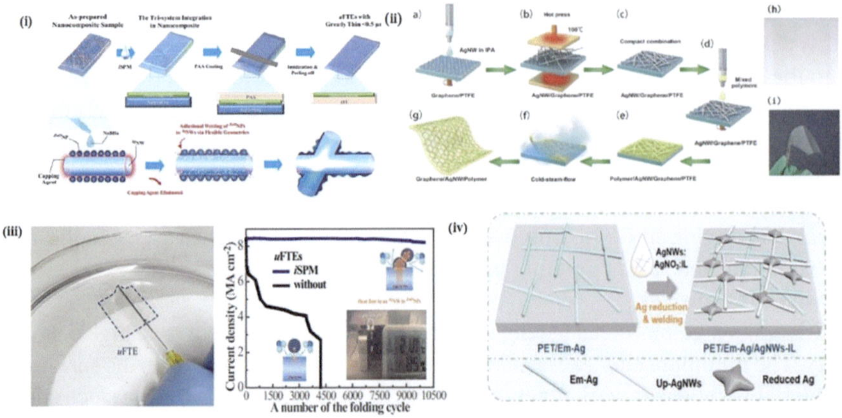

Fig. 6 Schematic images and structure representation of metal nanowire-based flexible transparent electrode. **(i)** A schematic diagram exhibiting the fabrication process of the substrate-integrated uFTEs with greatly thin < 0.5 μm containing the tri-system integration in the nanocomposite via the iSPM reproduced with permission [80] Copyright 2024, Nature communications. **(ii)** Schematic representation of typical thin and fully flexible hybrid electrode developed by integrating the encapsulation of silver nanowires (AgNWs) network between a monolayer graphene and polymer film as a sandwich structure (a–g) (h) optical image of MG-A-P hybrid electrode (i) curved exhibition of MG-A-P hybrid electrode [125] Copyright 2016 American Chemical Society. **(iii)** (a) The photo of the uFTE (4 cm^2) floating on the surface of the water getting lifted by a needle and Real-time monitoring of the operational current density of uFTEs according to with (blue) and without (black) the tri-system integration against the simultaneous multi-loading under the humid condition of relative humidity 85% [80] Copyright 2024, Nature communications (**iv**) Schematic representation of preparation of Ag NWs atomic level chemical welding on PET using ionic liquid [124] Copyright 2022, American Chemical Society

Au Nanowires

Gold (Au), known for its excellent conductivity, is widely used as an efficient electrode material in high-end devices, typically as a bottom electrode in solar cells [129]. Thin Au layers have also been explored as top electrodes; recently, an Au textile-based top electrode in an OSC achieved a PCE of 2.91% [130] via physical lamination. Additionally, a yarn-based OSC [131] using stainless wire as the main electrode and Au as the subordinate electrode attained a PCE of 4.6%. Similarly, low-cost Al [132], high-conductivity Ni [133], and high-work function Pt (5.65) [134] nanowires have also been used as FTEs. Metal alloys were also explored as FTEs in SCs [135].

3.2.2 Ultrathin Metals (UTMs)

Ultrathin and uniform metal films, such as those made from Ag, Au, Cu, or Al, with thicknesses reaching from 10 to 20 nm, offer a blend of good electrical conductivity, optical transparency, and mechanical flexibility. However, their electrical and optical properties are depending on film thickness. Higher the thickness, conductivity can suggestively improve, impending that of bulk metal, but their optical transmittance declines abruptly. To achieve an equilibrium between transmittance and conductivity, a film thickness of nearly 10–20 nm is ideal. Ultrathin metal films show great potential as replacements for ITO in transparent electrodes for flexible electronics. Due to the Volmer-Weber growth mode in thin films can result in island-like morphology, which often results in deprived optical transmittance and electrical conductivity [136]. Many research works have been conducted to discourse this issue.

Ultrathin Metal-Based Films

Silver and copper are the recurrently active materials for ultrathin m-FTEs. The metal layer exhibits high sheet resistance until it reaches the percolation threshold. Extra deposition (greater than ideal thickness) of metal above the percolation threshold will lessening the sheet resistance significantly and across multiple scales of magnitude, the optical transmittance will reduce dramatically [137]. Thus, enlightening the wettability of metals on designated substrates that can proficiently diminish the percolation threshold, which is the crucial to gaining highly-efficient m-FTEs composed of ultrathin metallic layers. Guo's group [138] in 2008, done fabrication using highly transparent metal electrodes (Au, Ag, and Cu) for rigid device through nanoimprinted lithography approach. Considering all metals, Ag is regarded as the optimal option for UTMFs such as nanowires, owing to its low resistivity (1.62 μΩ·cm), outstanding ductility, and superior stability [139]. Nevertheless, it still presents certain inevitable boundaries. It gets oxidized and corroded easily, that will suddenly reduce its conductivity and transparent nature. To tackle this problem, Formica et al. developed a high-performance FTE using sputtering technique and formed Ag–Ni bilayer onto a treated PET substrate with a 1 nm Ni capping layer. The resulting PET/Ag–Ni electrode demonstrated superior stability, conductivity, and flexibility, maintaining its sheet resistance subsequently 1200 bending cycles within a radius of 4–12 mm [140]. Later on, Wang's group developed a hybrid FTE (S-1805/Ag (0.6 nm)/Au (4.4 nm)) by suppressing the growth of 3D Island on Au UTMF [141].

OMO and DMD Electrodes

Oxide-metal-oxide (OMO) and dielectric-metal-dielectric (DMD) structures were introduced to enhance the performance of the FTEs. It involves a metal film with highly indexed dielectric layer and a multi layered highly indexed dielectric layer. It features outstanding transmittance, low sheet resistance, and remarkable flexibility. Explicitly, the high-refractive-index dielectric layers in the multilayer enhance transmittance by minimizing the metal film's reflection through destructive interference [142]. This prevents them from readily oxidizing under atmospheric conditions, enhancing their stability and robustness. Zou and group developed ultrathin Ag films on a PEN substrate by employing a self-assembled monolayer-modified ZnO/Ag/ZnO construction. The resulting electrodes demonstrated lesser sheet resistance (8.61 Ω/sq), improved transmittance (> 80% between 400 and 600 nm), and minimal surface roughness [142]. Afterward, Yun's group developed a ZnO/AgOx/

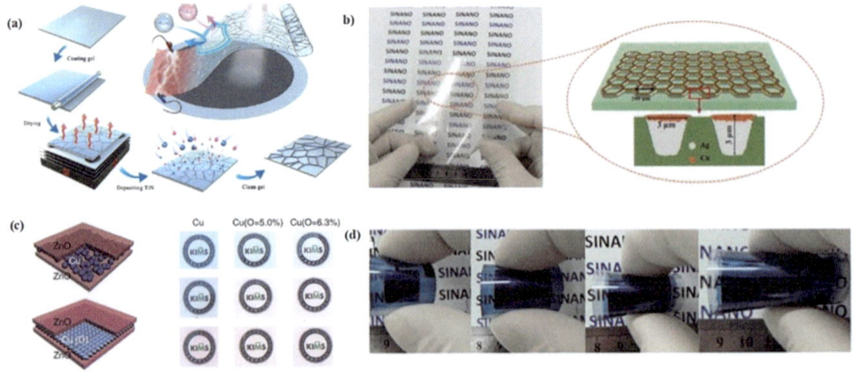

Fig. 7 Schematic images and structure representation of ultrathin metal-based electrodes. **a** Schematic representation of systematic preparation of TiN nano mesh transparent electrode [155] Copyright 2023, American Chemical Society. **b** Photograph and the schematic diagram of the large area PET/Ag/Cu electrode [152] Copyright 2019, Wiley VCH. **c** Schematic illustrations on behalf of the different morphologies of Cu and Cu(O) films sandwiched between 60-nm-thick ZnO films in an oxide-metal-oxide configuration (right) and optical photographs of ZnO/Cu/ZnO FTCEs with varying Cu and ZnO thicknesses (left) [144] copyright 2015, Nature communications. **d** Photographs of the large area flexible OSCs based on uTMs [152] copyright 2019, Wiley VCH

ZnO electrode through oxygen doping of an ultrathin Ag layer [143]. The AgOx layer exhibited a smoother and more continuous morphology at an identical thickness (6 nm). The ZnO/AgO*x*/ZnO electrode have shown better transmittance (on an average of 91%), lower sheet resistance (20 Ω sq^{-1}). Flexible devices displayed excellent long-term stability, retentive more than 85% of their original PCEs after 30 days in dark ambient conditions. Ultrathin Cu electrodes have grown attention as a low-cost, ample alternative to Ag owing to their similar resistivity (1.68 μΩ cm) (Fig. 7c) [144] PET/Cu electrodes [145] demonstrated low sheet resistance (lesser than 1), smooth surfaces, and excellent adhesion, with additional advancements Cu–Ni comprising layers, ZnO/Cu(O)/ZnO, and ZnO/Cu(N)/ZnO structures [146].

Ultrathin Metal-Based Grids

Metal grid-based flexible electrodes are used in flexible OSCs as an alternative to difficult-to-grow metal films. These grids can be patterned in various geometries, with their properties optimized by adjusting parameters like aspect ratio, line width, depth, and pitch size [147]. Key fabrication techniques include evaporative deposition, nanoimprint lithography, gravure offset printing, inkjet printing, and flexo-printing [148–151]. The composite electrode method is used to rectify the above issue which showed a lesser sheet resistance of 6.1 Ω sq^{-1} [146]. Lately, Han et al. created a 3 μm wide hexagonal Ag/Cu grid using nanoimprinting fabrication and an electrodeposition process, with highly conductive PEDOT: PSS (E100) applied on the upper part of the Ag/Cu grid. Further, many reports have been made using PET/Ag/Cu/E100 as an electrode that exhibited, long-term stability and superior mechanical durability, and preserved 95% of the original efficiency next 2000 bending cycles by a radius of 7.5 mm [152]. Ma et al. fabricated large area OSCs using metal grids. An Ag/Cu composite grid (Fig. 7b, d) with < 1 Ω/sq resistance and 84% transparency serves as a flexible OSC electrode, achieving 12.26% efficiency for 1 cm^2. For larger areas, it achieved efficiencies of 7.79 and 7.35% for 4 and 9 cm^2 devices, respectively [152]. In addition to the conductive polymer PEDOT: PSS, carbon materials were integrated with Ag grids to create high-conductivity FTEs for OSCs. Ag grids were hybrid with graphene, CP, and CNT for the betterment of the performance.

Like silver, Kang et al. made Cu grid electrodes utilizing a transfer printing method with a PDMS stamp, achieving 78% average transmittance, 22 Ω/sq sheet resistance, and excellent flexibility. Bending tests showed no conductivity decrease until the radius reached 3 mm [153]. The flexible device made using a Cu grid with P3HT: PC61BM achieved a PCE of 2.1%, similar to the conventional devices with ITO (2.24%).

Ultrathin Metal-Based Meshes and Foils

Employing metal mesh or foils will help the FOPV to be fabricated in a high conductivity in a large area at a low cost. The ease of fine-tuning the line width and pitch to have a better transparent nature with optimal sheet resistivity, transmittance, controlled morphology, and ease of fabrication is advantageous. The metal mesh-based electrode can display desirable electrical properties and surface topography. Jiang et al. recently testified a reverse-offset printing method to generate homogeneous, ultrathin (100 nm) silver mesh on flexible substrates. The conducting and transparent nature of the Ag mesh-based electrodes were optimized by adjusting the spacing between lines, achieving a sheet resistivity of 17 Ω/sq and 93.2% transmittance, outperforming sputtered ITO electrodes [154]. Recently, 2023 TiN nanomesh (Fig. 7a) electrode, created by sputtering TiN onto a cracked gel film, offers exceptional durability, chemical resistance, and performance. It maintains stable electrical properties through repeated bending and resists strong acids and alkalis, with 88% light transmission at 550 nm and ~ 200 Ω/sq sheet resistance. Successfully applied in electric heaters and thermochromic devices, it holds promise for diverse applications [155].

3.3 *Carbon-Based Material Electrode*

Carbon material have received enormous attention as a highly promising electrode material for flexible solar cells. Due to its high conductivity, excellent mechanical, chemical, and electrical stability, low interfacial resistance, superior transparency, high flexibility, lightweight, low cost, and large surface area. Graphene and CNTs are the commonly used carbon materials as FTE for OSCs [156].

3.3.1 Graphene

After the initial discovery of 2D graphene materials using mechanical exfoliation method, it gained enormous attention for various application due to its promising features such as high transmittance and conductivity [157, 158]. To fabricate graphene electrodes, huge effects were made on techniques such as mechanical exfoliation, liquid-phase exfoliation, semi-induced exfoliation, chemical vapor deposition (CVD), and reduced graphene oxide (rGO). The rGO and CVD grown Gr films have been the most widely studied candidates for FTE [159–161].

To improve the work function of pristine graphene ($\approx$ 4.4 eV) and to execute the sheet resistance ($\approx 6\ \Omega\ sq^{-1}$) various doping and transparent strategies for graphene transparent electrodes having been studied extensively [162, 163] Various doping and treatment strategies of graphene electrode can be classified as (1) Acid doping, (2) Organic component doping, (3) Metal Chloride/metal doping and (4) Molybdenum oxide doping.

Acid Doping

Among the mineral acids HNO_3 (chemical doping [119] and transfer doping [164]) and H_2SO_4 doping of graphene electrodes were studied to improve conductivity. Lee et al., and Chen et al., have experimented with CVD multilayer graphene (MLG) electrodes and CVD grown. Graphene films copper foils (thermal annealing) doped with HNO_3 resulting in enhanced conductivity and smoothness and wettability [165, 166]. To overcame degradation of cogently due to high temperature in HNO3 doping. Kwon et al. have doped graphene with macromolecular acid (Perfluorinated polymeric sulfonic acid (PFSA)), resulting in high stability, low Rsh (~ 91 sq^1), high transmittance (~ 89%), and increased work function (by 0.71 eV) [167].

Organic Component Doping

To improve hydrophilicity for solution-processed OCS, in 2009, Wang et al. have demonstrated CVD grown Gr electrode modified by pyrene butanoic acid succinimidyl ester (PBASE). Through non-covalent interactions in increased work function and conductivity [168]. Ni et al., and Kim eta al, have fabricated graphene electrode with CVD grown and electronically doped CVD grown with ferroelectric polymer of poly(vinylidenefluoride-trifluoroethylene-chlorotrifluoroethylene) (PVDF-TrFE)) resulting in large-scale single-layer Gr electrode and flexible. Graphene electrode-based OSC, respectively, and with high transmittance, low Rsh and high mechanical flexibility (Fig. 8d) [159, 163] Kim et al. [169] have employed bis(trifluoromethanesulfonyl)amide (TFSA [CF3SO2]2NH) doped treatment graphene electrode to improve optical transmittance, reduce sheet resistance and to shift work function film 4.1 to 5.1 eV, due to efficient hole transport layer from PEDOT: PSS HTL to graphene electrode. Recently, Koo et al. [170] have demonstrated the direct integration of PI on CVD grown graphene electrodes, resulting in flexible, durable, and thermally stable graphene electrodes. The resultant PI-assisted transferred graphene band FOSC showed an ultra-clean surface with a low sheet resistance of 83 $\Omega\ sq^1$, high optical transmittance (~ 92%), high PCE of 15.2%, and

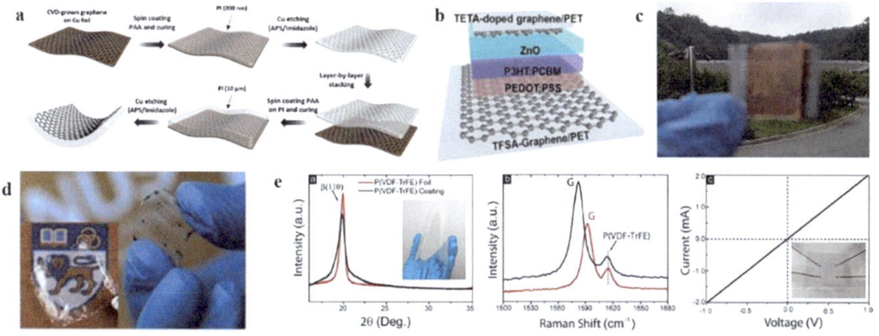

Fig. 8 Schematic images and structure representation of graphene flexible transparent electrode. **a** Schematic describing the Two-Step Fabrication Process of PI@GR. Reproduced with permission [170] Copyright 2017, Elsevier. **b** Schematic of the device architecture for a typical solution-processed semitransparent FOSC. **c** Photograph of a typical semitransparent FOSC. Reproduced with permission [73] Copyright 2018, American Chemical Society. **d** Optical image of GFeTC samples on the transparent PET substrate. **e** XRD results of both P(VDF-TrFE) foil and house-produced P(VDF-TrFE) film. Inset shows the optical image of large-scale prepolarized P(VDF-TrFE) foil on a PET substrate. Reproduced with permission [163] Copyright 2012, American Chemical Society

high mechanical flexibility for 1000 cycles of radius 2 mm. Ren et al. [171] have recently reported, highly flexible OSC, with graphene anode doped with benzimidazole and have demonstrated 3-layer graphene band 1.6 cm^{-1} FOSC with excellent mechanical robustness and small change in PCE with 250 flexing cycles.

Metal Chloride/Metal Doping

Many researchers have reported metal chlorides doped graphene electrodes, to improve the electrical conductivity [44, 83, 172]. However, reduction of metal cation leading to leakage of current from device is drawback. In, 2012, Kim et al., have used various metal ($AuCl_3$, $IrCl_3$, $MoCl_3$, $OsCl_3$, $PdCl_2$, and $RhCl_3$) as dopants to few layer graphene sheets to increase the work function upto 5.14 eV, which is due to the change transfer for graphene to metal ions [173]. Recently, Su et al., have developed a novel approach by doping layer by layer (LBL) stacked graphene with $AuCl_3$ resulting in improves performance of electrodes (Rsh of 40 Ω sq^1), transmittance 89.5%, wide range of work function tunability and good stability [174].

Molybdenum Oxide Doping

To improve wettability of hydrophilic graphene films, Wang et al. have developed as strategy by evaporating thin layer of MoO_x onto graphene film, thus improving charge separation efficiency and work function (5.47 eV) [175]. Later other researchers and Du's group [171] have introduced MoO_x interlayer on graphene electrode, to improve wettability and reduce energy barrier between active layer and graphene, leading uniform deposition of active layer.

All Graphene-Based FOSC

To fabricate FOSC, graphene electrodes were used as both anode and cathode, for instance Kang et al., in 2016, have developed dry transfer method to transfer graphene on organic layer with ethylene vinyl acetate as addition for adhesion [176].

Various researchers, have developed all graphene electrode-based semitransparent OPV device with PEDOT: PSS doping in graphene anode and metal oxide/CsF doping in graphene cathode resulting in improved PCE, optical transmittance and reduced sheet resistance [177, 178]. Shin et al., have demonstrated TFSA doped graphene electrodes as the anode (TFSA-G) and triethylenetetramine (TETA)-doped graphene cathodes for OPV, resulting in low electrical resistance, high transmittance, long-term stability and superior mechanical flexibility Kymakiset al, have developed a simple laser-based reduction method and later fabricated rGr micro mesh on PET substrate resulting in improved transparency, mechanical flexibility with good retention of efficiency under extreme bending conditions [178, 179]. In 2017, Blom and coworkers have developed electrochemical exfoliation method to fabricate flexible graphene electrode on glass and PEN substrates [180]. To fabricate low-cost FOSC, based on graphene electrode other methods such as spin coating and vacuum filtration have been reported [181].

Hybride Graphene Electrode

To realize high-performance FTE, hybrid composite electrodes have been explored to enhance adhesion and interlayer conduction. Various research groups have added perfluorinated ionomers (PFI) and nonionic small molecule surfactant (Triton X-100) into PEDOT: PSS to improve the wettability of the PEDOT: PSS films on CVD grown graphene, and consequently PCE [45]. Further, GO/PEDOT: PSS macro composite as TCE was explored, which second as hole caring material, resulting in enhances optical properties and experimental flexibilities, suitable for large-scale production methods [182]. In 2019, Chen et al. have employed composite electrode of PH1000/graphene on SV-8 polymer, resulting in 88.5% transparency at 550 nm and reduced Rsh of 147.5 Ω sq^{1}. The flexible OSC maintained PCE mostly after 5oo bending cycles with a radius of 3–8 mm [183].

3.3.2 Carbon NanoTube

Among the carbon materials, CNT have gained great interest among the academic and industry due to their appealing features such electrical, optical mechanical and thermal properties, tunable band gap, flexibility, large specific surface area, low weight, low cost, and mass production possibilities [45, 52, 184, 185] After the first report of needle like CNTs by Iigima in 1991 [186], Du Pasquier et al. [187] have employed for the first time in 2005 CNTs as bottom TCE. CNTs are hollow cylindrically tubes of graphite carbon atoms with ~ 1 nm diameter [188]. Based on the number of layer, CNTs we classified into single walled CNTs (SWCNTs), double walled CNTs (DWCNTS) and multiwall CNT (MWCNTs). Among the CNTs, SWCNTs

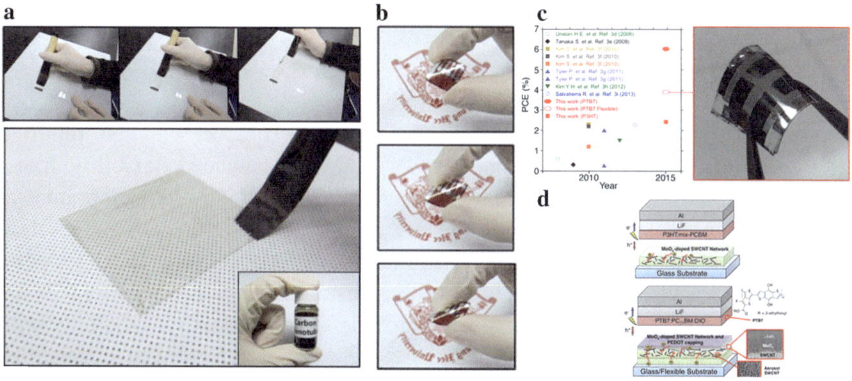

Fig. 9 Schematic representation of flexible carbon nanotube electrodes. **a** Picture of brush printing process (upper panels) to fabricate transparent CNT network electrodes and transparent 3-brush CNT electrode. **b** Flexibility of the FOSCs with a brush-painted CNT network electrode. Reproduced with permission [195] Copyright 2013, Elsevier. **c** Picture of the present PET-based flexible SWCNT OSC. **d** The most optimized device that gave high efficiency. Reproduced with permission [71] Copyright 2015, American Chemical Society

[185, 189] are the mostly widely studied due to their higher transparency and conductivity than the DWCNTs and MWCNTs [190]. In 2005, Chhowalla et al. at Rutgers have for the first time reported SWCNTS-based electrode in OSC, through solution processing [187] later Rowell et al. printed SWCNTs on PET substrate [191]. Various researchers groups have reported different techniques for fabricating CNT-based electrodes such as Bian's group for CVD method for ST-OPV [192], filtration-transferred method (Chhowalla group) [193], transfer printing method of Grunes et al. [194] and spin coating on PET [191] Apart from these methods, Kim's group have reported simple brush printing method for CNT electrodes on PET with excellent flexibility. Figure 9-a, b shows brush printing process of transparent CNT on substrate [195].

3.3.3 Doping

Various groups reported MoOx doped SWCNTs electrodes for FOSC by dry deposition on PET/PI substrates [71] by thermal annealing of MoOx on CNT in anaerobic condition [196] resulting in enhanced transmittance and efficiency. In 2017, Maruyana et al. [197] have reported for the first time MoOx doped SWCNT as electrode for small molecule (diketopyrrodopyrrole) based OSCs. In chemical doping, acid doping of CNT with HNO3 improves the transmittance and conductivity [72, 85]. Recently Jeon et al. [198] have reported chemical doping with TFMS, TFES and HNO_3 PCEs of 8.0, 8.3, and 8.5%, with good efficiency and excellent long-term stability, which have potential for sustainable devices.

Composites of CNT

On the other hand, a composite of CNT electrodes shows excellent optical, electrical, and mechanical properties. Organic composites, especially, easily deposit flexible substrates and having good mechanical properties. Researchers reported the PANI and PDMS organic composite CNTs having enhanced flexibility, transparency, high thermal stability, and mechanical robustness [199, 200]. Subramanya et al. [201] reported PEDOT: PSS with CNT composite using the solution method which enhanced electrode properties with PCE 7.35%. Huang et al. 2011 reported graphene with SWCNT as the electrode with adjustable WF (5.1–3.4 eV). [202] the porous network of carbon-based materials provides electron transport mobility with mechanical deformation, low interfacial resistance, and lightweight. These properties are suitable for making flexible electrodes.

3.4 MXene

MXenes are a class of two-dimensional materials derived from transition metal carbides, nitrides, or carbonitrides, known for their exceptional electrical conductivity, mechanical strength, and tunable surface chemistry. These properties make MXenes promising candidates for energy storage, sensors, and electromagnetic interference shielding applications [203]. In 2011 Gogotsi et al. first reported 2D Mxnen nanosheets based on transition metal carbide and nitride materials [204]. The Mxene is synthesized by a top-down selective etching procedure of their parent hexagonal MAX phase. MAX was first reported in 1996 by Borsoum et al. [205], which is represented as M_{n-1} AX_n, where "M" represents transition material (V, Ti, Mo, Z, etc.), "A" represents the group "A" elements, and "X" represents nitrogen or groups of two elements [206]. In 2008, the Mxene (Ti_3C_2Tx) was used as a conductive electrode in MAPbI3-based perovskite solar cells for the first time [207]. Mxene possesses excellent properties, including good electrical conductivity and high optical transmittance, mechanical flexibility, low-cost sustainable band alignment and uniform surface termination, which allow it to tune their electrical properties, significant carrier mobility, magnificent metallic conductivity, and tunable work function. Mxene also has high sheet resistance which is due to the percolation effect, this signifies that developing suitable and efficient thin film fabrication techniques can give the solution through multi-step spin coating on the substrate [83, 206].

Taylor et al. reported ultrathin transparent conducting electrode with low sheet resistance 437 Ω sq^{-1}, transmittance 77%, 550 nm, the result films deposited at a spin rate of 400 nm with the 7.40 $\pm$ 0.5 nm thickness due to the ultra-thickness of the electrode (< 10 nm) the sheet resistance was reduced [208]. In 2020, Zhang et al. reported that all solution processes have flexible semitransparent electrodes with different active layers PM_6: 76 and PTB7-Th: PC_7, BM, resulting in the PCE respective 13.15 and 7.37% on PET substrate. The pristine WF of the electrode is 4.36 eV after coating of PEDOT: PSS and polyethyleneimine on the surface of the films; the MXene (Ti_3C_2Tx) is employed as the device's cathode and anode. The MXene film possesses good mechanical flexibility and durability under blending stress [209]. At the same time, hybrid MXene electrodes also achieve excellent properties. Tang group as well as Zhang group recently developed AG NW/MXene hybrid electrode with high electrical conductivity, high surface area, low sheet resistance, high transmittance, and high mechanical robustness [75].

Very recently, Son et al. 2023 reported a hybrid material of Mxene nanosheet and Ag nanowires with idiosyncratic properties; the MXene/AgNw/CPI (colorless polyamide) exhibits significantly improved conductivity and transmittance compared with positive Mxene sheet resistance R_{sh} (13.08 Ω sq^{-1}) transmittance 79% at 550 nm. And lower work function (− 4.67 eV) with a higher PCE 10.3% J_{oc} of 22.3 mA cm^{-2} was reported. The results show great potential as a flexible transparent electrode [210]. Similarly, Kim et al., first reported high-efficiency Mxene (FTC) flexible transparent conducting electrode based on a multilayer electrode fabricated by spine and sputtering coating technology, the results show both binary and ternary active layers examined by respective hybrid electrode Ag@ MAM, Ni@ MAM, and Zn@ MAM PCE. The electrode in binary active layer (PBDB-T-2F: Yb) shows 8.44, 10.42, and 11.78%, than the PCE of ternary active layer (PBDB:T:Yc: $PC_{71}BM$) which shows 10.68, 11.92, and 13.86%. The Zn@ MAM is highly efficient with low sheets resistance Rsh (9% Ω sq^{-1}) and passed the 2000 cycles bending test. These results indicate that the hybrid layers of the Mxene electrode give great attention to large area flexible OSCs via potential fabrication methods and long-term stability [211] (Fig. 10).

4 Properties

The intriguing properties of devices based on organic semiconductors make them highly suitable for commercial applications, enabling large area fabrication and versatile functionalities, all while keeping costs low. It includes maximum to minimal sized devices, its all possible only through the exclusive properties of organic materials. To develop a comprehensive understanding of flexible organic photovoltaics, one must first grasp the fundamental properties of OSCs and then examine the specific attributes that enable mechanical flexibility, a critical aspect of these devices.

4.1 Fundamental Properties of OSCs

OSCs are premeditated to absorb sunlight and translate it into electrical energy thus a broad absorption spectrum for efficient light capture across various wavelengths to improve energy conversion. The performance of organic photovoltaics (OPVs) is primarily assessed by higher values of **power conversion efficiency (PCE)**, **open-circuit voltage (Voc)**, **short-circuit current density (Jsc)**, and **fill factor (FF)**, as

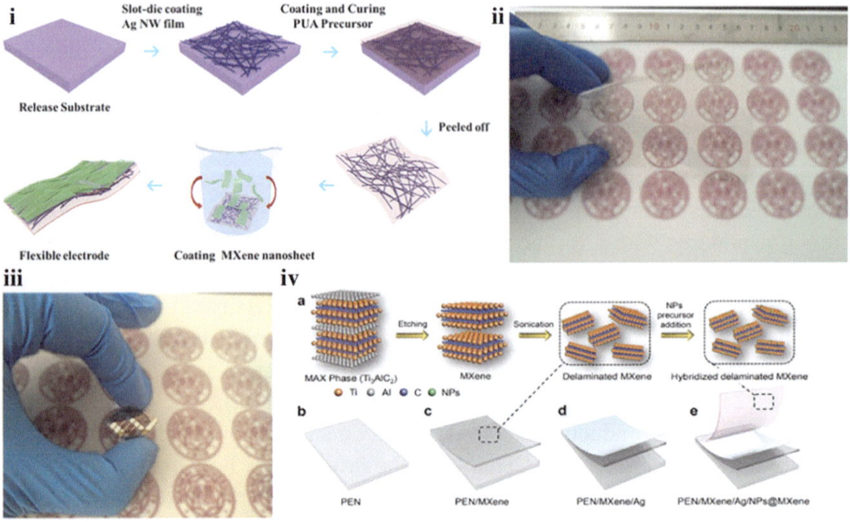

Fig. 10 Schematic representation of synthetic process and flexible photography's of flexible Mxene. (**i**) Fabrication Process of MXene-Based Flexible Transparent Electrode. (**ii**) Optical images of the optimized MXene/AgNW-PUA films. (**iii**) Optical photograph of flexible PSCs. Reproduced with permission [75] Copyright 2019, American Chemical Society. (**iv** Synthesis process of liquid exfoliation and delamination utilized for fabrication of a multilayer electrode. Step-by-step protocol for the fabrication of multilayer electrode. Reproduced with permission [211] Copyright 2023, Wiley-VCH GmbH

these metrics directly influence the efficiency and effectiveness of energy conversion. Device should be durable and stable over a long time and is possible with organic materials which can be tailor made to absorb specific wavelengths of light, influencing the range and efficiency of light absorption for high mobility of charges.

4.2 *Properties Essential for Flexibility*

The main property for hold the flexibility is the mechanical properties. Personified in the molecular structure are the reestablishing forces of covalent bonds, but also the intramolecular forces established in electrostatic interactions between distal units and dihedral potential energies between adjacent ones. The structure of OSCs must be intrinsically flexible to withstand repeated bending, stretching, or folding without important performance degradation. Flexibility requires ultrathin layers in both the active layer and electrodes to decrease stiffness while upholding efficiency (Fig. 11). Instead, the properties that seem to be of amplified importance in thin-film organic semiconductors are low tensile modulus in the elastic regime (Thin-film deposition techniques, such as roll-to-roll processing, are usually laboring to attain this). Flexibility allows for bending, while stretchability enables the material to

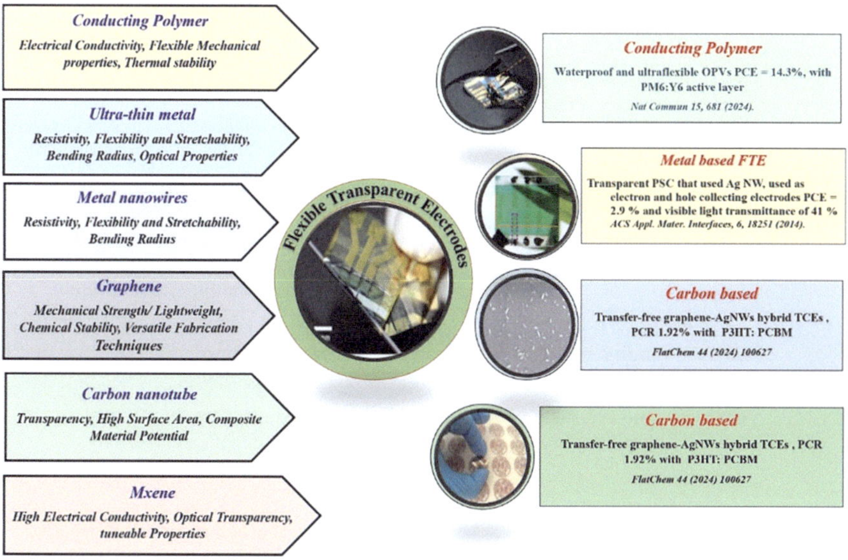

Fig. 11 Properties and performance metrics of flexible metal electrodes used in organic photovoltaic (OPV) devices. Reproduced with permission [76] Copyright 2020, Elsevier. Reproduced with permission [212] Copyright 2024, Nature. Reproduced with permission [213] Copyright 2014, American Chemical Society. Reproduced with permission [81] Copyright 2024, Elsevier. Reproduced with permission [75]. Copyright 2019, American Chemical Society

endure strain. This often involves using materials with elastic properties or including specific microstructures to dissipate strain effectively. In practical applications, OSCs are typically laminated onto flexible substrates, such as PET or PEN. Strong adhesion to these substrates is crucial for maintaining device integrity during flexing. To improve durability, flexible OSCs are frequently encapsulated with protective layers that protector against environmental influences while outstanding flexible. Additionally, according to the specific application, some more properties will be highlighted, i.e., for a wearable device washability, stretchability, and breathability should be optimal. According to the method of preparation the active layer molecules should have desired properties like solubility, good film forming nature, stability, and high conductivity, etc. The essential properties FOPVs, along with their impact on achieving high performance, are systematically outlined in a Table 1.

5 Stretchable Organic Solar Cell

Throughout the past few decades, stretchable electronics have drawn a lot of attention from both academia and industry because they can withstand repeated stretching and severe deformations without losing their electrical characteristics [214–216].

Table 1 List of flexible organic photovoltaic (FOPV) properties and their optimal relationship to performance

S. No.	Properties	Optimal relationship to performance
1	**Sheet resistance**	**Low**—This is crucial to prevent reductions in the fill factor or current density, particularly in large area organic solar cells (OSCs)
2	**Wettability**	**High—**This is particularly significant for the fabrication process, as good wettability permits for better uniformity, enhanced charge transport, and improved device performance
3	**Surface roughness**	**Low**—Ensure strong adhesion between the substrate and upper layers
4	**Cohesion and adhesion properties**	**High**—Cohesion within the material guarantees the strength and integrity of individual layers, preventing cracking or degradation
5	**Tensile strength**	**High—**Desirable and ensures that the device can endure mechanical stress, such as bending, stretching, or folding, without breaking or cracking
6	**Elastic or Young's modulus**	**Moderate to Low**—Generally preferred. While high stiffness can prevent bending or stretching, it can lead to brittleness, especially in FOPVs
7	**Crack onset strain**	**High—**It indicates that the material can endure significant deformation before cracks begin to form
8	**Transmittance**	**High**—This ensures that the incident light reaches the active layer without significant optical losses, thereby improving the overall device efficiency

Stretchable OPV devices fall into one of two categories: intrinsically or structurally stretchable OPVs. The former uses structural engineering to achieve stretchability and consists of fiber-shaped devices as well as ultrathin devices with buckling structures [78, 217]. Stretchability is even more significant than conformability and flexibility for some applications like wearable electronics and portable applications [218]. But these devices' limited flexibility prevents such OSCs from being successfully placed onto textiles or other wearables, which needs both stretchability and flexibility [219]. There are some strategies to follow for stretchability to electronic devices a) intrinsically stretchable materials should used for the device fabrication [220]. The use of pre-stretched elastomers in combination of the ultrathin device [119, 221–223]. The use of stretchable substrates and conductors including rigid-islands [224]. Mesh construction in which comfortable stretching, the devices' strain can be significantly decreased [225, 226]. The development of high-performance stretchable

OSCs is extremely demanding due to the upcoming wearable electronics. Moreover, buckle-based non-intrinsically stretchable OSCs and intrinsically stretchable OSCs are developed in order to scope with the environment for next generation wearable electronic devices.

5.1 Buckle-Based Non-intrinsically Stretchable Organic Solar Cells

Wearable electronic devices should be lightweight and able to tolerate different levels of stretching, compression, and deformability. Stretchability may be effectively achieved by employing ultrathin films having wrinkled architecture [227]. One of the most remarkable examples of stretchable OSCs was documented by Bao and colleagues in 2011 and they developed a stretchable OSCs that can withstand stresses of 27% by using mechanical buckling [223]. Bauer and his group in 2012, developed thick blend of P3HT with PC61BM around 200 nm of ultrathin flexible OSCs [224]. In a recent study, Chen et al. [228] found that combining the optical interference effect with an improved transparent Ag electrode's morphology enhanced the device's photovoltaic capabilities and it also achieved an efficiency of ~ 17% and a power-per-weight ratio of ~ 40 W g^{-1} (Fig. 12).

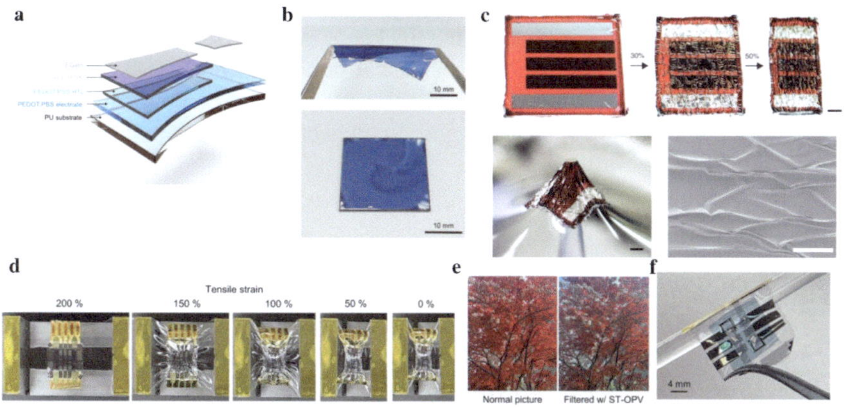

Fig. 12 Schematic representation of the ultrathin/stretchable OPV structure. a Schematic of the intrinsically stretchable organic photovoltaics. **b** Photographs showing elongation of the free-standing composite films with the active systems under tensile strain. Reproduced with permission [229] Copyright 2024, Nature. **c** Stretchable solar cells made by attaching the ultrathin solar cell to a pre-stretched elastomer and device attached to an elastomeric support under 3D conditions. Reproduced with permission [224] Copyright 2012, Springer Nature. **d** Optical images of PEI/Ag electrodes from pre-stretched condition. **e** Images captured without obstruction (left) and through parylene (right). **f** Photograph of ultraflexible ST-OPV. Reproduced with permission [79] Copyright 2023, Springer Nature

5.2 *Intrinsically Stretchable Organic Solar Cells*

Furthermore, an intrinsically stretchable (IT-SOSC) electronic materials have a comparable inherent stretchability to soft tissues, which can allow intrinsically stretchable electronic devices to mechanically mimic the human body. Stretchability of the devices is achieved by utilizing a structure based on a buckle or fiber, which also prevents tensile fracture of the functional layer material. This gadget has a number of drawbacks, including expensive, complex operations, and unstable structure. Further limiting its practical applicability is the fact that those devices were only extended in one direction [223, 230, 231]. Thus, for the purpose to eliminate the drawbacks, it is imperative to create naturally stretchable OSCs. An intrinsically stretchable organic photovoltaic system (OPV) comprises of many components, including the substrate, bottom transparent electrode, electron transport layer, photovoltaic layer, hole transport layer, and top electrode [45].

5.3 *Stretchable Substrates*

The development of substrates that are inherently flexible, or substrates that do not displays the limited mechanical compliance of current substrates like plastic-based PET, PEN, PDMS and PI, or stiff glassware and elastomer substrates of f-SOSCs, such as PDMS or polyurethane, TPU, acrylic-based 3 M acrylic foam tape and SEBS R1-10. A support substrate with a thickness ranging from 10 to 100 μm has a significant impact on the overall mechanical performance of an f-SOSC [232, 233]. Properties of stretchable substrate such as low elastic modulus, high optical transmittance, low surface roughness, good stretchability, and low moisture/oxygen permeability.

5.4 *Stretchable Electrodes*

For usage as flexible transparent electrodes, some potential materials are being investigated such as semiconducting polymers (i.e., PEDOT:PSS), metal-based inorganics (i.e., CuNWs/AgNWs), and patterned metal meshes/grids and carbon materials (i.e., CNT and 2D graphene) (Table 2).

Table 2 Summary of stretchability and strain values of recently reported IT-SOSC device structure

Year	Device structure	Stretchability (%)	Strain at PCE80% (%)	References
2024	Strem-AT/AI4083/PBQx-TCl:PYF-IT:N2200/PNDIT-F3N/EGaIn	51	12.6	[234]
2024	TPU/PH1000/ZnO/PM6:L8-BO/PNDITF3N-Br/liquid metal	63	14.6	[235]
2024	PU/PH1000/PEDOT:PSS HTL//Ter D18:Y6/EGaIn	52	14.2	[229]
2023	TPU/PH1000/AI4083/PM7-Thy10:L8-BO/PNDIT-F3N-Br/EGaIn	13.7	43	[236]
2023	TPU-AgNWs/PEDOT:PSS/PM6:BTP-eC9/Azo-AgNWs-AZO	10.9	9.9	[237]
2022	TPU/PH1000/AI4083/Active Layer/PNDIT-F3N-Br/EGaIn	12.7	31.6	[238]
2022	TPU/PH1000/AI4083/PM6-b-PDMS19k:L8-BO/PNDIT-F3N-Br/EGaIn	36.5	11.34	[239]
2021	Ag/PFN-Br/PM6:BTP-4C/PEDOT:PSS/patterned PDMS	10	~ 17	[240]

6 Coating and Printing Techniques

The higher efficiency of OSCs was generally developed by the spin-coating technique. The primary benefit of the spin-coating technique is its ease of use in producing an organic active layer with excellent performance, distinct compositions, and well-defined thicknesses. Even though spin-cast lab-scaling devices exhibit good characteristics, this technology has not yet been beneficial for upscale manufacturing [241].

6.1 Roll-to-Roll Printing

Roll-to-Roll (R2R) printing is used for fabricating flexible OPV devices on a large scale. It enables the high-speed continuous processing of flexible substrates in roll form, including metal or plastic foils [242]. Spin coating is a favorable and general technique that is often used in laboratories but R2R technique cannot be used because of its solution wasted and quick solvent evaporation. Large-scale deposition techniques are essential for the upscaling of OPVs such as inkjet printing, blade coating/

doctor-blade, screen printing, spray coating, slot die coating, and gravure printing. As a result, researchers must concentrate on developing more scalable coating techniques for OSCs, particularly those that can be used for R2R [243].

6.2 *Blade Coating/Doctor-Blade*

The blade coating technique was potentially used to fabricate large-scale films and also used as a scale-testing printing in laboratories. The factors such as viscosity, ink concentration, substrate temperature, blading speed, and surface energy can all have an impact on the quality of the film [244]. The process of blade coating is shown in Fig. 13a. The method of blade coating approach has been used to achieve an efficiency of ~ 17% on the small-area device with little material waste it shows a lower value compared to the spin coating. Min et al. group developed PM6:Y6-based LBL devices by blade coating technique and obtained a ~ 16% higher efficiency in a small area (around 0.04 cm^{-1}) [245].

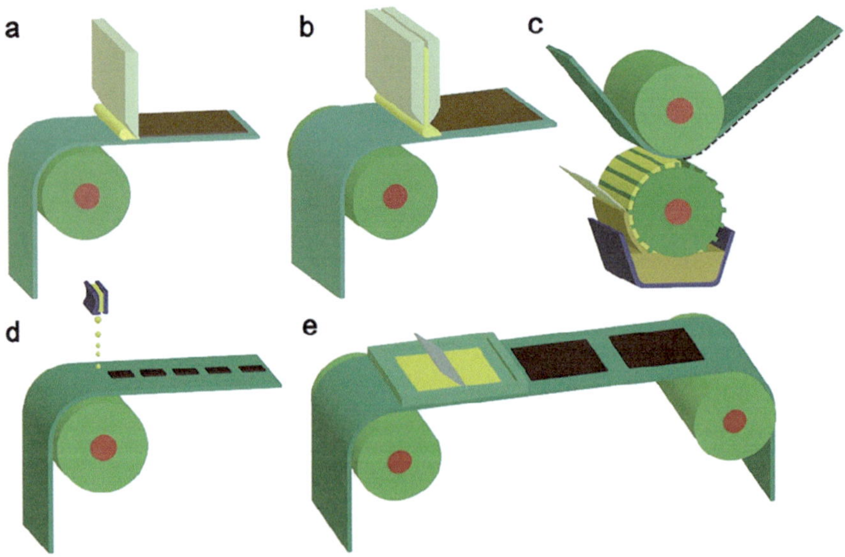

Fig. 13 Schematic representations of scalable solution deposition methods for large area flexible OSCs that are compatible with roll-to-roll manufacturing. **a** Blade coating. **b** Slot die coating. **c** Gravure printing. **d** Inkjet printing. **e** Screen printing. Reproduced with permission [32] Copyright 2021, Springer Nature

6.3 Slot Die Coating

The slot die coating method is an alternate printing process that was shown to be the best methodology for scaling large area OSCs by using the meniscus between the substrates and the slot die head [242]. During this process the ink is repeatedly pumped through the head of the slot die which is similar to the applicator of the doctor blade. The meniscus creation is also changed by capillary force and the solution viscosity between substrate and applicator. The substrate or slot die head speed as well as the ink feeding rate may be accurately adjusted to specify the film's thickness and the process of slot die coating is shown in Fig. 13b. In 2021, Zang et al. [246] developed PM6:Y6 films with two different solvents such as chlorobenzene and chloroform. The PM6:Y6 film shows a PCE of up to 14.42%. Recently Yang Group [243] developed 30 cm^{-1} large area OSC modules by slot die coating which showed ~ 12% of PCE and exceptional stability of storage exceeding 250 days.

6.4 Gravure Printing

Gravure printing is a technique for producing higher-resolution patterns and it consists of two cylinders a gravure cylinder and an impression cylinder [247]. During the process of printing, the gravure cylinder creates patterns, and the cylinder regulates the essential area for ink transfer and the process of gravure printing is shown in Fig. 13c. After filling the patterns with ink, the extra ink is removed by a gravure cylinder by rotating it with a blade. Meanwhile, by making contact with the impression cylinder, the substrate forms the pattern and completes the ink transfer process. The thickness of the film is managed depth of the mesh hole in the gravure cylinder [248]. The inks usually have low viscosity to help printing at a high speed of over 1000 m/min, and a low boiling point for fast drying on wet film [249].

6.5 Inkjet Printing

Inkjet printing is an electronically operated contact-less printing method that expels the ink in finely regulated droplet trajectory with the advantage of excellent resolution, no material loss, and patternable preparation [250]. Furthermore, heating or mechanical compression can be used to precisely adjust the size of the droplets. Additives such as 1,8-octanedithiol, 1-chloronaphthalene, *o*-dichlorobenzene are used as the solvents in inkjet printing because of their high boiling points [251]. The process of inkjet printing is shown in Fig. 13d. To prepare the wet film, the printing head is moved transversely as the substrate moves forward [252]. To achieve the thickness of the printed film ($1-10$ μm), the ink should have high viscosity and the solvent should have low volatility. Corzo group fabricated a PTB7-Th: IEICO-4F system was prepared by inkjet printing which showed a result of PCE up to 12.4% for opaque devices and 9.5% for semitransparent devices [253]. Chen et al. have developed a device of PBDB-T-2F: BTP-BO-4Cl with non-fullerene OSCs for optimization of the temperature and morphology exciton dissociation, effective suppression of molecular aggregation. This showed the result of highest value of PCE 13.09% [247].

6.6 Screen Printing

Screen printing is a versatile process that allows full 2D printing of films and it requires high-viscosity ink for high-conductivity electrodes and reproducibility is affected by paste properties [254] as shown in Fig. 13e. Screen printing can be classified into two methods such as flatbed screen printing and rotary screen printing [255]. In flatbed screen printing, the ink is applied to the screen continuously, followed by the screen's contact with the substrate's top and the movement of a squeegee over it. To create a particular design, the ink is shear thin and is to be deposited on the substrate. Rotary screen printing has a greater printing speed (over 100 m min^{1}) and higher resolution than flatbed screen printing. In rotary screen printing, the screen includes ink, squeegee, and patterns to be assembled as a cylinder, which rotates with the substrate.

7 Applications of Flexible OPVs

OPVs have grown significant attraction due to their incomparable properties [211, 256–260]. Recent advancements have led to portable and multiple applications constructed devices using FOPVs [178, 261–265]. It has proved that it can uphold a relentless level of PCE irrespective of the course of illumination. This made FOPVs can be employed for indoor organic photovoltaics [266–268] (IOPVs), building integrated organic photo voltaic [262, 263, 269] (BIPV), wearable electronics [8], agri-voltaics and sensor-based applications. Further, FOSCs can be coupled with bags, cloths, shelters, air crafts, and automobiles, etc. They can give fantastic power-per-weight ratios, with satisfactory manufacture costs compared with other common techniques [270].

7.1 IOPV

Unlike outdoor IOPV's should possess work with low power, predominantly function within artificial illuminations LED's, EL's. This will possess lesser interesting than natural sunlight this selection of active materials with narrower adsorption band

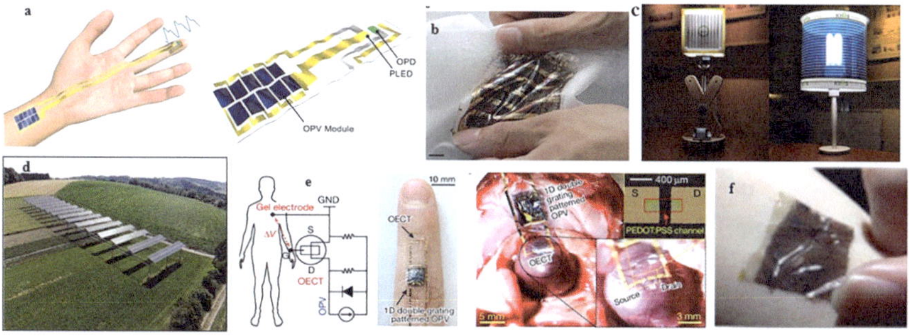

Fig. 14 Promising applications of OSCs. **a** Schematic diagram of the self-powered PPG sensor with PLED, OPD, and OPV module (right). Reproduced with permission [279] Copyright 2021, Springer Nature. **b** Stretchable and waterproof elastomer-coated OPV for washable electronic textile applications. Reproduced with permission [212] Copyright 2024, Springer Nature. **c** Photographs of indoor OSCs under LED illumination. Reproduced with permission [11] copyright 2021 Wiley–VCH GmbH. **d** Scientific agrivoltaics power plant of semitransparent OSCs in greenhouses. Reproduced with permission [280] Copyright 2020, Elsevier. **e** Photography of self-powered integrated electronics with flexible OSCs in biomedical devices. Reproduced with permission [9] Copyright 2018, Nature. **f** A real photo portrayed a fabricated stretchable device attached onto the skin. Reproduced with permission [281] Copyright 2018, Elsevier

and elevated (Fig. 14c). Absorbance coefficients become crucial [267, 271]. Selection of donor and acceptor should have energy level matching to work with indoor PV with high charge transfer capacities. Polymer donor and NF acceptors-based IOPV's or doing better than fullerene acceptor. In 2019, Hou et al. developed NFA semiconductor which has A-D-A structure. Several studies have shown that OPVs convert indoor illumination into energy more effectively than other conventional techniques [272]. In 2019, Yasuda and coworkers reported small molecule-based donor. Acceptor have resulted with PCE over 20% and white light illumination along with Voc of 4W [273]. Recently Yang et al. reported large bandgap NF's (1.75 eV). Modification in their side chains which influences in their morphological properties. This has reported over 29% PCE with various indoor lighting spectra which proved structure property relationship between molecular structure and IPV performance [267].

7.2 BIPV

Building integrated photovoltaics including photovoltaic windows, skylights and photovoltaic greenhouses need to employ semitransparent organic photovoltaic (ST-OPV) cells, with their transparency and aesthetic characteristics. They use solar system balancing both PCE and AVT. Also intrinsic brittleness is the disadvantage of Si-PVs, flexibility is the unique characteristics of OSCs compound. They have ability to convert sunlight into electricity while letting visible light transmissions allowing buildings to generate power while upholding natural lighting. Key parameters to evaluate the performance of ST-OPVs cells with AVT, LUE, CRI, and IPR [178].

Among various flexible transparent electrodes AgNWs are doing better without sacrificing their properties. Two research groups, Li et al. [274, 275] have reported employing multifunctional ST-OSCs for BIPVs simultaneously have the best performance in all the requires parameters. It has been shown that multifunctional active layer with ITO/PEDOT:PSS/PTBT-Th:I FIC-4F/BIS-FIMG/UTAg and ITO/PEDOT:PSS/TA/PM6: BO-4Cl/BiS-FIMG/Ag shown PCE upto 16.4%. Conversely, if Distributed Bragg Reflector (DBR), Ag (16 nm) has been integrated with (D-STOSCs) resulted with 32.07 AVT. Employing anti-reflective covering layer, 1D photonic crystals consisting of high and low refractive index films on devices will substantially improve ST-OPV properties to a high extent. Ultrathin Ag transparent electrodes are found to be suitable candidates to improve the flexibility and appearance of OPVs with AVT 86.4% [8].

7.3 Sensors and Self-powered Devices

Compared to traditional electronic sensors, flexible sensors (Fig. 14a) offer enhanced biocompatibility, real-time monitoring capabilities, and additional benefits. They are the utmost broadly employed in human health monitoring (Fig. 14e) and human–machine interaction, which converts physical motion, as well as compression, bending, tension, shear, stress, and strain, into electrical signals [276]. Self-powered devices will be highly required for telecommunication, transportation and advanced sensors [237, 277, 278] As a result, noteworthy research efforts have focused on developing flexible sensors that associate high sensitivity with mechanical flexibility. This has led to the emergence of FOSC-based sensors. In 2023, Dahiya et al. developed a self-powered e-Skin integrated touch sensors. e-Skin (Fig. 14f) that can together "sense" shadows of approaching objects and "feel" physical interactions. Along with pressure sensing (sensitivity of up to 0.26 kPa1 in the range of 1–10 kPa and the OPV can provide power density of higher than 0.32 μW cm^{-1} at 1500 lx [10]. FOSC based sensor-based integrated devices are highly applicable through wearable devices.

7.4 Wearable Electronics

For health monitoring and smart clothing wearable electronics will be the key (Fig. 14b). Two main requirements of WEs are highly flexible to face mechanical deformation and harvest energy from the environment while concurrently charging the device. Many fiber-based solar cells have been reported for wearable applications [52].

A textile-based OSC was established by integrating plasmonic nanostructures onto woven fabrics that are taking spatial opacity, irregularity and physical porosity which displayed an impressive PCE of 8.71% and waterproof OSC attached to the shirt obtained a PCE of 7.9% [282]. By employing polyurethane fibers in the Ag-ink has been deposited in electric device which can FTE be integrated with gloves and kneepads to assist as heat source [283]. For the low-cost integration of PV devices into textiles is fully spray-coated OSCs, which the process of coating and fabrication less expensive than other techniques. The FOSCs not only can easily incorporated into cloths but also can be woven into the textile to charge the smart watches [284]. For example tailor made smart watch produced by Kolon industries integrating OSC module for energy harvesting. Few observations and inferences can be made from the literature that Ag-wire-based FOSCs suggests that the suitable substitute counter electrode material for the CNT. Graphene electrodes will have high sheet resistance and low work function [8].

7.5 Agrivoltaics

Flexible ST-OSCs to a greenhouse construction (Fig. 14d) permits concurrent plant cultivation and electricity generation, thus dropping the green house energy demand [232]. The OSCs can thus mark plant growth, yield, power generation, and thermal load of the green houses. The ST-OSCs for agricultural applications must be compliant to the solar range in answer to plant photomorphogenesis [18, 285].

8 Future and Conclusion

Organic photovoltaics (OPVs) are on the cusp of revolutionizing the solar energy landscape, particularly for flexible, lightweight, and low-cost devices. Future progress in OPVs hinges on advancing material science, device architecture, and scalable manufacturing techniques. Non-fullerene acceptors and novel donor materials are likely to dominate, pushing power conversion efficiencies beyond 20%, while tandem OPV configurations could further enhance performance by broadening the absorption spectrum. The replacement of traditional electrodes, such as indium tin oxide (ITO), with alternatives like silver nanowires, graphene, or conductive polymers, will significantly reduce costs while improving flexibility and mechanical durability. Additionally, roll-to-roll processing and inkjet printing are expected to enable the large-scale production of OPVs at an economically competitive level. Stability remains a crucial challenge, as the intrinsic degradation of organic materials under environmental stress limits their operational lifetimes. Research into advanced encapsulation technologies, the development of intrinsically stable materials, and strategies for mitigating degradation mechanisms will be pivotal for ensuring long-term performance. The integration of OPVs into emerging markets, such as wearable electronics, smart packaging, and off-grid systems, represents a transformative opportunity, particularly in resource-constrained regions where low-cost and portable energy solutions are critical. Eventually, the future of OPVs lies in their ability to balance efficiency, cost, and durability while catering to a diverse range of applications. Multidisciplinary research, combining material innovation, advanced fabrication techniques, and sustainable practices, will be essential to unlocking the full potential of this technology. In conclusion, these challenges are addressed; OPVs could play a significant role in driving the transition to renewable energy, and meeting the growing demand for flexible, portable, and low-cost solar power solutions are critical.

Acknowledgements We wish to extend our gratitude to Dr. R. Prasanth, Department of Green Energy Technology, Pondicherry University, for offering us the opportunity and extending support to contribute to this book chapter. Dr. D. S. Sharada, M. Premkumar, J. Jayapratha, are thankful for the facilities and the support extended from Department of Green Energy Technology, Pondicherry Univeristy. Dr. B. Balambiga would wish to acknowledge all the support extended from the Department of Chemistry, Pachaiyappa's College, Pachaiyappa's Educational Trust, Chennai.

References

1. N. Kant, P. Singh, Review of next generation photovoltaic solar cell technology and comparative materialistic development. Mater. Today Proc. **56**, 3460–3470 (2022)
2. Y. Lin et al., Printed nonfullerene organic solar cells with the highest efficiency of 9.5%. Adv. Energy Mater. **8**(13), 1701942 (2018)
3. N. Shah et al., A review of third generation solar cells. Processes **11**(6), 1852 (2023)
4. G. Li, R. Zhu, Y. Yang, Polymer solar cells. Nat. Photonics **6**(3), 153–161 (2012)
5. S. Günes, H. Neugebauer, N.S. Sariciftci, Conjugated polymer-based organic solar cells. Chem. Rev. **107**(4), 1324–1338 (2007)
6. M. Riede, D. Spoltore, K. Leo, Organic solar cells—the path to commercial success. Adv. Energy Mater. **11**(1), 2002653 (2021)
7. Y. Li, Molecular design of photovoltaic materials for polymer solar cells: toward suitable electronic energy levels and broad absorption. Acc. Chem. Res. **45**(5), 723–733 (2012)
8. J. Wan, R. Zhang, Y. Li, Y. Li, Applications of organic solar cells in wearable electronics. Wearable Electron. **1**, 26–40 (2024)
9. S. Park et al., Self-powered ultra-flexible electronics via nano-grating-patterned organic photovoltaics. Nature **561**(7724), 516–521 (2018)
10. N.M. Nair, D. Shakthivel, K.M. Panidhara, V. Adiga, P.C. Ramamurthy, R. Dahiya, Self-powered e-skin based on integrated flexible organic photovoltaics and transparent touch sensors. Adv. Intell. Syst. **5**(10), 2300103 (2023)
11. M. Jahandar, S. Kim, D.C. Lim, Indoor organic photovoltaics for self-sustaining IoT devices: progress, challenges and practicalization. Chemsuschem **14**(17), 3449–3474 (2021)
12. X. Zhou, W. Cao, Flexible and stretchable carbon-based sensors and actuators for soft robots. Nanomaterials **13**(2), 316 (2023)
13. T. Qu et al., Biomimetic electrodes for flexible organic solar cells with efficiencies over 16%. Adv. Opt. Mater. **8**(17), 2000669 (2020)

14. Y. Hu, J. Wang, C. Yan, P. Cheng, The multifaceted potential applications of organic photovoltaics. Nat. Rev. Mater. **7**(11), 836–838 (2022)
15. J. Wang et al., Binary organic solar cells with 19.2% efficiency enabled by solid additive. Adv. Mater. **35**(25), 2301583 (2023)
16. O. Almora et al., Device performance of emerging photovoltaic materials (version 2). Adv. Energy Mater. **11**(48), 2102526 (2021)
17. A. Chakraborty et al., Photovoltaics for indoor energy harvesting. Nano Energy, 109932 (2024)
18. W. Song et al., Semi-transparent organic photovoltaics for agrivoltaic applications. Nano Energy, 108805 (2023)
19. H. Samadi, G. Ala, V. Lo Brano, P. Romano, F. Viola, Investigation of effective factors on vehicles integrated photovoltaic (VIPV) performance: a review. World Electr. Veh. J. **14**(6), 154 (2023)
20. C.W. Tang, Two-layer organic photovoltaic cell. Appl. Phys. Lett. **48**(2), 183–185 (1986)
21. D. Kearns, M. Calvin, Photovoltaic effect and photoconductivity in laminated organic systems. J. Chem. Phys. **29**(4), 950–951 (1958)
22. G. Yu, J. Gao, J.C. Hummelen, F. Wudl, A.J. Heeger, Polymer photovoltaic cells: enhanced efficiencies via a network of internal donor-acceptor heterojunctions. Science **270**(5243), 1789–1791 (1995)
23. P.W.M. Blom, V.D. Mihailetchi, L.J.A. Koster, D.E. Markov, Device physics of polymer: fullerene bulk heterojunction solar cells. Adv. Mater. **19**(12), 1551–1566 (2007)
24. J. Fu et al., Rational molecular and device design enables organic solar cells approaching 20% efficiency. Nat. Commun. **15**(1), 1830 (2024)
25. Z. Chen et al., 20.2% efficiency organic photovoltaics employing a π-extension quinoxaline-based acceptor with ordered arrangement. Adv. Mater. **36**(33), 2406690 (2024)
26. Z. Wang et al., Manipulating the macroscopic and microscopic morphology of large-area gravure-printed ZnO films for high-performance flexible organic solar cells. Energy Environ. Mater. **7**(2), e12592 (2024)
27. H. Wang et al., Large-area gravure-printed AgNWs electrode on water/oxygen barrier substrate for long-term stable large-area flexible organic solar cells. Chinese J. Chem. **42**(5), 478–484 (2024)
28. W. Wang et al., Indoor organic photovoltaic module with 30.6% efficiency for efficient wireless power transfer. Nano Energy, 109893 (2024)
29. C. Zhu, X. Wang, W. Liu, Y. Liu, X. Zhan, Organic interlayer materials for non-fullerene solar cells. Trends Chem. (2024)
30. M. Stephen, K. Genevičius, G. Juška, K. Arlauskas, R.C. Hiorns, Charge transport and its characterization using photo-CELIV in bulk heterojunction solar cells. Polym. Int. **66**(1), 13–25 (2017)
31. K. Fukuda, K. Yu, T. Someya, The future of flexible organic solar cells. Adv. Energy Mater. **10**(25), 2000765 (2020)
32. Y. Fu, Y. Huang, Y. Li, and Y. Li, Large-area flexible organic solar cells. NPJ Flex. Electron. **5**(1) (2021)
33. B.D. Choudhury et al., The photon absorber and interconnecting layers in multijunction organic solar cell. Sol. Energy **201**, 28–44 (2020)
34. J. Fu et al., 19.31% binary organic solar cell and low non-radiative recombination enabled by non-monotonic intermediate state transition. Nat. Commun. **14**(1), 1760 (2023)
35. D. Qian et al., Design, application, and morphology study of a new photovoltaic polymer with strong aggregation in solution state. Macromolecules **45**(24), 9611–9617 (2012)
36. Y. Li et al., Recent progress in organic solar cells: a review on materials from acceptor to donor. Molecules **27**(6), 1800 (2022)
37. S.M. Patel, K. Pal, P.N. Kumar, M. Deepa, D.S. Sharada, Design and synthesis of novel indole and carbazole based organic dyes for dye sensitized solar cells: theoretical studies by DFT/TDDFT. ChemistrySelect **3**(6), 1623–1628 (2018)

38. X. Li et al., Mapping the energy level alignment at donor/acceptor interfaces in non-fullerene organic solar cells. Nat. Commun. **13**(1), 2046 (2022)
39. V.V. Sharma, A. Landep, S.-Y. Lee, S.-J. Park, Y.-H. Kim, G.-H. Kim, Recent advances in polymeric and small molecule donor materials for Y6 based organic solar cells. Next Energy **2**, 100086 (2024)
40. G. Li et al., High-efficiency solution processable polymer photovoltaic cells by self-organization of polymer blends. Nat. Mater. **4**(11), 864–868 (2005)
41. M.M. Wienk et al., Efficient methano [70] fullerene/MDMO-PPV bulk heterojunction photovoltaic cells. Angew. Chemie Int. Ed. **42**(29), 3371–3375 (2003)
42. Y. Lin et al., An electron acceptor challenging fullerenes for efficient polymer solar cells. Adv. Mater. **27**(7), 1170–1174 (2015)
43. X. Li, P. Li, Z. Wu, D. Luo, H.-Y. Yu, Z.-H. Lu, Review and perspective of materials for flexible solar cells. Mater. Reports Energy **1**(1), 100001 (2021)
44. S. Li, Z. Li, X. Wan, Y. Chen, Recent progress in flexible organic solar cells. Escience **3**(1), 100085 (2023)
45. J. Qin et al., Recent progress in flexible and stretchable organic solar cells. Adv. Funct. Mater. **30**(36), 2002529 (2020)
46. L. Gao et al., All-polymer solar cells based on absorption-complementary polymer donor and acceptor with high power conversion efficiency of 8.27%. Adv. Mater. **28**(9), 1884–1890 (2015)
47. X. Zhan et al., A high-mobility electron-transport polymer with broad absorption and its use in field-effect transistors and all-polymer solar cells. J. Am. Chem. Soc. **129**(23), 7246–7247 (2007)
48. S.E. Shaheen, C.J. Brabec, N.S. Sariciftci, F. Padinger, T. Fromherz, J.C. Hummelen, 2.5% efficient organic plastic solar cells. Appl. Phys. Lett. **78**(6), 841–843 (2001)
49. G. Zhao, Y. He, Y. Li, 6.5% efficiency of polymer solar cells based on poly (3-hexylthiophene) and indene-C60 bisadduct by device optimization. Adv. Mater. **22**(39), 4355–4358 (2010)
50. S. Banerjee, S.K. Gupta, A. Singh, A. Garg, Buffer layers in inverted organic solar cells and their impact on the interface and device characteristics: an experimental and modeling analysis. Org. Electron. **37**, 228–238 (2016)
51. X. Xu, Q. Peng, Hole/electron transporting materials for nonfullerene organic solar cells. Chem. Eur. J. **28**(25), e202104453 (2022)
52. I. Ali et al., Advances in smart photovoltaic textiles. ACS Nano **18**(5), 3871–3915 (2024)
53. Y. Bai et al., Interfacial engineering and optical coupling for multicolored semitransparent inverted organic photovoltaics with a record efficiency of over 12%. J. Mater. Chem. A **7**(26), 15887–15894 (2019)
54. C. Hou, H. Yu, ZnO/Ti_3C2Tx monolayer electron transport layers with enhanced conductivity for highly efficient inverted polymer solar cells. Chem. Eng. J. **407**, 127192 (2021)
55. Z. Zheng et al., A highly efficient non-fullerene organic solar cell with a fill factor over 0.80 enabled by a fine-tuned hole-transporting layer. Adv. Mater. **30**(34), 1801801 (2018)
56. Y. Wang, J. Han, L. Cai, N. Li, Z. Li, F. Zhu, Efficient and stable operation of nonfullerene organic solar cells: retaining a high built-in potential. J. Mater. Chem. A **8**(40), 21255–21264 (2020)
57. N.-W. Teng et al., Highly efficient nonfullerene organic photovoltaic devices with 10% power conversion efficiency enabled by a fine-tuned and solution-processed hole-transporting layer. Sol. RRL **4**(9), 2000223 (2020)
58. R. Ma et al., Rational anode engineering enables progresses for different types of organic solar cells. Adv. Energy Mater. **11**(23), 2100492 (2021)
59. Y. Sun et al., Flexible organic photovoltaics based on water-processed silver nanowire electrodes. Nat. Electron. **2**(11), 513–520 (2019)
60. U. Shankar, A. Bandyopadhyay, Plastics in high chemical resistant applications (2022)
61. D. Lee, D.-Y. Youn, Z. Luo, I.-D. Kim, Highly flexible transparent electrodes using a silver nanowires-embedded colorless polyimide film via chemical modification. RSC Adv. **6**(36), 30331–30336 (2016)

62. J.-W. Lim et al., Mechanical integrity of flexible Ag nanowire network electrodes coated on colorless PI substrates for flexible organic solar cells. Sol. Energy Mater. Sol. Cells **105**, 69–76 (2012)
63. K.-W. Seo et al., Simple brush painted Ag nanowire network on graphene sheets for flexible organic solar cells. J. Vac. Sci. Technol. A **32**(6) (2014)
64. J.B. Brolly, D.I. Bower, I.M. Ward, Diffusion and sorption of CO_2 in poly (ethylene terephthalate) and poly (ethylene naphthalate). J. Polym. Sci. Part B Polym. Phys. **34**(4), 769–780 (1996)
65. J. Clatot, G. Campet, A. Zeinert, C. Labrugère, M. Nistor, A. Rougier, Low temperature Si doped ZnO thin films for transparent conducting oxides. Sol. Energy Mater. Sol. Cells **95**(8), 2357–2362 (2011)
66. S.K. Sethi, U. Shankar, G. Manik, Fabrication and characterization of non-fluoro based transparent easy-clean coating formulations optimized from molecular dynamics simulation. Prog. Org. Coatings **136**, 105306 (2019)
67. S.K. Sethi, L. Soni, U. Shankar, R.P. Chauhan, G. Manik, A molecular dynamics simulation study to investigate poly (vinyl acetate)-poly (dimethyl siloxane) based easy-clean coating: an insight into the surface behavior and substrate interaction. J. Mol. Struct. **1202**, 127342 (2020)
68. T. Akter, W.S. Kim, Reversibly stretchable transparent conductive coatings of spray-deposited silver nanowires. ACS Appl. Mater. Interfaces **4**(4), 1855–1859 (2012)
69. X. Yang et al., Large-scale stretchable semiembedded copper nanowire transparent conductive films by an electrospinning template. ACS Appl. Mater. Interfaces **9**(31), 26468–26475 (2017)
70. H. Park, S. Chang, X. Zhou, J. Kong, T. Palacios, S. Gradečak, Flexible graphene electrode-based organic photovoltaics with record-high efficiency. Nano Lett. **14**(9), 5148–5154 (2014)
71. I. Jeon et al., Direct and dry deposited single-walled carbon nanotube films doped with MoOx as electron-blocking transparent electrodes for flexible organic solar cells. J. Am. Chem. Soc. **137**(25), 7982–7985 (2015)
72. I. Jeon, C. Delacou, A. Kaskela, E.I. Kauppinen, S. Maruyama, Y. Matsuo, Metal-electrode-free window-like organic solar cells with p-doped carbon nanotube thin-film electrodes. Sci. Rep. **6**(1), 31348 (2016)
73. D.H. Shin, C.W. Jang, H.S. Lee, S.W. Seo, S.-H. Choi, Semitransparent flexible organic solar cells employing doped-graphene layers as anode and cathode electrodes. ACS Appl. Mater. Interfaces **10**(4), 3596–3601 (2018)
74. G. Zhao et al., Nitrogen-mediated growth of silver nanocrystals to form ultrathin, high-purity silver-film electrodes with broad band transparency for solar cells. ACS Appl. Mater. Interfaces **10**(47), 40901–40910 (2018)
75. H. Tang, H. Feng, H. Wang, X. Wan, J. Liang, Y. Chen, Highly conducting MXene–silver nanowire transparent electrodes for flexible organic solar cells. ACS Appl. Mater. Interfaces **11**(28), 25330–25337 (2019). https://doi.org/10.1021/acsami.9b04113
76. W. Huang et al., Efficient and mechanically robust ultraflexible organic solar cells based on mixed acceptors. Joule **4**(1), 128–141 (2020)
77. J. Wan et al., Solution-processed transparent conducting electrodes for flexible organic solar cells with 16.61% efficiency. Nano-micro Lett. **13**, 1–14 (2021)
78. S. Xiong et al., Ultrathin and efficient organic photovoltaics with enhanced air stability by suppression of zinc element diffusion. Adv. Sci. **9**(8), 2105288 (2022)
79. H. Lee et al., Ultra-flexible semitransparent organic photovoltaics. npj Flex. Electron. **7**(1), 27 (2023)
80. J.J. Kim et al., Tri-system integration in metal-oxide nanocomposites via in-situ solution-processed method for ultrathin flexible transparent electrodes. Nat. Commun. **15**(1), 2070 (2024). https://doi.org/10.1038/s41467-024-46243-6
81. M.S.A. Kamel, M. Oelgemöller, M.V. Jacob, Influence of deposition temperature and hydrogen on sustainable and transfer-free graphene transparent electrode for organic solar cells. FlatChem **44**, 100627 (2024)

82. W. Song, Q. Ye, Z. Chen, J. Ge, L. Xie, Z. Ge, Advances in stretchable organic photovoltaics: flexible transparent electrodes and deformable active layer design. Adv. Mater. **36**(37), 2311170 (2024). https://doi.org/10.1002/adma.202311170
83. X. Fan, Doping and design of flexible transparent electrodes for high-performance flexible organic solar cells: recent advances and perspectives. Adv. Funct. Mater. **31**(8), 2009399 (2021)
84. Y. Sun et al., Flexible high-performance and solution-processed organic photovoltaics with robust mechanical stability. Adv. Funct. Mater. **31**(16), 2010000 (2021)
85. F. Qiao, H. Chu, Y. Xie, Z. Weng, Recent progress of transparent conductive electrodes in the construction of efficient flexible organic solar cells. Int. J. Energy Res. **46**(4), 4071–4087 (2022)
86. K. Saranya, M. Rameez, A. Subramania, Developments in conducting polymer based counter electrodes for dye-sensitized solar cells—an overview. Eur. Polym. J. **66**, 207–227 (2015)
87. Y. Xu et al., Recent progress of electrode materials for flexible perovskite solar cells. Nano-Micro Lett. **14**(1), 117 (2022)
88. T. Niederhoffer, A. Vanhoestenberghe, H.T. Lancashire, Methods of poly (3, 4)-ethylenedioxithiophene (PEDOT) electrodeposition on metal electrodes for neural stimulation and recording. J. Neural Eng. **20**(1), 11002 (2023)
89. X. Guo, A. Facchetti, The journey of conducting polymers from discovery to application. Nat. Mater. **19**(9), 922–928 (2020)
90. V. Van Tran, S. Lee, D. Lee, T.-H. Le, Recent developments and implementations of conductive polymer-based flexible devices in sensing applications. Polymers (Basel) **14**(18), 3730 (2022)
91. N. Singh, U. Riaz, Recent trends on synthetic approaches and application studies of conducting polymers and copolymers: a review. Polym. Bull. **79**(12), 10377–10408 (2022)
92. C.K. Chiang et al., Electrical conductivity in doped polyacetylene. Phys. Rev. Lett. **40**(22), 1472 (1978)
93. L. Zhang et al., The role of mineral acid doping of PEDOT: PSS and its application in organic photovoltaics. Adv. Electron. Mater. **6**(1), 1900648 (2020)
94. X. Fan et al., PEDOT: PSS for flexible and stretchable electronics: modifications, strategies, and applications. Adv. Sci. **6**(19), 1900813 (2019)
95. R. Chen et al., Sequential solution polymerization of poly (3, 4-ethylenedioxythiophene) using V2O5 as oxidant for flexible touch sensors. IScience **12**, 66–75 (2019)
96. Q. Qiao, *Organic Solar Cells: Materials, Devices, Interfaces, and Modeling*. CRC Press (2017)
97. F. Zhang, M. Johansson, M.R. Andersson, J.C. Hummelen, O. Inganäs, Polymer photovoltaic cells with conducting polymer anodes. Adv. Mater. **14**(9), 662–665 (2002)
98. T. Aernouts et al., Printable anodes for flexible organic solar cell modules. Thin Solid Films **451**, 22–25 (2004)
99. S. Na, S. Kim, J. Jo, D. Kim, Efficient and flexible ITO-free organic solar cells using highly conductive polymer anodes. Adv. Mater. **20**(21), 4061–4067 (2008)
100. Y.H. Kim, C. Sachse, M.L. Machala, C. May, L. Müller-Meskamp, K. Leo, Highly conductive PEDOT: PSS electrode with optimized solvent and thermal post-treatment for ITO-free organic solar cells. Adv. Funct. Mater. **21**(6), 1076–1081 (2011)
101. D.A. Mengistie, P.-C. Wang, C.-W. Chu, Effect of molecular weight of additives on the conductivity of PEDOT: PSS and efficiency for ITO-free organic solar cells. J. Mater. Chem. A **1**(34), 9907–9915 (2013)
102. H. Park et al., Retarding ion exchange between conducting polymers and ionic liquids for printable top electrodes in semitransparent organic solar cells. ACS Appl. Mater. Interfaces **12**(2), 2276–2284 (2019)
103. W. Song et al., Foldable semitransparent organic solar cells for photovoltaic and photosynthesis. Adv. Energy Mater. **10**(15), 2000136 (2020)
104. W. Song et al., Over 14% efficiency folding-flexible ITO-free organic solar cells enabled by eco-friendly acid-processed electrodes. iScience **23**(4) (2020)

105. Y. Wang et al., A highly stretchable, transparent, and conductive polymer. Sci. Adv. **3**(3), e1602076 (2017)
106. X. Hu et al., A mechanically robust conducting polymer network electrode for efficient flexible perovskite solar cells. Joule **3**(9), 2205–2218 (2019)
107. X. Zhang, B. Zhang, X. Ouyang, L. Chen, H. Wu, Polymer solar cells employing water-soluble polypyrrole nanoparticles as dopants of PEDOT: PSS with enhanced efficiency and stability. J. Phys. Chem. C **121**(34), 18378–18384 (2017)
108. Y. Xia, K. Sun, J. Ouyang, Solution-processed metallic conducting polymer films as transparent electrode of optoelectronic devices. Adv. Mater. **18**(24), 2436–2440 (2012)
109. N. Kim et al., Highly conductive PEDOT: PSS nanofibrils induced by solution-processed crystallization. Adv. Mater. **26**(14), 2268–2272 (2014)
110. K. Sun, P. Li, Y. Xia, J. Chang, J. Ouyang, Transparent conductive oxide-free perovskite solar cells with PEDOT: PSS as transparent electrode. ACS Appl. Mater. Interfaces **7**(28), 15314–15320 (2015)
111. W. Meng et al., Conductivity enhancement of PEDOT: PSS films via phosphoric acid treatment for flexible all-plastic solar cells. ACS Appl. Mater. Interfaces **7**(25), 14089–14094 (2015)
112. W. Song et al., All-solution-processed metal-oxide-free flexible organic solar cells with over 10% efficiency. Adv. Mater. **30**(26), 1800075 (2018)
113. J. Wan et al., All annealing-free solution-processed highly flexible organic solar cells. J. Mater. Chem. A **9**(9), 5425–5433 (2021)
114. N. Kim, H. Kang, J. Lee, S. Kee, S.H. Lee, K. Lee, Highly conductive all-plastic electrodes fabricated using a novel chemically controlled transfer-printing method. Adv. Mater. **27**(14), 2317–2323 (2015)
115. J. Wan et al., Metal oxide-free flexible organic solar cells with 0.1 M perchloric acid sprayed polymeric anodes. J. Mater. Chem. A **8**(40), 21007–21015 (2020)
116. E. Bihar et al., Fully inkjet-printed, ultrathin and conformable organic photovoltaics as power source based on cross-linked PEDOT: PSS electrodes. Adv. Mater. Technol. **5**(8), 2000226 (2020)
117. X. Wang et al., Flexible ITO-free organic photovoltaics on ultra-thin flexible glass substrates with high efficiency and improved stability. Sol. RRL **3**(4), 1800286 (2019)
118. X. Meng et al., A general approach for lab-to-manufacturing translation on flexible organic solar cells. Adv. Mater. **31**(41), 1903649 (2019)
119. A. Kasry, M.A. Kuroda, G.J. Martyna, G.S. Tulevski, A.A. Bol, Chemical doping of large-area stacked graphene films for use as transparent, conducting electrodes. ACS Nano **4**(7), 3839–3844 (2010)
120. H. Zhu, Y. Shen, Y. Li, J. Tang, Recent advances in flexible and wearable organic optoelectronic devices. J. Semicond. **39**(1), 11011 (2018)
121. J.-Y. Lee, S.T. Connor, Y. Cui, P. Peumans, Solution-processed metal nanowire mesh transparent electrodes. Nano Lett. **8**(2), 689–692 (2008)
122. J.H. Seo et al., Cold isostatic-pressured silver nanowire electrodes for flexible organic solar cells via room-temperature processes. Adv. Mater. **29**(30), 1701479 (2017)
123. Y. Wang et al., Polyimide/ZnO composite cooperatively crosslinked by Zn2+ Salt-Bondings and hydrogen bondings for ultraflexible organic solar cells. Chem. Eng. J. **451**, 138612 (2023). https://doi.org/10.1016/j.cej.2022.138612
124. G. Zeng et al., Realizing 17.5% efficiency flexible organic solar cells via atomic-level chemical welding of silver nanowire electrodes. J. Am. Chem. Soc. **144**(19), 8658–8668 (2022). https://doi.org/10.1021/jacs.2c01503
125. H. Dong et al., A flexible and thin graphene/silver nanowires/polymer hybrid transparent electrode for optoelectronic devices. ACS Appl. Mater. Interfaces **8**(45), 31212–31221 (2016)
126. I.E. Stewart et al., Solution-processed copper–nickel nanowire anodes for organic solar cells. Nanoscale **6**(11), 5980–5988 (2014)
127. C. Sachse, N. Weiß, N. Gaponik, L. Müller-Meskamp, A. Eychmüller, K. Leo, ITO-free, small-molecule organic solar cells on spray-coated copper-nanowire-based transparent electrodes. Adv. Energy Mater. **4**(2), 1300737 (2014)

128. J. Chen et al., Solution-processed copper nanowire flexible transparent electrodes with PEDOT: PSS as binder, protector and oxide-layer scavenger for polymer solar cells. Nano Res. **8**, 1017–1025 (2015)
129. M.S.R. Khan, A. Reza, Optical and electrical properties of optimized thin gold films as top layer of MIS solar cells. Appl. Phys. A **54**, 204–207 (1992)
130. S. Lee, Y. Lee, J. Park, D. Choi, Stitchable organic photovoltaic cells with textile electrodes. Nano Energy **9**, 88–93 (2014)
131. A. Ehrmann, T. Blachowicz, Recent coating materials for textile-based solar cells. AIMS Mater. Sci. **6**(2) (2019)
132. D.E. Carlson, C.R. Wronski, Amorphous silicon solar cell. Appl. Phys. Lett. **28**(11), 671–673 (1976)
133. D. Bryant et al., A transparent conductive adhesive laminate electrode for high-efficiency organic-inorganic lead halide perovskite solar cells. Adv. Mater. **26**(44), 7499–7504 (2014)
134. D. Gu, S.K. Dey, P. Majhi, Effective work function of Pt, Pd, and Re on atomic layer deposited HfO2. Appl. Phys. Lett. **89**(8) (2006)
135. Z. Pang, Y. Zhao, Y. Duan, J. Duan, Q. Tang, L. Yu, Well-aligned NiPt alloy counter electrodes for high-efficiency dye-sensitized solar cell applications. J. energy Chem. **30**, 49–56 (2019)
136. J. Huang et al., Seed-layer-free growth of ultra-thin Ag transparent conductive films imparts flexibility to polymer solar cells. Sol. Energy Mater. Sol. Cells **184**, 73–81 (2018)
137. X. Lu, Y. Zhang, Z. Zheng, Metal-based flexible transparent electrodes: challenges and recent advances. Adv. Electron. Mater. **7**(5), 2001121 (2021). https://doi.org/10.1002/aelm.202001121
138. M.-G. Kang, L.J. Guo, M.-S. Kim, J. Kim, Organic solar cells using nanoimprinted transparent metal electrodes. Adv. Mater. **20** (2008)
139. J. Zou, C. Li, C. Chang, H. Yip, A.K. Jen, Interfacial engineering of ultrathin metal film transparent electrode for flexible organic photovoltaic cells. Adv. Mater. **26**(22), 3618–3623 (2014)
140. N. Formica, D.S. Ghosh, T.L. Chen, C. Eickhoff, I. Bruder, V. Pruneri, Highly stable Ag–Ni based transparent electrodes on PET substrates for flexible organic solar cells. Sol. energy Mater. Sol. cells **107**, 63–68 (2012)
141. X. Wen, F. Yi, J. Ji, Y. Bi, W. Zhang, W. Wang, S-1805/Ag/Au hybrid transparent electrodes for ITO-free flexible organic photovoltaics. Chem. Res. Chinese Univ. **35**(3), 509–513 (2019)
142. X. Guo, X. Liu, F. Lin, H. Li, Y. Fan, N. Zhang, Highly conductive transparent organic electrodes with multilayer structures for rigid and flexible optoelectronics. Sci. Rep. **5**(1), 10569 (2015)
143. W. Wang et al., Transparent ultrathin oxygen-doped silver electrodes for flexible organic solar cells. Adv. Funct. Mater. **24**(11), 1551–1561 (2014)
144. G. Zhao et al., Stable ultrathin partially oxidized copper film electrode for highly efficient flexible solar cells. Nat. Commun. **6**(1), 8830 (2015)
145. K. Li et al., Full-solution processed flexible organic solar cells using low-cost printable copper electrodes. Adv. Mater. **26**(42), 7271–7278 (2014)
146. D.S. Ghosh, R. Betancur, T.L. Chen, V. Pruneri, J. Martorell, Semi-transparent metal electrode of Cu–Ni as a replacement of an ITO in organic photovoltaic cells. Sol. Energy Mater. Sol. Cells **95**(4), 1228–1231 (2011)
147. H.B. Lee, W.-Y. Jin, M.M. Ovhal, N. Kumar, J.-W. Kang, Flexible transparent conducting electrodes based on metal meshes for organic optoelectronic device applications: a review. J. Mater. Chem. C **7**(5), 1087–1110 (2019)
148. M. Hösel, R.R. Søndergaard, M. Jørgensen, F.C. Krebs, Fast inline roll-to-roll printing for indium-tin-oxide-free polymer solar cells using automatic registration. Energy Technol. **1**(1), 102–107 (2013)
149. J.H. Park et al., Flexible and transparent metallic grid electrodes prepared by evaporative assembly. ACS Appl. Mater. Interfaces **6**(15), 12380–12387 (2014)
150. M. Song et al., ITO-free highly bendable and efficient organic solar cells with Ag nanomesh/ZnO hybrid electrodes. J. Mater. Chem. A **3**(1), 65–70 (2015)

151. J.-S. Yu et al., Silver front electrode grids for ITO-free all printed polymer solar cells with embedded and raised topographies, prepared by thermal imprint, flexographic and inkjet roll-to-roll processes. Nanoscale **4**(19), 6032–6040 (2012)
152. Y. Han et al., Efficiency above 12% for 1 cm^2 flexible organic solar cells with Ag/Cu grid transparent conducting electrode. Adv. Sci. **6**(22), 1901490 (2019)
153. M.-G. Kang, H.J. Park, S.H. Ahn, L.J. Guo, Transparent Cu nanowire mesh electrode on flexible substrates fabricated by transfer printing and its application in organic solar cells. Sol. Energy Mater. Sol. Cells **94**(6), 1179–1184 (2010)
154. Z. Jiang et al., Reverse-offset printed ultrathin Ag mesh for robust conformal transparent electrodes for high-performance organic photovoltaics. Adv. Mater. **30**(26), 1707526 (2018)
155. C. Li et al., Highly flexible and acid-alkali resistant TiN nanomesh transparent electrodes for next-generation optoelectronic devices. ACS Nano **17**(24), 24763–24772 (2023). https://doi.org/10.1021/acsnano.3c05211
156. L.M. González, D. Ramirez, F. Jaramillo, Current status and trends of carbon-based electrodes for fully solution-processed perovskite solar cells. J. Energy Chem. **68**, 222–246 (2022)
157. K.S. Novoselov et al., Electric field effect in atomically thin carbon films. Science (80-.). **306**(5696), 666–669 (2004)
158. Y. Chen, Y. Yue, S. Wang, N. Zhang, J. Feng, H. Sun, Graphene as a transparent and conductive electrode for organic optoelectronic devices. Adv. Electron. Mater. **5**(10), 1900247 (2019)
159. D. Konios et al., Reduced graphene oxide micromesh electrodes for large area, flexible, organic photovoltaic devices. Adv. Funct. Mater. **25**(15), 2213–2221 (2015)
160. Z. Liu, S.P. Lau, F. Yan, Functionalized graphene and other two-dimensional materials for photovoltaic devices: device design and processing. Chem. Soc. Rev. **44**(15), 5638–5679 (2015)
161. S. Park, J. An, J.R. Potts, A. Velamakanni, S. Murali, R.S. Ruoff, Hydrazine-reduction of graphite-and graphene oxide. Carbon N. Y. **49**(9), 3019–3023 (2011)
162. S. Bae et al., Roll-to-roll production of 30-inch graphene films for transparent electrodes. Nat. Nanotechnol. **5**(8), 574–578 (2010)
163. G.-X. Ni et al., Graphene–ferroelectric hybrid structure for flexible transparent electrodes. ACS Nano **6**(5), 3935–3942 (2012)
164. L. D'Arsié et al., Stable, efficient p-type doping of graphene by nitric acid. RSC Adv. **6**(114), 113185–113192 (2016)
165. S. Lee et al., Flexible organic solar cells composed of P3HT: PCBM using chemically doped graphene electrodes. Nanotechnology **23**(34), 344013 (2012)
166. Y. Chen et al., A two-step thermal annealing and HNO3 doping treatment for graphene electrode and its application in small-molecule organic solar cells. Org. Electron. **38**, 35–41 (2016)
167. S.-J. Kwon et al., Extremely stable graphene electrodes doped with macromolecular acid. Nat. Commun. **9**(1), 2037 (2018)
168. Y. Wang, X. Chen, Y. Zhong, F. Zhu, K.P. Loh, Large area, continuous, few-layered graphene as anodes in organic photovoltaic devices. Appl. Phys. Lett. **95**(6) (2009)
169. D. Kim, D. Lee, Y. Lee, D.Y. Jeon, Work-function engineering of graphene anode by bis (trifluoromethanesulfonyl) amide doping for efficient polymer light-emitting diodes. Adv. Funct. Mater. **23**(40), 5049–5055 (2013)
170. D. Koo et al., Flexible organic solar cells over 15% efficiency with polyimide-integrated graphene electrodes. Joule **4**(5), 1021–1034 (2020)
171. J. Du et al., Extremely efficient flexible organic solar cells with a graphene transparent anode: dependence on number of layers and doping of graphene. Carbon N. Y. **171**, 350–358 (2021)
172. M.S.A. Kamel, M. Oelgemöller, M.V. Jacob, Chemical vapor deposition-grown graphene transparent conducting electrode for organic photovoltaics: Advances towards scalable transfer-free synthesis. Renew. Sustain. Energy Rev. **203**, 114740 (2024)
173. K.C. Kwon, K.S. Choi, S.Y. Kim, Increased work function in few-layer graphene sheets via metal chloride doping. Adv. Funct. Mater. **22**(22), 4724–4731 (2012)

174. K.K. Kim et al., Enhancing the conductivity of transparent graphene films via doping. Nanotechnology **21**(28), 285205 (2010)
175. J. Schneider, P. Rohner, D. Thureja, M. Schmid, P. Galliker, D. Poulikakos, Electrohydrodynamic nanodrip printing of high aspect ratio metal grid transparent electrodes. Adv. Funct. Mater. **26**(6), 833–840 (2016)
176. Y. Song, S. Chang, S. Gradecak, J. Kong, Visibly-transparent organic solar cells on flexible substrates with all-graphene electrodes. Adv. Energy Mater. **6**(20), 1600847 (2016)
177. E. Muchuweni, B.S. Martincigh, V.O. Nyamori, Organic solar cells: current perspectives on graphene-based materials for electrodes, electron acceptors and interfacial layers. Int. J. Energy Res. **45**(5), 6518–6549 (2021)
178. Z. Wu, H. Yin, G. Li, Z. Ji, Recent progress in semitransparent organic solar cells. Org. Electron. 107060 (2024)
179. E. Kymakis, K. Savva, M.M. Stylianakis, C. Fotakis, E. Stratakis, Flexible organic photovoltaic cells with in situ nonthermal photoreduction of spin-coated graphene oxide electrodes. Adv. Funct. Mater. **23**(21), 2742–2749 (2013)
180. A.G. Ricciardulli, S. Yang, X. Feng, P.W.M. Blom, Solution-processable high-quality graphene for organic solar cells. ACS Appl. Mater. Interfaces **9**(30), 25412–25417 (2017)
181. A. Satharasinghe, T. Hughes-Riley, T. Dias, A review of solar energy harvesting electronic textiles. Sensors **20**(20), 5938 (2020)
182. M. Zafar et al., Graphene-based polymer nanocomposites for energy applications: Recent advancements and future prospects. Results Phys. 107655 (2024)
183. Y. Chen, Y. Yue, S.-R. Wang, N. Zhang, J. Feng, H.-B. Sun, Thermally-induced wrinkles on PH1000/graphene composite electrode for enhanced efficiency of organic solar cells. Sol. Energy Mater. Sol. Cells **201**, 110075 (2019)
184. M.A. Deshmukh, S.-J. Park, B.S. Hedau, T.-J. Ha, Recent progress in solar cells based on carbon nanomaterials. Sol. Energy **220**, 953–990 (2021)
185. L. Wieland, H. Li, C. Rust, J. Chen, B.S. Flavel, Carbon nanotubes for photovoltaics: from lab to industry. Adv. Energy Mater. **11**(3), 2002880 (2021)
186. S. Iijima, Helical microtubules of graphitic carbon. Nature **354**(6348), 56–58 (1991)
187. A. Du Pasquier, H.E. Unalan, A. Kanwal, S. Miller, M. Chhowalla, Conducting and transparent single-wall carbon nanotube electrodes for polymer-fullerene solar cells. Appl. Phys. Lett. **87**(20) (2005)
188. M.F.L. De Volder, S.H. Tawfick, R.H. Baughman, A.J. Hart, Carbon nanotubes: present and future commercial applications. Science (80-.). **339**(6119), 535–539 (2013)
189. I. Jeon, R. Xiang, A. Shawky, Y. Matsuo, S. Maruyama, Single-walled carbon nanotubes in emerging solar cells: synthesis and electrode applications. Adv. Energy Mater. **9**(23), 1801312 (2019)
190. E. Muchuweni, E.T. Mombeshora, B.S. Martincigh, V.O. Nyamori, Recent applications of carbon nanotubes in organic solar cells. Front. Chem. **9**, 733552 (2022)
191. M.W. Rowell et al., Organic solar cells with carbon nanotube network electrodes. Appl. Phys. Lett. **88**(23) (2006)
192. X. Xia et al., Infrared-transparent polymer solar cells. J. Mater. Chem. **20**(39), 8478–8482 (2010)
193. H.E. Unalan, P. Hiralal, D. Kuo, B. Parekh, G. Amaratunga, M. Chhowalla, Flexible organic photovoltaics from zinc oxide nanowires grown on transparent and conducting single walled carbon nanotube thin films. J. Mater. Chem. **18**(48), 5909 (2008)
194. Y. Zhou, L. Hu, G. Grüner, A method of printing carbon nanotube thin films. Appl. Phys. Lett. **88**(12) (2006)
195. D.-Y. Cho, K. Eun, S.-H. Choa, H.-K. Kim, Highly flexible and stretchable carbon nanotube network electrodes prepared by simple brush painting for cost-effective flexible organic solar cells. Carbon N. Y. **66**, 530–538 (2014)
196. S.L. Hellstrom et al., Strong and stable doping of carbon nanotubes and graphene by MoOx for transparent electrodes. Nano Lett. **12**(7), 3574–3580 (2012)

197. C. Delacou et al., Indium tin oxide-free small molecule organic solar cells using single-walled carbon nanotube electrodes. ECS J. Solid State Sci. Technol. **6**(6), M3181 (2017)
198. I. Jeon et al., Polymeric acid-doped transparent carbon nanotube electrodes for organic solar cells with the longest doping durability. J. Mater. Chem. A **6**(30), 14553–14559 (2018)
199. R.V. Salvatierra, C.E. Cava, L.S. Roman, A.J.G. Zarbin, ITO-free and flexible organic photovoltaic device based on high transparent and conductive polyaniline/carbon nanotube thin films. Adv. Funct. Mater. **23**(12), 1490–1499 (2013)
200. Y. Zhang et al., Polymer-embedded carbon nanotube ribbons for stretchable conductors. Adv. Mater. **22**(28), 3027 (2010)
201. B. Subramanyam, P.C. Mahakul, K. Sa, J. Raiguru, P. Mahanandia, Investigation of improvement in stability and power conversion efficiency of organic solar cells fabricated by incorporating carbon nanostructures in device architecture. J. Phys. Mater. **3**(4), 45004 (2020)
202. J.-H. Huang, J.-H. Fang, C.-C. Liu, C.-W. Chu, Effective work function modulation of graphene/carbon nanotube composite films as transparent cathodes for organic optoelectronics. ACS Nano **5**(8), 6262–6271 (2011)
203. L. Yin et al., MXenes for solar cells. Nano-Micro Lett. **13**, 1–17 (2021)
204. M. Naguib et al., *Two-Dimensional Nanocrystals Produced by Exfoliation of* Ti_3AlC_2 *in MXenes*. Jenny Stanford Publishing (2023) pp. 15–29
205. M.W. Barsoum, T. El-Raghy, Synthesis and characterization of a remarkable ceramic: Ti_3SiC_2. J. Am. Ceram. Soc. **79**(7), 1953–1956 (1996)
206. M.A. Saeed, A. Shahzad, K. Rasool, F. Mateen, J. Oh, J.W. Shim, 2D MXene: A potential candidate for photovoltaic cells? A critical review. Adv. Sci. **9**(10), 2104743 (2022)
207. C. Dall'Agnese, Y. Dall'Agnese, B. Anasori, W. Sugimoto, S. Mori, Oxidized Ti_3C_2 MXene nanosheets for dye-sensitized solar cells. New J. Chem. **42**(20), 16446–16450 (2018)
208. M. Mariano et al., Solution-processed titanium carbide MXene films examined as highly transparent conductors. Nanoscale **8**(36), 16371–16378 (2016)
209. L. Qin et al., A flexible semitransparent photovoltaic supercapacitor based on water-processed MXene electrodes. J. Mater. Chem. A **8**(11), 5467–5475 (2020)
210. E.-Y. Shin et al., An inverted layer-by-layer process to enable ultrasmooth MXene–Ag nanowire hybrid electrode for organic photovoltaics. Sol. RRL **7**(9), 2201130 (2023)
211. K.A. Nirmal, W. Ren, A.C. Khot, D.Y. Kang, T.D. Dongale, T.G. Kim, Flexible memristive organic solar cell using multilayer 2D titanium carbide MXene electrodes. Adv. Sci. **10**(19), 2300433 (2023)
212. S. Xiong et al., Waterproof and ultraflexible organic photovoltaics with improved interface adhesion. Nat. Commun. **15**(1), 681 (2024)
213. F. Guo et al., Fully solution-processing route toward highly transparent polymer solar cells. ACS Appl. Mater. Interfaces **6**(20), 18251–18257 (2014)
214. Y. Chen et al., Integrated ionic-additive assisted wet-spinning of highly conductive and stretchable PEDOT: PSS fiber for fibrous organic electrochemical transistors. Adv. Electron. Mater. **7**(8), 2100231 (2021)
215. J. Kim, N. Kim, M. Kwon, J. Lee, Attachable pulse sensors integrated with inorganic optoelectronic devices for monitoring heart rates at various body locations. ACS Appl. Mater. Interfaces **9**(31), 25700–25705 (2017)
216. Y. Li et al., A stretchable-hybrid low-power monolithic ECG patch with microfluidic liquid-metal interconnects and stretchable carbon-black nanocomposite electrodes for wearable heart monitoring. Adv. Electron. Mater. **5**(2), 1800463 (2019)
217. M. Kaltenbrunner et al., Flexible high power-per-weight perovskite solar cells with chromium oxide–metal contacts for improved stability in air. Nat. Mater. **14**(10), 1032–1039 (2015)
218. N. Matsuhisa et al., Printable elastic conductors with a high conductivity for electronic textile applications. Nat. Commun. **6**(1), 7461 (2015)
219. D.J. Lipomi, H. Chong, M. Vosgueritchian, J. Mei, Z. Bao, Toward mechanically robust and intrinsically stretchable organic solar cells: evolution of photovoltaic properties with tensile strain. Sol. Energy Mater. Sol. Cells **107**, 355–365 (2012)

220. N. Matsuhisa, X. Chen, Z. Bao, T. Someya, Materials and structural designs of stretchable conductors. Chem. Soc. Rev. **48**(11), 2946–2966 (2019)
221. L. Li et al., A solid-state intrinsically stretchable polymer solar cell. ACS Appl. Mater. Interfaces **9**(46), 40523–40532 (2017)
222. Y.-Y. Yu et al., Intrinsically stretchable nanostructured silver electrodes for realizing efficient strain sensors and stretchable organic photovoltaics. ACS Appl. Mater. Interfaces **9**(33), 27853–27862 (2017)
223. D.J. Lipomi, B.C. Tee, M. Vosgueritchian, Z. Bao, Stretchable organic solar cells. Adv. Mater. **15**(23), 1771–1775 (2011)
224. M. Kaltenbrunner et al., Ultrathin and lightweight organic solar cells with high flexibility. Nat. Commun. **3**(1), 770 (2012)
225. A. Lamoureux, K. Lee, M. Shlian, S.R. Forrest, M. Shtein, Dynamic kirigami structures for integrated solar tracking. Nat. Commun. **6**(1), 8092 (2015)
226. K. Yamagishi et al., Elastic kirigami patch for electromyographic analysis of the palm muscle during baseball pitching. NPG Asia Mater. **11**(1), 80 (2019)
227. M. Karl et al., Flexible and ultra-lightweight polymer membrane lasers. Nat. Commun. **9**(1), 1525 (2018)
228. X. Zheng et al., Versatile organic photovoltaics with a power density of nearly 40 W g^{-1}. Energy Environ. Sci. **16**(5), 2284–2294 (2023)
229. J. Wang et al., Intrinsically stretchable organic photovoltaics by redistributing strain to PEDOT: PSS with enhanced stretchability and interfacial adhesion. Nat. Commun. **15**(1), 4902 (2024)
230. R. Ma, J. Feng, D. Yin, H.-B. Sun, Highly efficient and mechanically robust stretchable polymer solar cells with random buckling. Org. Electron. **43**, 77–81 (2017)
231. D. Yin et al., Mechanically robust stretchable organic optoelectronic devices built using a simple and universal stencil-pattern transferring technology. Light Sci. Appl. **7**(1), 35 (2018)
232. C. Liu, C. Xiao, C. Xie, W. Li, Flexible organic solar cells: Materials, large-area fabrication techniques and potential applications. Nano Energy **89**, 106399 (2021)
233. J.A. Rogers, T. Someya, Y. Huang, Materials and mechanics for stretchable electronics. Science (80-.) **327**(5973), 1603–1607 (2010)
234. D. Han et al., Intrinsically stretchable organic solar cells and sensors enabled by extensible composite electrodes. Adv. Funct. Mater. **34**(44), 2407392 (2024)
235. S. Lee et al., Advancing high-efficiency, stretchable organic solar cells: novel liquid metal electrode architecture. Energy Environ. Sci. **17**(22), 8915–8925 (2024)
236. Q. Wan et al., High-performance intrinsically stretchable polymer solar cell with record efficiency and stretchability enabled by thymine-functionalized terpolymer. J. Am. Chem. Soc. **145**(22), 11914–11920 (2023)
237. J. Huang et al., Intrinsically stretchable, semi-transparent organic photovoltaics with high efficiency and mechanical robustness via a full-solution process. Energy Environ. Sci. **16**(3), 1251–1263 (2023)
238. J. Lee et al., Intrinsically stretchable, highly efficient organic solar cells enabled by polymer donors featuring hydrogen-bonding spacers. Adv. Mater. **34**(50), 2207544 (2022)
239. L. Zhu et al., Single-junction organic solar cells with over 19% efficiency enabled by a refined double-fibril network morphology. Nat. Mater. **21**(6), 656–663 (2022)
240. J. Huang et al., Stretchable ITO-free organic solar cells with intrinsic anti-reflection substrate for high-efficiency outdoor and indoor energy harvesting. Adv. Funct. Mater. **31**(16), 2010172 (2021)
241. Z. Bi, H.B. Naveed, Y. Mao, H. Yan, W. Ma, Importance of nucleation during morphology evolution of the blade-cast PffBT4T-2OD-based organic solar cells. Macromolecules **51**(17), 6682–6691 (2018)
242. L. Sun, K. Fukuda, T. Someya, Recent progress in solution-processed flexible organic photovoltaics. npj Flex. Electron. **6**(1), 89 (2022)
243. F. Yang, Y. Huang, Y. Li, Y. Li, Large-area flexible organic solar cells. npj Flexible Electron **5** (2021)

244. P.G.V. Sampaio et al., Overview of printing and coating techniques in the production of organic photovoltaic cells. Int. J. Energy Res. **44**(13), 9912–9931 (2020)
245. R. Sun et al., A layer-by-layer architecture for printable organic solar cells overcoming the scaling lag of module efficiency. Joule **4**(2), 407–419 (2020)
246. Y. Yang et al., Layer-by-layer slot-die coated high-efficiency organic solar cells processed using twin boiling point solvents under ambient condition. Nano Res. **14**, 4236–4242 (2021)
247. X. Chen et al., Balancing the molecular aggregation and vertical phase separation in the polymer: nonfullerene blend films enables 13.09% efficiency of organic solar cells with inkjet-printed active layer. Adv. Energy Mater. **12**(12), 2200044 (2022)
248. C. Kapnopoulos et al., Fully gravure printed organic photovoltaic modules: a straightforward process with a high potential for large scale production. Sol. Energy Mater. Sol. Cells **144**, 724–731 (2016)
249. D. Angmo, J. Sweelssen, R. Andriessen, Y. Galagan, F.C. Krebs, Inkjet printing of back electrodes for inverted polymer solar cells. Adv. Energy Mater. **3**(9), 1230–1237 (2013)
250. S.H. Eom et al., High efficiency polymer solar cells via sequential inkjet-printing of PEDOT: PSS and P3HT: PCBM inks with additives. Org. Electron. **11**(9), 1516–1522 (2010)
251. G. Wang, M.A. Adil, J. Zhang, Z. Wei, Large-area organic solar cells: material requirements, modular designs, and printing methods. Adv. Mater. **31**(45), 1805089 (2019)
252. S.K. Karunakaran et al., Recent progress in inkjet-printed solar cells. J. Mater. Chem. A **7**(23), 13873–13902 (2019)
253. D. Corzo, E. Bihar, E.B. Alexandre, D. Rosas-Villalva, D. Baran, Ink engineering of transport layers for 9.5% efficient all-printed semitransparent nonfullerene solar cells. Adv. Funct. Mater. **31**(7), 2005763 (2021)
254. G. Hu et al., Functional inks and printing of two-dimensional materials. Chem. Soc. Rev. **47**(9), 3265–3300 (2018)
255. Z. Wang et al., High power conversion efficiency of 13.61% for 1 cm2 flexible polymer solar cells based on patternable and mass-producible gravure-printed silver nanowire electrodes. Adv. Funct. Mater. **31**(4), 2007276 (2021)
256. S. Sun et al., Solution processed semi-transparent organic solar cells over 50% visible transmittance enabled by silver nanowire electrode with sandwich structure. Adv. Mater. **35**(46), 2305092 (2023)
257. C. Guan et al., Non-covalent interactions between polyvinyl chloride and conjugated polymers enable excellent mechanical properties and high stability in organic solar cells. Angew. Chemie **135**(44), e202312357 (2023)
258. W. Song et al., Ultra robust and highly efficient flexible organic solar cells with over 18% efficiency realized by incorporating a linker dimerized acceptor. Angew. Chemie **135**(41), e202310034 (2023)
259. Q. Ye et al., Ductile oligomeric acceptor-modified flexible organic solar cells show excellent mechanical robustness and near 18% efficiency. Adv. Mater. **35**(44), 2305562 (2023)
260. P. Subudhi, D. Punetha, Progress, challenges, and perspectives on polymer substrates for emerging flexible solar cells: a holistic panoramic review. Prog. Photovoltaics Res. Appl. **31**(8), 753–789 (2023)
261. M.A. Ansari, G. Ciampi, S. Sibilio, Novel materials for semi-transparent organic solar cells. Energies **17**(2), 333 (2024)
262. Z. Zhou, Z. Yuan, Z. Yin, Q. Xue, N. Li, F. Huang, Progress of semitransparent emerging photovoltaics for building integrated applications. Green Energy Environ. **9**(6), 992–1015 (2024)
263. Y. Li, X. Huang, H.K.M. Sheriff Jr., S.R. Forrest, Semitransparent organic photovoltaics for building-integrated photovoltaic applications. Nat. Rev. Mater. **8**(3), 186–201 (2023)
264. P. Kumar, S. You, A. Vomiero, Recent progress in materials and device design for semitransparent photovoltaic technologies. Adv. Energy Mater. **13**(39), 2301555 (2023)
265. H. Yu et al., Semi-transparent organic photovoltaics. Chem. Soc. Rev. **52**(13), 4132–4148 (2023)

266. X. Liu, S. Xu, B. Tang, X. Song, Indoor organic photovoltaics for low-power internet of things devices: recent advances, challenges, and prospects. Chem. Eng. J. 154944 (2024)
267. S. Biswas, Y. Lee, H. Choi, H.W. Lee, H. Kim, Progress in organic photovoltaics for indoor application. RSC Adv. **13**(45), 32000–32022 (2023)
268. M.-Z. Sha, Y.-J. Pu, H. Yin, X.-T. Hao, Recent progress of indoor organic photovoltaics—from device performance to multifunctional applications. Org. Electron. **114**, 106736 (2023)
269. K.A. Castro Morales, R. Corrao, Recent advancements of semi-transparent photovoltaic technologies for innovative BIPV products, in International Conference of Ar. Tec. (Scientific Society of Architectural Engineering). Springer (2024), pp. 180–195
270. S.A. Hashemi, S. Ramakrishna, A.G. Aberle, Recent progress in flexible–wearable solar cells for self-powered electronic devices. Energy Environ. Sci. **13**(3), 685–743 (2020)
271. R. Steim et al., Organic photovoltaics for low light applications. Sol. Energy Mater. Sol. Cells **95**(12), 3256–3261 (2011)
272. Y. Cui et al., Wide-gap non-fullerene acceptor enabling high-performance organic photovoltaic cells for indoor applications. Nat. Energy **4**(9), 768–775 (2019)
273. R. Arai, S. Furukawa, N. Sato, T. Yasuda, Organic energy-harvesting devices achieving power conversion efficiencies over 20% under ambient indoor lighting. J. Mater. Chem. A **7**(35), 20187–20192 (2019)
274. D. Wang et al., High-performance see-through power windows. Energy Environ. Sci. **15**(6), 2629–2637 (2022)
275. Y. Li et al., Color-neutral, semitransparent organic photovoltaics for power window applications. Proc. Natl. Acad. Sci. **117**(35), 21147–21154 (2020)
276. X. Liu, Y. Wei, Y. Qiu, Advanced flexible skin-like pressure and strain sensors for human health monitoring (2021). https://doi.org/10.3390/mi12060695
277. X. Bai et al., Boosting the photovoltaic performance of doctor-bladed organic solar cells using a low-boiling solvent additive. Org. Electron. **118**, 106794 (2023)
278. D. Lv, Q. Jiang, Y. Shang, D. Liu, Highly efficient fiber-shaped organic solar cells toward wearable flexible electronics. npj Flex. Electron. **6**(1), 38 (2022). https://doi.org/10.1038/s41528-022-00172-w
279. H. Jinno et al., Self-powered ultraflexible photonic skin for continuous bio-signal detection via air-operation-stable polymer light-emitting diodes. Nat. Commun. **12**(1), 2234 (2021)
280. S. Schindele et al., Implementation of agrophotovoltaics: techno-economic analysis of the price-performance ratio and its policy implications. Appl. Energy **265**, 114737 (2020)
281. Y.-T. Hsieh, J.-Y. Chen, C.-C. Shih, C.-C. Chueh, W.-C. Chen, Mechanically robust, stretchable organic solar cells via buckle-on-elastomer strategy. Org. Electron. **53**, 339–345 (2018)
282. S.H. Cho, J. Lee, M.J. Lee, H.J. Kim, S.-M. Lee, K.C. Choi, Plasmonically engineered textile polymer solar cells for high-performance, wearable photovoltaics. ACS Appl. Mater. Interfaces **11**(23), 20864–20872 (2019)
283. C. Zhang, L. Zhang, Z. Pu, B. Bao, W. Ouyang, D. Li, Fabricating 1D stretchable fiber-shaped electronics based on inkjet printing technology for wearable applications. Nano Energy **113**, 108574 (2023)
284. J. Huang, High-performed flexible organic photovoltaics towards wearable applications (2023)
285. A. Sharma, S. Masoumi, D. Gedefaw, S. O'Shaughnessy, D. Baran, A. Pakdel, Flexible solar and thermal energy conversion devices: organic photovoltaics (OPVs), organic thermoelectric generators (OTEGs) and hybrid PV-TEG systems. Appl. Mater. Today **29**, 101614 (2022). https://doi.org/10.1016/j.apmt.2022.101614

Combined Photovoltaic–Electrochemical Systems for Integrated Energy Storage and Conversion

Mousami Barman and Prasanth Ravindran

Abstract Integrating photovoltaic (PV) and electrochemical (EC) systems has emerged as a promising renewable energy utility by combining solar energy harvesting with efficient storage and conversion technologies. PV systems generate electricity by converting sunlight, while EC systems, including batteries, supercapacitors, and electrolyzers, store energy or produce clean fuels like hydrogen. This synergy guarantees energy reliability, minimizes energy loss during peak solar output, and supports smart-grid transitions. Revolutions in PV technologies, such as high-efficiency perovskite cells, and progressions in EC systems, including vigorous electrolyzers, have paved the way for scalable applications. Various integration configurations, such as PV–electrolyzer and PV–battery systems, outfit diverse energy storage and usage needs. Direct coupling offers simplicity but faces efficiency limitations, while indirect coupling with power management units enhances energy conversion efficiency—challenges like energy losses, high material costs, and long-term durability impact large-scale adoption. However, research into low-cost catalysts and stable materials continues to address these limitations. Integrated PV–EC systems offer key benefits, including sustainable hydrogen production, enhanced grid resilience, and improved energy independence. With ongoing technological advancements, these systems are poised to play a pivotal role in achieving carbon-neutral energy goals and fostering a sustainable future.

Keywords Hybrid systems · Integrated energy storage · PV–electrolyzer · Direct coupling · Indirect coupling

M. Barman · P. Ravindran (✉)
Department of Green Energy Technology, UNESCO Madanjeet School of Green Energy Technologies, Pondicherry University, Puducherry, India
e-mail: prasanth.get@pondiuni.edu.in

P. Ravindran et al. (eds.), *Photon to Power*, Progress in Optical Science and Photonics 32, https://doi.org/10.1007/978-981-96-5914-2_11

1 Introduction

The rapid expansion of renewable energy sources, particularly solar energy, has led to significant research in novel energy harvesting and storage technologies [1–5]. Among these, the integration of PV cells with electrochemical (EC) storage or conversion technologies offers energy generation and storage within a single framework which can enhance the efficiency and reliability of solar energy utilization [2, 4]. These systems capitalize on the strengths of both PV and EC components. PV systems are designed to capture sunlight and convert it into electrical energy. At the same time, electrochemical devices, such as batteries, supercapacitors, and electrolyzers, store this energy for later use or convert it into alternative forms like hydrogen fuel [6, 7]. For instance, during times of peak solar output, excess electricity from PV systems can be used to charge batteries or run electrolyzers for hydrogen generation, ensuring that surplus energy is efficiently stored for use during periods of low solar availability, preventing waste, and enhancing energy reliability [8]. These combined systems offer key benefits such as boosting energy independence by storing solar power for later use and enhancing grid resilience by supplying dispatchable energy during high demand [9]. This adaptability is crucial for supporting the transition to smart, flexible energy grids [10, 11]. The Future of PV–EC systems is heavily influenced by advancements in related technologies, including breakthroughs in solar cell efficiencies, storage capacities of batteries, and the development of novel electrochemical materials and processes [6]. For instance, the adoption of perovskite solar cells has shown promise due to their potential for high efficiency at lower manufacturing costs compared to traditional silicon solar cells, thus making solar energy more accessible to a wider audience [12–15]. Likewise, ongoing research into more stable and efficient electrolyzers contributes to the economically viable production of hydrogen, which can serve as a clean fuel for various applications.

However, integrated PV–EC systems face numerous challenges in material stability, energy transfer efficiency, high costs, power management, and scalability, all of which limit their large-scale adoption [7]. The efficiency of PV–EC systems hinges on the compatibility of PV and EC materials and the effectiveness of power electronics, which regulate energy flow within the system. As far as emerging materials like perovskites and solid-state electrolytes offer promising solutions to improve both the performance and longevity of PV–EC systems [15, 16]. By leveraging the strengths of both technologies, these systems not only enhance the reliability of energy supply but also contribute to a sustainable future with lower carbon emissions and energy independence [17].

2 Fundamentals of Photovoltaic and Electrochemical System

2.1 Photovoltaic Systems

Photovoltaic cells or solar cells are devices that convert sunlight directly into electricity using semiconducting materials through the photovoltaic effect [1]. The process begins when sunlight hits the cell, and photons excite electrons, creating electron–hole pairs which are separated by the electric field within the p–n junction, which initiates the flow of electric current [11]. The core structure of a PV cell consists of a p–n junction formed by doping silicon with elements like boron and phosphorus, creating a positive (p-type) and negative (n-type) layer. PV systems, often built from silicon, thin-film materials, or organic compounds, play a crucial role in solar energy collection. Silicon-based PV cells are available in monocrystalline, polycrystalline, and thin-film forms like cadmium telluride (CdTe) and copper indium gallium selenide (CIGS), each offering specific benefits in efficiency, cost, and application suitability [18]. However, challenges like energy loss and temperature effects are managed using anti-reflective coatings and cooling systems, making PV cells a cornerstone of sustainable energy [16, 19].

2.2 Electrochemical Systems

The electrochemical system involves energy storage and conversion processes including batteries, supercapacitors, fuel cells, and electrolyzers. Energy can be stored as chemical fuels, such as hydrogen (produced by proton exchange membrane (PEM) electrolyzers), or in batteries for later use [20]. PEM electrolyzers enable efficient hydrogen production, which acts as renewable fuel [21]. While batteries and supercapacitors provide fast charge or discharge cycles, ideal for grid support and rapid energy release. These systems include two important components that operate by moving electrons through redox reactions to store or convert energy, such as:

2.2.1 Energy Storage Systems (ESS)

ESS units store electrical energy as chemical energy for later use. These include batteries and supercapacitors, each with unique advantages based on application requirements [22]. Batteries like lithium-ion, sodium-ion, and metal–air, hold energy in chemical bonds offering high density and are ideal for long-term storage. Lithium-ion batteries are known for efficiency and rechargeability, while sodium-ion batteries are valued for their lower cost, and metal–air batteries promise higher density [23].

Supercapacitors, on the other hand, store energy through electrostatic charge accumulation, allowing rapid charge–discharge cycles, making them suitable for short-term, high-power applications, and grid stability support [12, 23].

2.2.2 Energy Conversion Systems (ECS)

ECS units are designed to transform chemical energy into electrical energy or vice versa, for playing a key role in storing renewable energy for later use. These include fuel cells and electrolyzers. A typical hydrogen fuel cell produces electricity by combining hydrogen and oxygen, yielding water as their byproduct and providing clean, efficient energy. However, proton exchange membrane (PEM) fuel cells are especially valued in PV–EC systems for their compact design and high efficiency [24]. Similarly, electrolyzers convert electrical energy into hydrogen and oxygen via a water-splitting process, storing energy in hydrogen form for later utilization in fuel cells or as a versatile fuel source, with PEM electrolyzers being highly compatible with PV systems for efficient hydrogen production, especially during peak solar output.

3 Integration of Photovoltaic and Electrochemical Systems

The combination of PV and EC systems creates a holistic energy solution, enabling efficient energy capture, storage, and utilization as shown in Fig. 1 [17]. The integration of these two systems provides numerous benefits, particularly in addressing the challenges posed by the intermittency of solar power [25]. Combined systems can efficiently capture sunlight, store energy, and offer flexibility for both electricity generation and chemical production (such as hydrogen) [26]. Different types of integration serve distinct roles based on energy storage duration and delivery speed, contributing to a more resilient and sustainable energy future.

3.1 Types of PV–EC Integration

3.1.1 PV–Electrolyzer Systems

These systems, utilize solar electricity generated by photovoltaic cells to power electrolyzers, which perform water-splitting reactions to produce hydrogen and oxygen [27]. The hydrogen generated through this method is a transportable fuel and storable energy carrier, later converted back into electricity via fuel cells or utilized in various applications, including industrial processes, especially in scenarios where demand is seasonal or intermittent [28]. PV–electrolyzer systems are particularly suited for

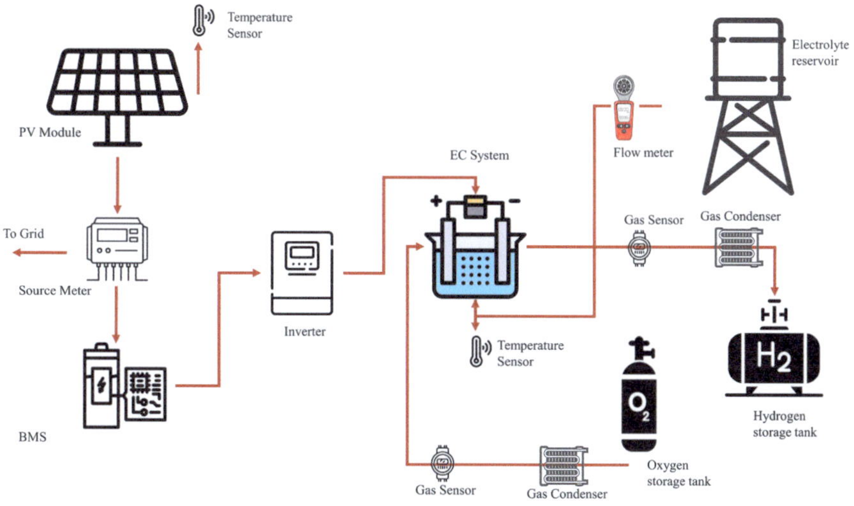

Fig. 1 Schematic representation of combined PV–EC system for integrated energy storage and conversion. (Images are from Freepik)

large-scale solar farms where substantial hydrogen production can occur, providing a scalable solution for renewable energy storage and utilization.

3.1.2 PV–Battery Systems

PV–battery systems directly store electricity from PV cells into batteries, allowing for energy use at times when sunlight isn't available [6]. It includes common battery choices like lithium-ion and flow batteries, which provide flexible storage durations and capacities, making PV–battery systems highly adaptable for on-demand energy release and peak load management [29]. These systems are widely used in residential solar installations, enabling homeowners to maximize their use of self-generated solar energy and reduce reliance on grid electricity.

3.1.3 PV–Supercapacitor Systems

Similar to PV–battery configurations, PV–supercapacitor systems store solar energy but are tailored for applications requiring rapid charge and discharge cycles. Supercapacitors can store energy electrostatically for a short duration and offer high power density, also well-suited for managing short-term energy fluctuations, such as those in microgrids or hybrid energy systems [30]. It effectively smooths out the power output by absorbing energy during peak production times and releasing it during short periods of low generation. PV–supercapacitor systems are especially valuable

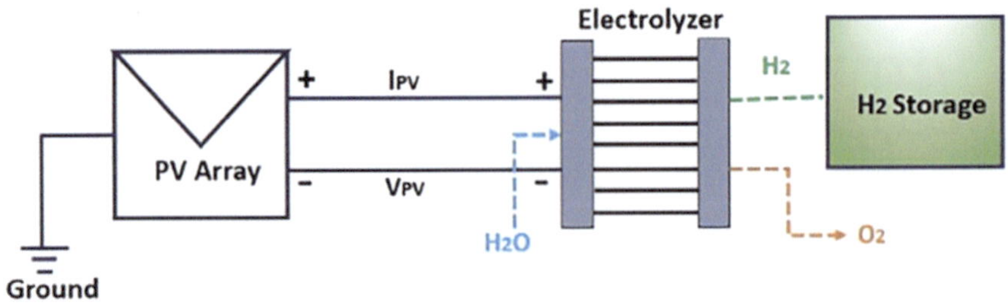

Fig. 2 Direct coupling configuration

in applications such as grid frequency regulation, electric vehicles, and backup power for critical systems, where quick energy delivery is essential.

3.2 Coupling Mechanisms

3.2.1 Direct Coupling

In a direct coupling configuration, the output of the PV generator is connected directly to the input of the electrolyzer without any intermediate power stage [31]. This setup is straightforward as it avoids additional electronic components, thereby reducing capital costs and minimizing conversion losses. Since direct coupling relies entirely on available sunlight, the EC device must match PV output, meaning both must operate effectively under similar voltage and current conditions to achieve optimal performance, typically near its maximum power point (MPP). The system's efficiency is significantly affected by varying irradiance and temperature conditions, as the electrolyzer may not always operate near its MPP [31, 32]. This can lead to less efficient energy transfer and reduced hydrogen production. Electrolyzers and some batteries work well in this setup if their input voltage range can tolerate solar variability. However, this setup may suffer from some mismatch issues, as the output power from the PV system fluctuates with sunlight intensity, making it prone to interruptions and energy losses during low-light conditions [32] (Fig. 2).

3.2.2 Indirect Coupling

An indirect setup uses an intermediate power stage that incorporates a maximum power point tracker (MPPT) and an electronics unit such as a DC–DC converter to manage the power transfer between PV and EC systems [33]. The MPPT ensures that the PV module operates at its maximum power regardless of changing meteorological conditions. Despite sunlight variability, these DC–DC converters regulate voltage and current, creating a stable energy flow that matches the optimal output

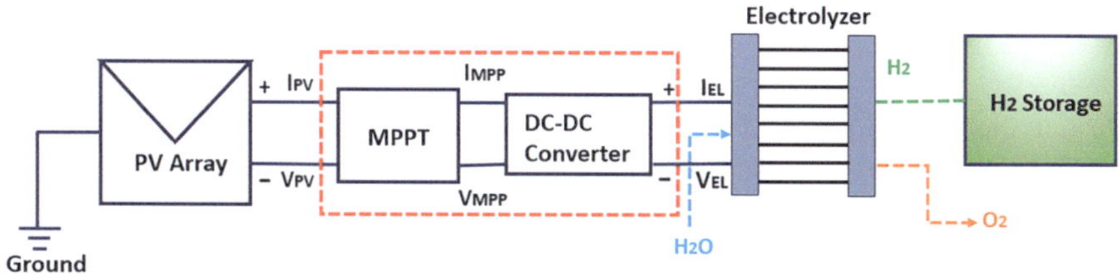

Fig. 3 Indirect coupling configuration

power of the PV to the input requirements of the electrolyzer, maximizing overall system efficiency. The strategy helps to optimize energy conversion by matching the operational requirements of both systems [34]. These setups support a variety of EC devices, such as batteries or hydrogen electrolyzers, across different operational ranges, accommodating large-scale grid-connected applications and smart-grid systems where reliability is essential [35] (Fig. 3).

4 Challenges in PV–EC Systems

PV–EC systems face several key challenges, including energy storage and conversion mechanisms, which affect their overall efficiency, cost-effectiveness, material stability, and scalability.

4.1 Energy Efficiency

Significant energy losses arise at various stages of a PV–EC system, reducing overall efficiency. For instance, in PV cells, losses occur due to reflection, heat dissipation, and recombination of the charge carriers, which limit the energy captured from sunlight [21]. In the context of electrolyzers which are essential for hydrogen production, face efficiency challenges largely due to overpotentials, an additional voltage required to drive the water-splitting reaction [36]. This need for extra voltage leads to energy losses, as a portion of the electrical energy input is not fully utilized for hydrogen production. Additionally, overpotentials result in heat generation, further decreasing efficiency and demanding a robust cooling mechanism [20, 36]. These inefficiencies are problematic especially when aiming for large-scale hydrogen production, where every increment in efficiency can greatly impact overall energy cost and sustainability.

4.2 Cost and Scalability

Despite the technological advances, the material cost for both PV and EC systems remains a huge challenge. Materials like platinum in fuel cells and lithium-ion batteries are quite expensive and subject to supply constraints. Research into alternative materials, such as low-cost catalysts for electrolyzers and abundant battery materials like sodium or zinc, has led to cost-effective large-scale deployment [6]. While the cost of solar panels also significantly decreased over the past decade, by about 90% according to recent trends in the renewable energy market, however, integrating photovoltaic (PV) systems with efficient electrolyzers, inverters, and energy storage systems still presents a considerable financial challenge.

4.3 Durability and Stability

Long durability and stability are critical for the widespread adoption of PV–EC systems. However, these systems face notable degradation over time, primarily due to environmental stresses and repeated usage cycles. PV modules due to prolonged exposure to UV radiation, humidity, and temperature, cause degradation of the material, leading to discoloration, efficiency loss of solar cells, and overall output reduction. Materials in PV cells, like traditional silicon or emerging perovskite, can lose their integrity and suffer from chemical breakdowns that impact performance [12, 14]. Conversely, electrochemical devices (e.g., batteries, electrolyzers) experience performance deterioration due to repeated charging–discharging cycles. Over time, this cycling can lead to capacity loss, chemical instability, and increased internal resistance which together lower the overall energy storage and conversion efficiency.

5 Benefits and Advances in Integrated PV–EC Systems

Combined PV–EC systems offer numerous significant benefits that make them an appealing choice for sustainable energy solutions. Primarily, it facilitates sustainable hydrogen production by harnessing renewable solar energy, thus supporting carbon-neutral goals and greenhouse gas emissions. The integration also provides an effective energy capability by transforming solar energy into hydrogen, allowing for storage and utilization during periods when no sunlight is available, which enhances energy reliability and stability [37]. Furthermore, now advancements in technology have led to improvements in solar-to-hydrogen (STH) efficiencies, achieving efficiencies above 10% and even higher in optimized configurations. Reduction in overpotentials in integrated PV–EC systems through advanced materials development and innovative designs, making the overall performance versatile, thus allowing for a wide range of applications in configuring to accommodate both large-scale and small-scale energy needs [38] (Table 1).

6 Applications and Future Prospects

6.1 Off-grid Power Solutions

Off-grid power solutions leverage combined photovoltaic–electrochemical systems to provide a transformative approach to decentralized energy, especially in remote regions, ensuring a continuous power supply, even during periods without sunlight [43]. These benefit from the coupling of solar energy generation and electrochemical energy storage mechanisms to create a self-sufficient energy system. PV–EC systems, especially PV–battery and PV–fuel cell combinations, are ideal for off-grid applications, to offer a sustainable solution for powering remote areas, telecommunications infrastructure, and emergency backup systems, where regular power supply is required. Enhanced development of integrated systems allows better synergy between PV, batteries, and electrolyzers, potentially leading to innovations in energy management. Advancements in technology will facilitate the scaling of these systems for broader applications, from small domestic units to larger village-scale setups. Ongoing research on materials and configurations will drive efficiency improvements and lower costs, making off-grid power solutions more accessible globally.

6.2 Grid Stabilization

Grid Stabilization provides a promising solution for balancing supply and demand, enhancing grid resilience, and supporting the integration of renewable energy sources on a large scale. It involves managing and maintaining the reliable operation of power systems, particularly as more intermittent energy sources are connected to the grid [44]. Stability is an essential parameter to prevent outages and ensure that supply matches demand at all times. However, grid stability alone faces challenges which include variability of renewable energy generation, influenced by weather conditions and time of day. This variability can create significant fluctuations that must be managed to maintain a reliable power supply. Additionally, voltage fluctuations also pose a challenge for rapid change in generation, which can affect the voltage levels critical for the proper functioning of electrical equipment. Combined PV and electrochemical systems offer effective solutions to the challenges of energy storage systems (ESS), which provide frequency regulation by absorbing excess energy during peak generation and releasing it during low-generation times, thus helping to stabilize grid frequency [10]. Conversely, batteries can respond rapidly to fluctuations, discharging

Table 1 This expanded timeline highlights significant milestones and developments in combined PV–EC research

Design concept	Materials used	System efficiency	Challenges	References
Single junction PV cells with traditional electrolyzer	**PV Cells**: Silicon (Si) **Catalysts**: Platinum (Pt), Iridium Oxide (IrO_2) **Electrolytes**: Liquid, acidic or alkaline	~ 5–10%	High cost, high energy losses, stability issues, minimum light absorption	[39]
Introduction of multi-junction cells, Initial research on tandem cells, novel catalyst development	**PV Cells**: Silicon-based multi-junction, GaAs **Catalysts**: Nickel-based alloys, cobalt phosphides **Electrolytes**: Alkaline, some exploration of polymer	10–15%	Catalyst stability, high degradation rate in electrolyzers	[40]
Advancements in multi-junction tandem cells, basic solid-state electrolytes	**PV Cells**: InGaP/GaAs with silicon-based tandem **Catalysts**: Non-precious metals like Ni-Mo **Electrolytes**: Solid-state (experimental)	12–16%	High cost of multi-junction cells, electrolyte limitations	[41]
Multi-junction tandem cells, focus on efficiency and catalyst durability	**PV Cells**: GaAs/Si, InGaP **Catalysts**: Nickel–molybdenum, cobalt phosphides, Fe–Ni alloys **Electrolytes**: Advanced alkaline, solid-state electrolytes	15–20%	Cost, complex manufacturing, catalyst degradation over time	[5]
Perovskite tandem cells; advanced non-PGM catalysts	**PV Cells**: Perovskite-silicon tandems, GaAs **Catalysts**: Iron–nickel sulfides, cobalt oxide, MoS_2 **Electrolytes**: Polymer-based solid-state	18–22%	Stability of perovskites, scalability of catalyst production	[14, 42]
Highly efficient perovskite/silicon tandems, advanced PV–fuel cell integration focused on non-PGM and stable catalysts	**PV Cells**: Perovskite-silicon tandem, III–V materials **Catalysts**: Iron–nickel, cobalt-nitride, MoS_2 **Electrolytes**: Polymer-based solid-state, PEM for improved durability	20–25%+	Scaling for grid-level storage, long-term catalyst durability	[7, 10, 12, 20]

stored energy when generation capacity decreases, and supporting overall grid flexibility. Integration with smart-grid technology also allows PV–EC systems to adjust energy storage and release in real-time responding to grid conditions automatically. As these technologies mature, the PV–EC system may be foundational in a resilient, low-carbon grid, supporting a global shift toward sustainable and stable renewable energy.

6.3 Decarbonizing Industry

Decarbonizing the industrial sector is a critical goal in addressing climate change and combined PV–EC systems, provide a promising pathway for energy storage and conversion to reduce carbon emissions on high-temperature processes, chemical production, and other energy-intensive operations [45]. For instance, the electroreduction of carbon dioxide and carbon monoxide using renewable electricity from PV systems can produce essential chemicals such as ethylene, ethanol, and methanol, contributing to sustainable industrial practices [46]. Prospects for PV–EC systems in industrial decarbonization focus on scaling up production, reducing the costs of catalysts, and improving efficiency to make the technology more feasible for widespread use. Moreover, as industries seek to improve energy efficiency and reduce their carbon footprints, the integration of PV–EC systems aligns well with broader electrification trends [47]. Additionally, the development of more robust energy storage solutions will bolster the reliability of these systems, ensuring that energy generated during peak sunlight hours can be effectively utilized when demand is high or solar generation is low [46, 47]. Overall, the integration of PV–EC systems with batteries could support both continuous and flexible energy demand toward a low-carbon industrial landscape, aligning with global efforts to achieve net-zero emissions, and ensuring a reliable, clean energy supply.

7 Conclusion

Combining the strengths of solar energy generation with effective electrochemical processes offers a pathway to greater energy efficiency, and reliability for renewable energy storage and conversion, particularly as demand grows for efficient, scalable, and low-carbon energy options. Advancements in tandem solar cells, solid-state electrolytes, and non-precious metal catalysts have significantly improved the efficiency and durability of PV–EC systems, moving closer to commercial viability. These innovations not only enhance energy conversion efficiency but also lower production costs, making clean hydrogen and long-term energy storage more accessible for both grid and industrial applications. Since, PV–EC systems are poised to play a critical role in a sustainable energy future, supporting grid resilience, enabling off-grid applications, and decarbonizing heavy industries, it is necessary to optimize materials for stability, refine system integration to ensure seamless integration, as well as enhancing grid compatibility, for large-scale deployment. With ongoing research and investment in these areas, the full potential of integrated photovoltaic–electrochemical systems can be realized, facilitating a transition toward a more sustainable and resilient energy future.

References

1. P. Hollmuller, J. Joubert, Evaluation of a 5 kwp photovoltaic hydrogen production and storage installation for a residential home in Switzerland. Int. J. Hydrogen Energy **25**, 97–109 (2000)
2. A. Szyszka, Neunburg vorm wald-test centre for solar hydrogen technology. Int. J. Hydrogen Energy **19**(10), 823–841 (1994)
3. J.P. Vanhanen, P.D. Lund, J.S. Tolonen, Electrolyser-metal hydride-fuel cell system for seasonal energy storage. Int. J. Hydrogen Energy **23**(4), 267–271 (1998). https://doi.org/10.1016/s0360-3199(97)00065-7
4. M. Stefanoni, Development of a solar-hydrogen cycle hydride storage. Int. J. Hydrogen Energy **22**(5) (1997)
5. J.W. Ager, M.R. Shaner, K.A. Walczak, I.D. Sharp, S. Ardo, Experimental demonstrations of spontaneous, solar-driven photoelectrochemical water splitting. Energy Environ. Sci. **8**(10), 2811–2824 (2015). https://doi.org/10.1039/c5ee00457h
6. O. Astakhov, S.N. Agbo, K. Welter, V. Smirnov, U. Rau, T. Merdzhanova, Storage batteries in photovoltaic–electrochemical device for solar hydrogen production. J. Power Sources **509** (2021). https://doi.org/10.1016/j.jpowsour.2021.230367
7. J. Kim, S. Jeong, M. Beak, J. Park, K. Kwon, Performance of photovoltaic-driven electrochemical cell systems for CO_2 reduction. Chem. Eng. J. **428** (2022). https://doi.org/10.1016/j.cej.2021.130259
8. N. Armaroli, V. Balzani, The future of energy supply: challenges and opportunities. Angew. Chem. Int. Ed. **46**(1–2), 52–66 (2007). https://doi.org/10.1002/anie.200602373
9. J. Kim, S. Jeong, M. Beak, J. Park, K. Kwon, Performance of photovoltaic-driven electrochemical cell systems for CO_2 reduction. Chem. Eng. J. **428,** 1–13 (2022). https://doi.org/10.1016/j.cej.2021.130259
10. S.G. Nnabuife, A.K. Hamzat, J. Whidborne, B. Kuang, K.W. Jenkins, Integration of renewable energy sources in tandem with electrolysis: a technology review for green hydrogen production. Int. J. Hydrogen Energy **107** (2025). https://doi.org/10.1016/j.ijhydene.2024.06.342
11. F. Obeidat, A comprehensive review of future photovoltaic systems. Sol. Energy **163**, 545–551 (2018). https://doi.org/10.1016/j.solener.2018.01.050
12. A.S.R. Bati, Y.L. Zhong, P.L. Burn, M.K. Nazeeruddin, P.E. Shaw, M. Batmunkh, Next-generation applications for integrated perovskite solar cells. Commun. Mater. **4**(1), 1–24 (2023). https://doi.org/10.1038/s43246-022-00325-4
13. H. Zhang, Y. Lu, W. Han, J. Zhu, Y. Zhang, W. Huang, Solar energy conversion and utilization: towards the emerging photo-electrochemical devices based on perovskite photovoltaics. Chem. Eng. J. **393**, 124766 (2020). https://doi.org/10.1016/j.cej.2020.124766
14. J.J. Yoo et al., Efficient perovskite solar cells via improved carrier management. Nature **590**(7847), 587–593 (2021). https://doi.org/10.1038/s41586-021-03285-w
15. J. Ekspong et al., Solar-driven water splitting at 13.8% solar-to-hydrogen efficiency by an earth-abundant electrolyzer. ACS Sustain. Chem. Eng. **9**(42), 14070–14078 (2021). https://doi.org/10.1021/acssuschemeng.1c03565
16. T. Rahman et al., Investigation of degradation of solar photovoltaics: a review of aging factors, impacts, and future directions toward sustainable energy management. Energies **16**(9) (2023). https://doi.org/10.3390/en16093706

17. M. Lee, S. Haas, V. Smirnov, T. Merdzhanova, U. Rau, Scalable photovoltaic-electrochemical cells for hydrogen production from water—recent advances. ChemElectroChem **9**(24) (2022). https://doi.org/10.1002/celc.202200838
18. A.C. Lazaroiu, M. Gmal Osman, C.V. Strejoiu, G. Lazaroiu, A comprehensive overview of photovoltaic technologies and their efficiency for climate neutrality. Sustainability **15**(23) (2023). https://doi.org/10.3390/su152316297
19. J. Kim, M. Rabelo, S.P. Padi, H. Yousuf, E.C. Cho, J. Yi, A review of the degradation of photovoltaic modules for life expectancy. Energies **14**(14), 1–21 (2021). https://doi.org/10.3390/en14144278
20. J. Jia et al., Solar water splitting by photovoltaic-electrolysis with a solar-to-hydrogen efficiency over 30%. Nat. Commun. **7**, 1–6 (2016). https://doi.org/10.1038/ncomms13237
21. S. Calnan et al., Development of various photovoltaic-driven water electrolysis technologies for green solar hydrogen generation. Sol. RRL **6**(5) (2022). https://doi.org/10.1002/solr.202100479
22. F. Bourry, J. Martin, M. Boyra, Conversion architectures for combined photovoltaic storage systems: modeling and efficiency analysis, in *2013 IEEE Grenoble Conference, POWERTECH 2013* (2013). https://doi.org/10.1109/PTC.2013.6652471
23. D. Rekioua, Energy storage systems for photovoltaic and wind systems: a review. Energies **16**(9) (2023). https://doi.org/10.3390/en16093893
24. F. Xiao et al., Recent advances in electrocatalysts for proton exchange membrane fuel cells and alkaline membrane fuel cells. Adv. Mater. **33**(50), 1–38 (2021). https://doi.org/10.1002/adma.202006292
25. Y. Lu, M. Chen, G. Zhu, Y. Zhang, Recent progress in the study of integrated solar cell-energy storage systems. Nanoscale **16**(18), 8778–8790 (2024). https://doi.org/10.1039/d4nr00839a
26. I. Holmes-Gentle, S. Tembhurne, C. Suter, S. Haussener, Kilowatt-scale solar hydrogen production system using a concentrated integrated photoelectrochemical device. Nat. Energy **8**(6), 586–596 (2023). https://doi.org/10.1038/s41560-023-01247-2
27. J. Li, N. Wu, Semiconductor-based photocatalysts and photoelectrochemical cells for solar fuel generation: a review. Catal. Sci. Technol. **5**(3), 1360–1384 (2015). https://doi.org/10.1039/c4cy00974f
28. T.L. Gibson, N.A. Kelly, Optimization of solar powered hydrogen production using photovoltaic electrolysis devices. Int. J. Hydrogen Energy **33**(21), 5931–5940 (2008). https://doi.org/10.1016/j.ijhydene.2008.05.106
29. M. Amir et al., Energy storage technologies: an integrated survey of developments, global economical/environmental effects, optimal scheduling model, and sustainable adaption policies. J. Energy Storage **72** (2023). https://doi.org/10.1016/j.est.2023.108694
30. J. Zhang, M. Gu, X. Chen, Supercapacitors for renewable energy applications: a review. Micro Nano Eng. **21** (2023). https://doi.org/10.1016/j.mne.2023.100229
31. L.G. Arriaga, W. Martínez, U. Cano, H. Blud, Direct coupling of a solar-hydrogen system in Mexico. Int. J. Hydrogen Energy **32**(13), 2247–2252 (2007). https://doi.org/10.1016/j.ijhydene.2006.10.067
32. R.E. Clarke, S. Giddey, F.T. Ciacchi, S.P.S. Badwal, B. Paul, J. Andrews, Direct coupling of an electrolyser to a solar PV system for generating hydrogen. Int. J. Hydrogen Energy **34**(6), 2531–2542 (2009). https://doi.org/10.1016/j.ijhydene.2009.01.053
33. A. González del Valle, P. García-Linares, A. Martí, Optimizing hydrogen production: a comparative study of direct and indirect coupling between photovoltaics and electrolyzer. Energy Convers. Manag. **315** (2024). https://doi.org/10.1016/j.enconman.2024.118751
34. F. Alhaj Omar, A new approach for improving the efficiency of the indirectly coupled photovoltaic-electrolyzer system. Int. J. Hydrogen Energy **48**(24), 8768–8782 (2023). https://doi.org/10.1016/j.ijhydene.2022.11.327
35. R. Mas, D. Pichilingue, A. Berastain, C. Celis, On the application of sliding mode control to indirectly coupled photovoltaic-electrolyzer systems used in the production of clean energy. Int. J. Thermofluids **23** (2024). https://doi.org/10.1016/j.ijft.2024.100747

36. S. Licht, B. Wang, S. Mukerji, T. Soga, M. Umeno, H. Tributsch, Efficient solar water splitting, exemplified by RuO_2-catalyzed AlGaAs/Si photoelectrolysis. J. Phys. Chem. B **104**(38), 8920–8924 (2000). https://doi.org/10.1021/jp002083b
37. H. Ouabi, R. Lajouad, M. Kissaoui, A. El Magri, Hydrogen production by water electrolysis driven by a photovoltaic source: a review. e-Prime - Adv. Electr. Eng. Electron. Energy **8** (2024). https://doi.org/10.1016/j.prime.2024.100608
38. H. Song, S. Luo, H. Huang, B. Deng, J. Ye, *Solar-Driven Hydrogen Production: Recent* (2022)
39. A. Brinner, *The 10 I < Jf7 Test and Research Facility** (2015)
40. O. Khaselev, J.A. Turner, A monolithic photovoltaic-photoelectrochemical device for hydrogen production via water splitting. Science **280**(5362), 425–427 (1998). https://doi.org/10.1126/science.280.5362.425
41. M.M. May, H.J. Lewerenz, D. Lackner, F. Dimroth, T. Hannappel, Efficient direct solar-to-hydrogen conversion by in situ interface transformation of a tandem structure. Nat. Commun. **6**, 4–10 (2015). https://doi.org/10.1038/ncomms9286
42. A. Rothschild, H. Dotan, Beating the efficiency of photovoltaics-powered electrolysis with tandem cell photoelectrolysis. ACS Energy Lett. **2**(1), 45–51 (2017). https://doi.org/10.1021/acsenergylett.6b00610
43. D. Akinyele, *Design and Optimization of Off-Grid Solar Power Systems for Rural Electrification* (2024)
44. M. Dada, P. Popoola, Recent advances in solar photovoltaic materials and systems for energy storage applications: a review. Beni-Suef Univ. J. Basic Appl. Sci. **12**(1) (2023). https://doi.org/10.1186/s43088-023-00405-5
45. M. Sameti, E. Syron, 100% renewable industrial decarbonization: optimal integration of solar heat and photovoltaics. Smart Energy **8**, 2022 (2021). https://doi.org/10.1016/j.segy.2022.100087
46. R. Xia, S. Overa, F. Jiao, Emerging electrochemical processes to decarbonize the chemical industry. JACS Au **2**(5), 1054–1070 (2022). https://doi.org/10.1021/jacsau.2c00138
47. D.S. Mallapragada et al., Decarbonization of the chemical industry through electrification: barriers and opportunities. Joule **7**(1), 23–41 (2023). https://doi.org/10.1016/j.joule.2022.12.008

Advancements Beyond Conventional Solar Conversion Techniques

Photovoltaic Assisted Water Electrolysis

S. P. Aravind and Prasanth Ravindran

Abstract This chapter outlines an integrated approach to water electrolysis by photovoltaic (PV) technology for sustainable green hydrogen generation. A foundational overview pertaining to the operational principles of photovoltaic systems and water electrolysis, this chapter further discusses the diversity of solar cells and electrolyzer, highlighting their respective efficiencies, proceeding to an in-depth knowledge of how the two technologies work and the synergy resulting from their combination. Key concepts discussed include principles of photovoltaics and electrochemistry relative to the splitting of water. An exploratory advancement in optimization and control strategies by advanced power electronics for lessening power fluctuations owing to the variability of solar energy and the dynamic requirements of the electrolytic systems are provided. These parts accentuate the importance of integrating smart management systems to improve the efficiency and stability of PV-assisted electrolysis setups. In the optimization context of the system, the chapter emphasizes techniques set to maximize hydrogen production. These include various control algorithms and system configurations designed in such a way that they can be adapted to different inputs of solar while maintaining an optimum output of hydrogen. This chapter analyzes the practical cases and real-world applications from which to extract the scalability and performance of existing systems by investigating the economic viability and market analysis for PV-assisted electrolysis, especially as a renewable energy source. Finally, it presents future challenges: the research needs that already exist and potential technological advances. It demands continuous innovation and collaboration to surpass barriers that exist currently and increase the feasibility of producing solar-powered hydrogen. By charting out these pathways, the chapter is intended to contribute toward the broader goal of transitioning toward a sustainable and economically viable hydrogen economy.

Keywords Photovoltaic (PV) technology · Water electrolysis · Green hydrogen generation · Optimization · Hydrogen production

S. P. Aravind · P. Ravindran (✉)
Nano Photonics Laboratory, Department of Green Energy Technology, UNESCO Madanjeet School of Green Energy Technologies, Pondicherry University, Kalapet, Puducherry, India
e-mail: prasanth.get@pondiuni.edu.in

P. Ravindran et al. (eds.), *Photon to Power*, Progress in Optical Science and Photonics 32, https://doi.org/10.1007/978-981-96-5914-2_12

1 Introduction

The global imperative underscores the urgency of decarbonization to mitigate climate change and transition from fossil-based energy systems to sustainable, renewable energy sources. In this context, green hydrogen emerges as a pivotal solution, offering a clean energy alternative with the potential to significantly reduce carbon emissions across various hard-to-abate sectors, including transportation, industrial heating, steel production, and power generation [1]. Unlike carbon-intensive methods such as steam methane reforming, which emits approximately 12 kg of CO_2 per kilogram of hydrogen produced, green hydrogen is generated via water electrolysis powered by renewable energy sources like wind and solar, thereby avoiding carbon emissions during production [2, 3]. This method not only supports the decarbonization of the energy sector but also addresses the intermittency of renewable energy sources by providing a means of storing surplus energy [4].

The growing global momentum toward a green hydrogen economy is evidenced by the emergence of national and regional roadmaps, policy initiatives, and investment programs to facilitate the large-scale deployment of green hydrogen technologies. Governments and businesses worldwide are recognizing the strategic importance of green hydrogen in achieving the climate goals set forth by the Paris Agreement. They are actively pursuing its development and integration into their energy systems. Despite the promising potential of green hydrogen, significant technological and economic challenges remain to be addressed [1, 5].

As demand for hydrogen is expected to rise significantly, projections indicate that it could constitute 18% of total energy demand by 2050, so the development and deployment of green hydrogen technologies are imperative. The economic feasibility of large-scale hydrogen production and storage is contingent on technological progress and cost reductions [6–8]. A promising approach to sustainable hydrogen production is integrating photovoltaic (PV) energy with water electrolysis, which harnesses abundant solar energy. This pathway addresses the intermittency of renewable energy sources and provides a clean and stable energy carrier, making it a critical component in the global shift toward decarbonization [9–11]. Direct and indirect coupling are the two main methods of integrating PV systems with water electrolysis. An indirect photovoltaic-water electrolyzer is predominantly used, which deploys maximum power point tracking (MPPT) devices and DC–DC converters for a consistent operation wherein direct coupling operates varyingly according to the climatic constraints [12–19].

Recent advancements in PV-electrolysis systems have demonstrated remarkable improvements in solar-to-hydrogen (STH) efficiency, achieving up to 30% efficiency by optimizing the match between PV cell voltage characteristics and electrolyzer performance. Advanced materials, such as Nafion membranes and noble metal catalysts in polymer electrolyte membrane (PEM) electrolyzers, have further enhanced system durability and performance [20–23].

The selection of electrodes plays a crucial role in water dissociation as the desired potential in the PV setup increases the cost. The electrode selection should meet the requisite of reducing the overpotential by prompting high catalytic active cathode materials. Less expensive, high catalytic active electrodes will make hydrogen production more feasible and efficient. The noble metals such as Ir, Ru, Pd, Pt as well as the d-block elements such as Fe, Co, Mo, Mn, Ni and their respective modified nano compounds has been proven better electrochemical activity, durability, and low over potential and thereby prompted hydrogen production. Despite that the corrosive nature of the transition elements and the dearth of availability of the noble metals drives the researchers to explore some alternatives. Recent studies have been reported the effectiveness of Ni-based alloys as catalysts for hydrogen generation [24–29].

Integrating PV energy with water electrolysis boosts hydrogen production efficiency and contributes to broader goals of renewable energy storage and grid stabilization. Moreover, integrating water treatment processes, such as desalination and waste heat for water distillation, enhances system sustainability, particularly for offshore applications. This holistic approach, combining electrolysis with renewable energy and advanced control systems, underscores the pivotal role of hydrogen in future energy strategies, offering a scalable and environmentally friendly solution for green hydrogen production [30–34].

Green hydrogen's versatility as an energy carrier further enhances its integration into various applications, including as a feedstock for industrial processes, fuel for electric vehicles, and a buffer for renewable energy, boosting energy security and sustainability. For instance, hydrogen is being injected into natural gas networks in Italy and used in Finland's energy system to reduce CO_2 emissions and fossil energy consumption despite a consequent increase in electricity demand. The adoption of hydrogen in maritime transportation and industrial port areas further underscores its role in decarbonizing operations and heavy transport systems [35–37].

The integration of hydrogen with other renewable technologies, such as battery energy storage systems, can enhance economic competitiveness and reduce environmental impacts. Coordinated policy efforts and continued research and development are essential to strategically deploy green hydrogen and achieve a sustainable, carbon–neutral future [38, 39].

However, realizing the full potential of green hydrogen requires overcoming several challenges, including technological advancements in electrolyzer design, prioritizing the development of robust control mechanisms by advanced power electronics to enhance the reliability and efficiency of systems for adapting into various solar conditions, infrastructure upgrades for hydrogen transmission and combustion [40–45].

2 Fundamentals

2.1 *Photovoltaics (PV)*

Photovoltaics is a field of study that has captivated the scientific community for decades, wherein photovoltaic cells act as the workhorses of solar energy conversions by exploiting the photovoltaic effect to generate electricity. This fundamental principle expedites certain materials, such as silicon, to absorb photons from sunlight and generate electron–hole pairs, which can then be harnessed to produce an electric current. These photons that impact the PV cell create an excitation in energy, building a potential difference in the circuit and producing voltage to operate it. The efficiency of photovoltaic cells depends on the material composition and the design and engineering of the overall PV system [46].

Generally, Si-based semiconductor materials are the most commonly preferred owing to the unique properties of Silicon, such as higher efficiency with minimal loss, durability, suitable bandgap, and technological maturity [47]. In addition to Si-based PVs, researchers have developed various promising alternatives and technologies, such as thin film-based PVs, Perovskite solar cells, Dye-sensitized, and Organic PVs. The choice depends upon distinctive applications, efficiency, availability, and material cost.

2.1.1 Silicon-Based Photovoltaic Cells

Monocrystalline Silicon Photovoltaic Cells

Monocrystalline silicon photovoltaic cells, made using the Czochralski process, offer high efficiency and long life. With thin wafers, Fresnel lenses, and bandgap engineering, they reach about 22% efficiency. Using wxAMPS software to optimize absorber thickness while maintaining performance by adjusting dopant levels to enhance charge mobility. This led to improvements in current density–voltage, energy band structure, and external quantum efficiency, achieving a 28% power conversion efficiency. Chemical wet etching improved light absorption, and a conductive distributed Bragg reflector (DBR) at the rear of the cell further enhanced light capture, especially for near-infrared photons (900–1200 nm), boosting efficiency [48–50].

Polycrystalline Silicon Photovoltaic Cells

Polycrystalline silicon photovoltaic cells have historically been less efficient, with a maximum recorded efficiency of 21.9%. Ongoing research aims to improve this through Metal-Induced Layer Exchange Crystallization, which reduces grain boundary and intragrain defects, enhancing poly-Si material quality. However, the high defect density in poly-Si layers remains a challenge, making optimization crucial for achieving more efficient polycrystalline silicon solar cells [51–53].

2.1.2 Thin Film Photovoltaic Cells

Amorphous Silicon Solar Cells

Amorphous silicon solar cells offer a cost-effective alternative to traditional crystalline silicon cells, with efficiencies suitable for terrestrial applications. Their p-i-n structure, with intrinsic amorphous silicon between p-type and n-type layers, makes them ideal for indoor power generation. These cells are typically produced using plasma-enhanced chemical vapor deposition (PE-CVD). However, the Staebler–Wronski effect, which increases dangling bonds under light exposure, reduces efficiency over time. To mitigate this, intermediate layers with optimized refractive indices improve light absorption. Adding a window layer, such as a-SiC:H or graphene oxide, has further enhanced short-circuit voltage and overall efficiency [54–57].

Cadmium Telluride (CdTe)

Cadmium telluride (CdTe) thin film solar cells are a highly promising technology, with efficiencies reaching 22.1%. By integrating a thin CuInTe2 (CIT) current booster layer and optimizing its parameters, CdTe devices have achieved an impressive efficiency of 37.46%, a VOC of 1.102 V, JSC of 38.50 mA/cm^2, and a fill factor of 88.30%. Additionally, photon recycling with high-reflectance Bragg reflectors could push efficiency to around 44.3%. Research also shows that excess tellurium alters the crystalline structure, affecting electrical properties, and a new CdTe powder synthesis method has been developed to improve thin film solar cell quality [58, 59].

Copper Indium Gallium Selenide (CIGS)

Copper indium gallium selenide (CIGS) solar cells are efficient thin film technologies with a high efficiency of 22.8%, comparable to crystalline silicon (c-Si) solar cells. Recent advancements, such as silver alloying and gallium grading at the back contact, have pushed CIGS efficiency to 23.6%. CIGS cells are favored for their high absorption coefficient, tunable bandgap, and overall efficiency. This study seeks to address the limitations of traditional vacuum and high-temperature processes by using scalable methods. The process includes depositing a CIGS absorber using nanoparticle ink, followed by printing a buffer layer and a conductive ZnO/[AgNW/ZnO] electrode as the window layer [60–63].

Gallium Arsenide (GaAs) Solar Cells

GaAs-based solar cells have demonstrated great potential for achieving over 30% efficiency in space applications, thanks to GaAs's optimal 1.42 eV bandgap, lightweight structure, and high photon absorption. By fine-tuning the absorber layer thickness to 200 μm, GaAs cells have reached a peak efficiency of 30.16%. Additionally, GaAs nanostructure-based solar cells show improved light-harvesting efficiency compared to conventional silicon cells [64–66].

2.1.3 Advanced Photovoltaic Cells

Perovskite Solar Cells

These types of cells are most suitable for photoelectrochemical water-splitting applications due to their usually high open-circuit voltage of roughly 1.1–1.4 V, which easily exceeds the thermodynamic potential needed to drive the solar-assisted water-splitting reactions with a single photo absorber [67]. The continuous improvement in STH efficiency, now achieving over 20%, has been through the development of perovskite/silicon tandem cells by bandgap engineering and using thin solid electrolytes with enhanced resistivity to decrease losses [68]. However, large-scale deployment still faces great challenges related to energy losses, fabrication complexity, and cost-effectiveness. Despite these daunting challenges, fast growth in power conversion efficiency and the prospect of high STH efficiency make the PSC a promising candidate for prospective solar energy applications related to sustainable hydrogen production [23, 69–71].

Dye-Sensitized

Dye-sensitized solar cells (DSSCs) consist of five key components: a transparent conducting substrate, dye layer, semiconductor film, electrolyte, and electrodes. The transparent conducting substrate is a transparent material with a conductive backing that supports the dye layer, which absorbs sunlight and initiates electron transfer. DSSC working electrodes are typically made from a conductive substrate coated with a porous semiconductor film, which is composed of semiconductor particles of various sizes. These particles are sensitized by dye molecules to enhance light absorption. A unique feature of DSSCs is the separation of light absorption and charge transport processes—dye molecules handle light harvesting, while semiconductors and electrolytes manage charge transport. This separation allows for optimization of each function, improving efficiency and reducing costs compared to traditional photovoltaics. The electrolyte in DSSCs, which can be liquid, gel, or polymer solid, contains a redox couple that facilitates electron transfer between electrodes [72–76].

Organic Solar Cells

Organic solar cells (OSCs) are emerging as a potential alternative to traditional inorganic solar cells, primarily because they use polymers instead of semiconductors like silicon or gallium arsenide. This makes OSCs lightweight, flexible, and cost-effective, which is appealing for sustainable energy solutions. Typical OSCs consist of multiple functional layers, including a substrate, an organic solar cell device, hydrophobic polymer layers, and metal layers. The hydrophobic polymer and metal layers are critical in preventing water and oxygen penetration, thereby enhancing durability and performance. Additionally, incorporating a conductive substrate with electrodes and strategically placed metal layers improves electron flow, boosting efficiency. OSC performance can also be enhanced by using multiple electron acceptors in the active layers, enabling better light absorption and electron transport, which increases the overall energy conversion rate [77, 78].

2.2 *Types of Electrolyzers*

Water electrolysis systems, essential for green hydrogen production, differ in their electrolytes, working conditions, and ion transport mechanisms. The main types of electrolyzers are:

2.2.1 Alkaline Electrolyzers (AWE)

Alkaline electrolyzers use a liquid alkaline solution, commonly potassium hydroxide (KOH), as the electrolyte. The electrochemical reactions involved are:

Anode:

$$4OH^- \rightarrow O_2 + 2H_2O + 4e^-$$

Cathode:

$$2H_2O + 2e^- \rightarrow H_2 + 2OH^-$$

The hydroxide ions (OH^-) produced at the cathode move toward the anode, oxidizing them, generating oxygen gas and water. A diaphragm separates the electrodes to prevent gas mixing. AWE is a well-established and economical technology. However, it operates at lower current densities than other systems and requires regular maintenance due to issues such as electrolyte degradation and gas crossover [79–81] (Fig. 1).

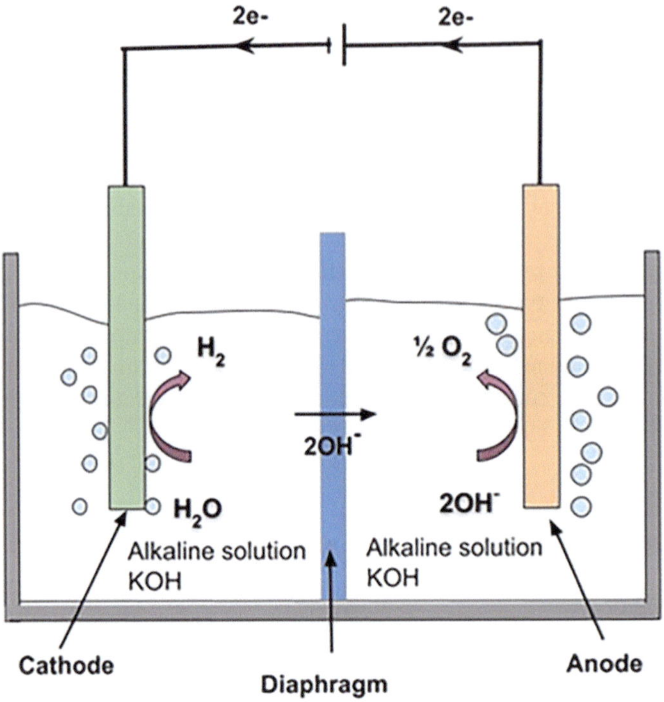

Fig. 1 Alkaline electrolyzer

2.2.2 Polymer Electrolyte Membrane Electrolyzers (PEM)

PEM electrolyzers use a solid proton-conducting polymer membrane as the electrolyte, allowing only protons to pass through. The reactions are:

Anode:

$$2H_2O \rightarrow O_2 + 4H + +4e^-$$

Cathode:

$$4H + +4e^- \rightarrow 2H_2$$

Water is oxidized at the anode to produce oxygen, protons, and electrons. The protons migrate through the membrane to the cathode, recombining with electrons to form hydrogen gas. PEM electrolyzers operate at temperatures between 30 and 80 °C, and are known for high current densities, rapid response to power fluctuations, and high efficiency. However, they require expensive platinum-group catalysts [82] (Fig. 2).

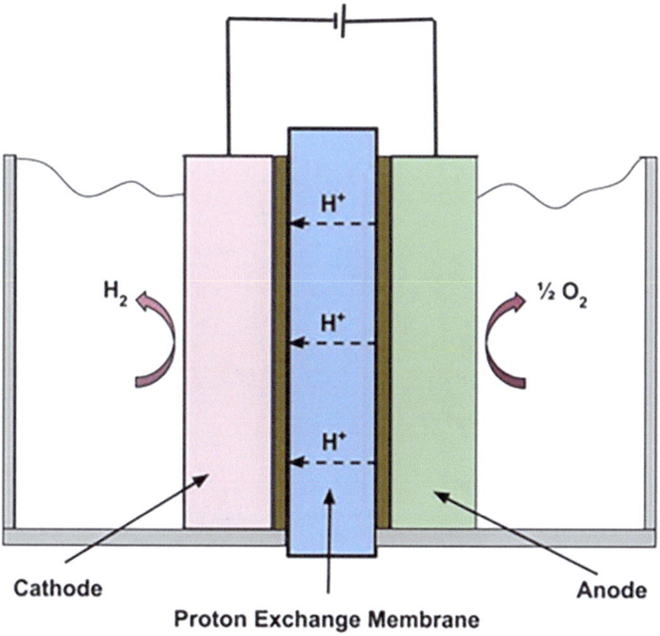

Fig. 2 Proton exchange membrane electrolyzer

2.2.3 Anion Exchange Membrane Electrolyzers (AEM)

AEM electrolyzers use a durable anion exchange membrane that facilitates the movement of hydroxide ions (OH^-) from the cathode to the anode. The electrochemical reactions involved are similar to alkaline electrolysis:

Anode:

$$4OH- \rightarrow O_2 + 2H_2O + 4e^-$$

Cathode:

$$2H_2O + 2e^- \rightarrow H_2 + 2OH^-$$

Hydroxide ions generated at the cathode move through the anion exchange membrane to the anode, oxidizing them, producing oxygen gas and water. AEM combines the advantages of both AWE and PEM systems, such as eliminating the need for liquid electrolytes and allowing non-precious metal catalysts, making it a promising technology for future green hydrogen production [84, 85] (Fig. 3).

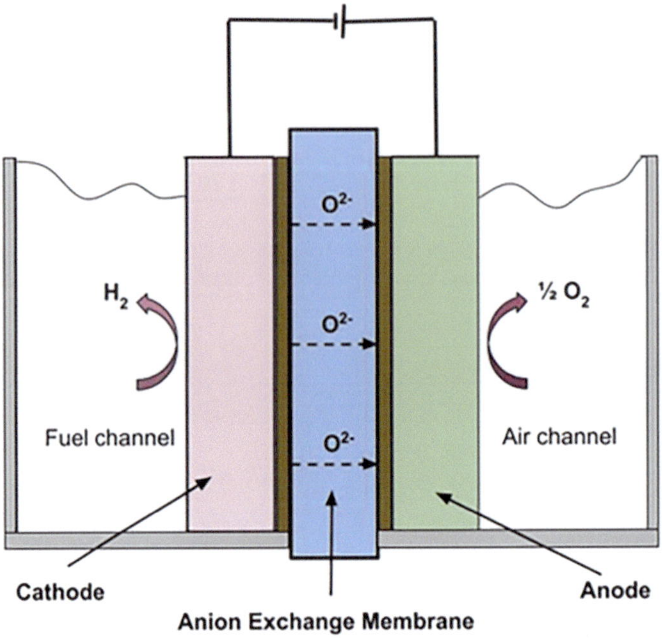

Fig. 3 Anion exchange membrane electrolyzer

2.2.4 Solid Oxide Electrolyzers (SOEC)

SOECs operate at elevated temperatures, typically between 700 and 1000 °C, using a solid ceramic electrolyte of yttria-stabilized zirconia (YSZ). The reactions are:

Anode:

$$O_2^- \rightarrow O_2 + 2e^-$$

Cathode:

$$H_2O + 2e^- \rightarrow H_2 + O_2^-$$

Water is reduced at the cathode to produce hydrogen gas and oxygen ions (O^{2-}). The oxygen ions travel through the ceramic electrolyte to the anode, where they are oxidized to release oxygen gas. The high operating temperature enhances reaction kinetics and enables the use of waste heat from industrial processes, potentially increasing energy efficiency. However, challenges like material degradation and maintaining durability at such high temperatures remain [86–89] (Fig. 4).

The development of advanced electrocatalysts and novel cell designs has significantly improved the performance and efficiency of water electrolysis systems.

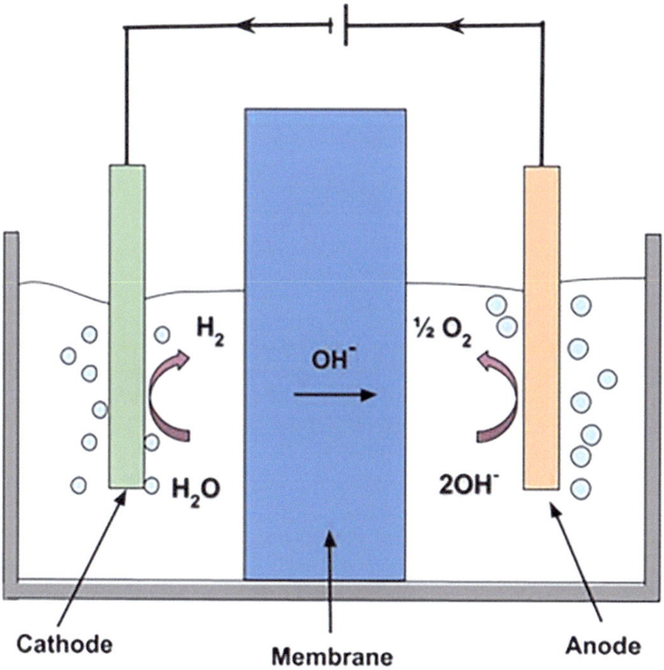

Fig. 4 Solid oxide electrolyzer

Research focuses on enhancing electrocatalyst activity, selectivity, and stability to reduce costs, increase durability, and improve overall hydrogen production efficiency [81, 89–93] (Table 1).

2.3 *Types of Membranes and Electrodes Used*

The types of membranes and electrodes used in electrolysis are crucial for optimizing the efficiency and cost-effectiveness of hydrogen production. Electrolysis technologies primarily include Alkaline Water Electrolysis (AWE) and Polymer Electrolyte Membrane Electrolysis (PEME), each utilizing distinct membranes and electrodes to facilitate the electrochemical reactions necessary for hydrogen generation. Below, we explore the various types of membranes and electrodes used in these processes.

Table 1 Comparison of AWE, PEM, AEM, and SOEC

Technology	Alkaline water electrolysis (AWE)	Proton exchange membrane (PEM)	Anion exchange membrane (AEM)	Solid oxide electrolysis cell (SOEC)
Operating temperature (°C)	65–100	50–80	45–80	500–1000
Operating pressure (bar)	2–10	30–80	30–80	< 30
Efficiency (%)	62–84	55–80	70–80	> 90
Current density (A/Cm^2)	0.2–0.5	0.5–10	1.0	0.2–1.3
Advantages	Low capital cost, mature technology, long lifespan, no need for precious metals	High efficiency, fast startup, high hydrogen purity, compact design	Low cost, uses non-precious metals, high gas purity, high current density	Very high efficiency, uses waste heat, high hydrogen production capacity
Disadvantages	Lower current density, slow response to load changes, sensitivity to impurities	High capital cost, use of precious metals, lower durability in acidic conditions	Low maturity, stability, and durability issues	High operating temperature, high cost, lower durability
Applications	Large-scale hydrogen production	Renewable energy integration	Emerging technology with cost reduction potential	Industrial applications with high-temperature steam

2.3.1 Types of Membranes

Alkaline Water Electrolysis (AWE) Membranes

Porous Separators: Traditionally, AWE uses diaphragms to separate the anode and cathode, allowing ion conduction while preventing gas crossover. Modern alternatives include polymeric diaphragms and ion-solvating membranes, which offer high conductivity even in lower KOH concentrations [94].

Anion Exchange Membranes (AEMs): These membranes facilitate anion conduction and are increasingly used due to their ability to operate with lower KOH concentrations, balancing conductivity and membrane lifespan [94, 95].

Thin Film Composite (TFC) Membranes: A new category in AWE, TFC membranes consist of a quaternary ammonium selective layer on a porous polyethylene support, offering high anion conductivity and mechanical stability, thus reducing gas crossover and enhancing performance [96].

Polymer Electrolyte Membrane Electrolysis (PEME) Membranes

Proton Exchange Membranes (PEMs): These membranes are used in PEME for their high proton conductivity and efficiency, though they require expensive catalysts like IrO_2 and Pt due to the acidic environment [97, 98].

Nanocomposite Membranes: Incorporating reinforcing polymer nanofibrous webs into the PFSA matrix, these membranes enhance mechanical robustness and reduce gas crossover while maintaining high proton conductivity [97].

2.3.2 Types of Electrodes

Alkaline Water Electrolysis (AWE) Electrodes

Nickel-Based Electrodes: Nickel and nickel–cobalt alloys are commonly used due to their corrosion resistance and catalytic activity. Nickel mesh electrodes, in particular, show excellent adhesion to catalytic coatings and high durability [99].

Zero-Gap Electrode Configuration: This involves pressing electrodes directly onto the separator surface, reducing ionic resistance and enhancing efficiency. Porous silicon separators with metal electrodes in a zero-gap configuration have shown promising results [100].

Polymer Electrolyte Membrane Electrolysis (PEME) Electrodes

Iridium Oxide Supported on Metal Oxides: To reduce iridium loading while maintaining high activity and durability, iridium oxide is supported on doped tin oxide materials, enhancing corrosion resistance and electronic interactions [97].

Nanocomposite Electrocatalysts: These are used to prepare membrane-electrode assemblies, offering improved performance and stability in PEME cells [97].

Advancements in membrane and electrode technologies have greatly enhanced the efficiency and cost-effectiveness of electrolysis, yet challenges persist. High costs of precious metal catalysts in Proton Exchange Membrane Electrolysis (PEME) and the mechanical and chemical stability of anion exchange membranes (AEMs) in Alkaline Water Electrolysis (AWE) remain as ongoing concerns. Furthermore, the development of membranes that reduce gas crossover without sacrificing ionic conductivity is vital for both AWE and PEME systems. These issues underscore the necessity for ongoing research and innovation in materials science to further improve the feasibility of electrolysis for large-scale hydrogen production. Selecting the appropriate membranes and electrodes is critical to the efficiency and sustainability of the electrolysis process. Despite considerable advancements, especially in the creation of advanced membranes and electrode configurations, continuous research is imperative to overcome current challenges and refine these technologies for future use [101–105].

2.4 Thermodynamics of Water Electrolysis

Water electrolysis involves splitting water into hydrogen and oxygen gases by applying electrical energy. This process is guided by thermodynamic principles that determine the energy required to drive the reaction and the efficiency limitations.

2.4.1 Electrochemical Reaction

The overall reaction for water electrolysis is as follows:

$$H_2O(l) \rightarrow H_2(g) + O_2(g)$$

This reaction can be broken down into two half-reactions at the anode and cathode:

Anode (oxidation):

$$2H_2O \rightarrow O_2 + 4H^+ + 4e^-$$

Cathode (reduction):

$$4H^+ + 4e^- \rightarrow 2H_2$$

The energy required to drive these reactions is dictated by the Gibbs free energy (ΔG) and enthalpy (ΔH) of the reaction [40].

2.4.2 Gibbs Free Energy and Enthalpy

Gibbs free energy (ΔG) represents the minimum electrical energy required for the reaction, under standard conditions, which is 237.13 kJ/mol. This is the amount of energy that must be supplied in the form of electrical energy.

Enthalpy (ΔH) represents the total energy required, accounting for both electrical and thermal energy. For water electrolysis, the enthalpy is 285.83 kJ/mol.

The thermodynamic efficiency of the system is defined by the ratio:

$$\text{Efficiency} = \frac{\Delta G}{\Delta H} \approx 83\%$$

This means that, ideally, about 83% of the energy for the reaction must be supplied electrically, while 17% can be supplied thermally [40, 106, 107].

2.4.3 Standard Electrode Potential

The relationship between Gibbs free energy and the theoretical minimum voltage for electrolysis is given by:

$$\Delta G = -nFE$$

where

n number of moles of electrons involved (2 for water electrolysis)
F Faraday constant (96,485 C/mol).

The theoretical standard electrode potential is:

$$E = \frac{\Delta G}{nF} \approx 1.23\text{V}$$

Thus, the minimum voltage required to split water is 1.23 V under ideal conditions [40, 107, 108].

2.4.4 Overpotentials and Practical Voltage

In practical electrolysis systems, the actual voltage required is higher than the theoretical minimum due to overpotentials. These arise from various energy losses, such as:

Activation overpotential: Due to the kinetic barriers at the electrodes.

Ohmic overpotential: Due to resistance in the electrolyte, electrodes, and other system components.

Concentration overpotential: Arising from mass transport limitations as reactants and products are transported to and from the electrode surfaces.

As a result, practical operational voltages range from 1.6 to 2.0 V [40, 107].

2.4.5 Temperature Effects on Electrolysis

The Gibbs free energy (ΔG) decreases as temperature increases, reducing the electrical energy required. However, the enthalpy (ΔH) remains relatively constant since the extra energy can be supplied as heat. In high-temperature electrolysis, such as solid oxide electrolyzer cells (SOECs), external heat can improve efficiency by reducing the electrical energy demand. This can be advantageous when utilizing waste heat from industrial processes [40, 107].

2.4.6 Thermal Management

Proper thermal management is crucial in electrolysis systems to optimize energy efficiency. The difference between ΔH and ΔG (i.e., the thermal energy) can be supplied as heat, which decreases the need for electrical energy input. This makes the process more efficient, particularly in high-temperature systems where heat can be harvested from external sources, such as waste heat from other processes [40, 107, 109, 110].

By understanding the thermodynamics of water electrolysis, researchers and engineers can design more efficient and cost-effective systems for large-scale hydrogen production.

2.5 *Integration of PV and Electrolysis*

The increasing demand for clean and sustainable energy sources has spurred significant interest in developing green hydrogen production technologies. Hydrogen, as a versatile energy carrier, holds immense potential to decarbonize various sectors, including transportation, industry, and power generation. Photovoltaic (PV) systems, with their ability to harness abundant solar energy, offer a promising pathway for sustainable hydrogen production through water electrolysis. This process involves splitting water molecules into hydrogen and oxygen using electricity, and PV systems can provide the necessary electrical energy for this process [10, 22, 33].

The integration of PV systems with electrolysis systems presents both opportunities and challenges. The efficiency, cost-effectiveness, and overall performance of such integrated systems are influenced by various factors, including the type of coupling employed, the power electronics used, and the control strategies implemented. This section explores the integration of PV systems with electrolysis, focusing on direct and indirect coupling methods, the role of power electronics, maximum power point tracking (MPPT) techniques, and the advantages and disadvantages associated with each approach [18, 23].

2.5.1 Solar-to-Hydrogen (STH)

Solar-to-hydrogen (STH) efficiency measures the effectiveness of converting solar energy into hydrogen through processes such as photovoltaic assisted water electrolysis or photoelectrochemical (PEC) water splitting. It is calculated as the ratio of the chemical energy in the produced hydrogen to the total solar energy input.

$$\text{STH efficiency} = \frac{\text{Energy content of hydrogen produced}}{\text{Total solar energy input}} \times 100$$

Key factors influencing STH efficiency include the photovoltaic efficiency, electrolyzer efficiency, optical losses, and overpotentials in the electrolysis process. In PV-electrolysis systems, typical STH efficiencies range from 10 to 20%, depending on the efficiencies of both the solar panels and the electrolyzer. PEC systems generally exhibit lower efficiencies, ranging from 1 to 15%, as they are more affected by material and optical losses. Enhancing STH efficiency is crucial for reducing the cost of solar-based hydrogen production, making it a key focus in the development of renewable hydrogen technologies [20, 111–114].

2.5.2 Direct Coupling of PV Systems with Electrolysis

Direct coupling involves connecting PV modules to an electrolyzer, eliminating the need for intermediate power conversion stages. This approach offers several advantages, including simplicity, cost-effectiveness, and adaptability. The absence of additional components, such as DC–DC converters or inverters, simplifies the system and reduces energy losses associated with power conversion. This can increase overall system efficiency, particularly under favorable operating conditions. Eliminating power conversion stages can reduce the initial investment and maintenance costs, making direct coupling an attractive option for small-scale or decentralized hydrogen production systems. Direct coupling allows the system to respond quickly to changes in solar irradiance, ensuring that the electrolyzer operates within its optimal range of current and voltage. This adaptability is particularly important for maximizing hydrogen production under varying solar conditions.

Despite its advantages, direct coupling also presents certain challenges and limitations, including variability of solar power, limited operating hours, and scalability challenges. The output of PV systems is inherently variable, depending on factors like solar irradiance, temperature, and shading. This variability can affect the electrolyzer's operation, potentially leading to inefficiencies or even damage if the voltage and current levels are not properly managed. Without energy storage, direct coupling systems can only produce hydrogen during daylight hours when solar energy is available. This limits the overall hydrogen production capacity and may necessitate additional energy sources or storage systems to ensure continuous operation. Scaling up direct coupling systems can be challenging due to the need to precisely match the PV output with the electrolyzer's input requirements. This can become complex and expensive for larger systems, limiting their scalability [13, 15, 40, 115–117].

2.5.3 Indirect Coupling of PV Systems with Electrolysis

Indirect coupling involves using power electronics, such as DC–DC converters or inverters, to interface the PV system with the electrolyzer. This approach offers several advantages over direct coupling, including improved control and flexibility, and integration of energy storage and scalability. Indirect coupling provides greater control over the voltage and current supplied to the electrolyzer, allowing for optimization of the electrolysis process under varying solar conditions. This can lead to improved system efficiency and hydrogen production. Indirect coupling systems can readily incorporate energy storage components, such as batteries, to store excess solar energy and enable hydrogen production even during periods of low or no solar irradiation. This enhances the system's reliability and flexibility. Indirect coupling is more scalable than direct coupling, as it allows for better power flow management and adaptation to larger and more complex systems.

While indirect coupling offers several advantages, it also has certain limitations, including increased complexity and cost, energy losses, and control complexity. Including power electronics increases the system's complexity and initial investment cost. This can be a significant consideration, particularly for smaller or cost-sensitive applications. Power conversion processes in indirect coupling systems can lead to energy losses, potentially reducing the overall system efficiency compared to direct coupling. Managing the power flow and optimizing the system's performance in indirect coupling systems can be more complex than in direct coupling systems, requiring sophisticated control strategies and algorithms [40, 118–121].

2.5.4 Calculation of the Operating Point in Directly and Indirectly Coupled Systems

In directly coupled systems, the photovoltaic (PV) panels are directly connected to the electrolyzer without any intermediate power electronics, such as a DC–DC converter. The operating point in this system is determined graphically by the intersection of the current–voltage (I–V) curve of the PV generator and the I–V curve of the electrolyzer. The intersection of these two curves represents the equilibrium where the power supplied by the PV panel matches the power consumed by the electrolyzer.

The conditions for calculating the operating point in direct coupling are:

The voltage of the PV generator must equal the voltage of the electrolyzer.

$$V_{\text{PV}} = V_{\text{Electrolyzer}}$$

The current supplied by the PV module must equal the current required by the electrolyzer:

$$I_{\text{PV}} = I_{\text{Electrolyzer}}$$

The non-linear equation for the voltage-current relationship is given as:

$$V_{\text{th}} \cdot \log\left(\frac{(I_{\text{ph}} - I_{\text{dc}})}{I_{\text{s}}} + 1\right) - R_{\text{s}} \cdot I_{\text{dc}} = R_{\text{electrolyzer}} \cdot I_{\text{dc}} + V_{\text{stack}}$$

where V_{th} is the thermal voltage, R_s is the series resistance, I_{dc} is the DC current, $R_{electrolyzer}$ is the resistance of the electrolyzer, and V_{stack} is the stack voltage.

In indirectly coupled systems, a DC–DC converter is inserted between the PV generator and the electrolyzer. This power electronics device helps decouple the PV panel's operating point from that of the electrolyzer, allowing the system to operate the PV panels at their maximum power point (MPP), where the PV generator achieves its highest efficiency.

In this configuration:

The PV generator operates at its MPP, optimized by a maximum power point tracker (MPPT).

The DC–DC converter ensures that the power from the PV panels is transferred to the electrolyzer while converting the voltage and current levels to match the electrolyzer's requirements.

The relationship for power matching in this case is:

$$P_{\text{PV,MPP}} \cdot \eta_{\text{converter}} = P_{\text{Electrolyzer}}$$

where

$P_{\text{PV,MPP}}$ power generated at the maximum power point.
$\eta_{\text{converter}}$ efficiency of the DC–DC converter, typically between 90 and 95%.

This approach allows for much more efficient operation, as the PV panels are always running at their optimal point regardless of the electrolyzer's characteristics.

In directly coupled systems, the operating point is determined by the natural intersection of the PV and electrolyzer characteristics, leading to simpler design but lower efficiency. In contrast, indirectly coupled systems, using a DC–DC converter and MPPT, offer more flexibility and higher efficiency by optimizing the PV generator's performance independently of the electrolyzer [121].

2.5.5 Choosing the Right Coupling Method

The choice between direct and indirect coupling depends on various factors, including the scale of the system, cost constraints, desired flexibility, and the specific application requirements.

Direct coupling is suitable for smaller, cost-sensitive systems with well-matched PV and electrolyzer characteristics, where simplicity and efficiency are prioritized. Indirect coupling is more appropriate for larger, more complex systems that require greater control, flexibility, and the ability to integrate energy storage.

2.5.6 The Role of Power Electronics

Power electronics, including DC–DC converters, inverters, and maximum power point tracking (MPPT) systems, are critical in integrating PV systems with electrolysis. These components enable efficient energy transfer, voltage and current regulation, and system performance optimization.

DC–DC Converters

DC–DC converters are used to match the voltage and current output of the PV system to the requirements of the electrolyzer. They can step up or down the voltage, ensuring the electrolyzer operates at its optimal point for maximum hydrogen production efficiency [122–125].

Inverters

Inverters are used in indirect coupling systems to convert the DC output of the PV system to AC, which can then be converted back to DC at the electrolyzer's required voltage and current levels. Inverters also enable grid connection, allowing for the export of excess energy or the import of energy during periods of low solar irradiation [16, 126, 127].

Maximum Power Point Tracking (MPPT)

MPPT systems are used to continuously track the maximum power point of the PV system, ensuring that the maximum possible energy is harvested from the solar panels. This is crucial for maximizing the hydrogen production capacity of the system [128–132].

MPPT techniques are essential for optimizing the energy output from PV systems and maximizing the efficiency of PV-electrolysis systems. Various MPPT algorithms have been developed, each with its advantages and limitations. Perturb and Observe (P&O) is a simple and widely used MPPT algorithm that involves perturbing the operating point of the PV system and observing the resulting change in power output. The algorithm then adjusts the operating point in the direction that increases the power output until the MPP is reached. Incremental Conductance (INC) is like P&O but uses the incremental change in conductance to determine the direction of the operating point adjustment. INC is generally considered more accurate and stable than P&O, especially under rapidly changing environmental conditions [133–135].

Several advanced MPPT techniques have been developed, including those based on fuzzy logic, artificial neural networks, and other optimization algorithms. These techniques offer improved accuracy, adaptability, and robustness compared to traditional methods like P&O and INC [136–139].

Integrating PV systems with electrolysis for hydrogen production offers a promising pathway to sustainable and decarbonized energy systems. Both direct and indirect coupling methods have their own advantages and disadvantages, and the choice of method depends on the specific application requirements. Power electronics, including DC–DC converters, inverters, and MPPT systems, are critical in optimizing energy transfer, ensuring efficient operation, and maximizing hydrogen production. Advancements in MPPT techniques, such as those based on artificial intelligence and machine learning, further enhance the efficiency and adaptability of PV-electrolysis systems. As research and development in this field continue, it is expected that the integration of PV systems with electrolysis will become increasingly efficient, cost-effective, and scalable, contributing significantly to the global transition toward a clean energy future [45, 118, 140–142].

2.6 Intermittent Solar Energy and Electrolyzer Demand: Challenges and Solutions

The increasing global demand for clean energy and the urgent need to decarbonize various sectors have placed significant emphasis on the utilization of renewable energy sources. Among these, solar energy has emerged as a frontrunner due to its abundance and potential for large-scale deployment. Photovoltaic (PV) technology has witnessed rapid advancements, making it an attractive option for generating electricity. However, the intermittent nature of solar energy, caused by factors such as diurnal cycles and weather variability, poses significant challenges to its integration into the existing energy infrastructure.

One promising approach to address the intermittency of solar energy is to utilize it for hydrogen production through water electrolysis. Electrolysis, a process that uses electricity to split water molecules into hydrogen and oxygen, offers a sustainable and carbon–neutral method for storing and utilizing energy. The hydrogen produced can be used in various applications, including transportation, industrial processes, and energy storage. However, the fluctuating nature of solar energy creates a dynamic demand on electrolyzers, affecting their efficiency, lifespan, and overall system stability.

This section explores the intricate relationship between intermittent solar energy and electrolyzer demand. It delves into the challenges posed by the variability of solar power and analyzes the solutions implemented to ensure efficient and sustainable hydrogen production. By examining various aspects, including the impact of weather patterns, the role of energy storage solutions, the integration of smart grid technology, and the optimization of PV and electrolysis systems, this review provides valuable insights into the current state and prospects of solar-powered hydrogen production [4, 34, 150–153].

2.6.1 Challenges Posed by Intermittent Solar Energy

The primary challenge associated with utilizing solar energy for hydrogen production is its inherent variability. Solar energy generation fluctuates significantly throughout the day and is influenced by weather conditions, including solar irradiance, cloud cover, and ambient temperature. These fluctuations create a dynamic demand on electrolyzers, which can lead to several operational and efficiency-related issues.

Fluctuating Demand

The intermittent availability of solar energy creates a fluctuating power supply to the electrolyzers. This necessitates the system to handle frequent start-stop cycles and rapid load changes, which can negatively impact the performance and lifespan of the electrolyzer components. Alkaline water electrolyzers (AWEs), a commonly used technology for hydrogen production, are particularly susceptible to degradation caused by frequent start-stop cycles and fluctuations in power input [40, 90].

Efficiency Losses

The variability in solar power output often results in electrolyzers operating at partial loads. This suboptimal operation can significantly impact the efficiency of hydrogen production, affecting parameters such as hydrogen flow rate, pressure, temperature, and purity. The dynamic operation requirements imposed by intermittent solar energy necessitate rapid startup capabilities, which are not always compatible with the traditional start/stop procedures of electrolysis systems [40, 116].

Weather Dependency

The output of PV systems is heavily influenced by weather patterns. Changes in solar irradiance, cloud cover, and ambient temperature directly impact the power generated by PV panels. For instance, high temperatures can reduce the efficiency of PV cells, while cloud cover can significantly decrease the amount of sunlight reaching the panels. This weather dependency creates uncertainty and unpredictability in the power supply to electrolyzers, making it challenging to maintain stable and efficient hydrogen production [40, 116, 154, 155].

2.6.2 Solutions for Mitigating the Impact of Intermittent Solar Energy

Several strategies and solutions have been developed to mitigate the challenges posed by its intermittent nature to harness the potential of solar energy for hydrogen production. These solutions focus on optimizing the integration of PV systems with electrolyzers, managing energy storage, and enhancing the overall efficiency and stability of the system.

Energy Storage Systems

Energy storage solutions, such as batteries and hydrogen storage, are pivotal in stabilizing the electrolyzer demand and ensuring a consistent hydrogen production process. Batteries can store excess energy generated during high solar irradiance periods and release it during periods of low solar irradiance or high demand. This allows for a smooth and continuous power supply to the electrolyzers, maximizing their efficiency and preventing damage caused by frequent start-stop cycles. Hydrogen storage, on the other hand, offers a longer-duration energy storage solution. The hydrogen produced through electrolysis can be stored and used as a clean energy carrier or reconverted back to electricity through fuel cells when needed. This approach not only enhances grid stability but also contributes to the development of a hydrogen economy, where hydrogen is utilized as a primary energy source for various applications [2, 82, 119, 156, 157].

Advanced Control Strategies

The integration of advanced control algorithms, such as model predictive control (MPC) and fuzzy logic, can significantly enhance the performance and efficiency of electrolyzer systems powered by solar energy. MPC algorithms utilize predictive models to anticipate changes in solar power output and make real-time adjustments to the operation of the electrolyzers. This proactive approach ensures that the electrolyzers operate optimally even under fluctuating power conditions, thereby minimizing energy wastage and improving overall system efficiency.

Fuzzy logic, on the other hand, is well-suited to handle the nonlinearities and uncertainties inherent in the electrolysis process. By incorporating various input variables, such as the state of charge of batteries and predicted future power values, fuzzy logic controllers can make informed decisions about the operation of electrolyzers, ensuring a balance between energy production and consumption [158–161].

Hybrid Systems

Combining PV systems with other energy sources, such as wind or grid power, can enhance system efficiency and reliability. Hybrid systems can utilize the grid to provide backup power during periods of low solar irradiance or high demand, thereby stabilizing the energy input to the electrolyzers. This integration allows for continuous hydrogen production and reduces the reliance on energy storage solutions [118, 140, 162].

System Design

The design and operation of electrolyzer systems also play a crucial role in addressing the challenges posed by intermittent solar energy. An optimal system design incorporates sufficient gas volumes to dilute gas contamination, liquid and solid volumes to maintain system temperature during part-load operations, and the ability to switch between separated and combined electrolyte cycles to manage gas impurities effectively [163].

Smart Grid Technology

Innovative grid technology is pivotal in managing solar energy and electrolyzer demand. By facilitating real-time monitoring and control of energy flow, intelligent grids enable dynamic adjustments to the operation of electrolyzers based on energy availability and demand. This ensures a steady and reliable hydrogen production process while maximizing the utilization of renewable energy sources [41].

2.6.3 Optimizing PV and Electrolysis Systems

Both PV and electrolysis systems need to be optimized to further enhance the efficiency and sustainability of solar-powered hydrogen production. This involves maximizing the power output of PV systems and ensuring the efficient operation of electrolyzers under varying conditions [40, 113, 146].

Optimizing PV Output

The efficiency of PV systems can be maximized through various strategies, including:

Maximum Power Point Tracking (MPPT)

MPPT algorithms ensure that PV panels operate at their maximum power point, maximizing energy conversion efficiency under varying solar irradiance and temperature conditions.

System Modeling

Accurate mathematical models can be used to predict the current voltage (I–V) characteristics of PV panels, allowing for precise adjustments to maintain optimal performance.

Advanced Optimization Algorithms

These algorithms can aid in developing novel operating strategies that consider technical and economic objectives.

High-Efficiency PV Cells

The use of high-efficiency PV cells, such as multi-junction designs, can significantly improve the overall system efficiency [164–167].

Optimizing Electrolysis Operation

Several factors influence the optimal operating conditions of electrolysis, including system design, temperature, current density, and the integration of renewable energy sources.

System Design

An optimal system design minimizes gas impurities and maintains system temperature during part-load operations.

Temperature Management

Maintaining the system temperature within an optimal range is essential for efficient operation and preventing the degradation of electrolyzer components.

Gas Purity

Ensuring high gas purity is critical for preventing explosive mixtures and ensuring the quality of the produced hydrogen.

Current Density

Optimizing current density can help mitigate gas contamination and improve efficiency, especially during low power availability.

Integration of Renewable Energy Sources

Efficient AC/DC converters are required to integrate renewable energy sources with electrolyzers, ensuring optimal operation under varying power conditions [41, 122, 123, 162, 168–171].

2.6.4 The Role of Energy Storage in Enhancing System Reliability

Energy storage systems play a crucial role in enhancing the reliability and stability of renewable energy systems, including PV-electrolysis systems for hydrogen production. Batteries and hydrogen storage offer different but complementary energy storage approaches, each with advantages and challenges.

Battery Storage

Batteries provide a direct and immediate storage solution, capturing excess solar energy for later use. They can be integrated with PV systems through DC-coupling or AC-coupling technologies, allowing for the efficient storage and utilization of renewable energy. Batteries also contribute to grid stability by providing ancillary services such as frequency regulation and voltage support [172, 173].

Hydrogen Storage

Hydrogen storage offers a long-duration energy storage solution, enabling the storage of excess energy generated from renewable sources over extended periods. Hydrogen can be stored in various forms, such as compressed gas, liquid hydrogen, or metal hydrides. The choice of storage method depends on the specific application and the desired storage capacity [174].

Hybrid Storage Solutions

Combining batteries and hydrogen storage can offer a more comprehensive and flexible approach to energy storage. Batteries can provide short-term energy storage and grid support services, while hydrogen storage can handle longer-duration storage requirements, especially during extended low solar irradiance periods. By combining these two technologies, the system can achieve a balanced and reliable energy supply, ensuring uninterrupted operation of the electrolyzer and maximizing the utilization of solar energy.

Intermittent solar energy poses challenges and opportunities for integrating electrolyzers in hydrogen production. Solar power variability impacts electrolyzer efficiency, lifespan, and system stability. However, advancements in energy storage, control algorithms like MPC, fuzzy logic, and hybrid systems are addressing these issues. Energy storage, particularly batteries, and hydrogen storage ensures a stable power supply, while hybrid systems increase flexibility and reliability. Optimizing PV-electrolysis systems with advanced strategies enhances efficiency and cost-effectiveness, positioning solar-powered hydrogen production as a critical player in achieving a sustainable, decarbonized energy future [2, 111, 118, 153, 175–179].

2.7 *Real-World Applications and Case Studies: Operational Projects Around the World*

Integrating photovoltaic (PV) systems with electrolysis for hydrogen production is increasingly recognized as a critical element in the global shift toward a low-carbon economy. This approach offers a sustainable alternative to traditional fossil-fuel-based hydrogen production by utilizing clean, renewable energy sources that can be applied across various sectors. However, scaling these systems to meet industrial demands involves significant technical, economic, and operational challenges. This section examines real-world applications and case studies of operational projects globally, focusing on lessons learned concerning scalability, performance, and the challenges encountered in these pioneering efforts (Fig. 5).

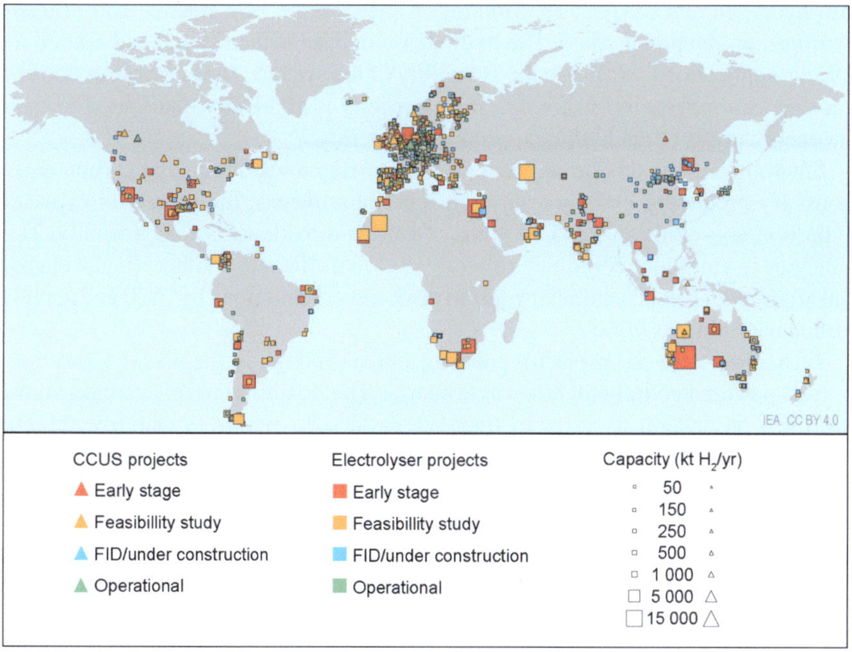

Fig. 5 Map of announced low-emission hydrogen production projects. This map displays various stages of hydrogen production projects globally, differentiated by project type and capacity. *Source* International Energy Agency (IEA) [36]

3 Global Overview of Operational Projects

3.1 Sinopec Xinjiang Kuqa Green Hydrogen Project (China)

The Sinopec Xinjiang Kuqa Green Hydrogen Project is a landmark development in the renewable energy sector, particularly in producing green hydrogen. With an annual capacity of 20,000 tons of hydrogen and an investment of $470.77 million, it is the largest photovoltaic-powered green hydrogen project globally. Located in Kuqa County, Xinjiang Uygur Autonomous Region, China, the project successfully integrates renewable solar energy with hydrogen production via water electrolysis.

The project's infrastructure includes a 300 MW photovoltaic power generation system, which provides the necessary electricity to drive hydrogen production. Through water electrolysis, solar-generated electricity is used to split water molecules into hydrogen and oxygen, contributing to the creation of a flexible and efficient hydrogen production process. The hydrogen produced is then stored and refined for various applications, adding to the versatility of the system. Power transmission and transformation systems ensure that the electricity is efficiently distributed to meet the energy needs of the hydrogen production facility.

Environmentally, this project is a significant step toward reducing carbon emissions. By producing hydrogen through renewable means, the project is expected to prevent approximately 485,000 tons of carbon dioxide emissions annually. This reduction in emissions contributes directly to global efforts to combat climate change and aligns with China's national goals to peak carbon emissions by 2030 and achieve carbon neutrality by 2060.

Technologically, the project represents a breakthrough in the use of water electrolysis powered entirely by renewable energy. The flexibility of this method allows hydrogen production to adapt to fluctuations in solar power availability, thereby increasing the system's efficiency and reliability. By demonstrating this scalable integration of renewable energy with hydrogen production, Sinopec reinforces its position as a global leader in the energy transition.

Despite its successes, the project faces challenges, particularly in managing the inherent variability of renewable energy sources like solar power. Maintaining grid stability and power quality requires continuous innovation and coordination among government bodies, utilities, and research institutions. The integration of renewable energy with hydrogen production also demands sophisticated technologies to ensure smooth operation and consistent hydrogen output.

As the largest hydrogen producer in China, Sinopec's role in this project highlights its leadership in developing a green hydrogen economy. The Sinopec Xinjiang Kuqa Green Hydrogen Project is not only a testament to technological innovation but also a strategic initiative in China's broader plan to shift toward a low-carbon economy. This project exemplifies the potential of green hydrogen to become a cornerstone of future energy systems, reinforcing Sinopec's commitment to sustainability and global climate goals [180–182].

3.1.1 Fukushima Hydrogen Energy Research Field (FH2R) (Japan)

The Fukushima Hydrogen Energy Research Field (FH2R) is one of the largest hydrogen production facilities powered by renewable energy, marking a significant advance in the pursuit of a sustainable, carbon-free energy future. Situated in Namie Town, Fukushima Prefecture, Japan, FH2R was inaugurated in March 2020. Its primary goal is to demonstrate the feasibility and efficiency of producing green hydrogen on a large scale using solar power.

FH2R operates with an installed solar power generation capacity of 20 MW, which powers a 10 MW alkaline electrolyzer capable of producing 1200 Nm^3 of hydrogen per hour. The facility also uses a hydrogen demand and supply forecasting system, which optimizes both production and storage, enhancing overall operational efficiency. By utilizing on-site solar energy and supplementing it with grid electricity when necessary, FH2R showcases an advanced model of renewable energy integration for hydrogen production.

The project's significance extends beyond technological innovation. It highlights the potential of solar power as a viable source for large-scale hydrogen production, contributing to Japan's renewable energy goals and advancing the global transition toward sustainable energy. The hydrogen produced at FH2R is entirely green, meaning it is generated without emitting carbon dioxide, providing an environmentally friendly alternative to fossil fuels.

Technologically, FH2R represents a step forward in reducing the costs and improving the efficiency of hydrogen production. By pushing the boundaries of what is possible in green hydrogen generation, the facility plays a key role in advancing the technology and contributing to the overall reduction of hydrogen production costs. Additionally, FH2R enhances Japan's energy security by diversifying the country's energy sources and decreasing its reliance on fossil-fuel imports.

Beyond energy production, FH2R contributes to regional revitalization in Fukushima. Following the region's economic challenges after the 2011 disaster, the project fosters job creation and spurs innovation in the green energy sector, supporting the recovery and transformation of the local economy.

However, challenges remain. Green hydrogen must achieve cost competitiveness with conventional hydrogen derived from fossil fuels, necessitating continued technological advances and scaling up production to leverage economies of scale. Infrastructure development is also essential, particularly in hydrogen storage, transportation, and distribution, to facilitate the expansion of green hydrogen markets. The scalability of facilities like FH2R will be crucial in meeting the increasing demand for green hydrogen as countries around the world move toward decarbonization [183].

3.1.2 Ningxia Solar Hydrogen Project (China)

The Ningxia Solar Hydrogen Project, developed by Ningxia Baofeng Energy Group, is the largest green hydrogen production facility in the world. Located in the Ningxia Hui Autonomous Region of China, the project marks a significant step forward in the integration of renewable energy with advanced electrolyzer technology. With an electrolyzer capacity of 150 MW (alkaline electrolyzer) and a solar power capacity of 200 MW, the facility is capable of producing 27,000 tons of green hydrogen annually. Since becoming fully operational in December 2022, the project has demonstrated the feasibility of large-scale hydrogen production powered by solar energy.

This project underscores China's leadership in the global green hydrogen sector. By deploying advanced electrolyzer technology, the Ningxia facility has contributed 50% of the global electrolyzer capacity, positioning China as a major player in the renewable hydrogen market. A key objective for Baofeng Energy Group is to reduce hydrogen production costs to $1.2 per kilogram, which would make green hydrogen a competitive alternative to fossil fuels. The technological advancements realized in this project are essential for achieving cost reductions and improving the efficiency of green hydrogen production.

Strategically, the Ningxia project aligns with China's broader energy and industrial goals. Hydrogen has been identified as a key industry for the future in China's Five-Year Plan, and the country's hydrogen market is expected to grow significantly, from 1 million tons annually to hundreds of megatons by 2050. The competitive landscape for green hydrogen in China is rapidly evolving, with other projects, such as Sinopec's 260 MW facility powered by solar and wind energy, aiming to surpass the capacity of the Ningxia project. China's dominance in electrolyzer manufacturing, which accounts for a third of global capacity, further strengthens its competitive position in the green hydrogen market.

The economic and environmental impacts of the Ningxia project are significant. By relying on renewable energy for hydrogen production, the facility drastically reduces carbon emissions compared to conventional methods that rely on fossil fuels. In addition to its environmental benefits, the project is driving local economic growth through job creation and investments in the green hydrogen sector.

However, challenges remain for the future development of the project. Scaling up the electrolyzer and solar power capacities is a priority to further increase hydrogen production. Efficiency improvements, such as adopting PEM electrolyzers or advanced hydrogen storage solutions, could also enhance the facility's overall performance. Infrastructure development, particularly in hydrogen storage, transportation, and distribution, will be essential to support the growing production and utilization of green hydrogen.

Continued policy support and incentives will be crucial for the long-term success of the Ningxia project and the broader green hydrogen industry. By addressing these challenges, the Ningxia Solar Hydrogen Project could play a pivotal role in advancing China's renewable energy goals and promoting global decarbonization efforts [184, 185].

3.2 Economic Viability and Market Analysis of Photovoltaic-Electrolysis Systems for Hydrogen Production

The transition to renewable energy is a cornerstone of global efforts to mitigate climate change and reduce dependence on fossil fuels. Among the various renewable energy technologies, photovoltaic-electrolysis (PV–E) systems have emerged as a promising approach for producing green hydrogen, which is seen as a key enabler for a sustainable energy future. Green hydrogen, produced by splitting water using renewable electricity, has applications across multiple sectors, including transportation, industry, and energy storage. However, the widespread adoption of PV–E systems hinges on their economic viability and market competitiveness [1, 7, 39].

This section provides an in-depth analysis of the economic viability of PV–E systems, exploring the key cost components, the levelized cost of hydrogen (LCOH), and the impact of government incentives and market trends. By examining the current state of the technology and future projections, this analysis aims to offer insights into the potential of PV–E systems to compete with other hydrogen production methods and their role in the global energy transition.

3.2.1 Cost Components of PV–E Systems

The economic viability of PV–E systems is largely determined by three primary cost components: the photovoltaic (PV) modules, the electrolyzers, and the balance-of-system (BoS) costs. Each of these components plays a critical role in influencing the overall cost of hydrogen production [116, 118].

Photovoltaic (PV) Modules

PV modules are the heart of the PV–E system, converting sunlight into electricity that powers the electrolysis process. The cost of PV modules has seen a dramatic decline over the past decade due to technological advancements, economies of scale, and increased production efficiencies. This reduction in cost has been a key driver in lowering the overall LCOH for PV–E systems.

There are several types of PV technologies available, each with varying costs and efficiencies. For instance, traditional silicon-based PV modules are widely used due to their maturity and lower cost, but they offer moderate efficiency levels. In contrast, concentrator photovoltaic (CPV) systems, which use lenses or mirrors to focus sunlight onto high-efficiency cells, offer higher efficiencies but at a higher cost. The choice of PV technology significantly impacts the overall cost-effectiveness of the PV–E system. For example, CPV-electrolysis systems, despite their high initial costs, can achieve a competitive LCOH due to their superior solar-to-hydrogen (STH) efficiency [111, 153, 177].

Electrolyzers

Electrolyzers are the devices that split water into hydrogen and oxygen using electricity. The choice of electrolyzer technology, such as proton exchange membrane (PEM), alkaline, or solid oxide, affects both the efficiency and cost of hydrogen production. PEM electrolyzers, for instance, are known for their high efficiency and compact design but are more expensive than alkaline electrolyzers. Alkaline electrolyzers, while less expensive, typically offer lower efficiency and are bulkier.

The capacity of the electrolyzer also plays a role in determining the LCOH. Larger electrolyzers benefit from economies of scale, reducing the cost per unit of hydrogen produced. However, the optimal sizing of the electrolyzer relative to the PV array is crucial for minimizing costs. Oversizing the PV array, as some studies suggest, can lead to cost savings by ensuring the electrolyzer operates efficiently even during periods of lower solar irradiance [21, 118, 179, 186].

Balance-of-System (BoS) Costs

BoS costs encompass all other components and infrastructure required for the operation of the PV–E system. This includes inverters, wiring, mounting structures, energy storage systems, and control systems. BoS costs can be a significant portion of the total system cost, particularly in large-scale installations.

The integration of energy storage systems, such as batteries, can stabilize the energy supply and allow for continuous hydrogen production, even when solar power is intermittent. However, energy storage adds to the overall cost and may not always be economically viable in regions with abundant solar resources. Optimizing the BoS design and minimizing these costs is critical for enhancing the economic viability of PV–E systems [187, 188].

Levelized Cost of Hydrogen (LCOH)

The LCOH is a key metric for assessing the economic viability of hydrogen production technologies. It represents the per-unit cost of hydrogen over the lifetime of the production system, considering capital expenditures, operational costs, and maintenance. The LCOH for PV–E systems is influenced by several factors, including the cost of PV modules, the efficiency and cost of electrolyzers, and BoS costs [116, 118, 176, 177, 188].

Comparative Analysis

The LCOH for PV–E systems varies widely depending on the configuration, location, and technological advancements. For instance, in regions with excellent solar resources, PV–E systems can achieve lower LCOH values, making them competitive with other renewable and conventional hydrogen production methods. In contrast, systems in regions with less favorable solar conditions may face higher LCOH values, limiting their competitiveness.

A comparative analysis of different PV–E configurations shows that grid-supplemented systems can achieve lower LCOH by leveraging cheaper grid electricity during periods of low solar output. However, fully off-grid systems, while more expensive, offer greater energy independence and are often preferred in remote locations [142, 189–191].

Impact of Technological Advancements

Technological advancements play a crucial role in reducing the LCOH for PV–E systems. Innovations such as high-efficiency PV modules, advanced electrolyzer technologies, and optimized BoS designs can lead to significant cost reductions. For example, the integration of Concentrated Solar Power (CSP) with PV systems can enhance the operational hours of electrolyzers, reducing the LCOH by increasing the overall efficiency of the system.

Furthermore, advancements in control strategies, such as maximum power point tracking (MPPT) and improved electrolyzer control algorithms, can optimize the performance of the PV–E system, reducing energy losses and further lowering the LCOH [165, 192].

Future Projections

Future projections suggest that the LCOH for PV–E systems could decline significantly with continued R&D efforts and scaling up of production. Achieving a hydrogen production cost of $2/kg, often cited as a competitive target with fossil-fuel-derived hydrogen, is considered feasible with aggressive technological advancements and cost reductions in PV and electrolyzer technologies. In locations with optimal solar resources, this target could be achieved by 2030, positioning PV–E systems as a cornerstone of the green hydrogen economy [7, 176].

3.2.2 Government Incentives and Market Trends

Government incentives and market trends play a pivotal role in shaping the economic viability and adoption of PV–E systems. Policies aimed at reducing the cost of renewable energy technologies, promoting hydrogen as an energy carrier, and supporting R&D are crucial for accelerating the deployment of PV–E systems.

Government Incentives

Government incentives, such as subsidies, feed-in tariffs (FiTs), tax rebates, and R&D funding, have been instrumental in driving the growth of the PV industry. These incentives reduce the upfront capital costs for investors, shorten the payback period, and enhance the overall return on investment for PV–E systems.

For instance, in regions like Europe, FiTs have been successful in promoting the adoption of PV systems by guaranteeing a fixed price for the electricity generated, thus providing a stable revenue stream. In the context of PV–E systems, similar incentives could be applied to encourage the production of green hydrogen, particularly in the early stages of market development.

R&D investments are also critical for advancing the technology and reducing costs. Government-funded research initiatives can accelerate the development of high-efficiency PV modules, more cost-effective electrolyzers, and optimized BoS components, all of which contribute to lowering the LCOH [179, 188, 191].

Market Trends

The market for PV–E systems is influenced by several trends, including the decreasing cost of PV modules, the growing demand for green hydrogen, and the increasing focus on sustainability. The rapid decline in PV module costs has been a key enabler for the economic viability of PV–E systems. As the cost of PV modules continues to fall, driven by technological advancements and economies of scale, the LCOH for PV–E systems is expected to decrease further.

The demand for green hydrogen is also on the rise, driven by its potential to decarbonize hard-to-abate sectors such as heavy industry, transportation, and chemical production. Governments and industries are increasingly recognizing the role of green hydrogen in achieving carbon neutrality, leading to growing investments in hydrogen infrastructure and technologies.

Furthermore, sustainability is becoming a key consideration in energy markets. Companies and governments are under increasing pressure to reduce their carbon footprint and transition to renewable energy sources. PV–E systems, with their ability to produce green hydrogen with zero direct emissions, are well-positioned to capitalize on this trend [1, 39, 45, 116, 193].

3.2.3 Global Examples and Lessons Learned

Several countries and regions have made significant strides in deploying PV–E systems, providing valuable lessons for future projects. For example, China has emerged as a global leader in PV manufacturing and deployment, driven by strong government support and strategic investments in R&D. The country's focus on scaling up production and reducing costs has led to significant cost reductions in

PV technologies, which can be leveraged to enhance the competitiveness of PV–E systems [194].

In Australia, large-scale PV–E projects are being developed to produce green hydrogen for both domestic use and export. The country's abundant solar resources and supportive policy environment make it an ideal location for such projects. The experience in Australia highlights the importance of site selection and resource availability in determining the economic viability of PV–E systems.

Europe, particularly Germany and Spain, has also been at the forefront of PV–E deployment. These countries have implemented various policy mechanisms, such as FiTs and subsidies, to promote the adoption of PV systems. However, the experience in Spain, where a boom-bust cycle occurred due to policy uncertainty, underscores the need for stable and predictable policy frameworks to support the long-term growth of the PV–E market [195].

3.2.4 Technological Advancements

Ongoing R&D is critical for addressing the technological challenges facing PV–E systems. Advancements in PV module efficiency, electrolyzer technology, and BoS components are necessary to reduce costs and improve performance. Additionally, the development of hybrid systems that combine PV with other renewable energy sources, such as wind or CSP, could further enhance the economic viability of PV–E systems.

The economic viability of PV-electrolysis systems is a key determinant of their potential to contribute to the global transition to renewable energy. While significant progress has been made in reducing costs and improving efficiency, challenges remain that must be addressed through continued innovation, supportive government policies, and strategic market development.

PV–E systems have the potential to produce green hydrogen at competitive costs, particularly in regions with abundant solar resources. As technological advancements continue and the market for green hydrogen expands, PV–E systems are likely to play an increasingly important role in the energy landscape.

The future of PV–E systems is bright, but realizing their full potential will require a coordinated effort between governments, industry, and research institutions. By addressing the challenges and seizing the opportunities presented by this emerging technology, PV–E systems can become a cornerstone of the sustainable energy future [195].

3.2.5 Future Perspectives

The future of solar-powered hydrogen production through electrolysis holds immense promise. Continued research and development efforts are focused on improving the efficiency, cost-effectiveness, and scalability of these systems. Advancements in PV technology, such as the development of higher efficiency solar cells and the integration of perovskite-based solar cells, are expected to further boost the performance and affordability of PV systems.

Similarly, innovations in electrolyzer technology, including the development of advanced electrocatalysts and novel cell designs, will lead to increased efficiency and reduced costs of hydrogen production. The integration of artificial intelligence and machine learning algorithms for real-time optimization and control of PV-electrolysis systems will also enhance their performance and adaptability to varying conditions.

Furthermore, the exploration of hybrid systems that combine solar energy with other renewable energy sources, such as wind and biomass, holds potential for further improving the reliability and sustainability of hydrogen production. The development of decentralized and microgrid-based systems that incorporate solar-powered electrolysis can also contribute to a more resilient and localized energy infrastructure.

As the demand for clean energy continues to rise, solar-powered hydrogen production through electrolysis is poised to play a pivotal role in meeting this demand. With ongoing advancements in technology and increasing investment in research and development, we can expect a significant expansion in the deployment of these systems, contributing to a cleaner and more sustainable energy future [7, 11, 34, 102, 196–198].

References

1. M. Wappler, D. Unguder, X. Lü, H. Ohlmeyer, H. Teschke, W. Lueke, Building the green hydrogen market—current state and outlook on green hydrogen demand and electrolyzer manufacturing. Int. J. Hydrog. Energy **47**(79), 33551–33570 (2022)
2. E.B. Agyekum, C. Nutakor, A.M. Agwa, S. Kamel, A critical review of renewable hydrogen production methods: factors affecting their scale-up and its role in future energy generation. Membranes **12**(2), 173 (2022)
3. D. Çelik, M. Yıldız, Investigation of hydrogen production methods in accordance with green chemistry principles. Int. J. Hydrog. Energy **42**(36), 23395–23401 (2017)
4. M. Nasser, T.F. Megahed, S. Ookawara, H. Hassan, A review of water electrolysis–based systems for hydrogen production using hybrid/solar/wind energy systems. Environ. Sci. Pollut. Res. **29**(58), 86994–87018 (2022)
5. L.L. Guarieiro, J.P.D. Anjos, L.A.D. Silva, A.Á. Santos, E.E. Calixto, F.L. Pessoa, J.L.D. Almeida, M. Andrade Filho, F.S. Marinho, G.O.D. Rocha, J.B.D. Andrade, Technological perspectives and economic aspects of green hydrogen in the energetic transition: challenges for chemistry. J. Br. Chem. Soc. **33**(8), 844–869 (2022)
6. IEA, *The Future of Hydrogen*(2019). License: CC BY 4.0. https://www.iea.org/reports/the-future-of-hydrogen
7. C. Idrissov, A. Holland, C. O'Brien, *Hydrogen Economy 2023–2033: Production, Storage, Distribution & Applications* (2023). https://www.idtechex.com/en/research-report/hydrogen-economy-2023-2033-production-storage-distribution-and-applications/946
8. F.J. De Sisternes Jimenez, C.P. Jackson, *Green Hydrogen in Developing Countries* (World Bank Group, 2020). http://documents.worldbank.org/curated/en/953571597951239276/Green-Hydrogen-in-Developing-Countries
9. E. Wilhelm, M. Fowler, A technical and economic review of solar hydrogen production technologies. Bull. Sci. Technol. Soc. **26**(4), 278–287 (2006)

10. J. Turner, G. Sverdrup, M.K. Mann, P.C. Maness, B. Kroposki, M. Ghirardi, R.J. Evans, D. Blake, Renewable hydrogen production. Int. J. Energy Res. **32**(5), 379–407 (2008)
11. T. Terlouw, L. Rosa, C. Bauer, R. McKenna, Future hydrogen economies imply environmental trade-offs and a supply-demand mismatch. Nat. Commun. **15**(1), 7043 (2024)
12. S. Cheng, Y. Li, Z. Yu, Y. Sun, B. Jin, Z. Mao, In situ thermal and electricity utilization of photovoltaic devices by membrane distillation and electrochemical advanced oxidation for desalination and degradation of wastewater. Adv. Sustain. Syst. **5**(5), 2000278 (2021)
13. Y.F. Liu, J.L. Ma, M.J. Zhou, Experimental study on direct coupling in a photovoltaic-electrolyte hydrogen generation system. Adv. Mater. Res. **860**, 18–21 (2014)
14. A. Khalilnejad, A. Sundararajan, A.I. Sarwat, in *IEEE/IAS 52nd Industrial and Commercial Power Systems Technical Conference (I&CPS)*. Performance Evaluation of Optimal Photovoltaic-Electrolyzer System with the Purpose of Maximum Hydrogen Storage (IEEE, 2016), pp. 1–9
15. R. García-Valverde, C. Miguel, R. Martínez-Béjar, A. Urbina, Optimized photovoltaic generator–water electrolyser coupling through a controlled DC–DC converter. Int. J. Hydrog. Energy **33**(20), 5352–5362 (2008)
16. S. Dahbi, R. Aboutni, A. Aziz, N. Benazzi, M. Elhafyani, K. Kassmi, Optimised hydrogen production by a photovoltaic-electrolysis system DC/DC converter and water flow controller. Int. J. Hydrog. Energy **41**(45), 20858–20866 (2016)
17. M.E. Şahin, A photovoltaic powered electrolysis converter system with maximum power point tracking control. Int. J. Hydrog. Energy **45**(16), 9293–9304 (2020)
18. A. Kumar, S.Y. Pan, Opportunities and challenges of electrochemical water treatment integrated with renewable energy at the water-energy nexus. Water Energy Nexus **3**, 110–116 (2020)
19. S. Assahout, H. Elaissaoui, A. El Ougli, B. Tidhaf, H. Zrouri, A neural network and fuzzy logic based MPPT algorithm for photovoltaic pumping system. Int. J. Power Electron. Drive Syst. **9**(4), 1823–1833 (2018)
20. J. Jia, L.C. Seitz, J.D. Benck, Y. Huo, Y. Chen, J.W.D. Ng, T. Bilir, J.S. Harris, T.F. Jaramillo, Solar water splitting by photovoltaic-electrolysis with a solar-to-hydrogen efficiency over 30%. Nat. Commun. **7**(1), 13237 (2016)
21. T.L. Gibson, N.A. Kelly, Predicting efficiency of solar powered hydrogen generation using photovoltaic-electrolysis devices. Int. J. Hydrog. Energy **35**(3), 900–911 (2010)
22. T.L. Gibson, N.A. Kelly, Optimization of solar powered hydrogen production using photovoltaic electrolysis devices. Int. J. Hydrog. Energy **33**(21), 5931–5940 (2008)
23. W.J. Chang, K.H. Lee, H. Ha, K. Jin, G. Kim, S.T. Hwang, H.M. Lee, S.W. Ahn, W. Yoon, H. Seo, J.S. Hong, Design principle and loss engineering for photovoltaic–electrolysis cell system. ACS Omega **2**(3), 1009–1018 (2017)
24. Y. Li, X. Tan, R.K. Hocking, X. Bo, H. Ren, B. Johannessen, S.C. Smith, C. Zhao, Implanting Ni-O-VOx sites into Cu-doped Ni for low-overpotential alkaline hydrogen evolution. Nat. Commun. **11**(1), 2720 (2020)
25. S. Sharma, Metal-based catalysts for hydrogen production reactions. Mater. Today Proc. **81**, 1068–1072 (2023)
26. D.L. Stojić, B.D. Cekić, A.D. Maksić, M.P.M. Kaninski, ŠS. Miljanić, Intermetallics as cathode materials in the electrolytic hydrogen production. Int. J. Hydrog. Energy **30**(1), 21–28 (2005)
27. R. Ciriminna, M. Pagliaro, Enhanced nickel catalysts for producing electrolytic hydrogen. RSC Sustain. **1**(6), 1386–1393 (2023)
28. K. Kannan, D. Chanda, M.M. Meshesha, B.L. Yang, Impressive efficiency of zinc oxide-manganese oxide/MAX composite in two-electrode system for photovoltaic-electrolyzer water splitting. Colloids Surf. A Physicochem. Eng. Aspects **689**, 133599 (2024)
29. M.E. Mert, C. Edis, Ş Akyıldız, B.N. Demir, H. Nazligul, Y. Gurdal, B.D. Mert, Design and performance analysis of a PV-assisted alkaline electrolysis for hydrogen production: an experimental and theoretical study. Fuel **355**, 129497 (2024)

30. A. Singlitico, J. Østergaard, S. Chatzivasileiadis, Onshore, offshore or in-turbine electrolysis? Techno-economic overview of alternative integration designs for green hydrogen production into offshore wind power hubs. Renew. Sustain. Energy Trans. **1**, 100005 (2021)
31. G. Calado, R. Castro, Hydrogen production from offshore wind parks: current situation and future perspectives. Appl. Sci. **11**(12), 5561 (2021)
32. J. Li, W. Liu, W. Qi, Hydrogen production technology by electrolysis of water and its application in renewable energy consumption. E3S Web Conf. **236**, 02001 (2021) (EDP Sciences)
33. D.M. Santos, C.A. Sequeira, J.L. Figueiredo, Hydrogen production by alkaline water electrolysis. Quim. Nova **36**, 1176–1193 (2013)
34. L. Vidas, R. Castro, Recent developments on hydrogen production technologies: state-of-the-art review with a focus on green-electrolysis. Appl. Sci. **11**(23), 11363 (2021)
35. N. Parpulova, Green hydrogen-the missing piece in the European energy mix. Yearbook UNWE **2**(2), 87–93 (2022)
36. IEA, *Global Hydrogen Review 2023* (2023). https://www.iea.org/reports/global-hydrogen-review-2023
37. M. Ali, J. Ekström, M. Lehtonen, Sizing hydrogen energy storage in consideration of demand response in highly renewable generation power systems. Energies **11**(5), 1113 (2018)
38. M. Wanner, Transformation of electrical energy into hydrogen and its storage. Eur. Phys. J. Plus **136**(5), 593 (2021)
39. H. Ma, Z. Sun, Z. Xue, C. Zhang, Z. Chen, A systemic review of hydrogen supply chain in energy transition. Front. Energy **17**(1), 102–122 (2023)
40. V.M. Lopez, H. Ziar, J.W. Haverkort, M. Zeman, O. Isabella, Dynamic operation of water electrolyzers: a review for applications in photovoltaic systems integration. Renew. Sustain. Energy Rev. **182**, 113407 (2023)
41. B.L. Nguyen, M. Panwar, R. Hovsapian, K. Nagasawa, T.V. Vu, in *IECON 2021—47th Annual Conference of the IEEE Industrial Electronics Society*. Power Converter Topologies for Electrolyzer Applications to Enable Electric Grid Services (IEEE, 2021), pp. 1–6
42. M.D. Cabezas, J.I. Franco, H.J. Fasoli, Optimization of self-regulated hydrogen production from photovoltaic energy. Int. J. Hydrog. Energy **45**(17), 10391–10397 (2020)
43. B.B. Choi, J.H. Jo, T. Lee, S.Y. Jeon, J. Kim, Y.S. Yoo, Operational characteristics of high-performance kW class alkaline electrolyzer stack for green hydrogen production. J. Electrochem. Sci. Technol. **12**(3), 302–307 (2021)
44. W. Jiang, Y.K. Wu, T. Yang, F.Y. Yu, W. Wang, S. Hashimoto, in *2012 Asia-Pacific Power and Energy Engineering Conference*. Identification and Power Electronic Module Design of a Solar Powered Hydrogen Electrolyzer (IEEE, 2012), pp. 1–4
45. S. Calnan, R. Bagacki, F. Bao, I. Dorbandt, E. Kemppainen, C. Schary, R. Schlatmann, M. Leonardi, S.A. Lombardo, R.G. Milazzo, S.M. Privitera, Development of various photovoltaic-driven water electrolysis technologies for green solar hydrogen generation. Solar RRL **6**(5), 2100479 (2022)
46. V.R. Pannase, H.B. Nanavala, in *Proceedings of the 2017 International Conference on Inventive Systems and Control (ICISC)*. A review of PV technology power generation, PV material, performance and its applications (IEEE, 2017), pp. 1–5
47. M. Okil, M.S. Salem, T.M. Abdolkader, A. Shaker, From crystalline to low-cost silicon-based solar cells: a review. SILICON **14**(5), 1895–1911 (2022)
48. R.S. Dubey, S. Saravanan, S. Kalainathan, Performance enhancement of thin film silicon solar cells based on distributed Bragg reflector & diffraction grating. AIP Adv. **4**(12) (2014)
49. S. Mitra, H. Ghosh, H. Saha, S.K. Datta, P. Chaudhuri, C. Banerjee, Improvement of photon management in partial rear contact solar cells using a combination of DBR and Mie scatterers. Opt. Commun. **397**, 1–9 (2017)
50. R.S. Dubey, S. Saravanan, S. Kalainathan, Performance evaluation of thin film silicon solar cell based on dual diffraction grating. Nanoscale Res. Lett. **9**, 1–5 (2014)
51. N. Budini, R.D. Arce, R.H. Buitrago, J.A. Schmidt, in *Alternative Energy and Shale Gas Encyclopedia*. Polycrystalline Silicon for Thin Film Solar Cells (2016), pp. 226–232

52. C. Becker, D. Amkreutz, T. Sontheimer, V. Preidel, D. Lockau, J. Haschke et al., Polycrystalline silicon thin-film solar cells: status and perspectives. Solar Energy Mater. Solar Cells **119**, 112–123 (2013)
53. Y. Kuo, C.-C. Lin, S. Verkhoturov, Crystallization of a-Si Thin Film Using an Ultra Thin n+ Poly-Si Seed Layer for Solar Cell Applications, in *2012 38th IEEE Photovoltaic Specialists Conference.* (IEEE, Austin, TX, USA, 2012), pp.000342–000345
54. X. Sheng, J. Liu, I. Kozinsky, A.M. Agarwal, J. Michel, L.C. Kimerling, Efficient light trapping structure in thin film silicon solar cells, in *2010 35th IEEE Photovoltaic Specialists Conference.* (IEEE, Honolulu, HI, USA, 2010), pp.001575–001576
55. L. Zhou, L. Xiao, H. Yang, J. Liu, X. Yu, Greatly enhanced photovoltaic performance of crystalline silicon solar cells via metal oxide. Nanomaterials **8**(7), 505 (2018)
56. T. Matsuoka, Y. Kuwano, Quality improvement in a-Si films and their application to a-Si solar cells. IEEE Trans. Electron Dev. **37**(2), 397–402 (1990)
57. A. Feltrin, T. Meguro, E. Van Assche, T. Suezaki, M. Ichikawa, T. Kuchiyama et al., Advanced light trapping designs for high efficiency thin film silicon solar cells. Solar Energy Mater. Solar Cells **119**, 219–227 (2013)
58. K. Zweibel, The impact of tellurium supply on cadmium telluride photovoltaics. Science **328**(5979), 699–701 (2010)
59. D.B. Li, S.S. Bista, R.A. Awni, S. Neupane, A. Abudulimu, X. Wang et al., 20%-efficient polycrystalline Cd (Se, Te) thin-film solar cells with compositional gradient near the front junction. Nat. Commun. **13**(1), 7849 (2022)
60. J.S. Britt, A.E. Delahoy, Z.J. Kiss, in *National Renewable Energy Laboratory and Sandia National Laboratories Photovoltaics Program Review Meeting*, vol. 394. Technologies for Thin Film CIGS-Based Photovoltaics (1997), pp. 115–122
61. S. Aksu, S. Pethe, A. Kleiman-Shwarsctein, S. Kundu, M. Pinarbasi, in *2012 38th IEEE Photovoltaic Specialists Conference.* Recent Advances in Electroplating Based CIGS Solar Cell Fabrication (IEEE, 2012), pp. 003092–003097
62. S. Aksu, S. Pethe, A. Kleiman-Shwarsctein, S. Kundu, M. Pinarbasi, in *ECS Meeting Abstracts*, no. 38. Overview of Electrodeposition Based Copper Indium Gallium Selenide (CIGS) Solar Cell Fabrication. (IOP Publishing, 2012), p. 2910
63. K. Ramanathan, R. Noufi, B. To, D.L. Young, R. Bhattacharya, M.A. Contreras et al., in *2006 IEEE 4th World Conference on Photovoltaic Energy Conference*, vol. 1. Processing and Properties of Sub-Micron CIGS Solar Cells. (IEEE, 2006), pp. 380–383
64. Y. Fan, J. Wang, Z. Yang, Z. Cheng, H. Hu, R. Zhang et al., in *2017 16th International Conference on Optical Communications and Networks (ICOCN).* Design and Optimization of GaAsP/Si Dual-Junction Solar Cells. (IEEE, 2017), pp. 1–3
65. M. Khizar, M.Y.A. Raja, in *8th International Conference on High-capacity Optical Networks and Emerging Technologies.* Design and Development of High Efficiency Solar Cells Using Thin Film GaAs on Nanostructured Silicon. (IEEE, 2011), pp. 42–45
66. A. Belghachi, N. Limam, Effect of the absorber layer band-gap on CIGS solar cell. Chin. J. Phys. **55**(4), 1127–1134 (2017)
67. A.J. Gurudayal, N. Mathews, in *Wiley-VCH Verlag GmbH & Co. KGaA.* Perovskite Solar Cells for Photoelectrochemical Water Splitting and CO_2 Reduction (2018)
68. H. He, Q. Zhang, Z. Wang, S. Pan, Y. Zhao, X. Zhang, Advances and practical prospects for bias-free photovoltaic-driven electrochemical water splitting systems. Adv. Energy Mater. **14**(11), 2303713 (2024)
69. Y. Rong, Y. Hu, A. Mei, H. Tan, M.I. Saidaminov, S.I. Seok et al., Challenges for commercializing perovskite solar cells. Science, **361**(6408), eaat8235 (2018)
70. C.J. Yu, Advances in modelling and simulation of halide perovskites for solar cell applications. J. Phys. Energy **1**(2), 022001 (2019)
71. G. Perrakis, A.C. Tasolamprou, G. Kakavelakis, K. Petridis, M. Graetzel, G. Kenanakis et al., Transparent multispectral photonic electrode for all-weather stable and efficient perovskite solar cells (2023). arXiv preprint arXiv:2308.03386

72. M. Akhtaruzzaman, V. Selvanathan, A.M. & Hassan, in *Comprehensive Guide on Organic and Inorganic Solar Cells*. Dye-Sensitized Solar Cells. (Academic Press, 2022), pp. 195–244
73. H. Iftikhar, G.G. Sonai, S.G. Hashmi, A.F. Nogueira, P.D. Lund, Progress on electrolytes development in dye-sensitized solar cells. Materials **12**(12), 1998 (2019)
74. J. Li, Y. Shi, in *ASME International Mechanical Engineering Congress and Exposition*, vol. 54976. Electron Transport and Recombination in TiO_2 Nanofiber Dye Sensitized Solar Cell (2011), pp. 361–364
75. I. Benesperi, H. Michaels, M. Freitag, The researcher's guide to solid-state dye-sensitized solar cells. J. Mater. Chem. C **6**(44), 11903–11942 (2018)
76. J.F. Kang, X. Wang, T.S. Zhang, Z.W. Fan, H.Y. Yin, Z.Q. Lu, L.F. Liu, in *2010 Photonics Global Conference*. Design and Optimization of Larger-Sized Dye Sensitized Solar Cell (DSSC) (IEEE, 2010), pp. 1–3
77. L. Sun, Y. Chen, M. Sun, Y. Zheng, Organic solar cells: physical principle and recent advances. Chem. Asian J. **18**(5), e202300006 (2023)
78. Y.Q. Wong, H.Y. Wong, C.S. Tan, H.F. Meng, Performance optimization of organic solar cells. IEEE Photon. J. **6**(4), 1–26 (2014)
79. A. Manabe, H. Domon, J. Kosaka, T. Hashimoto, T. Okajima, T. Ohsaka, Study on separator for alkaline water electrolysis. J. Electrochem. Soc. **163**(11), F3139 (2016)
80. A. Manabe, M. Kashiwase, T. Hashimoto, T. Hayashida, A. Kato, K. Hirao et al., Basic study of alkaline water electrolysis. Electrochim. Acta **100**, 249–256 (2013)
81. M.A. Khan, H. Zhao, W. Zou, Z. Chen, W. Cao, J. Fang et al., Recent progresses in electrocatalysts for water electrolysis. Electrochem. Energy Rev. **1**, 483–530 (2018)
82. R. Maric, H. Yu, Proton exchange membrane water electrolysis as a promising technology for hydrogen production and energy storage. Nanostruct. Energy Gen. Trans. Storage **13** (2019)
83. R. Vinodh, S.S. Kalanur, S.K. Natarajan, B.G. Pollet, Recent advancements of polymeric membranes in anion exchange membrane water electrolyzer (AEMWE): a critical review. Polymers **15**(9), 2144 (2023)
84. D. Li, A.R. Motz, C. Bae, C. Fujimoto, G. Yang, F.Y. Zhang et al., Durability of anion exchange membrane water electrolyzers. Energy Environ. Sci. **14**(6), 3393–3419 (2021)
85. A. Khataee, A. Shirole, P. Jannasch, A. Krüger, A. Cornell, Anion exchange membrane water electrolysis using Aemion™ membranes and nickel electrodes. J. Mater. Chem. A **10**(30), 16061–16070 (2022)
86. J. Milewski, M. Wołowicz, J. Lewandowski, Solid oxide electrolysis cell systems—variant analysis of the structures and parameters. Appl. Mech. Mater. **459**, 106–112 (2014)
87. S. Vaidya, J.H. Kim, in *Continuum Mechanics: Progress in Fundamentals and Engineering Applications*, ed. by Y. Gan. Continuum Mechanics of Solid Oxide Fuel Cells Using Three-Dimensional Reconstructed Microstructures (2012), pp. 73–88
88. Y. Xu, B. Zhang, Recent advances in electrochemical hydrogen production from water assisted by alternative oxidation reactions. Chem Electro Chem **6**(13), 3214–3226 (2019)
89. H. Li, J. Guo, Z. Li, J. Wang, Research progress of hydrogen production technology and related catalysts by electrolysis of water. Molecules **28**(13), 5010 (2023)
90. H.G. Xu, X.Y. Zhang, Y. Ding, H.Q. Fu, R. Wang, F. Mao et al., Rational design of hydrogen evolution reaction electrocatalysts for commercial alkaline water electrolysis. Small Struct. **4**(8), 2200404 (2023)
91. J. Zhang, G. Chen, K. Müllen, X. Feng, Carbon-rich nanomaterials: fascinating hydrogen and oxygen electrocatalysts. Adv. Mater. **30**(40), 1800528 (2018)
92. U. Babic, M. Suermann, F.N. Büchi, L. Gubler, T.J. Schmidt, Critical review—identifying critical gaps for polymer electrolyte water electrolysis development. J. Electrochem. Soc. **164**(4), F387 (2017)
93. S. Wang, A. Lu, C.J. Zhong, Hydrogen production from water electrolysis: role of catalysts. Nano Converg. **8**(1), 4 (2021)
94. D. Henkensmeier, W.C. Cho, P. Jannasch, J. Stojadinovic, Q. Li, D. Aili, J.O. Jensen, Separators and membranes for advanced alkaline water electrolysis. Chem. Rev. **124**(10), 6393–6443 (2024)

95. R. Gupta, in *Electrochemical Society Meeting Abstracts 243*, no. 36. Development of membranes for alkaline water electrolysis (The Electrochemical Society, Inc, 2023), pp. 2037–2037
96. J. Choi, H. Kim, S. Jeon, M.G. Shin, J.Y. Seo, Y.I. Park et al., Thin film composite membranes as a new category of alkaline water electrolysis membranes. Small **19**(37), 2300825 (2023)
97. S. Cavaliere, I. Jiménez-Morales, J. Al Cheikh, D.J Jones, J. Rozière, In *Electrochemical Society Meeting Abstracts 243*, no. 36. Nanocomposite Electrodes and Membranes for PEM Electrolysis (The Electrochemical Society, Inc., 2023), pp. 2074–2074
98. G. Lindquist, S.W. Boettcher, in *Electrochemical Society Meeting Abstracts 243*, no. 36. Membrane Electrolyzers for Hydrogen Production: From Catalyst Fundamentals to Devices (The Electrochemical Society, Inc., 2023), pp. 2061–2061
99. V.N. Kuleshov, S.V. Kurochkin, N.V. Kuleshov, A.A. Gavrilyuk, I.V. Pushkareva, M.A. Klimova, O.Y. Grigor'eva, Alkaline water electrolysis with anion-exchange membranes and nickel-based catalysts. Russian J. Electrochem. **59**(11), 915–929 (2023)
100. A. Raman, S. van der Werf, H. Gardeniers, D. Fernandez Rivas, A. Susarrey Arce, in *Electrochemical Society Meeting Abstracts 244*, no. 41. Zero-gap porous silicon membrane electrodes for alkaline electrolysis. (The Electrochemical Society, Inc., 2023), pp. 2020–2020
101. K. Zeng, D. Zhang, Recent progress in alkaline water electrolysis for hydrogen production and applications. Prog. Energy Combust. Sci. **36**(3), 307–326 (2010)
102. S.A. Grigoriev, V.N. Fateev, D.G. Bessarabov, P. Millet, Current status, research trends, and challenges in water electrolysis science and technology. Int. J. Hydrog. Energy **45**(49), 26036–26058 (2020)
103. K. Ayers, The potential of proton exchange membrane–based electrolysis technology. Curr. Opin. Electrochem. **18**, 9–15 (2019)
104. M.F. Ahmad Kamaroddin, N. Sabli, T.A. Tuan Abdullah, S.I. Siajam, L.C. Abdullah, A. Abdul Jalil, A. Ahmad, Membrane-based electrolysis for hydrogen production: a review. Membranes **11**(11), 810 (2021)
105. C. Santoro, A. Lavacchi, P. Mustarelli, V. Di Noto, L. Elbaz, D.R. Dekel, F. Jaouen, What is next in anion-exchange membrane water electrolyzers? Bottlenecks, benefits, and future. Chem Sustain Chem **15**(8), e202200027 (2022)
106. R.L. LeRoy, C.T. Bowen, D.J. LeRoy, The thermodynamics of aqueous water electrolysis. J. Electrochem. Soc. **127**(9), 1954 (1980)
107. C. Xiang, K.M. Papadantonakis, N.S. Lewis, Principles and implementations of electrolysis systems for water splitting. Mater. Horiz. **3**(3), 169–173 (2016)
108. X. Li, L. Zhao, J. Yu, X. Liu, X. Zhang, H. Liu, W. Zhou, Water splitting: from electrode to green energy system. Nano-Micro Lett. **12**, 1–29 (2020)
109. A. Brisse, J. Schefold, M. Zahid, High temperature water electrolysis in solid oxide cells. Int. J. Hydrog. Energy **33**(20), 5375–5382 (2008)
110. E.M. Leal, J. Brouwer, in *ASME International Mechanical Engineering Congress and Exposition*, vol. 42118. Thermodynamic Analysis of Production of Hydrogen Using High-Temperature Fuel Cells (2005), pp. 481–490
111. M.A. Khan, I. Al-Shankiti, A. Ziani, H. Idriss, Demonstration of green hydrogen production using solar energy at 28% efficiency and evaluation of its economic viability. Sustain. Energy Fuels **5**(4), 1085–1094 (2021)
112. D.P. Pham, S. Lee, J. Yi, Photovoltaic partner selection for high-efficiency photovoltaic-electrolytic water splitting systems: brief review and perspective. SILICON **14**, 753–760 (2022)
113. S.M. Bashir, M.A. Nadeem, M. Al-Oufi, M. Al-Hakami, T.T. Isimjan, H. Idriss, Sixteen percent solar-to-hydrogen efficiency using a power-matched alkaline electrolyzer and a high concentrated solar cell: effect of operating parameters. ACS Omega **5**(18), 10510–10518 (2020)
114. J. Chen, D. Yang, D. Song, J. Jiang, A. Ma, M.Z. Hu, C. Ni, Recent progress in enhancing solar-to-hydrogen efficiency. J. Power Sour. **280**, 649–666 (2015)

115. R.E. Clarke, S. Giddey, F.T. Ciacchi, S.P.S. Badwal, B. Paul, J. Andrews, Direct coupling of an electrolyser to a solar PV system for generating hydrogen. Int. J. Hydrog. Energy **34**(6), 2531–2542 (2009)
116. A. Liponi, A. Baccioli, L. Ferrari, U. Desideri, in *E3S Web of Conferences,* vol. 334. Techno-economic analysis of hydrogen production from PV plants (EDP Sciences, 2022), p. 01001
117. S. Watson, D. Bian, N. Sahraei, T. Buonassisi, I.M. Peters, Advantages of operation flexibility and load sizing for PV-powered system design. Sol. Energy **162**, 132–139 (2018)
118. A. Rosenstiel, N. Monnerie, J. Dersch, M. Roeb, R. Pitz-Paal, C. Sattler, Electrochemical hydrogen production powered by PV/CSP hybrid power plants: a modelling approach for cost optimal system design. Energies **14**(12), 3437 (2021)
119. W. Xiaochen, L. Mengxuan, F. Yifan, in *E3S Web of Conferences*, vol. 248. Analysis of key issues in electric-hydrogen coupled operation (EDP Sciences, 2021), p. 01055
120. D. Ferrero, M. Santarelli, Investigation of a novel concept for hydrogen production by PEM water electrolysis integrated with multi-junction solar cells. Energy Convers. Manage. **148**, 16–29 (2017)
121. A. Djafour, M. Matoug, H. Bouras, B. Bouchekima, M.S. Aida, B. Azoui, Photovoltaic-assisted alkaline water electrolysis: basic principles. Int. J. Hydrog. Energy **36**(6), 4117–4124 (2011)
122. D. Concha, H. Renaudineau, M.S. Hernandez, A.M. Llor, S. Kouro, Evaluation of DCX converters for off-grid photovoltaic-based green hydrogen production. Int. J. Hydrog. Energy **46**(38), 19861–19870 (2021)
123. B. Yodwong, D. Guilbert, M. Phattanasak, W. Kaewmanee, M. Hinaje, G. Vitale, AC-DC converters for electrolyzer applications: state of the art and future challenges. Electronics **9**(6), 912 (2020)
124. V.T. Vinh, in *Energy Conversion—Current Technologies and Future Trends.* The Solutions of DC-DC Converters for Renewable Energy System (IntechOpen, 2019).
125. S. Lu, K. Sun, H. Shi, S. Jiang, Y.W. Li, in *2018 IEEE International Power Electronics and Application Conference and Exposition (PEAC).* Comparison of High Power DC-DC Converters for Photovoltaic Generation Integrated into Medium Voltage DC Grids (IEEE, 2018), pp. 1–6.
126. A. Jemmali, H. Abid, K. Elleuch, A. Toumi, in *2020 17th International Multi-Conference on Systems, Signals & Devices (SSD).* Control Structure for a Photovoltaic Chain Connected to the Three-Phase Grid. (IEEE, 2020), pp. 525–529
127. Ö. Çelik, A. Teke, A. Tan, Overview of micro-inverters as a challenging technology in photovoltaic applications. Renew. Sustain. Energy Rev. **82**, 3191–3206 (2018)
128. M.L. Katche, A.B. Makokha, S.O. Zachary, M.S. Adaramola, A comprehensive review of maximum power point tracking (MPPT) techniques used in solar PV systems. Energies **16**(5), 2206 (2023)
129. A. Ibrahim, R. Aboelsaud, S. Obukhov, Maximum power point tracking of partially shading PV system using cuckoo search algorithm. Int. J. Power Electron. Drive Syst. **10**(2), 1081–1089 (2019)
130. M. Bahrami, R. Gavagsaz-Ghoachani, M. Zandi, M. Phattanasak, G. Maranzana, B. Nahid-Mobarakeh et al., Hybrid maximum power point tracking algorithm with improved dynamic performance. Renew. Energy **130**, 982–991 (2019)
131. M.A. Jusoh, M.F.N. Tajuddin, S. Ayob, M.A. Roslan, Maximum power point tracking charge controller for standalone PV system. TELKOMNIKA **16**(4), 1413–1426 (2018)
132. S. Bandri, Z. Zulkarnaini, A. Sofian, in *MATEC Web of Conferences*, vol. 215. Implementation of maximum power point tracking to improving the solar cell efficiency. (EDP Sciences, 2018), p. 01016
133. C. Sun, J. Ling, J. Wang, Research on a novel and improved incremental conductance method. Sci. Rep. **12**(1), 15700 (2022)
134. L. Shang, H. Guo, W. Zhu, An improved MPPT control strategy based on incremental conductance algorithm. Protect. Control Modern Power Syst. **5**(2), 1–8 (2020)

135. E. Kim, M. Warner, I. Bhattacharya, in *2020 47th IEEE Photovoltaic Specialists Conference (PVSC)*. Adaptive Step Size Incremental Conductance-Based Maximum Power Point Tracking (MPPT). (IEEE, 2020), pp. 2335–2339
136. R. Shalaby, H.H. Ammar, A.T. Azar, M.I. Mahmoud, Optimal fractional-order fuzzy-MPPT for solar water pumping system. J. Intell. Fuzzy Syst. **40**(1), 1175–1190 (2021)
137. M.Y. Heelan, F.A.M. Al-Qrimli, in *2020 International Conference on Electrical, Communication, and Computer Engineering (ICECCE)*. Design and Simulation of Neuro-Fuzzy Based MPPT Controller for PV Power System. (IEEE, 2020), pp. 1–6
138. A.M. Farayola, A.N. Hasan, A. Ali, in *2018 9th International Renewable Energy Congress (IREC)*. Use of MPPT Techniques to Reduce the Energy Pay-Back Time in PV Systems (IEEE, 2018), pp. 1–6
139. M.A. Ghasemi, H.M. Foroushani, M. Parniani, Partial shading detection and smooth maximum power point tracking of PV arrays under PSC. IEEE Trans. Power Electron. **31**(9), 6281–6292 (2015)
140. C.K. Kim, H.S. Cho, C.H. Kim, W. Cho, H.G. Kim, A feasibility study of photovoltaic—electrolysis—PEM hybrid system integrated into the electric grid system over the Korean Peninsula. Front. Chem. **9**, 732582 (2021)
141. E. Crespi, P. Colbertaldo, G. Guandalini, S. Campanari, Design of hybrid power-to-power systems for continuous clean PV-based energy supply. Int. J. Hydrog. Energy **46**(26), 13691–13708 (2021)
142. F. Gutiérrez-Martín, L. Amodio, M. Pagano, Hydrogen production by water electrolysis and off-grid solar PV. Int. J. Hydrog. Energy **46**(57), 29038–29048 (2021)
143. S. Preet, S.T. Smith, A comprehensive review on the recycling technology of silicon-based photovoltaic solar panels: challenges and future outlook. J. Clean. Prod. 141661 (2024)
144. M.K. Abid, M.V. Kumar, V.A. Raj, M. Dhas, Environmental impacts of the solar photovoltaic systems in the context of globalization. Ecol. Eng. Environ. Technol. **24** (2023)
145. S.O. Ganiyu, C.A. Martinez-Huitle, The use of renewable energies driving electrochemical technologies for environmental applications. Curr. Opin. Electrochem. **22**, 211–220 (2020)
146. N. Mathur, S. Singh, J.W. Sutherland, Promoting a circular economy in the solar photovoltaic industry using life cycle symbiosis. Resour. Conserv. Recycl. **155**, 104649 (2020)
147. N.M. Kumar, S.S. Chopra, P. Rajput, in *Photovoltaic Solar Energy Conversion*. Life Cycle Assessment and Environmental Impacts of Solar PV Systems (Academic Press, 2020), pp. 391–411.
148. A.P. Borole, A.L. Greig, in *Biohydrogen*. Life-Cycle Assessment and Systems Analysis of Hydrogen Production (Elsevier, 2019), pp. 485–512
149. V. Aryan, M. Font-Brucart, D. Maga, A comparative life cycle assessment of end-of-life treatment pathways for photovoltaic backsheets. Prog. Photovolt. Res. Appl. **26**(7), 443–459 (2018)
150. S.K. Dash, S. Chakraborty, D. Elangovan, A brief review of hydrogen production methods and their challenges. Energies **16**(3), 1141 (2023)
151. W.H. Cheng, A. De La Calle, H.A. Atwater, E.B. Stechel, C. Xiang, Hydrogen from sunlight and water: a side-by-side comparison between photoelectrochemical and solar thermochemical water-splitting. ACS Energy Lett. **6**(9), 3096–3113 (2021)
152. M. Martino, C. Ruocco, E. Meloni, P. Pullumbi, V. Palma, Main hydrogen production processes: an overview. Catalysts **11**(5), 547 (2021)
153. S.E. Hosseini, M.A. Wahid, Hydrogen from solar energy, a clean energy carrier from a sustainable source of energy. Int. J. Energy Res. **44**(6), 4110–4131 (2020)
154. M. Alqudah, T. Dokic, M. Kezunovic, Z. Obradovic, Prediction of solar radiation based on spatial and temporal embeddings for solar generation forecast (2022). arXiv preprint arXiv:2206.08832
155. V. Paduani, Q. Xiao, B. Xu, D. Lubkeman, N. Lu, Optimal control design for operating a hybrid PV plant with robust power reserves for fast frequency regulation services (2022). arXiv preprint arXiv:2212.03803

156. FCHEA, *Road Map to a US hydrogen economy* [PDF]. Fuel Cell and Hydrogen Energy Association (2024). https://fchea.org/wp-content/uploads/2024/07/RoadMaptoaUSHydrogenEconomyFullReport.pdf
157. V. Dias, M. Pochet, F. Contino, H. Jeanmart, Energy and economic costs of chemical storage. Front. Mech. Eng. **6**, 21 (2020)
158. G.I. Giurgi, D. Petreus, D.V. Giurgi, L. Szolga, in *IOP Conference Series: Earth and Environmental Science*, vol. 897, no. 1. Power conversion performance of a single-phase PV inverter with fuzzy logic algorithm. (IOP Publishing, 2021), p. 012018
159. C. Conker, M.K. Baltacioglu, Fuzzy self-adaptive PID control technique for driving HHO dry cell systems. Int. J. Hydrog. Energy **45**(49), 26059–26069 (2020)
160. A. Emad, M. El-Shimy, G. Amer, I. Ihoume, in *2020 International Conference on Electrical and Information Technologies (ICEIT)*. Power Management Control of Hydrogen-Based System Using Fuzzy Logic Method. (IEEE, 2020), pp. 1–6
161. M.L. Anjulo, Power management of standalone hybrid PV-micro hydro battery energy storage system using fuzzy logic controller. Int. J. Eng. Res. Technol. **8** (2019)
162. B.L. Nguyen, M. Panwar, R. Hovsapian, Y. Agalgaonkar, T. Vu, in *2022 IEEE 1st Industrial Electronics Society Annual On-Line Conference (ONCON)*. Hierarchical Control of Grid-Connected Hydrogen Electrolyzer Providing Grid Services (IEEE, 2022), pp. 1–6
163. R.E. Clarke, S. Giddey, S.P.S. Badwal, Stand-alone PEM water electrolysis system for fail-safe operation with a renewable energy source. Int. J. Hydrog. Energy **35**(3), 928–935 (2010)
164. Z.A. Memon, M.A. Akbari, M. Zare, An improved cheetah optimizer for accurate and reliable estimation of unknown parameters in photovoltaic cell and module models. Appl. Sci. **13**(18), 9997 (2023)
165. B.V. Hien, T.V. Anh, N.T. Linh, P.Q. Khanh, Rapidly determine the maximum power point in the parallel configuration of the photovoltaic system. Sensors **23**(17), 7503 (2023)
166. Y. Son, S. Mukherjee, R. Mallik, B. Majmunović, S. Dutta, B. Johnson et al., Levelized cost of energy-oriented modular string inverter design optimization for PV generation system using geometric programming. IEEE Access **10**, 27561–27578 (2022)
167. K.B.D. Melo, H.S. Moreira, M.G. Villalva, Influence of solar position calculation methods applied to horizontal single-axis solar trackers on energy generation. Energies **13**(15), 3826 (2020)
168. D. Henkensmeier, M. Najibah, C. Harms, J. Žitka, J. Hnát, K. Bouzek, Overview: state-of-the-art commercial membranes for anion exchange membrane water electrolysis. J. Electrochem. Energy Convers. Storage **18**(2), 024001 (2021)
169. D. Guilbert, G. Vitale, Improved hydrogen-production-based power management control of a wind turbine conversion system coupled with multistack proton exchange membrane electrolyzers. Energies **13**(5), 1239 (2020)
170. J. Brauns, T. Turek, Alkaline water electrolysis powered by renewable energy: a review. Processes **8**(2), 248 (2020)
171. S. Bintz, D. Häfele, V. Ketchedjian, J. Roth-Stielow, in *2019 21st European Conference on Power Electronics and Applications (EPE'19 ECCE Europe)*. Parallel-Serial-Rectifier for Power-to-Hydrogen Applications. (IEEE, 2019), p. P-1
172. Y.J. Kim, H. Lee, S. Hwang, W. Kim, S. Kim, S.Y. Kim, S. Bae, Research needs for realization of zero-carbon power grids with selected case studies. Appl. Sci. **12**(5), 2533 (2022)
173. P.T. Krein, J.A. Galtieri, Active management of photovoltaic system variability with power electronics. IEEE J. Emerg. Selected Topics Power Electron. **9**(6), 6507–6523 (2020)
174. P.C. Rao, M. Yoon, Potential liquid-organic hydrogen carrier (LOHC) systems: a review on recent progress. Energies **13**(22), 6040 (2020)
175. M.A. Hannan, V.L.Z. Dao, Y.A. Ahmed, Hydrogen production from water electrolysis: Current practice vs sustainable alternatives. ASEAN Eng. J. **12**(3), 129–136 (2022)
176. D.S. Mallapragada, E. Gençer, P. Insinger, D.W. Keith, F.M. O'Sullivan, Can industrial-scale solar hydrogen supplied from commodity technologies be cost competitive by 2030? Cell Rep. Phys. Sci. **1**(9) (2020)

177. A. Grimm, W.A. de Jong, G.J. Kramer, Renewable hydrogen production: a techno-economic comparison of photoelectrochemical cells and photovoltaic-electrolysis. Int. J. Hydrog. Energy **45**(43), 22545–22555 (2020)
178. T. Grube, J. Reul, M. Reuß, S. Calnan, N. Monnerie, R. Schlatmann et al., A techno-economic perspective on solar-to-hydrogen concepts through 2025. Sustain. Energy Fuels **4**(11), 5818–5834 (2020)
179. R. Boudries, Techno-economic assessment of solar hydrogen production using CPV-electrolysis systems. Energy Proc. **93**, 96–101 (2016)
180. H. Xu, Hydrogen development in China and the EU: a recommended Tian Ji's horse racing strategy (2024). arXiv preprint arXiv:2408.08874
181. H. Tang, Y. Yu, Q. Sun, Progress, challenges, and strategies for China's natural gas industry under carbon-neutrality goals. Processes **12**(8), 1683 (2024)
182. Q. Wang, Hydrogen production from renewable energy: current status, prospects, and challenges. Ann. Rep. China's Petrol. Gas New Energy Indus. **2022–2023**, 231–252 (2024)
183. Fukushima Hydrogen Energy Research Field in Japan ready for green hydrogen production. Fuel Cells Bull. **2020**(3), 1 (2020)
184. M. Zhu, D. Xiang, H. Cao, L. Liu, C. Guo, Techno-economic analysis of green hydrogen production using a 100 MW photovoltaic power generation system for five cities in North and Northwest China. Sol. Energy **269**, 112312 (2024)
185. H. Yang, A. Wu, Z. Liu, Y. Su, X. Hu, K. Świerczek et al., A power-to-hydrogen nearby consumption system based on a flat-tube rSOC coupled with local photovoltaics and Yellow River water. Int. J. Hydrog. Energy **57**, 1111–1117 (2024)
186. M. Modestino, C. Rodriguez, D. Psaltis, C. Moser, in *2014 IEEE Photonics Conference*. Design Principles of Deployable Solar-Hydrogen Generators. (IEEE, 2014), pp. 441–442
187. I.M. Peters, S. Watson, N. Sahraei, T. Buonassisi, In *2017 IEEE 44th Photovoltaic Specialist Conference (PVSC)*. Global Design Aspects of Persistent and Autonomous PV Powered Systems (IEEE, 2017), pp. 648–651
188. E. Vartiainen, C. Breyer, D. Moser, E. Román Medina, C. Busto, G. Masson et al., True cost of solar hydrogen. Solar RRL **6**(5), 2100487 (2022)
189. A. Bilich, K. Langham, R. Geyer, L. Goyal, J. Hansen, A. Krishnan et al., Life cycle assessment of solar photovoltaic microgrid systems in off-grid communities. Environ. Sci. Technol. **51**(2), 1043–1052 (2017)
190. A.S. Mundada, K.K. Shah, J.M. Pearce, Levelized cost of electricity for solar photovoltaic, battery, and cogen hybrid systems. Renew. Sustain. Energy Rev. **57**, 692–703 (2016)
191. M.R. Shaner, H.A. Atwater, N.S. Lewis, E.W. McFarland, A comparative technoeconomic analysis of renewable hydrogen production using solar energy. Energy Environ. Sci. **9**(7), 2354–2371 (2016)
192. W. Hu, L. Zhu, Z. Dou, C. Zhang, K. Wang, in *E3S Web of Conferences*, vol. 406. Modeling and Control Strategy for Hydrogen Production Systems Coupled with PV and Battery Storage (EDP Sciences, 2023), p. 02048
193. C. Ogbonnaya, C. Abeykoon, A. Nasser, A. Turan, C.S. Ume, Prospects of integrated photovoltaic-fuel cell systems in a hydrogen economy: a comprehensive review. Energies **14**(20), 6827 (2021)
194. A. Chadly, K. Moawad, K. Salah, M. Omar, A. Mayyas, State of global solar energy market: Overview, China's role, challenges, and opportunities. Sustain. Horiz. **11**, 100108 (2024)
195. H. Song, S. Luo, H. Huang, B. Deng, J. Ye, Solar-driven hydrogen production: recent advances, challenges, and future perspectives. ACS Energy Lett. **7**(3), 1043–1065 (2022)
196. R. Xia, S. Overa, F. Jiao, Emerging electrochemical processes to decarbonize the chemical industry. JACS Au **2**(5), 1054–1070 (2022)
197. S. Aeschlimann, C. Charalambous, Breaking barriers in carbon dioxide removal with electrochemistry. *RMI* (2024). https://rmi.org/insight/breaking-barriers-in-carbon-dioxide-removalwith-electrochemistry
198. MIT Technology Review Insights, Scaling green hydrogen technology for the future. MIT Technol. Rev. (2024). https://www.technologyreview.com/2024/06/18/1092956/scaling-green-hydrogen-technology-for-the-future/

Biohybrid Solar Cells: Integrating Biological Components for Enhanced Performance

C. Clement Raj

Abstract Biohybrid solar cells are one of the fast-emerging photovoltaic technologies that integrates biological components replacing traditional solar cell materials to enhance the performance of the cells. The biohybrid solar cell aims to utilize the unique properties of biological molecules to improve the stability and functionality of the solar cells. Natural light absorbers such as photosynthetic proteins, dyes and pigments can be integrated with photoelectrodes for improved light absorption efficiency. DNA and enzymes can be used as effective charge transport materials in solar cells. Biohybrid solar cells can be designed by mimicking natural systems to optimize the orientation of solar cells. Biomaterials can also be used with nanomaterials as hybrid structures for better light absorption, charge separation and charge transport. In summary, biohybrid solar cells provide a cleaner, cost effective, and sustainable alternative to conventional solar cells. However, challenges remain in the integration, compatibility and stability of the biomaterials. Research in biohybrid solar cells is still in the early stages, with a huge potential to revolutionize solar energy by leveraging the natural efficiencies of biological systems.

Keywords Photosynthetic proteins · Biohybrid photovoltaics · Dye-sensitized solar cells (DSSC) · Protein integration · Photosystem I (PSI) · Photosystem II (PSII) · Biomimetic solar energy · Electron transfer pathways

1 Introduction

Energy has become an integral part of human life as it makes life easier. Affordable and clean energy is crucial for sustainability. Global energy demand is expected to rise at a faster rate over the next three years, growing by an average of 3.4% annually through 2026 [1]. In the past few years, global energy demand has increased unprecedentedly which has resulted in the increased use of fossil fuels and nuclear power.

C. C. Raj (✉)
Department of Electronics and Communication Engineering, Easwari Engineering College, Chennai, India
e-mail: clementraj.c@eec.srmrmp.edu.in

P. Ravindran et al. (eds.), *Photon to Power*, Progress in Optical Science and Photonics 32, https://doi.org/10.1007/978-981-96-5914-2_13

Depletion of fossil fuels, global warming and other environmental concerns have led to extensive research in the development of renewable energy technologies in the search for alternatives. Solar energy, wind power, biomass, geothermal energy, and hydropower are the most widely available renewable energy sources. The improvement in the performance characteristics of the systems has led to declining capital costs and shorter energy payback times [2]. Among the renewable energy sources solar and wind are the most prominent as witnessed by the growth in the respective technologies over the past decades. At present worldwide need of approximately 17 TW energy is substantially exceeded by the nearly 120,000 TW of solar energy that is received by the Earth. So, one extremely appealing area of current research into alternative energy technologies is the study and creation of innovative solar or photovoltaic devices for the conversion of photonic energy from solar light to electricity [3].

Solar cells are generally categorized into three generations based on the materials used [4, 5]. Conventional wafer-based cells made of crystalline silicon make up the first generation. Thin film solar cells made of amorphous silicon, cadmium telluride, and copper indium gallium selenide are part of the second generation. The third generation of developing solar cells uses many topologies, including quantum dot solar cells, dye-sensitized solar cells, and organic photovoltaics (OPV). There are limits to the efficiency, design advantages, and resource availability of each form of solar cell. The inherent high-energy input needs of present technologies, the geopolitical and natural resource limitations, and the environmental toxicity of certain fabrication-process-related components are only a few of their shortcomings. Owing to these challenges, a growing number of scientists are investigating more organic approaches, like the process of solar energy conversion through biohybrid solar cells. Traditional silicon based photovoltaic technologies have limitations such as high manufacturing costs, material scarcity, and environmental concerns regarding their disposal [6].

In this context, biohybrid solar cells emerge as a promising alternative by integrating biological molecules with synthetic materials, harnessing the efficiency, and elegance of natural processes to improve solar energy harvesting. Biohybrid solar cells combine the unique attributes of biological components, such as proteins or photosynthetic systems, with inorganic materials typically used in photovoltaic (PV) technology. The goal of biohybrid solar cells is to mimic and improve upon the natural process of photosynthesis to convert sunlight into electrical energy more efficiently.

This chapter will delve into the fundamental concepts of biohybrid solar cells, their working principles, the biological and synthetic components used, recent developments in the field, and the future potential of this innovative technology.

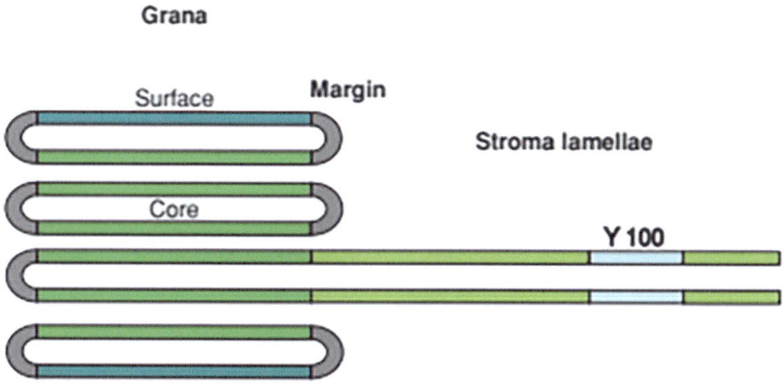

Fig. 1 Schematic model of domains of the thylakoid membrane. Reproduced with permission from [8]. Copyright 2001. Science Direct

1.1 Working Principles of Biohybrid Solar Cells

Biohybrid solar cells are inspired by nature, particularly by the photosynthetic machinery of plants, algae, and certain bacteria. Photosynthesis is one of the most efficient natural processes for capturing solar energy, converting it into chemical energy via light-absorbing pigments like chlorophyll and reaction centers that drive electron transfer processes. Early biochemical studies showed that chloroplast thylakoid membranes oxidize H_2O, reduce NADP, and synthesize ATP. These reactions are catalyzed by two photosystems [photosystem I (PSI) and photosystem II (PSII)], an ATP synthase (F-ATPase) that produces ATP at the expense of the proton motive force (pmf) formed by light-driven electron transfer reactions, and the cytochrome (cyt) b6f complex, which mediates electron transport between PSII and PSI and converts the redox energy into part of the proton gradient used for ATP formation [7] (Figs. 1 and 2).

1.2 Biohybrid Solar Cell Architecture

There are several approaches to constructing biohybrid solar cells, each with unique advantages and challenges. The operational principle of biohybrid solar cells closely mirrors that of traditional photovoltaic cells but with a few key differences due to the incorporation of biological materials. Here's a step-by-step breakdown of how biohybrid solar cells work:

1. **Light Absorption**: The biohybrid cell's biological component (such as PSI or PSII) absorbs photons from sunlight. This light energy excites the chlorophyll molecules, generating high-energy electrons.

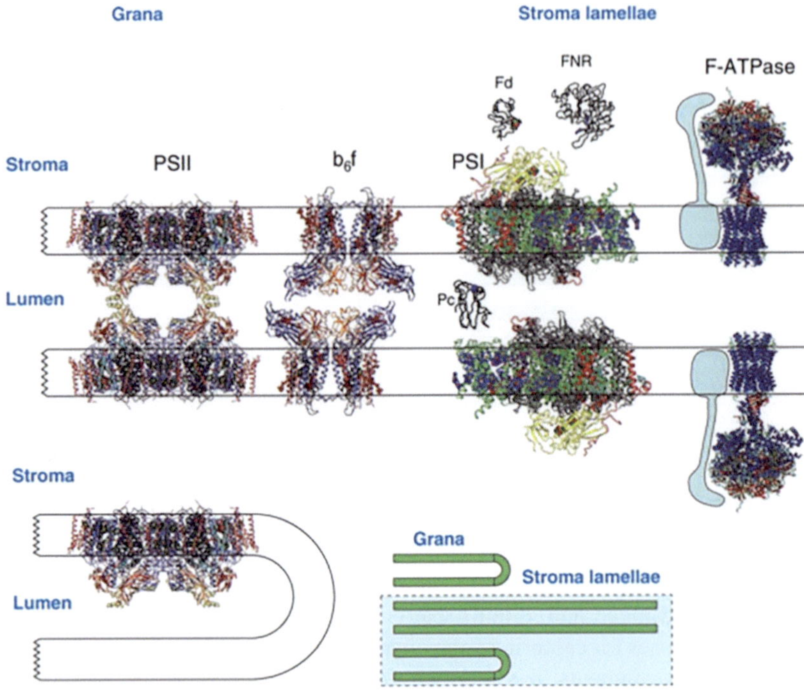

Fig. 2 Architecture of thylakoid membranes and soluble proteins. Reproduced with permission from [8]. Copyright 2001. Science Direct

2. **Electron Transfer**: Once the electrons are excited, they are transferred through a series of electron carriers. In nature, this electron transport chain generates chemical energy (ATP and NADPH); in biohybrid solar cells, it is instead used to drive an electrical current.
3. **Electron Collection**: The synthetic material in the biohybrid cell, often a semiconductor like silicon or a transparent conducting oxide, captures the electrons and transports them to an external circuit, where they can be used to power devices or stored in batteries.
4. **Energy Output**: Just like traditional solar cells, biohybrid solar cells can be integrated into larger solar panels to scale energy production. The potential advantage lies in their ability to perform efficiently in low-light conditions and at lower manufacturing costs (Fig. 3).

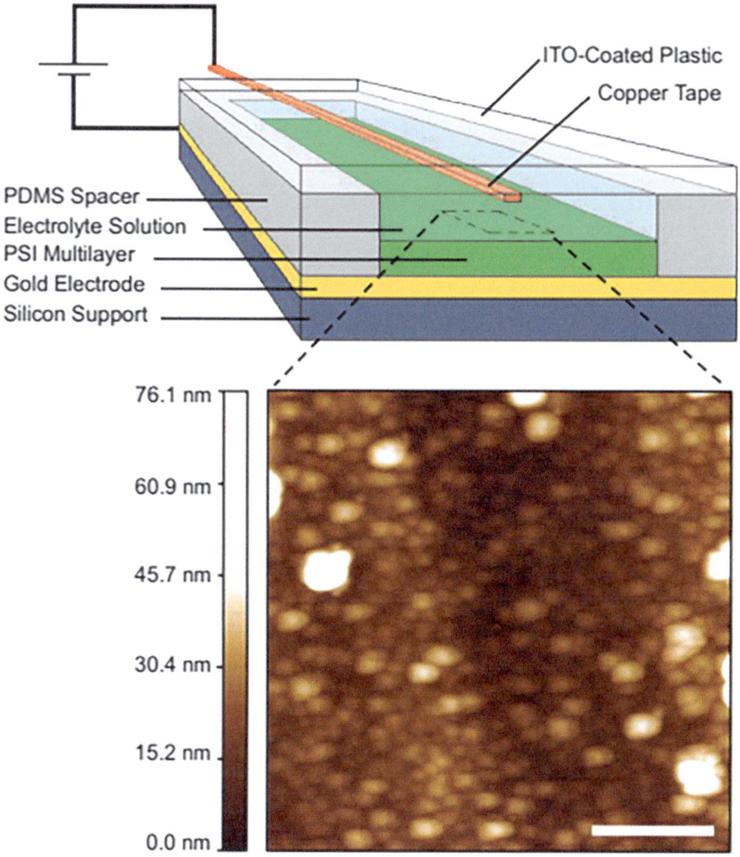

Fig. 3 Schematic of a PSI catalyzed photoelectrochemical cell. Inset: AFM image of PSI multilayer. Reproduced with permission from [9]. Copyright 2010. Science Direct

1.3 Integration of Biological Components and Synthetic Components

Biohybrid solar cells are unique because they incorporate biological molecules to harvest and convert sunlight into electricity. These components, primarily derived from photosynthetic organisms like plants, algae, and bacteria, are highly efficient at capturing light and facilitating electron transfer, making them attractive for solar cell applications [10].

Thylakoids form a physically continuous three-dimensional network enclosing an aqueous space called the lumen and are differentiated into two distinct physical domains: cylindrical stacked structures (called grana) and interconnecting single membrane regions (stroma lamellae). The protein complexes that catalyze electron transfer and energy transduction are unevenly distributed in thylakoids: PSI is located in the stroma lamellae, PSII is found almost exclusively in the grana, the F-ATPase is located mainly in the stroma lamellae, and the cyt b6f complex is found in grana and grana margins [11].

2 Biohybrid Electrodes

2.1 Thylakoid as Sensitizer

Biohybrid electrodes that used thylakoid membranes as the sensitizer produced the highest photocurrent value. Without the entire native lipid bilayer, the related membrane photosynthetic complexes PSII, b6f complex, and PSI functioned as sensitizers. In this work, the natural photosynthesis circuit was re-engineered using single step aerosol techniques to effectively manufacture an ecologically friendly, mediator-free, electron donor, and linker-free photobioreactor. Nanostructured TiO_2 was the substrate used in the membrane preparation process. Water served as the main donor; no exogenous electron donors were added. DDM was used to solubilize the membrane proteins in thylakoid membranes. The electrospinning technique was used to apply the resultant solution on the TiO_2 surface. PSI, PSII, and cytochrome b6/f congregate on the surface and stick to the sensitized TiO_2 once it is exposed to the electrolyte, whereas the surfactant diffuses into the electrolyte. Although this solar cell has a fully functional Z-scheme, it is not a thylakoid-based solar cell according to technical standards. The maximum photocurrent density was 6.7 mA cm^{-2} when exposed to UV and visible light, and 12 μA cm^{-2} when exposed to visible light [12].

One of the central concepts in biohybrid solar cells is the use of photosynthetic proteins, such as **photosystem I (PSI)** and **photosystem II (PSII)**, which play a crucial role in natural photosynthesis. These systems are integral in converting light energy into chemical energy, but in biohybrid solar cells, they are re-engineered to produce electrical energy. **Photosystem I (PSI)** and **Photosystem II (PSII)** are protein complexes involved in the natural process of photosynthesis. Found in the thylakoid membranes of plants, algae, and cyanobacteria, these systems play crucial roles in converting light energy into chemical energy (Fig. 4).

Photosystem I (PSI): PSI absorbs light at wavelengths around 700 nm (red light) and is responsible for the final steps of the light-dependent reactions of photosynthesis. It drives electron transfer from plastocyanin to ferredoxin, ultimately contributing to the generation of NADPH (a reducing agent in plants). In biohybrid solar cells, PSI can be incorporated into an electrode to transfer excited electrons into the external circuit, producing electrical current.

The protein complex known as Photosystem I (PSI) is present in the thylakoid membranes of cyanobacteria, algae, and plants. It is the first stage in light-driven electron transport and absorbs light with a wavelength of about 700 nm. Mershin et al. [13] claim that the self-assembly of dry PSI complexes occurred on nanostructured semiconductors [13]. Ac-AAAAAAK-NH_2 (A6K), a 2.4 nm long cationic peptide

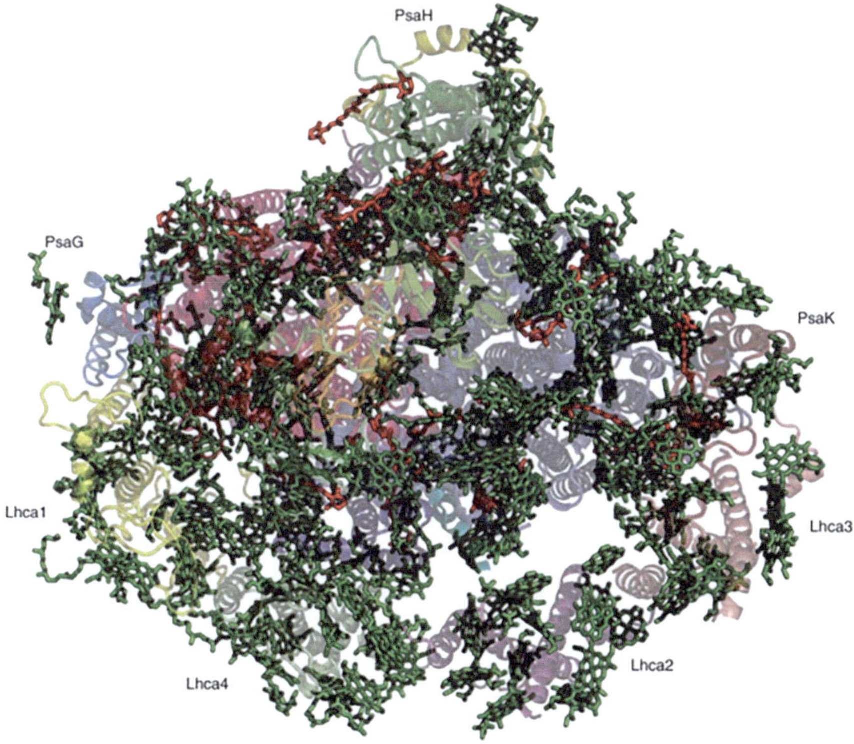

Fig. 4 View of the structure of plant photosystem I from the stromal side. Reproduced with permission from [7]. Copyright 2004. Nature

surfactant with six alanine residues and a lysine at the amidated C-terminus, stabilized these complexes.

PSI combined with this peptide in solution indicated increased stability. Using PSI as a sensitizer, researchers constructed the entire solar cell in the manner of a dye-sensitized solar cell. PSI complexes functioned as the charge separator and light collector in solar cells. To offer a sizable effective surface area for PSI adsorption and light harvesting, they employed nanocrystalline TiO_2 and ZnO-nanowires. Mutant PSI complexes are necessary for the effective immobilization of the ZnO substrate. The engineering of native PSI complexes was not necessary for TiO_2. The process of screen printing was used to deposit TiO_2 paste.

The TiO_2-based BPEC achieved electrical power outputs of up to 81 $\mu W\ cm^{-2}$ and an area-normalized short-circuit current density of up to 362 $\mu A\ cm^{-2}$ in typical sunlight. The efficiency of the ZnO-based BPEC was less than seen for the older device. Results of light absorption are shown for electron transfer to primary and then secondary electron acceptors when the primary electron donor (P700) is excited. The Co(II)/Co(III) redox couple electrolyte, known as Z813, filled the role of plastocyanin in their BPEC, and ferredoxin was swapped out for nanocrystalline TiO_2 or ZnO-nanowires to provide a large surface area for taking in electrons [14].

The most efficient PSI and LHCII-based solar cells to date were reported by Yu et al. in 2015 [15]. They constructed a number of solar cells in their work using different mesoporous layer architectures. A scattering layer's effect was also investigated. The nanoparticles in the scattering layer are around 200 nm in size, while the nanoparticles in the transparent mesoporous layer are about 22 nm in size. Layer-by-layer screen printing of TiO_2 pastes was applied to the conductive FTO-glass, and the coated glass was then sintered at 500 °C. The first layer to form was the translucent mesoporous layer. Next, the transparent layer was covered with a layer of TiO_2 scattering.

To prevent multilayer formation on the TiO_2 surface and to maintain PSI from A. platensis or LHCII from spinach suspended in solution, the surfactant n-Dodecyl-b-D-maltoside (DDM) was utilized. The pigment-protein complexes and DDM solution in MES buffer were applied to the electrode. Whereas the smaller LHCII complex can permeate both the transparent layer and the scattering layer, PSI complexes can only penetrate the scattering layer. As a result, LHCII loaded the TiO_2 substrate more thoroughly than PSI complexes alone. The performance of these watermark achievement solar cells was as follows: For the PSI-based solar cell, JSC is equal to 1.31 mA cm^{-2} and η is 0.47%; for the LHCII-based solar cell, these values are 1.51 mA cm^{-2} and η is 0.52%.

However, the results indicate that the efficiency of electron transfer from PSI to TiO_2 was higher than that of LHCII. These solar cells were built using FTO-glass and an electrolyte based on iodine. Additionally, the particular light condition used in these tests was provided via an AM1.5 filter. A potentiometric analysis of PSI/TiO_2, a biohybrid electrode, was reported by Shah et al. in 2014. In this work, an aerosol-chemical vapor deposition (ACVD) method was used to create columnar TiO_2 nanostructured films onto ITO-glass.

In this procedure, PSI complexes were electrospray onto the TiO_2 surface after titanium tetraisopropoxide had been purged with nitrogen. In short, a syringe was used to apply a voltage to the electrode surface while a solution of PSI and DDM in ethanol was deposited. The complexes were then left on the surface after the electrode was submerged in the electrolyte solution, the surfactant was removed, and the electrode was dried. By adjusting the deposition time, one can regulate the thickness of the sensitizer layer. To remove surfactant from the electrode with electrolyte solution quickly, the concentration of DDM must be kept within a certain range.

The amount of PSI complexes adsorbed onto the electrode increased with the speed at which the DDM concentration dropped below the threshold micelle concentration. Conversely, adequate DDM is required to stop the PSI complexes from aggregating. Dichlorophenol and sodium ascorbate functioned as electron donors in the solution. The electrode was subjected to both visible and UV + visible light to evaluate titania's role in the creation of photocurrent. For the PSI/TiO_2 electrode, photocurrent under UV + visible light was about twice as high as under visible light alone, and more than three times higher for the bare TiO_2. The photocurrent from PSI/TiO_2 was around

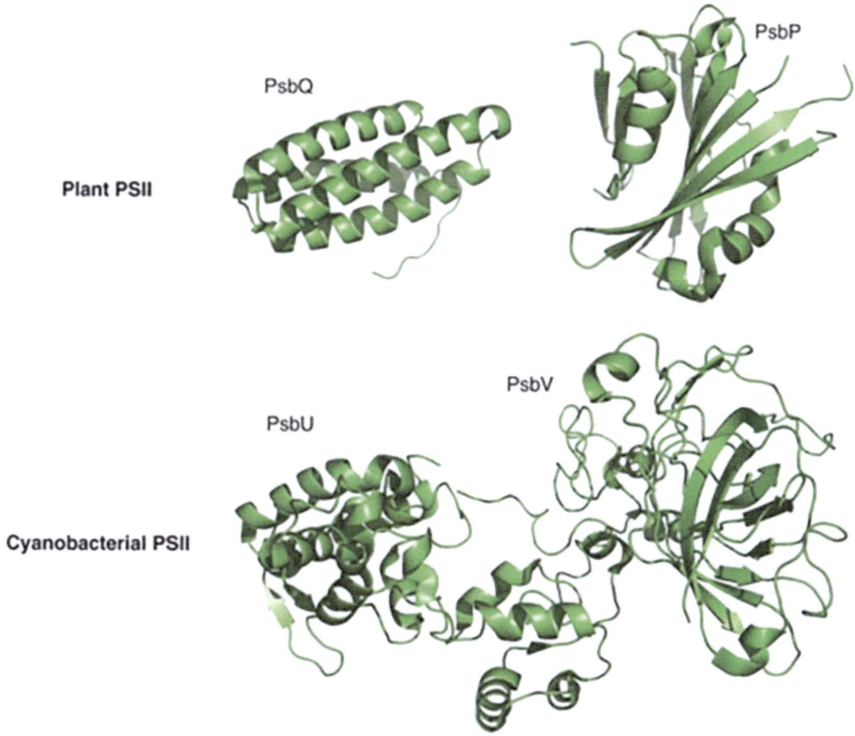

Fig. 5 Structures of the smaller extrinsic polypeptides of photosystem II. Reproduced with permission from [16]. Copyright 2009. Science

two times higher than that from bare TiO_2 when compared to each other. The best result for this work is about 4.15 mA cm^{-2} for PSI/TiO_2 at UV + visible light, with an optimal deposition time of 10 min [15] (Figs. 5, 6, and 7).

Photosystem II (PSII): PSII is the first protein complex in the light-dependent reactions, absorbing light at around 680 nm (red light). It is responsible for splitting water molecules into oxygen, protons, and electrons, which are used to drive the electron transport chain. PSII provides a source of high-energy electrons in biohybrid solar cells, which can then be directed to synthetic materials for electron capture and electricity generation.

To create a TiO_2 nanotube network, the Li group incorporated photosystem II from spinach thylakoid membranes in 2016 [17]. They looked into the relationship between the structure of TiO_2 and the efficiency of light-to-current conversion, as well as the differences between mediated and direct electron transfer (MET). TiO_2 nanotubes were created from Titanium (IV) n-butoxide solution using the surface sol–gel technique. The surface sol–gel approach was carried out using cellulose as the scaffold. The filter paper was the source of cellulose fibers. After that, a single

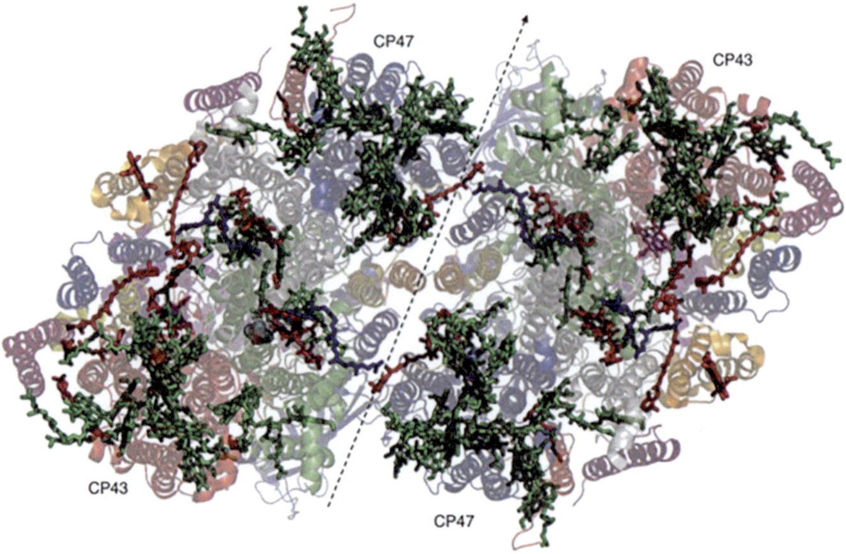

Fig. 6 Stromal side view of the structure of the cyanobacterial photosystem II (PSII) dimer. Reproduced with permission from [7]. Copyright 2004. Nature

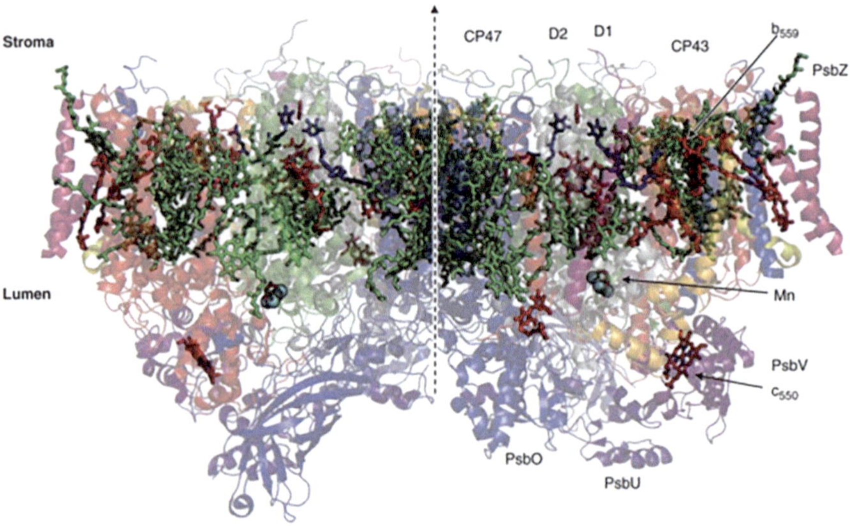

Fig. 7 Side view of the structure of cyanobacterial photosystem II. Reproduced with permission from [7]. Copyright 2004. Nature

deposition cycle was used to coat each cellulose nanofiber of the filter paper with a thin layer of titania gel, around 0.5 nm in thickness.

Layers of varying thicknesses were created by applying varying numbers of deposition cycles. To get rid of the cellulose, this film was calcined. Commercialized TiO_2 nanopowder or TiO_2 nanotubes made in-lab were dissolved in ethanol and sonicated to create the paste. Without sintering, the paste was applied to ITO-glass and allowed to air dry. Additionally, the researchers tested electrodes lacking PSII complexes (bare TiO_2) and electrodes without TiO_2 as controls. PSII complexes were immobilized on the ITO surface in the later case.

Two different illumination conditions were tested: red and white. The mediator utilized in the experiments was 2,5-dichloro-1,4-benzoquinone (DCBQ). Under red light, electrodes with bare TiO_2 produce no current in either the DET or MET scenarios. Under red light, the PSII/TiO_2/ITO-glass current increased with the thickness of the TiO_2 layer without the need for a mediator. However, it was demonstrated that in all other circumstances (red light with MET and white light), the less current was created, the thicker the TiO_2 layer. It's fascinating to see how the mediator affects the current flowing from the exposed TiO_2 electrode. Their findings imply that DCBQ and TiO_2 can interact. MET was consistently far more successful than DET.

The explanation for these results was that the TiO_2 was not involved in the DCBQ-mediated electron transfer from PSII to the ITO. The total current values of PSII/ITO and bare TiO_2/ITO were less than the photocurrent from PSII/TiO_2/ITO-glass. It was proposed that electrons from PSII travel more quickly to the valence band of TiO_2, where they are subsequently attracted to ultraviolet light and migrate to the conductive band. If DCBQ is present, it can take an electron from TiO_2's conduction band and move it to the ITO. Thus, electrons can move from PSII to the electrode in two different ways: directly from QA to the TiO_2 and through the DCBQ.

About 10.6 $\mu A\ cm^{-2}$ was the best current value for PSII/TiO_2/ITO-glass systems, which is slightly less than twice as much as from bare TiO_2. Compared to TiO_2 nanotube paste, the TiO_2 paste made from commercial nanopowder offered less current [17]. The channels for electron transfer in a hybrid PSII/TiO_2 system were also studied by Brinkert et al. PSII was isolated from T. elongatus for this investigation.

The outcomes exhibit a good correlation with the findings from the earlier research. The compact TiO_2 layer between sensitized TiO_2 and ITO was applied to verify the DCBQ-mediated electron transfer to ITO right away. In the presence of DCBQ, the electrode containing this layer did not show a rise in photocurrent. Given that the compact layer is impermeable to the mediator species, this was anticipated. As a result, the electron transfer pathway PSII quinone—TiO_2—DCBQ—ITO is suggested.

This could be because the electrons from quinone PSII don't have enough energy to get into the TiO_2 conduction band; instead, they stay at the surface in lower energy states where they can interact with mediators again and proceed slowly to the conducting electrode. Inhibitor research revealed the surprising fact that direct electron transfer happens from the QA site in addition to the QB site. For complexes that are not in direct contact with TiO_2, DCBQ can take electrons from the acceptor side of PSII. Because of the energy mismatch between the PSII acceptor side and the conduction band edge of TiO_2, electron diffusion in TiO_2 is the rate-limiting phase in current production for PSII/TiO_2 hybrid electrodes [18].

2.2 Chlorophyll and Other Pigments

Chlorophyll is the primary pigment found in plants and algae responsible for absorbing light during photosynthesis. It absorbs light most efficiently in the blue and red regions of the electromagnetic spectrum and plays a central role in transferring energy to reaction centers in PSI and PSII (Fig. 8).

In biohybrid solar cells, chlorophyll or its derivatives can be used as light-harvesting pigments. They are often immobilized on electrodes or embedded in dye-sensitized systems, where they absorb photons and generate excited electrons that can then be captured and transferred to a conductive layer.

Other Pigments: In addition to chlorophyll, biohybrid solar cells can utilize other natural pigments such as **carotenoids**, **phycobilins**, and **xanthophylls**, which also play roles in light absorption and energy transfer in photosynthetic organisms. These pigments extend the range of light absorption, allowing biohybrid solar cells to harness a broader spectrum of sunlight.

Carotenoids: These pigments absorb blue and green light and transfer energy to chlorophyll molecules. Their wide absorption spectrum and stability under light exposure make them useful for enhancing light capture in biohybrid solar cells. **Phycobilins**: Found in cyanobacteria and red algae, phycobilins are pigments that absorb light in the green and orange-red regions of the spectrum. Their ability to absorb wavelengths that chlorophyll cannot makes them valuable additions to biohybrid solar cells, improving their efficiency in capturing light energy across a broader spectrum.

Carotenoids, unlike chlorophylls, are also found in many other types of species, therefore their evolutionary history may reflect many additional activities in addition to photosynthesis [19, 20]. Phycobiliproteins, which are found in cyanobacterial light-harvesting structures termed phycobilisomes, work similarly to previously identified antennae complexes in that they absorb photonic energy and direct it toward the photosystem. Phycobiliproteins have maxima in the green, yellow, and red portions of their absorption spectra. Therefore, cyanobacteria were originally called blue-green algae because many of them reflected blue light [21].

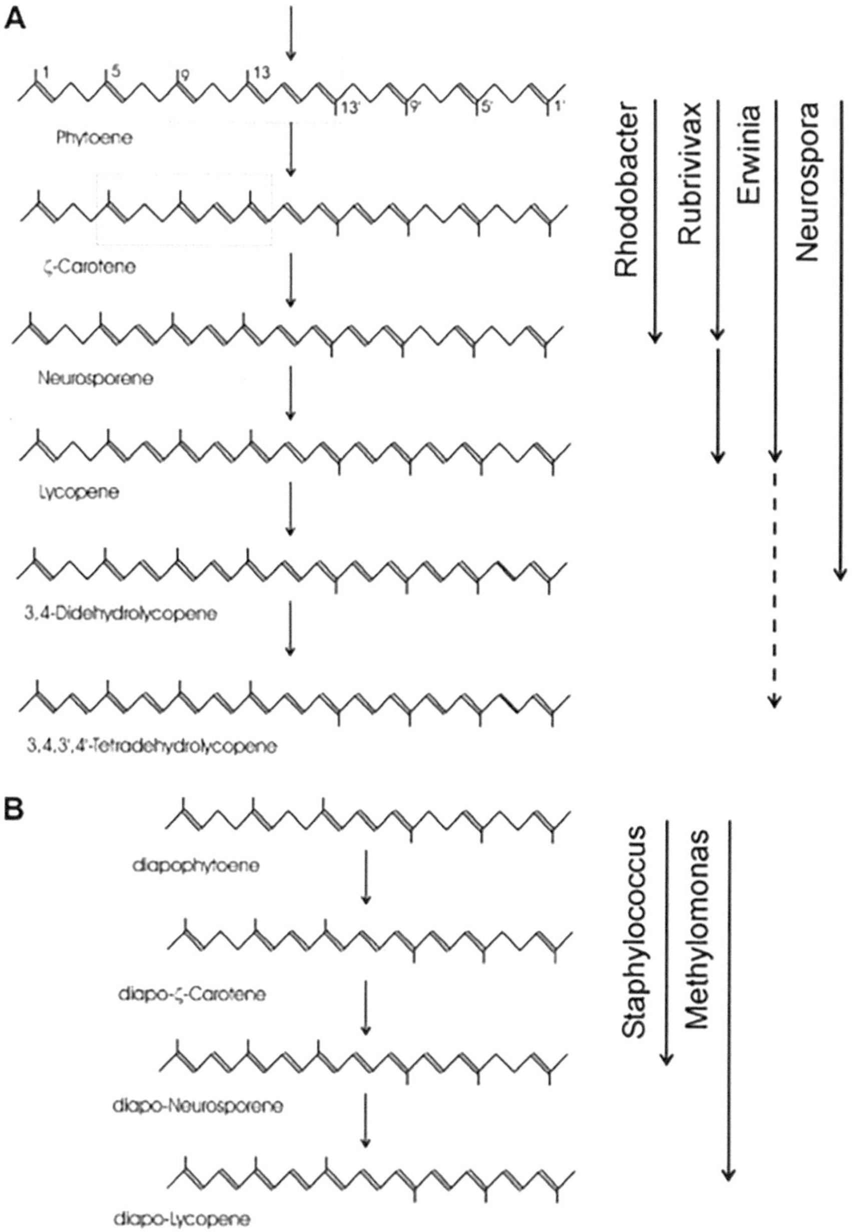

Fig. 8 Reaction sequences of CrtI-type phytoene (**a**) and diapophytoene (**b**) desaturases from different bacteria and fungi. Arrows cover all the desaturation steps catalyzed by the individual enzymes. The proposed recognition sites for the desaturase are boxed. Reproduced with permission from [21]. Copyright 2007. Science Direct

2.3 Reaction Centers

Photosynthetic Reaction Centers (RCs) are specialized protein complexes in photosynthetic organisms where the primary charge separation occurs, leading to the conversion of light energy into chemical energy. The most common RCs used in biohybrid solar cells are derived from photosystem I and photosystem II (Fig. 9).

Purple bacterial reaction centers were cultivated on WO_3–TiO_2 mesoporous film that had been specially pore-sized by Lu et al. [22]. Their design was based on research suggesting that, in comparison with pure TiO_2 or WO_3 alone, the TiO_2-doped amorphous WO_3 and nanocluster WO_3–TiO_2 exhibit better performance of light-induced charge separation. Tungsten hexachloride and titanium butoxide were the reactants used in the sol–gel process to create the WO_3–TiO_2 paste. Using the spin coating approach, an even layer of the paste was coated onto ITO glass, and the electrode was then subjected to high-temperature sintering. The WO_3–TiO_2 structures and a pure TiO_2 film made by anodic electrodeposition are contrasted by the authors.

The freshly made WO_3–TiO_2 films were immersed in a pH 8.0 Tris–HCl buffer solution containing RCs (at 4 °C) for two to three days to immobilize the protein. The substrate exhibits the best RC loading, with pores measuring roughly 7.1 nm. The researchers hypothesize that WO_3–TiO_2's greater hydrophilicity explains why there

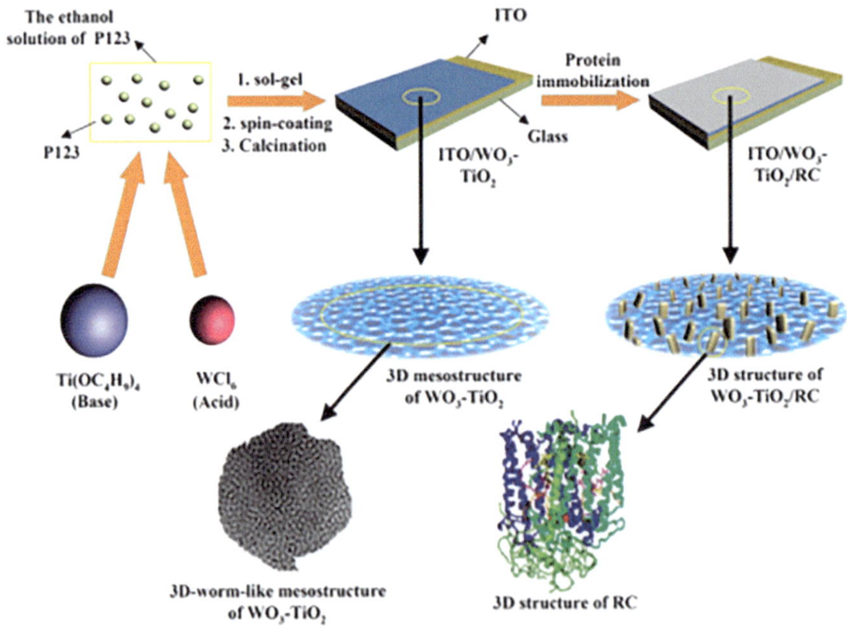

Fig. 9 Experimental Setup for Preparation of a Bio-PE Based on the Tailored Mesoporous WO_3–TiO_2 Film. Reproduced with permission from [22]. Copyright 2005. American Chemical Society

is superior RC loading onto it as opposed to the TiO_2 layer. The more electrostatically stable orientation of the RC was such that its more hydrophilic H-subunit domains faced the substrate before the addition of this hydrophobic semiconductor. Based on differential absorption spectra, the maximum loading of RC on these WO_3–TiO_2 films is determined to be 0.63 μmol/g. The photoelectric response was evaluated by the authors using the three-electrode scheme.

During these tests, the electrode bias was roughly -0.3 V relative to the Ag/AgCl electrode. The electrochemical cell's buffer contains 8 mM sodium dithionite. At 5 mW cm^{-2} white light and 0.1 mW cm^{-2} red light ($\lambda > 600$ nm), the electrode generated approximately 30 μA cm^{-2} and 0.9 μA cm^{-2}, respectively. When sensitizer (RC) was added to WO_3–TiO_2 films, the photocurrent practically quadrupled. The electrode with bare WO_3–TiO_2 films produced 6.7 μA cm^{-2} under white light and almost produced no current at red light. The ITO/WO_3–TiO_2/RC electrode's stability was also assessed, and it was found that after one hour of continuous light, photocurrent decreased by about 15% [22].

Almost two years after the Lu group conducted a very similar study, a group of Russian scientists published a method of immobilizing RC onto nanostructured TiO_2 [23]. Due to hydrophobic interactions, RC complexes in this work were fixed in a porous substrate of TiO_2 alone without the presence of tungsten (VI) oxide. The H subunit should face the buffer in this work and not come into touch with the TiO_2 substrate. As the antecedents for mesoporous TiO_2 layer, these researchers compared various forms of TiO_2 nanopowder. The size of the particles separates them from one another. Every nanopowder was prepared using the same method for TiO_2 paste. TiO_2 nanopowder, polyethylene glycol, and detergent Triton X-100 were combined in an aqueous solution to create TiO_2 paste.

To keep the TiO_2 nanoparticles from aggregating, different acids were introduced. By rolling on the paste, films were stacked onto the ITO-glass, creating a layer that was about 4 μm thick. A mesoporous layer with the most effective surface is produced from the nanopowder containing the smallest nanoparticles; as a result, this film has the potential to adsorb the greatest number of RC particles. To prepare the best film, acetic acid was utilized as an anti-aggregation agent. For the photocurrent measurement, the three-electrode setup was once more employed.

For TiO_2 without RC, the bias supplied to the electrode was around − 0.3 V against Ag/AgCl, while for TiO_2-RC, it was approximately − 0.25 V. With an incoming white light intensity of around 80 mW cm^{-2}, the TiO_2-RC generated roughly 50 μA cm^{-2}. Bare TiO_2 electrodes produced a current that was just two times smaller under the same lighting conditions. Bare TiO_2 produced negligible current under red light illumination (45 mW cm^{-2}), but TiO_2-RC produced approximately 1.125 μA cm^{-2}). The last acceptor in this ETC is quinone (Q), and there is an energetically unfavorable electron transfer from Q to TiO_2. Researchers used sodium dithionite to artificially lower all Qs to transfer electrons to the electrode successfully.

Plasmodiolin thus functioned as the electron transductor connecting the protein to the electrode. Based on these findings, the authors concluded that the P840's excited triplet form might contribute electrons to the TiO_2 conductive band [23]

The first RC complex found is related to the photosynthetic purple sulfur bacteria and has been biochemically studied for over 20 years. As per the previously mentioned protocols, this is a kind II RC. The core domain of all purple bacterial RCs is composed of the H-, L-, and M-polypeptides. Cofactors involved in energy or electron transfer are encased in L- and M-polypeptides [24]. These cofactors consist of two monomers of bacteriochlorophyll a (BChl), two ubiquinone molecules (QA and QB), two bacteriopheophytin a (BPhe), two bacteriochlorophyll a special pair (P840 or simply P), and carotenoids.

With its broad cytoplasmic domain, the H subunit is more peripheral [25]. The unique pair of bacteriochlorophyll is called P870 because of its longwave absorption maxima, which is roughly 10 nm in length. The redox potential of QB in the purple bacterial RC is approximately 0 V versus NHE, and the redox potential of this unique pair is E o ($P870^+/P870$) $\cong + 0.45$ V. Charge separation thus leads to a steady-state redox differential of roughly $- 0.5$ to $- 0.4$ V. There are two primary benefits that purple bacteria have when it comes to fabricating biohybrid electrodes.

They are the smallest known complexes that can separate charges in response to light. As a result, compared to other photosystem complexes, more active purple bacterial RC complexes can be immobilized in the same space. Secondly, compared to photosystem complexes from oxygenic species, purple bacterium RCs are simpler to extract. In electrochemical cells, RC-biohybrid electrodes can achieve cathodic or anodic current depending on how these RCs are oriented onto the substrate. Trammel et al. [26] describe the construction of two varieties of RC-based solar cells using carbon-based electrodes. The first of these devices used N-(1-pyrene)-iodoacetamide as a linker to produce light-induced anodic current, and the second used His6-$Ni2^+$-nitrilotriacetic acid (NTA) as a linker to produce light-induced cathode current [26].

The unique orientation of RC complexes to the carbon substrate varies between these electrodes. To produce anodic and cathodic current, respectively, the following artificial electron carriers were utilized as the donor: cytochrome c (reduced by $Na_2S_2O_4$) and ubiquinone-10 (2,3-dimethoxy-5-methyl-6-geranio-1,4-benzoquinone) (Fig. 7a). In these studies, the apparatus that generated cathodic current was more effective. This is because, in this particular device, the distance between P840 and carbon is smaller than the distance between quinone and the electrode in the scenario where anodic current is produced. The length of the linker plus the size of the RC complex are added to determine the distance [27].

The reaction centers contain special pairs of chlorophyll molecules that are excited by photons, which triggers electron transfer. These reaction centers are engineered into biohybrid solar cells to capture light and generate charge separation, which is necessary for producing electrical current.

Reaction centers are attractive because of their ability to achieve near-perfect energy conversion efficiency in natural systems. By embedding them into synthetic materials, researchers hope to replicate this efficiency in solar cells.

2.4 Bacteriorhodopsin and Other Light-Activated Proteins

Bacteriorhodopsin is a light-activated protein found in certain archaea (such as *Halobacterium salinarum*) that functions as a proton pump. It absorbs light and uses the energy to transport protons across the membrane, generating a proton gradient used for ATP synthesis.

In biohybrid solar cells, bacteriorhodopsin can be used as a light-harvesting material due to its ability to absorb green light (around 570 nm) and convert it into a proton gradient. This proton gradient can be utilized to drive electron transport processes in conjunction with synthetic materials to generate electrical current.

The membrane protein bacteriorhodopsin (bR), which is photoactive and contains retinal, is one of the biosensitizers that exhibits promise in terms of simplicity, stability, and quantum efficiency. In the current work, we have integrated low-cost, carbon-based components, such as a hydroquinone/benzoquinone (HQ/BQ) redox electrolyte and a cathode composed of PEDOT (poly(3,4-ethylenedioxythiophene) functionalized with multi-walled carbon nanotubes (CNT), in a photoanode-sensitized TiO_2 solar cell. We have employed a mutant of bR, D96N. Using SEM, TEM, and Raman, the photoanode and cathode were morphologically and chemically investigated. Impedance spectroscopic analysis (EIS), open circuit potential decay (VOC), and linear sweep voltammetry (LSV) were used to examine the electrochemical performance of the bR-BSCs. The champion device produced a power conversion efficiency (PCE) of 0.16%, a fill factor of about 24%, a current density (JSC) of 1.0 mA/cm^2, and a voltage output current (VOC) of -669 mV [28].

2.4.1 Enzymes and Redox Proteins

Enzymes play a crucial role in biohybrid solar cells, particularly those involved in electron transfer processes. For example, **ferrodoxins** and **cytochromes** are redox-active proteins that can shuttle electrons between the biological light-harvesting systems and the synthetic conductive components of the cell.

Ferrodoxins are iron-sulfur proteins that facilitate electron transfer in photosynthesis, especially between PSI and other components of the electron transport chain. In biohybrid solar cells, ferrodoxins can help bridge the gap between biological systems and synthetic electrodes, ensuring smooth electron flow (Fig. 10).

Using the native electron transport protein ferredoxin (Fd) as a scaffold for the binding of a ruthenium photosensitizer (PS) and a molecular cobaloxime catalyst (Co), this unique solar fuel biohybrid produces H_2 under light. A long-lived (> 1.5 ms) Ru(III)–Fd–Co(I) charge separated state created by an electron relay across the Fd [2Fe–2S] cluster is directly demonstrated by EPR and transient optical measurements. This state starts the catalytic cycle for $2H^+ + 2e^- \rightarrow H_2$ [29].

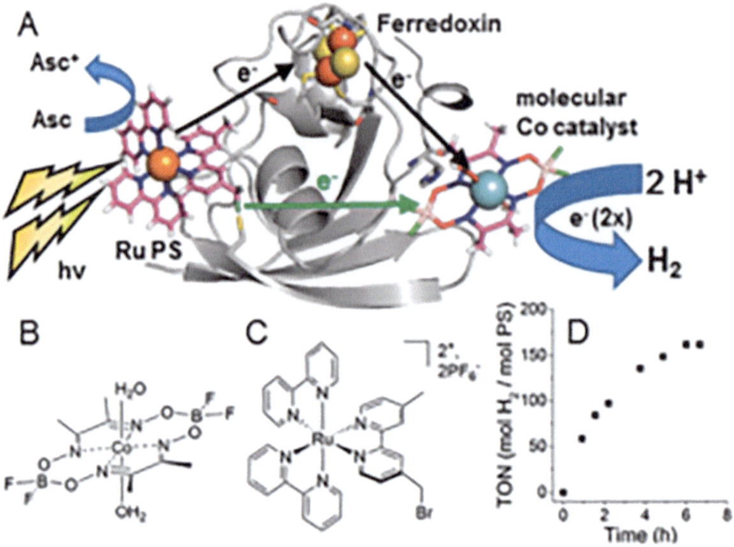

Fig. 10 **a** Photocatalytic H_2 production scheme in Ru–Fd–Co biohybrids with two potential pathways for electron transfer from PS to catalyst; via the [2Fe–2S] cluster of Fd (1A70), black arrows; directly from PS to catalyst, green arrow. **b** and **c** Chemical structures of **b** cobaloxime catalyst (Co), Co(dmgBF2)22H_2O and **c** ruthenium photosensitizer (Ru PS), [Ru(4—CH_2Br-40—CH_3-2,20 -bpy)(bpy)2] 2PF6 used in the study. **d** Representative time course profile of H_2 production from Ru–Fd–Co hybrid upon visible light illumination. The assay conditions are 2 mM Ru–Fd–Co hybrid (0.6 Ru/Fd, 0.9 Co/Fd) in 10 mM MES pH 6.3, 100 mM sodium ascorbate. Reproduced with permission from [29]. Copyright 2007. Royal Society of Chemistry

Cytochromes are another class of proteins that contain heme groups and participate in electron transport in various biological processes. They can be engineered into biohybrid solar cells to enhance electron mobility and improve overall efficiency.

The membrane-bound cytochrome bc1 or b6f complex is a component of the ETC. Quinols generated by photochemistry in type II RCs or during cyclic electron transfer involving type I RCs are oxidized by these complexes. Protons are pumped across the thylakoid membrane by these proteins in order to create a proton gradient and reduce the lumen's pH. This is their second function. A cytochrome bc1 or b6f type complex with a generally comparable architecture is present in almost all phototrophic organisms [30].

2.5 DNA- or Peptide-Based Scaffolds

DNA and peptides have also been explored as scaffolding materials in biohybrid solar cells. They can provide a flexible framework to stabilize biological components (such as reaction centers or enzymes) on synthetic electrodes, ensuring optimal orientation and spacing for efficient electron transfer.

DNA scaffolds can be engineered to precisely position photosynthetic proteins or light-harvesting pigments in a way that maximizes their interaction with synthetic components. DNA's ability to form highly ordered structures makes it an excellent candidate for creating biohybrid systems with well-organized light-harvesting complexes.

A DNA origami platform with a programmable antenna array that allows the creation of logically constructed antenna shapes. Using ensemble and single-molecule fluorescence spectroscopy and in-depth Förster modeling, we systematically examine the light-harvesting efficiency of a ring-like antenna with regard to the number of donors and interdye distances. This extensive study highlights DNA origami as a highly versatile platform for evaluating design principles in artificial light-harvesting networks and shows excellent and dependable structural control over multichromophoric geometries [31].

Peptide scaffolds are short chains of amino acids that can be designed to bind specific biological components to conductive surfaces. They offer the advantage of biocompatibility and flexibility in design, enabling researchers to tailor the biohybrid cell architecture for optimal energy conversion.

2.6 Whole Cells and Microorganisms

In some cases, **whole cells** of photosynthetic organisms (such as algae or cyanobacteria) have been integrated into biohybrid solar cells. These organisms naturally contain all the machinery necessary for capturing light and converting it into energy. By incorporating living cells, biohybrid solar cells can potentially harness the full power of natural photosynthesis in an artificial system. For example, cyanobacteria can be immobilized on electrodes, where they can carry out photosynthesis and generate electrons, which are then harvested by the biohybrid solar cell. This approach offers the potential for self-repairing and self-regenerating solar cells, as living organisms can naturally replace damaged components.

3 Challenges in Biohybrid Solar Cell Development

Despite the potential of biohybrid solar cells, several challenges must be addressed before they can achieve widespread use. These include:

- **Stability of Biological Materials**: The biological components, such as photosynthetic proteins, are inherently fragile and can degrade over time when exposed to oxygen, temperature fluctuations, or intense light. Stabilizing these proteins in a way that preserves their functionality is a significant challenge.
- **Integration with Synthetic Components**: Efficiently coupling biological molecules with synthetic materials without losing electron transfer efficiency is another technical hurdle. The interface between the biological and synthetic layers must be optimized to ensure maximum energy capture and transfer.
- **Scaling Production**: While biohybrid solar cells offer a potential for lower-cost production, scaling these technologies to the levels needed for global energy demands remains a challenge. The sourcing, purification, and integration of biological molecules into large-scale solar panels must be streamlined for commercial viability.

4 Recent Developments and Future Directions

In recent years, researchers have made significant strides in improving the efficiency, stability, and scalability of biohybrid solar cells. Innovations in nanotechnology, such as the use of nanomaterials to stabilize biological components, have shown promise in extending the lifespan of biohybrid solar cells. Additionally, the use of new synthetic materials, such as conductive polymers and graphene, has improved the efficiency of electron transport.

Future research may focus on optimizing the interaction between biological and synthetic components, improving the longevity of these cells, and finding ways to integrate them into existing solar technologies.

5 Conclusion: The Promise of Biohybrid Solar Cells

Biohybrid solar cells represent an exciting convergence of biology and technology. By mimicking the processes found in nature and integrating them with modern materials, these cells offer a promising route to more sustainable and efficient solar energy harvesting. While there are still challenges to overcome, the progress made so far points to a future where biohybrid solar cells could become an important part of the global renewable energy landscape.

In the chapters that follow, we will explore the detailed mechanisms of various biological components, the latest advancements in material science for biohybrid cells, and the environmental and economic impact of this technology. Biohybrid solar cells stand at the forefront of a new era in solar energy, where nature and innovation work together to power a sustainable future.

References

1. M. Asif, T. Muneer, Energy supply, its demand and security issues for developed and emerging economies. Renew. Sustain. Energy Rev. **11**(7), 1388–1413 (2007)
2. M. Grätzel, Recent advances ins sensitized mesoscopic solar cells. Acc. Chem. Res. **42**, 1788–1798 (2009). https://doi.org/10.1021/ar900141y
3. N.S. Lewis, D.G. Nocera, Powering the planet: chemical challenges in solar energy utilization. Proc. Natl. Acad. Sci. U.S.A. **103**, 15729–15735 (2006). https://doi.org/10.1073/pnas.0603395103
4. T.V. Arjunan, T.S. Senthil, Review: dye sensitised solar cells. Mater. Technol. **28**, 9–14 (2013). https://doi.org/10.1179/1753555712Y.0000000040
5. M. Grätzel, Photovoltaic and photoelectrochemical conversion of solar energy. Philos. Trans. A. Math. Phys. Eng. Sci. **365**, 993–1005 (2007). https://doi.org/10.1098/rsta.2006.1963
6. N. Sekar, R.P. Ramasamy, Recent advances in photosynthetic energy conversion. J. Photochem. Photobiol. C Photochem. Rev. **22**, 19–33 (2014). https://doi.org/10.1016/j.jphotochemrev.2014.09.004
7. N. Nelson, A. Ben-Shem, The complex architecture of oxygenic photosynthesis. Nat. Rev. Mol. Cell Biol. **5**, 971–982 (2004)
8. N. Nelson, C.F. Yocum, Structure and function of photosystem I and II. Annu. Rev. Plant Biol. **57**(1), 521–625 (2006). https://doi.org/10.1146/annurev.arplant.57.032905.105350
9. P.N. Ciesielski, F.M. Hijazi, A.M. Scott, C.J. Faulkner, L. Beard, K. Emmett, S.J. Rosenthal, D. Cliffel, G. Kane Jennings, Photosystem I—based biohybrid photoelectrochemical cells. Bioresour. Technol. **101**(9), 3047–3053 (2010)
10. J.M. Anderson, Insight into consequences of grana stacking of thylakoid membranes in vascular plants: a personal perspective. Aust. J. Plant Physiol. **26**, 625–639 (1999)
11. J.M. Anderson, Changing concepts about the distribution of photosystems I and II between grana-appressed and stroma exposed thylakoid membranes. Photosynth. Res. **73**, 157–164 (2002)
12. R.F. Simmerman, T. Zhu, D.R. Baker, L. Wang, S.R. Mishra, C.A. Lundgren, B.D. Bruce, Engineering photosystem I complexes with metal oxide binding peptides for bioelectronic applications. Bioconjug. Chem. **26**, 2097–2105 (2015). https://doi.org/10.1021/acs.bioconjchem.5b00374
13. A. Mershin, K. Matsumoto, L. Kaiser, D. Yu, M. Vaughn, M.K. Nazeeruddin, B.D. Bruce, M. Gratzel, S. Zhang, Self-assembled photosystem-I biophotovoltaics on nanostructured TiO(2) and ZnO. Sci. Rep. **2**, 234 (2012). https://doi.org/10.1038/srep00234
14. D. Yu, M. Wang, G. Zhu, B. Ge, S. Liu, F. Huang, Enhanced photocurrent production by bio-dyes of photosynthetic macromolecules on designed TiO_2 film. Sci. Rep. **5**, 9375–9384 (2015). https://doi.org/10.1038/srep09375
15. V.B. Shah, W.R. Henson, T.S. Chadha, G. Lakin, H. Liu, R.E. Blankenship, P. Biswas, Linker-free deposition and adhesion of photosystem i onto nanostructured TiO_2 for biohybrid photoelectrochemical cells. Langmuir **31**, 1675–1682 (2015). https://doi.org/10.1021/la503776b
16. K.N. Ferreira, T.M. Iverson, K. Maghlaoui, J. Barber, S. Iwata, Architecture of the photosynthetic oxygen-evolving center. Science **303**, 1831–1838 (2004)
17. J. Li, X. Feng, J. Fei, P. Cai, J. Huang, J. Li, Integrating photosystem II into a porous TiO_2 nanotube network toward highly efficient photo-bioelectrochemical cells. J. Mater. Chem. A **4**, 12197–12204 (2016). https://doi.org/10.1039/C6TA04964H

18. K. Brinkert, F. Le Formal, X. Li, J. Durrant, A.W. Rutherford, A. Fantuzzi, Photocurrents from photosystem II in a metal oxide hybrid system: electron transfer pathways. Biochim. Biophys. Acta Bioenerg. **2016**, 1497–1505 (1857). https://doi.org/10.1016/j.bbabio.2016.03.004
19. B.R. Green, W.W. Parson, *Light-Harvesting Antennas in Photosynthesis* (Kluwer Academic, 2003)
20. S.I. Beale, Enzymes of chlorophyll biosynthesis. Photosynth. Res. **60**, 43–73 (1999). https://doi.org/10.1023/A:1006297731456
21. G. Sandmann, Evolution of carotene desaturation: the complication of a simple pathway. Arch. Biochem. Biophys. **483**, 169–174 (2009). https://doi.org/10.1016/j.abb.2008.10.004
22. Y. Lu, M. Yuan, Y. Liu, B. Tu, C. Xu, B. Liu, D. Zhao, J. Kong, Photoelectric performance of bacteria photosynthetic proteins entrapped on tailored mesoporous WO_3-TiO_2 films. Langmuir **21**, 4071–4076 (2005). https://doi.org/10.1021/la0470129
23. E.P. Lukashev, V.A. Nadtochenko, E.P. Permenova, O.M. Sarkisov, A.B. Rubinc, Dokl. Biochem. Biophys. **415**, 211–216 (2007). https://doi.org/10.1134/S1607672907040138
24. T. Mizoguchi, M. Isaji, J. Harada, H. Tamiaki, Isolation and pigment composition of the reaction centers from purple photosynthetic bacterium Rhodopseudomonas palustris species. Biochim. Biophys. Acta Bioenerg. **2012**, 395–400 (1817). https://doi.org/10.1016/j.bbabio.2011.12.001
25. R. Das, P.J. Kiley, M. Segal, J. Norville, A.A. Yu, L. Wang, S.A. Trammell, L.E. Reddick, R. Kumar, F. Stellacci, N. Lebedev, J. Schnur, B.D. Bruce, S. Zhang, M. Baldo, Integration of photosynthetic protein molecular complexes in solid-state electronic devices. Nano Lett. **4**, 1079–1083 (2004). https://doi.org/10.1021/nl049579f
26. S.A. Trammell, L. Wang, J.M. Zullo, R. Shashidhar, N. Lebedev, Orientated binding of photosynthetic reaction centers on gold using Ni NTA self-assembled monolayers. Biosens. Bioelectron. **19**, 1649–1655 (2004). https://doi.org/10.1016/j.bios.2003.12.034
27. J.P. Allen, G. Feher, T.O. Yeates, H. Komiya, D.C. Rees, Structure of the reaction center from Rhodobacter sphaeroides R-26: the cofactors. Proc. Natl. Acad. Sci. U.S.A. **84**, 5730–5734 (1987). https://doi.org/10.1073/pnas.84.16.5730
28. C. Espinoza-Araya, R. Starbird, E. S. Prasad, V. Renugopalakrishnan, A. Mulchandani, B.D. Bruce, C.C. Villarreal, A bacteriorhodopsin-based biohybrid solar cell using carbon-based electrolyte and cathode components, Biochim. Biophys. Acta Bioenerg. **1864**(4), 148985 (2023). https://doi.org/10.1016/j.bbabio.2023.148985
29. S.R. Solitau, J. Niklas, P.D. Dahlberg, O.G. Poluektov, D.M. Tiede, K.L. Mulfort, L.M. Utschig, Aqueous light driven hydrogen production by a Ru–ferredoxin–Co biohybrid. Chem. Comm. **51**, 10628–10631 (2015)
30. M.F. Yanyushin, M.C. del Rosario, D.C. Brune, R.E. Blankenship, New class of bacterial membrane oxidoreductases. Biochemistry **44**, 10037–10045 (2005). https://doi.org/10.1021/bi047267l
31. E.A. Hemmig, C. Creatore, B. Wunsch, L. Hecker, M.A. Parker, S. Emmott, P. Tinnefeld, U.F. Keyser, A.W. Chin, Nano Lett. **16**(4), 2369–2374 (2016)

Artificial Photosynthesis: Mimicking Nature for Solar Fuel Production

C. G. Jothi Prakash and Jeong-Won Lee

Abstract Unique capabilities of natural living beings aid its survival in harsh environment and inspired scientists toward biomimicry. Biomimicry enables adaptation of heritable traits and behavior of those natural elements. Mimicking nature to solve human needs gained significant attention in addressing sustainability over technological innovations. These innovations are highly interdisciplinary are not limited to architectures, clothing, aerodynamics, medicine, and so on. Globalization and increasing energy utilization, demand alternative resources for energy harvesting, conversion, and storage. Artificial photosynthesis is a technology that attempts to mimic natural process of photosynthesis in converting light energy to usable chemical energy. This process utilizes a catalyst that facilitate light energy conversion to storable hydrogen and other carbon-based fuels. The use of various types of catalyst including homogeneous, heterogeneous, and biocatalyst offer selectivity in the conversion processes. The conversion efficiency of natural photosynthesis range between 3 and 6% and the artificial photosynthesis struggles to achieve similar conversion efficiencies. Being a sustainable alternative for energy conversion and storage, artificial photosynthesis is hardly realized in application due to its limited conversion efficiency. Hence, the technology is still under investigation and development to help address the sustainability and future energy security. This chapter aims at summarizing the progress and prospects of nanoscience and technology in the field of artificial photosynthesis. Specifically, the principles and strategies adopted to engineer nanomaterial-based catalysts, and its performance improvement in energy conversion has been given special attention. The chapter could provide valuable insights into efficient catalyst design and bring in integration to develop renewable fuel production from solar energy.

Keywords Biomimicry · Artificial photosynthesis · Nanomaterials · Hydrogen · CO_2 reduction · Solar fuel

C. G. J. Prakash (✉) · J.-W. Lee
Advanced Nano Surface Engineering Laboratory, Department of Mechanical Engineering, Chosun University, Gwangju, South Korea
e-mail: prakash@chosun.ac.kr

P. Ravindran et al. (eds.), *Photon to Power*, Progress in Optical Science and Photonics 32, https://doi.org/10.1007/978-981-96-5914-2_14

1 Introduction

Globalization and increased energy consumption reverberate in production and mostly rely on fossil fuels. It expected that, the energy demand could increase from 33,000 TWh to 40,800 TWh by 2030 and will increase to 70,000 TWh by 2050 [1]. The primarily available source of fossil fuels will become scarcely available to address the energy demands. The use of fossil fuel for energy production is associated with emission of green-house gases and the levels of CO_2 that implicates global climatic changes [2, 3]. Hence, it is vital to look for alternates that minimize emissions and further develop technological support to capture CO_2 efficiently. The primary energy use of world with respect to different fuels was projected by U.S. Energy Information Administration as is represented as a chart (Fig. 1).

Based on the projected energy requirements and dependence of fuels, the possible shift to renewables is driven by resources, energy policies, and associated technological costs. The introduction of policy incentives over battery cost, efficiency standards and electricity price could boost up the sales of light duty electric vehicles by 75%. It is projected that, improvement in renewable resources could decrease the CO_2 intensity to 19% by the year 2050 [1]. Thus, the energy security hastens a potential renewable alternative for the sustainable development.

Certainly, technological innovations on non-fossil fuel-based sources for energy use are improving. Limitless light energy from sun is a sole natural resource to fulfill the needs of humanity. The received energy of ~ 3×10^{18} MJ from sun in a year is over 10,000 folds greater than the humankind consumption. Thus, new protocols toward solar energy harvesting and transformation into alternative energy resources must be developed and deployed [4]. The focus on using solar energy to capture light and produce electricity by photovoltaic (PV) systems are improving daily [5–7]. However, the existing price of PV systems comparing with fossil fuels are expensive due to high cost of fabrication. In progress with time, the blending principles of PV systems using less expensive organic and inorganic materials could

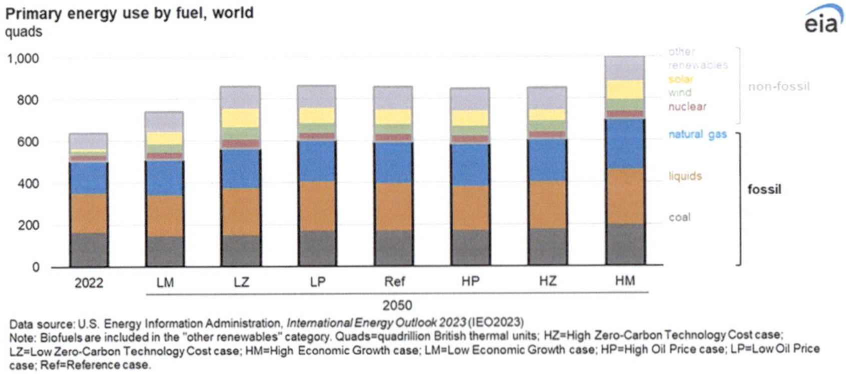

Fig. 1 Projected energy use by fuel [reproduced from IEO2023 release, EIA] [1]

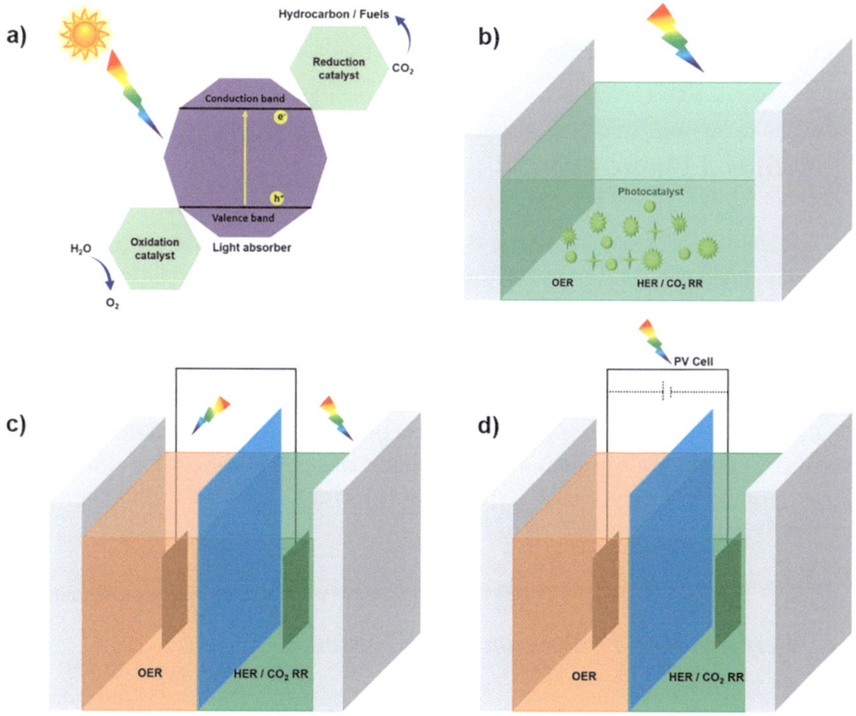

Fig. 2 **a** Principle of photocatalysis, **b–d** photochemical, photoelectrochemical and photovoltaic-electrochemical catalytic systems

relatively increase the employability and ultimately provide long-term solution to energy harvesting, conversion, and storage [5, 8].

In nature, plants, algae, and certain bacteria converts the typical light energy from sun into useful chemical energy through natural photosynthesis process. The solar energy storage in light reaction (NADPH/H^+) during photosynthesis is achieved by water splitting for production of carbohydrates, biomass and other organic molecules [2]. The energy conversion in such photosystems (chloroplast) trigger electron transfer to generate ATP and NADPH. During the Calvin cycle, the generated ATP and NADPH is further utilized to harness glucose in the dark reaction from H_2 and CO_2 [9–12]. This simple $2e^-/2H^+$ process referring to hydrogen evolution reaction (HER), is more facile in carbon dioxide reduction. Thus, the biological photosynthesis process stands effective to reduce CO_2 with H_2 equivalents (NADPH/H^+) and not with H_2O. In the objective of artificial photosynthesis development, the system demands two-step process: 1. Solar-driven water splitting for hydrogen production. 2. Dark reaction to reduce CO_2 using produced H_2 [13]. The process of natural photosynthesis is noteworthy on water oxidation and CO_2 reduction, which functions independently at different times and at different spaces and is depicted as schematic.

Inspired from nature, we could design and build similar systems to produce chemicals and fuels based on the underlying fundamental scientific principles [14].

2 Artificial Photosynthesis

Artificial photosynthesis (AP) attempts to utilize energy from the sun and convert into useful chemical energy. The process and functions of AP, mimic natural photosynthesis and has gained reasonable attention in the recent decades. This refers to solar water splitting or carbon dioxide reduction with proton and electron as the source that could be light, electricity or a combination of both. Artificial photosynthesis opens up specific energy sources classified as photochemical (PC), photovoltaic-electrochemical (PV-EC), and photoelectrochemical (PEC) catalytic systems. In such systems, the catalyst determines the performance and energy conversion efficiencies. Hence, an efficient catalyst must be capable of reducing CO_2 and further oxidize water for hydrogen production [14].

2.1 *Performance Evaluation of AP Systems*

The performance evaluation of AP systems can be deduced with various parameters that include solar to hydrogen conversion efficiency (STH), external quantum efficiency (EQE) or incident photon to current conversion efficiency (IPCE), apparent quantum efficiency (AQE), applied bias photon to current conversion efficiency (ABPE), and absorbed photon to current conversion efficiency (APCE) or internal quantum efficiency (IQE).

STH is defined by the ratio of output chemical energy to input photon energy as expressed in Eqs. (1) and (2). The chemical energy obtained as production rate (mmol/s) and change in Gibbs free energy (237.2 kJ/mol) to produce one mol of hydrogen can be calculated. Similarly, solar to CO and other fuel conversion efficiencies; where, P_{total} can be obtained from total intensity of incident light in a given area.

$$\text{STH} = \left[\frac{\left(\text{mmol H}_2\text{s}^{-1}\right) \times \left(237.2\ \text{kJ mol}^{-1}\right)}{P_{\text{total}}\left(\text{mW cm}^{-2}\right) \times \text{Area}\left(\text{cm}^2\right)}\right]_{\text{AM 1.5G}} \tag{1}$$

$$\text{STH} = \left[\frac{j_{\text{sc}}\left(\text{mA cm}^{-2}\right) \times (1.23\ \text{V}) \times \text{FE}}{P_{\text{total}}\left(\text{mW cm}^{-2}\right)}\right]_{\text{AM 1.5G}} \tag{2}$$

Faradaic efficiency (FE) with respect to specific product is defined by the ratio of charge of formed chemical to total charge transfer in the reaction process (Q). This can be determined by number of transferred electrons (N), amount of chemical (n) and Faraday constant ($F = 96{,}485$ C mol^{-1}) as shown in Eq. (3).

$$\mathrm{FE} = \left[\frac{N \times n(\mathrm{mol}) \times F\left(\mathrm{C\ mol}^{-1}\right)}{Q(C)}\right] \tag{3}$$

IPCE describes the current generated per incident photon for a given wavelength of light and is defined as the ratio of number of electrons used to produce H_2 to the incident photons. Rather than being "power out/ power in" for STH, the efficiency in IPCE gives rise to "electron out/photon in" as shown in Eq. (4). Where P_{total} ($\mathrm{mW\ cm^{-2}}$) and λ (nm) are incident power and wavelength of monochromatic light respectively. The value 1239.8 V nm is obtained by the product of speed of light with Planck's constant. Alternatively, AQE can be determined using Eqs. (5) and (6) in which N, N_A, n, t, c, h is number of incident photons, Avogadro constant (6.022 $\times 10^{23}$ $\mathrm{mol^{-1}}$), amount of H_2 evolved, reaction time, speed of light (3×10^8 $\mathrm{ms^{-1}}$) and Planck constant (6.626×10^{-34} $\mathrm{Js^{-1}}$).

$$\mathrm{IPCE}(\lambda) = \mathrm{EQE}(\lambda) = \left[\frac{\left|j_{ph}\left(\mathrm{mA\ cm}^{-2}\right)\right| \times 1239.8(\mathrm{V\ nm})}{P_{total}\left(\mathrm{mW\ cm}^{-2}\right) \times \lambda(\mathrm{nm})}\right]_{\mathrm{AM1.5G}} \tag{4}$$

$$\mathrm{AQE} = \left[\frac{N_A \times n \times 2}{N}\right] \tag{5}$$

$$N = \left[\frac{P \times t \times \lambda}{h \times c}\right] \tag{6}$$

Notably, STH, EQE/IPCE includes photon losses as a result of transmission and reflection. Hence, quantum efficiency excluding the losses are essential to determine the intrinsic performance of energy conversion and can be deduced using APCE as shown in Eqs. (7) and (8). This describes the photocurrent obtained per absorbed photon in which I and I_0 is the intensity of incident light and transmitted light, respectively. While ABPE differs from STH and gives the efficiency (Eq. 9) where a bias is applied (V_b) between counter and working electrodes.

$$\mathrm{APCE}(\lambda) = \mathrm{IQE}(\lambda) = \left[\frac{\left|j_{ph}\left(\mathrm{mA\ cm}^{-2}\right)\right| \times 1239.8(\mathrm{V\ nm})}{P_{mono}\left(\mathrm{mW\ cm}^{-2}\right) \times \lambda(\mathrm{nm}) \times \left(1 - 10^{-A}\right)}\right]_{\mathrm{AM1.5G}} \tag{7}$$

$$A = -\log\left(\frac{I}{I_0}\right) \tag{8}$$

$$\mathrm{ABPE} = \left[\frac{\left|j_{ph}\left(\mathrm{mA\ cm}^{-2}\right)\right| \times (1.23 - |V_b|)(V)}{P_{total}\left(\mathrm{mW\ cm}^{-2}\right)}\right]_{\mathrm{AM1.5G}} \tag{9}$$

Though AP is under development, it could be a potential alternative for sustainable energy harvesting and can address challenges in energy security and climate change.

2.2 Photoelectrochemical Cells (PECs)

The discovery of photoelectric effect by French scientist Edmond Becquerel in 1839, infatuated engineers and scientists to convert light into electricity and/or chemicals [15]. This was inspired by the photographic images acquired by Daguerre in 1837 [16]. The pioneering experiments by illuminating electrolyte solution containing metal halide produced electric current between the platinum electrodes. In the fact of energy conversion from light to electricity, films were inefficient to absorb red light; which was recognized as semiconductor nature of silver halide grains that ranged between 2.7 and 3.2 eV with insignificant absorption beyond 460 nm. In 1883, Vogel discovered that the halide solutions could be sensitized with dyes to improve photosensitivity at higher wavelengths photo [16].

The modern photo-electrochemistry laid its foundation from the support of Brattain and Garret [17]. Later, Gerischer [18] detailed the first electrochemical and photoelectrochemical studies at the semiconductor-electrolyte interface. Immediately after the oil crisis in 1973, the worldwide quest for alternative energy sources stimulated research on photoelectrochemical cells resulting in over a thousand publications [19].

Energy photons exceeding band gap generates electron–hole pairs. In a regenerative, this light energy conversion into electrical energy do not leave any chemical change. The work on regenerative cells were given attention in electron doping of III/V or II/VI semiconductors with I_2/I^-, vanadium (II)/vanadium (III) or sulfide/polysulfide redox couples as electrolytes. For a multijunction regenerative cell, a conversion efficiency of up to 19.6% is reported [20]. In a similar principle, photosynthetic cells operate with bi-redox systems; the one reacts with holes at semiconducting electrode and the other reacting with electrons at counter electrode. This refers to the sunlight driven cleavage of water molecules. In 1972, Fujishima and Honda [21] employed TiO_2 for water photolysis which showed low conversion efficiency resulting from the large band gap (3–3.2 eV) that absorbs UV region of solar spectrum. Numerous efforts to improve spectral absorption region of TiO_2 have failed and created disillusionment toward PECs. Moreover, these semiconductors with narrow visible spectrum absorption are prone to photocorrosion. Hence the use of sensitizer, absorbs photons in visible spectrum and inject the charge at semiconductor-electrolyte interface and bring in stability to corrosion.

3 Nanostructured Materials for Solar Fuel Production

The need for efficient incident light absorber, drive the development of nanostructured semiconductor materials that possess high internal surface area. Utilizing nanostructures with high internal surface parallelly increases carrier kinetics at the interface [22]. Additionally, the advantage over direction of light absorption and diffusion path length of photogenerated carriers enhances the efficiency of energy conversion [23].

Chalcogenides such as CdSe, CdTe, CdS, and metal oxides such as TiO_2, Fe_2O_3, Nb_2O_5, V_2O_5, SnO_2, Cu_2O WO_3, and $BiVO_4$ have gained significant attention for engineering the materials at nanoscale. Applications of TiO_2 in water splitting is hindered by the large band gap. Strategies including introduction of heterojunction (TiO_2/CdS, TiO_2/CdSe) and doping (N-TiO_2) narrows the bandgap of semiconductor and further be utilized for water splitting [24–26]. Nanostructures in form of tubes (NT), wires (NW), and rods (NR) of TiO_2 improves the catalytic activity of OER with higher photocurrent (2–3.5 mA/cm^2) at the standard potential (1.23 V vs. RHE) under irradiation equivalent to 1 and 0.5 sun [27, 28].

The nanostructured semiconductor materials and its performance over photosystems strongly rely on size, thickness and dimension (0D, 1D, 2D). Thus, by reducing the size of materials to nanoscale, the exposure of atoms/ions to surface increases leading to enhanced surface-volume ratio [29, 30]. Size of nanomaterial is a critical factor in charge recombination that accounts optimal thickness $L_D + W$, where L_D and W refers to carrier diffusion length and width of depletion layer, respectively. The carrier diffusion length in defined by $\sqrt{D_m \tau_m}$, where D_m and τ_m refers to diffusion coefficient and carrier lifetime [31].

To better understand the importance of thickness, Kay et al. [32] demonstrated a photocurrent density of 2.3 mA/cm^2 at 1.23 V versus RHE in a 7 nm width depletion layered Fe_2O_3 photoanode. The cauliflower shaped photoanode performance in energy conversion is evidenced without any cocatalyst. Thus, the charge carriers are subject to recombination beyond this critical width and thickness [33, 34]. Also, surface and interface charge recombination could be influenced by specific surface area [35]. However, the demonstrated surface modification strategies effectively reduced these interfacial recombination losses and charge carrier deficits on the surface [36–39].

3.1 1D Nanomaterials

One dimensional nanostructured material for photoelectrochemical cells have shown good progress in the recent decades [40, 41]. Nanowire morphology as 1D structure with Bi_2O_3/BiAl oxide generated 696 μmol/cm^2 hydrogen contributing to a faradaic efficiency of 93%. In a similar other composite material, gold nanoparticles were decorated over ZnO/CdS nanotubes and demonstrated a conversion efficiency of 3.45% under visible light ($\lambda > 420$ nm) illumination [42]. Irrespective of the synthesis and fabrication methodologies, 1D nanostructures such as tubes, wires, rods, and fibers are being employed as photocatalysts to produce sustainable fuels and chemicals [43, 44]. Recently, ceria (CeO_2) has gained attention in water splitting due to its strong redox capability. Nanowire CeO_2 demonstrated higher photocatalytic water splitting with hydrogen evolution rate of 1290 μmol g^{-1} h^{-1} under white light irradiation. However, CeO_2 NW as a composite with CdS nanoparticles produced 473 μmol g^{-1} h^{-1} with visible light as photon source [45]. Similarly, Au nanoparticle loaded TiO_2 NWs present higher production of 1436 μmol g^{-1} h^{-1} for hydrogen and is enhanced to 11.5 times to bare TiO_2 NWs [46].

Similar to nanowires and nanotubes, one dimensional nanorod structures also gained attention due to the advantage over length and could be fabricated as metals, non-metals and mixed compounds as well. Being photoactive, ZnO is least preferred for hydrogen due to high recombination rate and photocorrosion that adversely affects catalytic activity [47]. To overcome such difficulties, Liu [48] and group attempted to address the drawbacks by fabricating ZnO nanorods as a composite with CuO. The composite surface presented H_2 evolution of 1700 μmol g^{-1} h^{-1} and is marked by specific surface area and reduction of recombination. With improvements in materials science research, the exploration of new materials and its composites as core–shell nanostructures are interesting in photoabsorber. These core–shell nanostructures are aimed at improving the photocorrosion resistance and efficient charge transfer processes. Molybdenum carbide (Mo_2C) in the presence of carbon as Mo_2C–C shell and CdS nanorods as core produced 17.24 mmol g^{-1} h^{-1} of hydrogen. However, the 1D CdS nanorod alone produced 410 μmol g^{-1} h^{-1} of hydrogen. The large difference in efficiency is attributed to the efficient charge carrier transport between core and shell materials. Additionally, the C based Mo_2C shell material can protect the CdS core nanorod from light induced corrosion as shown in Fig. 3 [49].

Similarly, carbon supported Cu_2O nanowire arrays present 93.08 μmol cm^{-2} as phenol oxidized hydrogen produced over a time period of 8 h [50]. The attribution of structural materials in photo-assisted charge transfer systems pave the way for construction of efficient sustainable energy conversion devices.

Irrespective of the nanostructure dimensionality, semiconductors are prone to recombination of charge carriers thus limiting the conversion efficiencies. These limitations are overcome by heterostructure construction, doping and composite material synthesis that includes graphene/carbon as additives. In such an approach, graphitic carbon nitride was used as composite with CdS nanowires and evaluated water splitting efficiency with respect to percentage of g-C_3N_4 [45]. Similar other methodologies involve cobalt hydroxide additive as composite with CdS nanowires that report 14,430 μmol g^{-1} h^{-1} at wavelength $\geq$ 420 nm. This corresponds to 206 time's higher production rate compared with pristine CdS nanowires. Additionally, it is reported that the conversion efficiency is threefold higher than the 1 wt% Pt/CdS catalyst used as benchmark [51]. To boost the efficiency of conversion, plasmonic mechanism is being introduced with nanofibers of titanium dioxide (TiO_2). Under visible light irradiation, a total of 70% removal of nitrogen, carbon is achieved with estrogen (17-β-ethynyl estradiol) and Cu^{2+} mixed waste-water electrolyte. Additionally, H_2 conversion of 98% was achieved with the plasmonic system [52]. The plasmonic mechanism further extends to use of gold (Au) nanoparticles in TiO_2 nanowires that exhibit 1436 μmol g^{-1} h^{-1} as hydrogen evolution; while, the pristine TiO_2 nanowires show rate of 125 μmol g^{-1} h^{-1}. SEM images of nanowires at different magnifications are presented in Fig. 4 below. The incorporation of plasmonic effect with similar photo-semiconductor (ZnO) nanostructures shows lower hydrogen production rate compared with TiO_2 [47]. The lower conversion efficiencies are associated

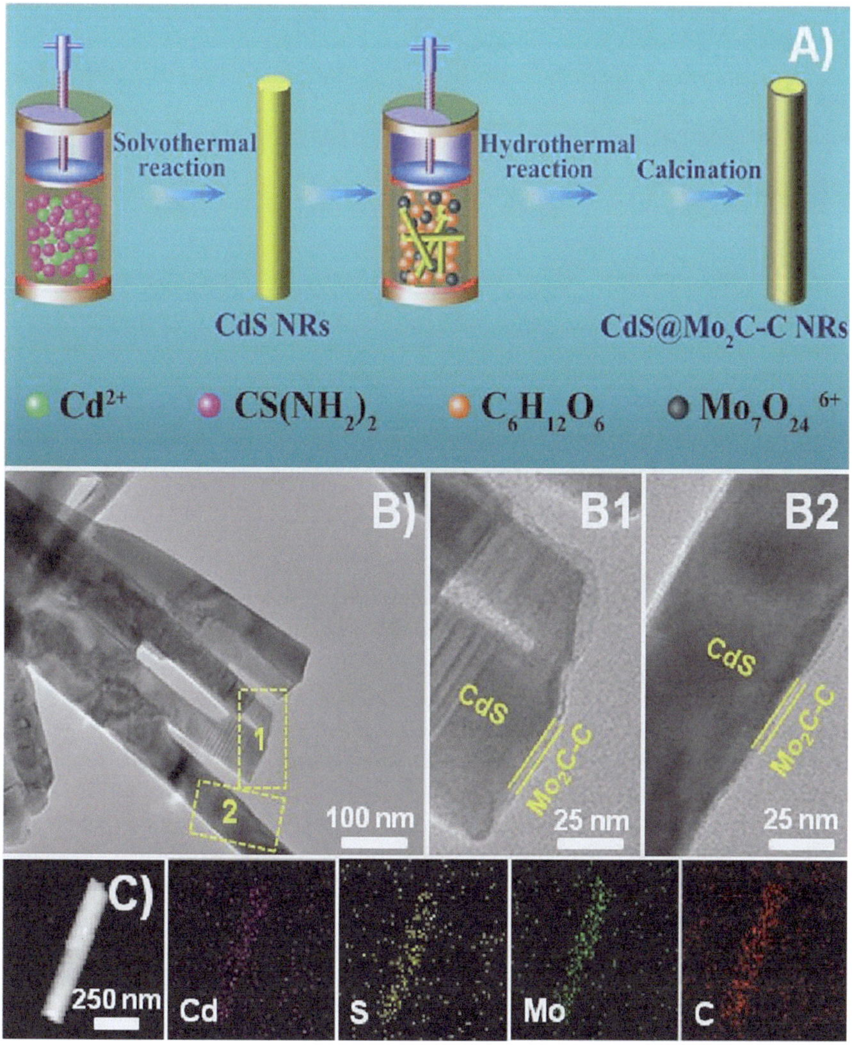

Fig. 3 Schematic of synthesis and respective TEM images of CdS/Mo_2C–C core–shell nanostructure. Reproduced with permission [49]. Copyright 2017. The Royal Society of Chemistry

with similar recombination losses and photocorrosion of semiconductors that seek specific additional attention.

The use of 1D nanomaterials in hydrogen production have been explored and reported to have significant enhancement as discussed earlier. Similar 1D nanostructured materials are being used in photosystems to reduce CO_2 to other fuels such as CO, CH_4, C_2H_6 [53]. Nanotube structures of TiO_2, SiO_2 and TiN with diameter and length ranging nano to micro and millimeter scale are able to achieve high CO_2 reduction. The performance enhancement are attributed to specific surface area, improved

Fig. 4 SEM images of TiO_2 nanowires. Reproduced with permission from [46]. Copyright 2017

light absorption and improved catalytic sites for redox reactions [54]. The supporting material $ZnIn_2S_4$ with TiO_2 nanotubes improved the reduction capacity. However, it also demonstrated disadvantages with smaller range of absorption wavelength and consequent recombination of charge carriers [55]. The advantages of tubular structures are further extended to bismuth oxyhalide ($Bi_{12}O_{17}C_{12}$) that showed a 17-fold increase in solar catalytic activity [56]. In a similar nanotube halide system ($Bi_{12}O_{17}Br_2$) with gold nanoparticles showed selectivity as high as 90% for C_2H_6 [57]. The improvements in activity are attributed to higher adsorption of CO_2 to tube walls and associated sites for further redox reactions.

Besides the investigation with tubular structures, nanowires are also being investigated as pristine materials and as composites. In a cobalt supported $W_{18}O_{49}$ nanowire catalyst, production rates of 16 μmol g^{-1} h^{-1} and 22 μmol g^{-1} h^{-1} as rate of H_2 and CO production, respectively [58]. In an S-scheme heterojunction system, 25 wt% $AgVO_3$/Ag supported TiO_2 nanowires exhibited 9.5 mmol g^{-1} h^{-1} methanol production rate marked by the formation of internal electric field with enhanced light capture and transfer of photogenerated charge carriers. The efficiency enhancement is additionally supported by palsmonic resonance of Ag that greatly increased the photoreduction of CO_2 to methanol [59]. Vacancy rich materials are equally important to enhance the active sites that enhance redox reactions. In a study conducted by Jun Li et al. [60], vacancy rich metallic V-$Bi_{19}Br_3S_{27}$ nanowires showed roughly 2.3-fold higher methanol generation in NIR region and attributed to unique interfacial engineering. Tian and coworkers investigated ethylenediamine (EDA) supported CdS hybrid nanowires which showed 115.6 and 959.4 μmol g^{-1} h^{-1} as CO and H_2 generation rates, respectively [61]. The high sulfur vacancy in CdS/EDA are capable of trapping photogenerated electrons and holds higher lifetime. This further promotes higher hydrogen adsorption and CO_2 reduction.

Nanobelt and nanoribbon structures constructed with Zn_2GeO_4 material demonstrated a reasonably high selectivity in CO_2 reduction. The high reduction is due to high specific surface area, reduced recombination rate, spatial transport channel for charge separation and improved vacancy sites that enable higher redox reactions [62, 63]. Similar nanobelt heterostructure fabricated using TiO_2 and $ZnFe_2O_4$ shows 180 μmol g^{-1} for cyclohexanone (CH) and cyclohexyl formate (CF), respectively for 8 h of irradiation [64]. The availability of high redox active sites increased the CO_2 reduction and associated fuel production.

Nanorods are commonly investigated 1D nanostructures marked by its advantages of controllable aspect ratio and enabling high redox active sites. Nickel as supporting catalyst with CeO_2 forms a p–n junction at Ni–CeO_2 interface. The advantages of oxygen vacancy of CeO_2 boost the charge transfer kinetics of and enhance the life time resulting in formaldehyde product formation (421 μmol g^{-1} h^{-1}) and 7039 μmol g^{-1} h^{-1} yield of methanol production [65, 66]. ZnSe nanorods as composite with $CsSnCl_3$ is highly efficient in charge separation and high active site for CO_2 reduction. The suppression in charge carrier recombination resulted with CO_2 reduction to CO and CH_4 by the rate of 57 μmol g^{-1} h^{-1}, respectively [67]. The physicochemical advantages of 1D nanostructured materials exhibit suppressed recombination rates owing to the heterojunction formation. Moreover, the demonstrated specific surface area, improved light absorption and vacancy sites show high selectivity to solar fuel production.

3.2 2D Nanomaterials

Among the nanomaterial systems (0D, 1D, 2D, and 3D) and their functional composites, the ultrathin structure and high surface-volume ratio of 2D materials stand advantageous in offering low material usage, faster processing and high efficiency [68]. The distinct morphological features of 2D nanomaterials unlocked new prospects photocatalysis [69]. The critical steps of absorption, separation of charges and interfacial redox reactions can considerably be enhanced with 2D nanomaterials due to their unique atomic properties. The advantages over ultrathin nature have the possibility to reduce recombination due the shorter diffusion length. Furthermore, the density of active sites can significantly enhance the redox reactions contributing to increased conversion efficiencies [70]. However, the advantages of 2D materials have not been realized yet in practical application demands.

Peng et al. [71] evaluated H_2 generation with vacancy induced MoS_2 incorporated on $ZnIn_2S_4$. The discovery over synergistic structural and electronic properties dramatically enhanced the photo-assisted hydrogen evolution. The composite exhibits 23.2 mmol g^{-1} h^{-1} as hydrogen generation accounting to apparent quantum yield of $\approx 5\%$. The electron transfer properties and mechanism of H_2 production in $ZnIn_2S_4@MoS_2$ is shown in Fig. 5.

A strategy proposed by Zhang et al. [72] replaced sulfur vacancies with MoS_2 quantum dots (QDs) in ZnI_2S_4. It is said that photogenerated electrons quickly migrates to QDs resulting in hydrogen generation activity of 6.88 mmol g^{-1} h^{-1}. In case of carbon nitride nanosheets, the vacancies over carbon can improve the conversion efficiency by 20-fold compared with the bulk C_3N_4 [73]. In similar other case, a defective $K_4Nb_6O_{17}$ nanosheets can produce 1661 μmol g^{-1} $h^{-1}$of hydrogen and is 21 times higher than defect-free nanosheets [74]. These studies show that defective sites in nanosheet structure can significantly improve donor density and kinetics of charge leading to enhanced photoconversion efficiencies. The concept proposed by Lei et al. to enabled visible light photocatalytics [75]. In that study, $Ca_2Nb_3O_{10}$

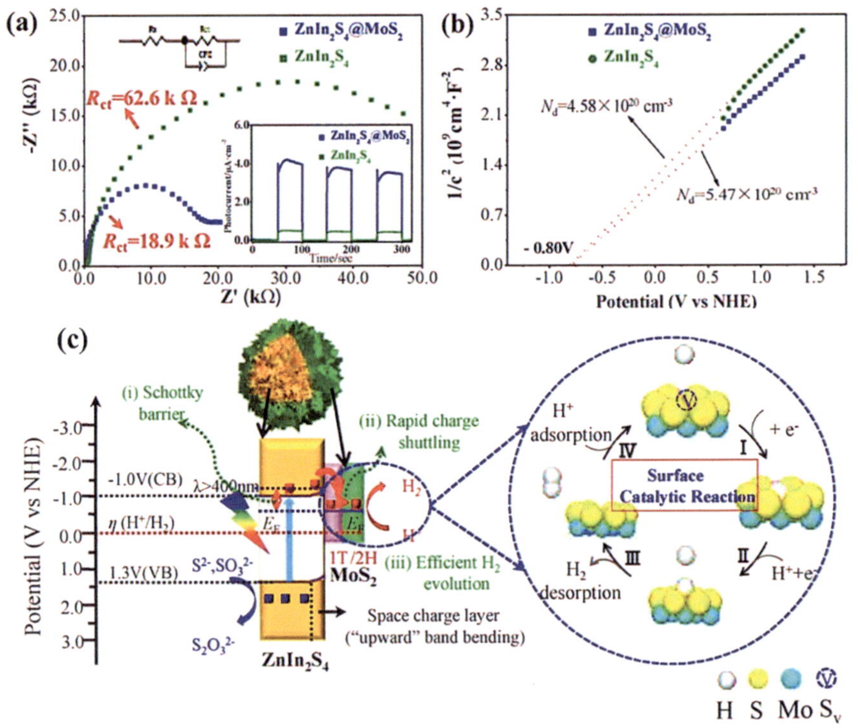

Fig. 5 **a** Nyquist plot of MoS_2 composite, **b** Mott–Schottky of MoS_2 composite and **c** mechanism for hydrogen generation in composite. Reproduced with Permission from [71] Copyright 2021 Elsevier B.V

nanosheet with RGO as photosensitizer showed 0.8 mmol g^{-1} h^{-1} as hydrogen generation rate. The theoretical and experimental results prove the photogenerated electron transfer process from RGO to $Ca_2Nb_3O_{10}$ and involve in photocatalytic reactions to produce hydrogen. This further proves the ability of perovskite catalytic nanohybrids for its employability in visible light harvesting for solar fuel production. The reported works discussed above are aimed at facilitating recombination less transfer of charge carriers and to enhance active sites for the catalytic reactions. This also accounts the heterostructure fabrication that generates strong electric field within the nanomaterial system [76]. Thermal annealing can help to attain such heterostructure in MoS_2 system that changes the ratio of its phases (1T/2H). The in-plane heterostructure demonstrated 1.5 mmol g^{-1} h^{-1} hydrogen generation and further shows better stability with 100 mM lactic acid solution.

The use of semiconductor photocatalyst in CO_2 reduction can produce chemical fuels such as CH_3OH, CH_2O, CH_4, CH_2O_2, and CO [77, 78]. However, the CO_2 conversion requires more electrons which is complex and challenging. Since the work pioneered by Inoue et al. [79] in 1979, the number of catalyst capable of converting CO_2 into fuels are being published. The Bi_2WO_6 powder system, demonstrated 75 μmol g^{-1} h^{-1} for CH_3OH production that was estimated for an operational time of 5 h. Similar to hydrogen production, vacancy systems are equally important to reduce CO_2 into fuels. In a Zn vacancy rich $ZnIn_2S_4$ nanosheet, a 33 μmol g^{-1} h^{-1} CO formation rate is recorded which is roughly fourfold higher than vacancy deficit $ZnIn_2S_4$ nanomaterial [80]. The g-C_3N_4/NiAl-LDH 2D/2D nanosheet heterojunction formed an efficient interface for electron transfer that resulted with higher CO formation comparing with pristine nanosheets. Furthermore, this heterojunction presented nearly 82% selectivity for CO formation and even improved H_2 and O_2 evolution [79].

Tri-material Z-Schemes are being investigated in parallel and has considerable attention in CO_2 reduction systems. In this particular study, Bi_2WO_6/RGO/g-C_3N_4 is constructed to enhance the electric field at heterojunctions thus further improved the migration of charge carriers. This hybrid Z-scheme heterojunction showed 92% of selectivity in CO_2 reduction and H_2, O_2 generation and kinetic process in Z-scheme is presented in Fig. 6a [81]. A 0.157 mmol g^{-1} h^{-1} CH_3OH is formed with functionalized Bi_2WO_6 accounting to 60 h of reaction time and present no evident loss in CO_2 reduction implying the employability in solar fuel production (232 of 2D materials). The targets of reducing charge carrier recombination and improving charge carrier separation also demands enhancement in adsorption capacity of CO_2. The excellent electrical conductivity of metallic sites of titanium carbide MXene's are choice of extensively researched material in catalysis. The combination of two dimensional Bi_2WO_6 with 2D Ti_3C_2 established a strong interface for charge transfer kinetics that aid excellent photocatalytic activity [82]. Furthermore, increased specific surface in the form of pores promoted CO_2 adsorption, which is crucial for efficient reduction to fuels. However, the data on electrons involved in CO_2 reduction is unclear. The electron transfer process between Ti_3C_2 and Bi_2WO_6 is demonstrated in Fig. 6b. In a similar study, commercial P25 nanoparticles have been used as cocatalyst with Ti_3C_2 MXene that resulted with improved photocatalytic CO_2 reduction [83]. Moreover,

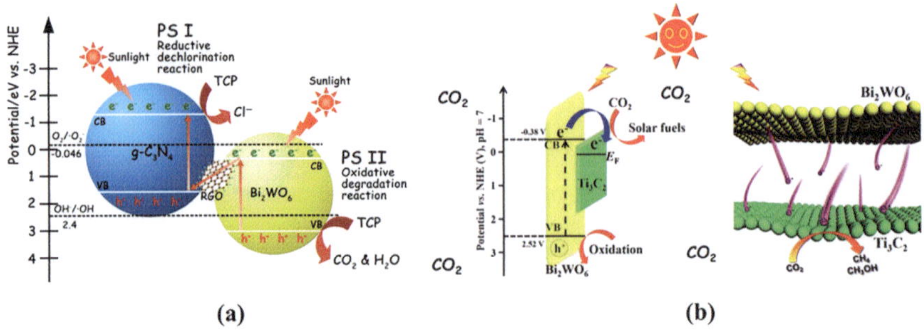

Fig. 6 **a** Electron transfer process in Z-Scheme. Reproduced with permission from [81] Copyright 2016. Elsevier B.V **b** electron transfer process in Bi_2WO_6/Ti_3C_2 heterojunction Reproduced with permission from [82] Copyright 2018. WILEY–VCH

the DFT analysis reported in this study shows higher CO_2 adsorption energy on TC-OH thus improving the adsorption of CO_2 molecules. The findings demonstrate improved electrical conductivity, charging isolation, CO_2 adsorption and activation sites by using metal-free cocatalyst for photocatalytic process. Xu et al. performed a similar study to utilize TiO_2 nanoparticles with rice crust morphology and produced CH_4 by reducing CO_2 [84]. The efficient electron transfer between TiO_2 and Ti_3C_2 is marked by the morphology that offered more active sites and promoted photocatalytic CO_2 reduction. However, straightforward evaluation of CH_4 production seem difficult due to the presence of multiphasic TiO_2.

Similar architectural nanostructured materials are being investigated for ammonia (NH_3) and hydrogen peroxide (H_2O_2) production. Ammonia is essential for fertilizers and also an important carrier in energy and fuels [85, 86]. 2D nanomaterials as photocatalyst including V_O-BiOBr, Bi_5O_7I and MoS_2 are promising in N_2 fixation which is essential in ammonia production [87–89]. However, these materials are less efficient in energy conversion and demand more improvements. Nanosheet MoS_2 material with sulfur vacancies presented 325 μmol g^{-1} as NH_3 production rate for the tested time period of 10 h while the commercial MoS_2 does not show any photocatalytic activity [88]. The reports discussed above are aimed at tuning light absorption, limiting recombination and improving active sites. Therefore, heterojunction construction and proper interface engineering of 2D nanomaterials could pave the way for efficient solar fuel production.

3.3 3D Nanomaterials

Robustness, conversion efficiency of any photocatalytic system involves materials engineering in micro to nanoscale and specifically, how these materials are structured. The photocatalytic solar fuel production efficiency is tuned by controlling size, morphology, and crystallinity of structural nanomaterials [90, 91]. The dependence of morphology and dimensionality on conversion efficiency of semiconducting photoactive materials are being investigated till date. Emerging aspects of 3D architectures and its associated benefits in artificial photosynthesis systems to produce fuels are discussed here.

In natural leaves, nanolayered thylakoid granules are involved in photosynthesis process where the pigments, charge carriers and functional proteins are periodically and precisely arranged [92]. Those layered thylakoid structures possess large surface area and are high light absorbing with total internal reflections [93]. Inspired by the nature, Zhou et al. [94] demonstrated 3D leaf architecture in a perovskite titanate system for CO_2 photoreduction. The architectural concept of leaf prototype demonstrated hydrocarbon fuel production (CO and CH_4). The demonstrated prototype could open broad pathways for the development of network that improves gas diffusion and improved light harvesting. A schematic differentiating natural and artificial photosynthetic system is exemplified in Fig. 7.

Light harvesting is more important in artificial photosynthesis systems as it influences the efficiency of energy conversion. Various 3D photonic structures with unique absorption characteristics are being explored and used as advantageous over antireflection and light trapping [95]. 3D structures are primarily targeted to antireflection that is attributed to the refractive index of propagative medium. The layer composed

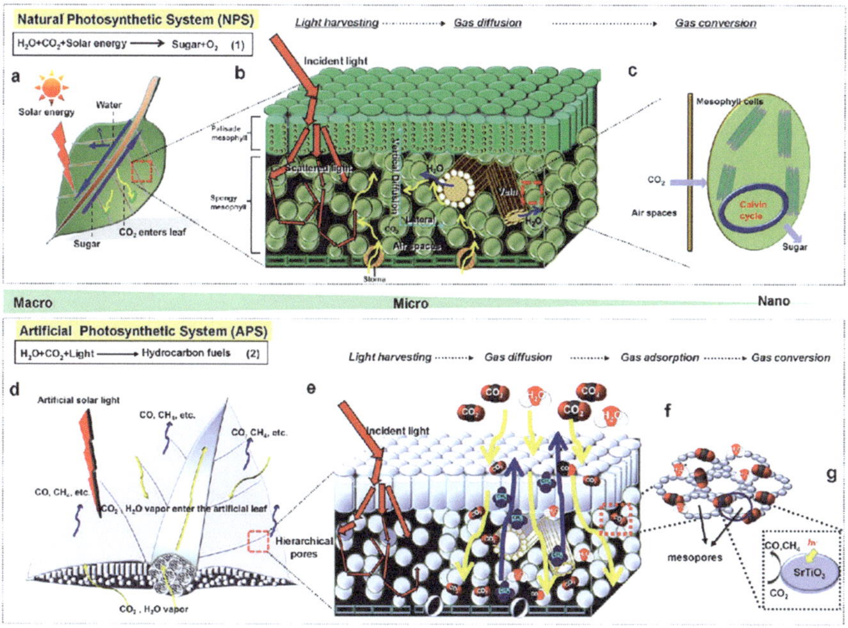

Fig. 7 Basic natural photosynthesis process in leaf structure demonstrated at macro, micro and nanoscale (**a–c**); artificial photosynthesis process at macro, micro and nanoscale (**d–f**). Reproduced with permission from [94]. Copyright 2013. Springer nature

of low refractive index (n) film and a high refractive indexed surface (n_s) makes the best antireflective surface; here, the light propagation follows thin film law of interference. However, minimal loss of reflection is ascertained only while the incident light is received at a specific polarization, wavelength, and angle when $n^2 = n_s n_{air}$ as defined by Fresnel [96, 97]. Wang et al. [98] nanospheres of g-C_3N_4 that mimic membrane structures of thylakoid. This biomimetic system exhibited 7.5% hydrogen precipitation quantum yield thus presenting high artificial photosynthesis performance. Similar photosystem with Co_3O_4 cocatalyst was used to enhance overall water splitting [99]. Jiang et al. synthesized multi-shell g-C_3N_4 structures to promote light trapping and charge separation. An optical simulation-based model was used to fabricate TiO_2 diatoms with periodic structures using lithography technique. Photoreduction to CO and CH_4 was 2 to threefold higher compared with powders of same composition [100].

Similar to thylakoid structures in leaf, hierarchical skeleton structure of butterfly wings has inspired scientists to develop CdS/Au/TiO_2 Z-scheme system [101]. Such skeleton structures have the capability of arresting light without any reflection thus enhancing the photoconversion efficiencies [102, 103]. From the experimental demonstration, it is reported that the reflectivity in skeleton structure in minimized by 40% that enhanced the hydrogen generation by 200% [101]. The inspirations not only stopped by butterfly wings and further extended to Papilio nephelus wings architecture constructed using graphitic carbon with higher hierarchical porosity for promotive CO_2 reduction [104]. These investigations on biomimetic nanomaterial structures have contributed tremendously to artificial photosynthetic systems. However, controlling the nanosized structures for scalable production has remained a great challenge.

As refractive index of materials is crucial in arresting the absorbed light energy, embedding 3D photonic crystals could also be promising in AP. Photonic crystal (PC) is composed of two refractive index components that could produce photon energy bands and forbidden band structures [105]. The slow photon effect of PCs can effectively increase the lifetime of photons of inside the material and thus enhance the efficiency of capture and energy conversion [105–107]. The energy delivery rate and group velocity of non-absorbing dielectric materials can be expressed as shown in Eq. (10) [108].

$$\vartheta_{\text{group}} = \frac{\partial \omega}{\partial k} \tag{10}$$

The lifetime of photons propagating in Bloch mode in photonic crystal of length L can be described as Eq. (11) [109].

$$1/\tau = \frac{\vartheta_{\text{group}}}{L} + \frac{\alpha}{(4\pi)(2\pi/L)^2} + \cdots \tag{11}$$

where α is defined as the curvature of dispersion curve $\left(\alpha = d^2\omega/d^2k\right)$ in Bloch mode [109]. Any slowdown in group velocity is related to increase in intensity, photon density and lifetime of photons [108]. It is important to achieve slow photon effect to improve light absorption efficiency and hence overall conversion efficiency in AP. Furthermore, it is also important to use high liquid wettable structures that is capable of high mass transfer as quickly as possible [110, 111]. However, PC architectures were not widely used due to the limitations of high cost in fabrication technique [105]. To date, electrochemical deposition, spray pyrolysis, vapor deposition and colloidal self-assembly methods are being adopted [112–115]. As a result of surface plasmon polariton, use of metallic and dielectric materials as composite can have high absorption capacity. Furthermore, dielectric materials such as transition metal oxides and wide band gap semiconductors are also good candidates for composites that has made great progress in the recent years [110, 116, 117].

The inverse opal structure investigated for platinum loaded TiO_2 does not report a huge increase in hydrogen generation marked by the easy recombination of photo-generated charge carriers [118]. A porous 3D rod-on-rod hierarchical structure constructed with CdS–TiO_2 showed 51.6 and 11.9 μmol h^{-1} as hydrogen production rate in photoelectrochemical and photocatalytic systems, respectively [119]. While, hierarchical inverse opal structures investigated by Liu and coworkers show a 2.5-fold increase in hydrogen generation compared to the conventional Pt–TiO_2 and the life time of charge carrier still does not show any remarkable improvements. PC being advantageous in life time of charge carriers, the theoretical values are not realized in experimental levels. But, the demonstration of PCs in artificial photosynthesis are promising and realization in large-scale precision synthesis are still challenging.

A 3D sponge-like microporous thin film CdS increased the light absorption that resulted with 47.6% photoelectric conversion efficiency. The light absorption capacity and small work function of CdS efficiently transferred charge carriers contributed to such increased conversion efficiency. Using noble metals as cocatalyst with 3D architecture of CdS can further promote photocatalytic activity by localized surface plasmon resosnance [LSPR]. Fang et al. [120] fabricated hierarchical CdS/Au wing scales inspired from butterfly wings. The LSPR generated by rib structures and good charge transfer interface between CdS and Au boosted the hydrogen production rate to 221 μmol h^{-1} under visible light irradiation. On the other side, Soto et al. [121] propose the contributing factor that affect photocatalytic hydrogen production efficiency. Firstly, the crystallinity of CdS, the existence of smaller nanocrystals of CdS, then the presence of Ag_2S species that can have proper p–n heterojunction resulting with increased activity marked by the confinement effect.

Transition metals are also being investigated as cocatalyst with 3D CdS nanosystems. However, the conversion efficiencies in photocatalytic hydrogen generation rate are significantly smaller compared to noble metal cocatalysts which are limited by cost and resources. A highly conductive β-NiS film on CdS nanoflowers benefited the transfer of photogenerated e^- and resulted with 30 mmol g^{-1} h^{-1} as hydrogen production rate [122]. The exploration of photocatalyst engaged with utilizing organic materials which are advantageous over inorganic materials. The advantage of workability, control, purity and flexibility are typical characteristics of organic materials. Combination of diethylenetriamine (DETA) with CdS exhibit 8 mmol g^{-1} h^{-1} as hydrogen generation with satisfactory durability [123]. In a similar structure, the composite of CdS with PANI is capable of producing 310 μmol g^{-1} h^{-1} of H_2 without degradation for a period of 30 h. The CdS based 3D nanomaterials and its composites are capable of exhibiting solar to hydrogen efficiencies from 1.5 to 43% [124]. This accommodates composites of hierarchical structures, transition metals, LSPR, and organic materials.

As an extension to 3D nanomaterials, additive manufacturing technology is accurate to some extent with an advantage of faster fabrication over a wide range of functional materials in micro and nanoscale. Hierarchical 3D porous structures of titania exhibited large porous structure comparable to natural leaves that enhance the mass transfer performance [125]. This system shows twofold and sixfold enhancement in CO and CH_4 evolution activity, respectively. Moreover, the numerical simulation conducted for this 3D printed system also demonstrate the advantage of scale-up methodologies for AP. He et al. [126] proposed a 3D printing methodology to fabricate carbon nitride-based aerogels and is valid in versatile materials such as transition metal carbides, boron nitride, and carbon nitride for seawater desalination, AP etc. [127]. Lee and coworkers pioneered the 3D printing methodology to fabricate metal-based photoelectrodes which was further used for TiO_2 nanotube array formation [128]. The enhancement in photocatalytic water splitting performance and improved quantum efficiency of photoelectrode was assisted by the high specific surface area.

The broad prospects of structure controllability, complex design and hybrid structural design makes additive manufacturing as a potential candidate for material fabrication. However, limitations over reliability due to surface roughness still persist. The diversified advantages of 3D printing enabled researchers to explore surface modification strategies that aim to minimize the roughness so as to enhance the employability and to reduce the frictional wear damages [129, 130].

3.4 Facet–Engineering

Besides tuning nanostructure and its morphology to improve photocatalytic activity, facet engineering of semiconductor with appropriate control over morphology have gained parallel attention. The facets have capability to manipulate performance in sensors, photodetectors, photovoltaic cells, and photocatalysis [131–134]. The semiconductors discussed in earlier sections have the potential of being controlled in their exposed facets. Use of different chemical agents and its thermophysical properties can lead to tailored facets with precise atomic arrangements [135, 136]. Crystal facet engineering follows Gibbs–Wulff theorem; higher surface energy facets have the potential to grow faster and vanish without forming final crystal. Whereas, low surface energy crystal forms at slower rate and are more stable in constructing final crystal structure [137]. Therefore, the crystals facets obtained under equilibrium conditions are less reactive and are not favorable for artificial photosynthesis. In TiO_2 semiconductor, (001) facets are desirable than (101) for solar energy conversion. By employing organic and inorganic additives as capping agents, the high energy of facets is reduced to achieve higher percentage of preferred facets. By adopting such chemical modifications, nearly 100% (001) facets of TiO_2 are synthesized [138, 139]. Another strategy to achieve desirable facet accommodate selective etching technique in other words defined as top-down route. Notably, corrosion resistance of crystal facets may change accordingly with the etching agents (i.e.,) facets having high corrosion resistance are least etched while letting out other crystal facets. Hydrofluoric (HF) acid can selectively etch (001) of anatase TiO_2 without altering (101) facets and the same is evidenced in theoretical calculations [57 of fa4]. However, similar chemical approached can be utilized to protect preferential facets as demonstrated with Ag_2O crystals. The strong interaction of hydroxide ions on (111) facet of silver atoms protects from etching [140].

Generally, semiconductor crystals are composed of diverse facets rather than holding to one isolated facet. Therefore, the performance of semiconducting crystal in photosystems are a synergistic effect of coexisting facets. Density Functional Theory (DFT) simulations show similarly positioned conduction bands of (101) and (001) facets of TiO_2; while, the conduction band of (101) is higher than (001). Thus, the coexistence and exposure of (101) and (001) facets are beneficial in effective charge separation resulting with enhanced photocatalytic activity [141]. Other semiconducting facet oriented crystals favorable for photocatalytic process are explored and that accounts to $BiOIO_3$—(010)/(100), $BiVO_4$—(010)/(110), CeO_2—(100)/(111), $SrTiO_3$—(001)/(110), $ZnGa_2O_4$—(110)/(111), and Bi_3TiNbO_9—(001)/(110) [134]. Therefore, tailoring the exposure of crystal facets can effectively improve charge separation and transfer in semiconducting materials that is beneficial in photocatalytic energy conversion devices. Apart from TiO_2, facet control of $BiVO_4$ is widely and equally investigated as photoanodes for solar fuel production [142]. The design and development of $BiVO_4$ controlled architectures, facet engineering and cocatalyst are highlighted to enhance the absorption efficiency, charge separation efficiency and charge transfer efficiency [143, 144]. Photoanode with (040) crystal facet engineering demonstrated 42.1% as the photon to current conversion efficiency [142]. The improved charge separation and injection efficiency driven by (040) facet produced 0.02 mmol g^{-1} h^{-1}.

Photocorrosion is equally important in employability of photocatalytic energy conversion devices. As discussed earlier for selective etching, Cu_2O nanocrystal with (100), (110), and (111) facets were employed for MO dye degradation. Photocorrosion in nanocrystal completely removed the (100) and (110) facets and thus decreasing the catalytic performance limiting the longevity. This photocorrosion of (100) and (110) facet even affected the stability in water splitting electrocatalysis [134]. In case of Cu_2WS_4 photocatalytic hydrogen production, photocorrosion was observed on (101) with no changes in (001) facets [134].

Wang et al. [145] demonstrated more reactive [0001] facets of CdS highly efficient photocatalytic hydrogen production. This is achieved by injecting Cd^{2+} ions along [0001] and [10 $\bar{1}$. 1] directions thus lead to tunable morphologies such as nanocones, nanofrustums and nanoplates. In Cu_2O nanocrystal with facets (100), (110), and (111) orientation showed largest driving force for electron transfer at interface for efficient solar fuel generation [146]. The role of crystal facets in band alignment is exemplified in Fig. 8.

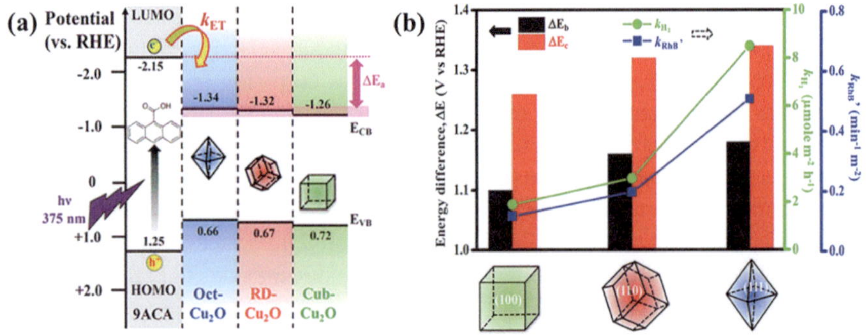

Fig. 8 **a** Energy diagram of facet dependent Cu_2O, **b** correlation of change in energy band with RhB photodegradation rate constant (k_{RhB}') and hydrogen production rate constant (k_{H_2O}). Reproduced with permission [146]. Copyright 2022. American Chemical Society

Li et al. [147] explored the career mobility of CeO_2 crystal facets (111) and (100) in CH_4 generation capacity. In general, probability of carriers to reach reactive sites within their lifetime rely on effective mass of it. The gradient approximation functional calculations showed 1.140 m_o and 1.031 m_o (m_o mass of free electron) as effective mass of holes for (111) and (100), respectively. While the effective mass of electron reads 3.33 m_o and 8.37 m_o for (111) and (100), respectively. It is seen that (111) facet has significantly smaller mass than (100) favoring straightforward electron transfer. However, (100) facet favor efficient hole transport is much efficient than (111). This depicts the role of crystal facet to effectively improve photocatalytic CH_4 generation [147]. In noble metal (Pt and Au) supported (001)—TiO_2 showed 61.3 mmol g^{-1} h^{-1} hydrogen generation which is attributed to effective separation and transport kinetics of electrons and holes [148]. In a similar fashion, Zhu et al. [148] demonstrated 71 mmol g^{-1} h^{-1} as hydrogen generation on (001) and (101) faceted TiO_2. Specifically, the study is conducted with control over Pt loading that changes the overall conversion efficiency. Hence, the hydrogen generation reported in this study could be devoted to the role of Pt loading and the particular role of facets is not detailed. Besides the control of faceted nanomaterial species, noble metal loading contributes more even in CO_2 reduction. Iron-based MOFs with (111) exposed facet is further deposited with PtO species that resulted with 285.6 and 18.1 µmol g^{-1} as CO and CH_4 yield. A notable selectivity of 94% for CO and is attributed to efficient charge carrier separation, transfer with effective adsorption/desorption intermediates at the interface [149].

Engineering the nanostructures with unaltered crystal facets do affect performance in photoelectrocatalyis and are marked by charge recombination. As discussed in the previous sections, structuring of nanomaterials play a significant role in migration of charge carriers and in turn minimize the recombination. These advantages are coexisting properties of active surface area, vacancy defects and improved active sites. Nanoplate based $BiVO_4$, highly reactive (001) facets demonstrate fivefold enhancement in photocurrent density compared to nanorod structures with (100) exposed facets [134]. However, the specific contribution of facet could not be derived as nanorod structure has different exposed facets. A clear information on role of crystal facet of WO_3 nanomultilayers with 60% exposed (002) facets and nanorods with 20% exposed (002) facets is reported [134]. A threefold enhancement is observed for photocatalytic hydrogen production. The research on facet dependence is further extended to heterojunction construction with (001) and (111) exposed facets of In_2O_3 and AlGaN. This construction accounted to a 6 times higher photocurrent density. The combination of semiconductor facets BiOI (001)/BiOCl (010), TiO_2 (001)/ $SrTiO_3$ (001) showed higher photo current density attributed to the optimized electron transfer pathways [134]. Such heterojunctions are being investigated in quantum dots (101) as loading on TiO_2 (001) which presents 2.05% as photoconversion efficiency [134].

Till date, creating vacancy defects in nanomaterials are appealing since they have the potential to reduce charge carrier recombination. Specifically, CO_2 adsorption catalytic reaction demand vacancies and has proved to enhance activation, charge carrier separation and light absorption [150]. Approaches such as ferroelectric induced polarization have also prove to inhibit electron–hole recombination. Additionally, the selectivity of carbon-based reduction reactions strongly relies on dynamic and thermodynamic factors. Dynamic factors include charge separation, active sites, adsorption and desorption of reactants or intermediates; while, thermodynamic factors include photon energy and band position. Facet engineered spongy nickel organic catalyst exhibited 100% selectivity toward CO production [150]. Moreover, the reaction selectivity can be tuned by metal additives such as Ag and Rh. The access to Ni^{2+} active sites enhanced capture and stabilization of CO_2^- intermediates resulting in enhanced CO production with limited H_2 evolution. In a different approach, an isolated single atom cobalt is fabricated with Bi_3O_4Br that improved CO_2 reduction with 32-fold higher CO yield than bulk Bi_3O_4Br [150]. Porous atomic layers of CoN are shown to improve CO production with 100% selectivity [151]. The demonstration over high CO is a result of increase in IR absorption, reduced recombination and further promotion of COOH* intermediate radicals that are very essential in CO production.

The same kinetics are favorable for CO_2 reduction to produce CH_4. The CO_2 reduction to produce CH_4 can take two different pathways; 1. Deoxygenation, $CO_2 \rightarrow CO \rightarrow C^{\cdot} \rightarrow CH_3 \rightarrow CH_4$ or 2. Hydrogenation $CO_2 \rightarrow HCOOH \rightarrow HCHO \rightarrow CH_3OH \rightarrow CH_4$. Vacancies of oxygen and silver are the primary drivers for the formation of intermediates either through deoxygenation or hydrogenation and promote charge transfer for higher CH_4 selectivity [152, 153]. Moreover, CO_2 to CH_3OH conversion also demand strong intermediate formation that includes CHO, CH_2O, and CH_3O [154]. It is reported that a high selectivity of 95% is achieved with graphitic carbon composite [154]. In addition to other carbon fuels, production of C_{2+} products are attracting more interest. But the product formation is more challenging as it involves complex reaction pathways [150]. Titanium dioxide with (101) facet decorated onto Au–Pd alloy is found to aid the formation of C_2H_4 and C_2H_6 compared to mono metal/TiO_2 composites [155]. To aid our discussion, the coexistence of structure, material, morphology with controlled facets is contribute

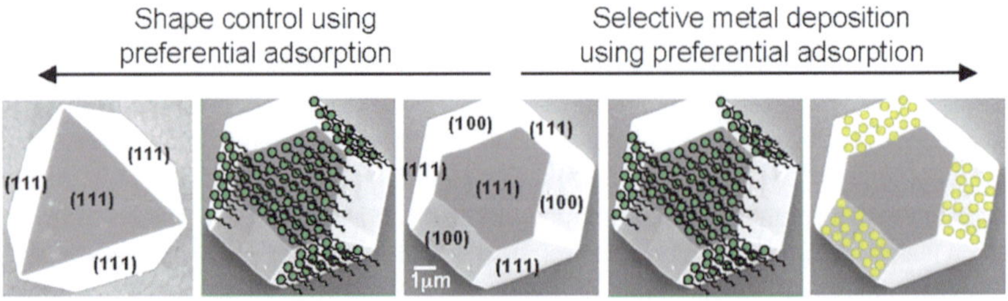

Fig. 9 Preferential adsorption technique for shape control (left) and selective metal deposition (right) Reproduced with Permission from [156]. Copyright 2009. American Chemical Society

to efficient catalytic systems. Therefore, higher percentage of controlled facets with controlled morphology and its associated performance assessment could reveal the sole critical contribution in photoconversion efficiencies. A schematic demonstrating shape control and deposition using preferential adsorption technique is exemplified in Fig. 9.

Facet engineering strategies reported above are primarily focused on carrier separation efficiency. However, emphasis on transfer kinetics is essential to deduce a proper design strategy for efficient catalytic processes. Moreover, reported studies have structural contributions that could participate in charge transfer process and there exist inconsistency in morphological optimization. Hence, facet engineering of nanomaterials needs more attention to be effectively employed in artificial photosynthesis.

4 Durability of Engineered Materials and Future Prospects

Materials science in macro, micro, and nanoscale is being explored to understand the artificial photosynthesis process. Efficient modification methodologies are still proposed to achieve higher conversion efficiencies in solar fuel production. In real-world system conditions, an energy conversion device is expected to operate for several thousands of hours. Despite the fabrication techniques and efficiency boosting strategies, durability analysis and its associated degradation mechanism are not well explored. The fundamental physicochemical and thermophysical properties of material promote degradation and further compromise the primary function of photocatalytic energy conversion. However, these fundamental properties involved in durability is least understood.

Primarily, photocorrosion is a failure mechanism that significantly affects the interfacial process and hence the reaction sites lose the potential to convert energy. Quantitative methods and approaches have been explored to understand the corrosion process that includes mass analysis and in-situ microscopy techniques. The combination of theoretical models and experimental data reveals the underlying mechanism of photocorrosion. Moreover, the mechanism involved in photo conversion process is understood further with DFT calculations that increased the emphasis on materials research. Methodologies to inhibit corrosion process is simultaneously investigated to come up with a common selection criterion that could open up effective strategies to design corrosion free materials. Additionally, the employability of efficient materials in longer operational hours are much needed to explore the scalability and further commercialization.

The inefficiency and degradation of materials puts forth hurdles for implementation and no manufacturing industry exist for standalone solar fuel production. This further limits the human force to achieve net-zero carbon sustainable economy. Hence, it is expected that researchers could be inspired by the technological advances of photovoltaics, and other hydrogen production systems, which have undergone decades of expansion to achieve the current production capacity. It is expected that the pathway for building a sustainable renewable energy technological environment could be realized with further investigations on efficient materials discovery.

The so far developed prototype system in the scale of 100 m^2 with 1 GW hydrogen plant produces 220,000 metric tons/year of hydrogen with the phtotoelectrochemical generator design [157]. Therefore, photovoltaic supported energy systems can be realized in large-scale which is additionally supported by the appropriate infrastructure, supply chain management and associated distribution networks. Such a large-scale realization is a result of initial studies focused on materials, energy requirements with cost analysis linked to panel-lifetime and solar to hydrogen efficiency [158]. The key findings related to implementation has led to significant advances in water splitting and CO_2 reduced solar fuels.

Over the last decades, artificial photosynthetic materials research has seen improvements and is more crucial in addressing energy challenges associated with conversion efficiency. Reported methodologies and techniques to enhance efficiency are still in lab scale trial and error stage. It is important to consider environmental properties which are crucial to be employed in commercial applications.

One other big challenge in catalyst development, is the standardization of test procedures that involves different liquids, reactants and reactor properties that change with every individual reported works. Additionally, the performance efficiency reported in each study could not be compared directly. More specifically, the activity, selectivity, and conversion rate of CO_2 reduction to products have diversified dependence. These inconsistencies limit the possibilities of optimization and hence a unified methodical method for catalyst development could not be ascertained. Therefore, future work in catalyst development could be focused to enhance quantum efficiency and relatively reduce the limiting factors that inhibit conversion efficiencies.

5 Conclusion

The chapter pays attention to mechanism of artificial photosynthesis and explored various nanomaterials in the form of structure engineering and facet engineering. Specifically, reaction mechanism for hydrogen production and CO_2 reduced fuel production were focused. The progress in materials science to discover efficient photocatalytic materials for solar fuel production is discussed. The process involved in artificial photosynthesis such as light absorption, charge separation, recombination and transfer kinetics are investigated accordingly with physicochemical properties of materials were explored and detailed here. The role of dimensional nanostructure, morphology, and facets in efficiency of photocatalytic fuel production is explored.

Despite the advantages of engineered nanomaterials, the limitations in quantum efficiency are undeniable and remains a bottleneck to enhance the large-scale implementation of catalytic materials in artificial photosynthesis. Secondly, the synthesis methodologies involved are limited to lab scale and the possibilities for large-scale production are limited by the capability of repetitive geometries and dimensions. Thirdly, the localized temperature effects on photocatalytic fuel production are not properly understood and explored. The in-situ measurements and effect of temperature must be given special attention to unveil the shortcomings of conversion efficiency. The final big consideration is about standardization of test methodologies which is more crucial to develop efficient sustainable energy systems.

References

1. U.S Department of Energy Information Administration (EIA) (International Energy Outlook 2023, 2023)
2. J. Barber, P.D. Tran, From natural to artificial photosynthesis. J. R. Soc. Interf. **10** (2013). https://doi.org/10.1098/rsif.2012.0984
3. World energy assessment: energy and the challenge of sustainability (UNDP, 2000)
4. H. Li, R. Li, G. Liu et al., Noble-metal-free single- and dual-atom catalysts for artificial photosynthesis. Adv. Mater. **36** (2024)
5. N.S. Lewis, G. Crabtree, A.J. Nozik et al., *Basic Research Needs for Solar Energy Utilization. Report of the Basic Energy Sciences Workshop on Solar Energy Utilization* (United States, 2005)
6. D.P. Hagberg, J.-H. Yum, H. Lee et al., Molecular engineering of organic sensitizers for dye-sensitized solar cell applications. J. Am. Chem. Soc. **130**, 6259–6266 (2008). https://doi.org/10.1021/ja800066y
7. S.E. Shaheen, D.S. Ginley, G.E. Jabbour, Organic-based photovoltaics: toward low-cost power generation. MRS Bull. **30**, 10–19 (2005). https://doi.org/10.1557/mrs2005.2
8. N.S. Lewis, D.G. Nocera, Powering the planet: Chemical challenges in solar energy utilization. Proc. Natl. Acad. Sci. **103**, 15729–15735 (2006). https://doi.org/10.1073/pnas.0603395103
9. A. Machín, K. Fontánez, J.C. Arango et al., One-dimensional (1d) nanostructured materials for energy applications. Materials **14** (2021). https://doi.org/10.3390/ma14102609
10. J.H. Kim, D. Hansora, P. Sharma et al., Toward practical solar hydrogen production—an artificial photosynthetic leaf-to-farm challenge. Chem. Soc. Rev. **48**, 1908–1971 (2019). https://doi.org/10.1039/C8CS00699G

11. X. Wang, Q. Wu, H. Ma et al., Fabrication of PbO_2 tipped Co_3O_4 nanowires for efficient photoelectrochemical decolorization of dye (reactive brilliant blue KN-R) wastewater. Solar Energy Mater Solar Cells **191**, 381–388 (2019). https://doi.org/10.1016/j.solmat.2018.12.005
12. A. Machín, M. Cotto, J. Ducongé, F. Márquez, Artificial photosynthesis: current advancements and future prospects. Biomimetics **8** (2023)
13. D.K. Dogutan, D.G. Nocera, Artificial photosynthesis at efficiencies greatly exceeding that of natural photosynthesis. Acc. Chem. Res. 3143–3148 (2019). https://doi.org/10.1021/acs.accounts.9b00380
14. L. Zhang, Y. Wang, Decoupled artificial photosynthesis. Angewandte Chemie—International Edition **62** (2023)
15. A.E. Becquerel, Recherches sur les effets de la radiation chimique de la lumiere solaire au moyen des courants electriques. Comptes Rendus de L'Academie des Sci. **9** (1839)
16. M. Grätzel, Photoelectrochemical cells. Nature **414**, 338–344 (2001). https://doi.org/10.1038/35104607
17. W.H. Brattain, C.G.B. Garrett, Experiments on the interface between germanium and an electrolyte. Bell Syst. Tech. J. **34**, 129–176 (1955). https://doi.org/10.1002/j.1538-7305.1955.tb03766.x
18. H. Gerischer, Electrochemical behavior of semiconductors under illumination. J. Electrochem. Soc. **113**, 1174 (1966). https://doi.org/10.1149/1.2423779
19. K. Kalyanasundaram, Photoelectrochemical cell studies with semiconductor electrodes—a classified bibliography (1975–1983). Solar Cells **15**, 93–156 (1985). https://doi.org/10.1016/0379-6787(85)90011-0
20. S. Licht, Multiple band gap semiconductor/electrolyte solar energy conversion. J. Phys. Chem. B **105**, 6281–6294 (2001). https://doi.org/10.1021/jp010552j
21. A. Fujishima, K. Honda, Electrochemical photolysis of water at a semiconductor electrode. Nature **238**, 37–38 (1972). https://doi.org/10.1038/238037a0
22. M.G. Walter, E.L. Warren, J.R. McKone et al., Solar water splitting cells. Chem. Rev. **110**, 6446–6473 (2010). https://doi.org/10.1021/cr1002326
23. Z. Han, F. Qiu, R. Eisenberg et al., Robust photogeneration of H_2 in water using semiconductor nanocrystals and a nickel catalyst. Science (1979) **338**, 1321–1324 (2012). https://doi.org/10.1126/science.1227775
24. K. Shankar, J.I. Basham, N.K. Allam et al., Recent advances in the use of TiO_2 nanotube and nanowire arrays for oxidative photoelectrochemistry. J. Phys. Chem. C **113**, 6327–6359 (2009). https://doi.org/10.1021/jp809385x
25. S. Hoang, S. Guo, N.T. Hahn et al., Visible light driven photoelectrochemical water oxidation on nitrogen-modified TiO_2 nanowires. Nano Lett. **12**, 26–32 (2012). https://doi.org/10.1021/nl2028188
26. V. Srimurugan, C.G. Jothiprakash, V. Souparnika, R. Prasanth, Photocorrosion-less stable heterojunction photoanode for efficient visible-light driven solar hydrogen generation. Int. J. Hydrog. Energy **47**, 12515–12527 (2022). https://doi.org/10.1016/j.ijhydene.2022.02.013
27. V. Srimurugan, C.G. Jothiprakash, R. Prasanth, Efficiency enhancement in a stoichiometrically stable CdS/TiO_2 nanotube heterostructure electrode for sunlight-driven hydrogen generation. New J. Chem. **45**, 12838–12847 (2021). https://doi.org/10.1039/D1NJ00735A
28. I.S. Cho, J. Choi, K. Zhang et al., Highly efficient solar water splitting from transferred TiO_2 nanotube arrays. Nano Lett. **15**, 5709–5715 (2015). https://doi.org/10.1021/acs.nanolett.5b01406
29. X. Chen, S.S. Mao, Titanium dioxide nanomaterials: synthesis, properties, modifications, and applications. Chem. Rev. **107**, 2891–2959 (2007). https://doi.org/10.1021/cr0500535
30. F.E. Osterloh, Inorganic nanostructures for photoelectrochemical and photocatalytic water splitting. Chem. Soc. Rev. **42**, 2294–2320 (2013). https://doi.org/10.1039/C2CS35266D
31. J. Li, N. Wu, Semiconductor-based photocatalysts and photoelectrochemical cells for solar fuel generation: a review. Catal. Sci. Technol. **5**, 1360–1384 (2015). https://doi.org/10.1039/C4CY00974F

32. A. Kay, I. Cesar, M. Grätzel, New benchmark for water photooxidation by nanostructured α-Fe_2O_3 films. J. Am. Chem. Soc. **128**, 15714–15721 (2006). https://doi.org/10.1021/ja0643801
33. J. Li, F. Meng, S. Suri et al., Photoelectrochemical performance enhanced by a nickel oxide–hematite p–n junction photoanode. Chem. Commun. **48**, 8213–8215 (2012). https://doi.org/10.1039/C2CC30376K
34. S.D. Tilley, M. Cornuz, K. Sivula, M. Grätzel, Light-induced water splitting with hematite: improved nanostructure and iridium oxide catalysis. Angewandte Chemie Int. Ed. **49**, 6405–6408 (2010). https://doi.org/10.1002/anie.201003110
35. M.A. Grela, A.J. Colussi, Kinetics of stochastic charge transfer and recombination events in semiconductor colloids. relevance to photocatalysis efficiency. J. Phys. Chem. **100**, 18214–18221 (1996). https://doi.org/10.1021/jp961936q
36. F. Le Formal, M. Grätzel, K. Sivula, Controlling photoactivity in ultrathin hematite films for solar water-splitting. Adv. Funct. Mater. **20**, 1099–1107 (2010). https://doi.org/10.1002/adfm.200902060
37. S. Tanriseven, B. Corbett, Low effective surface recombination in In(Ga)As/GaAs quantum dot diodes. J. Appl. Phys. **110**, 034508 (2011). https://doi.org/10.1063/1.3611387
38. B. Atomsa Gonfa, H. Zhao, J. Li et al., Air-processed depleted bulk heterojunction solar cells based on PbS/CdS core–shell quantum dots and TiO_2 nanorod arrays. Solar Energy Mater. Solar Cells **124**, 67–74 (2014). https://doi.org/10.1016/j.solmat.2014.01.037
39. F. Le Formal, N. Tétreault, M. Cornuz et al., Passivating surface states on water splitting hematite photoanodes with alumina overlayers. Chem. Sci. **2**, 737–743 (2011). https://doi.org/10.1039/C0SC00578A
40. J. Li, X. Jin, R. Li et al., Copper oxide nanowires for efficient photoelectrochemical water splitting. Appl. Catal. B **240**, 1–8 (2019). https://doi.org/10.1016/j.apcatb.2018.08.070
41. W. Feng, L. Lin, H. Li et al., Hydrogenated TiO_2/ZnO heterojunction nanorod arrays with enhanced performance for photoelectrochemical water splitting. Int. J. Hydrog. Energy **42**, 3938–3946 (2017). https://doi.org/10.1016/j.ijhydene.2016.10.087
42. P.E.A. Salomão, D.S. Gomes, E.J.C. Ferreira et al., Photoelectrochemical hydrogen production from water splitting using heterostructured nanowire arrays of Bi_2O_3/BiAl oxides as a photocathode. Solar Energy Mater. Solar Cells **194**, 276–284 (2019). https://doi.org/10.1016/j.solmat.2018.12.037
43. D.Y.C. Leung, X. Fu, C. Wang et al., Hydrogen production over titania-based photocatalysts. ChemSusChem **3**, 681–694 (2010). https://doi.org/10.1002/cssc.201000014
44. H. Ahmad, S.K. Kamarudin, L.J. Minggu, M. Kassim, Hydrogen from photo-catalytic water splitting process: a review. Renew. Sustain. Energy Rev. **43**, 599–610 (2015). https://doi.org/10.1016/j.rser.2014.10.101
45. J. Zhang, Y. Wang, J. Jin et al., Efficient visible-light photocatalytic hydrogen evolution and enhanced photostability of core/shell CdS/g-C_3N_4 nanowires. ACS Appl. Mater. Interf. **5**, 10317–10324 (2013). https://doi.org/10.1021/am403327g
46. A. Machin, M. Cotto, J. Ducong&eacute et al., Synthesis and characterization of Au@TiO_2 NWs and their catalytic activity by water splitting: a comparative study with Degussa P25. Am. J. Eng. Appl. Sci. **10**, 298–311 (2017). https://doi.org/10.3844/ajeassp.2017.298.311
47. A. Machín, M. Cotto J. Duconge et al., Hydrogen production via water splitting using different Au@ZnO catalysts under UV–vis irradiation. J. Photochem. Photobiol. A Chem. **353**, 385–394 (2018). https://doi.org/10.1016/j.jphotochem.2017.11.050
48. Z. Liu, H. Bai, S. Xu, D.D. Sun, Hierarchical CuO/ZnO "corn-like" architecture for photocatalytic hydrogen generation. Int. J. Hydrog. Energy **36**, 13473–13480 (2011). https://doi.org/10.1016/j.ijhydene.2011.07.137
49. S.-S. Yi, J.-M. Yan, B.-R. Wulan, Q. Jiang, Efficient visible-light-driven hydrogen generation from water splitting catalyzed by highly stable CdS@Mo_2C–C core–shell nanorods. J. Mater. Chem. A Mater. **5**, 15862–15868 (2017). https://doi.org/10.1039/C7TA03191B
50. Z. Zhou, Z. Wu, Q. Xu, G. Zhao, A solar-charged photoelectrochemical wastewater fuel cell for efficient and sustainable hydrogen production. J. Mater. Chem. A Mater. **5**, 25450–25459 (2017). https://doi.org/10.1039/C7TA08112J

51. X. Zhou, J. Jin, X. Zhu et al., New $Co(OH)_2$/CdS nanowires for efficient visible light photocatalytic hydrogen production. J. Mater. Chem. A Mater. **4**, 5282–5287 (2016). https://doi.org/10.1039/C6TA00325G
52. D. Wang, Y. Li, G. Li Puma et al., Dye-sensitized photoelectrochemical cell on plasmonic Ag/AgCl @ chiral TiO_2 nanofibers for treatment of urban wastewater effluents, with simultaneous production of hydrogen and electricity. Appl. Catal. B **168–169**, 25–32 (2015). https://doi.org/10.1016/j.apcatb.2014.11.012
53. C. Perdomo, N.T. Nguyen, Recent advances in 1D nanostructured catalysts for photothermal and photocatalytic reduction of CO_2. Curr. Opin. Colloid Interf. Sci. **61** (2022)
54. J. Low, S. Qiu, D. Xu et al., Direct evidence and enhancement of surface plasmon resonance effect on Ag-loaded TiO_2 nanotube arrays for photocatalytic CO2 reduction. Appl. Surf. Sci. **434**, 423–432 (2018). https://doi.org/10.1016/j.apsusc.2017.10.194
55. G. Zhang, H. Wu, D. Chen et al., A mini-review on ZnIn2S4-Based photocatalysts for energy and environmental application. Green Energy Environ. **7**, 176–204 (2022). https://doi.org/10.1016/j.gee.2020.12.015
56. J. Di, C. Zhu, M. Ji et al., Defect-rich $Bi_{12}O_{17}C_{l2}$ nanotubes self-accelerating charge separation for boosting photocatalytic CO_2 Reduction. Angewandte Chemie Int. Ed. **57**, 14847–14851 (2018). https://doi.org/10.1002/anie.201809492
57. Y. Wang, J. Zhao, Y. Liu et al., Synergy between plasmonic and sites on gold nanoparticle-modified bismuth-rich bismuth oxybromide nanotubes for the efficient photocatalytic CC coupling synthesis of ethane. J. Colloid Interf. Sci. **616**, 649–658 (2022). https://doi.org/10.1016/j.jcis.2022.02.109
58. H. Zhang, Y. Wang, S. Zuo et al., Isolated cobalt centers on $W_{18}O_{49}$ nanowires perform as a reaction switch for efficient CO_2 photoreduction. J. Am. Chem. Soc. **143**, 2173–2177 (2021). https://doi.org/10.1021/jacs.0c08409
59. Y.-M. Dai, C.-Y. Li, W.-H. Ting, J.-M. Jehng, Plasmon Ag/$AgVO_3$/TiO_2-nanowires S-scheme heterojunction photocatalyst for CO_2 reduction. J. Environ. Chem. Eng. **10**, 108045 (2022). https://doi.org/10.1016/j.jece.2022.108045
60. J. Li, W. Pan, Q. Liu et al., Interfacial engineering of $Bi_{19}Br\ 3S_{27}$ nanowires promotes metallic photocatalytic CO_2 reduction activity under near-infrared light irradiation. J. Am. Chem. Soc. **143**, 6551–6559 (2021). https://doi.org/10.1021/jacs.1c01109
61. T. Tian, X. Jin, N. Guo et al., CdS/ethylenediamine nanowires 3D photocatalyst with rich sulfur vacancies for efficient syngas production from CO_2 photoreduction. Appl. Catal. B **308**, 121227 (2022). https://doi.org/10.1016/j.apcatb.2022.121227
62. J. Zhu, W. Shao, X. Li et al., Asymmetric triple-atom sites confined in ternary oxide enabling selective CO_2 photothermal reduction to acetate. J. Am. Chem. Soc. **143**, 18233–18241 (2021). https://doi.org/10.1021/jacs.1c08033
63. Q. Liu, Y. Zhou, J. Kou et al., High-yield synthesis of ultralong and ultrathin Zn_2GeO_4 nanoribbons toward improved photocatalytic reduction of CO_2 into renewable hydrocarbon fuel. J. Am. Chem. Soc. **132**, 14385–14387 (2010). https://doi.org/10.1021/ja1068596
64. G. Song, F. Xin, X. Yin, Photocatalytic reduction of carbon dioxide over $ZnFe_2O_4$/TiO_2 nanobelts heterostructure in cyclohexanol. J. Colloid Interf. Sci. **442**, 60–66 (2015). https://doi.org/10.1016/j.jcis.2014.11.039
65. H.R. Park, A.U. Pawar, U. Pal et al., Enhanced solar photoreduction of CO_2 to liquid fuel over rGO grafted NiO-CeO_2 heterostructure nanocomposite. Nano Energy **79**, 105483 (2021). https://doi.org/10.1016/j.nanoen.2020.105483
66. P.K. Prajapati, A. Malik, N. Nandal et al., Morphology controlled Fe and Ni-doped CeO_2 nanorods as an excellent heterojunction photocatalyst for CO_2 reduction. Appl. Surf. Sci. **588**, 152912 (2022). https://doi.org/10.1016/j.apsusc.2022.152912
67. N. Li, X. Chen, J. Wang et al., ZnSe nanorods–$CsSnC_{l3}$ perovskite heterojunction composite for photocatalytic CO_2 reduction. ACS Nano **16**, 3332–3340 (2022). https://doi.org/10.1021/acsnano.1c11442
68. M. Ikram, A. Raza, S.O.A. Ahmad et al., Solar-triggered engineered 2D-materials for environmental remediation: status and future insights. Adv. Mater. Interf. **10** (2023)

69. F. Haque, T. Daeneke, K. Kalantar-zadeh, J.Z. Ou, Two-dimensional transition metal oxide and chalcogenide-based photocatalysts. Nanomicro Lett **10**, 23 (2017). https://doi.org/10.1007/s40820-017-0176-y
70. J. Di, J. Xia, H. Li, Z. Liu, Freestanding atomically-thin two-dimensional materials beyond graphene meeting photocatalysis: opportunities and challenges. Nano Energy **35**, 79–91 (2017). https://doi.org/10.1016/j.nanoen.2017.03.030
71. N. Li, J. Ma, W. Wang et al., Dual S-scheme MoS_2/$ZnIn_2S_4$/Graphene quantum dots ternary heterojunctions for highly efficient photocatalytic hydrogen evolution. J. Colloid Interf. Sci. **676**, 496–505 (2024). https://doi.org/10.1016/j.jcis.2024.07.144
72. S. Zhang, X. Liu, C. Liu et al., MoS_2 quantum dot growth induced by S vacancies in a $ZnIn_2S_4$ monolayer: atomic-level heterostructure for photocatalytic hydrogen production. ACS Nano **12**, 751–758 (2018). https://doi.org/10.1021/acsnano.7b07974
73. Q. Liang, Z. Li, Z.-H. Huang et al., Holey graphitic carbon nitride nanosheets with carbon vacancies for highly improved photocatalytic hydrogen production. Adv. Funct. Mater. **25**, 6885–6892 (2015). https://doi.org/10.1002/adfm.201503221
74. W. Bi, C. Ye, C. Xiao et al., Spatial location engineering of oxygen vacancies for optimized photocatalytic H_2 evolution activity. Small **10**, 2820–2825 (2014). https://doi.org/10.1002/smll.201303548
75. D. Li, H. Zhao, L. Li et al., Graphene-sensitized perovskite oxide monolayer nanosheets for efficient photocatalytic reaction. Adv. Funct. Mater. **28**, 1806284 (2018). https://doi.org/10.1002/adfm.201806284
76. R. Peng, L. Liang, Z.D. Hood et al., In-Plane heterojunctions enable multiphasic two-dimensional (2D) MoS_2 nanosheets as efficient photocatalysts for hydrogen evolution from water reduction. ACS Catal. **6**, 6723–6729 (2016). https://doi.org/10.1021/acscatal.6b02076
77. Z. Sun, H. Wang, Z. Wu, L. Wang, G-C_3N_4 based composite photocatalysts for photocatalytic CO_2 reduction. Catal. Today **300**, 160–172 (2018). https://doi.org/10.1016/j.cattod.2017.05.033
78. J. Ran, M. Jaroniec, S.-Z. Qiao, Cocatalysts in semiconductor-based photocatalytic CO_2 reduction: achievements, challenges, and opportunities. Adv. Mater. **30**, 1704649 (2018). https://doi.org/10.1002/adma.201704649
79. S. Tonda, S. Kumar, M. Bhardwaj et al., G-C_3N_4/NiAl-LDH 2D/2D hybrid heterojunction for high-performance photocatalytic reduction of CO_2 into renewable fuels. ACS Appl. Mater. Interf. **10**, 2667–2678 (2018). https://doi.org/10.1021/acsami.7b18835
80. X. Jiao, Z. Chen, X. Li et al., Defect-mediated electron-hole separation in one-unit-cell $ZnIn_2S_4$ layers for boosted solar-driven CO_2 reduction. J. Am. Chem. Soc. **139**, 7586–7594 (2017). https://doi.org/10.1021/jacs.7b02290
81. D. Ma, J. Wu, M. Gao et al., Fabrication of Z-scheme g-C_3N_4/RGO/Bi_2WO_6 photocatalyst with enhanced visible-light photocatalytic activity. Chem. Eng. J. **290**, 136–146 (2016). https://doi.org/10.1016/j.cej.2016.01.031
82. S. Cao, B. Shen, T. Tong et al., 2D/2D heterojunction of ultrathin MXene/Bi_2WO_6 nanosheets for improved photocatalytic CO_2 reduction. Adv. Funct. Mater. **28** (2018). https://doi.org/10.1002/adfm.201800136
83. M. Ye, X. Wang, E. Liu et al., Cover feature: boosting the photocatalytic activity of P25 for carbon dioxide reduction by using a surface-alkalinized titanium carbide MXene as cocatalyst (ChemSusChem 10/2018). ChemSusChem **11**, 1577 (2018). https://doi.org/10.1002/cssc.201800959
84. C. Liu, Q. Xu, Q. Zhang et al., Layered BiOBr/Ti_3C_2 MXene composite with improved visible-light photocatalytic activity. J. Mater. Sci. **54**, 2458–2471 (2019). https://doi.org/10.1007/s10853-018-2990-0
85. C. Ling, X. Niu, Q. Li et al., Metal-free single atom catalyst for N_2 fixation driven by visible light. J. Am. Chem. Soc. **140**, 14161–14168 (2018). https://doi.org/10.1021/jacs.8b07472
86. A.J. Medford, M.C. Hatzell, Photon-driven nitrogen fixation: current progress, thermodynamic considerations, and future outlook. ACS Catal. **7**, 2624–2643 (2017). https://doi.org/10.1021/acscatal.7b00439

87. X. Xue, R. Chen, H. Chen et al., Oxygen vacancy engineering promoted photocatalytic ammonia synthesis on ultrathin two-dimensional bismuth oxybromide nanosheets. Nano Lett. **18**, 7372–7377 (2018). https://doi.org/10.1021/acs.nanolett.8b03655
88. S. Sun, X. Li, W. Wang et al., Photocatalytic robust solar energy reduction of dinitrogen to ammonia on ultrathin MoS_2. Appl. Catal. B **200**, 323–329 (2017). https://doi.org/10.1016/j.apcatb.2016.07.025
89. Y. Bai, L. Ye, T. Chen et al., Facet-dependent photocatalytic N_2 fixation of bismuth-rich Bi_5O_7I nanosheets. ACS Appl. Mater. Interf. **8**, 27661–27668 (2016). https://doi.org/10.1021/acsami.6b08129
90. Y.-J. Yuan, D. Chen, Z.-T. Yu, Z.-G. Zou, Cadmium sulfide-based nanomaterials for photocatalytic hydrogen production. J. Mater. Chem. A Mater. **6**, 11606–11630 (2018). https://doi.org/10.1039/C8TA00671G
91. X. Kong, F. Yu, H. Zhang et al., Synthesis and study of morphology regulation, formation mechanism and photocatalytic performance of CdS. Appl. Surf. Sci. **576**, 151817 (2022). https://doi.org/10.1016/j.apsusc.2021.151817
92. P. Kumar, N. Saxena, V. Gupta et al., Correlation between surface phonon mode and luminescence in nanocrystalline CdS thin films: an effect of ion beam irradiation. J. Appl. Phys. **116**, 043517 (2014). https://doi.org/10.1063/1.4891452
93. X. Yang, Y. Yang, B. Wang et al., Synthesis and photocatalytic property of cubic phase CdS. Solid State Sci. **92**, 31–35 (2019). https://doi.org/10.1016/j.solidstatesciences.2019.04.004
94. H. Zhou, J. Guo, P. Li et al., Leaf-architectured 3D hierarchical artificial photosynthetic system of perovskite titanates towards CO_2 photoreduction into hydrocarbon fuels. Sci. Rep. 3 (2013). https://doi.org/10.1038/srep01667
95. H. Zhou, J. Xu, X. Liu et al., Bio-inspired photonic materials: prototypes and structural effect designs for applications in solar energy manipulation. Adv. Funct. Mater. **28**, 1705309 (2018). https://doi.org/10.1002/adfm.201705309
96. J. Cai, L. Qi, Recent advances in antireflective surfaces based on nanostructure arrays. Mater. Horiz. **2**, 37–53 (2015). https://doi.org/10.1039/C4MH00140K
97. S. Chattopadhyay, Y.F. Huang, Y.J. Jen et al., Anti-reflecting and photonic nanostructures. Mater. Sci. Eng. R. Rep. **69**, 1–35 (2010). https://doi.org/10.1016/j.mser.2010.04.001
98. J. Sun, J. Zhang, M. Zhang et al., Bioinspired hollow semiconductor nanospheres as photosynthetic nanoparticles. Nat. Commun. **3**, 1139 (2012). https://doi.org/10.1038/ncomms2152
99. D. Zheng, X.-N. Cao, X. Wang, Precise formation of a hollow carbon nitride structure with a janus surface to promote water splitting by photoredox catalysis. Angewandte Chemie Int. Ed. **55**, 11512–11516 (2016). https://doi.org/10.1002/anie.201606102
100. P. Xie, Z. Chen, J. Xu et al., Artificial ceramic diatoms with multiscale photonic architectures via nanoimprint lithography for CO_2 photoreduction. J. Am. Ceram. Soc. **102**, 4678–4687 (2019). https://doi.org/10.1111/jace.16334
101. L. Ding, H. Zhou, S. Lou et al., Butterfly wing architecture assisted $CdS/Au/TiO_2$ Z-scheme type photocatalytic water splitting. Int. J. Hydrog. Energy **38**, 8244–8253 (2013). https://doi.org/10.1016/j.ijhydene.2013.04.093
102. S. Tadepalli, J.M. Slocik, M.K. Gupta et al., Bio-optics and bio-inspired optical materials. Chem. Rev. **117**, 12705–12763 (2017). https://doi.org/10.1021/acs.chemrev.7b00153
103. D. Zhang, W. Zhang, J. Gu et al., Inspiration from butterfly and moth wing scales: characterization, modeling, and fabrication. Prog. Mater. Sci. **68**, 67–96 (2015). https://doi.org/10.1016/j.pmatsci.2014.10.003
104. J. Xu, H. Zhou, K. Shi et al., Bio-directed morphology engineering towards hierarchical 1D to 3D macro/meso/nanoscopic morph-tunable carbon nitride assemblies for enhanced artificial photosynthesis. J.. Mater. Chem. A Mater. **5**, 2195–2203 (2017). https://doi.org/10.1039/C6TA08691H
105. J. Liu, H. Zhao, M. Wu et al., Slow photons for photocatalysis and photovoltaics. Adv. Mater. **29**, 1605349 (2017). https://doi.org/10.1002/adma.201605349

106. X. Li, J. Yu, M. Jaroniec, X. Chen, Cocatalysts for selective photoreduction of CO_2 into solar fuels. Chem. Rev. **119**, 3962–4179 (2019). https://doi.org/10.1021/acs.chemrev.8b00400
107. X. Zheng, Y. Yang, S. Chen, L. Zhang, Slow photons for solar fuels. Chin. J. Catal. **39**, 379–389 (2018). https://doi.org/10.1016/S1872-2067(17)62930-9
108. G. von Freymann, V. Kitaev, B.V. Lotsch, G.A. Ozin, Bottom-up assembly of photonic crystals. Chem. Soc. Rev. **42**, 2528–2554 (2013). https://doi.org/10.1039/C2CS35309A
109. D. Duché, L. Escoubas, J.-J. Simon et al., Slow Bloch modes for enhancing the absorption of light in thin films for photovoltaic cells. Appl. Phys. Lett. **92**, 193310 (2008). https://doi.org/10.1063/1.2929747
110. A. Stein, B.E. Wilson, S.G. Rudisill, Design and functionality of colloidal-crystal-templated materials—chemical applications of inverse opals. Chem. Soc. Rev. **42**, 2763–2803 (2013). https://doi.org/10.1039/C2CS35317B
111. C.G. Jothi Prakash, R. Prasanth, Approaches to design a surface with tunable wettability: a review on surface properties. J. Mater. Sci. **56**, 108–135 (2021). https://doi.org/10.1007/s10853-020-05116-1
112. L. Sun, M. Yang, J. Huang et al., Freestanding graphitic carbon nitride photonic crystals for enhanced photocatalysis. Adv. Funct. Mater. **26**, 4943–4950 (2016). https://doi.org/10.1002/adfm.201600894
113. J. Kaschke, L. Blume, L. Wu et al., A helical metamaterial for broadband circular polarization conversion. Adv. Opt. Mater. **3**, 1411–1417 (2015). https://doi.org/10.1002/adom.201500194
114. L. Zhou, Z. Zhuang, H. Zhao et al., Intricate hollow structures: controlled synthesis and applications in energy storage and conversion. Adv. Mater. **29**, 1602914 (2017). https://doi.org/10.1002/adma.201602914
115. K. Cui, P. Lemaire, H. Zhao et al., Tungsten-carbon nanotube composite photonic crystals as thermally stable spectral-selective absorbers and emitters for thermophotovoltaics. Adv. Energy Mater. **8**, 1801471 (2018). https://doi.org/10.1002/aenm.201801471
116. S. Nasiri Khalil Abad, J. Moghaddam, M. Mozammel et al., Growth mechanism and charge transport properties of hybrid Au/ZnO nanoprisms. J. Alloys Compd. **777**, 1386–1395 (2019). https://doi.org/10.1016/j.jallcom.2018.11.084
117. X. Zheng, L. Zhang, Photonic nanostructures for solar energy conversion. Energy Environ. Sci. **9**, 2511–2532 (2016). https://doi.org/10.1039/C6EE01182A
118. F. Sordello, C. Minero, Photocatalytic hydrogen production on Pt-loaded TiO2 inverse opals. Appl. Catal. B **163**, 452–458 (2015). https://doi.org/10.1016/j.apcatb.2014.08.028
119. Y. Zhong, S. Yang, S. Zhang et al., CdS branched TiO_2: rods-on-rods nanoarrays for efficient photoelectrochemical (PEC) and self-bias photocatalytic (PC) hydrogen production. J. Power Sour. **430**, 32–42 (2019). https://doi.org/10.1016/j.jpowsour.2019.04.116
120. J. Fang, J. Gu, Q. Liu et al., Three-dimensional CdS/Au butterfly wing scales with hierarchical rib structures for plasmon-enhanced photocatalytic hydrogen production. ACS Appl. Mater. Interf. **10**, 19649–19655 (2018). https://doi.org/10.1021/acsami.8b03064
121. E. Soto, F. Vaquero, N. Mota et al., CdS photocatalysts modified with Ag: effect of the solvothermal temperature on the structure and photoactivity for hydrogen production. Catalysts **2**, 110 (2019)
122. Y. Zhang, Z. Peng, S. Guan, X. Fu, Novel β-NiS film modified CdS nanoflowers heterostructure nanocomposite: extraordinarily highly efficient photocatalysts for hydrogen evolution. Appl. Catal. B **224**, 1000–1008 (2018). https://doi.org/10.1016/j.apcatb.2017.11.043
123. J. Lv, J. Liu, J. Zhang et al., Construction of organic–inorganic cadmium sulfide/diethylenetriamine hybrids for efficient photocatalytic hydrogen production. J. Colloid Interf. Sci. **512**, 77–85 (2018). https://doi.org/10.1016/j.jcis.2017.10.052
124. J. Prakash, P. Kumar, N. Saxena et al., CdS based 3D nano/micro-architectures: formation mechanism, tailoring of visible light activities and emerging applications in photocatalytic H_2 production, CO_2 reduction and organic pollutant degradation. J. Mater. Chem. A Mater.. **11**, 10015–10064 (2023)
125. L. Chen, X. Tang, P. Xie et al., 3D printing of artificial leaf with tunable hierarchical porosity for CO_2 photoreduction. Chem. Mater. **30**, 799–806 (2018). https://doi.org/10.1021/acs.chemmater.7b04313

126. P. He, X. Tang, L. Chen et al., Patterned carbon nitride-based hybrid aerogel membranes via 3D printing for broadband solar wastewater remediation. Adv. Funct. Mater. **28**, 1801121 (2018). https://doi.org/10.1002/adfm.201801121
127. X. Tang, H. Zhou, Z. Cai et al., Generalized 3D printing of graphene-based mixed-dimensional hybrid aerogels. ACS Nano **12**, 3502–3511 (2018). https://doi.org/10.1021/acsnano.8b00304
128. C.-Y. Lee, A.C. Taylor, S. Beirne, G.G. Wallace, 3D-printed conical arrays of TiO_2 electrodes for enhanced photoelectrochemical water splitting. Adv. Energy Mater. **7**, 1701060 (2017). https://doi.org/10.1002/aenm.201701060
129. J. Prakash, C. Gunarasan C, 안동규 et al., 전해연마된 SLM 기반 316L Steel의 질량 손실 및 표면 피복 특성 연구 Mass loss and surface passivation characteristics of electropolished SLM 316L. Steel **33** (2024). https://doi.org/10.5228/KSTP.2024.33.4.255
130. J.U. Jeong, K.K. Lee, D.G. Ahn, J.W. Lee, Study of surface roughness change of AlSi10Mg based on selective laser melting (SLM) using electro-polishing technique. J. Mech. Sci. Technol. (2024). https://doi.org/10.1007/s12206-024-2412-y
131. C. Wang, H. Tissot, J.H. Stenlid et al., High-density isolated Fe_1O_3 sites on a single-crystal $Cu_2O(100)$ surface. J. Phys. Chem. Lett. **10**, 7318–7323 (2019). https://doi.org/10.1021/acs.jpclett.9b02979
132. Y. Li, K. Luo, R. Tao et al., A new concept and strategy for photovoltaic and thermoelectric power generation based on anisotropic crystal facet unit. Adv. Funct. Mater. **30**, 2002606 (2020). https://doi.org/10.1002/adfm.202002606
133. Z. Li, Z. Li, Z. Shi, X. Fang, Facet-dependent, fast response, and broadband photodetector based on highly stable all-inorganic $CsCu_2I_3$ single crystal with 1D electronic structure. Adv. Funct. Mater. **30**, 2002634 (2020). https://doi.org/10.1002/adfm.202002634
134. S. Wang, G. Liu, L. Wang, Crystal facet engineering of photoelectrodes for photoelectrochemical water splitting. Chem. Rev. **119**, 5192–5247 (2019). https://doi.org/10.1021/acs.chemrev.8b00584
135. S. Karthikeyan, S. Kumar, L.J. Durndell et al., Size-dependent visible light photocatalytic performance of Cu_2O nanocubes. ChemCatChem **10**, 3554–3563 (2018). https://doi.org/10.1002/cctc.201800439
136. Y.-C. Pu, H.-Y. Chou, W.-S. Kuo et al., Interfacial charge carrier dynamics of cuprous oxide-reduced graphene oxide (Cu_2O-rGO) nanoheterostructures and their related visible-light-driven photocatalysis. Appl. Catal. B **204**, 21–32 (2017). https://doi.org/10.1016/j.apcatb.2016.11.012
137. N. Li, W. Yan, Y. Niu et al., Photoinduced in situ spontaneous formation of a reduced graphene oxide-enwrapped Cu–Cu_2O nanocomposite for solar hydrogen evolution. ACS Appl. Mater. Interf. **13**, 9838–9845 (2021). https://doi.org/10.1021/acsami.0c20636
138. Q. Xiang, K. Lv, J. Yu, Pivotal role of fluorine in enhanced photocatalytic activity of anatase TiO_2 nanosheets with dominant (001) facets for the photocatalytic degradation of acetone in air. Appl. Catal. B **96**, 557–564 (2010). https://doi.org/10.1016/j.apcatb.2010.03.020
139. J.S. Chen, Y.L. Tan, C.M. Li et al., Constructing hierarchical spheres from large ultrathin anatase TiO_2 nanosheets with nearly 100% exposed (001) facets for fast reversible lithium storage. J. Am. Chem. Soc. **132**, 6124–6130 (2010). https://doi.org/10.1021/ja100102y
140. L.-M. Lyu, M.H. Huang, Investigation of relative stability of different facets of Ag_2O nanocrystals through face-selective etching. J. Phys. Chem. C **115**, 17768–17773 (2011). https://doi.org/10.1021/jp2059479
141. Z. Zheng, B. Huang, J. Lu et al., Hierarchical TiO_2 microspheres: synergetic effect of {001} and {101} facets for enhanced photocatalytic activity. Chem. Eur. J. **17**, 15032–15038 (2011). https://doi.org/10.1002/chem.201101466
142. C.W. Kim, Y.S. Son, M.J. Kang et al., (040)-crystal facet engineering of $BiVO_4$ plate photoanodes for solar fuel production. Adv. Energy Mater. **6** (2016). https://doi.org/10.1002/aenm.201501754
143. R. Li, H. Han, F. Zhang et al., Highly efficient photocatalysts constructed by rational assembly of dual-cocatalysts separately on different facets of BiVO4. Energy Environ. Sci. **7**, 1369–1376 (2014). https://doi.org/10.1039/C3EE43304H

144. T.W. Kim, K.-S. Choi, Nanoporous $BiVO_4$ photoanodes with dual-layer oxygen evolution catalysts for solar water splitting. Science (1979) **343**, 990–994 (2014). https://doi.org/10.1126/science.1246913
145. X. Wang, M. Liu, Z. Zhou, L. Guo, Toward facet engineering of CdS nanocrystals and their shape-dependent photocatalytic activities. J. Phys. Chem. C **119**, 20555–20560 (2015). https://doi.org/10.1021/acs.jpcc.5b07370
146. S.J. Yang, Y.K. Lin, Y.C. Pu, Y.J. Hsu, Crystal facet dependent energy band structures of polyhedral Cu_2O nanocrystals and their application in solar fuel production. J. Phys. Chem. Lett. **13**, 6298–6305 (2022). https://doi.org/10.1021/acs.jpclett.2c01632
147. P. Li, Y. Zhou, Z. Zhao et al., Hexahedron prism-anchored octahedronal CeO_2: crystal facet-based homojunction promoting efficient solar fuel synthesis. J. Am. Chem. Soc. **137**, 9547–9550 (2015). https://doi.org/10.1021/jacs.5b05926
148. C. Zhu, Y. Shen, F. Yang et al., Engineering hydrogenated TiO_2 nanosheets by rational deposition of Ni clusters and Pt single atoms onto exposing facets for high-performance solar fuel production. Chem. Eng. J. **466**, 143174 (2023). https://doi.org/10.1016/j.cej.2023.143174
149. X.Y. Zhang, P. Wang, X.Y. Lu et al., Synergistic effect on photocatalytic CO2 reduction of facet-engineered Fe-soc-MOFs with photo-deposited PtO species. Chem. Eng. J. **476**, 146560 (2023). https://doi.org/10.1016/j.cej.2023.146560
150. W. Peng, T.H. Chuong Nguyen, D.L.T. Nguyen et al., A roadmap towards the development of superior photocatalysts for solar- driven CO_2-to-fuels production. Renew. Sustain. Energy Rev. **148**, 111298 (2021)
151. L. Liang, X. Li, J. Zhang et al., Efficient infrared light induced CO_2 reduction with nearly 100% CO selectivity enabled by metallic CoN porous atomic layers. Nano Energy **69**, 104421 (2020). https://doi.org/10.1016/j.nanoen.2019.104421
152. X. Yang, S. Wang, N. Yang et al., Oxygen vacancies induced special CO_2 adsorption modes on Bi_2MoO_6 for highly selective conversion to CH_4. Appl. Catal. B **259**, 118088 (2019). https://doi.org/10.1016/j.apcatb.2019.118088
153. X. Li, Y. Sun, J. Xu et al., Selective visible-light-driven photocatalytic CO_2 reduction to CH_4 mediated by atomically thin $CuIn_5S_8$ layers. Nat. Energy **4**, 690–699 (2019). https://doi.org/10.1038/s41560-019-0431-1
154. J.-Y. Liu, X.-Q. Gong, A.N. Alexandrova, Mechanism of CO_2 photocatalytic reduction to methane and methanol on defected anatase TiO_2 (101): a density functional theory study. J. Phys. Chem. C **123**, 3505–3511 (2019). https://doi.org/10.1021/acs.jpcc.8b09539
155. Q. Chen, X. Chen, M. Fang et al., Photo-induced Au–Pd alloying at TiO_2 101 facets enables robust CO_2 photocatalytic reduction into hydrocarbon fuels. J. Mater. Chem. A Mater. **7**, 1334–1340 (2019). https://doi.org/10.1039/C8TA09412H
156. C.G. Read, E.M.P. Steinmiller, K.S. Choi, Atomic plane-selective deposition of gold nanoparticles on metal oxide crystals exploiting preferential adsorption of additives. J. Am. Chem. Soc. **131**, 12040–12041 (2009). https://doi.org/10.1021/ja9036884
157. R. Sathre, C.D. Scown, W.R. Morrow et al., Life-cycle net energy assessment of large-scale hydrogen production via photoelectrochemical water splitting. Energy Environ. Sci. **7**, 3264–3278 (2014). https://doi.org/10.1039/C4EE01019A
158. H.A. Atwater, Artificial photosynthesis: pathway to solar fuels. Phys. Today **76**, 32–39 (2023). https://doi.org/10.1063/PT.3.5360

Ferroelectric Photovoltaic Effect: Past, Present, and Future Perspectives

Subhajit Pal and Durga Sankar Vavilapalli

Abstract Ferroelectric (FE) materials, a subclass of non-centrosymmetric piezoelectric crystals, exhibit spontaneous and switchable polarization responses. These materials are being used to convert one form of energy into another or to preserve a specific form of energy in the system. Among them, the photon-to-electric energy conversion phenomenon by the FE materials has made them an important class of materials in recent days due to their anomalous photovoltaic (PV) effect, which allows them to convert photon energy into electric energy with a power conversion efficiency that exceeds the Shockley–Queisser limit. By combining PV properties with strain, temperature, and polarization, these materials demonstrate fascinating properties like photostriction, flexo-PV effect, photocaloric effect, photothermal effect, and photoferroelectric effect. Therefore, the observed PV effect in the ferroelectric system and its tunable characteristics with the intrinsic and extrinsic parameters make them potential next-generation optoelectronic materials in terms of both fundamental and technological aspects. However, the development of FE optoelectronic devices is hindered due to their lack of absorption capability and bad electrical conductivity. To address these challenges, researchers have been exploring methods such as interface engineering, bandgap tuning, surface charge manipulation, and plasmonics enhancement. In this chapter, the historical development, detailed understanding of the mechanism, recent progress, and future outlooks of the FE–PV effect will be described.

Keywords Ferroelectrics · Photovoltaics · Bulk photovoltaic effect · Shockley–Queisser limit · Polarization

---

S. Pal
School of Engineering and Materials Science, Queen Mary University of London, London, UK
e-mail: subhajit.pal@qmul.ac.uk

D. S. Vavilapalli (✉)
Materials Design Division, Department of Physics, Chemistry and Biology (IFM), Linköping University, Linköping, Sweden
e-mail: durga.sankar.vavilapalli@liu.se

P. Ravindran et al. (eds.), *Photon to Power*, Progress in Optical Science and Photonics 32, https://doi.org/10.1007/978-981-96-5914-2_15

1 Introduction

Currently, several materials are being studied for photovoltaic (PV) applications, including semiconductor-based p–n junction solar cells, dye-sensitized solar cells, quantum dot dye-sensitized solar cells, hybrid solar cells, multi-junction solar cells, perovskite solar cells, and organic solar cells [1]. The primary limitation of conventional solar cells is their bandgap, which limits the open-circuit voltage and restricts efficiency to below the Shockley–Queisser limit [1, 2]. In contrast, ferroelectric materials, which belong to the perovskite solar cell category, are being studied for PV applications due to their significantly large open-circuit voltage that exceeds the bandgap limit [3]. Being a subclass of pyroelectric materials, ferroelectric materials exhibit spontaneous polarization responses that can also be switched under an electric field. This class of materials is well known for using pressure sensors, actuators, memories, capacitors, nonlinear optics elements, and so on [4]. Recently, there has been an increasing focus on their potential for photovoltaic applications. Unlike the conventional PV effect driven by the built-in field at the junction, the PV effect in ferroelectric materials arises from the internal electric field present throughout the entire sample. The internal bias field generated during the poling process in ferroelectric material aligns ferroelectric dipoles, which leads to the separation of photo-generated charge carriers when exposed to light [5, 6]. It was only in the 1970s that the first quantitative estimation and model for the ferroelectric PV effect were proposed [7]. In 1974, a breakthrough in the ferroelectric PV effect was observed, which demonstrated the photocurrents in Fe-doped $LiNbO_3$ crystals with saturation voltages in excess of 1000 V (~ 10^5 V/cm) [8]. Importantly, the observed efficiency exceeding the Shockley–Queisser limit in $BaTiO_3$ thin film makes ferroelectrics a promising next-generation PV material [9]. Several mechanisms, such as the bulk PV effect, domain wall theory, Schottky theory, and depolarization field model, are proposed to explain such an anomalous PV effect. Overall, understanding the physics behind the ferroelectric PV effect and the scope for it to be the next-generation solar cell material makes the ferroelectric system an important class of materials. Apart from the observation of the above bandgap photovoltaic effect, the photoinduced ferroelectric phenomena have captured the attention of the physics community due to the coupling effect with the ferroic order parameter [10]. Here, several factors are highlighted that could potentially influence the outcome of the PV effect in ferroelectric materials , along with their challenges and future prospects.

2 Mechanism of Ferroelectric PV Effects

2.1 Depolarization Field Model

The surface charges generated by polarization are not perfectly screened by the free charges in the metal or semiconductor when the ferroelectric material is sandwiched between two metal or semiconductor electrodes [11]. Imperfect screening of surface charges happens when the centres of gravity for both the polarization charge and the free compensation charges are not in alignment. This misalignment can lead to uneven distribution of electric fields at the surface, affecting the overall performance of the material and potentially impacting its electrical properties. Additionally, the distribution of screening charges relies on the characteristics of ferroelectric materials as well as the metal or semiconductor, including factors such as remnant polarization, free charge density, and dielectric constant. The unscreened polarization charges create a significant electric field within the ferroelectric layer, which opposes the direction of polarization. This electric field is referred to as the depolarization field (E_{dp}), as shown in Fig. 1a. The strength of the E_{dp} is influenced by the thickness of the ferroelectric material. Specifically, a thinner ferroelectric layer generates a larger depolarization field. It has been observed that the E_{dp} remains significant when the thickness is less than 100 nm, while it becomes negligible when the thickness exceeds 100 μm. The estimated E_{dp} in the 10–30 nm thick BTO film, sandwiched between metal $SrRuO_3$ electrodes, is approximately 25–50 V/mm [12]. The presence of a strong E_{dp} creates a driving force that helps separate the electron–hole pairs produced by light. Additionally, it is believed that the unusual photovoltage observed in ferroelectric materials is closely linked to the amount of spontaneous polarization present in those materials.

2.2 Schottky Barrier Effect

When a metal is connected to an *n*/*p*-doped semiconductor, the electrochemical potential of the electrons must equalize on both sides of the junction [13]. As a result, a built-in electric field develops at the interface, causing band bending in both the valence and conduction bands. For example, when *n*-type ferroelectric is in contact with the metal electrode, electrons moving into the semiconductor from metal will encounter a potential barrier height known as the Schottky barrier height. The schematic representation of Schottky barrier form due to the contact between the *n*-type semiconductor and metal is displayed in Fig. 1b. The barrier leads to an increased free electron density in the conduction band while the valence band becomes depleted of holes at the interface. Consequently, the recombination times for holes are faster than those for electrons. Under illumination, there is a pronounced increase in the density of minority carriers, specifically electrons. This notable change results in a divergence between the electrochemical potential of the electrons and that of the

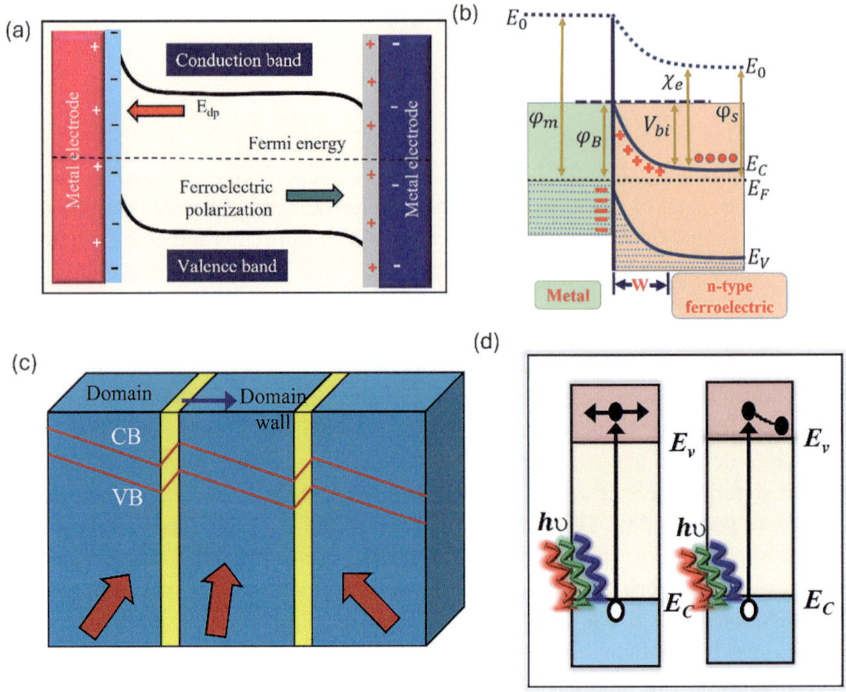

Fig. 1 Schematic describing the PV mechanisms in ferroelectrics. **a** Depolarization model, **b** Schottky barrier effect, **c** domain wall theory, and **d** Bulk photovoltaic effect (BPVE) [18]

holes, thereby establishing a potential difference. This potential difference ultimately leads to the generation of photovoltage, a critical phenomenon in the operation of various semiconductor devices, facilitating the conversion of light energy into electrical energy. The presence of the Schottky barrier typically has a diode-like I–V characteristic, very much like the p–n junction.

2.3 Domain Wall Contribution

According to domain wall theory, the separation of photo-generated charge carriers occurs exclusively at domain walls. This phenomenon is attributed to the periodic potential generated by the varying orientations of adjacent domains [3]. The schematic representation of the domain and the electric field in the domain walls are shown in Fig. 1c. The domain wall theory, proposed by Yang et al., arises from observations of the photovoltaic effect in $BiFeO_3$ (BFO) with a strip domain configuration. It was observed that the photovoltage in the BFO film increased linearly when the PV measurements were taken parallel to the direction of the domain walls. However, the PV effect disappears when the measurements are performed along the direction perpendicular to the domain walls. The potential drop at domain walls occurs due to changes in the polarization components between neighbouring domains. The potential drop across the domain walls is approximately 10 mV, which creates a significant electric field of around 5×10^6 V/m within the narrow domain wall [3]. This electric field at the domain walls drives the dissociation of photo-generated charge carriers. In this model, the high electric field at the domain walls is said to reduce the recombination rate of charge carriers. Upon exposure to illumination, each domain wall serves as a nanoscale generator of photovoltage. When conducting photovoltaic measurements parallel to the orientation of the domain walls, the generated photovoltage is connected in series, leading to an accumulation of voltage in the direction of the net polarization. This process enhances the overall efficiency of energy conversion in the system. Consequently, the material exhibits anomalous photovoltage, not limited by their bandgap.

2.4 Bulk PV Effect

The PV effect on several ferroelectric oxide systems is demonstrated to yield a giant PV response, and the origin is mostly correlated to the bulk photovoltaic effect (BPVE). As per the phenomenological theory of BPVE, the uniform illumination on the non-centrosymmetric crystal produced the PV current, and the same can be expressed as [15],

$$J_i^{pv} = \beta_{ijk} e_j e_k^* I_0 + i\beta_{ik} \left[ee^*\right]_k I_0$$

where I_0 is the light intensity, e_j and e_k^* are components of the incident light polarization vector. β_{ijk} and β_{ik} are third- and second-rank tensors, respectively. It was proposed that the total current in the non-centrosymmetric crystal involves linear and circular parts. Additionally, these two current components can also be separated from each other depending on the crystal structure of the materials. Linear PV current can be determined under the illumination of linearly polarized light in non-centrosymmetric crystals. On the other hand, circular PV current can be determined under the illumination of circularly polarized light in the gyrotropic crystals. From the phenomenological approach of BPVE, 27 components of β_{ijk} and 9 components of β_{ik} of non-centrosymmetric crystals can be determined based on crystal symmetry [3]. In general, the phenomenological theory of BPVE helps compare the PV responses of different crystals based on the magnitude of these tensor components. However, it does not explain the origins of the overall PV outcomes. In this regard, the microscopic theory of BPVE helps to find the origin of BPVE in the system. The microscopic theory of BPVE consists of two important phenomena: (i) ballistic and (ii) shift current [7, 14]. The ballistic mechanism relates to the non-thermal excitation of photo-generated charge carriers in a non-centrosymmetric system, resulting in an asymmetric momentum distribution of the carriers within the conduction band (CB). After light excitation, charge carriers are excited from the valence band (VB) to the CB. After being photo-excited, the carriers lose energy and transition to the conduction band minima (CBM) through non-radiative decay. The ballistic mechanism involves the transport of carriers in a band, thus depending on the mobility of the charge carriers. On the other hand, the shift current mechanism is a quantum mechanical phenomenon related to the displacement of wave packets during photoexcitation process [16]. In this process, charge carriers shift asymmetrically during inter-band transitions which causes a phase change in the wave packet. Consequently, resultant current is generated in the material. The shift current mechanism is based on inter-band transitions, which means that the shift current is independent of the mobility of photo-generated charge carriers. The shift current can be calculated using the tight-binding Hamiltonian approach by projecting atomic basis wave functions into reduced symmetric atomic orbitals, such as Wannier functions. In 2016, it was also demonstrated that the BPVE can also be combined with domain wall theory and exhibits combined outcomes [14, 17]. The single domain BFO film alone cannot explain the overall PV behaviour; however, combined BVPE with domain wall PV effect considerably described the observed PV outcomes in BFO thin film.

3 Bandgap Engineering:

Most FE oxides are insulators with wide band gap energy, typically between 3 and 5 eV, limiting light absorption to the ultraviolet (UV) region, accounting for just 3–5% of the entire solar spectrum [19–21]. Developing ferroelectric materials with a narrow bandgap to maximize solar energy absorption is critical to increasing FEPV efficiency. Researchers are rapidly discovering narrow bandgap ferroelectrics, which were once rare but still face numerous synthesis challenges. For example, the ferroelectric phase of Bi_2O_2Te (Eg = 0.15 eV) is metastable with a propensity to decompose into binary phases [22]. Recent studies have demonstrated the ferroelectricity of many narrow bandgap materials, including In_2Se_3, SnS, SnSe, and GeS, but they still need to address issues with surface instabilities and phase competition [23–26]. Bandgaps below 2 eV can be found in antimony sulphur iodide (SbSI), methylammonium lead iodide ($[CH_3NH_3]PbI_3$), and hexagonal rare-earth ferrites like $LuFeO_3$ [27–30]. A layered growth approach via molecular beam epitaxy can stabilize the hexagonal ferrites, a promising set of multiferroic materials that also compete with multiple other stable polymorphs [30]. There are many more potentially ferroelectric materials that are predicted but not yet experimentally realized. There were several attempts to reduce the bandgap in the FE material, but in most cases, the polarization was greatly compromised. An alternative approach to stabilizing materials with an intrinsically lower Eg is to decrease the bandgap in existing ferroelectric materials through doping-induced polarization rotation, transition metal-induced defect levels, and structural modification.

3.1 Structural Modification

Tetrahedrally coordinated compounds, which are intrinsically polar with an inverted t_{2g}/e_g orbital and have a smaller coordination number than octahedral compounds, can offer low bandgap energy. Zhanga et al. reported a new high-TC multiferroic $KBiFe_2O_5$ compound. This interesting multiferroic compound is in the brownmillerite family and is made up of tetrahedra that are coordinate [31]. It has weak ferromagnetic and ferroelectric properties. The structure of $KBiFe_2O_5$ incorporates tetrahedral Fe^{3+} within a Fe_2O_3 block, which alternates with a K, BiO_2 block. $KBiFe_2O_5$ is an extraordinarily interesting material from the point of view of having a very low bandgap value of 1.59 eV (direct bandgap), which is a lesser bandgap than BFO (Eg = 2.7 eV, theoretical photoconversion efficiencies to 7%), enabling high visible light absorption and a higher concentration of carriers for energy transportation with a theoretical photoconversion efficiency of 30% (Fig. 2a and b). The robust photoelectric effects in the polar, orthorhombic $KBiFe_2O_5$ have become a revival in multiferroic PV research because of the high photovoltage (V_{oc} = 8.8 V) and photocurrent density (Jsc = 15 μA cm^{-2}) (Fig. 2c–e). Like other oxide multiferroics, $KBiFe_2O_5$ also possesses strong magnetodielectric coupling around 780 K [32]. The resistivity of the material is three orders less than that corresponding to the well-known BFO. The value of the absorption coefficient is > 10^4 cm^{-1}, which is the optimal value for solar cell applications. Together, these results leave no doubt that there is a strong photoelectric effect in polar, orthorhombic $KBiFe_2O_5$. However, the reported high leakage current obscures the evidence for improved FE characteristics.

Nechache et al. demonstrated the development of double perovskite Bi_2FeCrO_6 with varying bandgaps, achieving conversion efficiencies of 3.3% in single junction solar cells and 8.1% in tandem solar cells [6]. An adequate comprehension of the correlation among growth conditions, Fe/Cr ordering in Bi_2FeCrO_6, and the bandgap is essential. In another workform, Quattropani et al. demonstrated bandgap tuning in BCFO double perovskite by adjusting the Fe/Cr cationic order parameters [33]. The influence of long-range ordering on the optical characteristics is examined in films formed on an STO (111) substrate (Fig. 3c). The long-range ordering was defined by two parameters, *R* and *D*, where *R* indicates the Fe–Cr ordering and *D* represents the size of the ordered domain. Single-layer and multilayer films are fabricated by manipulating the cationic ordering parameters by controlling the deposition rate and temperature. Figure 3a and b illustrates the fluctuation of the bandgap and polarization with respect to the *R* and *D* parameters. In BFCO, polarization is controlled by Bi^{3+} cations, whereas the bandgap is determined by the ordered interaction of Fe and Cr at the B-site via oxygen. Figure 3c and d illustrates the schematic diagram representing the measuring geometry and the photovoltaic performance of BCFO. They have reported a substantial PCE of 8.1% under AM1.5G irradiation.

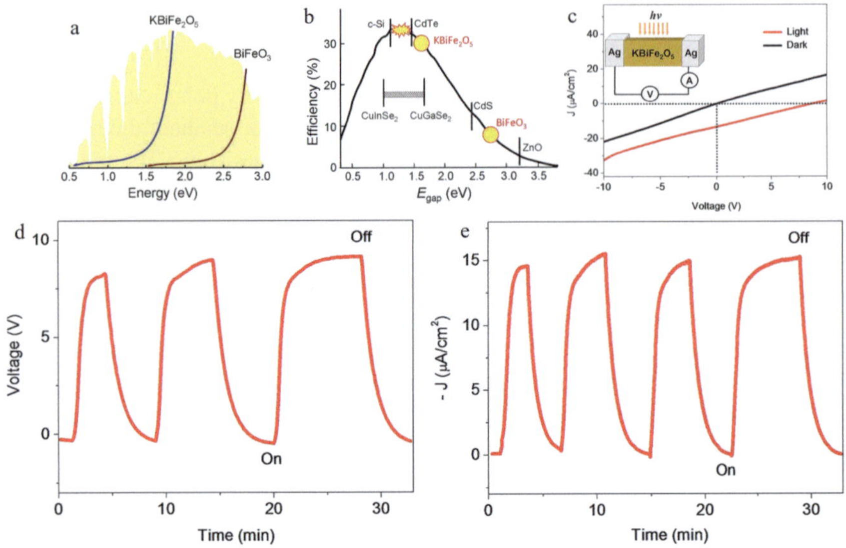

Fig. 2 **a** Spectra of solar radiance and UV–Vis-near IR absorption of $BiFeO_3$ and $KBiFe_2O_5$. **b** maximum theoretical efficiency vs. band gap *J–V* curves in the dark and under UV illumination at room temperature. Inset shows the measurement setup schematically. **c** Zero-voltage photocurrent density following on–off UV exposure and **d** open-circuit photovoltaic voltage following on–off UV exposure [31]

3.2 *Transition Metal-Induced Defect Levels*

Das et al. reported an efficient way to reduce the band gap of a classical ferroelectric material, $BaTiO_3$, without compromising its ferroelectric polarization by co-doping with bivalent Mn^{3+} John–Teller active ion at the Ti^{4+}-site of $BaTiO_3$ along with Nb^{5+} ions. The Jahn–Teller distortion of Mn^{3+} ion plays an important role in stabilizing the ferroelectric phase. Replacing Mn^{3+} with non-Jahn–Teller Fe^{3+} deteriorates the ferroelectric property of $BaTiO_3$ rapidly. Additionally, the Mn^{3+}-doped sample exhibits a greater bandgap reduction than the Fe^{3+}-doped sample, and the theoretical calculations corroborate that the bandgap reduction is a result of the Mn–3d mid-gap states. The Mn–Nb co-doped samples are particularly appealing for low bandgap applications due to these characteristics, which do not substantially compromise polarization. In $BaTi_{1-x}(Mn_{1/2}Nb_{1/2})_xO_3$ (BTMN), the bandgap variations and polarization can be inferred from Fig. 4a and b, respectively, in relation to the Mn concentration and polarization variation. The $BaTi_{0.925}(Mn_{1/2}Nb_{1/2})_{0.075}O_3$ sample is reported to have a minimum bandgap of 1.66 eV and maintain a substantial polarization of nearly 70% of BTO [34].

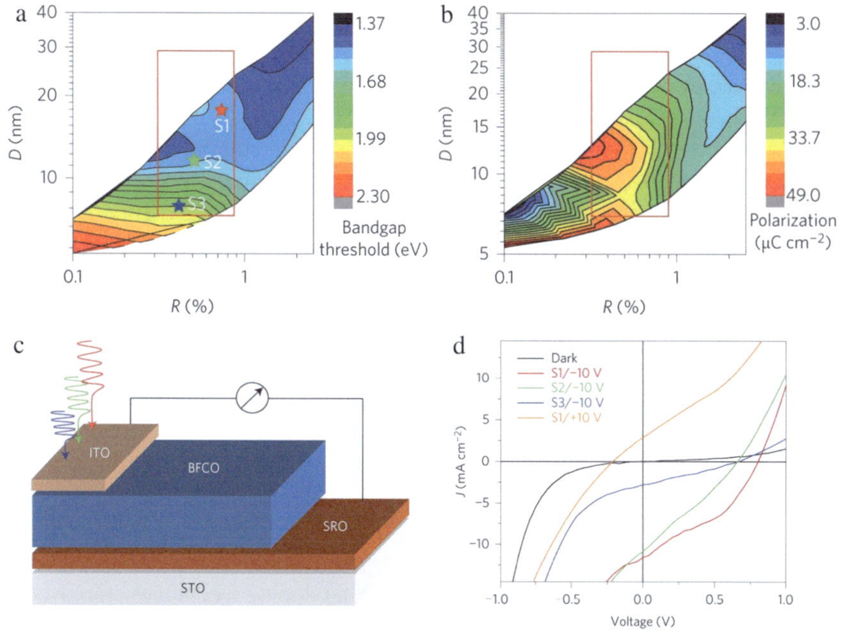

Fig. 3 Mapping of absorption peak showing the variation of order parameters with **a** bandgap threshold and **b** polarization of BFCO thin films. **c** Schematic diagram and **d** *J*–*V* curves of BFCO single layer under AM 1.5G illumination [6]

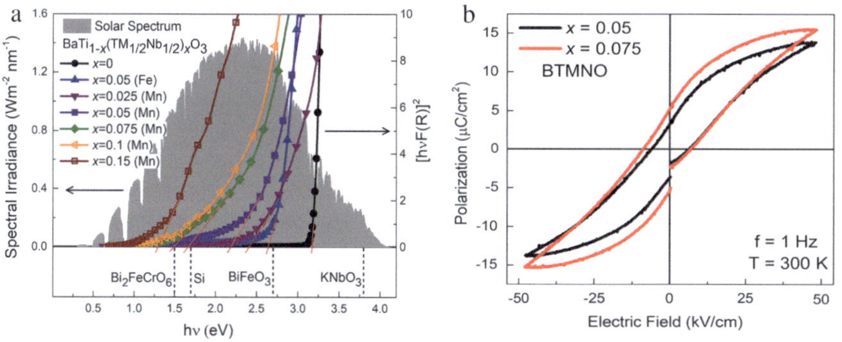

Fig. 4 **a** Kubelka–Munk plot to find the bandgaps of BTMN, $x = 0$–0.15 and $BaTi_{1-x}(Fe_{1/2}Nb_{1/2})_xO_3$, $x = 0.05$. **b** Polarization hysteresis loop for two compositions (with $x =$ 0.05 and 0.75) measured at 300 K [34]

The photovoltaic efficiency of ferroelectric perovskite oxides that have been engineered with a bandgap can be further improved by high-field polarization. Nevertheless, the mechanism of the high-field polarization-enhanced ferroelectric photovoltaic effect in terms of bandgap evolution is not well understood. Su et al. reported a material with good ferroelectricity and four absorption bandgaps, three of which have sub-bandgaps generated by Ni doping, $[KNbO_3]_{1-x}-[SrNi_{0.5}Hf_{0.5}O_{3-\delta}]_x$ [35]. The influence of elevated defects resulting from high-field polarization on the evolution of the bandgap and the photovoltaic properties was investigated. The increase in defect concentration, which is induced by high-field polarization, is the cause of the bandgap structure variation in $KNSNH_2$. The photovoltaic performance is further improved by the production of additional carriers, which is a consequence of the bandgap structure evolution in $KNSNH_2$ after polarization. Additionally, the assimilation of visible light is improved. The short-circuit photocurrent density (Jsc) is increased by twofold under simulated solar irradiation as a consequence of polarization in the $KNSNH_2$. This research provides new insights into the mechanism of bandgap evolution and the influence of defects caused by polarization on the enhancement of ferroelectric photovoltaic performance.

3.3 Doping-Induced Polarization Rotation

In addition to structural effects and transition metal doping techniques employed to reduce the bandgap, the literature reports another mechanism for bandgap reduction through polarization instability or rotation. The initial theoretical investigation by Wang et al. shown that distortions in BO_6 octahedra caused by the off-centring of B-site transition metals significantly influence the bandgap of the ferroelectric material [36]. Their research establishes an efficient method to reduce the FE material's bandgap without substantially disrupting its polarization via structural transitions. The structural transition from the rhombohedral phase to the tetragonal phase, accompanied with polarization rotation, results in a bandgap reduction of 1.2 eV. You et al. exhibited the improved photovoltaic effects through polar engineering utilizing La-doped BFO. The isovalent La^{3+} doping at the Bi^{3+} site significantly affects the stability of the polar phase, altering polar ordering or inducing polarization rotation [37]. A significant photovoltaic response has been noted at the interface between polar and non-polar phases. The 20% La-doped sample exhibits an enhancement in photocurrent by at least an order of magnitude, with photovoltage values double those of undoped BFO. Optical investigations and theoretical forecasts have validated that the polarization rotation caused by chemical substitution results in direct–indirect transitions and extended carrier lifetimes. Thus, the boost in the PV performance was due to the connection between polarization, lattice, and orbital degrees of freedom.

4 Bulk Photovoltaic Effect (BPVE) in 2D Ferroelectrics

Two-dimensional materials have developed a research focus because they can tune their bandgaps using a top-down strategy to scale down to a monolayer and a bottom-up strategy to scale up to a bulk form [38]. The investigation of bulk photovoltaic effect (BPVE) for multiple 2D materials has revived interest in solar energy to electricity conversions. Researchers have conducted several studies on the growth of exfoliated 2D heterogeneous materials and p–n junction geometries, including lateral, vertical, and mixed-dimensional materials based on 2D p–n junctions. However, very few studies are focused on BPVE in 2D materials since the BPVE mechanism differs from standard p–n junction-based photovoltaics. The Shockley–Queisser (S–Q) limit, the maximum theoretical efficiency for a solar cell with a single p–n junction, is 30% at 1.1 eV [39]. The bulk PV effect in 2D ferroelectrics could give a solar cell a higher efficiency than the S–Q limit. Atomically thin layers of 2D materials with van der Waals (vdW) interaction have demonstrated the ability to replace cutting-edge silicon-based solar cell technology. Transition metal dichalcogenides (TMDs) are being studied for use in next-generation transparent conducting materials [40, 41]. They have higher visible and infrared light absorption, ultrahigh mobility, transparency, flexibility, stability, electrical conductivity, and thermal conductivity than traditional Si and GaAs. Both 2D materials and non-centrosymmetric crystals can exhibit significant photovoltage due to the lack of inversion symmetry in their structures. In recent years, the ferroelectricity in 2D materials has been predicted and observed in materials like $MoTe_2$, $Sn_2P_2S_6$, and $CuInP_2S_6$ [42–44]. In this context, researchers have proposed low-dimensional materials like layered van der Waals (vdW) structures, which lack inversion symmetry, as potential candidates for high-efficiency photocurrent collection through BPVE. The intriguing physics in the two-dimensional (2D), such as magnetism, hyperferroelectricity, and exotic topological phases, have been widely explored, but very few direct experimental studies have been conducted on BPVE in 2D materials.

Molybdenum ditelluride ($MoTe_2$) demonstrated room-temperature ferroelectricity, and it is regarded as the most promising 2D layered nanomaterial in regards to enabling phase transition from 2H-to-1T', which is due to a significantly small energy gap (Eg < 50 meV) among TMDs. 2D TMDs exhibit strong absorption from the near-infrared (IR) to visible light region, indicating the efficient utilization of solar energy due to intrinsic crystal anisotropy. The smaller band gaps and more delocalized valence band states in bulk 2D TMDs cause the PV effect to be significantly greater than in monolayers. The bulk PV effect of 2D TMDs can be improved through the use of strain engineering, which has also been demonstrated to be a successful strategy. A promising polar 2D material candidate without centrosymmetry, $MoTe_2$ with metastable phase (1T'), is available for studying strong bulk PV effects. The $MoTe_2$-based bulk PV device exhibits a dominant transformation with both the open-circuit voltage (V_{oc}) and the direct-current (DC) photocurrent (I_{sc}) values without an external bias. More band-to-band transitions occur when illuminated with the lowest wavelength of the incident light. A large bulk PV response at laser power 90.4 W/cm^2, with a $V_{oc} = 16$ mV, an $I_{sc} = 60$ μA, FF = 0.30, and an efficiency of 0.4 was achieved [42]. The PV response of $MoTe_2$ before and after distortion is shown in Fig. 5a and b.

BPVE in 2D ferroelectric $CuInP_2S_6$ was a magnitude greater than most materials made from ferroelectric perovskite oxides [43]. The room-temperature polar ordering in 2D $CuInP_2S_6$ inextricably links the BPVE. The incident photons create

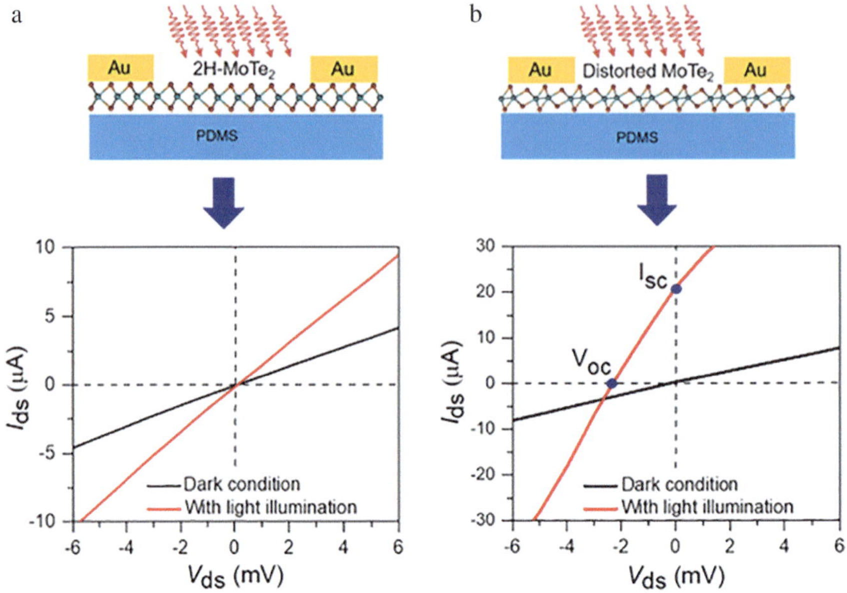

Fig. 5 **a** Before stretching. *I–V* characteristics of 2H-$MoTe_2$ nanoflake demonstrating photoconductive effect. **b** *I–V* characteristics of distorted 2H-$MoTe_2$ (1T′-$MoTe_2$) demonstrating induced photocurrent under visible light illumination (50.3 W cm^{-2}, 600 nm) [42]

electron–hole pairs that are separated by the built-in electric field and drift to create a considerable short-circuit current (I_{sc}) when the device is exposed to light. Furthermore, the direction of the electric polarization can affect the sign of the asymmetric energy potential barrier between the top and bottom surfaces, leading to a switchable photocurrent (as shown in Fig. 6a). When the poling voltage sweeps back and forth, the J_{SC} follows an obvious hysteresis loop, as seen in Fig. 6b. Figure 6c displays the typical output *I–V* characteristics of a device. In order to keep the electric polarization from changing, the bias in the DC measurement is set to ± 0.1 V, which is comparable to a field strength of 105 V cm^1. A large and positive J_{sc} (the short-circuit current density) as roughly ~ 10 mA cm^{-2} using a laser spot area of ~ 20 μm^2. The current density at 0.1 V is around 1 μA cm^{-2}. The V_{oc} is calculated to be − 0.8 V using a linear fit to the *I–V* curve of the output characteristic when compared to bulk ferroelectric photovoltaic cells. Figure 6d shows the temperature-dependent open-circuit voltage, with the V_{oc} decreasing as the temperature rises and disappearing entirely when the temperature is greater than 315 K. The absence of photovoltage above Curie temperature (T_c) shows that 2D ferroelectricity is responsible for the increased BPVE. They also investigate the photocurrent-photocurrent relationship upon illumination. Figure 6e displays the conductivity and J_{sc} corresponding values, as the laser power increases from 1 to 20 W/cm^2. This study underscores the potential application of ultrathin 2D ferroelectric materials in third-generation solar cells. Solar cells made using just monolayers in the literature have relatively low efficiencies.

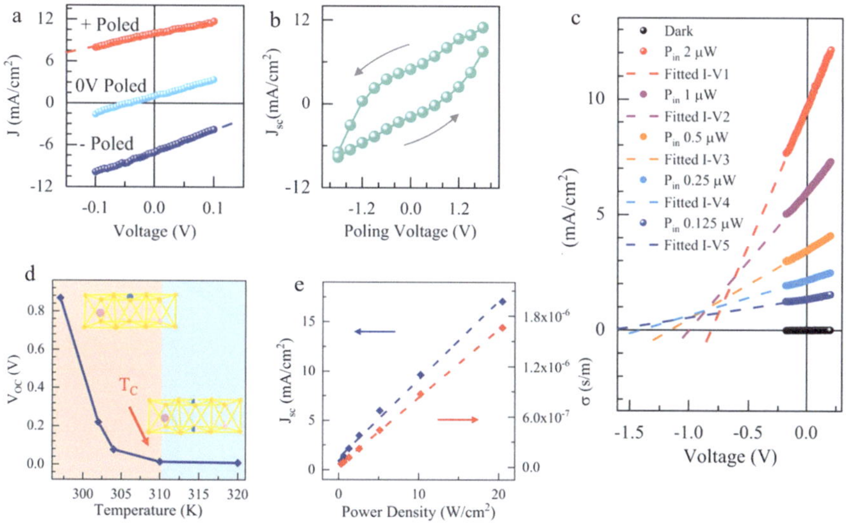

Fig. 6 **a** *I–V* characteristics of output at various poling voltages. **b** J_{sc} versus poling voltage, **c** *I–V* plots of typical output in the dark and under 405 nm illumination. **d** The temperature influences the open-circuit voltage. **e** Short-circuit current density versus power density J_{sc} and power density have a linear relationship with excitation power [43]

The robust ferroelectricity of an atomic-thin $Sn_2P_2S_6$ nanosheet at room temperature was experimentally demonstrated in semiconducting $Sn_2P_2S_6$ crystals with 2D morphology through the use of a chemical vapour transport approach with the assistance of space confinement [44]. A multidirectional electrical control mechanism enables a tunable bulk photovoltaic (BPV) effect by intercorrelated programming of ferroelectric order along out-of-plane (OOP) and in-plane (IP) directions (Fig. 7a and b). A $Sn_2P_2S_6$ optoelectronic device that is highly integrated and vertically sandwiched with graphene electrodes achieves a polarization-dependent open-circuit photovoltage with a dichroic ratio of 2.0 under 405 nm light illumination by combining the capability of anisotropic in-plane optical absorption. The migration of Sn^{2+} ions off the lattice symmetry centre is the cause of the spontaneous polarization of the $Sn_2P_2S_6$ crystal. The piezoresponse force microscopy (PFM) tip was used to apply positive and negative voltage bias (~ + 3 and − 3 V) to the nanosheet vertically, thereby accomplishing the artificial writing of ferroelectric domains. The PFM phase and amplitude images of an 18 nm thick $Sn_2P_2S_6$ nanosheet after the construction of box-in-box graphics with reversed DC bias inside and outside the centre. The vertical BPV response can be modulated through electric polarization in multiple orientations due to the intercorrelated ferroelectricity in $Sn_2P_2S_6$. The short-circuit photocurrent and voltage were effectively modulated by the application of an IP bias of ± 10 V. The high coercive field strength along the [001] crystallographic orientation is highly associated with the relatively larger critical voltage for IP ferroelectric switching, in contrast to OOP polarization. At a light density of

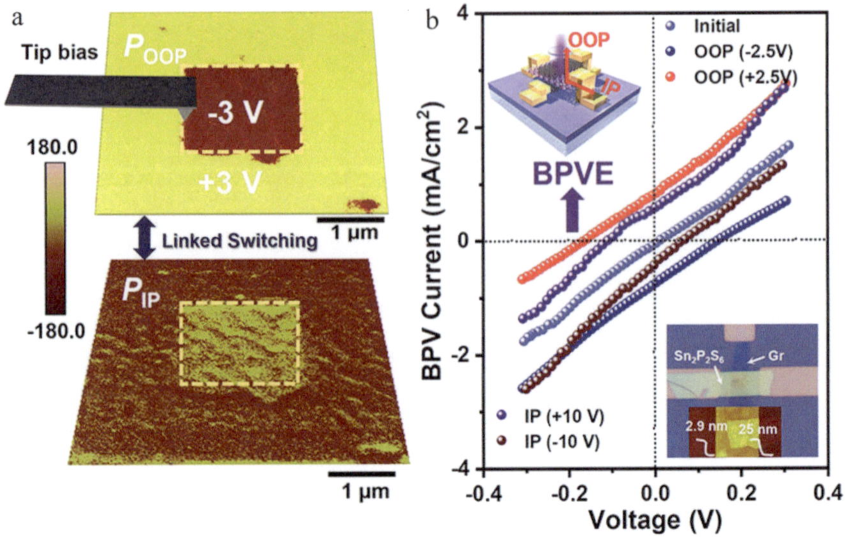

Fig. 7 **a** OOP and IP phase images for 2D $Sn_2P_2S_6$ with box-in-box patterns written by reverse DC bias (+ 3 and − 3 V), **b** *I*–*V* characteristics with the inversion of polarization direction with OOP (± 2.5 V) and IP (± 10 V) bias poling [44]

9.6 W/cm^2, both J_{sc} and V_{oc} would increase monotonically as the incident intensity changes, resulting in the maximum values of 1.1 mA/cm^2 and 0.16 V. The values are identical to those of previously reported 2D ferroelectric $CuInP_2S_6$ with a thickness of 20 nm (Jsc of 0.8 mA/cm^2, $P = 0.3$ W/cm^2) and are higher than those of reported BPV devices based on bulk ferroelectrics $BiFeO_3$.

5 Conclusion and Future Perspective

Here it is discussed about the technological and fundamental aspect of ferroelectric PV effect. Although it was shown that power conversion efficiency in ferroelectrics is not limited by the bandgap of the materials, achieving efficiency beyond the Shockley–Queisser limit occurs only at the nanoscale. Therefore, current research on the ferroelectric PV effect is focused on increasing its efficiency by incorporating various strategies such as bandgap engineering, nano-laying, strain engineering, incorporating plasmonic effect with PV, electrode modification, and materials engineering. Importantly, the advantage of the polarization angle-dependent photovoltaic effect is utilized in spintronic applications by selectively choosing spin-polarized current using circularly polarized light. The circular PV effect observed in ferroelectrics is also applied in circular photodetection. Above all, the most important aspect of ferroelectrics is that the interplay of light with the degrees of freedom in ferroic materials in general leads to a richness of properties extending beyond the mere generation of electricity. This gives rise to photoinduced random access memory devices, photoelectrocatalysis for water splitting, piezophototronic devices, photosensitive actuators, self-powered photodetectors, photovoltaic devices, photoinduced microsensing, microactuator, photostrictive-magnetostrictive magnetometers, light control gas storage applications, and so on. For example, controlling the ferroelectric polarization with illumination gives rise to optical control resistance state manipulation, which can be used in energy-efficient memory applications. This can be used for future optical and electrical control in neuromorphic computing. There are numerous promising applications related to the optical coupling of ferroelectric properties that have gained increased attention from the research community. Focusing on these areas could lead to significant advancements and innovations in the field.

References

1. A. Polman, M. Knight, E.C. Garnett, B. Ehrler, W.C. Sinke, Photovoltaic materials: present efficiencies and future challenges. Science **352**(6283), aad4424 (2016). https://doi.org/10.1126/science.aad4424
2. K.T. Butler, J.M. Frost, A. Walsh, Ferroelectric materials for solar energy conversion: photoferroics revisited. Energy Environ. Sci. **8**(3), 838–848 (2015). https://doi.org/10.1039/C4EE03523B.DOI:10.1039/C4EE03523B
3. S.Y. Yang, J. Seidel, S.J. Byrnes, P. Shafer, C.H. Yang, M.D. Rossell, P. Yu, Y.H. Chu, J.F. Scott, J.W. Ager et al., Above-bandgap voltages from ferroelectric photovoltaic devices. Nat. Nanotechnol. **5**(2), 143–147 (2010). https://doi.org/10.1038/nnano.2009.451
4. J.F. Scott, C.A. Paz de Araujo, Ferroelectric memories. Science **246**(4936), 1400–1405 (1989). https://doi.org/10.1126/science.246.4936.1400
5. C. Paillard, X. Bai, I.C. Infante, M. Guennou, G. Geneste, M. Alexe, J. Kreisel, B. Dkhil, Photovoltaics with ferroelectrics: current status and beyond. Adv. Mater. **28**(26), 5153–5168 (2016). https://doi.org/10.1002/adma.201505215
6. R. Nechache, C. Harnagea, S. Li, L. Cardenas, W. Huang, J. Chakrabartty, F. Rosei, Bandgap tuning of multiferroic oxide solar cells. Nat. Photon. **9**(1), 61–67 (2015). https://doi.org/10.1038/nphoton.2014.255
7. J.E. Spanier, V.M. Fridkin, A.M. Rappe, A.R. Akbashev, A. Polemi, Y. Qi, Z. Gu, S.M. Young, C.J. Hawley, D. Imbrenda et al., Power conversion efficiency exceeding the Shockley-Queisser limit in a ferroelectric insulator. Nat. Photon. **10**(9), 611–616 (2016). https://doi.org/10.1038/nphoton.2016.143
8. A.M. Glass, D. von der Linde, D.H. Auston, T.J. Negran, Excited state polarization, bulk photovoltaic effect and the photorefractive effect in electrically polarized media. J. Electron. Mater. **4**(5), 915–943 (1975). https://doi.org/10.1007/BF02660180
9. A.M. Glass, D. von der Linde, T.J. Negran, High-voltage bulk photovoltaic effect and the photorefractive process in $LiNbO_3$. Appl. Phys. Lett. **25**(4), 233–235 (1974). https://doi.org/10.1063/1.1655453
10. B. Kundys, Photostrictive materials. Appl. Phys. Rev. **2**(1) (2015). https://doi.org/10.1063/1.4905505
11. J. Zhang, X. Su, M. Shen, Z. Dai, L. Zhang, X. He, W. Cheng, M. Cao, G. Zou, Enlarging photovoltaic effect: combination of classic photoelectric and ferroelectric photovoltaic effects. Sci. Rep. **3**(1), 2109 (2013). https://doi.org/10.1038/srep02109
12. Y. Yuan, Z. Xiao, B. Yang, J. Huang, Arising applications of ferroelectric materials in photovoltaic devices. J. Mater. Chem. A **2**(17), 6027–6041 (2014). https://doi.org/10.1039/C3TA14188H.DOI:10.1039/C3TA14188H

13. D.A. Neamen, *Semiconductor Physics and Devices: Basic Principles* (McGraw-Hill, 2012)
14. A.M. Burger, R. Agarwal, A. Aprelev, E. Schruba, A. Gutierrez-Perez, V.M. Fridkin, J.E. Spanier, Direct observation of shift and ballistic photovoltaic currents. Sci. Adv. **5**(1), eaau5588 (2019). https://doi.org/10.1126/sciadv.aau5588
15. B. Sturman, V. Fridkin, *Photovoltaic and Photo-refractive Effects in Noncentrosymmetric Materials*, 1st edn. (Routledge, 1992). https://doi.org/10.1201/9780203743416
16. S.M. Young, A.M. Rappe, First principles calculation of the shift current photovoltaic effect in ferroelectrics. Phys. Rev. Lett. **109**(11), 116601 (2012). https://doi.org/10.1103/PhysRevLett.109.116601
17. H. Matsuo, Y. Kitanaka, R. Inoue, Y. Noguchi, M. Miyayama, T. Kiguchi, T.J. Konno, Bulk and domain-wall effects in ferroelectric photovoltaics. Phys. Rev. B **94**(21), 214111 (2016). https://doi.org/10.1103/PhysRevB.94.214111
18. S. Pal, N.V. Sarath, K.S. Priya, P. Murugavel, A review on ferroelectric systems for next generation photovoltaic applications. J. Phys. D Appl. Phys. **55**(28), 283001 (2022). https://doi.org/10.1088/1361-6463/ac52f4
19. I. Grinberg, D.V. West, M. Torres, G. Gou, D.M. Stein, L. Wu, G. Chen, E.M. Gallo, A.R. Akbashev, P.K. Davies et al., Perovskite oxides for visible-light-absorbing ferroelectric and photovoltaic materials. Nature **503**(7477), 509–512 (2013). https://doi.org/10.1038/nature12622
20. D.S. Vavilapalli, K. Srikanti, R. Mannam, B. Tiwari, K. Mohan Kant, M.S.R. Rao, S. Singh, Photoactive brownmillerite multiferroic $KBiFe_2O_5$ and its potential application in sunlight-driven photocatalysis. ACS Omega **3**(12), 16643–16650 (2018). https://doi.org/10.1021/acsomega.8b01744
21. H. Liu, J. Chen, Y. Ren, L. Zhang, Z. Pan, L. Fan, X. Xing, Large photovoltage and controllable photovoltaic effect in $PbTiO_3$-$Bi(Ni^{2/3+}Nb^{1/3-})O^{3-}$ Ferroelectrics. Adv. Electron. Mater. **1**(4), 1400051 (2015). https://doi.org/10.1002/aelm.201400051
22. X. Zou, F. Tian, H. Liang, Y. Li, Y. Sun, C. Wang, Coexistence of anisotropic large magnetoresistance and ferroelectricity in two-dimensional narrow-bandgap Bi_2O_2Te. ACS Nano **16**(11), 19543–19550 (2022). https://doi.org/10.1021/acsnano.2c09997
23. C. Cui, W.-J. Hu, X. Yan, C. Addiego, W. Gao, Y. Wang, Z. Wang, L. Li, Y. Cheng, P. Li et al., Intercorrelated in-plane and out-of-plane ferroelectricity in ultrathin two-dimensional layered semiconductor In_2Se_3. Nano Lett. **18**(2), 1253–1258 (2018). https://doi.org/10.1021/acs.nanolett.7b04852
24. N. Higashitarumizu, H. Kawamoto, C.-J. Lee, B.-H. Lin, F.-H. Chu, I. Yonemori, T. Nishimura, K. Wakabayashi, W.-H. Chang, K. Nagashio, Purely in-plane ferroelectricity in monolayer SnS at room temperature. Nat. Commun. **11**(1), 2428 (2020). https://doi.org/10.1038/s41467-020-16291-9
25. W. Shi, M. Gao, J. Wei, J. Gao, C. Fan, E. Ashalley, H. Li, Z. Wang, Tin selenide (SnSe): growth, properties, and applications. Adv. Sci. **5**(4), 1700602 (2018). https://doi.org/10.1002/advs.201700602
26. Y. Yan, Q. Deng, S. Li, T. Guo, X. Li, Y. Jiang, X. Song, W. Huang, J. Yang, C. Xia, In-plane ferroelectricity in few-layered GeS and its van der Waals ferroelectric diodes. Nanoscale **13**(38), 16122–16130 (2021). https://doi.org/10.1039/D1NR03807A.DOI:10.1039/D1NR03807A
27. I. Cho, B.-K. Min, S.W. Joo, Y. Sohn, One-dimensional single crystalline antimony sulfur iodide, SbSI. Mater. Lett. **86**, 132–135 (2012). https://doi.org/10.1016/j.matlet.2012.07.050
28. J.S. Manser, J.A. Christians, P.V. Kamat, Intriguing optoelectronic properties of metal halide perovskites. Chem. Rev. **116**(21), 12956–13008 (2016). https://doi.org/10.1021/acs.chemrev.6b00136
29. B.S. Holinsworth, D. Mazumdar, C.M. Brooks, J.A. Mundy, H. Das, J.G. Cherian, S.A. McGill, C.J. Fennie, D.G. Schlom, J.L. Musfeldt, Direct band gaps in multiferroic h-$LuFeO_3$. Appl. Phys. Lett. **106**(8) (2015). https://doi.org/10.1063/1.4908246
30. L. Lin, H.M. Zhang, M.F. Liu, S. Shen, S. Zhou, D. Li, X. Wang, Z.B. Yan, Z.D. Zhang, J. Zhao et al., Hexagonal phase stabilization and magnetic orders of multiferroic $Lu_{1-x}Sc_xFeO_3$. Phys. Rev. B **93**(7), 075146 (2016). https://doi.org/10.1103/PhysRevB.93.075146

31. G. Zhang, H. Wu, G. Li, Q. Huang, C. Yang, F. Huang, F. Liao, J. Lin, New high Tc multiferroics $KBiFe_2O_5$ with narrow band gap and promising photovoltaic effect. Sci. Rep. **3**(1), 1265 (2013). https://doi.org/10.1038/srep01265
32. D.S. Vavilapalli, A.A. Melvin, S. Kavita, A.K. Yadav, S.N. Jha, D. Bhattacharyya, S.C. Sarma, S.C. Peter, M.S. Ramachandra Rao, S. Singh, Multifunctional brownmillerite $KBiFe_2O_5$: Structural, magneto-dielectric, optical, photoelectrochemical studies and enhanced photocatalytic activity over perovskite $BiFeO_3$. Sol. Energy Mater. Sol. Cells **200**, 109940 (2019). https://doi.org/10.1016/j.solmat.2019.109940
33. A. Quattropani, D. Stoeffler, T. Fix, G. Schmerber, M. Lenertz, G. Versini, J.L. Rehspringer, A. Slaoui, A. Dinia, S. Colis, Band-gap tuning in ferroelectric Bi_2FeCrO_6 double perovskite thin films. J. Phys. Chem. C **122**(2), 1070–1077 (2018). https://doi.org/10.1021/acs.jpcc.7b10622
34. S. Das, S. Ghara, P. Mahadevan, A. Sundaresan, J. Gopalakrishnan, D.D. Sarma, Designing a lower band gap bulk ferroelectric material with a sizable polarization at room temperature. ACS Energy Lett. **3**(5), 1176–1182 (2018). https://doi.org/10.1021/acsenergylett.8b00492
35. M. Su, D. Yang, C. Yuan, L. Meng, B. Zhu, J. Xu, C. Zhou, G. Rao, The effect of defects induced by polarization on bandgap evolution and photovoltaic performance in KN-based semiconducting ferroelectric ceramics. Mater. Sci. Semicond. Process. **165**, 107680 (2023). https://doi.org/10.1016/j.mssp.2023.107680
36. F. Wang, I. Grinberg, A.M. Rappe, Band gap engineering strategy via polarization rotation in perovskite ferroelectrics. Appl. Phys. Lett. **104**(15) (2014). https://doi.org/10.1063/1.4871707
37. L. You, F. Zheng, L. Fang, Y. Zhou, L.Z. Tan, Z. Zhang, G. Ma, D. Schmidt, A. Rusydi, L. Wang et al., Enhancing ferroelectric photovoltaic effect by polar order engineering. Sci. Adv. **4**(7), eaat3438 (2018). https://doi.org/10.1126/sciadv.aat3438
38. S. Aftab, M.Z. Iqbal, Z. Haider, M.W. Iqbal, G. Nazir, M.A. Shehzad, Bulk photovoltaic effect in 2D materials for solar-power harvesting. Adv. Opt. Mater. **10**(23), 2201288 (2022). https://doi.org/10.1002/adom.202201288
39. S. Rühle, Tabulated values of the Shockley-Queisser limit for single junction solar cells. Sol. Energy **130**, 139–147 (2016). https://doi.org/10.1016/j.solener.2016.02.015
40. S.N. Shirodkar, U.V. Waghmare, Emergence of ferroelectricity at a metal-semiconductor transition in a 1*T* monolayer of MoS_2. Phys. Rev. Lett. **112**(15), 157601 (2014). https://doi.org/10.1103/PhysRevLett.112.157601
41. W. Ding, J. Zhu, Z. Wang, Y. Gao, D. Xiao, Y. Gu, Z. Zhang, W. Zhu, Prediction of intrinsic two-dimensional ferroelectrics in In_2Se_3 and other III2-VI3 van der Waals materials. Nat. Commun. **8**(1), 14956 (2017). https://doi.org/10.1038/ncomms14956
42. S. Aftab, M.A. Shehzad, H.M. Salman Ajmal, F. Kabir, M.Z. Iqbal, A.A. Al-Kahtani, Bulk photovoltaic effect in two-dimensional distorted $MoTe_2$. ACS Nano **17**(18), 17884–17896 (2023). https://doi.org/10.1021/acsnano.3c03593
43. Y. Li, J. Fu, X. Mao, C. Chen, H. Liu, M. Gong, H. Zeng, Enhanced bulk photovoltaic effect in two-dimensional ferroelectric $CuInP_2S_6$. Nat. Commun. **12**(1), 5896 (2021). https://doi.org/10.1038/s41467-021-26200-3
44. D. Li, J.-K. Qin, B. Zhu, L.-Q. Yue, P.-Y. Huang, C. Zhu, F. Zhou, L. Zhen, C.-Y. Xu, Intercorrelated ferroelectricity and bulk photovoltaic effect in two-dimensional $Sn_2P_2S_6$ semiconductor for polarization-sensitive photodetection. ACS Nano **18**(13), 9636–9644 (2024). https://doi.org/10.1021/acsnano.4c00382

Solar Optics

Geometric Optics-Based Solar Light Management Strategies

Adhithya S. Thampi, M. S. Sangeetha, Devika Mahesh, Mohamed Shaheer, and Adersh Asok

Abstract Geometric optics (GO)-based solar light management strategies use optical phenomena like specular reflection, refraction, and total internal reflection to optimize solar energy utilization across buildings, industry, and agriculture. These strategies improve energy efficiency in buildings by using optics systems for renewable energy generation, such as integrated photovoltaics, photothermal applications, or direct utilization of solar energy for daylighting applications. In agriculture, GO can provide precision-controlled greenhouse lighting and optics-enabled renewable energy systems for improved light management in agrivoltaics. In industry applications, solar light management is crucial for many sectors. The broad adoption of GO in industries can be exemplified by applications like solar photoreactors for chemical processing, solar tracking, biofuel production, carbon capture, hydrogen generation, and generating high-value compounds like pharmaceuticals, food additives, pigments, etc. This chapter explores the importance of GO in these technologies, their design principles, and the challenges of integrating GO-based solutions, highlighting their role in advancing sustainable development goals.

1 Introduction

The escalating need for sustainable energy solutions has catalyzed considerable advancements in effectively utilizing solar energy, a renewable and readily accessible resource. Among the diverse methodologies employed, geometric optics (GO)-based strategies have emerged as a fundamental technology, facilitating enhanced control and optimization of solar light for various applications. By leveraging the principles of reflection, refraction, total internal reflection, and concentration, these approaches maximize solar energy capture and offer tailored solutions that cater to

A. S. Thampi · M. S. Sangeetha · D. Mahesh · M. Shaheer · A. Asok (✉)
Centre for Sustainable Energy Technologies, CSIR-National Institute for Interdisciplinary Science and Technology, Thiruvananthapuram, India
e-mail: adersh.asok@niist.res.in

A. S. Thampi · M. S. Sangeetha · A. Asok
Academy of Scientific and Innovative Research (AcSIR), Ghaziabad, India

P. Ravindran et al. (eds.), *Photon to Power*, Progress in Optical Science and Photonics 32, https://doi.org/10.1007/978-981-96-5914-2_16

application-specific requirements. The purposes of using GO are diverse, focusing on maximizing light utilization, improving efficiency, and reducing costs. Applications benefit from precise and controlled light delivery, enabling scalability, reliability, and safety in operations. Additionally, it allows customization of light properties for specific processes, supports innovation in product design, and enhances sustainability by improving energy efficiency.

The pursuit of energy efficiency in buildings has driven innovations in sustainable design, and one of the most promising approaches is integrating GO into solar management systems. By strategically designing and utilizing optical devices such as lenses, mirrors, and concentrators, GO offers a solution to enhance solar energy harvesting, reduce unwanted heat gain, and improve building thermal comfort. This approach is integral to developing energy-efficient, solar-powered buildings that minimize reliance on mechanical systems and contribute to sustainable architecture. In today's rapidly evolving industrial landscape, the significance of incorporating innovative solar energy solutions cannot be overstated. Among these solutions, photoreactors, photobioreactors, photovoltaics, and solar trackers stand out as pivotal technologies that enhance energy efficiency and contribute to sustainable practices. GO plays a crucial role in optimizing solar energy utilization across these industrial domains driving advancements in functionality and economic viability. In recent years, the application of GO has extended beyond buildings and industrial sectors and into agriculture, where it plays a critical role in managing solar light for enhanced crop production. As the global population continues to rise, the demand for efficient agricultural practices becomes increasingly urgent. By harnessing the principles of GO, farmers and agricultural engineers can optimize light distribution within greenhouses and open fields, leading to improved photosynthesis and crop yields. This strategic manipulation of light not only supports healthier plant growth but also allows for the cultivation of crops in regions with suboptimal sunlight conditions.

As we delve deeper into this chapter, we will explore the multifaceted applications of geometric optics in buildings, industries, and agriculture, discussing various technological advancements and their implications for future practices. By understanding how to effectively manage solar light, we can pave the way for sustainable solar energy utilization techniques that align with global efforts to promote energy efficiency and security.

2 Fundamentals of Geometric Optics

GO provides a critical framework for the effective management of solar light by conceptualizing light as linear rays. This perspective allows for precise manipulation of light through fundamental optical principles, including reflection, refraction, total internal reflection (TIR), and concentration of light. Reflection occurs when light bounces off surfaces, enabling systems like mirrors to direct sunlight efficiently. In contrast, refraction involves the bending of light as it passes through different mediums, which is crucial for lenses that focus or disperse light in various applications. TIR is particularly important in fiber optics and solar concentrators, where light is kept within a medium without escaping, thereby enhancing efficiency. Light concentration techniques, such as the use of parabolic reflectors or prism arrays, focus sunlight onto a smaller area, increasing energy density and improving the overall capture of solar energy. Together, these principles are vital in developing advanced technologies that enhance the absorption of solar energy, ensure optimal distribution of natural daylight in buildings, and boost the overall effectiveness of renewable energy solutions.

Following is the comprehensive overview of the theory, its fundamental principles, and the governing equations.

2.1 Reflection

The law of reflection outlines the principles governing the behavior of light when it strikes reflective surfaces, such as mirrors. This principle of reflection states that the angle formed by light as it strikes a surface called the angle of incidence, is equal to the angle it forms as it bounces off, known as the angle of reflection as shown in Fig. 1.

$$\boldsymbol{\theta_i} = \boldsymbol{\theta_r} \tag{1}$$

Here, $\boldsymbol{\theta_i}$ is the angle of incidence where $\boldsymbol{\theta_r}$ is the angle of reflection.

Mirror-based solar energy systems rely on the principles of reflection and concentration to direct sunlight effectively for solar applications. These systems utilize various configurations, such as parabolic troughs, parabolic dishes, heliostats, Fresnel reflectors, and compound parabolic collectors each designed to optimize sunlight focus and energy efficiency.

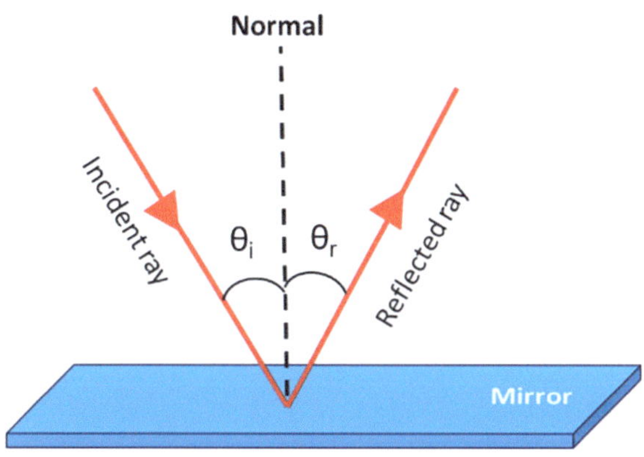

Fig. 1 Schematic diagram showing the principle of reflection

A parabolic trough (PT) consists of a curved, reflective mirror shaped like a parabola. The geometry ensures that parallel sunlight rays converge along a focal line, where a receiver captures and converts the concentrated energy. These troughs are commonly fabricated from reflective materials like polished aluminum or reflective silver. The parabolic dish (PD) employs a segment of a paraboloid surface, directing sunlight to a singular focal point. This configuration enhances energy concentration, making it suitable for high-intensity solar applications. A heliostat system features an array of flat mirrors, each precisely oriented to reflect sunlight onto a central receiver, typically mounted atop a tower. This design facilitates large-scale energy collection. Fresnel reflectors (FL), consisting of multiple narrow, flat reflectors arranged in series that reflect sunlight toward a shared focal line, utilize a receiver to capture energy, making this arrangement more compact and space-efficient than parabolic systems. The Compound Parabolic Collector (CPC) system is a fixed tubular reactor that features two semi-involute, highly polished aluminum reflectors positioned beneath the reactor tube. These reflectors are oriented at an optimized tilt angle, thereby enhancing solar light absorption while eliminating the necessity for tracking systems. A schematic representation of the overview of mirror-based GO is shown in Fig. 2.

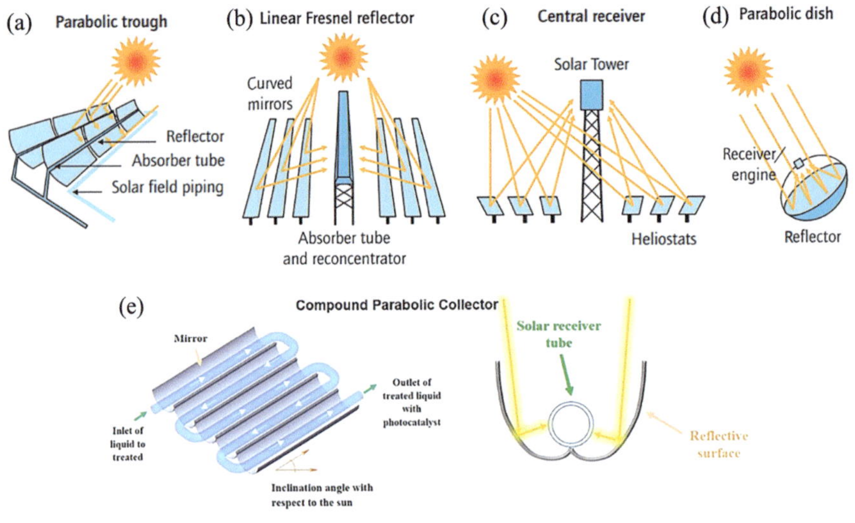

Fig. 2 An overview of mirror-based GO. *Source* Reused with permission from [1] Copyright © 2023 De Gruyter Open Access

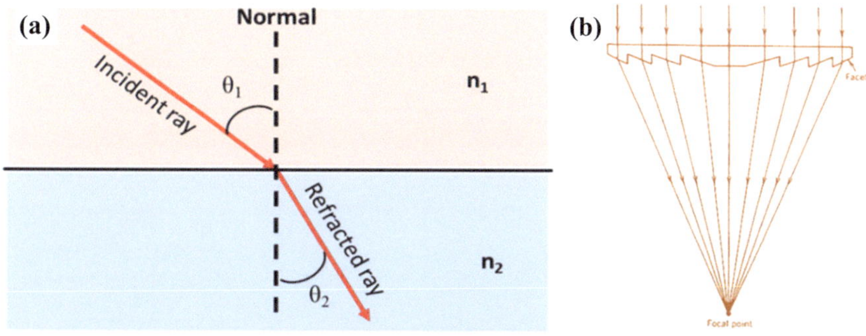

Fig. 3 Schematic representation of **a** the principles of refraction and **b** the working principle of FL. *Source* Reused with permission from [2] Copyright© 2014 Maxwell Scientific Publication Corp.

2.2 Refraction

Refraction refers to the bending of light as it passes from one material to another with a different refractive index. This phenomenon is governed by Snell's law, which mathematically defines the relationship between the angles of incidence and refraction based on the refractive indices of the materials involved (Fig. 3a).

$$n_1 sin(\theta_1) = n_2 sin(\theta_2) \tag{2}$$

where n_1 and n_2 are the refractive indices of 2 different media, θ_1 and θ_2 are the angle of incidence and angle of refraction, respectively. Fresnel lenses (FL) and prismatic lenses (PL) are the common examples adopted in actual practices. PL redirects and diffuses sunlight to enhance indoor daylighting. These are commonly used in skylights and windows to improve lighting conditions while minimizing glare.

FL works by utilizing a series of concentric grooves that are molded into a flat surface, which helps to reduce the overall thickness compared to traditional glass lenses. Each groove acts like a small prism, bending light toward a common focal point (Fig. 3b). This design allows Fresnel lenses to capture and focus light efficiently, making them ideal for applications like lighthouses, projectors, and solar energy systems. Their lightweight and compact nature provides significant advantages in various optical applications.

2.3 Total Internal Reflection (TIR)

TIR happens when light traveling through a denser medium is completely reflected at the boundary with a less dense medium. This occurs only when the angle of incidence surpasses the critical angle (Fig. 4), which is mathematically defined as follows:

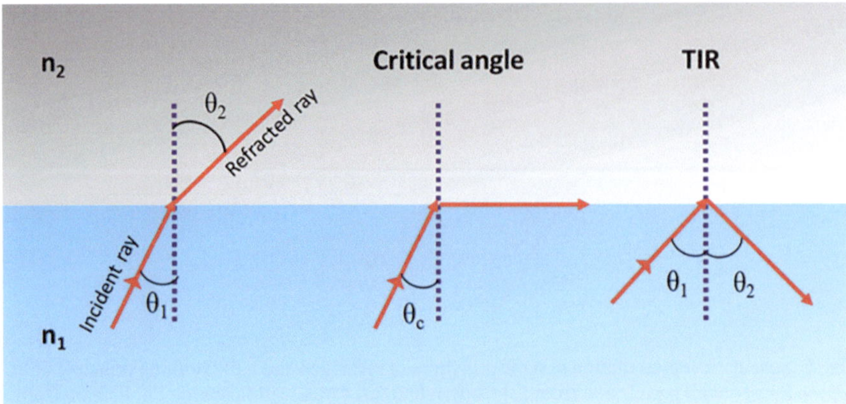

Fig. 4 Schematic representation of the principle of TIR

$$\boldsymbol{sin(\theta_c)} = \frac{n_2}{n_1}. \tag{3}$$

Here, $n_1 > n_2$ where n_1, n_2 are the refractive indices of the two media. This principle is essential for optical fibers (OF), light pipes (LP), and concentrators, ensuring efficient light transport with minimal loss.

OF relies on TIR to transport light over long distances with minimal loss. In solar-based systems, they are used to direct sunlight into areas not directly exposed to it, such as the deep interiors of buildings. LPs are hollow or solid structures that guide sunlight using TIR. Light pipes are often employed in daylighting systems to illuminate spaces naturally while reducing artificial lighting demands.

2.4 Light Concentration

Geometric light concentrators are optical arrangements that focus sunlight onto smaller areas, enhancing energy density (Fig. 5). The concentration ratio (C_g) is expressed as:

$$\mathbf{Cg} = \frac{A_{\text{aperture}}}{A_{\text{receiver}}}. \tag{4}$$

Here, the A_{aperture} and A_{receiver} are the aperture area of the concentrator and the area of the receiving surface, respectively. GO systems concentrate sunlight onto the target area accompanied by either reflection or refraction. PT, PD, FR, CPC, and heliostat are reflection-based geometric optics, which concentrate light on the focal point or line while FL and PL concentrate light followed by refraction.

Fig. 5 Schematic representation of light concentration

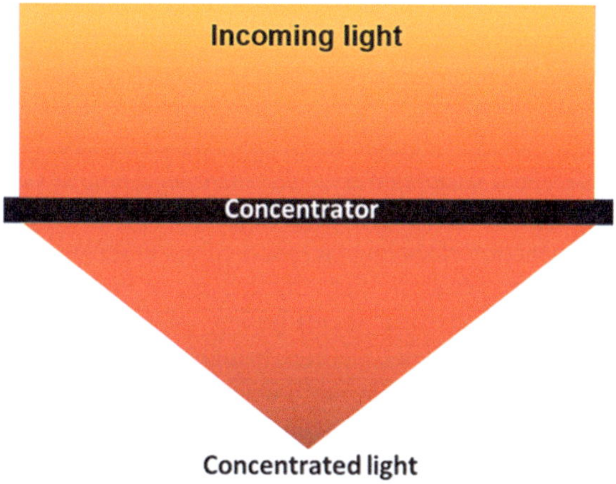

A comprehensive schematic illustration of GO is given in Fig. 6. In conclusion, geometric optics-based strategies for solar light management offer a scientifically grounded approach to optimize solar energy capture. By leveraging principles such as reflection, refraction, and concentration, these methods efficiently direct sunlight to desired focal points, enhancing energy conversion processes. Such strategies underpin the design of advanced solar systems, enabling greater energy efficiency and sustainability. As the demand for renewable energy grows, these techniques provide a critical foundation for innovative solutions that harness the power of sunlight with precision and effectiveness.

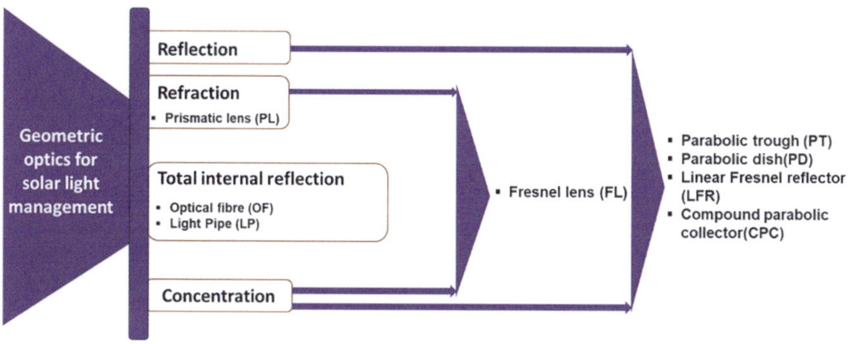

Fig. 6 A comprehensive schematic illustration of GO

3 Solar Light Management Opportunities in Buildings Through GO

Contemporary architecture increasingly emphasizes the importance of energy efficiency and sustainability, with the integration of GO serving a crucial function in managing solar energy in buildings. This innovative optical technology can be effortlessly incorporated into various architectural features, such as windows, facades, and skylights, enhancing the effective use of natural light while reducing the harmful impacts of solar heat gain (Fig. 7).

The fundamental theories behind the optical processes are applied to regulate the amount, direction, and intensity of sunlight entering and interacting with the building's interior. In the context of solar energy systems, CPCs, and other concentrator designs use parabolic or hyperbolic shapes to focus light onto Building Integrated Photovoltaic (BIPV) panels or other energy-capturing surfaces, increasing their efficiency by enhancing light intensity [3–6]. Furthermore, optical systems that incorporate TIR principles, such as LPs or segmented prismatic façades, can direct sunlight into areas far from direct windows, providing natural illumination while minimizing the reliance on artificial lighting. Adjustable optical elements within building façades can dynamically alter the amount of light and heat entering the space throughout the day, optimizing solar gain during cooler months and minimizing heat load in warmer months. Collectively, these applications of GO enhance

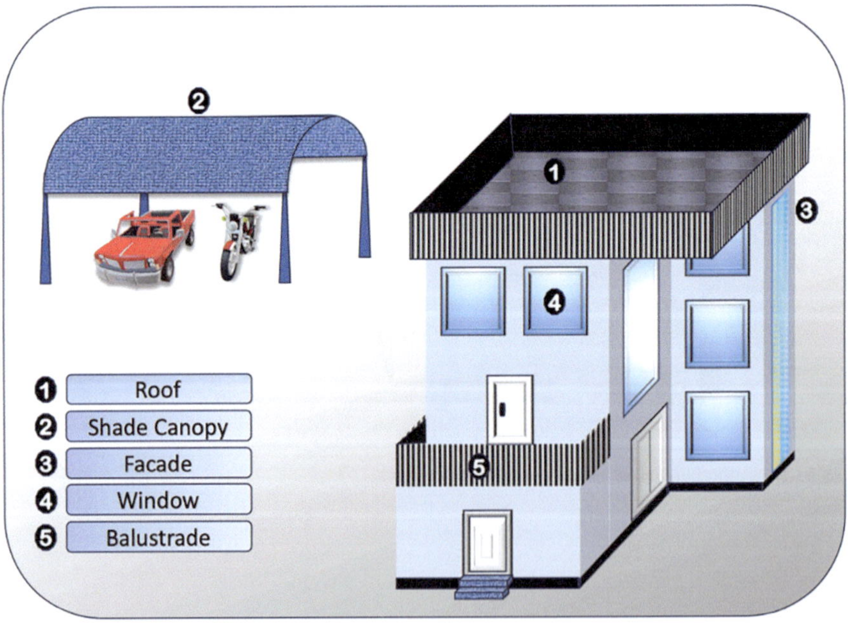

Fig. 7 Diagram illustrating the architectural zones for the integration of GO

building energy performance by balancing natural light utilization, thermal comfort, and energy generation, while reducing the need for mechanical heating, cooling, and lighting systems. This section highlights the key strategies based on GO applied to building roofs, windows, facades, awnings, and canopies.

4 GO for Roofing Elements

Integrating GO principles in rooftops presents significant opportunities for improving solar energy management, thermal efficiency, and daylight utilization. One of the pivotal applications of GO in rooftop design is solar concentrators, which play a crucial role in harnessing solar energy efficiently. These systems employ surfaces that utilize TIR to focus sunlight onto photovoltaic (PV) cells, significantly enhancing energy capture while reducing the material footprint typically associated with conventional solar arrays.

In building rooftops, the PT geometry can be scaled down into smaller parabolic concentrators or reflectors that concentrate sunlight onto smaller, highly efficient PV cells. Reflectors positioned at certain angles on a building's rooftop can enhance the energy capture of PV systems. The Ring Array Concentrator (RAC) developed by SVV Technology Innovations uses this principle, where reflectors emulate an aspherical point focus lens, increasing the intensity of light reaching solar cells and thus improving overall efficiency [7]. This technology is especially useful for urban buildings with limited rooftop space, allowing for increased power output with fewer panels.

Such innovations highlight the potential to harness solar energy more effectively, aligning with global sustainability goals. The application of TIR extends beyond solar energy to daylighting systems, such as light pipes and tubular daylighting devices. These technologies channel natural light from rooftop collectors deep into building interiors, effectively reducing the need for artificial lighting and enhancing overall energy efficiency in both commercial and residential spaces. Further, daylighting systems utilizing FLs and OFs have emerged as an ideal solution for solar energy concentration due to their compact design, affordability, scalability, and ability to enhance energy density efficiently. While the lens directs light to a concentrated area, it produces a diffused image, transmitting only visible light into the fibers. The system incorporates a hybrid solar tracking mechanism that combines a photosensitive sensor with a GPS-based astronomical algorithm, achieving precise solar alignment through a dual-axis rotational platform, as illustrated in Fig. 8a [8]. In the heliostat daylighting system, the primary mirror, crafted from circular aluminized glass, serves to track the sun, collect sunlight, and deliver high-performance reflectance and durability metrics, as depicted in Fig. 8b and c. A secondary mirror, derived from the reflective surface of the primary mirror, focuses the collected sunlight and channels it into the building's interior. The building's central area features a cavity structure designed to guide the redirected sunlight downward, while an additional mirror positioned at the room's forward-facing opening further directs the sunlight to the lower parts of the structure. This system includes a rotatable supporting mechanism that allows the mirror to adjust vertically and horizontally, tailoring the sunlight's angle and intensity based on the room's layout [8].

In addition to all these optical systems, LPs are designed to channel sunlight from a building's roof to its interior using the principle of TIR (Fig. 9). Typically made from highly reflective materials like aluminum or silver, these pipes achieve reflectivity rates of 85–94%. Comprising a light collector, the pipe itself, and a luminaire, the system effectively captures sunlight at the roof and distributes it throughout the building, with mechanisms in place to regulate brightness and prevent overheating or glare [8].

Another critical advancement in rooftop design involves refractive technologies, particularly prismatic skylights (Fig. 10a). These skylights disperse sunlight uniformly across indoor spaces, mitigating glare and ensuring a consistent and comfortable daylighting experience [9]. Additionally, the beta ray spherical solar collector is an innovative photovoltaic system that utilizes a transparent glass sphere to concentrate sunlight onto high-efficiency solar cells (Fig. 10b). By employing advanced optics, the design enhances light capture and power generation, even under diffuse light conditions. Its unique geometry allows for multi-prismatic skylight. Beta ray type of solar collector directional energy collection, making it highly effective in urban or limited-space installations [10].

Mu et al. (2019) introduced a refraction-based approach to enhance clean water production in traditional solar stills by incorporating an FL into a single-basin, single-slope design. The lens concentrated sunlight onto a focal point at the basin's bottom, leading to two notable outcomes: a "high hourly productivity window" (HHPW) characterized by consistent and elevated water production rates, and the occurrence of nucleate boiling, which significantly improved heat transfer during the water's phase change process. This innovative design which is integrated in building rooftops achieved a remarkable 467% increase in water yield (L/m^2/day) and an 84.7% boost in daily efficiency compared to systems without the FL. Experimental findings further revealed that optimal performance was influenced by basin water depth and forced air cooling, which enhanced distillate output and system efficiency [11].

Liang et al. created and examined an annular Fresnel solar concentrator equipped with a central circular Fresnel lens (AFSCFL) (Fig. 11). This setup includes an annular Fresnel solar concentrator (AFSC), a circular Fresnel lens positioned in the center, and a receiver. Simulation studies showed that the AFSCFL receiver outperformed the AFSC in solar energy generation, average radiation levels, and peak radiation levels, with enhancements of 9.49%, 10.70%, and 9.49%, respectively. Experimental evaluations further showed that the AFSCFL could achieve up to 20% higher thermal efficiency than the AFSC under conditions of low solar radiation. This adaptability not only maximizes energy production but also enhances the overall sustainability of building designs. Further complementing these technologies are daylight redirection systems, which employ rooftop mirrors to direct sunlight into shaded or darker areas of a building. This capability ensures even illumination in deep or irregularly shaped spaces that might otherwise lack adequate natural light [12].

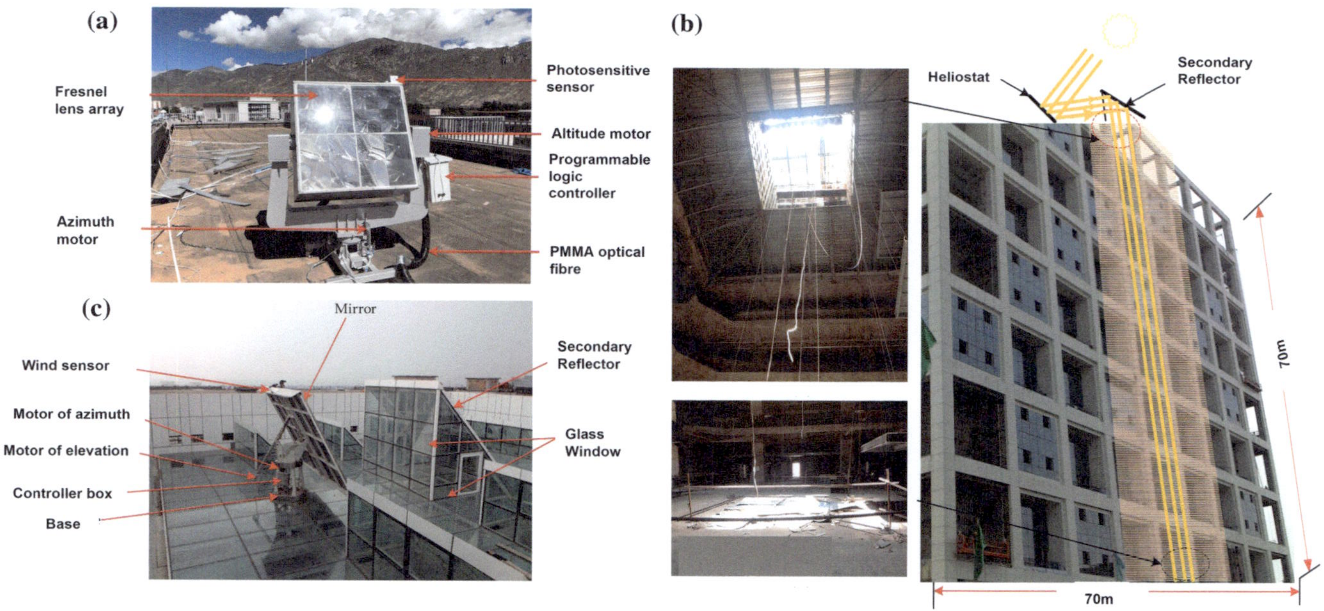

Fig. 8 Photographs of **a** the configuration of a Fresnel lens-based daylighting system utilizing OF, **b** roof-mounted heliostat daylighting system, and **c** heliostat daylighting system implemented in a large structure: light trajectory from the eighth floor (top left), illumination on the ground level (bottom left), overall design layout (right). *Source* Reused with permission from [8] Copyright©2023 MDPI, Basel, Switzerland

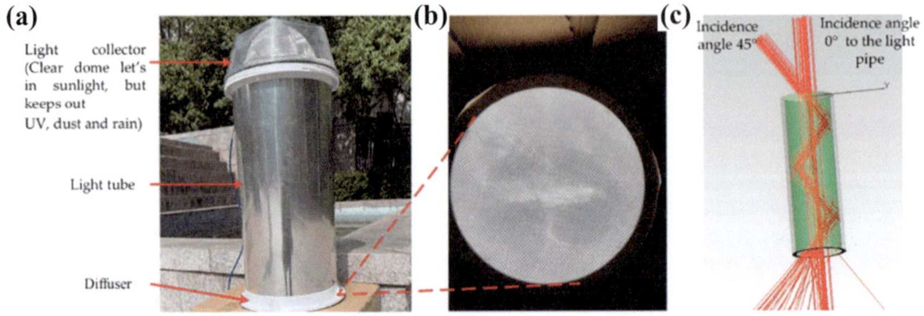

Fig. 9 Overview of a typical daylighting system using a light pipe: **a** image of the light pipe, **b** diffusing element, **c** explanation of the operational principle. *Source* Reused with permission from [8] Copyright©2023 MDPI, Basel, Switzerland

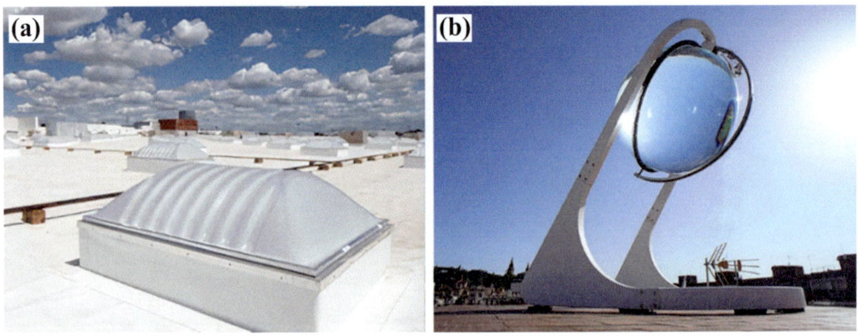

Fig. 10 **a** Prismatic skylight and **b** beta ray solar collector

Fig. 11 Picture of AFSCFL. *Source* Reused with permission from [12] Copyright © 2020 Elsevier Ltd

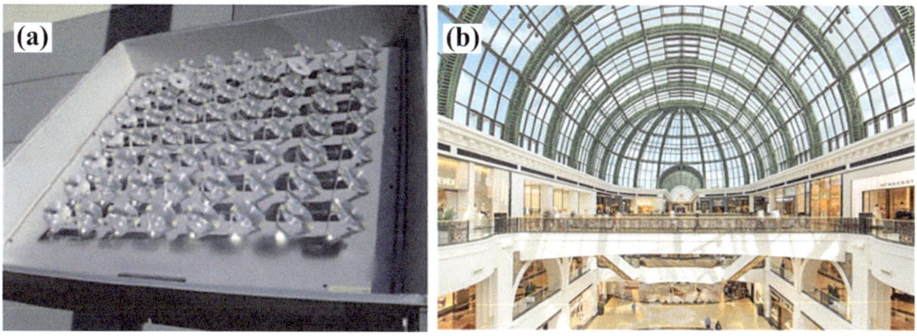

Fig. 12 Photographs of **a** Parans daylighting system at the rooftop and **b** the Mall of the Emirates in Dubai. *Source* Reused with permission from [15] © The Chartered Institution of Building Services Engineers 2011

The Parans system uses advanced technology to bring natural sunlight into interior spaces that lack direct access to daylight. Solar collectors with high-precision lenses are installed on roofs or walls to capture and focus sunlight efficiently (Fig. 12a). These lenses track the sun's movement throughout the day, ensuring consistent light collection. The captured sunlight is transmitted indoors via fiber optic cables, which preserve its brightness and color spectrum. Specially designed fixtures then distribute the light inside, creating a natural daylight experience. By harnessing sunlight directly, the system reduces energy consumption and enhances indoor environments with the health benefits of natural light [13]. The Mall of the Emirates in Dubai building architecture have dome shaped glass roof structure which facilitating the optimization of natural light entering the atrium of the mall (Fig. 12b) [14].

The advancements in GO are transforming rooftops into multifunctional systems that play a vital role in modern green building design and renewable energy strategies. Among the notable innovations, the Cassegrain optics-based concentrator successfully commercialized by SolFocus exemplifies the potential of advanced optical technologies for flat roof installations [16].

Furthermore, Morgan Solar Inc. has made significant strides with its Light-Guide Solar Optics (LSO) system, which optimizes solar collection and minimizes installation complexities. The introduction of the Sun Simba in 2011 marks a key milestone for Morgan Solar (Fig. 13b), providing a modular concentrating solar panel optimized for high heat, wind loads, and moisture, thus facilitating reduced maintenance costs [17]. Solar concentrators are integral to the application of solar thermal systems in buildings, enhancing their efficiency by focusing sunlight onto a smaller area to generate heat. This heat can be used directly for heating water, air, or other fluids, or it can drive thermodynamic cycles for electricity generation. The prototype of the hybrid CPV/T ANU-Chromasun Micro-Concentrator (MCT) was set up on the rooftop of the Australian National University in Australia (Fig. 13c) [18].

Fig. 13 Images of **a** Morgan solar CPV modules, **b** thermal hybrid concentrator installed roof of Bruce Hall, Australia. *Source* Reused with permission from [18]

4.1 GO for Building Facades and Windows

As sustainable architecture continues to gain momentum, the integration of GO into building facades and windows has emerged as a key strategy for enhancing both energy efficiency and indoor comfort. Prismatic glass, for instance, can bend light at precise angles, allowing it to penetrate deeper into the building's interior with the advantage of lowering glare and limiting heat gain from sunlight. This approach not only optimizes daylighting but also minimizes the need for artificial lighting, contributing to overall energy savings.

Incorporating more advanced optical technologies, such as FLs solar collection systems into building facades, has shown considerable promise in reducing energy consumption. For example, in studies conducted in the Netherlands, a FL system for heating achieved a 16% reduction in energy usage compared to traditional modulating boilers with the same heat pump. When compared to a boiler system connected directly to floor heating, energy consumption decreased by 43%, illustrating the potential for optical systems to significantly reduce heating demands [19]. Similarly, the development of low-concentration photovoltaic (PV) integrated prismatic segmented facades designed to capture direct normal irradiance (DNI) during peak sunlight hours has demonstrated substantial improvements in energy capture (Fig. 14a). Detailed simulations have shown that prismatic TIR systems, with segment head angles ranging between 17° and 25°, offer flexibility in energy capture, where smaller angles are optimal for midday radiation, and larger angles ensure consistent energy generation throughout the day. This adaptability in energy collection strategies makes these systems highly effective for buildings in regions with varying sunlight angles [20].

In addition to these innovations, technologies such as Photovoltaic Reflective Integrated Design for Efficiency (PRIDE) (Fig. 14b) have integrated optical concentrators with photovoltaic systems, enabling significant improvements in energy generation.

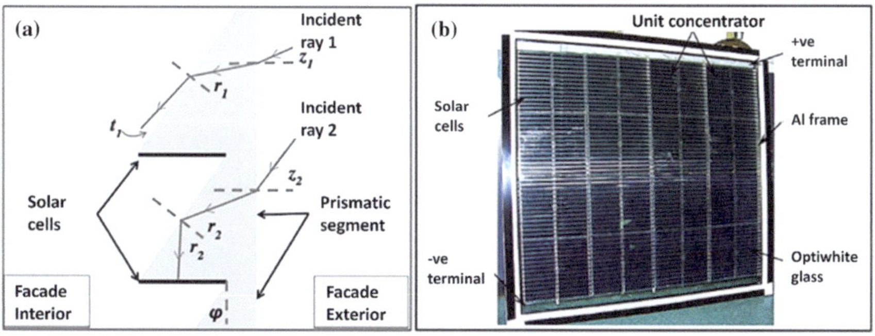

Fig. 14 **a** Schematic representation of optical trajectories of two light rays interacting with the proposed segments of the prismatic façade and **b** image of PRIDE solar collector. *Source* Reused with permission from [20] Copyright © 2016 Elsevier Ltd. and [6] Copyright © 2014 Elsevier Ltd

The PRIDE system, which uses single-crystal silicon solar cells and optical concentrators, outperforms traditional non-concentrating photovoltaic systems by concentrating light onto high-efficiency cells, thus improving overall energy production. Indoor tests using flash and continuous solar simulators have confirmed the superior performance of these integrated systems in building façades and windows applications. However, long-term studies have raised concerns about the durability of the dielectric materials used in the concentrators, suggesting that further development is needed to ensure their longevity and reliability [6, 21]. Furthermore, researchers have explored various concentrator geometries for concentrating photovoltaic (CPV) systems, with the Square Elliptical Hyperboloid (SEH) concentrator (Fig. 15c) showing particular promise for use in building facades. This design, which features an entry aperture of elliptical geometry and an exit aperture of square geometry joined by hyperbolic branches, concentrates light using TIR. While this design improves light concentration, its performance diminishes as the incident angle increases, resulting in a lower short-circuit current and fill factor, both of which are critical parameters for optimizing system efficiency [22].

Another exemplary case is the Institut du Monde Arabe in Paris (Fig. 15a), where the building's façade draws inspiration from the traditional mashrabiya design, incorporating optical lenses and reflective surfaces to regulate light and heat entering the building. The façade features elliptical apertures that adjust throughout the day, tracking the movement of the sun to control the intensity of light entering through the windows. These apertures precisely modulate sunlight, reducing glare while ensuring that the interior is bathed in natural light. This dynamic response to solar movement improves thermal comfort and contributes to energy savings by reducing reliance on artificial lighting and mechanical cooling systems [24]. Beyond energy efficiency, GO can also elevate the aesthetic qualities of buildings, as demonstrated by the Mercedes-Benz Museum in Stuttgart, Germany (Fig. 15b). The museum's double helix structure of optics facade not only manages solar radiation but also contributes to the building's distinctive architectural identity, blending aesthetic innovation with sustainable functionality. This dual-purpose approach underscores the potential of

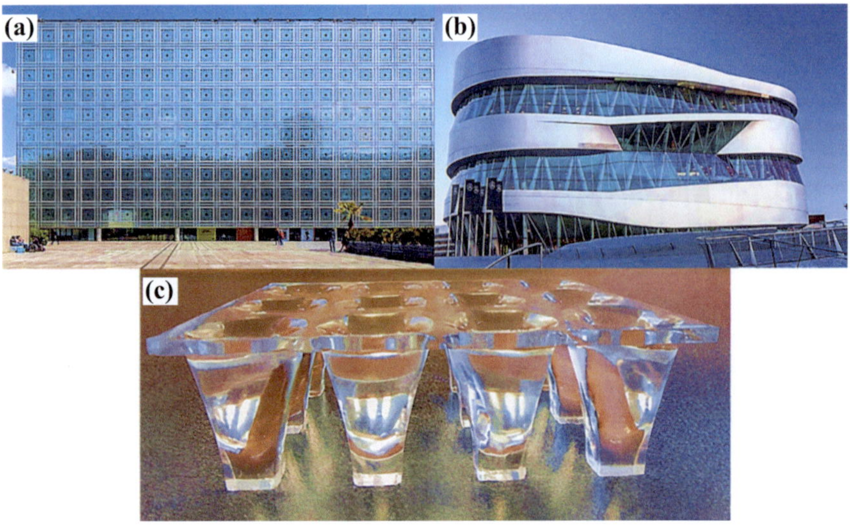

Fig. 15 Photographs of **a** Facade of Institut du Monde Arabe, **b** Mercedes-Benz museum, **c** an illustrative sample of the SEH concentrator array. *Source* Reused with permission from [26, 93, 94] Copyright © 2014 Elsevier B.V.

geometric optics to shape the future of architecture, where both beauty and energy efficiency coexist [25].

By incorporating lenses, mirrors, parabolas, prisms, and concave/convex reflectors, architects can design building components that not only control solar heat gain and glare but also enhance the amount of natural light entering the building. Notable examples like the Institut du Monde Arabe) illustrate how GO can be seamlessly integrated into modern architecture to optimize solar management, reduce energy consumption, and improve occupant comfort.

4.2 GO for Building Canopies and Balustrades

Geometric optics-based solar management solutions can be applied in several ways, from the reflective coatings used in awnings and canopies to the use of patterned surfaces in balustrades that manipulate light. Moreover, advanced techniques such as CSP systems and patterned optics can be incorporated to further optimize solar management and increase energy efficiency.

This approach not only contributes to reducing a building's carbon footprint but also aligns with the growing demand for environmentally responsible design solutions in architecture. As technology advances, the integration of dynamic, smart materials into building facades, shading devices, and sun-control elements—driven by the principles of GO—will further enhance the performance of these systems, making buildings more energy-efficient and better adapted to the challenges posed by climate change.

In balustrades or facades, TIR allows sunlight to be redirected along the surface of the optics, minimizing glare and enhancing indirect daylighting. For example, curved glass balustrades can employ TIR to ensure that sunlight reflects along the interior side of the glass, diffusing through the space rather than causing harsh direct lighting. Patterned optics involve using microstructures on transparent surfaces to control how light interacts with the surface. This can be particularly useful in architectural facades, canopies, or balustrades. For instance, glass panels with micro-patterns can be used to diffuse light more evenly across an interior, reducing direct glare while still allowing adequate daylight to enter. These patterned optics can also serve to improve the aesthetics of a building, as they can create intricate light patterns that change dynamically with the movement of the sun. One example of this innovative approach is seen in the glass canopies produced by Aparna Craft Exteriors Pvt. Ltd. which uses specially treated glass to reflect a portion of incoming solar radiation. This reduces heat gain and minimizes the need for mechanical cooling, while also refracting light to distribute it more evenly, thus enhancing the overall lighting quality inside the building [27, 95].

McDonald's unveiled its first Zero-Energy Building in the USA at Disney's theme park in Florida, featuring a PV glass canopy over the outdoor dining patio. This canopy not only provides essential shade, reducing heat gain and keeping the area cooler but also generates energy, supporting the building's sustainability goals. In Florida's intense sun, this innovative design creates a comfortable and eco-friendly space for visitors [28, 96].

5 Innovative Solar Light Management Opportunities in Farming Using GO

Agriculture plays a crucial role in the global economy, and the growing population demands innovative solutions to ensure sustainable food production. Smart farming technologies, such as precision agriculture (PA) and precision livestock farming, optimize resources while reducing waste. GOs like FL, parabolic concentrators, and OFs enhance solar light management, playing a vital role in advancing the technologies for efficient and sustainable agriculture. In terms of farming, solar light management can be deduced to making the spectral region (400–780 nm) encompassing Photosynthetically Active Radiation (PAR) available for plants and blocking the UV and IR regions. PAR plays a crucial role in crop growth, as it is directly utilized by plants for photosynthesis for photosynthesis. PAR intensity and quality directly influence photosynthetic rates, growth patterns, nutritional quality, stress resistance, and crop yield. Studies have shown that plants exposed to balanced PAR distribution exhibit enhanced growth compared to those under non-uniform or suboptimal light conditions [29]. Light quality and specific wavelengths play a pivotal role in determining the phytochemical composition and nutritional quality of crops [30]. The findings underscore the critical role of tailoring light spectra in controlled environments to balance growth, yield, and nutritional value, which can be done by spectral filters and light management technologies using GO, to maximize PAR while mitigating excess UV and NIR radiation that could be detrimental to plant health.

5.1 GO for Greenhouses

Greenhouses (GH) act as controlled environments to optimize crop growth by managing the solar radiation that enters their interior. Recent innovations in GO in GH have focused on enhancing PAR transmission and distribution, creating optimal lighting conditions for photosynthetic activity.

A theoretical study using computational fluid dynamics (CFD) to analyze heat transfer and fluid flow phenomena in a FL-Concentrated Photovoltaic Thermal (FL-CPVT) system integrated with a GH has provided valuable insights into how different GH configurations (closed or ventilated) affect the temperature and flow fields within the greenhouse [31]. Importantly, FL-CPVT systems can enhance temperature control and lighting conditions, which are critical for optimizing plant growth, particularly during warmer months and for sensitive crops such as pot plants. In systems like FL-CPVT, such simulation (CFD) tools can be further optimized to support sustainable agricultural practices while improving crop yields [32, 33]. An innovative application of FL in a GH system equipped with a glass raster featuring linear FL on the south-facing roof was proposed [34] where the design concentrated direct solar radiation onto collectors while permitting diffuse light to enter. For the climatic conditions of Central Europe, the collectors absorbed approximately 12% of the total incoming global solar radiation. This energy was converted into heat at 30–50 °C, reducing heating energy consumption by about 50% compared to traditional systems. Furthermore, the system demonstrated successful parallel production of vegetables and fish, achieving efficient recycling of water and nutrients.

An energy-saving agricultural GH was proposed by adding a solar energy collector to improve energy utilization efficiency without occupying extra land or significantly affecting illumination inside the greenhouse [35]. The system, as shown in Fig. 16, uses a sliding-type cylindrical FL to concentrate sunlight onto a receiver, which collects and stores heat for use in the greenhouse. The system shows a receiving efficiency of over 80% when the inclination incident angle of sunlight ranges from $-38°$ to $38°$, allowing it to work for about 5 h in heat collection. The light environment inside the GH is maintained, with ground illumination reaching up to 8.38×10^5 Lux and not dropping below 4.22×10^5 Lux. The system was seen to be preferable in economic terms also by providing energy, equivalent to approximately $6153.9 per year. Overall, integrating FL in GHs, particularly in combination with simple concentrating thermal systems, offers a cost-effective and efficient solution. This approach enables precise light and temperature control, supports energy-efficient operations, and contributes to sustainable food production systems.

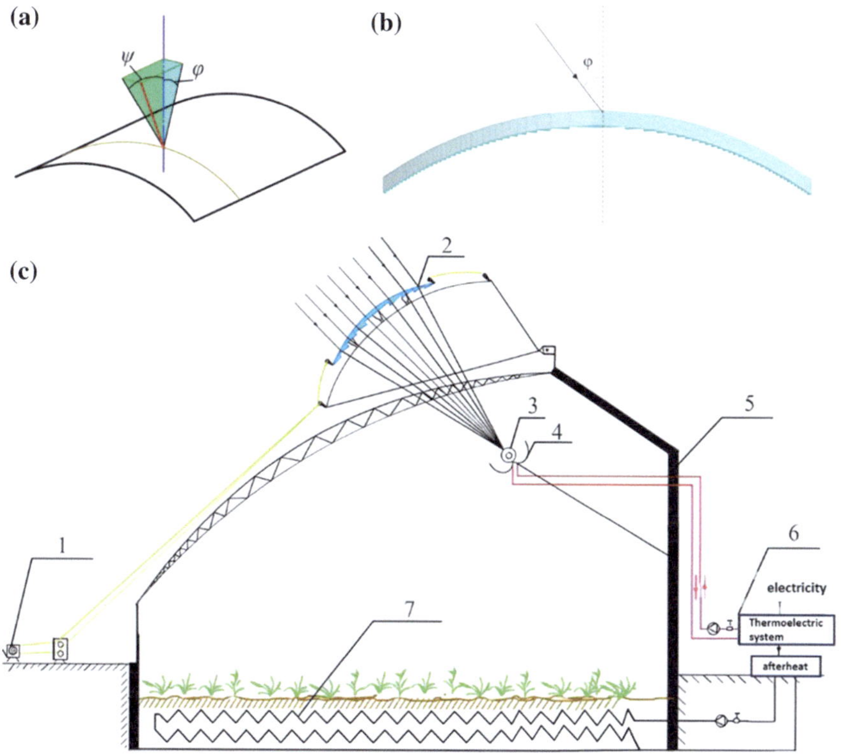

Fig. 16 Schematic representations of **a** decomposition of the incident angle in cylindrical FL, **b** cylindrical FL used, and **c** greenhouse with a sliding-type cylindrical lens as a solar collector for light and thermal requirements (1) sliding control, (2) FL, (3) vacuum tube, (4) secondary concentrator, (5) back wall, (6) equipment for power generation, (7) underground heat storage unit

High light diffusion in GHs has shown significant potential in enhancing plant growth and productivity by optimizing light distribution [36]. A study conducted at Wageningen University Research Greenhouse Horticulture in Bleiswijk in 2022, demonstrated that young tomato plants grown under high diffusion conditions (hortiscatter = 0.9) experienced improved biomass accumulation, increased leaf area, and higher light use efficiency compared to medium diffusion (hortiscatter = 0.65) and direct light (hortiscatter ~ 0) setups. These conditions also enhanced net photosynthesis, particularly in the middle crop layers, due to better light penetration and reduced shadowing. Furthermore, the light interception was notably higher in diffused light treatments, aligning with increased photosynthetic capacity. Advanced GH designs and coatings prioritize maximizing PAR transmission while mitigating non-beneficial UV and NIR components. Diffusive light technologies, for example, improve light uniformity within the GH, promoting consistent growth across plants. These approaches ensure efficient energy use and better crop outcomes.

5.2 GO for AV

Innovative and sustainable technologies like AV (AV) are gaining attraction since they double the farmers' income through energy generation along with farming. It maximizes the land use and contributes to achieving the renewable energy goals and food production. In AV, GO plays a pivotal role in shade mitigation and promoting uniform solar light distribution under photovoltaic panels by employing FL, waveguides, and OFs to enhance light management and crop productivity.

Traditional AV systems often produce uneven irradiance patterns, negatively affecting crop yield and quality. To address this, an innovative AV system employing a grooved glass plate is shown in Fig. 17 was developed to provide uniform lighting to crops [37]. A grooved glass plate is designed to scatter incident sunlight into three distinct paths via refraction. This scattering ensures the even illumination of the shadowed areas beneath the PV panels, creating a uniform light and providing optimal growth conditions for crops and photosynthesis. The grooved glass plate forms an integral part of the even-lighting AV system (EAS) structure, which also includes metal brackets and regular PV panels. The setup replaces one-third of the system's light-receiving area with the grooved glass plate. It maintains a PV array density comparable to that of conventional photovoltaic power stations, thus balancing energy generation without compromising crop light requirements. The parameters of the grooved glass plate were determined through precise calculations using Snell's law and GO principles. Advanced optical simulation tools, such as SolidWorks and Zemax, were used to validate these calculations, and the simulations confirmed the effectiveness of the grooved glass plate in achieving uniform light scattering, with minimal differences in the amount of light reaching previously shadowed areas. The study evaluated the system's effectiveness through two experiments: a small-scale trial involving lettuce (*Lactuca sativa L.*) (Fig. 18), while the large-scale study included diverse crops such as broccoli, shallot, garlic sprouts, rape, and Jerusalem artichoke. Results demonstrated that the EAS system significantly improved the light environment, increasing photosynthetic photon flux density (PPFD) by 47.38% compared to conventional systems. Crops under EAS exhibited faster growth, higher yield, and enhanced quality, including elevated soluble sugar and reduced nitrate content. Economic analysis showed that EAS increased farmers' income by 5.14 times on average and achieved a Land Equivalent Ratio (LER) of 1.64, underscoring efficient land utilization.

Several remarkable experimental models of AV systems integrated with FL for light and thermal management were brought up. A concentrated photovoltaic thermal (CPVT) greenhouse AV system incorporating FL with one-axis tracking showed that the removal of direct solar radiation reduced summer cooling demands by a factor of four [38] Concentrated radiation was directed onto a CPVT, achieving a thermal yield of 56%, an electrical yield of 11%, and a combined efficiency of 67%. The thermal energy collected was stored for winter heating, while electricity generated was either supplied to the grid or used for additional cooling or desalination systems. An innovative design for integrating concentrated photovoltaics (CPVs) into AV

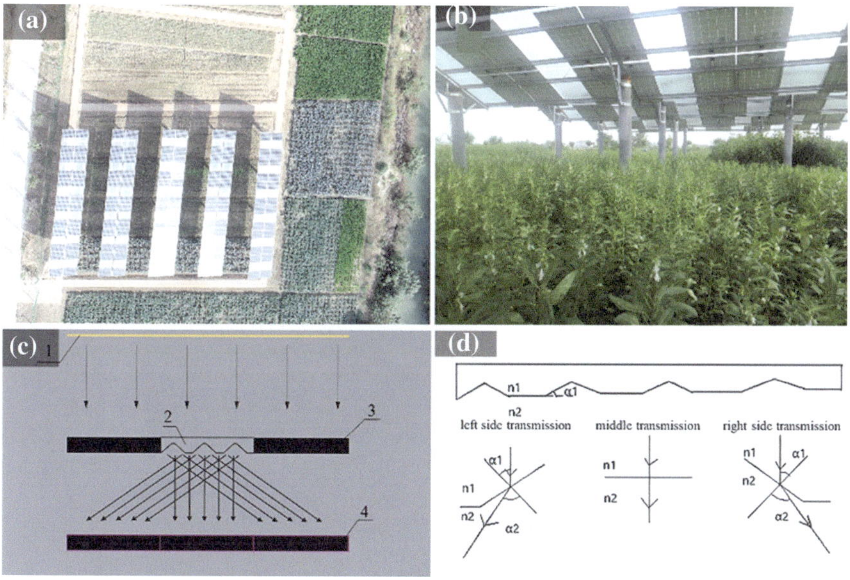

Fig. 17 **a**, **b** Photographs of the large-scale experimental facilities, **c** the light path to achieve even light distribution, and **d** the structure of the grooved glass plate

Fig. 18 Photograph showing the morphological difference of the lettuce grown at **a** open environment, **b** conventional AV system, **c** EAS system, **d** EAS with supplementary LED lamps for an extra 2 h of light daily

systems using curved linear FL is given in Fig. 19a [39]. These lenses were engineered to concentrate sunlight onto CPV modules, allowing diffused light to pass through for crop cultivation in greenhouse settings. Three distinct lens geometries—cylindrical, parabolic, and elliptical were tested for performance. The cylindrical FL emerged as the most efficient in concentrating light, outperforming the other designs.

The use of spectrum-splitting systems like dichroic mirrors along various GO like FL, OFs, parabolic concentrators, PTs optimize the availability of PAR while

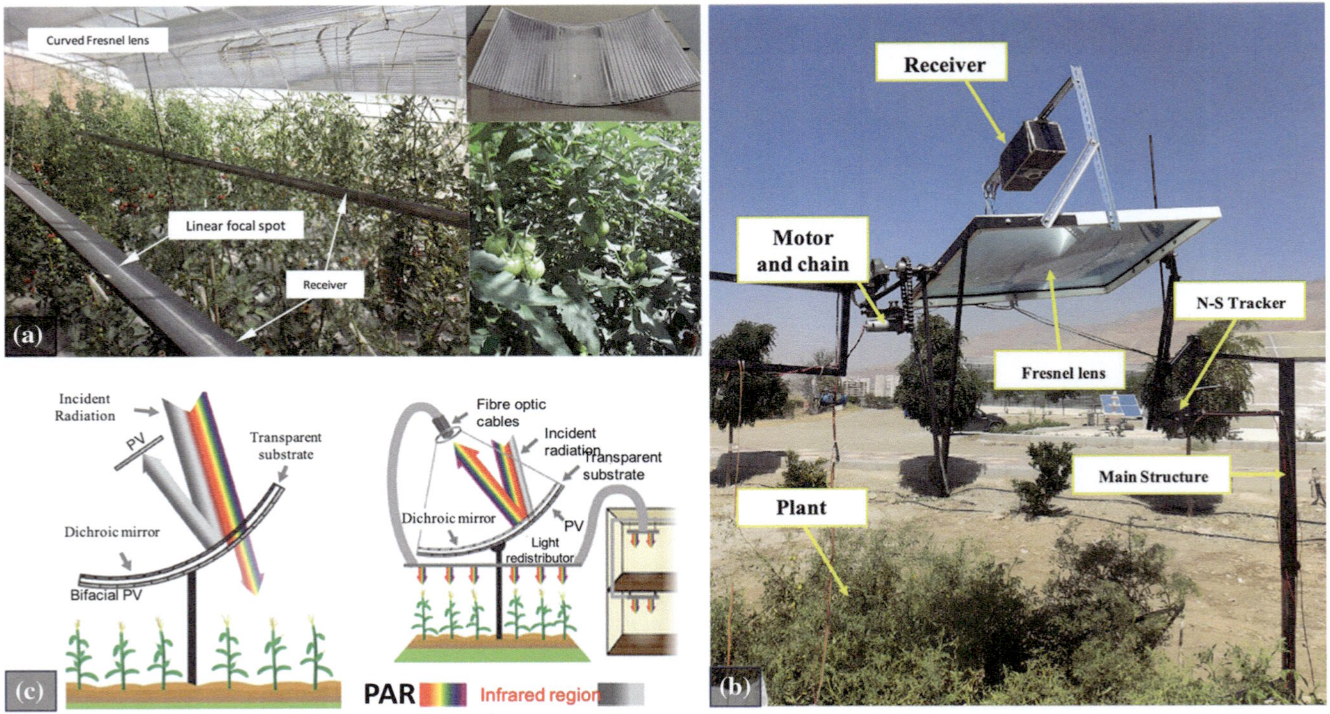

Fig. 19 Photographs of **a** the curved FL placed in the solar greenhouse (top left), curved FL (top right), and plant canopy and light distribution under the FL (bottom right), **b** photograph of the concentrating AV system, and **c** schematic representation of AV systems using parabolic troughs and OFs. *Source* Reused with permission from [39] Copyright © 2020 Elsevier Ltd [41] Copyright © 2024 Elsevier B.V., [42] Copyright © 2019 Springer Nature Limited

harnessing infrared (IR) light for energy production, thus enhancing both agricultural and photovoltaic outputs. For instance, PT concentrators coated with short-pass dichroic films [40] allow PAR to transmit to crops below while concentrating the IR spectrum onto solar cells. One such system (shown in Fig. 19b) uses a highly transparent concentrator AV setup with a FL coated with a dichroic thin film and a receiver made of silicon solar cells [41]. The lens focuses sunlight and splits it into two spectral ranges: PAR for plants and the remaining spectrum for electricity generation. The receiver reduces the intensity of focused radiation by 3.28 times, with an optical efficiency of 83.1% and electricity generation up to 18.1 W. Cherry tomatoes grown under the system were larger, heavier and had a deeper red hue, whereas mint plants had larger, greener, and longer leaves. The system provided cooler soil temperatures (3.66–9.53 °C cooler) and retained 5.7–8.4% more moisture, reducing water consumption. This dual functional system showed noteworthy techno-economic aspects as well. Fiber optic bundles attached to long-pass dichroic mirrors present another promising configuration (Fig. 19c). The OFs collect the PAR, reflected by the dichroic mirror, and redistribute it to the plants below the PV, supporting both horizontal and vertical farming. During off-season periods, these mirrors can be replaced with photovoltaic cells, ensuring year-round light utilization. Additionally, CPCs equipped with long-pass dichroic films can redistribute PAR to the plants via central waveguides while transmitting IR light to underlying PV cells or thermal systems. Heliostat-based systems with short-pass dichroic mirrors further expand the possibilities [42]. These systems reflect long-wavelength light to concentrated photovoltaic or solar thermal setups, enabling efficient energy conversion. Such modular systems demonstrate the flexibility and integration of geometrical optics like FL, compound parabolic concentrators, OFs, waveguides, and heliostats with dichroic optical designs for sustainable AV. A novel approach of Spectral-Splitting Concentrator Agrivoltaic (SCAPV) system [43] utilizes a concave mirror-shaped filter to manage sunlight selectively. The PAR required for photosynthesis are filtered and allowed to pass to the plants, whereas the unused wavelengths are taken for energy generation. The system was found to have several advantages such as improving the crop yields, enhancing soil-moisture retention, low installation costs, and improved Land Equivalent Ratio.

Luminescent solar concentrators (LSCs) offer a novel approach to AV systems by utilizing luminescent properties for solar light management. The PAR is selectively transmitted to crops, while the non-PAR wavelengths are redirected laterally through TIR for energy generation [44, 45]. Attempts have been made experimentally with the support of Monte Carlo ray tracing simulations [45] to integrate a CdSe/CdS nanocomposite layer over a Si-based bilayer LSC to improve optical efficiency. The system effectively transmits PAR wavelengths essential for plant growth, making it suitable for agrivoltaic applications. However, the relatively low levels of diffuse transmission and reflection of the bilayer LSC can limit the overall light uniformity available for crops. The ability of a spectral converting material like calcium strontium sulfide phosphor doped with divalent europium ($Ca_{0\cdot 4}Sr_{0\cdot 6}S:Eu^{2+}$) was investigated by applying it on a backlight foil of the PV panels. Spinach was grown in the AV systems, and a 25% increase in photosynthetic activity was observed, as compared to the external conditions [46].

Soliculture has developed a luminescent enhanced solar panel that integrates bifacial PV cells with a luminescent back sheet, specifically focusing on AV applications [47]. This design improves electricity generation and converts green light to red light. The enhanced red light is ensured for the crops through a combination of total internal reflection and luminescence. Though a model of even light-sharing greenhouse AV was developed with transparent luminescent PV panels, the efficiency was a major concern [48]. These findings suggest that such luminescent solar concentrators can be integrated into solar technologies to simultaneously improve solar energy capture and optimize light conditions for agricultural applications. This dual benefit highlights the potential for innovative materials in sustainable AV systems.

5.3 GO for Precision Farming

GO plays a crucial role in precision farming by optimizing light management for plant growth, improving resource efficiency, and enhancing crop health monitoring. Optical elements such as lenses, mirrors, and OFs, based on GO, can help control light paths and intensity, contributing to the optimization of crop growth environments, irrigation, and even pest control. This integration of optical principles supports more sustainable and productive agricultural practices, demonstrating the intersection of precision farming and GO for advancing modern agriculture. GO contributes significantly to disease detection and stress assessment in plants through spectral imaging and laser-based diagnostics. By analyzing how light interacts with plant tissues, these optical techniques allow for early detection of diseases and stress factors, enabling timely intervention to preserve plant health.

A key development in precision farming is the use of optical sensors, which monitor vital plant parameters such as soil moisture, plant health, and nutrient level. Soil-moisture sensors like frequency domain reflectometry (FDR) and time domain reflectometry (TDR) use electrodes to detect water content in soil, while optical sensors, such as photodiodes, evaluate key soil properties, including moisture, clay content, and organic matter. Advanced techniques like ground-penetrating radar (GPR) and gamma-ray spectrometry (GRS) offer further insights into soil nutrients and texture, enhancing decision-making for crop management [49, 50]. Innovations in seeding technologies, such as electric seeders with OF detection, allow for precise seed placement and real-time monitoring of seed quality, improving operational efficiency [51]. Innovations in GO have become indispensable in modern agriculture, offering practical solutions for improving crop growth and resource management. Optical elements like FL, OFs, and CPCs are being applied in AV systems and greenhouses to manage light more effectively. By integrating these optical solutions, more sustainable farming practices are being implemented, addressing critical challenges in food and energy production.

6 Advanced Solar Light Management Solutions for Industry Using GO

In solar light management applications for industrial purposes, GOs play a critical role in harnessing, concentrating, and directing sunlight to optimize energy generation and utilization. In this section, we cover four major specified solar energy utilization technologies, namely photoreactors (PR), photovoltaics (PV), photobioreactors (PBR), and solar trackers (ST), which have a variety of extended industrial applications. Industries currently exploring or utilizing solar photoreactors include chemical manufacturing, wastewater treatment plants, pharmaceutical companies, textile industries, energy production (hydrogen generation), and environmental remediation businesses. Any sector looking to generate electricity directly from sunlight using PV technology can be considered part of this industry; energy utilities, solar panel manufacturers, construction companies, commercial building owners, industrial facilities, telecommunications companies, agriculture, and transportation are come under this. Industries primarily working with solar tracking systems include: solar panel manufacturers, solar project developers, engineering, procurement, and construction (EPC) companies, electrical contractors, utility companies, agricultural operations (for solar-powered irrigation), and research institutions focused on renewable energy technologies. By precisely managing how sunlight is concentrated and absorbed, geometric optics enables scalable, efficient, and environmentally friendly solutions for industrial solar energy applications.

6.1 Solar Photoreactors

Solar photoreactors are specialized systems designed to utilize sunlight as an energy source to drive photochemical or photocatalytic processes. The process begins with the capture and concentration of sunlight using optical components such as parabolic reflectors, FLs, or mirrors. This concentrated light is directed onto a reaction chamber, typically made of transparent materials like quartz, allowing optimal light penetration. Within the reactor, the absorbed light provides the activation energy required for chemical reactions. These can be photochemical reactions, where the reactant molecules directly absorb light and undergo transformations, or photocatalytic reactions, where catalysts like titanium dioxide (TiO_2) absorb light to generate reactive species (e.g., hydroxyl radicals or hydrogen ions) that drive the reaction. Reactants, often circulated to ensure uniform exposure, interact with light and catalysts to form desired products. Solar PRs utilize GO to distribute or focus sunlight to maximize the exposure of reactants to sunlight, which is absorbed to initiate or enhance specific reactions.

PTs are the most popular and extensively adopted geometric configuration for PR with commonly used TiO_2 photocatalysts in suspended or deposited form in various matrices inside a tubular glass or acrylic located at the focal line [53, 54]. Considering various photocatalytic or photochemical applications, innovative ideas have been engineered to maximize the reactor performance. One latest report introduces a sunlight-adjustable parabolic trough reactor (SA-PTR) (Fig. 20a and b) that shows excellent weather adaptability for diurnal photocatalytic wastewater treatment [54]. The flexible nature of the PT allows it to change its size based on solar conditions. This geometric adaptability ensures that the amount of concentrated solar energy can be optimized throughout the day and across different seasons. For instance, the PT can be expanded during low sunlight conditions (e.g., winter or cloudy days) and rolled up during peak sunlight hours, maintaining optimal light conditions for photocatalysis (Fig. 20c).

The design variation of PT, known as CPC, allows to capture of the sunlight from a wider range of angles. Using a CPC in conjunction with photocatalytic monoliths represents a significant advancement. The CPC's parabolic geometry allows for effective concentration and collection of solar radiation, which is crucial for maximizing the efficiency of photocatalytic reactions (Fig. 21a) [55, 56]. A continuous flow photo reactor (CFPR) with the integration of CPC as shown in Fig. 21b improved the solar-to-photocatalyst interaction by providing increased solar radiation to the reactor [57]. While the CPC photoreactor offers several benefits, its design faces notable commercial challenges. These include limited treatment capacity relative to the solar footprint of CPC arrays and difficulties in scaling up to larger diameters due to pipe fragility. Moreover, larger end connections add to costs and are more prone to leakage.

Recent advancements in concentrator geometry, such as the innovative "winged" CPC (Fig. 22a) [58] and nature-inspired V-shaped reflectors (Fig. 22b), aim to improve operational efficiency. Research by Kant et al. shows that a V-shaped concentrated photoreactor can deliver four times the light exposure of a traditional non-concentrated tubular photoreactor, significantly boosting performance in applicable systems [59]. The high efficiency and compact design of PD technology for concentrating solar energy have led to the development of specialized PD-based photoreactors. Photocatalytic decomposition of a commercially developed detergent was executed via a PD-based batch-mode reactor under real-time solar illumination [60]. The system included a 1.0 m PD reflector with a 70-sun concentration ratio and a flask-shaped reaction vessel at its focal point (Fig. 23a). Concentrated sunlight enabled highly efficient photocatalytic degradation of refractory detergent, outperforming conventional biodegradation. A recently implemented pilot-scale photoreactor based on PD showed a solar-to-hydrogen conversion efficiency greater than 20% (Fig. 23b). The PD concentrates sunlight 1000-fold over 38.5 m^2 onto the reactor with PV and electrolysis components to generate electron–hole pairs and split water into oxygen and hydrogen at an optimal flow rate [61].

In cost-sensitive applications or where moderate light concentration is sufficient for the photochemical process, LFR-based photoreactors showcase novel technology by effectively controlling solar energy. An example of green wastewater treatment via an LFR-based photoreactor adjusted sunlight efficiently by managing six mirrors according to the sun path and weather characteristics. It ensures optimal light and temperature on sunny days and maximizes them under overcast conditions [62].

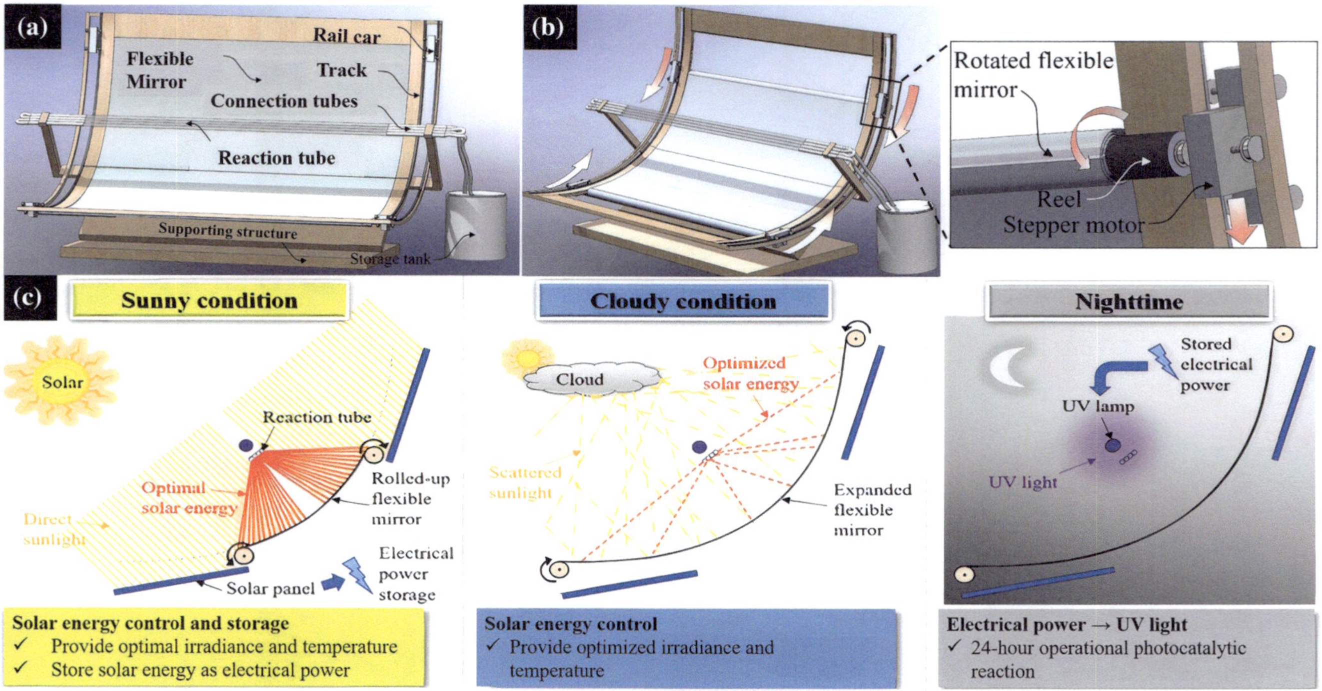

Fig. 20 **a** SA-PTR structural design, **b** control system for device operation, and **c** schematic representation of geometric adaptability of SA-PTR for achieving optimized solar energy. *Source* Reused with permission from [52] Copyright © 2024 Elsevier B.V.

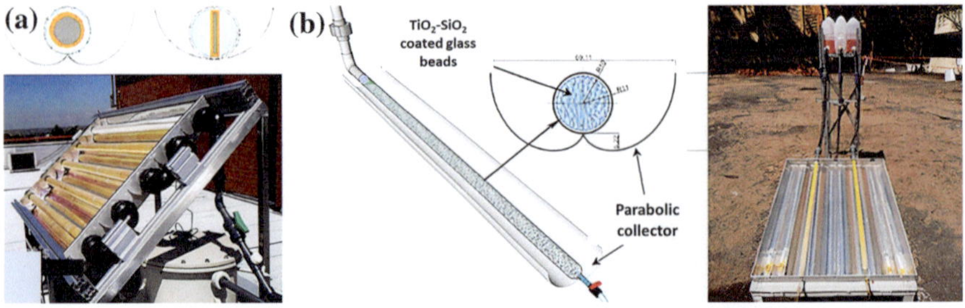

Fig. 21 **a** CPC-based photoreactor for solar water disinfection with two varying configurations (coaxial and fin type) shown above and **b** schematic representation (left) and actual photograph of CPC-based CFPR (right) for dye degradation. *Source* Reused with permission from [55] Copyright © 2023 Elsevier B.V [57] Copyright © 2021 Elsevier Ltd.

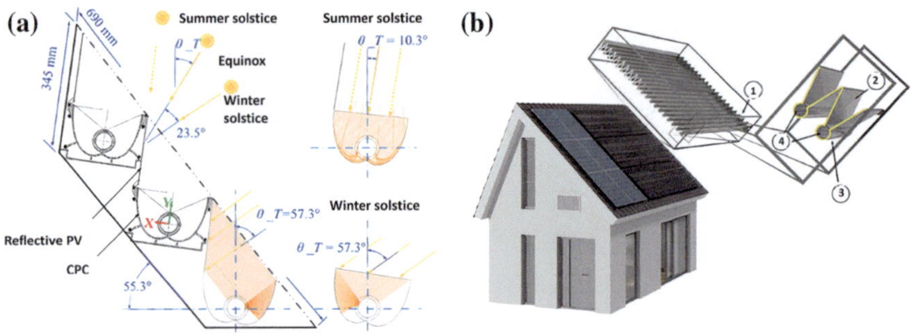

Fig. 22 Schematic representation of the geometric design of **a** winged CPC and **b** V-shaped reflector. *Source* Reused with permission from [58] Copyright © 2023 RSC publications [59]. Copyright © 2023 Elsevier Inc

Fig. 23 Photographs of reactor setup based on PD for **a** commercial detergent degradation, and **b** electrolytic hydrogen production. *Source* Reused with permission from [60, 97] Copyright © 2004 Elsevier Ltd [61] Copyright © Institute of Physics and IOP Publishing Limited 2024

With the potential use of efficiently concentrating sunlight onto a specific focal area while being lightweight and compact design, FLs are the best option for applications requiring high-intensity solar radiation in a small form factor. An experiment conducted to produce hydrogen via photocatalytic conversion of aqueous urea and urine with a Fresnel lens-based reactor underscored ultrahigh-level intensity of solar concentration up to 15,000 W/m^2 [63]. FL-based photoreactors are proven to be highly efficient in intensifying solar illumination and in turn, boosting photocatalytic efficiency and are widely employed in clean fuel production, hydrogen generation, and wastewater treatment (Fig. 24) [64, 71].

Coming to GO works by TIR mode, one potential approach is adopting an OF-integrated reactor design (Fig. 25). The OF coated with photocatalysts provides maximized exposure to sunlight and uniform illumination across the catalyst surface which are crucial for activating the photocatalytic process (Fig. 26a). Photocatalytic hydrogen production using ethanol–water mixtures in an OF-based reactor coated with Au/TiO_2, photoreduction of CO_2 for synthesizing methane and ethylene in OF-based reactor-coated TiO_2–SiO_2 mixed oxide are certain experimental studies demonstrating the advantages of fiber optics such as enhanced light efficiency and uniform light illumination with optimal mass transfer making it suitable for scaling up the photocatalytic process for practical applications [4, 65]. Additionally, the approach of OF coated with photocatalysts minimizes the issue of light scattering

Fig. 24 Photograph of FL-based reactor setup for real-time monitoring of solar photocatalytic process. *Source* Reused with permission from [64] Copyright© 2019 Elsevier B.V.

loss in the powdered photocatalysts resolving the challenge of industrial-level scaling up of the photocatalysis process. Augmenting the number of OF coated with Cu/TiO_2 outperformed its powdered counterpart showing an improved hydrogen production rate with more than one order in comparison with that resulted from photocatalysis using powdered Cu/TiO_2 [66].

The development of micro-structured OF canes (MOFCs) (Fig. 26b) showed boosted photocatalytic performance for hydrogen generation. MOFCs consist of

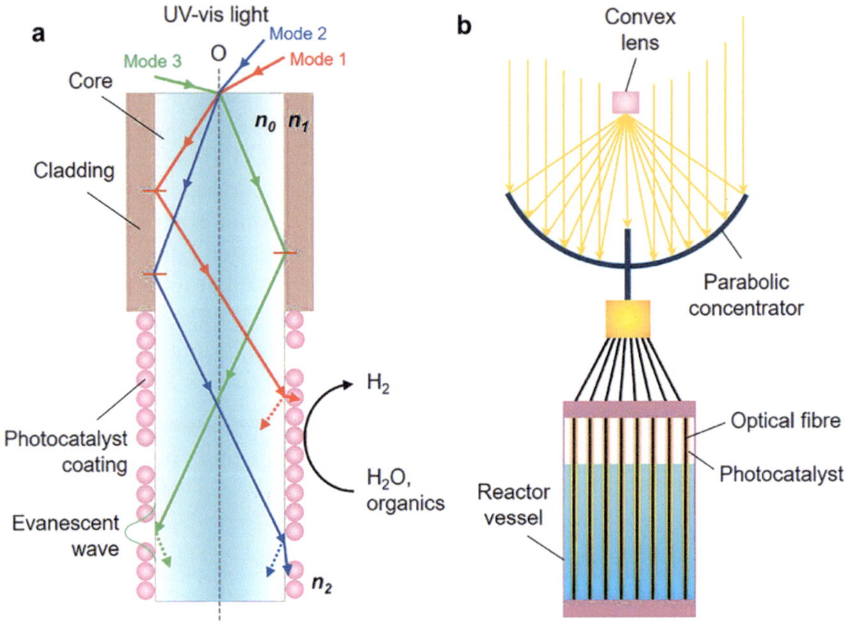

Fig. 25 **a** Schematic illustration of the working mechanism of OF coated with photocatalysts and **b** OF-based solar PR. *Source* Reused with permission from [67] Copyright © 1999–2024 John Wiley & Sons, Inc

Fig. 26 Images showing the photographs of **a** photocatalyst coated and **b** the hexagonally channeled OF-based solar photoreactor. *Source* Reused with permission from [70] Copyright © 2007 Elsevier B.V. and (Potter et al. 2020) Copyright © 2020 American Chemical Society

multiple capillaries running along the length of the fiber, which increases the internal surface area available for photocatalyst deposition [68]. The unique design of the channeled OF reactor (COFR) enhances light utilization, maintains high efficiency across varying conditions, and improves the interaction between reactants and the photocatalyst, making it a promising solution for air quality control applications. The hexagonally channeled structure of the COFR ensured complete confinement of light (all refracted light) within the OF coated with TiO_2, longer effective light propagation lengths, high surface area to volume ratio, and increased uniform light distribution, enhancing the gas-phase photodegradation of ethylene regardless of the incident angle of light [69].

A simulation study via Monte Carlo ray tracing (MCRT) methods to investigate the behavior of light interaction with the PT concentrator and the OF monolith reactor optimized the design for better light capture and utilization. The PT focuses sunlight on the OF increasing the irradiative flux, which is essential for driving the photocatalytic reduction of CO_2 effectively (Fig. 27). By keeping the maximum incident angle below 15 degrees, the system minimizes energy loss due to attenuation, thereby improving the efficiency of light transmission into the reactor [71].

So far, two types of OFs have been utilized as substrates for the photocatalyst coating: quartz and polymer OF. Quartz fiber is highly stable and offers excellent

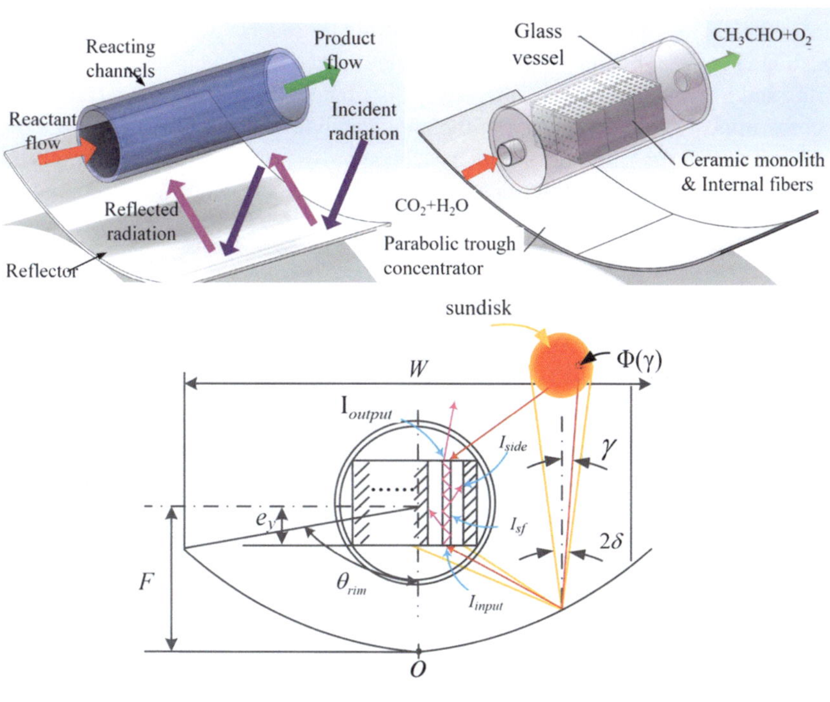

Fig. 27 Schematics of PT integrated OF monolith solar photoreactor. *Source* Reused with permission from [71] Copyright ©2021 MDPI (Basel, Switzerland)

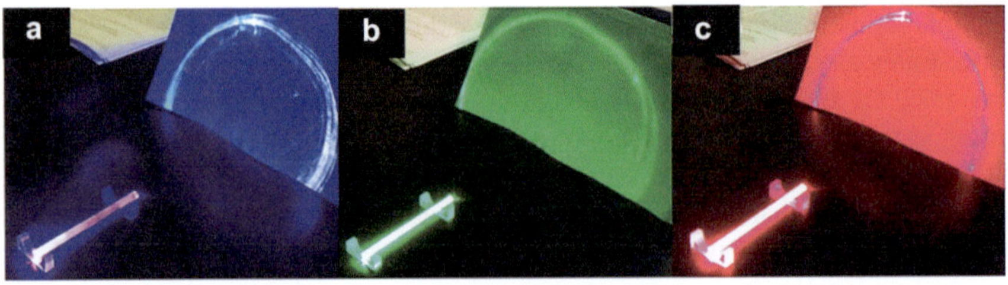

Fig. 28 Red (732 nm) and green (535 nm) laser light at an incident angle of 60° propagate and internally reflect through the coated optical rod and transmit through to the other end (**a**–**c**). Under blue laser illumination, the intensity diminishes along the length of the rod, indicating that the coating partially absorbs blue light as it undergoes internal reflection within the quartz rod. *Source* Reused with permission from [73]. Copyright © 2021 Springer Nature Limited

light transparency but is associated with high costs. In contrast, polymer OF, typically made from polymethylmethacrylate (PMMA) resin, is more affordable but lacks stability against photocatalytic oxidation due to its organic composition [72].

Waveguide photoreactor emerged as an effective alternative light management strategy using GO that optimizes photon utilization, leading to remarkable improvements in photocatalytic efficiency. It is reported that adopting an optical system involving a cylindrical quartz rod waveguide marked improvement in photon-to-yield efficiencies and CO production rates during the reverse water gas shift (RWGS) reaction (Fig. 28a, b, and c). The waveguide utilizes the principles of optical scattering and internal reflection to effectively distribute light across the catalyst surface, thereby improving the efficiency of the photocatalytic process [73].

Acrylic (PMMA) rods and sheets have been proven to be the potential waveguide for photoreactors for water purification and photo-oxidation of organic compounds [72, 74].

6.2 *Solar Photobioreactor*

A solar photobioreactor (PBR) is a system specifically designed to utilize sunlight for growing photosynthetic microorganisms, like algae or cyanobacteria, under controlled conditions. These microorganisms use solar energy to convert carbon dioxide (CO_2) and water into valuable products through photosynthesis. Solar PBRs are widely used for applications such as biofuel production, carbon capture, hydrogen generation, and generating high-value compounds like pharmaceuticals, food additives, or pigments.

The reactor typically consists of a transparent vessel made of materials like glass or acrylic to allow maximum sunlight penetration. Key components include a gas exchange system to supply CO_2 and remove excess oxygen, a nutrient delivery system, and mixing mechanisms to ensure even light distribution and nutrient availability. In certain designs, solar photobioreactors can also produce hydrogen gas through processes like biophotolysis, where specific algae or cyanobacteria split water molecules into oxygen and hydrogen under specific conditions.

A tubular serpentine PBR utilizing concentrated solar radiation has been installed at the Academic and University Centre of Nové Hardy in the Czech Republic. This system employs linear Fresnel lenses as solar concentrators, positioned within a climate-controlled greenhouse to focus direct sunlight onto the cultivation tubes. The cultivation unit, comprising a 24-m-long loop of six parallel horizontal glass tubes connected by U-bends, is mounted on a movable frame that automatically adjusts to optimize direct solar irradiance. The PBR has been utilized to investigate microalgae acclimation to extremely high solar irradiance, achieving light levels up to 3.5 times greater than those naturally available (Fig. 29a, c) [75].

Building on the initial design, an improved two-stage PBR was developed and tested, featuring two vertical and two inclined units with a total capacity of 450 L. It enables cultivation under much higher light intensities than naturally available. It is advantageous in high-latitude regions and provides ultrahigh irradiance to induce stress for accumulating high-value products like pigments and superoxide dismutase (Fig. 29b) [77]. The use of CPCs in photobioreactors (PBRs) could be advantageous for cultivating microalgae in high-latitude regions with lower solar irradiance (Fig. 30) [78]. For example, the comparative study of photosynthesis of *Chlorella vulgaris* cultures in a glass tube coupled with CPC and a plain tube without CPC demonstrated that the CPC system enhanced the culture's irradiance level by 351% and increased the average light intensity on the cells by 462% compared to the plain tube system [79].

Coupling with a light-dependent resistor (LDR), solar energy excited optical fiber-based (SEEOF) photobioreactor offered a feasible approach that can be adopted to achieve intensified light source efficiency. The combined light illumination from SEEOF and LDR demonstrated effective improvement in hydrogen production via phototropic reaction with the added benefit of reduced electrical consumption [80]. Creating additional rough surfaces on the optical fiber provides an alternative option to enhance the additional sites for the attachment of photosynthetic bacteria (PSB) in a photobioreactor. This increased surface area allows for more cells to adhere

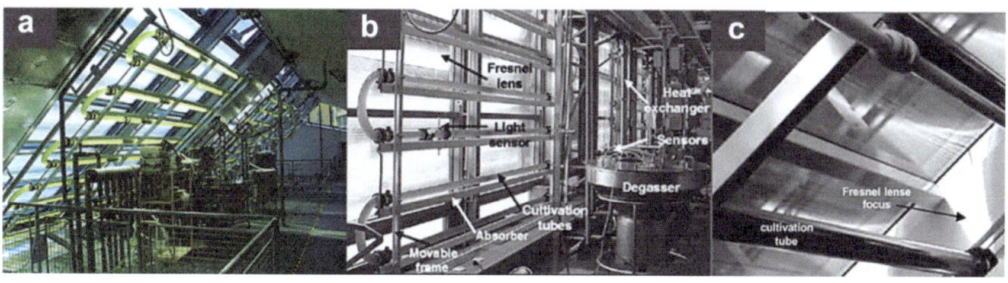

Fig. 29 Photographs of FL-based **a** solar PBR installed at the Academic and University Centre of Nové Hardy in the Czech Republic, **b** and **c** two-stage PBR developed. *Source* Reused with permission from [76]Copyright © 2015, Springer International Publishing Switzerland, [75] Copyright © 2003, Kluwer Academic Publishers, and [77] Copyright © 2008, Springer Science Business Media B.V.

Fig. 30 Photographs of the CPC-based PBR with inoculated culture (left) and the reactor front view (right). *Source* Reused with permission from [78] Copyright © MDPI (Basel, Switzerland)

to the fibers, facilitating the formation of a dense biofilm that leads to higher cell concentrations, which is beneficial for hydrogen production [81].

7 Photovoltaic Systems

As global industries increasingly prioritize renewable energy solutions, GO has emerged as a transformative technology in advancing photovoltaic (PV) systems. By focusing on optimizing light harvesting, minimizing material usage, and enhancing energy yield, GO is driving the scalability and efficiency of industrial-scale PV installations. These principles are not only revolutionizing PV technology but also catalyzing innovations that facilitate the integration of renewable energy into energy-intensive sectors, such as manufacturing, transportation, and urban infrastructure.

A prominent example of large-scale renewable energy innovation is the. Noor Solar Complex, located near Ouarzazate, Morocco, recognized as one of the largest concentrated solar power (CSP) plants globally, Noor represents a transformative step in Morocco's commitment to renewable energy. The project exemplifies the nation's strategic pivot from dependency on fossil fuels toward achieving energy independence and sustainability. With a planned capacity exceeding 580 MW, the Noor integrates advanced CSP technologies to meet the growing energy demands of the industrial sector. Beyond its impressive scale, Noor demonstrates the potential of combining renewable energy solutions with cutting-edge optical and thermal storage systems to ensure reliable and efficient energy output. By delivering a steady and sustainable energy supply, this project sets a benchmark for the global adoption of renewable energy technologies, reinforcing solar power's pivotal role in future energy systems [82, 98].

The integration of GO into solar technologies extends beyond CSP systems, significantly influencing the development of concentrated photovoltaic (CPV) systems. CPV systems utilize multijunction solar cells mounted on two-axis solar trackers and incorporate sophisticated optical geometries, such as refractive concentrators like FLs or reflective assemblies like Cassegrain systems. These designs achieve high solar concentration ratios, significantly boosting energy density, making them ideal for powering industrial processes, including hydrogen production via water electrolysis. CPV systems are further enhanced by the incorporation of maximum power point tracking (MPPT) devices and DC/DC converters, which ensure efficient energy conversion and stable power output, even in standalone industrial applications. These advancements underscore the versatility of GO in transforming renewable energy technologies to meet industrial-scale demand [83].

Notable advancements in CPV systems have been achieved by scientists at the National Renewable Energy Laboratory (NREL). By employing geometric concentrators, such as FLs, parabolic mirrors, and CPCs, NREL has demonstrated the ability to focus sunlight with exceptional precision, achieving the high-power densities required for hydrogen production through electrolysis (Fig. 31a). Additionally, NREL has explored hybrid CPV systems that integrate advanced thermal management strategies. These systems capture and repurpose excess heat for industrial applications, maximizing overall energy yield. This integration of thermal and PV technologies improves efficiency and expands the practical applications of CPV systems in industries seeking to adopt renewable energy solutions [84].

Further exemplifying the potential of GO, Palomino-Resendiz et al. developed an innovative active solar distiller that combines an FL concentrator with a PV energy collector (Fig. 31b). In this system, the PV component powers an electrical water preheating process, while a robotic mechanism with three degrees of freedom ensures precise sun tracking and optimal light concentration. The design, refined through concurrent engineering, involved comprehensive analyses of optical, thermal, structural, and dynamic factors to enhance performance. By efficiently utilizing solar radiation, this system significantly accelerates water distillation compared to passive methods, offering a self-sustaining, energy-independent solution for water purification [85].

Reflective surfaces play a vital role in PV systems by redirecting and concentrating light, forming multi-element waveguide-based photonic light concentrators

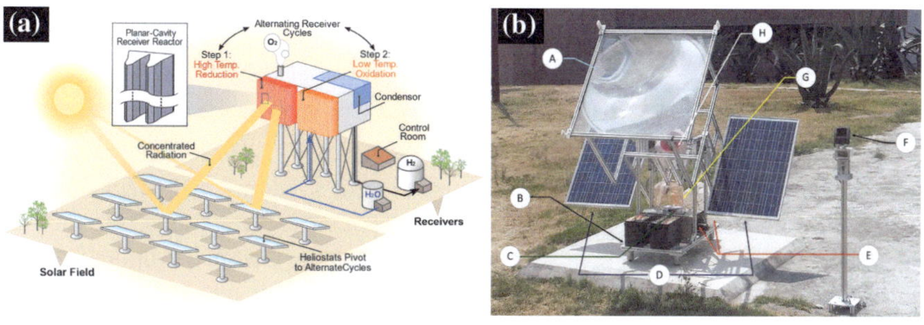

Fig. 31 **a** Schematic diagram of CPV-driven electrolytic hydrogen production proposed by NREL and **b** photograph of FL-based solar distiller. *Source* Reused with permission from [84] Copyright © 2022 Elsevier Ltd., and [85] Copyright © 2018 Elsevier Ltd.

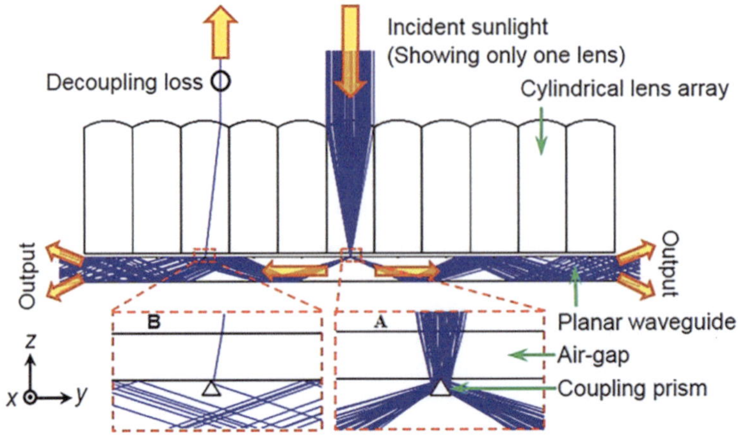

Fig. 32 Schematic representation of light propagation cylindrical lens-based planar light concentrator design. *Source* Reused with permission from [87] Copyright © 2014 Optical Society of America

with optically distant components. For instance, Kuo et al. introduced a two-stage planar solar concentrator featuring transparent waveguide sheets combined with arc segment structures. In the first stage, these arc segments act as TIR collectors, concentrating light onto the underlying waveguide, where it propagates with a skewed trajectory. The second stage incorporates movable light-guiding channels to collect and direct the concentrated light to a solar cell, where a CPC enhances energy capture [86].

Similarly, Peng Xie and colleagues explored a cylindrical lens-based planar light concentrator design (Fig. 32), evaluated using the Zemax ray tracing tool. This system features a light-guiding layer embedded with an air prism as a coupling element, separated from a lens array by an air gap. These innovative designs demonstrate advancements in light manipulation, achieving high concentration efficiency, wide angular tolerance, and improved energy output for PV applications [87].

8 Solar Tracking Systems

Solar trackers are devices that use sensors, control systems, and mechanical or motorized components to follow the sun's movement and maximize the amount of sunlight received by solar panels or solar concentrators. Geometric optics informs the development of solar tracking systems, ensuring panels and reflectors remain aligned with the sun to maximize exposure. A solar tracker based on GO essentially "tracks" the sun's position based on calculated geometric calculations; this allows for optimal sunlight capture and increased energy generation compared to stationary panels. These trackers often employ multiple mirrors or lenses arranged in a specific geometric configuration, allowing them to focus the sunlight onto a concentrated area. By calculating the sun's position based on time and location, the system adjusts the angles of the mirrors or lenses to always direct the sunlight toward the target point. Adoption of solar tracking can lead to achieving high solar concentration levels, leading to greater energy output, and precise control over the light path due to the geometric calculations involved.

A solar tracker utilizing motorized rotating mirrors is often employed to accurately direct light along the spectrometer axis, thereby compensating for the apparent rotation of the Sun. While it is feasible to calculate the Sun's azimuth and altitude for a specific time and location with high precision, various sources of angular offsets can manifest in practice when adjusting the positioning of the mirrors [88]. The altitude-azimuth design adopted in solar tracking has been installed at Lauder (45° S), New Zealand, and Arrival Heights (78° S), Antarctica proven to be of good reliability even in polar conditions. The custom coaxial power transformer of the system facilitates 360° azimuth rotation, thereby ensuring reliability and uninterrupted measurements. Additionally, the incorporation of four photodiodes enhances tracking accuracy by rectifying error offsets, thus maintaining a consistent pointing precision of 0.02°(Fig. 33) [89].

Beam-steering lens arrays offer a novel approach to solar tracking through millimeter-scale relative motion between paired lens arrays, which may reduce the mechanical complexity of traditional trackers. A research group has developed a design that directs sunlight with a divergence half-angle of less than 2° and achieves an average yearly efficiency of 73.4%. A simplified version attained 75.4% efficiency with a divergence half-angle under 3.5°(Fig. 34) [90]. These results highlight the potential of beam-steering lens arrays to enhance the cost-effectiveness of solar tracking systems in stationary solar concentrators.

An innovative semi-passive solar tracking concentrator incorporates two distinct arrays of polymethyl methacrylate (PMMA) prisms designed to track solar movement by refracting sunlight through the rotation of these prisms (Fig. 35a). The

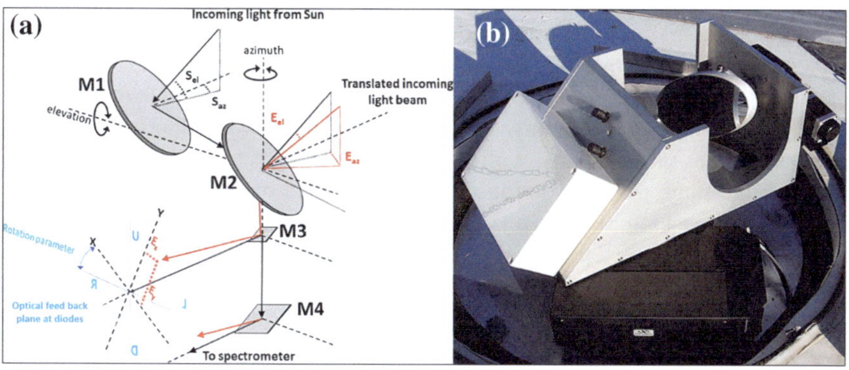

Fig. 33 **a** Scheme of the geometric configuration of the Lauder solar tracker optics and **a** photograph of altitude-azimuth design-based solar tracker

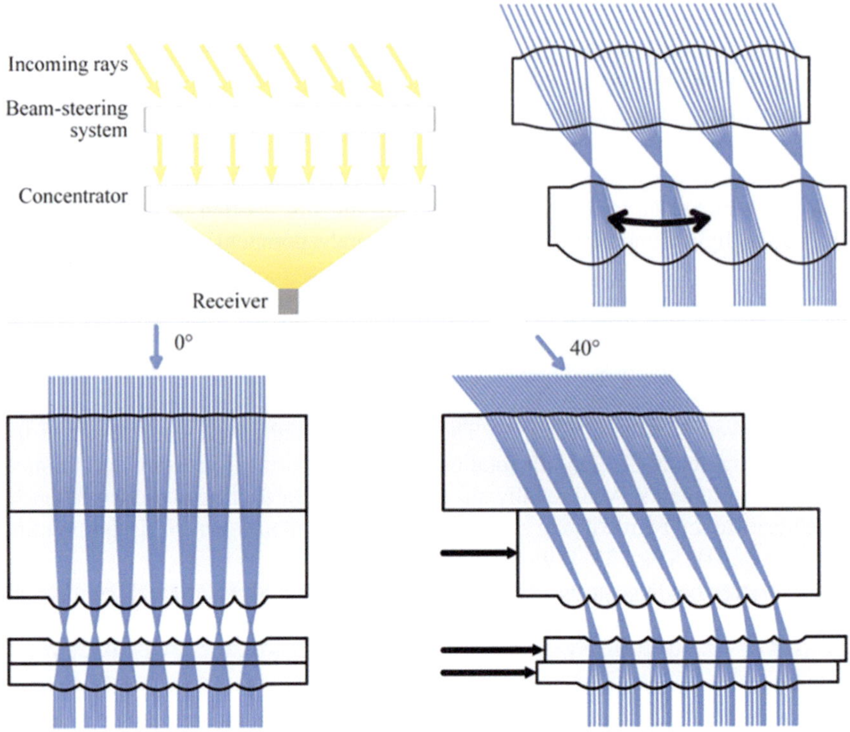

Fig. 34 Scheme of working mechanism of the beam-steering system coupled with conventional concentrator optics (top left) and the sunlight tracking using the beam-steering system. *Source* Reused with permission from [90]. Copyright © 2020 Optical Society of America

first array focuses on eliminating one directional component of solar radiation by rotating the prisms at specified angles. The second array addresses an additional directional component, adjusting its orientation to achieve a perpendicular alignment with the array. This configuration yields downward vertical radiation, facilitating the implementation of a stationary Fresnel lens to concentrate solar energy for diverse applications. The design has been thoroughly validated using simulation software, thereby confirming the feasibility of the concept (Fig. 35b). Although the system still incorporates moving parts, it offers several advantages over traditional tracking and concentrating systems: (i) reduced installation and maintenance requirements; (ii) simplified tracking mechanisms, as the only moving components are lightweight prisms that can be rotated using a single linear actuator per array, unlike conventional systems requiring motion of heavier structures; (iii) the Fresnel lens, serving as the concentrating element, remains stationary; (iv) as a result, the receiver does not need repositioning; and (v) decreased wind loads, as the gaps between prisms allow air to pass through freely [91].

Using prisms for large panels is impractical due to their weight and thickness. Instead, a Rotational Prismatic Director (RPD) employs two rotating Fresnel prisms,

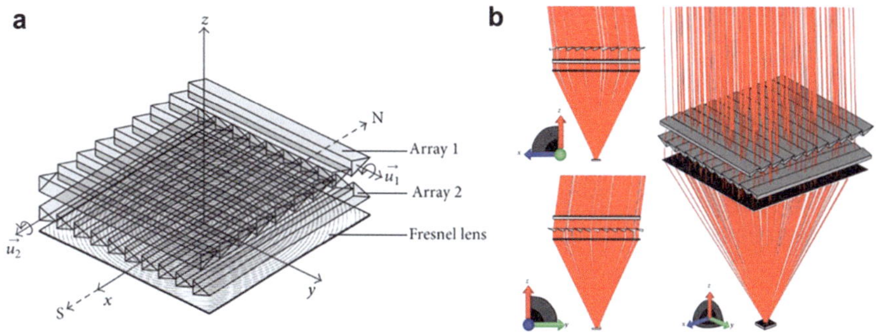

Fig. 35 **a** Geometric configuration of two prism arrays and a Fresnel lens and **b** simulation results by TracePro software 3.0

or "benders." The circular symmetry of the concentrating lens allows it to rotate with the second bender, combining their functions into a single optical component called an Inverse Off-Axis (IOA). This integration reduces the optical layers, minimizing light loss from reflections [92].

9 Conclusion

Geometric optics is at the forefront of technological advancements in solar light management, presenting transformative solutions across wide applications in building energy and architecture, industry, and agriculture sectors. By utilizing fundamental optical principles such as specular reflection, refraction, and TIR, GO facilitates the development of energy-efficient light management systems that seamlessly integrate sustainability, functionality, and aesthetics. In the building energy and architecture sector, the application of GO is revolutionizing the concept of energy efficiency and aesthetics. GO integration into building elements such as rooftops, facades, canopies, and balustrades can enable effective solar energy capture, optimized daylight distribution, and solar heat gain control with the prospectus of building aesthetics. In agriculture, the influence of GO has recently become profound, as it fundamentally supports resource management and sustainability strategies. Advanced optical systems ensure uniform light distribution and regulate light intensity, creating optimal photosynthesis and crop growth conditions. This approach enhances agricultural yield and minimizes adverse environmental impacts, promoting a greener, more resource-efficient framework within the farming sector. In the industrial context, GO drives innovation in renewable energy technologies, notably in solar photoreactors/bioreactors, CPV systems, and solar tracking technologies. The ongoing evolution of optical technologies signals the potential for even deeper incorporation of GO into sustainable practices. By regulating energy performance in buildings, adopting renewable energy systems in industrial settings, and enhancing energy and resource efficiency in agricultural production, GO stands as an effective solution to some of the most pressing global challenges. As these research and innovations mature, GO is poised to become a cornerstone of sustainable development, facilitating the creation of resilient future generations.

References

1. A. Aksnes, H.J.D. Johnsen, J. Torgersen, High-performance stationary solar tracking through multi-objective optimization of beam-steering lens arrays. Opt. Express **28**(14), 20503–20522 (2020). https://doi.org/10.1364/OE.396477
2. G. Alers, Identification and significance of problem, in *Luminescent Enhancement for Combined Solar and Commercial Agriculture*, (Final Report, Soliculture, Scotts Valley, CA, 2020)
3. A.H. Aziznezhad, S. Gorjian, H. Mokhtarzadeh, Design, development and experimental evaluation of a concentrator agrivoltaic system with integrated spectrally splitting Fresnel lens. Results Eng. **24** (2024). https://doi.org/10.1016/J.RINENG.2024.103119
4. H. Baig, N. Sarmah, D. Chemisana et al., Enhancing performance of a linear dielectric based concentrating photovoltaic system using a reflective film along the edge. Energy **73**, 177–191 (2014). https://doi.org/10.1016/J.ENERGY.2014.06.008
5. H. Baig, N. Sellami, T.K. Mallick, Performance modeling and testing of a building integrated concentrating photovoltaic (BICPV) system. Sol. Energy Mater. Sol. Cells **134**, 29–44 (2015). https://doi.org/10.1016/J.SOLMAT.2014.11.019
6. M. Burhan, M.W Shahzad, K.C Ng, Concentrated photovoltaic (cpv): hydrogen design methodology and optimization. Adv. Hydrog. Gener. Technol. (2018). https://doi.org/10.5772/INTECHOPEN.78055
7. R. Campbell, M. Machado, Low cost CPV = embedded CPV with internal tracker, in Conference Record of the IEEE Photovoltaic Specialists Conference, pp. 3003–3007. https://doi.org/10.1109/PVSC.2010.5614802
8. D. Chemisana, C. Lamnatou, Y. Tripanagnostopoulos, The effect of Fresnel lens—solar absorber systems in greenhouses. Acta Hortic **952**, 425–432 (2012). https://doi.org/10.17660/ACTAHORTIC.2012.952.53
9. C.Y. Chen, G.D. Saratale, C.M. Lee et al., Phototrophic hydrogen production in photobioreactors coupled with solar-energy-excited optical fibers. Int. J. Hydrogen Energy **33**, 6886–6895 (2008). https://doi.org/10.1016/J.IJHYDENE.2008.09.014
10. C. Corrado, S.W. Leow, M. Osborn, et al., Power generation study of luminescent solar concentrator greenhouse. J. Renew. Sustain. Energy **8** (2016). https://doi.org/10.1063/1.4958735
11. P. Denholm, M. Hand, M. Jackson, S. Ong, Land use requirements of modern wind power plants in the United States (2009). https://doi.org/10.2172/964608
12. F. Denny, J. Scott, G.D. Peng, R. Amal, Channelled optical fibre photoreactor for improved air quality control. Chem. Eng. Sci. **65**, 882–889 (2010). https://doi.org/10.1016/J.CES.2009.09.038
13. A.H. El-Ladan, M.H. Ruslan, H.Y. Chan et al., Comparative evaluation of reflective and refractive optical concentration systems in tropical climate. Res. J. Appl. Sci. Eng. Technol. **8**, 1319–1325 (2014). https://doi.org/10.19026/RJASET.8.1102
14. X. Fang, Z. Gao, H. Lu, Z. Zhang, Boosting CO_2 photoreduction activity by large Fresnel lens concentrated solar light. Catal. Commun. **125**, 48–51 (2019). https://doi.org/10.1016/J.CATCOM.2019.03.022
15. M.A. Fendrich, A. Quaranta, M. Orlandi, et al., Solar concentration for wastewaters remediation: a review of materials and technologies. Appl. Sci. **9**, 118 (2018). https://doi.org/10.3390/APP9010118

16. H. García, C. Ramírez, N. León, Innovative solar tracking concept by rotating prism array. Int. J. Photoenergy (2014). https://doi.org/10.1155/2014/807159
17. D. Gunawan, J. Zhang, Q. Li et al., Materials Advances in Photocatalytic Solar Hydrogen Production: Integrating Systems and Economics for a Sustainable Future. Adv. Mater. **36**, 2404618 (2024). https://doi.org/10.1002/ADMA.202404618
18. C.L. Guo, X. Zhu, Q. Liao et al., Enhancement of photo-hydrogen production in a biofilm photobioreactor using optical fiber with additional rough surface. Bioresour. Technol. **102**, 8507–8513 (2011). https://doi.org/10.1016/J.BIORTECH.2011.04.075
19. Q. He, H. Zheng, X. Ma, G. Wang, Optical analysis of a sliding-type cylindrical Fresnel lens concentrating collector for agricultural greenhouse. J. Daylighting **8**, 110–119 (2021). https://doi.org/10.15627/JD.2021.8
20. M. Herrando, G. Fantoni, A. Cubero et al., Numerical analysis of the fluid flow and heat transfer of a hybrid PV-thermal collector and performance assessment. Renew. Energy **209**, 122–132 (2023). https://doi.org/10.1016/J.RENENE.2023.03.125
21. N.T.T. Hoang, A.T.K. Tran, T.A. Le, D.D. Nguyen, Enhancing efficiency and photocatalytic activity of TiO_2-SiO_2 by combination of glycerol for MO degradation in continuous reactor under solar irradiation. J. Environ. Chem. Eng. **9**, 105789 (2021). https://doi.org/10.1016/J.JECE.2021.105789
22. X. Jin, Q. Li, K. Zhao, et al., Development and test of an electric precision seeder for small-size vegetable seeds. Int. J. Agric. Biol. Eng. **12**, 75–81 (2019). https://doi.org/10.25165/IJABE.V12I2.4618
23. V. Jirka, V. Kučeravý, M. Malý et al., Energy flow in a greenhouse equipped with glass raster lenses. Renew. Energy **16**, 660–664 (1999). https://doi.org/10.1016/S0960-1481(98)00247-X
24. M. Kacira, S. Sase, L. Okushima, Optimization of vent configuration by evaluating greenhouse and plant canopy ventilation rates under wind-induced ventilation. Trans. Am. Soc. Agric. Eng. **47**, 2059–2067 (2004). https://doi.org/10.13031/2013.17803
25. P. Kant, S. Liang, M. Rubin et al., Low-cost photoreactors for highly photon/energy-efficient solar-driven synthesis. Joule **7**, 1347–1362 (2023). https://doi.org/10.1016/J.JOULE.2023.05.006
26. J. Keil, Y. Liu, U. Kortshagen, V.E. Ferry, Bilayer luminescent solar concentrators with enhanced absorption and efficiency for agrivoltaic applications. ACS Appl. Energy Mater. **4**, 14102–14110 (2021). https://doi.org/10.1021/acsaem.1c02860
27. T. Lakhankar, H. Ghedira, M. Temimi, et al., Non-parametric methods for soil moisture retrieval from satellite remote sensing data. Remote Sensing **1**, 3–21 (2009). https://doi.org/10.3390/RS1010003
28. Q. Li, L. Jiang, G. Huang et al., A ternary system exploiting the full solar spectrum to generate renewable hydrogen from a waste biomass feedstock. Energy Environ. Sci. **16**, 3497–3513 (2023). https://doi.org/10.1039/D3EE00603D
29. Q. Li, C. Kubota, Effects of supplemental light quality on growth and phytochemicals of baby leaf lettuce. Environ. Exp. Bot. **67**, 59–64 (2009). https://doi.org/10.1016/J.ENVEXPBOT.2009.06.011
30. Y.-J. Li, T.-C. Teng, C.-H. Kuo (2020) Planar solar concentrator composed of stacked waveguides with arc-segment structures and movable receiving assemblies. Opt. Express **28**(23), 34362–34377. https://doi.org/10.1364/OE.405909
31. K. Liang, H. Zhang, H. Chen et al., Design and test of an annular Fresnel solar concentrator to obtain a high-concentration solar energy flux. Energy **214**, 118947 (2021). https://doi.org/10.1016/J.ENERGY.2020.118947
32. J.Y.Y. Loh, A. Mohan, A.G. Flood et al., Waveguide photoreactor enhances solar fuels photon utilization towards maximal optoelectronic—photocatalytic synergy. Nat. Commun. **12**, 1–11 (2021). https://doi.org/10.1038/s41467-020-20613-2
33. A.P. Lopes, F.M. Santos, T.F.C.V. Silva et al., Outdoor cultivation of the microalga chlorella vulgaris in a new photobioreactor configuration: the effect of ultraviolet and visible radiation. Energies **13**, 1962 (2020). https://doi.org/10.3390/EN13081962

34. R. Ma, H. Su, J. Sun et al., Concentrating photo-thermo-organized single-atom and 2D-raft Cu catalyst for full-spectrum solar harmonic conversion of aqueous urea and urine into hydrogen. Appl. Catal. B **315**, 121493 (2022). https://doi.org/10.1016/J.APCATB.2022.121493
35. Z. Ma, P. Davenport, G. Saur, System and technoeconomic analysis of solar thermochemical hydrogen production. Renew. Energy **190**, 294–308 (2022). https://doi.org/10.1016/j.renene.2022.03.108
36. A.X. Maldonado Pérez, J.D.J. Pérez Bueno, Solar lighting systems applied in photocatalysis to treat pollutants—a review. Rev. Adv. Mater. Sci. **62**, 20220293 (2023). https://doi.org/10.1515/RAMS-2022-0293
37. A.X. Maldonado Pérez, J.D.J. Pérez Bueno, Solar lighting systems applied in photocatalysis to treat pollutants—a review. Rev. Adv. Mater. Sci. **62**, 20220293 (2023). https://doi.org/10.1515/RAMS-2022-0293/ASSET/GRAPHIC/J_RAMS-2022-0293_FIG_009.JPG
38. T.K. Mallick, P.C. Eames, Design and fabrication of low concentrating second generation PRIDE concentrator. Sol. Energy Mater. Sol. Cells **91**, 597–608 (2007). https://doi.org/10.1016/J.SOLMAT.2006.11.016
39. J. Masojídek, Š Papáček, M. Sergejevová et al., A closed solar photobioreactor for cultivation of microalgae under supra-high irradiance: basic design and performance. J. Appl. Phycol. **15**, 239–248 (2003). https://doi.org/10.1023/A:1023849117102/METRICS
40. J. Masojídek, M. Sergejevová, K. Rottnerová et al., A two-stage solar photobioreactor for cultivation of microalgae based on solar concentrators. J. Appl. Phycol. **21**, 55–63 (2009). https://doi.org/10.1007/S10811-008-9324-6/METRICS
41. M.S. Mayhoub, D.J. Carter, Hybrid lighting systems: performance and design. Light. Res. Technol. **44**, 261–276 (2012). https://doi.org/10.1177/1477153511416324
42. A. Merlaud, Hermans C. de Mazière, A. Cornet, Equations for Solar Tracking. Sensors **12**, 4074–4090 (2012). https://doi.org/10.3390/S120404074
43. L.W. Miller, M.I. Tejedor-Tejedor, M.P. Moya et al., Photocatalyst-coated acrylic waveguides for oxidation of organic compounds. Stud. Surf. Sci. Catal. **130**, 1925–1930 (2000). https://doi.org/10.1016/S0167-2991(00)80483-8
44. C.K. Miskin, Y. Li, A. Perna et al., Sustainable co-production of food and solar power to relax land-use constraints. Nat. Sustain. **10**(2), 972–980 (2019). https://doi.org/10.1038/s41893-019-0388-x
45. A. Monteiro, S. Santos, P. Gonçalves, Precision agriculture for crop and livestock farming—brief review. Animals **11**, 2345 (2021). https://doi.org/10.3390/ANI11082345
46. L. Mu, X. Xu, T. Williams et al., Enhancing the performance of a single-basin single-slope solar still by using Fresnel lens: experimental study. J. Clean. Prod. **239**, 118094 (2019). https://doi.org/10.1016/J.JCLEPRO.2019.118094
47. N. Negishi, F. He, S. Matsuzawa et al., Wave-guide type photoreactor for water purification. C. R. Chim. **9**, 822–828 (2006). https://doi.org/10.1016/J.CRCI.2005.08.001
48. T.V. Nguyen, J.C.S. Wu, Photoreduction of CO_2 in an optical-fiber photoreactor: effects of metals addition and catalyst carrier. Appl. Catal. A: Gen. **335**, 112–120 (2008). https://doi.org/10.1016/J.APCATA.2007.11.022
49. T. Oyama, A. Aoshima, S. Horikoshi et al., Solar photocatalysis, photodegradation of a commercial detergent in aqueous TiO_2 dispersions under sunlight irradiation. Sol. Energy **77**, 525–532 (2004). https://doi.org/10.1016/J.SOLENER.2004.04.020
50. S.I. Palomino-Resendiz, D.A. Flores-Hernández, N. Lozada-Castillo et al., Design and implementation of a robotic active solar distiller based on a Fresnel concentrator and a photovoltaic system. Energy Convers. Manag. **166**, 637–647 (2018). https://doi.org/10.1016/J.ENCONMAN.2018.04.069
51. M.E. Potter, D.J. Stewart, A.E. Oakley et al., Combining photocatalysis and optical fiber technology toward improved microreactor design for hydrogen generation with metallic nanoparticles. ACS Photonics **7**, 714–722 (2020). https://doi.org/10.1021/ACSPHOTONICS.9B01577/ASSET/IMAGES/LARGE/PH9B01577_0007.JPEG
52. J.L. Reyes-García, C.A. Arancibia-Bulnes, F. Méndez-Arriaga et al., Optical and hydrodynamic performance of photocatalytic monoliths of different shapes in a solar photoreactor

with compound parabolic collector. Catal. Today **429**, 114498 (2024). https://doi.org/10.1016/J.CATTOD.2023.114498
53. J. Robinson, D. Smale, D. Pollard, H. Shiona, Solar tracker with optical feedback and continuous rotation. Atmos Meas Tech **13**, 5855–5871 (2020). https://doi.org/10.5194/AMT-13-5855-2020
54. M. Sabry, Prismatic TIR (total internal reflection) low-concentration PV (photovoltaics)-integrated façade for low latitudes. Energy **107**, 473–481 (2016). https://doi.org/10.1016/J.ENERGY.2016.04.057
55. N. Sarmah, T.K. Mallick, Design, fabrication and outdoor performance analysis of a low concentrating photovoltaic system. Sol. Energy **112**, 361–372 (2015). https://doi.org/10.1016/J.SOLENER.2014.12.019
56. N. Sarmah, B.S. Richards, T.K. Mallick, Design, development and indoor performance analysis of a low concentrating dielectric photovoltaic module. Sol. Energy **103**, 390–401 (2014). https://doi.org/10.1016/J.SOLENER.2014.02.029
57. N. Sellami, T.K. Mallick, Optical characterisation and optimisation of a static window integrated concentrating photovoltaic system. Sol. Energy **91**, 273–282 (2013). https://doi.org/10.1016/J.SOLENER.2013.02.012
58. J. Song, B.B. Dessie, L. Gao, Analysis and comparison of daylighting technologies: light pipe, optical fiber, and heliostat. Sustainability **15**, 11044 (2023). https://doi.org/10.3390/SU151411044
59. P.J. Sonneveld, G.L.A.M. Swinkels, J. Campen et al., Performance results of a solar greenhouse combining electrical and thermal energy production. Biosyst. Eng. **106**, 48–57 (2010). https://doi.org/10.1016/J.BIOSYSTEMSENG.2010.02.003
60. E.J. Stallknecht, C.K. Herrera, C. Yang, et al., Designing plant-transparent agrivoltaics. Sci. Rep. **13**, 1903 (2023) (123AD). https://doi.org/10.1038/s41598-023-28484-5
61. B.E. Stephens-Camacho, D. Sauceda-Carvajal, M.P. Sánchez-Saavedra, Growth, proximate composition, and photosynthesis of chlorella vulgaris cultures between a photobioreactor based in a compound parabolic concentrator and a plain tubular system for a biorefinery. Bioenergy Res 15, 230–241 (2022). https://doi.org/10.1007/S12155-021-10304-1/METRICS
62. K. Tong, L. Chen, L. Yang et al., Energy transport of photocatalytic carbon dioxide reduction in optical fiber honeycomb reactor coupled with trough concentrated solar power. Catalysts **11**, 829 (2021). https://doi.org/10.3390/CATAL11070829
63. G. Torzillo, G. Chini Zittelli (2015) Tubular photobioreactors, in *Algal biorefineries: volume 2: products and refinery design*, pp. 187–212. https://doi.org/10.1007/978-3-319-20200-6_5
64. S. Tsoutsou, C. Infante Ferreira, J. Krieg, M. Ezzahiri, Building integration of concentrating solar systems for heating applications. Appl. Therm. Eng. **70**, 647–654 (2014). https://doi.org/10.1016/J.APPLTHERMALENG.2014.05.079
65. S.V. Vasylyev, V.P. Vasylyev, V.A. Sergeev, Concept and design of "flat-plate" CPV module based on ring-array concentrator, in *Conference Record of the IEEE Photovoltaic Specialists Conference*, pp 3087–3091 (2010). https://doi.org/10.1109/PVSC.2010.5614474
66. N.G. Victoria, E.B. Romero, G. Franken, et al., Effect of high scattering lamellae on growth and photosynthesis of young tomato plants : smart materials crop experiments (2020). https://doi.org/10.18174/564877
67. L. Villén, F. Manjón, D. García-Fresnadillo, G. Orellana, Solar water disinfection by photocatalytic singlet oxygen production in heterogeneous medium. Appl. Catal. B: Environ. **69**, 1–9 (2006). https://doi.org/10.1016/J.APCATB.2006.05.015
68. M. Vivar, V. Everett, M. Fuentes et al., Initial field performance of a hybrid CPV-T microconcentrator system. Prog. Photovoltaics Res. Appl. **21**, 1659–1671 (2013). https://doi.org/10.1002/PIP.2229
69. P.S. Walko, R.N. Devi, Scalable optical fiber reactor for photocatalytic H2 production: addressing scattering issues. Int. J. Hydrogen Energy **48**, 17086–17096 (2023). https://doi.org/10.1016/J.IJHYDENE.2023.01.148
70. L. Wondraczek, A. Winnacker, M. Batentschuk et al., Enhanced photosynthetic activity in Spinacia Oleracea by spectral modification with a photoluminescent light converting material. Opt. Express **21**, A909–A916 (2013). https://doi.org/10.1364/OE.21.00A909

71. G. Wu, Q. Yang, Y. Zhang et al., Energy and optical analysis of photovoltaic thermal integrated with rotary linear curved Fresnel lens inside a Chinese solar greenhouse. Energy **197**, 117215 (2020). https://doi.org/10.1016/J.ENERGY.2020.117215
72. P. Xie, Y. Liu, H. Lin, B. Li, Total internal reflection-based planar waveguide solar concentrator with symmetric air prisms as couplers. Opt. Express **22**(S6), A1389–A1398 (2014). https://doi.org/10.1364/OE.22.0A1389
73. A. Yano, M. Cossu, Energy sustainable greenhouse crop cultivation using photovoltaic technologies. Renew. Sustain. Energy Rev. **109**, 116–137 (2019). https://doi.org/10.1016/J.RSER.2019.04.026
74. X. Yu, Y. Su, H. Zheng, S. Riffat, A study on use of miniature dielectric compound parabolic concentrator (dCPC) for daylighting control application. Build. Environ. **74**, 75–85 (2014). https://doi.org/10.1016/J.BUILDENV.2013.12.008
75. C. Zhang, G. An, Y. Zhu et al., Development of a sunlight adjustable parabolic trough reactor for 24-hour efficient photocatalytic wastewater treatment. Chem. Eng. J. **497**, 154582 (2024). https://doi.org/10.1016/J.CEJ.2024.154582
76. C. Zhang, N. Liu, J. Ming et al., Development of a novel solar energy controllable Linear Fresnel photoreactor (LFP) for high-efficiency photocatalytic wastewater treatment under actual weather. Water Res. **208**, 117880 (2022). https://doi.org/10.1016/J.WATRES.2021.117880
77. F. Zhang, M. Li, W. Zhang, et al., Large-scale and cost-efficient agrivoltaics system by spectral separation. iScience **26**, 108129 (2023). https://doi.org/10.1016/j.isci.2023.108129
78. J. Zheng, S. Meng, X. Zhang et al., Increasing the comprehensive economic benefits of farmland with Even-lighting Agrivoltaic Systems. PLoS ONE **16**, e0254482 (2021). https://doi.org/10.1371/JOURNAL.PONE.0254482
79. Technology Roadmap: Concentrating Solar Power (2010). https://doi.org/10.1787/9789264088139-EN
80. Home—Sunoptics. https://www.sunoptics.com/. Accessed 12 Dec 2024a
81. Spherical solar energy collector designed by Rawlemon works | The InfoGreenGlobal. https://infogreenglobal.com/spherical-solar-energy-collector-by-rawlemon/. Accessed 12 Dec 2024b
82. White roof and skylights on Romeoville, III. Walmart | Flickr. https://www.flickr.com/photos/walmartcorporate/5258813438. Accessed 14 Dec 2024c
83. Computer vision of optimal geometric concentration ratio for the solar ball lens. https://www.researchgate.net/publication/364232565_Computer_Vision_of_Optimal_Geometric_Concentration_Ratio_for_the_Solar_Ball_Lens. Accessed 14 Dec 2024d
84. Parans system: SP4 sunlight collector | Parans. https://parans.com/products/parans-system-sp4-sunlight-collector/. Accessed 12 Dec 2024e
85. Case study: Mall of the Emirates expansion—construction week online. https://www.constructionweekonline.com/projects-tenders/article-39722-case-study-mall-of-the-emirates-expansion. Accessed 12 Dec 2024f
86. File:Mall of the Emirates (3679338750).jpg—Wikimedia commons. https://commons.wikimedia.org/wiki/File:Mall_of_the_Emirates_%283679338750%29.jpg. Accessed 14 Dec 2024g
87. SolFocus vs. Concentrix Solar: battle of the solar concentrator PhotoVoltaic (CPV) start-ups | GUNTHER Portfolio. https://guntherportfolio.com/2006/08/solfocus-vs-concentrix-solar-battle-of-the-solar-concentrator-photovoltaic-cpv-start-ups/. Accessed 12 Dec 2024h
88. Home. https://beta.savanna.morgansolar.com/. Accessed 12 Dec 2024i
89. Morgan solar tracking units being tested at the Morgan Sol… | Flickr. https://www.flickr.com/photos/greenenergyfutures/10561400973. Accessed 14 Dec 2024j
90. Façade lighting: the illuminati residence, Gurugram. https://wfmmedia.com/facade-lighting-adding-depth-and-character-to-building-design/.. Accessed 12 Dec 2024k
91. Still standing: Arab World Institute, Paris, 1987—Architecture Today. https://architecturetoday.co.uk/still-standing-arab-world-institute-paris/. Accessed 12 Dec 2024l
92. Mercedes-Benz Museum. https://www.mercedes-benz.com/en/art-and-culture/museum/. Accessed 12 Dec 2024m

93. Institut Du Monde Arabe | Umberto | Flickr. https://www.flickr.com/photos/acquea/2610887359/. Accessed 14 Dec 2024
94. File:Mercedes-Benz Museum 201312 07 blue hour.jpg—Wikimedia commons. https://commons.wikimedia.org/wiki/File:Mercedes-Benz_Museum_201312_07_blue_hour.jpg. Accessed 14 Dec 2024o
95. Customizable Canopies manufacturing company in India—Aparna Craft Exteriors Pvt Ltd. https://aparna-craft.com/facades/glass-canopies/. Accessed 12 Dec 2024p
96. McDonald´s restaurant—Onyx solar's Projects. https://onyxsolar.com/mcdonalds-flagship-restaurant. Accessed 12 Dec 2024q
97. Concentrated solar reactor generates unprecedented amounts of hydrogen—Physics World. https://physicsworld.com/a/concentrated-solar-reactor-generates-unprecedented-amounts-of-hydrogen/. Accessed 13 Dec 2024r
98. Noor Ouarzazate: the world's largest concentrated solar power plant built in Morocco. https://esfccompany.com/en/articles/solar-energy/noor-ouarzazate-the-world-s-largest-concentrated-solar-power-plant-csp-built-in-morocco/. Accessed 13 Dec 2024s

Solar Concentrators Based on Inorganic Luminescent Materials Potential to Elevate the Efficiency

Monalisha Behera, Puja Kumari, Sudipta Som, and Subrata Das

Abstract The global quest for sustainable and efficient energy solutions has led to significant advancements in solar energy technologies. Among these, solar concentrators based on luminescent materials have emerged as a promising area of research, offering a unique combination of efficiency, versatility, and potential for integration into various environments. This review comprehensively analyzes the state-of-the-art luminescent solar concentrators (LSCs), exploring their fundamental principles, recent developments, and prospects. We begin by outlining the basic operating principles of LSCs, including the role of luminescent materials in absorbing sunlight and re-emitting it at longer wavelengths and the subsequent waveguide-based collection of this light to achieve concentration. The review then delves into the various luminescent materials used in LSCs, such as organic dyes, quantum dots, and perovskite nanocrystals, discussing their optical properties, performance metrics, and stability. Much of the review is dedicated to the latest advancements in LSC design, including integrating photonic structures and using spectral conversion layers to enhance light absorption and reduce re-absorption losses. We also explored the integration of LSCs with photovoltaic cells and other renewable energy systems to create hybrid systems with improved energy conversion efficiencies. The challenges facing the widespread adoption of LSCs are critically examined, including issues related to material stability, cost, scalability, and the trade-offs between efficiency and concentration ratios. Strategies for overcoming these challenges, such as developing new materials and optimizing device architectures, are discussed. Finally, the review concludes with an outlook on the future of LSCs, highlighting the potential for these devices to play a significant role in the next generation of solar energy harvesting

M. Behera · S. Som (✉)
Department of Physics, School of Science and Humanities, Shiv Nadar University Chennai, Chennai, Tamil Nadu, India
e-mail: sudiptasom@snuchennai.edu.in

P. Kumari
Department of Physics, New Government Polytechnic Patna, Bihar, India

S. Das (✉)
Materials Science and Technology Division, CSIR-National Institute for Interdisciplinary Science and Technology, Thiruvananthapuram, Kerala, India
e-mail: subratadas@niist.res.in

P. Ravindran et al. (eds.), *Photon to Power*, Progress in Optical Science and Photonics 32, https://doi.org/10.1007/978-981-96-5914-2_17

systems. Integrating LSCs with emerging technologies, such as building-integrated photovoltaics and bright windows, is discussed as a pathway to achieving sustainable and ubiquitous solar energy utilization.

1 Introduction

The rising perseverance of renewable energy sources has grown enormous advancements in various technologies related to solar energy. Solar energy, a plentiful and sustainable resource, is the foundation of global efforts in fighting climate change and reducing dependence on fossil fuels because of extraordinary solar radiation realized on the Earth's surface (140,000 TWh) [1–3]. In this regard, photovoltaic (PV) technology and devices have gained significant attention due to their ability to convert sunlight straight to electricity with enhanced efficiency and affordability [4, 5]. However, the extensive acceptance of PV technologies is quite delayed owing to challenges including costly raw materials, loss in efficiency because of unequal illumination, and restrictions in integrating old-style solar panels into urban locations [6, 7]. Solar concentrators can solve these issues by concentrating sunlight onto smaller-sized high-efficiency PV cells and lowering the overall cost per watt of electricity produced [8]. Conformist optical concentrators, which include parabolic mirrors and Fresnel lenses, depend on geometric optics to direct sunlight; however, they frequently need detailed tracking systems and occupy noteworthy space. These limits make them less appropriate for applications demanding compression, scalability, and artistic integration [9–12].

Luminescent solar concentrators (LSCs) that utilize luminescent materials to absorb and re-emit sunlight at greater wavelengths can create a transformative path. The released photons are internally reflected within a waveguide (classically a transparent slab) and directed to edges so that the PV cells transform it into electricity. This machinery allows LSCs to function proficiently without requiring mechanical tracking, which makes the system multipurpose, lightweight, and perfect for numerous applications such as building-integrated photovoltaics (BIPV), wearable electronics, and portable energy systems. Furthermore, LSCs can be engineered to be semi-transparent or colorful, enabling seamless integration into windows, facades, and decorative elements [13, 14].

The achievement of LSC technology centered on cutting-edge luminescent materials. Over the years, intense research on various types of fluorescent materials has been reported that count thin films, organic dyes, rare-earth phosphors, quantum dots, and perovskites. Each section of materials offers exceptional benefits and specific challenges [15, 16]. For example, organic dyes can provide tuneable emissive properties but frequently suffer from photodegradation. Quantum dots deliver incomparable luminescence efficiency and tunability, yet scalability and toxicity remain significant concerns [16, 17]. Hybrid perovskites, a newer class of materials, exhibit outstanding optical properties and potential for large-scale production, even though stability under operational conditions remains an area of active research [16–18].

Beyond materials development, revolutions in the innovative architecture of devices have further advanced LSC performance. Multi-layered strategies, tandem configurations, and a combination of light-management structures such as photonic crystals and dielectric mirrors have significantly enhanced light absorption and waveguiding efficiency [19]. Additionally, computational modeling and machine learning have enabled precise optimization of LSC designs, accelerating development and improving performance metrics [19].

This chapter delves into the significant potential of inorganic luminescent materials to enhance the efficiency of solar concentrators, providing an in-depth exploration of both foundational principles and the latest advancements in the field. It addresses the core mechanisms involved in light absorption, emission, and guiding within luminescent solar concentrators (LSCs), shedding light on the material science innovations that drive their functionality. The chapter examines how various luminescent materials, particularly inorganic ones, contribute to optimizing energy conversion and minimizing losses in these systems. Furthermore, the chapter explores recent breakthroughs that have made scaling LSC technologies more feasible for commercial applications. It emphasizes the importance of overcoming challenges such as material stability, scalability, and cost-effectiveness, which are critical for the widespread adoption of LSCs in renewable energy solutions. These breakthroughs highlight the role of LSCs in addressing key bottlenecks in the energy sector, such as improving solar energy harvesting efficiency while keeping costs competitive with other renewable technologies.

In addition to these technical considerations, the chapter underscores the aesthetic versatility of LSCs, a crucial factor for integrating solar energy systems into building materials and urban infrastructure. The potential for LSCs to be seamlessly incorporated into windows, facades, and other architectural elements makes them a promising solution for the creation of aesthetically pleasing, sustainable energy systems. As advancements in material design and fabrication techniques continue, the dream of highly efficient, cost-effective, and visually appealing solar energy systems moves ever closer to becoming a reality. Ultimately, this chapter aims to systematically review the state of LSC technology, providing a comprehensive understanding of its current capabilities and the exciting developments that lie ahead. By laying the groundwork for future research, it envisions a future where luminescent solar concentrators play a pivotal role in advancing solar energy technologies and contributing to the global shift toward sustainable, renewable energy sources. Through continued innovation and interdisciplinary collaboration, LSC-powered solar concentrators are poised to transform the way we harness the power of the sun.

2 Working Principals of LSC

Luminescent solar concentrators (LSCs) signify a unique and well-organized method of harvesting solar energy by utilizing the optical properties of luminescent materials to capture, convert, and project sunlight to photovoltaic (PV) cells, as shown in Fig. 1 [20].

Unlike conventional solar concentrators that depend on geometric optics, LSCs use photoluminescence and wave-guiding mechanisms, enabling efficient operation even under diffuse light conditions. This makes them particularly suitable for urban environments and applications where traditional systems face integration challenges. Below, the working principles of LSCs are elaborated in detail [20, 21].

2.1 Absorption of Light

The operation of a luminescent solar concentrator (LSC) begins with the absorption of sunlight by luminescent samples embedded within a transparent waveguide. These materials serve as the core components of the LSC system and are carefully selected based on their ability to effectively capture and manipulate solar radiation. The luminescent samples, such as organic dyes, quantum dots, inorganic phosphors, and perovskites, offer unique optical properties that contribute to the system's efficiency. Their primary function is to absorb specific portions of the solar spectrum and subsequently re-emit the energy at different wavelengths, enabling effective

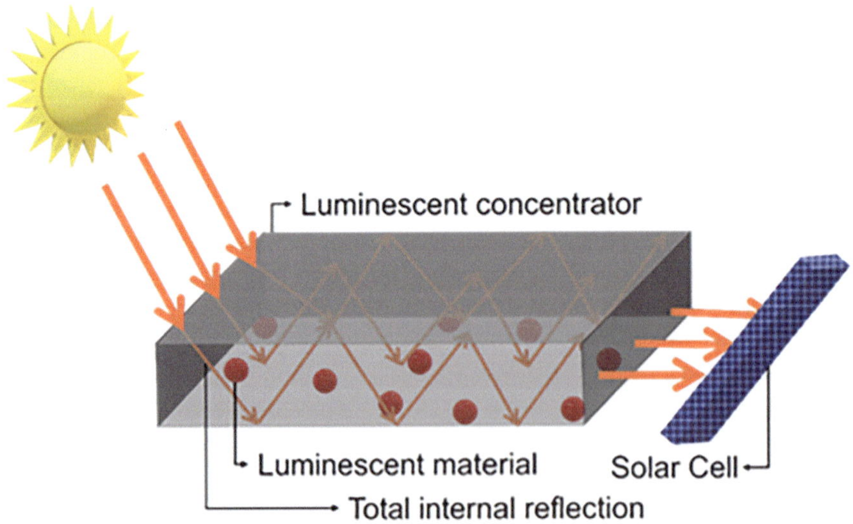

Fig. 1 A schematic of a luminescent solar concentrator

concentration and redirection of light toward photovoltaic cells. A crucial characteristic of luminescent materials in an LSC is their ability to absorb light across a wide range of the solar spectrum. This typically includes ultraviolet (UV), visible, and near-infrared (NIR) wavelengths, ensuring that a substantial portion of incident solar energy is captured and utilized. By efficiently harnessing a broad spectrum of sunlight, the LSC system maximizes the potential for energy conversion, reducing wasted radiation and enhancing overall performance.

The absorption characteristics of luminescent materials can be precisely tuned to target specific wavelengths of light. This selective absorption allows the system to optimize the process of energy collection and transfer, ensuring that the captured photons contribute effectively to photovoltaic conversion. By engineering the luminescent materials to absorb and emit light within desired spectral ranges, energy losses can be minimized, and the efficiency of the LSC system can be significantly improved. When photons from sunlight interact with the luminescent materials, their energy is absorbed, leading to the excitation of electrons within the material. This process includes promoting electrons from the ground state to a higher energy state. The excitation of electrons is a fundamental step in the operation of the LSC, as it forms the basis for subsequent energy conversion stages. Once in the excited state, the electrons eventually relax back to a lower energy level, during which they release energy in the form of re-emitted photons. These emitted photons are then guided through the transparent waveguide toward the LSC edges, where photovoltaic cells are positioned to capture and switch the intense light into electricity. By leveraging this process of absorption, excitation, and controlled emission, LSC systems efficiently concentrate diffuse solar radiation and direct it toward photovoltaic cells, enhancing solar energy harvesting while maintaining a lightweight and cost-effective design.

2.2 Emission of Light

Once the luminescent material in a luminescent solar concentrator (LSC) absorbs incoming photons, it undergoes a process known as radiative relaxation, during which it re-emits light at a longer wavelength. This shift in wavelength, referred to as the Stokes shift, is a critical mechanism that enhances the efficiency of the LSC system. By ensuring that the emitted photons are at a different wavelength than those initially absorbed, the system significantly reduces the likelihood of re-absorption losses, allowing more photons to propagate through the waveguide without being lost. This mechanism enhances light transport within the device and improves the overall energy conversion process.

3 Key Aspects of Photon Manipulation in LSCs

Down-Conversion of High-Energy Photons for Enhanced Utilization: One of the fundamental processes in an LSC is the down-conversion of photons, where high-energy photons (such as those in the ultraviolet (UV) or blue spectral range) are converted into lower-energy photons in the visible or near-infrared (NIR) spectrum. This transformation is advantageous because many photovoltaic (PV) cells have limited efficiency in directly utilizing high-energy photons. Instead, by converting them into wavelengths better suited for PV absorption, the LSC improves the overall light-to-electricity conversion efficiency. For instance, traditional silicon-based solar cells have peak efficiency in the red and near-infrared regions, making it beneficial to shift high-energy UV and blue photons into this range. This tailored spectral modification ensures that a greater proportion of the incident solar energy can be captured and effectively converted into electrical power.

Emission Spectral Tuning for Optimal Photovoltaic Compatibility: The efficiency of an LSC depends not only on its ability to capture light but also on its ability to emit photons at wavelengths that align with the optimal absorption range of the PV cells. This is achieved through precise emission spectral tuning, which involves designing the luminescent materials to emit light in a controlled manner. By carefully selecting the composition, molecular structure, and doping concentrations of luminescent materials, researchers can fine-tune the emission spectrum to match the spectral response of the solar cells placed at the edges of the concentrator. This optimization ensures that the maximum amount of emitted light is absorbed by the PV cells rather than being lost due to reflection or scattering. Moreover, different types of luminescent materials offer varying degrees of emission tunability.

Quantum dots, for instance, allow for highly customizable emission wavelengths based on their size and composition. Rare-earth ions provide sharp and well-defined emission peaks with minimal re-absorption losses. Perovskite-based emitters offer high quantum efficiency and broad tunability, making them promising for next-generation LSCs.

By leveraging these advanced materials, researchers can develop LSCs with tailored light-emission properties that maximize solar energy capture.

Ensuring Long-Term Photostability for Sustained Performance: A major challenge in implementing LSCs is photostability. That is its ability of luminescent materials to maintain their efficiency over long periods of sunlight exposure. Many traditional luminescent materials, such as organic dyes, are prone to photobleaching—a gradual degradation of their luminescent properties due to prolonged UV exposure. This degradation leads to a decline in light emission efficiency, ultimately reducing the effectiveness of the LSC system. To address this challenge, advanced luminescent materials are engineered to resist photobleaching and maintain their performance over extended operational lifetimes. Several strategies are employed to enhance photostability:

Encapsulation techniques: Protective coatings and polymer matrices shield luminescent materials from environmental degradation.

Use of inorganic phosphors and rare-earth-doped materials: These materials exhibit superior stability compared to organic dyes and retain their luminescence for longer durations.

Incorporation of UV-resistant additives: These additives help mitigate damage from high-energy UV exposure, prolonging the functional lifespan of the LSC.

By integrating these stability-enhancing techniques, modern LSC designs can achieve greater durability and reliability, making them viable for large-scale solar energy applications.

Efficient Light Guiding and Directional Control: In addition to converting and re-emitting photons, the luminescence process also facilitates effective light guiding within the waveguide. The transparent waveguide material ensures that emitted photons are internally reflected and efficiently transported toward the edges of the concentrator, where PV cells are positioned. This directional control is achieved through a combination of total internal reflection (TIR) that ensures that emitted photons remain trapped within the waveguide, minimizing light loss with low optical scattering materials that reduce dispersion and ensure that more light reaches the PV cells and optimized geometrical design (the shape and size of the LSC can be tailored to maximize photon collection and concentration at the edges). By ensuring that a high percentage of absorbed solar energy is successfully redirected toward PV cells, this process significantly boosts the effectiveness of LSC technology in harnessing sunlight.

3.1 Waveguiding and Total Internal Reflection (TIR)

Once photons are re-emitted by the luminescent materials embedded within the luminescent solar concentrator (LSC), they must be efficiently transported to the photovoltaic (PV) cells located at the device's edges. This process is facilitated by total internal reflection **(TIR)**, a fundamental optical mechanism that ensures light remains confined within the transparent waveguide, preventing unwanted escape and directing the photons toward the solar cells. The ability of the LSC to effectively guide light is crucial for maximizing energy capture and improving the system's efficiency.

4 Fundamental Principles of Total Internal Reflection in LSCs

Critical Angle and Internal Reflection: Total internal reflection occurs when light suffers an interface in a medium with lower refractive index at an angle higher than the material's critical angle. Instead of passing through the interface and escaping, the light is completely reflected into the waveguide, allowing it to continue traveling toward LSC's edges.

Role of the Waveguide's Refractive Index: The refractive index of the waveguide material is a key determinant of TIR efficiency. A higher refractive index results in a smaller critical angle, meaning that a larger fraction of the emitted light remains trapped inside the waveguide.

Materials with high refractive indices (e.g., glass or specialized polymers) improve photon confinement and reduce light leakage. Refractive index engineering allows optimization of the waveguide material to maximize light retention while minimizing unwanted absorption or scattering. Some hybrid materials, such as nanoparticle-doped polymers, can be used to further enhance optical properties without significantly increasing material costs or weight. By carefully selecting waveguide materials with the appropriate refractive index, LSC designers can achieve greater photon confinement and increase the overall efficiency of light transport to the PV cells.

Reducing Photon Losses for Maximum Light Transport: While TIR plays a dominant role in guiding photons, optical losses due to scattering, absorption, and imperfections in the waveguide material can reduce efficiency.

To counteract these losses, the following strategies are employed:

Minimizing Scattering Losses: Scattering occurs when light interacts with microscopic irregularities or impurities in the waveguide material, causing it to deviate from its intended path. To minimize scattering high-purity materials with minimal internal defects are used.

Smooth waveguide surfaces are engineered to reduce roughness and prevent unwanted light dispersion. Advanced fabrication techniques, such as nano-coatings, are applied to maintain optical clarity and structural integrity.

Reducing Absorption Losses: Absorption losses arise when waveguide materials inadvertently absorb some of the guided light before it reaches the PV cells. This issue is addressed by using low-absorption polymers or glass that are transparent in the wavelength range of interest. Employing fluorescent reabsorption mitigation techniques, such as using luminescent materials with large Stokes shifts, which reduce the chances of self-absorption within the LSC. Incorporating UV-stable coatings to prevent degradation over time, ensuring consistent optical performance. By optimizing both scattering and absorption properties, the LSC ensures that the majority of re-emitted photons successfully reach the solar cells, improving the system's overall energy conversion efficiency.

Waveguiding as a Passive, Cost-Effective Light Transport Mechanism: One of the major advantages of the LSC's waveguiding mechanism is that it is entirely passive, meaning it does not require any moving parts, sensors, or active tracking systems to function. Unlike traditional solar concentrators, which often rely on mechanical tracking systems to align with the sun's movement throughout the day, LSCs leverage internal light guiding principles to passively direct photons toward PV cells.

This results in several practical benefits:

Simplified design: LSCs do not require complex tracking mechanisms, making them easier to manufacture and integrate into various settings.

Lower maintenance: Since there are no moving parts, wear and tear are significantly reduced, increasing the system's longevity.

Versatile applications: LSCs can be integrated into windows, facades, and other architectural elements without the need for active orientation adjustment.

4.1 Photon Transport to Photovoltaic Cells

As photons travel through the waveguide, they are efficiently directed toward its edges, where high-performance photovoltaic (PV) cells are strategically positioned to capture and convert the concentrated light into electrical energy. This process significantly enhances the energy yield of the luminescent solar concentrator (LSC) by increasing the intensity of incident light on the PV cells while maintaining a compact and scalable design.

5 Key Aspects of Light Concentration and PV Energy Conversion in LSCs

Concentration Factor (Maximizing Light Intensity for Efficient Energy Harvesting):

One of the fundamental advantages of an LSC is its ability to concentrate diffuse solar radiation onto a much smaller area, thereby increasing the intensity of light reaching the PV cells. This is achieved by:

- **Collecting light over a large surface area:** The waveguide captures sunlight over a broad region, including both direct and diffuse sunlight, making LSCs effective even under cloudy conditions.
- **Guiding and concentrating light toward the edges:** Through **total internal reflection (TIR)** and controlled photon transport, the waveguide channels photons efficiently to its periphery.

- **Enhancing photon flux per unit area:** By funneling a large amount of light onto a small PV cell area, LSCs amplify the photon density, leading to improved power output.

The **concentration factor**—the ratio of the light intensity at the PV cell edges to the intensity of incident sunlight—plays a crucial role in determining overall efficiency. By increasing the concentration factor, fewer PV cells are needed, reducing costs while maintaining high energy output.

6 Edge-Mounted Photovoltaic Cells: High-Efficiency Energy Conversion

Unlike traditional solar panels, where photovoltaic material covers an entire surface, LSCs use edge-mounted PV cells, which are typically high-performance and optimized for maximum energy conversion.

Several advanced photovoltaic technologies are used in LSCs to ensure optimal efficiency:

- **Multi-junction solar cells**: These cells consist of multiple layers of semiconductor materials, each designed to absorb different wavelengths of light, significantly increasing efficiency (often exceeding 40% in laboratory conditions).
- **Perovskite solar cells**: Due to their high absorption coefficients and tunable bandgaps, perovskites are emerging as an ideal candidate for LSC applications, offering excellent efficiency even at low light intensities.
- **Silicon-based PV cells**: While traditional monocrystalline silicon cells can also be used, they are often combined with other materials to improve spectral response and overall energy yield.

The placement of these high-efficiency PV cells along the edges of the waveguide significantly reduces the amount of expensive photovoltaic material needed while maintaining strong energy output. This design not only improves cost-effectiveness but also enhances modularity, making it easier to integrate LSCs into buildings, vehicles, and urban infrastructure.

7 Enhanced Uniformity: Optimized Light Distribution for Higher Efficiency

A major advantage of LSC technology is its ability to deliver uniform light distribution to the PV cells, minimizing hotspots and ensuring consistent energy conversion. In traditional solar concentrator systems, uneven light distribution can lead to localized heating (hotspots), which negatively impacts PV cell performance and lifespan. LSCs, however, are designed to:

- **Evenly distribute re-emitted photons** across the waveguide, reducing intensity fluctuations.
- **Prevent thermal buildup** by spreading absorbed energy more uniformly, enhancing the durability of the PV cells.
- **Ensure steady electrical output**, improving reliability and reducing energy conversion losses.

By maintaining a homogeneous photon flux across the PV cell edges, LSCs optimize their efficiency while preventing performance degradation over time.

7.1 Mitigation of Re-absorption Losses

One of the primary loss mechanisms in luminescent solar concentrators (LSCs) is re-absorption, where the luminescent material absorbs its own emitted light before it can reach the photovoltaic (PV) cells. This process reduces the overall efficiency of the system by trapping photons within the waveguide rather than directing them toward energy conversion. To mitigate re-absorption losses and improve the performance of LSCs, several key strategies are employed.

Key Strategies for Reducing Re-absorption Losses

1. Significant Stokes Shift: Minimizing Spectral Overlap

The Stokes shift refers to the difference between the wavelengths of absorbed and emitted light. By engineering luminescent materials with a large Stokes shift, the overlap between absorption and emission spectra is minimized, reducing the probability of self-absorption.

- Materials with inherently large Stokes shifts, such as rare-earth-doped phosphors and certain quantum dots, are preferred for LSC applications.
- Organic dyes and inorganic phosphors can be chemically modified to increase the Stokes shift while maintaining a high luminescence efficiency.
- A well-separated absorption and emission spectrum ensures that emitted photons are efficiently transported through the waveguide with minimal re-absorption.

By increasing the Stokes shift, the system allows more emitted photons to travel to the PV cells, improving light collection efficiency.

2. Energy Transfer Mechanisms: Förster Resonance Energy Transfer (FRET) and Other Techniques

Advanced energy transfer mechanisms can be used to redirect absorbed energy to nearby luminescent molecules before it has a chance to be re-absorbed within the same material.

- Förster Resonance Energy Transfer (FRET) is a process where energy is non-radiatively transferred between closely spaced molecules via dipole interactions. This allows energy to "hop" from one molecule to another without emitting a photon, reducing the chance of re-absorption.
- Cascade energy transfer systems can be designed, where energy is transferred stepwise between different luminescent species, gradually shifting toward longer wavelengths that are less prone to re-absorption.
- Hybrid nanomaterial structures, such as energy-transfer-enhanced quantum dot pairs, optimize photon transport efficiency while minimizing losses.

By integrating these mechanisms, LSCs can improve photon harvesting efficiency while maintaining high optical transparency.

3. Material Optimization: High Quantum Yield and Low Self-Absorption

Selecting and engineering luminescent materials with optimal properties is crucial for reducing self-absorption and enhancing LSC performance.

- Quantum dots (QDs), particularly core–shell structures, exhibit high quantum yield (the efficiency of converting absorbed light into emitted light) and can be tuned to minimize self-absorption.
- Hybrid perovskites offer strong luminescence and tunable emission spectra, making them promising candidates for high-performance LSCs.
- Nanocomposite materials can be tailored to exhibit both high absorption efficiency and minimal re-absorption, enhancing photon transport through the waveguide.

By designing materials that balance high quantum efficiency, broad absorption spectra, and minimal spectral overlap, the re-absorption issue can be significantly reduced. Mitigating re-absorption losses is essential for maximizing light transport efficiency in LSCs. By employing materials with large Stokes shifts, incorporating energy transfer mechanisms like FRET, and optimizing luminescent material properties, LSCs can significantly reduce re-absorption losses and enhance photon delivery to PV cells. These innovations contribute to the development of next-generation solar concentrators with improved performance, scalability, and real-world applicability in building-integrated photovoltaics, transparent solar windows, and other renewable energy technologies.

7.2 *Photovoltaic Conversion*

The final step in the luminescent solar concentrator (LSC) process is the conversion of concentrated light into electrical energy through the photovoltaic (PV) effect. As the guided and amplified light reaches the edge-mounted PV cells, it is absorbed by semiconductor materials, generating electron–hole pairs that produce an electrical current. The enhanced light intensity and spectral optimization provided by the LSC significantly improve the efficiency of this process compared to traditional solar collection methods.

8 Key Factors Enhancing PV Performance in LSC Systems

1. **High-Efficiency PV Cells: Maximizing Energy Conversion with Minimal Material Use**

One of the primary advantages of LSC technology is its ability to concentrate light from a large area onto a much smaller PV cell area. This allows for the use of high-performance PV cells without the high costs associated with covering an entire surface with expensive semiconductor materials.

- **Multi-junction PV cells**: These advanced solar cells, consisting of multiple layers of semiconductor materials with varying bandgaps, can efficiently capture a broader spectrum of light, boosting overall conversion efficiency.
- **Perovskite and quantum dot-enhanced cells**: Due to their tunable optical properties, perovskite solar cells and quantum dot-based PV cells are well-suited for LSC applications, providing high efficiency even in low-light conditions.
- **III-V semiconductor PV cells**: These materials, commonly used in space and high-performance applications, can achieve superior conversion efficiencies when paired with LSCs, thanks to the concentrated photon flux.

By significantly reducing the required PV cell area, LSCs enable a cost-effective deployment of high-efficiency materials, making advanced solar technologies more economically viable.

2. **Optimized Energy Yield: Improved Light Utilization and Conversion Efficiency**

The uniform and concentrated illumination provided by the LSC improves PV cell performance by ensuring consistent light exposure, which enhances energy output compared to conventional solar panels.

- **Elimination of intensity fluctuations**: Unlike traditional solar setups that can experience uneven lighting conditions due to shading, dust, or angle variations, LSCs provide uniform photon delivery, reducing the risk of performance losses.

- **Spectral tuning for optimized absorption**: By selectively emitting light at wavelengths best suited for PV cells, LSCs minimize energy losses and ensure that the absorbed photons are efficiently converted into electricity.
- **Reduction in thermal losses**: Conventional solar panels suffer from thermalization losses, where excess energy from high-energy photons is dissipated as heat. LSCs mitigate this by shifting high-energy photons to lower-energy ones, better matching the PV cell's optimal operating range.

9 Additional Features and Benefits of LSC Technology

Beyond enhancing photovoltaic performance, LSCs offer several advantages that make them a promising solution for integrated and scalable solar energy applications.

1. **Diffuse Light Utilization: Effective Performance in Cloudy or Shaded Conditions**

One of the biggest challenges in traditional solar technology is the reliance on direct sunlight, which makes efficiency drop significantly under cloudy, shaded, or indirect light conditions. LSCs, however, are highly effective at capturing and converting diffuse sunlight, making them ideal for urban environments, overcast regions, and partially shaded locations.

- Unlike traditional concentrators, which rely on mirrors or lenses that work best with direct sunlight, LSCs efficiently harvest both direct and scattered sunlight, ensuring a more consistent power output.
- This ability makes LSCs especially useful for applications such as solar windows, building-integrated photovoltaics (BIPVs), and wearable solar devices, where light conditions can vary throughout the day.

2. **Aesthetic Integration: Customizable Designs for Architectural Applications**

LSCs can be engineered to be semi-transparent, colorful, or patterned, enabling their seamless integration into buildings and everyday structures without compromising aesthetics.

- **Transparent or tinted solar windows**: LSCs can be incorporated into glass facades, skylights, and windows, allowing buildings to generate solar energy while maintaining natural lighting and aesthetic appeal.
- **Decorative solar panels**: Unlike conventional panels, which are often bulky and visually intrusive, LSCs can be designed with various colors and artistic patterns, making them ideal for modern architectural designs, urban landscapes, and public installations.
- **Flexible and lightweight materials**: The versatility of LSC materials allows for integration into lightweight, bendable, and rollable solar applications, expanding the possibilities for portable energy solutions, vehicle-integrated photovoltaics (VIPVs), and wearable solar technology.

3. No Mechanical Tracking Required: Reduced Complexity and Maintenance Costs

Traditional solar concentrators often require mechanical tracking systems to follow the sun's movement, ensuring that light remains focused on the PV cells. This increases installation complexity, maintenance costs, and energy consumption. LSCs, however, offer a passive light concentration mechanism, eliminating the need for tracking systems.

- **Static operation**: LSCs efficiently capture sunlight from multiple angles, ensuring consistent energy generation throughout the day without repositioning.
- **Lower maintenance and durability**: Since LSCs contain no moving parts, they experience less wear and tear, reducing operational costs and enhancing long-term reliability.
- **Scalability and modular design**: The simplicity of LSCs allows for easy scalability, making them ideal for urban environments, industrial applications, and decentralized energy systems.

The LSC mechanism transforms how solar energy is captured and converted, enhancing photovoltaic efficiency while enabling versatile applications. By concentrating light onto high-efficiency PV cells, optimizing spectral matching, and eliminating the need for tracking systems, LSCs provide a cost-effective and architecturally adaptable solution for modern solar energy needs. With their ability to utilize diffuse sunlight, integrate into aesthetically pleasing designs, and reduce material costs, LSCs are paving the way for next-generation solar energy solutions that can be seamlessly embedded into everyday environments—from buildings and vehicles to personal electronics and smart infrastructure.

In summary, the operation of luminescent solar concentrators (LSCs) hinges on a sophisticated interplay of advanced material science and optical physics, resulting in a compact, highly efficient, and versatile system for solar energy harvesting. By harnessing the phenomena of luminescence, waveguiding, and the innovative design of materials, LSCs represent a promising pathway for producing scalable and cost-effective solutions to meet the growing global demand for renewable energy. The fundamental principle behind LSCs is their ability to absorb sunlight, convert it into lower-energy photons through luminescence, and then guide these photons to the edges of the concentrator where they can be efficiently converted into electricity by photovoltaic cells. This process relies on the careful selection and engineering of materials that can both efficiently absorb sunlight and ensure effective light trapping, reducing losses and maximizing the amount of energy that can be captured and utilized. The waveguiding aspect of LSCs is particularly significant. It involves the use of materials with specific optical properties that allow light to be confined within the concentrator, effectively channeling it toward the edges for conversion. This waveguiding mechanism is critical for optimizing the amount of light that interacts with the photovoltaic cells, enhancing overall efficiency even when the surface area of the collector is relatively small.

LSC materials are designed not only for their optical performance but also for aesthetic flexibility, enabling seamless integration into architectural and urban

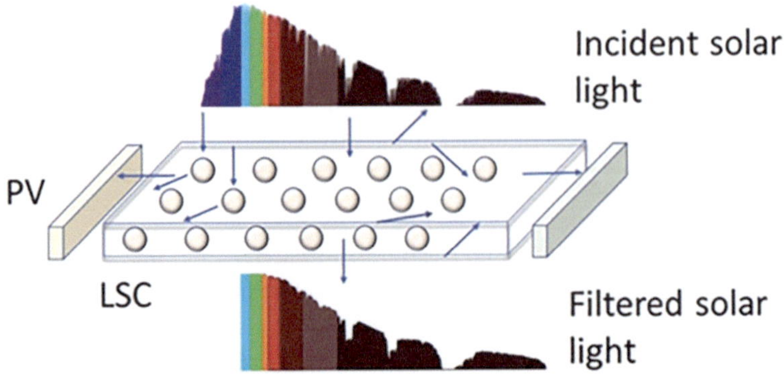

Fig. 2 Schematic model for luminescent-based solar concentrator [21]

settings. Transparent or semi-transparent LSCs can be incorporated into windows and facades, allowing solar energy harvesting without compromising design. These technologies address scalability and cost challenges in renewable energy by using low-cost materials and manufacturing methods, making them suitable for large-scale deployments in urban and commercial environments. Their compact size enables diverse applications, from solar farms to building-integrated systems. Figure 2 illustrates how LSCs absorb, convert, and guide light to photovoltaic cells. As LSC technology advances, it will play a crucial role in the transition to sustainable energy.

In light-guide-based solar concentrators, sunlight is coupled into a plate and then guided toward small PV cells by TIR. Such devices are gorgeous owing to their unique features, including low cost, thin geometry, ease of integration with appliances, etc. There are numerous methods to combine sunlight into a light guide, including non-luminescent methods such as refraction, scattering, diffraction, or luminescent strategies. The performance of non-luminescent concentrators suffers from limitations related to the Etendue conservation [9]. In the circumstance of non-tracking systems and light diffusion, the extreme achievable concentration equals n^2, where n is recognized as the refractive index of the light-guiding material. In the case of direct sunlight, only a limited angular part of the sky has to be covered, and the achievable concentration is greater [9].

LSCs do not undergo the above constraint. In an LSC, incident short-wavelength light is transformed into longer-wavelength photons by a luminescent source guided to the nearby PV cells. Suppose the energy difference among the incident short and emitted long-wavelength photons is $\Delta E = E_\mathrm{i} - E_\mathrm{o}$. In that case, the highest possible concentration equals $n^2(E_\mathrm{o}/E_\mathrm{i})^2 \exp(\Delta E/kT)$, where T is the concentrator's temperature. The heat ΔE produced during the luminescence procedure can be demoralized in lowering light entropy by an amount $\Delta E/T$ [9].

The fundamental perception of the device can be assumed by associating a scattering and a luminescent concentrator, as revealed in Fig. 3. Herein, in all the diagrams, a faultless bottom mirror in the device prevents light from absconding. In the scattering concentrator (Fig. 3a), the incident photons are scattered through

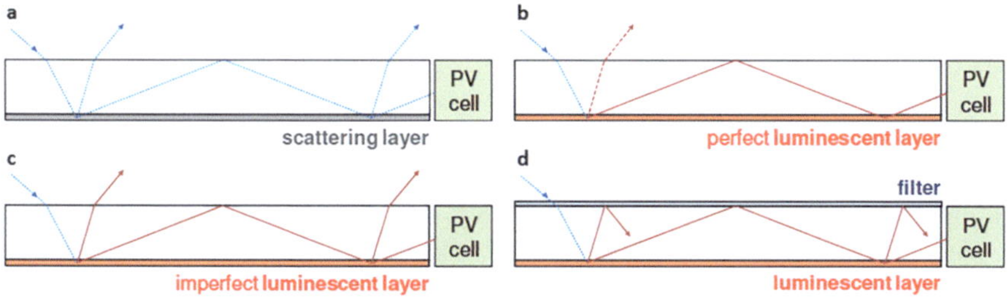

Fig. 3 **a** In a scattering concentrator, light primarily leaves the light guide through the process of scattering. **b** In a luminescent concentrator, the majority of the emitted luminescent light remains confined within the light guide. **c** If the luminescent material is imperfect, some of the luminescent light may escape. **d** A well-designed wavelength-selective filter prevents the loss of luminescent light [9]

a scattering film. A portion of the scattered light outflows, but a considerable part remains in the light guide. Nevertheless, if the light guide is lengthy, the remaining portion may reach the scattering layer and outflow. Hence, the realistic application of a scattering concentrator is restricted. In the LSC (Fig. 3b), the luminescent film absorbs the incident photons and transforms them into longer wavelength photons. A portion of this luminescent light outflows the surfaces without any supplementary actions [9].

A considerable amount of light will still remain within the guide. When this light interacts with the luminescent layer again, it will continue to propagate without restriction, as long as the luminescent layer is non-scattering and does not reabsorb the light. Under these ideal conditions, the system can achieve a high concentration of light, maximizing its efficiency. However, in practical scenarios, the luminescent film may not be perfect and could introduce some degree of scattering and reabsorption. If reabsorption occurs, a portion of the luminescent light may be lost, while another fraction could be reemitted at a different wavelength, as illustrated in Fig. 3c. This unwanted scattering and reabsorption can negatively impact the overall efficiency of the concentrator by reducing the amount of usable luminescent light directed toward the light guide. To enhance the performance of the devices shown in Figs. 3b and 3c, it is crucial to effectively manage the behavior of the luminescent light. One effective strategy involves the implementation of a wavelength-selective filter designed to optimize light retention within the system. Such a filter is engineered to reflect luminescent photons back into the light guide while allowing incident photons—those capable of being absorbed by the luminescent layer—to pass through. This ensures that the luminescent light remains confined within the system, minimizing losses due to scattering and reabsorption while maximizing the concentration of light within the guide, as illustrated in Fig. 3d. By employing this approach, the achievable concentration levels can be significantly increased, potentially reaching several thousand times the original intensity. This high concentration factor is essential for improving overall efficiency of the system, making it highly effective for applications requiring enhanced light capture and redirection, such as energy harvesting and optical signal processing [9].

A luminescent solar concentrator (LSC) is an advanced optical device designed to enhance solar energy harvesting by utilizing a transparent glass or plastic plate embedded with or coated in luminophores, such as organic dyes or phosphors. These luminophores play a crucial role in absorbing incoming sunlight and re-emitting it at longer wavelengths. Unlike conventional solar panels that require direct sunlight exposure, LSCs can efficiently capture diffuse sunlight, making them highly suitable for a variety of environments, including urban and shaded areas. Once the sunlight is absorbed and re-emitted, the emitted light is trapped within the plate through the principle of total internal reflection (TIR). This ensures that the light is guided toward the edges of the plate rather than escaping into the surroundings. At these edges, photovoltaic (PV) cells are strategically positioned to capture the concentrated light and convert it into electricity. This unique light-guiding mechanism significantly enhances energy conversion efficiency, as it enables the system to utilize both direct and indirect sunlight effectively. The concept of LSCs was first introduced over 30 years ago as a cost-effective alternative to traditional solar concentrators, which rely on bulky tracking systems to focus sunlight. Traditional LSCs typically consist of three key components: a transparent plate made of plastic or glass, luminescent dye molecules embedded within or applied as a coating, and photovoltaic cells attached to the edges. The dye molecules absorb incident sunlight, transfer the absorbed energy into the plate, and re-emit it as confined light, ensuring minimal loss. This guided light is then funneled toward the PV cells, where it is efficiently converted into electrical energy. One of the key advantages of LSCs is their potential for integration into various architectural and energy-harvesting applications. Unlike conventional solar panels, which are often bulky and require large surface areas, LSCs can be seamlessly incorporated into windows, facades, skylights, and even portable devices. Their ability to operate efficiently in low-light conditions further enhances their appeal for urban and indoor applications. Additionally, ongoing advancements in material science and nanotechnology have led to the development of more efficient and stable luminophores, further improving the performance and longevity of LSCs. Overall, luminescent solar concentrators offer a promising approach to sustainable energy generation by capturing and concentrating sunlight in a highly efficient manner. With continued research and technological advancements, LSCs have the potential to become a key player in the transition toward renewable energy solutions.

9.1 Efficient Photoluminescent Materials for the LSC

In LSCs, quantum dots (QDs) are advanced luminophores that enhance performance due to their unique optical properties. These nanoscale semiconductor particles absorb sunlight and emit light at longer wavelengths with high efficiency and tunable emission spectra. Quantum dots offer several advantages over traditional dyes or phosphors: [17, 22]

1. **Wide absorption range**: QDs can absorb a broad spectrum of sunlight, capturing more solar energy.
2. **Tunable emission**: By varying their size, the emission wavelength can be precisely controlled to match the optimal absorption range of photovoltaic (PV) cells.
3. **High quantum yield**: QDs exhibit excellent light conversion efficiency, reducing energy losses.
4. **Reduced reabsorption losses**: Carefully engineered QDs minimize self-absorption, enhancing light guidance to the edges.

In QD-based LSCs, the dots are embedded within or coated onto a transparent plate. They absorb sunlight, re-emit it at longer wavelengths, and confine it via total internal reflection, directing it to PV cells at the edges for electricity generation. This approach improves overall efficiency, making QD-LSCs a promising technology for cost-effective solar energy harvesting. Recently, quantum dots have gained good attention as lumiphores for LSCs because of their high quantum yield, tuneable absorption, and photoluminescent properties with high Stokes shift [17, 22]. LSCs based on QDs are reported by implanting diverse types of QDs and polymer composites, namely carbon dots in PMMA, and CuGaAlS/ZnS core/shell QDs in a polylauryl-methacrylate [23, 24]. However, QDs have serious toxicity issues. Furthermore, the visible light absorption nature of QDs make a typical color, which reduce their transparency and limits their use in several important PV devices including PV windows [22]. Zinc oxide (ZnO) QDs are looked more favorable compared to the other reported QDs for LSCs owing to their exclusive as well as tunable optical properties and high photoluminescent efficiency with outstanding stability [22]. Notwithstanding the imperfect absorption spectral region of ZnO because of the wider bandgap (3.3 eV), it is having high photon absorption ability in the longer energy zone of the solar spectrum, where a substantial solar energy is obtainable. Eventually, this high bandgap could allow absorption in the ultraviolet region, where silicon solar cells perform badly. Importantly, such band gap allows visible light transmission and enables the construction of smart transparent LSC devices without any color that can be used as PV windows in building integrated PV technologies. Other advantages of ZnO QDs for its application in LSCs comprise: the representative photoluminescent signal, situated in the visible spectral section, which is superior for the absorption of orthodox silicon solar cells; the eco-friendly synthetic procedure; and the comfort to acquire nanoparticles with desired shape, sizes, and surface properties, which can be a close match with the precise necessities of diverse LSCs [22].

M. J. Fimbres-Romero and coworkers fabricated transparent as well as colorless LSCs by implanting nontoxic and low-cost ZnO QDs with an average size of 5.5 nm in a PMMA matrix to utilize the QD's optical features and PMMA's transparency [22]. The synthesized colloidal ZnO QDs showed a broad yellow emission band under UV excitations and are transparent at the working concentrations (>95%

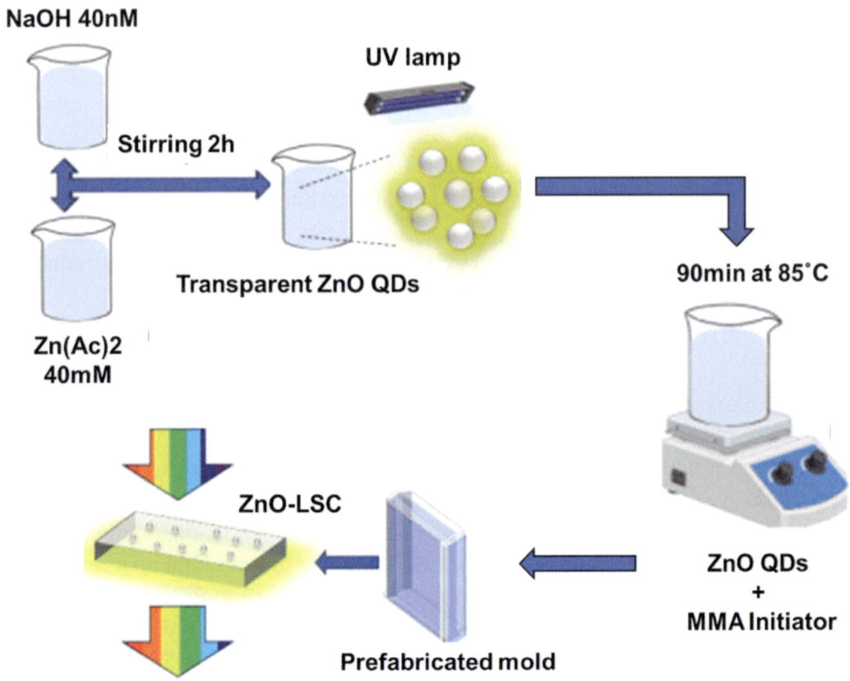

Fig. 4 Experimental techniques for producing colorless and transparent ZnO QD LSCs [22]

beyond 400 nm). The fabricated ZnO-LSCs presented a well-visible 80.3% average visible transmission. The optimized ZnO-LSC achieved a maximum power conversion efficiency of 5.45%. These outcomes specify that ZnO QD LSCs may be useful as transparent PV for their usage in BIPV. Overall experimental methodology of fabricating PMMA/ZnO-LSC fabrication is depicted in Fig. 4 [22].

It is crucial to emphasize the desirable optical properties that luminophores must possess for their effective application in luminescent solar concentrators (LSCs). These properties ensure optimal light absorption, emission, and guidance within the system, ultimately enhancing the efficiency of solar energy harvesting. The key optical features include: (1) minimal absorption in the visible spectral range to maintain high transparency and avoid undesired color tinting, (2) well-matched absorption and emission spectra with the specific photovoltaic (PV) cells used to maximize energy transfer, and (3) negligible spectral overlap between absorption and emission spectra to minimize self-absorption losses, which can significantly reduce device performance. As illustrated in Fig. 5, colloidal ZnO quantum dots (QDs) demonstrate strong photon absorption predominantly in the ultraviolet (UV) region, specifically below 375 nm, while exhibiting almost negligible absorption in the visible spectral range. This property is particularly advantageous because it ensures that ZnO QDs do not interfere with visible light transmission, maintaining the colorless and transparent nature of the material. This transparency is highly desirable for applications where LSCs need to be integrated into architectural elements such as windows, skylights, and building facades without altering their visual appearance. The inset of Fig. 5 visually confirms this property, showing that ZnO QDs maintain their transparency under ambient lighting conditions. In addition to their favorable absorption characteristics, ZnO QDs exhibit a broad and efficient emission spectrum. Figure 5 presents the emission spectrum of ZnO QDs in solution under UV excitation at 325 nm, revealing a wide emission range extending from 425 to 750 nm. The excitation spectrum peaks at approximately 360 nm, while the broad emission band reaches its highest intensity at around 540 nm. This emission is primarily attributed to defect-related states within the ZnO QD structure, a phenomenon that has been extensively documented in previous research. Such a broad emission spectrum is advantageous in LSC applications as it enables efficient spectral coverage, ensuring compatibility with a wide range of solar cells, particularly those optimized for visible light absorption. Furthermore, the inset of Fig. 5 demonstrates the optical behavior of ZnO QDs under different lighting conditions. Under normal visible light, the ZnO QDs appear transparent, reinforcing their suitability for applications requiring optical clarity. When exposed to UV light, they exhibit a distinctive yellowish emission, highlighting their strong luminescent properties. This feature is particularly useful in LSC systems, where effective photon emission and light guidance toward the photovoltaic cells are crucial for maximizing energy conversion efficiency. Overall, the unique optical characteristics of ZnO QDs align well with the fundamental requirements of transparent LSCs. Their ability to strongly absorb UV light while remaining highly transparent in the visible region allows them to function as efficient luminophores without compromising aesthetic transparency. Additionally, their broad emission spectrum enhances energy transfer to PV cells, improving overall device performance. These properties make ZnO QDs a promising candidate for the development of next-generation transparent LSCs, with potential applications in energy-efficient building-integrated photovoltaics, smart windows, and other innovative solar energy technologies. As research and material engineering continue to advance, further optimization of ZnO QDs may lead to even greater efficiency and stability, driving their widespread adoption in sustainable energy solutions [22–24].

Some of the desired properties of lumiphores to be used in transparent LSC technology are low absorption in the visible range of the electromagnetic spectrum, a luminescent spectrum in a wavelength range compatible with the attached solar cell, and a minimum overlap between their absorption and luminescent spectra. As shown in Fig. 5, the colloidal ZnO QDs show the characteristic strong absorption of photons below 375 nm, in the ultraviolet range, while a practically negligible absorption in the visible range, a property that makes the synthesized material colorless and highly transparent, as corroborated in the inset picture. Figure 4 also shows the photoluminescence spectrum of ZnO QDs in solutions under an excitation of 325 nm. Here, two signals are identified: the characteristics excitonic peak at around 360 nm, and the broad luminescent band located between 425 and 750 with a maximum at around 540 nm attributed to the collection of defects, as reported elsewhere. The inset picture displays the colloidal ZnO QDs under visible and ultraviolet excitation, showing high transparency and yellow luminescence produced by the defect band. The absorption

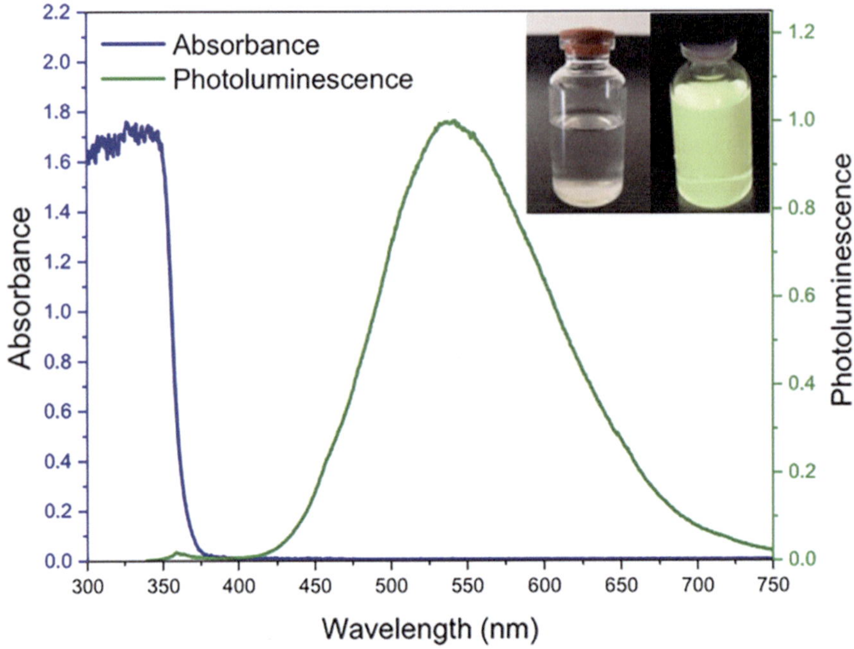

Fig. 5 Absorption and emission profiles of colloidal ZnO QDs [22]

and luminescence spectra of the synthesized ZnO QDs fulfill some of the most important properties to be considered as great candidates for luminophores in transparent LSC technology [25].

A major challenge with luminescent materials used in LSCs, such as organic dyes [26] and quantum dots [22], is significant reabsorption caused by overlap between their absorption and emission spectra. In contrast, rare-earth compounds offer greater potential due to their large spectral shifts and the forbidden nature of optical absorption between 4f levels. However, both inorganic phosphors [27] and organo-metallic rare-earth complexes [28] investigated so far absorb only a limited portion of the solar spectrum.

Optimizing phosphor properties requires balancing trade-offs. Reducing reabsorption necessitates a large spectral shift, which limits absorption at longer wavelengths. Additionally, common polymers used as light guides, such as polymethylmethacrylate (PMMA) [29] or polycarbonate, strongly absorb light above 850 nm due to C-H vibrational overtones. As a result, the ideal phosphor [30] for such light guides should have a broad absorption spectrum from the UV range to about 100 nm below its emission onset, with narrow emission centered around 800 nm. Phosphors based on Sm^{2+} fulfill most requirements for LSCs. As illustrated in Fig. 6, Sm^{2+} absorbs light via transitions from the 4f ground state to the 5d band. The position of the 5d band relative to the lowest 5D energy level is critical for optical performance [31]. For LSC applications, an optimal configuration occurs when the 5d band is not too

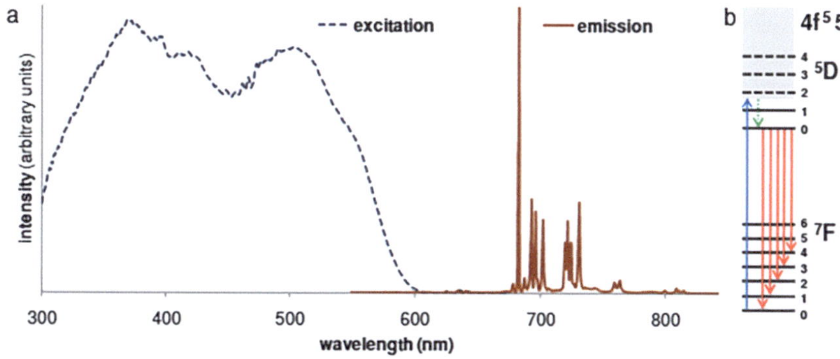

Fig. 6 **a** Excitation and **b** emission spectra of SrB_4O_7 doped with samarium and europium and corresponding energy level diagram [9]

close to the 5D level, allowing non-radiative relaxation into the 5D levels (Fig. 6, dashed line), followed by line emission. This configuration minimizes reabsorption due to the forbidden nature of the transitions, as seen in Sm^{2+} embedded in an SrB_4O_7 matrix.

Figure 6 shows the broad absorption and narrow emission spectra for this material, with a main emission peak at 685 nm [32] and smaller peaks extending to 820 nm. While the primary emission wavelength is shorter than ideal for LSCs, this system is the most effective phosphor we identified. Previous research indicates that the introduction of Eu^{2+} is necessary for significant absorption and emission, as Eu prevents the formation of Sm^{3+}, enabling efficient Sm^{2+} emission. Optimal results were achieved with equal concentrations of europium and samarium. Figure 6 confirms that no significant emission from Eu^{2+}, Eu^{3+}, or Sm^{3+} is observed, suggesting efficient energy transfer from Eu^{2+} to Sm^{2+} [32]. The phosphor SrB_4O_7: 5% Sm^{2+}, 5% Eu^{2+} was synthesized by mixing $SrCO_3$, Sm_2O_3, Eu_2O_3, and excess H_3BO_3. The mixture was fired at 800 °C in air, then at 935 °C in a CO atmosphere, and the excess boron oxide was dissolved in water. The resulting material was milled to an average particle size of ~ 5 μm. Figure 6 displays the material's absorption (excitation) and emission spectra, with Sm^{2+} emission lines at 685 nm, 695 nm, 720 nm, and 770 nm corresponding to transitions from $^5D_0$ to $^7F_0$, $^7F_1$, $^7F_2$, and $^7F_3$, respectively. The absorption coefficient is ~ 300 cm^{-1} at 500 nm and ~ 0.1 cm^{-1} at 685 nm. X-ray diffraction confirmed the absence of secondary phases in the material [32].

Efficient phosphors play a critical role in LSCs due to their ability to absorb sunlight, re-emit it at longer wavelengths, and guide it via total internal reflection to photovoltaic (PV) cells. The ideal phosphors for LSCs exhibit high quantum yield, broad absorption spectra, a large Stokes shift, and excellent stability under prolonged solar exposure. Efficient phosphors include rare-earth-doped materials specially the materials doped with Eu^{3+}, Tb^{3+}, Ce^{3+}, etc., transition metal ion-doped materials including the dopants as Cr^{3+}, Mn^{2+}, etc., upconverting elements doped with Er^{3+}, Yb^{3+}, Pb-based perovskite materials and its Pb free alternatives [33–41]. Jiurong

Li et al. investigated the fluorescence mechanism of CDs/RhB@SiO_2 composite nanoparticles by developing highly luminescent FRET-based materials. Their study emphasized the importance of carefully selecting an appropriate energy donor (CDs) and acceptor (RhB) to optimize energy transfer efficiency and enhance fluorescence performance [42]. The fundamental principle behind this selection is to ensure that the donor and acceptor possess well-matched spectral properties, allowing for efficient Förster Resonance Energy Transfer (FRET).

A key challenge in designing such composite materials is mitigating fluorescence quenching, which can significantly reduce the overall luminescence efficiency. During the composite formation process, quenching can occur if the CDs interact unfavorably with surrounding materials, leading to energy loss instead of efficient transfer. To overcome this issue, it is crucial to use CDs with appropriate surface functional groups, well-matched absorption and emission wavelengths, and large Stokes shifts. Large Stokes shifts minimize self-absorption by ensuring that the emission wavelength is sufficiently separated from the absorption wavelength, preventing reabsorption of emitted light. This careful selection process is essential for realizing the full potential of the FRET mechanism and maintaining high fluorescence efficiency in the final composite material. FRET, a non-radiative energy transfer mechanism, typically occurs between a donor and an acceptor through dipole–dipole interactions. As depicted in Fig. 7, the FRET process in CDs/RhB@SiO_2 composites involves the donor CDs transferring energy to the acceptor RhB molecules. When the CDs enter an excited state after absorbing energy, they interact with nearby RhB molecules via dipole–dipole coupling. This interaction leads to the deactivation of the CDs' excited state while simultaneously exciting multiple RhB molecules. Consequently, RhB emits energy at longer wavelengths, producing a characteristic fluorescence emission.

This well-orchestrated energy transfer mechanism significantly enhances the luminescence properties of the composite, making it a promising material for a wide range of applications. The ability to fine-tune the donor–acceptor pairing allows

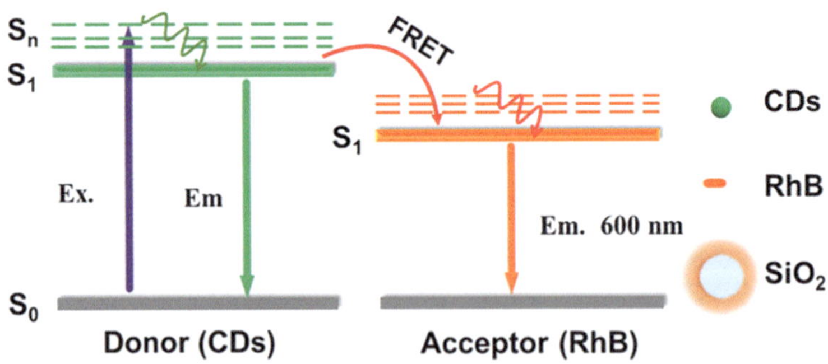

Fig. 7 Energy transfer takes place between a donor and an acceptor. The figure depicts FRET between donor CDs and acceptor RhB, where dipole–dipole interactions deactivate the CDs' excited state, exciting RhB and resulting in long-wavelength emission [42]

researchers to optimize emission wavelengths for specific uses, such as bioimaging, optoelectronic devices, and photovoltaic applications. Furthermore, the transparency and biocompatibility of silica (SiO_2) make these composite nanoparticles highly suitable for biomedical imaging, drug delivery, and sensor development. In summary, the study by Jiurong Li et al. highlights the crucial role of selecting suitable CDs to facilitate efficient FRET-based fluorescence enhancement. By ensuring optimal spectral alignment, minimizing fluorescence quenching, and leveraging dipole–dipole interactions, these composite nanoparticles achieve superior luminescence performance. Continued research in this area is expected to further refine the design and functionality of FRET-based nanomaterials, expanding their potential applications in both scientific and commercial domains.

Building on the research by Kantarod et al. [43], who developed a series of heptagon-embedded push–pull molecules with exceptionally high quantum yields (up to 90%) and small HOMO–LUMO energy gaps to enhance solar energy harvesting, Alberto Picchi et al. advanced this concept by fabricating luminescent solar concentrator (LSC) slabs using two of the most efficient compounds identified in that study [44]. The first fluorophore, designated as Nap2Car and illustrated in Fig. 8a, is a push–pull molecule where a naphthalene dicarboximide unit serves as the electron-accepting moiety, while a dibenzo-heptagon structure substituted with two carbazole groups' functions as the electron donor. This carefully designed molecular architecture enhances intramolecular charge transfer, improving its photophysical properties and overall efficiency in LSC applications. The second fluorophore, referred to as Peri2F and also shown in Fig. 8a, shares a similar structural framework but differs in its core composition. Instead of a naphthalene dicarboximide acceptor, Peri2F incorporates a perylene dicarboximide core embedded with a seven-membered ring. Additionally, its donor moiety—a di-benzoheptagon—contains two fluorine substituents, which influence its electronic properties and stability. Compared to Nap2Car, Peri2F exhibits unique bonding characteristics that facilitate the formation of stable radicals through homolytic cleavage, potentially contributing to improved photostability and charge transport efficiency.

A key advantage of incorporating a heptagonal ring into the molecular design is its impact on the overall shape and packing behavior of the fluorophore. The curved geometry induced by the seven-membered ring disrupts molecular aggregation, thereby reducing self-quenching effects that often limit the performance of organic luminophores in LSC applications. Moreover, this structural rigidity helps maintain a fixed molecular conformation, ensuring consistent optical performance and contributing to the exceptionally high quantum yield observed in these compounds. By integrating these advanced fluorophores into LSC slabs, Picchi et al. demonstrated the potential of heptagon-embedded push–pull molecules for enhancing solar energy collection and conversion. Their work paves the way for the development of highly efficient, stable, and scalable materials for next-generation photovoltaic applications, offering promising solutions for improving light harvesting efficiency in sustainable energy technologies.

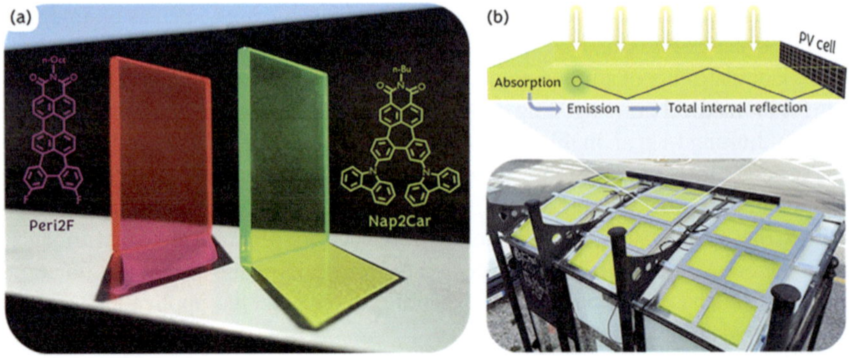

Fig. 8 **a** Molecular structures of Peri2F and Nap2Car, along with their respective LSC visual appearances; **b** representation of LSC operation [44]

Rare-earth-doped materials are among the most efficient and stable phosphors used in LSCs, offering sharp emission peaks and excellent durability. Europium-doped phosphors (e.g., Y_2O_3: Eu^{3+}) emitting sharp red light with high luminescence efficiency are in the forefront of this research. These materials are ideal for coupling with silicon-based PV cells that operate efficiently in the red spectrum and highly resistant to photobleaching, making them suitable for outdoor applications. Another kinds of materials are terbium-doped phosphors (e.g., $LaPO_4$: Tb^{3+}), which emit green light, complementing solar cells optimized for shorter wavelengths. These materials are known for their chemical and thermal stability [33, 34]. Cerium-doped phosphors (e.g., YAG: Ce^{3+}) also emit in the yellow range with a broad spectrum, enabling better solar spectrum coverage. These materials are widely used due to their durability and efficient energy conversion [33, 34].

Transition metal-doped phosphors exhibit tunable emission properties and are suitable for a wide range of LSC designs. Among all transition metal-doped phosphors, Mn^{2+}-doped phosphors (e.g., ZnS: Mn^{2+}), which can emit orange or yellow light with a relatively large Stokes shift, reducing reabsorption losses can be a cost-effective and versatile for integration into polymer matrices [35]. Cr^{3+}-doped phosphors (e.g., Al_2O_3: Cr^{3+}, Ruby) emitting in the red to near-infrared region, making them ideal for pairing with specific PV technologies like gallium arsenide or perovskite solar cells. These materials are extremely stable under high temperatures and UV radiation [36].

The development of high-efficiency luminescent solar concentrators (LSCs) is a complex process that requires selecting optimal phosphor materials, which play a pivotal role in determining the overall performance of these devices. Phosphor materials need to possess several key properties, including high quantum efficiency, excellent stability, and minimal reabsorption losses. These factors are essential for maximizing the energy conversion efficiency of the LSC. Among the wide range of phosphor materials available, perovskite-based, copper-doped, carbon dot-based, upconversion, and silicate-based phosphors have emerged as highly promising candidates. Each of these materials has distinct characteristics that can be tailored to enhance the performance of LSCs, making them highly valuable in next-generation solar energy technologies. Perovskite phosphors, particularly lead halide perovskites, have garnered significant interest in the LSC field due to their outstanding optoelectronic properties. These materials exhibit high quantum yield, which allows them to effectively absorb and re-emit sunlight. Lead halide perovskites such as $CsPbBr_3$ are capable of tunable emission across the visible spectrum, offering flexibility in adapting to the specific needs of various LSC designs. These properties, combined with their strong absorption characteristics, enable perovskite phosphors to efficiently capture sunlight and convert it into usable energy. The compatibility of perovskites with thin-film fabrication techniques further enhances their potential, enabling the development of flexible and versatile LSC designs that can easily integrate into modern solar technologies. Moreover, the ability to fine-tune the properties of perovskite materials has led to the development of doped perovskites. For instance, Mn^{2+}-doped $CsPbCl_3$ improves the stability of the material and reduces reabsorption losses, ensuring that the high emission efficiency is maintained over time. These modifications are particularly beneficial for applications requiring long-term environmental stability, which remains one of the main challenges with conventional perovskite materials [37, 38].

As the demand for sustainable technologies grows, there has been an increasing focus on environmentally friendly phosphors. Copper-doped phosphors such as ZnSe:Cu are emerging as a viable non-toxic alternative to traditional lead-based phosphors. These materials offer strong visible-light emission, making them effective for use in LSCs, while their environmentally friendly nature makes them highly attractive for large-scale applications. Copper-doped phosphors are also cost-effective, providing an affordable solution for solar energy systems. Carbon dot-based phosphors represent another promising alternative due to their low toxicity, tunable emission properties, and excellent photostability. Carbon dots are also easily synthesized from inexpensive and abundant precursors, which makes them an economically viable option for large-scale deployment. The ability of carbon dots to absorb and re-emit light across a wide spectrum enables them to effectively harness solar energy, contributing to high-efficiency LSCs. Additionally, their robust performance in outdoor environments ensures that these materials can maintain their luminescence over extended periods, making them suitable for long-term use in solar energy applications [39, 40].

One of the key challenges in traditional solar energy conversion systems is the inefficient utilization of low-energy (infrared) photons. Upconversion phosphors address this issue by converting low-energy photons into higher-energy emissions, which expands the usable portion of the solar spectrum. Materials such as $NaYF_4:Er^{3+},Yb^{3+}$ are capable of absorbing near-infrared (NIR) light and converting it into visible light, thereby enhancing the LSC's ability to capture and convert sunlight. The upconversion process allows LSCs to tap into previously underutilized regions of the solar spectrum, improving the overall energy conversion efficiency. By modifying the concentration of rare-earth dopants, such as Er^{3+}, Yb^{3+}, Tm^{3+}, and Ho^{3+}, researchers can further tailor the emission spectrum of upconversion phosphors to meet the specific requirements of different photovoltaic cells. This tunability ensures that upconversion phosphors can be optimized for hybrid LSC systems, where both visible and infrared light are harnessed to maximize energy capture [41].

For LSCs deployed in outdoor environments or in industrial applications, durability is a key consideration. Silicate-based phosphors have proven to be highly durable under harsh environmental conditions, making them an ideal choice for outdoor and industrial LSC applications. These phosphors offer high photoluminescence efficiency while maintaining their stability in the face of environmental stressors such as high temperatures, humidity, and chemical exposure. Unlike organic dyes or certain types of perovskites, silicate-based phosphors demonstrate excellent thermal and chemical stability, making them well-suited for long-term solar applications. Their broad emission spectrum allows them to be used in a variety of LSC designs, offering flexibility for different types of solar concentrators. Additionally, their robust structural integrity makes them compatible with both glass-based and polymer-based LSC designs, ensuring long-term durability and enhancing the overall performance of building-integrated photovoltaics (BIPVs), solar windows, and flexible solar panels [42].

The selection of the right phosphor material is critical for optimizing the efficiency, scalability, and cost-effectiveness of LSC technology. By utilizing perovskite phosphors for high quantum yield and spectral tunability, copper-doped and carbon dot-based materials for sustainable, non-toxic solutions, upconversion phosphors to extend the usable solar spectrum, and silicate-based phosphors for their durability and outdoor resilience, LSC technology is poised to continue its evolution toward higher performance, lower costs, and greater real-world applicability. As the field of LSC technology advances, ongoing research and material innovations will drive further improvements in efficiency, scalability, and cost reduction. This progress will make advanced LSCs a key player in the future of solar energy, helping to accelerate the transition to a renewable energy future. The continuous refinement of LSC materials will not only enhance their energy harvesting capabilities but also expand their potential applications, making them a crucial component of the growing renewable energy landscape.

10 Fluorescent Thin Films for LSC

Fluorescent thin films play a pivotal role in the development of highly efficient and adaptable luminescent solar concentrators (LSCs). These films are engineered to absorb sunlight, convert it into longer-wavelength emissions, and guide the emitted photons toward edge-mounted photovoltaic (PV) cells, where they are ultimately converted into electrical energy. The integration of luminophores—including organic dyes, quantum dots (QDs), phosphors, and perovskites—into these films enhances light absorption, optimizes emission properties, and improves overall solar energy conversion efficiency. Fluorescent thin films are essential for maximizing light-harvesting efficiency, minimizing reabsorption losses, and ensuring long-term stability, all of which contribute to the practical implementation of LSC technology in real-world applications. The following sections provide a comprehensive exploration of the types of fluorescent thin films, their material compositions, unique properties, and applications in LSCs [31–41].

10.1 Polymer-Based Fluorescent Thin Films

Poly(methyl methacrylate) (PMMA)-doped film is a widely used transparent polymer matrix for LSC applications. It is typically doped with fluorescent dyes like Rhodamine B, Coumarin, or Lumogen F. The properties of these films include high optical clarity and UV resistance, excellent compatibility with a wide range of organic dyes and mechanically robust, making it suitable for outdoor applications. These materials can be applied to window-mounted LSCs for urban environments and lightweight and flexible solar harvesting devices [45]. *Polycarbonate (PC) films* are another durable polymer often used with fluorescent dyes or quantum dots. These have high thermal stability compared to PMMA, making it suitable for high-temperature conditions. These are superior impact resistance, enabling use in rugged environments. These films can be used in rooftop LSC systems where durability is critical [45]. *Polyethylene terephthalate (PET) films* are coated with fluorescent dyes and laminated onto substrates. These films are flexible and lightweight, low-cost and scalable for large-area LSCs. These films can be applied to curved surfaces or portable LSC systems [45, 46].

10.2 Quantum Dot (QD)-Embedded Thin Films

Cadmium selenide (CdSe) quantum dots with a zinc sulfide (ZnS) shell are popular for their tunable optical properties and high quantum yield. These films are having absorption across a broad solar spectrum. The films are highly efficient re-emission with minimal reabsorption losses and have tunable emission wavelengths by controlling quantum dot size. These films can be applied to high-efficiency LSCs for rooftop or building-integrated systems [47]. *Carbon quantum dot films* (CQDs) are a sustainable, non-toxic alternative to traditional quantum dots. These films are having broad absorption range and adjustable emission properties and low environmental impact and ease of synthesis. These films can be applied to eco-friendly LSC designs for residential and urban use [45–47]. *Silicon quantum dot films* provide a non-toxic and abundant alternative to cadmium-based materials having high stability under UV light and emission in the near-infrared range, ideal for pairing with specialized PV cells. These films can be applied to advanced LSCs integrated with tandem solar cells [47].

10.3 Dye-Loaded Glass or Silica Thin Films

Organic dye-doped glass films or thin glass films doped with organic dyes such as Nile Red, Rhodamine 6G, or Fluorescein are in this list having durable and resistant to weathering, and high photoluminescence efficiency. These can be applied to transparent LSCs for architectural glazing systems [48]. *Silica thin films with dyes* are robust hosts for dyes, ensuring long-term stability and reducing photobleaching. These films have enhanced thermal and chemical stability, and compatibility with various dye molecules for long-lasting LSC modules for outdoor use [48].

10.4 Perovskite-Based Fluorescent Thin Films

Lead halide perovskites (e.g., $MAPbBr_3$, $CsPbBr_3$) thin films are a cutting-edge material for LSCs, offering high quantum yield and tunable emission spectra. The properties include broad absorption range and efficient re-emission, ease of fabrication into thin films using solution processing and potential for high efficiency in LSCs. The challenges are the stability under humidity and UV exposure remains a limitation, requiring encapsulation strategies. These materials can be applied to high-performance LSCs for commercial and industrial applications [49, 50]. *Non-toxic perovskites (e.g., $Cs_2AgBiBr_6$),* i.e., lead-free perovskites, are being developed to address toxicity concerns while retaining excellent optical properties. These perovskites have moderate quantum efficiency with improved environmental stability and potential for scaling to large-area applications. These can be applied to sustainable LSC solutions for eco-conscious markets [51].

10.5 Hybrid Thin Films

Combining organic dyes with inorganic nanoparticles like TiO_2 or ZnO to make *organic–inorganic composite films* to enhance light-harvesting and emission efficiency. These films can have broader absorption spectrum due to synergistic effects, improved stability and reduced reabsorption losses for advanced LSC designs for optimized performance [51, 52]. *Quantum dot-dye hybrid films* combine quantum dots and fluorescent dyes to capture and emit light more effectively with enhanced photostability and efficiency and ability to fine-tune emission properties by varying material combinations for high-efficiency, large-area LSCs for building facades [51, 52]. Building-integrated photovoltaics (BIPVs) are gaining significant traction worldwide as they combine renewable energy generation with functional building components like roofs, windows, and facades. This dual-purpose integration is a key driver in the rapid adoption of BIPVs within the photovoltaic (PV) industry. Among the innovative technologies enabling this trend is the luminescent solar concentrator (LSC), a system designed to enhance solar energy efficiency while maintaining architectural aesthetics [41, 45–47].

LSCs consist of transparent plates made from materials such as glass or plastic embedded with luminescent materials, including organic dyes, quantum dots, or nanostructured inorganic substances. These materials absorb sunlight, re-emit it at longer wavelengths, and guide the light through total internal reflection to the plate's edges, where solar cells convert it into electricity. By replacing the need for large, expensive solar cell arrays with a lightweight, cost-effective plate, LSCs make solar energy integration more accessible. They can also capture both direct and diffuse sunlight without requiring complex or expensive tracking systems, offering flexibility in terms of color, shape, and size [46–50]. Despite these advantages, LSCs face several challenges that hinder their widespread adoption. Reabsorption losses, where emitted light is absorbed again by the luminescent material, reduce efficiency. Additionally, dyes and quantum dots often lack stability when exposed to prolonged sunlight or high temperatures. Furthermore, highly efficient quantum dots are not yet commercially available, and existing options are costly, less efficient, and prone to degradation over time [49–51]. To address these issues, researchers have proposed an advanced LSC design featuring a thin-film coating of luminescent materials on a transparent plate as shown in Fig. 9. This approach confines absorption and emission processes to the thin film, minimizing reabsorption losses. Using high-purity materials with excellent optical clarity further enhances the system's efficiency. This thin-film LSC design represents a promising step toward overcoming current limitations, making LSCs a more viable and efficient option for BIPV applications while maintaining their architectural appeal [50–53].

For LSCs in building-integrated photovoltaic (BIPV) systems, suitable luminescent materials must exhibit high quantum yield (QY), long-term stability, and minimal reabsorption. Inorganic phosphors, commonly used in white light-emitting diodes (LEDs), show promise for LSC applications due to their advantages over organic dyes. For instance, YAG:Ce boasts a quantum efficiency exceeding 85% even at 200 °C and demonstrates excellent photostability and resistance to UV radiation and moisture, making it ideal for outdoor use [48–52]. Inorganic phosphors are also cost-effective, easy to handle, and more thermally stable compared to organic dyes and quantum dots. These qualities make them well-suited as luminescent materials for LSCs. Research should focus on assessing the optical performance of LSC structures under solar irradiation, particularly those incorporating a thin film of inorganic phosphor on a transparent plate [51–53]. However, few studies have focused on the use of inorganic phosphor in LSCs. de Boer et al. [54] demonstrated that Sm^{2+}-based phosphors meet most requirements for LSCs, showing that placing a thin phosphor layer at the bottom of the LSC with a wavelength-selective filter on top effectively reduces reabsorption and cone losses. Similarly, Desmet et al. [55] designed an LSC with an inorganic phosphor layer at the bottom, achieving an overall system efficiency of 2.5% using a microcellular polyethylene terephthalate (MCPET) reflector behind the phosphor layer. Hughes et al. [13]. explored LSCs with a phosphor layer, using a widely available phosphor to highlight their potential for commercialization.

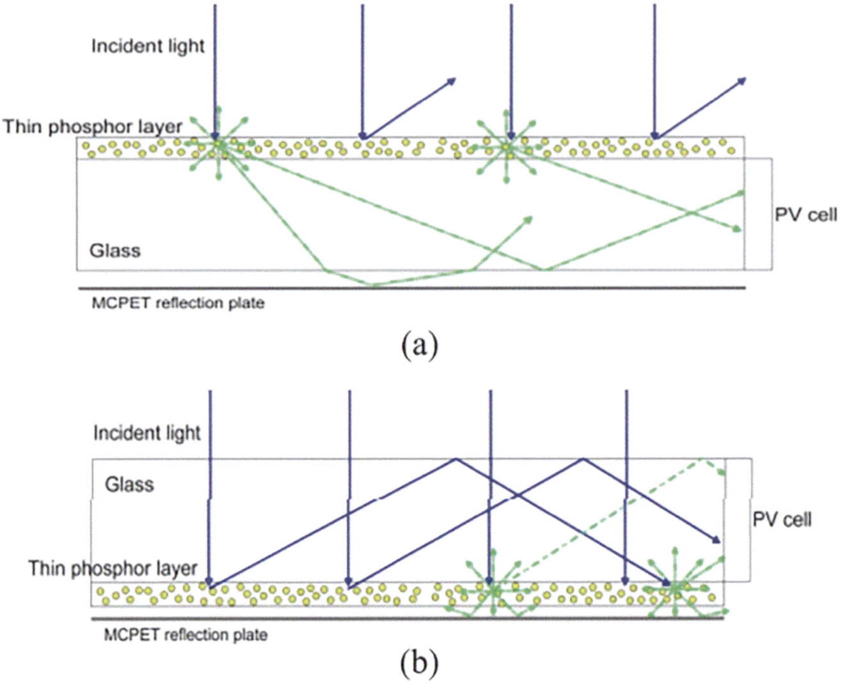

Fig. 9 Schematic of the thin film LSCs using inorganic phosphor with thin phosphor layer on **a** top surface and **b** bottom surface of the glass plate [56]

Through Monte Carlo simulations, they optimized the phosphor layer's location and concentration, finding that a high-concentration phosphor layer at the bottom of the LSC significantly enhances light concentration. Their work emphasized the importance of further research on thin-film LSCs incorporating various commercial phosphors to advance the technology for practical applications. Shang-Ping Ying and colleagues [56]. studied the performance of thin-film luminescent solar concentrators (LSCs) using commercial phosphors and photovoltaic (PV) cells. They explored various designs with different phosphor types, concentrations, and layer positions to optimize efficiency and provide guidelines for commercialization.

11 Effect of Geometry on Quantum Efficiency of LSC

Figure 11 illustrates how the length of the LSC influences both external and internal light transport processes, keeping all other parameters at their baseline values. As the LSC length increases significantly, the external photon efficiency experiences a slight decline of 8.0%, decreasing from 12.9% at 5 cm to 11.8% at 150 cm. This suggests that the performance of a lab-scale LSC can be largely maintained even

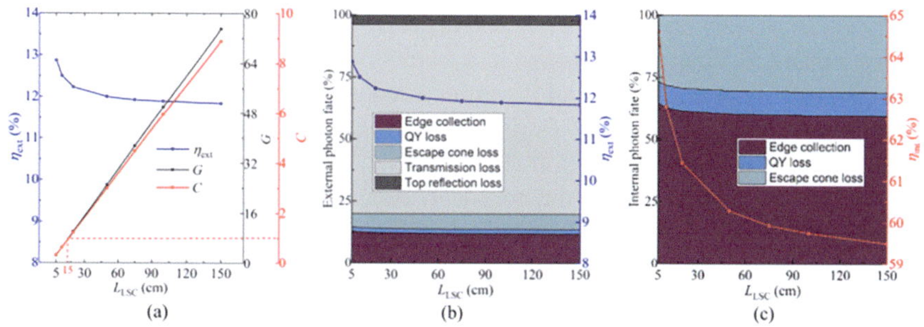

Fig. 10 Impact of LSC length on external photon efficiency, along with geometric gain and concentration factor [57]

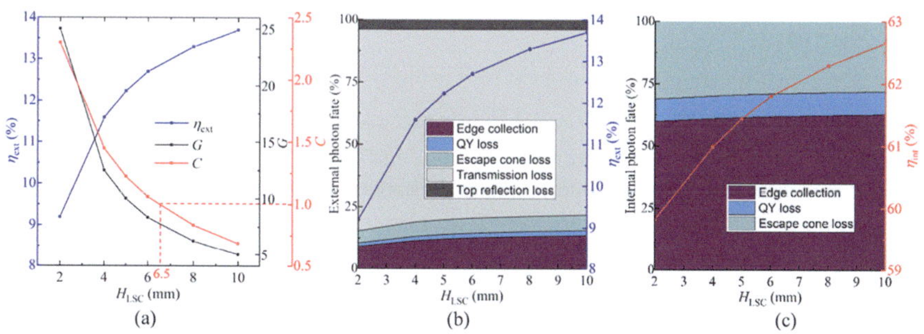

Fig. 11 Influence of LSC thickness on external photon efficiency, along with geometric gain and concentration factor [57]

when scaled up to a pilot-scale device, indicating promising potential for future commercial applications [21, 57]. As a result, the concentration factor increases consistently—from 0.32 at 5 cm to 8.9 at 150 cm—due to the predominant effect of the linear rise in geometric gain (Fig. 10a).

As depicted in Fig. 10b, the contributions from top reflection loss (3.9%) and transmission loss (76.2%) remain unchanged, meaning that LSC length primarily affects internal light transport. A longer LSC extends the light transport path, leading to more photoluminescence (PL) reabsorption events by downstream LFR305 particles. This results in increased escape cone loss and quantum yield (QY) loss (Fig. 10c). Consequently, internal photon efficiency declines steadily with increasing LSC length, dropping from 64.6% at 5 cm to 59.5% at 150 cm. Based on these findings, a meter-scale LSC is recommended to achieve a high concentration factor while maintaining external photon efficiency above 11%.

Figure 11 examines how LSC thickness affects performance. When thickness increases from 2 to 10 mm, geometric gain decreases by 80% (from 25 to 5), while external photon efficiency improves by 49% (from 9.2% to 13.7%). Since the decline in geometric gain has a greater impact, the overall concentration factor decreases from 2.3 at 2 mm to 0.7 at 10 mm (Fig. 11a). For external light transport (Fig. 11b), greater

LSC thickness means an extended path along the light incidence direction, enhancing absorption and reducing transmission loss from 80.8% at 2 mm to 74.3% at 10 mm. This results in more photons being PL emitted and guided to the edges through total internal reflection (TIR), though escape cone and QY losses also increase slightly. Regarding internal photon transport (Fig. 11c), η_{int} exhibits a sublinear increase as LSC thickness grows, primarily due to a decrease in escape cone loss—from 31.1% at 2 mm to 28.0% at 10 mm—among absorbed photons. The maximum η_{int} of 62.7% occurs at a thickness of 10 mm.

Since concentration factor and external photon efficiency exhibit opposite trends, a balance must be struck when determining LSC thickness. However, to prevent an undesirable concentration factor below one, the thickness should not exceed 6.5 mm.

12 Summary

Luminescent solar concentrators (LSCs) represent an innovative approach to harnessing solar energy, offering a promising alternative to traditional photovoltaic (PV) systems. Given the challenges associated with conventional solar panels, such as high costs, efficiency losses due to uneven illumination, and difficulties in integrating them into urban environments, LSCs provide a viable solution. These devices utilize luminescent materials to absorb sunlight, re-emit it at longer wavelengths, and guide the emitted photons toward PV cells through total internal reflection. Their ability to function efficiently without mechanical tracking makes them lightweight, versatile, and suitable for various applications, including building-integrated photovoltaics (BIPV), wearable electronics, and portable energy systems.

The working mechanism of LSCs involves four primary steps. Initially, luminescent materials embedded within a transparent waveguide absorb incoming sunlight. These materials, which include organic dyes, quantum dots, and perovskites, are selected based on their ability to capture specific wavelengths of light. Once absorbed, the light is re-emitted at longer wavelengths, a process known as the Stokes shift, which minimizes re-absorption losses and facilitates efficient photon transport. The emitted photons are then guided through the waveguide by total internal reflection, ensuring they reach the device's edges where PV cells convert the concentrated light into electricity. By employing materials with large Stokes shifts, advanced energy transfer mechanisms, and optimized device structures, LSCs can achieve high efficiency while mitigating energy losses.

Material selection plays a crucial role in determining the efficiency of LSCs. Quantum dots (QDs), for instance, have gained significant attention due to their tunable emission properties, high quantum yield, and reduced re-absorption losses. Among them, zinc oxide (ZnO) quantum dots have emerged as a particularly promising material due to their non-toxic nature, high stability, and strong absorption in the ultraviolet range, making them ideal for transparent photovoltaic windows. Rare-earth-doped phosphors, such as those containing Sm^{2+} and Eu^{2+}, are also being explored for their ability to exhibit large spectral shifts, which help minimize re-absorption losses. Additionally, transition metal-doped phosphors like Mn^{2+}-doped ZnS and Cr^{3+}-doped Al_2O_3 offer excellent optical properties for solar energy applications. Lead halide perovskites have shown exceptional light conversion efficiency and tunability, though challenges related to their long-term stability remain.

Advancements in LSC design have significantly improved their performance and integration potential. Thin-film LSCs, where luminescent coatings are applied to transparent plates, have been developed to minimize re-absorption and enhance efficiency. Hybrid LSCs, which combine organic dyes with inorganic materials such as quantum dots, further optimize light absorption and emission properties. Another key development is the use of LSCs in building-integrated photovoltaics, where they can be seamlessly incorporated into windows, facades, and rooftops, allowing for solar energy generation without compromising architectural aesthetics. The geometry of LSCs plays a crucial role in their overall efficiency. Studies have shown that increasing the length of an LSC results in a slight decline in external photon efficiency but significantly enhances the concentration factor, making longer LSCs more effective for energy harvesting. Similarly, adjusting the thickness of an LSC affects its performance. Thicker LSCs tend to have higher external photon efficiency but lower geometric gain, leading to reduced concentration factors. To maintain optimal efficiency, a thickness of less than 6.5 mm is generally recommended. Despite their advantages, LSCs still face several challenges that must be addressed to facilitate their widespread adoption. Issues such as material stability, re-absorption losses, and cost-effective scalability continue to hinder commercialization. Future research is focused on developing non-toxic and highly stable luminescent materials, improving device architectures through photonic structures and spectral conversion layers, and integrating LSCs with emerging solar technologies such as tandem solar cells and transparent PV systems. By addressing these challenges, LSCs have the potential to revolutionize solar energy harvesting, particularly in urban settings where space constraints and architectural considerations play a crucial role. With continued advancements, LSCs could become a key component of future solar energy solutions, making renewable energy more accessible, efficient, and aesthetically adaptable.

References

1. A.S. Al-Ezzi, M.N.M. Ansari, Photovoltaic solar cells: a review. Appl. Syst. Innovation **5**, 67 (2022)
2. F.J. Nijsse, J.F. Mercure, N. Ameli, F. Larosa, S. Kothari, J. Rickman, P. Vercoulen, H. Pollitt, The momentum of the solar energy transition. Nat. Commun. **14**, 6542 (2023)
3. G. Li et al., Solar energy utilisation: current status and roll-out potential. Appl. Therm. Eng. **209**, 118285 (2022)
4. D. Hao et al., Solar energy harvesting technologies for PV self-powered applications: A comprehensive review. Renew. Energy **188**, 678–697 (2022)
5. K. Sodiqjon, S. Begyor, K. Aleksandr, D. Farrukh, M. Mukhtasar, A. Akbarjon, Prospective aspects of using solar energy. J. New Century Innov. **18**, 142–148 (2022)

6. M. Dada, P. Popoola, Recent advances in solar photovoltaic materials and systems for energy storage applications: a review. J. Basic Appl. Sci. **12**, 66 (2023)
7. L. Alhmoud, Why does the pv solar power plant operate ineffectively? Energies **16**, 4074 (2023)
8. L. Li, B. Wang, R. Bader, T. Cooper, W. Lipiński, Chapter one—concentrating collector systems for solar thermal and thermochemical applications. Adv. Chem. Eng. Acad. Press. **58**, 1–53 (2021)
9. D.K.G. de Boer, D.J. Broer, M.G. Debije, W. Keur, A. Meijerink, C.R. Ronda, P.P.C. Verbunt, Progress in phosphors and filters for luminescent solar concentrators. Opt. Express **20**, A395–A405 (2012)
10. P.P.C. Verbunt, A. Kaiser, K. Hermans, C.W.M. Bastiaansen, D.J. Broer, M.G. Debije, Controlling light emission in luminescent solar concentrators through use of dye molecules aligned in a planar manner by liquid crystals. Adv. Funct. Mater. **19**, 2714–2719 (2009)
11. M.G. Debije, M.P. Van, P.P.C. Verbunt, M.J. Kastelijn, R.H.L. van der Blom, D.J. Broer, C.W.M. Bastiaansen, Effect on the output of a luminescent solar concentrator on application of organic wavelength-selective mirrors. Appl. Opt. **49**, 745–751 (2010)
12. D.K.G. de Boer, C.W. Lin, M.P. Giesbers, H.J. Cornelissen, M.G. Debije, P.P.C. Verbunt, D.J. Broer, Polarization-independent filters for luminescent solar concentrators. Appl. Phys. Lett. **98**, 021111 (2011)
13. M.D. Hughes, D.A.B. Tasciuc, D.A. Kaminski, Highly efficient luminescent solar concentrators employing commercially available luminescent phosphors. Sol. Energy Mater. Sol. Cells **171**, 293–301 (2017)
14. M. Cao, X. Zhao, X. Gong, Achieving high-efficiency large-area luminescent solar concentrators. JACS Au **3**, 25–35 (2023)
15. J. Choi, S.J. Kim, A highly transparent and efficient luminescent solar concentrator based on nanosized molybdenum clusters and quantum-cutting perovskite nanocrystals. J. Mater. Chem. C **12**, 12026–12036 (2024)
16. P. Moraitis, R.E.I. Schropp, W.G.J.H.M. van Sark, Nanoparticles for luminescent solar concentrators—a review. Opt. Mater. **84**, 636–645 (2018)
17. Y. Zhou, H. Zhao, D. Ma, F. Rosei, Harnessing the properties of colloidal quantum dots in luminescent solar concentrators. Chem. Soc. Rev. **47**, 5866–5890 (2018)
18. S.J. Ha, J.H. Kang, D.H. Choi, S.K. Nam, E. Reichmanis, J.H. Moon, Upconversion-assisted dual-band luminescent solar concentrator coupled for high power conversion efficiency photovoltaic systems. ACS Photonics **5**, 3621–3627 (2018)
19. J. Lin, L. Wang, Q. Jing, H. Zhao, Highly efficient and high color rendering index multilayer luminescent solar concentrators based on colloidal carbon quantum dots. J. Chem. Eng. **481**, 148441 (2024)
20. B.S. Richards, A. Ian, Howard, luminescent solar concentrators for building integrated photovoltaics: opportunities and challenges. Energy Environ. Sci. **16**, 3214–3239 (2023)
21. S. Castelletto, A. Boretti, Luminescence solar concentrators: A technology update. Nano Energy **109**, 108269 (2023)
22. M.J. de Fimbres-Romero, Á. Flores-Pacheco, M.E. Álvarez-Ramos, R. Lopez-Delgado, Transparent and colorless luminescent solar concentrators based on ZnO quantum dots for building-integrated photovoltaics. ACS Omega **9**, 28008–28017 (2024)
23. X. Gong et al., Fabrication of high-performance luminescent solar concentrators using N-doped carbon dots/PMMA mixed matrix slab. Org. Electron. **63**, 237–243 (2018)
24. Y. You et al., High-efficiency luminescent solar concentrators based on composition-tunable eco-friendly Core/shell quantum dots. Chem. Eng. J. **452**, 139490 (2023)
25. H.H. Kim et al., Realization of excitation wavelength independent blue emission of ZnO quantum dots with intrinsic defects. ACS Photonics **7**, 723–734 (2020)
26. F. Mateen, S.Y. Lee, S.K. Hong, Luminescent solar concentrators based on thermally activated delayed fluorescence dyes. J. Mater. Chem. A **8**, 3708–3716 (2020)
27. R. Reisfeld, Y. Kalisky, Improved planar solar converter based on uranyl neodymium and holmium glasses. Nature **283**, 281–282 (1980)

28. L.R. Wilson, B.C. Rowan, N. Robertson, O. Moudam, A.C. Jones, B.S. Richards, Characterization and reduction of reabsorption losses in luminescent solar concentrators. Appl. Opt. **49**, 1651–1661 (2010)
29. J. Zubia, J. Arrue, Plastic optical fibers: an introduction to their technological processes and applications. Opt. Fiber Technol. **7**, 101–140 (2001)
30. D.K.G. de Boer, Optimizing wavelength-selective filters for luminescent solar concentrators. Proc. SPIE **7725**, 77250Q-Q77259 (2010)
31. C.R. Ronda, S. de Brouwer, W. Keur, D.K.G. de Boer, A. Meijerink, A.J.J. Bos, P. Doorenbos, Luminescence of Sm^{2+} in strontium borates and aluminosilicates (2012)
32. C. Kulshreshtha, S.H. Cho, Y.S. Jung, K.S. Sohn, Deep red color emission in an Sm^{2+}-doped SrB4O7 phosphor. J. Electrochem. Soc. **154**, J86–J90 (2007)
33. M.G. Debije, P.P.C. Verbunt, Thirty years of luminescent solar concentrator research: solar energy for the built environment. Adv. Energy Mater. **2**, 12–35 (2011)
34. V.M. Granchak, T.V. Sakhno, S.Y. Kuchmy, Light-emitting materials-active components of luminescent solar concentrators. Theor. Exp. Chem. **50**, 1–20 (2014)
35. D.K.G. deBoer, D.J. Broer, M.G. Debije, W. Keur, A. Meijerink, C.R. Ronda, P.P.C. Verbunt, Progress in phosphors and filters for luminescent solar concentrators. Opt. Express **20**, A395 (2012)
36. W.G.J.H.M. van Sark, K.W.J. Barnham, L.H. Slooff et al., Luminescent solar concentrators–a review of recent results. Opt. Express **16**, 21773–21792 (2008)
37. van WGJHM, A. Meijerink, R.E.I. Schropp, Solar spectrum conversion for photovoltaics using nanoparticles, in *Third Gener. Photovoltaics*, (Springer Berlin Heidelberg, Netherlands, 2012) (pp. 1–28)
38. C. Feldmann, T. Jüstel, C.R. Ronda, P.J. Schmidt, Inorganic luminescent materials: 100 years of research and application. Adv. Funct. Mater. **13**, 511–516 (2003)
39. J. Levitt, W. Weber, Materials for luminescent greenhouse solar collectors. Appl. Opt. **16**, 2684–2689 (1977)
40. J.S. Batchelder, A.H. Zewail, T. Cole, Luminescent solar concentrators. 1: theory of operation and techniques for performance evaluation. Appl. Opt. **18**, 3090–3110 (1979)
41. M.G. Debije, P.P.C. Verbunt, Thirty years of luminescent solar concentrator research: solar energy for the built environment. Adv. Energy Mater. **2**, 12–35 (2012)
42. J. Li, S. Zheng, X. Zhao, A. Vomiero, X. Gong, Ultrahighly-efficient and stable luminescent solar concentrators enabled by FRET-based carbon nanodots. Nano Energy **134**, 110514 (2025)
43. Elisavet Tatsi et al. Thermoresponsive Host Polymer Matrix for Self-Healing Luminescent Solar Concentrators, ACS Appl. Energy Mater. https://doi.org/10.1021/acsaem.9b02196
44. A. Picchi et al., Sustainable luminescent solar concentrators with superior photodegradation resistance employing heptagon-embedded polycyclic aromatic dicarboximides as emitters. Mater. Today Energy **44**, 101646 (2024)
45. B. Norton et al., Enhancing the performance of building integrated photovoltaics. Sol. Energy **85**, 1629–1664 (2011)
46. D. Chemisana, Building integrated concentrating photovoltaics: a review. Renew. Sustain. Energy Rev. **15**, 603–611 (2011)
47. B. Rezaie, E. Esmailzadeh, I. Dincer, Renewable energy options for buildings: case studies. Energy Buildings **43**, 56–65 (2011)
48. Y. Li, P. Miao, W. Zhou, X. Gong, X. Zhao, N-doped carbon-dots for luminescent solar concentrators. J. Mater. Chem. A **5**, 21452–21459 (2017)
49. X. Gong, W. Ma, Y. Li, L. Zhong, W. Li, X. Zhao, Fabrication of high-performance luminescent solar concentrators using N-doped carbon dots/PMMA mixed matrix slab. Organic Electron. **63**, 237–243 (2018)
50. M. Zhu, Y. Li, S. Tian, Y. Xie, X. Zhao, X. Gong, Deep red emitting zinc and aluminium co-doped copper indium sulfide quantum dots for luminescent solar concentrators. J. Colloid Interface Sci. **534**, 509–517 (2019)
51. H. Zhao, Y. Zhou, D. Benetti, D. Ma, F. Rosei, Perovskite quantum dots integrated in large-area luminescent solar concentrators. Nano Energy **37**, 214–223 (2017)

52. H. Zhao, D. Benetti, L. Jin, Y. Zhou, F. Rosei, A. Vomiero, Absorption enhancement in 'Giant' core/alloyed-shell quantum dots for luminescent solar concentrator. Small **12**, 5354–5365 (2016)
53. R. Reisfeld, New developments in luminescence for solar energy utilization. Opt. Mater. **32**, 850–856 (2010)
54. D.K.G. de Boer et al., Progress in phosphors and filters for luminescent solar concentrators. Opt. Express **20**, A395–A405 (2012)
55. L. Desmet, A.J.M. Ras, D.K.G. deBoer, M.G. Debije, Monocrystalline silicon photovoltaic luminescent solar concentrator with 4.2% power conversion efficiency. Opt. Lett. **37**, 3087–3089 (2012)
56. S.P. Ying, B.M. Chen, W.L. Tseng, Thin-film luminescent solar concentrators using inorganic phosphors. IEEE Trans. Electron Devices **66**, 2290–2294 (2019)
57. S. Li, S. Haussener, Radiative transfer in luminescent solar concentrators. J. Quant. Spectrosc. Radiat. Transfer **319**, 108957 (2024)

Surface Texturing for Solar Light Trapping

B. S. Akila

Abstract One important parameter to consider in the fabrication of solar cell devices is the optical absorption coefficient of the semiconducting material used. Efficient absorption of incident light by a solar cell enhances the generation of carriers in the photovoltaic material, which, in turn, contributes to achieving high conversion efficiency in the device. To improve light trapping in photovoltaic devices, surface texturing is crucial. This process significantly reduces optical losses, such as reflection and unabsorbed radiation. Surface texturing minimizes the substantial mismatch in the refractive index between the air and the photovoltaic material. This chapter provides a detailed discussion of the optical principles related to tuning the refractive index of photovoltaic materials and the process of surface texturing.

Keywords Texturization · Etching · Refractive index · Light trapping · Reflection · Anisotropic etching · Isotropic etching · Upright pyramids · Inverted pyramids · Surface morphology

1 Introduction

Solar power is obtained from a non-polluting form of renewable energy known as solar energy. In this regard, solar cell which converts solar energy into electricity has a major scope in meeting the energy demands globally. As a consequence, invention and improvement in solar cells have achieved great heights leading to more than one generation of solar cells in order to meet the energy demand across the world at cheaper cost [1].

The manufacturing process of laboratory- and commercial-scale Si solar cells varies with their processing technique and material selection. The PERL and PERC solar cells introduced at the University of South Wales provide the basic design of the industrial Si solar cell [2]. The high-efficiency laboratory process reported for the fabrication of PERL Si solar cell by the University of South Wales comprises the

B. S. Akila (✉)
Department of Physics, MES College of ACS, Malleswaram, Bengaluru 560003, India
e-mail: akila.bspu@gmail.com

P. Ravindran et al. (eds.), *Photon to Power*, Progress in Optical Science and Photonics 32, https://doi.org/10.1007/978-981-96-5914-2_18

fabrication process of the saw damage etching followed by texturing with inverted pyramid structures and the diffusion of p–n junction buried contact deposition and passivation, introduction of SiO_2-based thin dielectric layer in the emitter region followed by double-layer antireflection coating of MgF_2 and ZnS. In addition, a further introduction of rear contact set the laboratory process condition to achieve the highest conversion efficiency of $25.0 \pm 0.5\%$ [3].

Reflection of light at the boundary or the interface of two different mediums is an optical phenomenon that is favorable in mirrors and unfavorable in glasses and lenses. The reflected light from the glass, lenses, or any optoelectronic device is considered an optical loss. Therefore, reduction of the reflected light is essential. Also, the reduction of optical loss in any optoelectronic devices such as photovoltaic cells is a promising approach for improving the efficiency of the device [4]. Given the above, it is reported that silicon wafers found reflecting the highest values of about 50–60% in the wavelength range between 315 and 400 nm, 60–70% in 280–315 nm, and more than 70% in the UV-C region 200–280 nm depending upon the quality of the wafer [5]. Furthermore, a reflectance percentage between 20 and 40% in the absorption region of 400–1100 nm has been reported depending upon the wafer quality, fabrication, and surface process conditions [6]. Therefore, the optical reflectance of silicon wafers has been reduced by modifying the surface using various patterns and techniques [7].

Antireflection is a phenomenon that plays a major role in increasing the incident light coupling into the device enhancing the efficiency of photovoltaic devices. In order to achieve this, the mismatch in refractive index between substrate and air medium has to be reduced. This refractive index mismatch is reduced by methods like (1) surface texturing, (2) antireflection coating, and (3) combining both surface texturing and antireflection coating [8]. Among the various steps in light trapping of solar cells, this chapter focuses on surface texturing.

2 Texturization

Texturization refers to the selective removal of atoms from the surface of the Si wafers, which creates less surface density of atoms at the surface which reduces the optical refractive index. The etching further creates an inward increase in the atomic density and establishes a gradual increase in the refractive index. Therefore, the incident light penetrates into the substrate instead of reflecting [8]. Therefore, surface texturization of Si wafer minimizes the reflection and enhances the refraction.

3 Optics of Textured Features

A flat and smooth surface, such as a silicon wafer, reflects a significant portion of the incident light. In contrast, a rough or textured surface scatters the incident light, diffusing it in various directions. The textured features on the surface of the Si wafer reduce the reflection due to the following phenomenon. (i) Multiple reflections between adjacent facets increase the likelihood of light penetrating the surface, (ii) the inclined facets create wide oblique angles, enhancing light trapping, and (iii) multiple internal reflections of transmitted low-energy photons facilitate their absorption [9]. Different ways of trapping incident light in the case of upright pyramidal textures are described below.

i. **Multiple light bounce** The incoming light bounces between the adjacent facets of the pyramids ensuring more than one chance for light to penetrate into the surface	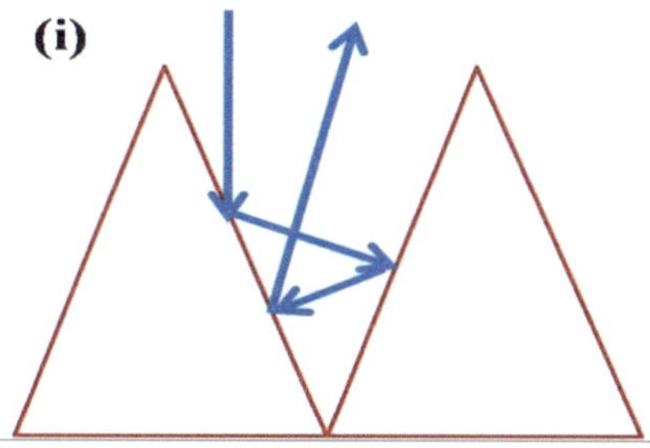
ii. **Wide oblique refracted angle** The path length of incoming light is increased inside the silicon substrate due to wide oblique refracted angle	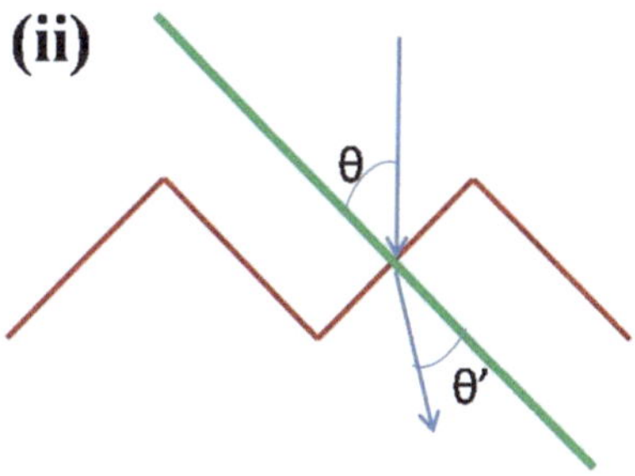
iii. **Multiple internal reflection** The transmitted low-energy photons are absorbed due to multiple internal reflections	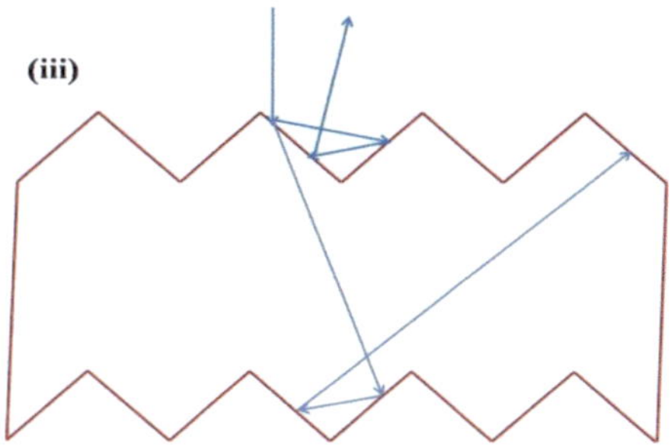

4 Correlation of Wavelength of Light and Dimension of Textured Feature

Depending upon the texture size and dimension of the wavelength of the incident light, the optical interaction of light over the textured surface differs. Various conditions and their corresponding optical phenomenon are discussed below.

4.1 λ (Wavelength of Incident Light) << Dimension of Textured Surface

In this case, the wavelength of the incident light is very small when compared to the dimension of the textured surface [10]. This can be explained by ray optics since the interacting feature size is larger in comparison with the wavelength of the interacting light. In this case, a large number of diffraction orders are generated, resulting in multiple reflections between adjacent facets. A similar light-trapping technique is employed in V-grooved surface textures produced through mechanical etching [11].

4.2 λ (Wavelength of Incident Light) ≈ Dimension of Textured Surface

In this case, the dimensions of both the interacting light and the textured feature are similar in size. Under this condition, the interference effect is dominant leading to specific diffraction orders. As a consequence, the reduction of reflection is strongly wavelength-dependent. Hence, the reduction of surface reflection happens only for a narrow spectral range [12].

4.3 λ (Wavelength of Incident Light) >> Dimension of the Textured Surface

In this case, the interacting wavelength of light is significantly larger than the dimensions of the textured features. As a result, the light cannot resolve the fine structure, causing the textured surface to behave as a medium with effective refractive indices. This light-trapping technique is utilized in moth-eye structures, as reported in [13, 14]. However, for large wavelengths of light interaction, these textures act as a continuous surface resulting in high surface reflectance.

5 Types of Texturization

The etching of valence electrons from the substrate surface is known as texturization. It is established by various methods which is broadly classified into two types they are:

i. dry etching;
ii. wet etching [15].

5.1 *Dry Etching*

The etching technique which uses plasma, metal blades, or laser is categorized as dry etching where no wet chemicals are involved. The dry etching process is very sophisticated since it requires vacuum conditions, gas lasers, etc. Also, there is a risk of toxicity of the gas used in the dry etching process. Unlike other etching techniques, it is capable of texturing of size < 100 nm. The dry etching is carried out using various techniques such as mechanical grooving, reactive ion etching, and laser etching [16].

5.1.1 Mechanical Grooving

Mechanical etching of Si wafers refers to a mechanical way of texturing using metal blades. The grooving tool consists of a metal body with a V-grooved abrasive layer and a diamond coating at the tip. Mechanical grooving is carried out on both multi-crystalline and mono-crystalline Si wafers. Mechanical grooving is performed on both multi-crystalline and mono-crystalline silicon wafers, creating periodic V-shaped grooves as the textured features. Technology has developed to texture more than 1 million Si wafers with a single abrasive layer coating [17].

5.1.2 Reactive Ion Etching (RIE)

RIE is commonly used to texture multi-crystalline Si wafers to form hillocks or random cones. Also, the technique is more commonly used in micro-machining and micro-electronics because the etch rate is more easily controllable in the RIE process [18]. The substrate to be etched is loaded into a glass chamber with an inlet and outlet for the etching gas. The glass chamber is fixed with a top and bottom electrode through which the current is passed. Further, the electric field formed in between the electrodes ionizes the gas molecules. These reactive ions react with the substrate and do the action of etching leaving behind the volatile product. $SF6/O_2$ or Cl_2 are commonly used plasma for etching Si substrates [19].

5.1.3 Laser Etching

The laser technique is not yet widely implemented because of its high cost. The pulsed laser radiation melts the silicon substrate and subsequently evaporates. Depending upon the pulse duration, laser power, and pulse repetition frequency, the textured features are controlled [20]. A post-etching step is employed to remove the amorphous silicon generated during laser texturing. Silicon wafers consisting of micro-/nanostructures resulted in 4.9% reflectance at 500 nm using an ultra-fast laser [21].

5.2 *Wet Chemical Etching*

In wet etching, the etchants used are liquid phase. The classifications under the wet chemical etching are given below [22].

1. **Electrochemical etching**
    - **Electroless etching**
    - **Anodic etching**
2. **Chemical etching**
    - **Anisotropic**
    - **Isotropic**

5.2.1 Electroless Etching

Electroless etching is the process of charge exchange between semiconductor and electrolyte without an external potential. In this process, a strong oxidizing agent that can extract electrons from the valence band of the semiconductor is used. An example of electroless etching is the dipping of Gallium Arsenide (GaAs) crystal in the ferricyanide ($Fe(CN)_6$) solution. The ferricyanide gets reduced to form h+ and reacts with GaAs and OH^-. The chemical reaction involved is given as an equation below [23].

$$GaAs + 10OH^- + 6h^+ \rightarrow GaO_2^- + AsO_3^{3-} + 5H_2O$$

5.2.2 Anodic Etching

Extraction of electrons from the valence band of a semiconductor material by applying an external potential at the semiconductor-electrolyte interface is known as anodic etching. An example of anodic etching is the dipping of a Si wafer in hydrazine solution along with a platinum rod. The external circuit is connected to apply a constant voltage. Also heater is used to keep the bath at a certain temperature. With the applied voltage, the hydrazine solution gets ionized releasing OH^- ions. These OH^- ions combine with the surface silicon to form orthosilicic acid and leave the solid surface. The overall reaction of anodic etching of silicon using hydrazine is given in the equation below [24].

$$2(NH_2 - NH_2) + 4H_2O + Si \rightarrow SiO_2(OH)_2^- + 2\left(NH_2 - NH_2^+\right) + 2H_2$$

5.3 *Chemical Etching*

The wet chemical etching can be performed using the assistance of the chemical activity of etchants. During the chemical etching, the etchants such as sodium hydroxide (NaOH), potassium hydroxide (KOH), tetramethylammonium hydroxide (TMAH), Na_3PO_4, Na_2CO_3, Na_2SiO_3, HF, HNO_3, CH_3COOH, and silicon exchange the electron and hence the Si dissolves into the etchant [25–29]. This dissolution of Si into the etchant differs with its crystalline orientation and the combination of etchants. Based on this, there are two types of etching processes that occur while etching Si.

They are:

i. anisotropic etching;
ii. isotropic etching.

5.3.1 Anisotropic Etching

Anisotropic etching is a process where the exchange of electrons occurs at the interface of substrate and etching agent. This process is also called as dissolution of surface material. An example of anisotropic etching is the texturing of a mono-crystalline Si wafer in the aqueous KOH solution. It consists of two-step processes. Initially, the Si wafers are cleaned for native oxide removal which leaves the surface Si atoms as H terminated. The first step of the chemical etching is nucleophilic attraction between OH^- and Si–H surface bond resulting in OH terminated Si atom at the surface. During the second step, the Si–OH surface bonds are attacked by the water molecules forming a complex molecule. Further, the back-bond of the Si is polarized by the presence of OH^- ion present in the etchant medium detaching the Si from the surface. The dissolution process for Si (100) in an alkaline solution is described below [30].

Step. 1

Step. 2

The overall reaction as given by H. Seidel is as follows [31]:

$$Si + 2H_2O + 2OH^- \rightarrow \left[Si(OH)_2O_2\right]^{2-} + 2H_2$$

5.3.2 Isotropic Etching

In the isotropic etching process, all the surface atoms are removed with a similar etch rate. Hence, the morphology of isotropically etched surface consists of round pits with equal etching in all directions. Accordingly, smooth surfaces of the etched patterns are established irrespective of crystalline orientation.

Extraction of electrons from the valence band of a semiconductor material by applying an external potential at the semiconductor-electrolyte interface and using it to remove Si atoms from the surface is known as anodic etching. In detail, a Si wafer is dipped into a hydrazine solution along with a platinum counterelectrode. The external circuit is connected to both the Si wafer and the 'Pt' counterelectrode and applied a constant voltage. This applied voltage ionizes the hydrazine and releases OH^- ions. This OH^- ion combines with the surface silicon and creates orthosilicic acid which dissolves and leaves the solid surface. The overall reaction of anodic etching of silicon using hydrazine is given in Eq. 1 [32]. This type of etching can be also carried out at elevated constant temperature using a controlled temperature bath.

$$2(NH_2 - NH_2) + 4H_2O + Si \rightarrow SiO_2(OH)_2^- + 2\left(NH_2 - NH_3^+\right) + 2H_2 \quad (1)$$

6 Type of Textured Structures

Various types of textured structures obtained on Si substrate for crystalline Si solar cell applications are random conical, micro-/nano-pyramids, regular inverted pyramids, porous upward pyramids, random upright pyramids, honeycomb, grooves, and moth-eye structures. Discussion on these textured structures and their optical reflectance is discussed below in detail.

6.1 Random Conical

Random conical structures are obtained generally on multi-crystalline wafers using techniques like reactive ion etching (RIE) or acidic etching [33]. Maskless texturing using RIE etching on multi-crystalline Si wafers has resulted in random conical textures with a hemispherical reflectance of less than 5%. [18]. While masked RIE has resulted in periodic conical frustum structures with a reflectance of 0.5%. [34]. Though the maskless and masked RIE etching process gives the least reflection by forming conical structures, they are costly and hence cannot be applicable to commercial production of Si solar cells. Random conical structures are also obtained by a simple acidic wet chemical etching process. Acidic etching has resulted in bringing down the reflectance from 34% to 27.6% [33].

6.2 Micro-/Nano-pyramids

Micro-/nano-pyramids on the silicon surface can be obtained by various techniques such as metal-assisted etching [35]. Near-zero reflectance has been achieved by incorporating micro/nanostructures by using the alkaline wet chemical method to form micro-sized pyramids and nanocones by the RIE method [36]. The other methods for obtaining micro/nanostructures are anisotropic etching for micro pyramids and metal-assisted etching for nanostructures with a reduced reflectance of 3.8% in the 300-1000 nm wavelength range [37]. Also, anisotropic etching combined with electroless etching for forming micro/nanostructures has proven 3% reflectance for the spectral range of 300–900 nm [38].

6.3 Regular Inverted Pyramids

Regular inverted pyramids are formed by masked anisotropic etching using the TMAH solution. The masks generally used are thermal oxides SiO_2. Crystalline Si wafer with regular inverted pyramids using PERL technology has obtained a reflectance value of 5% for the spectral range of 500–1000 nm with a conversion efficiency of 24.20% [39].

6.4 Porous Upright Pyramid

Porous upright pyramids consist of nanoporous structures formed over the micro-sized upright pyramids. Micro-sized upright pyramids are generally formed by anisotropic etching by alkaline solutions like KOH, NaOH, and TMAH. Whereas the nanoporous are formed by metal-assisted chemical etching (also called strain etching), acidic etching, and electrochemical etching. Porous pyramids formed by 3 wt% of KOH and 8 volume% for forming random upright pyramids and stain etching using HF and Fe $(NO_3)_3$ to form porous structures resulted in the mean reflectance of 4.18% in the 400–900 nm spectral range [40]. Another group formed porous upright pyramids involving alkaline etching for random upright pyramids and electrochemical etching for forming nano-sized porous structures and achieved a low reflectance of 1.9% in the 400–800 nm spectral range [41].

6.5 Random Upright Pyramids

Due to cost efficiency, the random upright pyramids formed by wet chemical etching of (100) silicon using the solution are broadly accepted as commercial standard methods [42]. Random upright pyramids are formed by anisotropic etching using various alkaline etchants such as Ethylenediamine pyrocatechol (EDP), hydrazine, NaOH, Na_2CO_3, Na_3PO_4, Na_2SiO_3, KOH, and TMAH have been tried to obtain surface texturing with an expectation of uniform distribution of pyramidal structures. Among them, KOH along with IPA is reported to have predominant anisotropy in etch rates along different orientations.

6.6 Grooves

Groove structures for reducing surface reflections are made by two methods: (i) mechanical grooving and (ii) laser grooving. In the laser grooving technique a pulsed laser beam is used to form 'V'-shaped grooves with constant spacing. Parallel grooves with 90 μm spacing were made using a laser with an output power of 50W, a speed of laser beam equal to 20 mm/s, and pulse repetition of 15 kHz frequency. Using the above-said conditions, surface reflectance was reduced to 13.96% [43]. Similar 'V'-shaped grooves were made by mechanical methods consisting of metal blades which resulted in the reflectance below 10%.

6.7 Characterization Techniques

Relevant characterization techniques used to evaluate the surface conditions of the etched substrate scanning electron microscope (SEM), field emission scanning electron microscope (FESEM), Zeta optical profiler, and UV–Visible-NIR spectrometer. The scanning electron microscope and field emission scanning electron microscope allow for determining the various surface morphologies obtained by surface texturing. Zeta optical profiler is an optical microscope that provides three-dimensional images of the surface of the sample under investigation from which the roughness and its dimensions are computed using the software. The UV–visible spectroscopy is used to measure the surface reflectance spectra of textured surfaces.

References

1. A.M. Bagher, M.M.A. Vahid, M. Mohsen, Types of solar cells and application. Am. J. Opt. Photonics **3**(5), 94–113 (2015)
2. T. Dullweber, J. Schmidt, Industrial silicon solar cells applying the passivated emitter and rear cell (PERC) concept—a review. IEEE J. Photovolt. 2156–338 (2016)
3. J. Zhao, A. Wang, P.P. Altermatt, S.R. Wenham, M.A. Green, Sol. Energy Mater. Sol. Cells **41**, 87–99 (1996)
4. B.S. Kim, D.H. Lee, S.H. Kim, G.H. An, K.J. Lee, N.V. Myung, Y.H. Choa, Silicon solar cell with nanoporous structure formed on a textured surface. J. Am. Ceram. Soc. **92**, 2415–2417 (2009)
5. C.W. Liu, C.L. Cheng, B.T. Dai, C.H. Yang, J.Y. Wang, Int. J. Photoenergy **2012**, 701809 (2012)
6. Y. Cao, A. Liu, H. Li, Y. Liu, F. Qiao, Z. Hu, Y. Sang, Appl. Surf. Sci. **257**, 7411–7414 (2011)
7. H. Park, S. Kwon, J.S. Lee, H.J. Lim, S. Yoon, D. Kim, Sol. Energy Mater. Sol. Cells **93**, 1773–1778 (2009)
8. F. Llopis, I. Tobias, J. Appl. Phys. **100**, 124504 (2006)
9. Y. Yang, M.A. Green, A. Ho-Baillie, H. Kampwerth, S. Pillai, H. Mehrvarz, Sol. Energy Mater. Sol. Cells **115**, 42–51 (2013)
10. C. Simeon, B. Finch, K.R. Mclntosh, Prog. Photovoltaic: Res. Appl. **21**, 960–971 (2013)
11. C. Cho, H. Kim, S. Jeong, S.W. Baek, J.W. Seo, D. Han, K. Kim, Y. Park, S. Yoo, J.Y. Lee, Sol. Energy Mater. Sol. Cells **115**, 36–41 (2013)
12. M.K. Hedayati, M. Elbahri, Materials **497**, 9 (2016)
13. F.M.M. Souren, J. Rentsch, M.C.M. Van de sanden, Prog. Photovoltaics: Res. Appl. **23**, 352–366 (2015)
14. F. Jiao, Q. Huang, W. Ren, W. Zhou, F. Qi, Y. Zheng, J. Xie, Microelectron. Eng. **103**, 126–130 (2013)
15. A.B. Anoop Prakash, J.G. Jency, M.C. Mathew, A review of various wet etching techniques used in micro fabrication for real estate consumption. Int. J. Comput. Appl.
16. M. Schnell, R. Ludemann, S. Schaefer, Plasma surface texturing for multicrystalline silicon solar cells, in *Proceeding of the 28th IEEE Photovoltaic Specialists Conference,*Anchorage, p. 367 (2000)
17. P. Fath, C. Borst, C. Zechner, E. Bucher, G. Willeke, S. Narayanan, Progress in a novel high throughput mechanical texturing technology for highly efficient multicrystalline silicon solar cells. Sol. Energy Mater. Sol. Cells **48**, 229–236 (1997)
18. A.R. Burgers, J.H. Bultman, Silicon solar cells textured by reactive ion etching with natural lithography, in *16th European Photovoltaic Solar Energy Conference* (2000) VA1/57

19. C. Cho, D. Kong, J.H. Oh, B. Kim, B. Lee, J. Lee, Surface texturing method for silicon solar cell using reactive ionetching with metal mesh. Phy. Status Solidi A **8**, 1844–1849 (2014)
20. B. Voisiat, S. Indrisiunas, G. Raciukaitis, I. Simkiene, A. Reza, R. Suzanoviciene, Application of laser texturing in silicon solar cell technology. Mater. Sci. **20** (2014)
21. B.K. Nayak, V.V. Iyengar, M.C. Gupta, Efficiency light trapping in silicon solar cells by ultrafast laser induced self assembled micro/nano structures. Prog. Photovolt.: Res. Appl. **19**, 631–639 (2011)
22. J.J. Kelly, H.G.G. Philipsen, Anisotropy in the wet etching of semiconductors. Curr. Opin. Solid State Mater. Sci. **9**, 84–90 (2005)
23. P.H.L. Notten, J.J. Kelly, Evidence for cathodic protection of crystallographic facets from GaAs etching profiles. J. Electrochem. Soc. **134**, 444–448 (1987)
24. K.B. Sundaram, H.W. Chang, Electrochemical etching of silicon by hydrazine. J. Electrochem. Soc. **140**, 6 (1993)
25. J.S. You, D. Kim, J.Y. Huh, H.J. Park, J.J. Pak, C.S. Kang, Sol. Energy Mater. Sol. Cells **66**, 37–44 (2001)
26. Y. Nishimoto, K. Namba, Sol. Energy Mater. Sol. Cells **61**, 393–402 (2000)
27. Z. Xi, D. Yang, W. Dan, C. Jun, X. Lin, D. Que, Renew. Energy **29**, 2101–2107 (2004)
28. U. Gangopadhyay, K. Kim, A. Kandol, J. Yi, H. Saha, Sol. Energy Mater. Sol. Cells **90**, 3094–3101 (2006)
29. Y. Nishimoto, T. Ishihara, K. Namba, J. Electrochem. Soc. **146**, 457–461 (1999)
30. T. Baum, D.J. Schiffrin, Mechanistic aspects of anisotropic dissolution of materials etching of single-crystal silicon in alkaline solutions. J. Chem. Soc. Faraday Trans. **94**, 691–694 (1998)
31. H. Seidel, L. Csepregi, A. Heuberger, H. Baumgartel, Anisotropic etching of crystalline silicon in alkaline solution. J. Electrochem. Soc. **137**, 3626–3632 (1990)
32. K.B. Sundaram, H.W. Chang, J. Electrochem. Soc. **140**, 6 (1993)
33. D.H. Macdonald, A. Cuevas, M.J. Kerr, C. Samundsett, D. Ruby, S. Winderbaum, A. Leo, Texturing industrial multicrysalline silicon solar cells. Sol. Energy **76**, 277–283 (2004)
34. Y. Lu, A. Lal, High efficiency ordered silicon nano conical frustum array solar cells by self powered parallel electron lithography. Nano Lett. **10**, 4651–4656 (2010)
35. Z. Huang, N. Geyer, P. Werner, J.de Boor, U. Gosele, Metal assisted chemical etching of silicon: a review. Asv. Mater. **23**, 285–308 (2011)
36. A. Ingenito, O. Isabella, M. Zeman, Nano-cone on micro pyramids: modulated surface tetures for maximal spectral response and high efficiency solar cells. Prog. Photovolt: Res. Appl. **23**, 1649–1659 (2015)
37. Y. Xiu, S. Zhang, V. Yelundure, A. Rohatgi, D.W. Hess, C.P. Wong, Superhydrophobic and low light reflectivity silicon surfaces fabricated by hierarchical etching. Langmuir **24**, 10421–10426 (2008)
38. D.Z. Dimitrov, C.H. Du, Crytalline silicon solar cells with micro/nano texture. Appl. Surf. Sci. **266**, 1–4 (2013)
39. A. Wang, J. Zhao, M. A. Green, 24% efficienct silicon solar cell. Appl. Phys. Lett. **57**, 602 (1990)
40. J. Xiao, L. Wang, X. Li, X. Pi, D. Yang, Reflectivity of porous pyramids structured silicon surface. Appl. Surf. Sci. **25**, 472–475 (2010)
41. H. Lv, H. Shen, Y. Jiang, C. Gao, H. Zhao, J. Yuan, Porous pyramids structured silicon surface with low reflectance over a broad band by electrochemical etching. Appl. Surf. Sci. **258**, 5451–5454 (2012)
42. P. Papet, O. Nichiporuk, A. Kaminski, Y. Rozier, J. Kraiem, J.F. Lelievre, A. Chaumartin, A. Fave, M. Lemiti, Pyramidal texturing of silicon solar cell with TMAH chemical anisotropic etching. Sol. Energy Mater. Sol. Cells **90**, 2319–2328 (2006)
43. L.A. Dobrzanski, A. Drygala, Surface texturing of multicrystallline silicon solar cells. J. Achiev. Mater. Manuf. Eng. **31**, 1 (2008)

Fabrication Techniques of Silicon Microwire Arrays and Realization of Radial Junction Silicon Microwire Solar Cells

Anjali Saini, Sanjay K. Srivastava, Prashant Misra, and Mrinal Dutta

Abstract Commercial Si photovoltaics (PV) occupied over 95% of the total PV market in 2019. Mono-crystalline Si PV dominates 65% of the Si PV market. Being, not cost-competitive mono-crystalline Si wafers' cost alone ascribes half of the total solar cell production cost. Si nanowire (SiNW) array cells, with excellent light confining and antireflection with short carrier transportation pathways, show daylight towards the realization of high efficiency using lower-quality Si wafers having short minority carrier diffusion lengths. All the benefits of SiNW arrays come to reality based on the realization of radial junctions. In spite of the advent of low-cost wafer-scale SiNW production technique, realization of radial junction SiNWs never happened. The focus of this research has been shifted towards the fabrication of higher diameter SiNWs, i.e. Si microwires (SiMWs) in wafer scale to realize radial junction Si solar cells. This chapter summarizes the recent development of the SiMW-based photovoltaic technologies. Different fabrication techniques developed till date for wafer-scale production of SiMWs are summarized. The challenges faced in this direction and the prevailing difficulties in the production of these kinds of devices have also been conferred.

Keywords Silicon · Microwire · Tapered · Light trapping · Radial junction · Solar cell

---

A. Saini · S. K. Srivastava
Photovoltaic Metrology Section, Advanced Materials and Device Metrology Division, CSIR-National Physical Laboratory, New Delhi, India

Academy of Scientific and Innovative Research (AcSIR), Uttar Pradesh, Ghaziabad, India

P. Misra · M. Dutta (✉)
National Institute of Solar Energy, Gurgaon, Haryana, India
e-mail: duttamrinal@nise.res.in

P. Ravindran et al. (eds.), *Photon to Power*, Progress in Optical Science and Photonics 32, https://doi.org/10.1007/978-981-96-5914-2_19

1 Introduction

The last few decades witnessed a rapid increase in global energy demand, pushing world energy consumption to double [1]. Four-fifths of this energy consumption come from non-renewable energy sources, consequential in a 6.2% surge in CO_2 emission. At present, renewable resources share nearly 36% of global energy consumption in 2021 and will continue to thrive strongly in the coming years [2]. Solar radiation is an immeasurable energy source available at practically no cost [3, 4]. The great advantage of this renewable source over other conventional non-renewable ones is that solar light can be directly harvested into electricity using photovoltaic (PV) cells [5, 6]. Currently, the per watt cost of solar cells has become invaluable in the last few years and has grown as a better technology in terms of both worth and applications [7, 8]. The conventional materials used for PV solar cell production are mainly silicon (Si) (single crystal, multi-crystal, poly-silicon), cadmium-telluride (CdTe), copper-indium-gallium-selenide (CIGSe) and copper-indium-gallium-sulphide (CIGS). Based on the use of these materials and their deposition and growth techniques, the PV cells are classified into three generations: (a) 1st generation (Si wafer-based cells), (b) 2nd generation (thin film cells) and (c) 3rd generation (new emerging technologies). Since the 1970s, Si has been under investigation for use in solar cell applications. The global PV marketplace is mostly captured (above 90%) by Si wafer-based solar cells (mainly mono-crystalline cells), due to various components: abundancy of Si (available on the earth's crust), non-toxic nature, well-matured and evolved production technology, favourable band gap for efficient PV conversion. The power conversion efficiency (PCE) of Si cells on a lab scale was recorded to be more than 25% in 1999 [9, 10] which stood unsurpassed for more than 10 years. The record PCE raised to 25.6% in 2014 [11] and to 26.7% in 2017 [12, 13] owing to the HJT technology that refers to heterojunction intrinsic thin (passivating a-Si) layer and interdigitated back contacts (IBCs) on *n*-type Si. At the production level, the efficiency of Si cells has already reached ~ 24% and is constantly increasing. In the past few years, after hydropower and wind power, solar power has become the third most used source of renewable energy. In several countries, solar power has started to compete with conventional power as the manufacturing cost of solar power is already tied to grid parity. This achievement is backed up not only by the advancement in technology but also by the encouragement of government policies and the geopolitical environment. However, for the additional growth of the solar industry to win the race with traditional electricity generation, further improvements in efficiency with lower production costs are necessary. However, the ample preponderance of the high efficiency PV panels has been based on c-Si at the expense of higher cost, mainly due to the costlier Si wafers [14, 15]. The basic needs to achieve high efficiency in these planar c-Si solar cells are: (1) a large amount of material to absorb a good amount of solar light, as the absorption of Si is low in visible to near-infrared regions. This requires additional antireflection coating that increases the cost. (2) Use of a large amount of material enunciates another requirement; the requirement of long minority carrier diffusion length that comes with the restriction to use very high-quality material. All these drawbacks could be resolved by using Si nanowire (SiNW) array radial junction solar cells. The main advantage of radial junction nanowire (NW) cells is the higher trapping of solar light in contrast to planar cells and producing a short carrier path for the photogenerated charge carriers to reach the electrode after separation at the p–n junction [16]. Moreover, trapping of the solar light by the SiNW arrays increases with an increase in NW length. These properties of the SiNW arrays have been reported by several groups. As reported by Garnet et al. [17] the path length of AM1.5G solar radiation in SiNW arrays increases by 73 times. Thus, these advantages dilute the need for large amounts of Si with long minority carrier diffusion length and also the need for extra antireflection coating. However, many problems remain to be solved in fabricating SiNW arrays-based PV cells with expected PCE. More than that fabricating radial junctions using SiNWs remains a challenge till now [18, 19] as the average diameter of such NWs is only around 100–150 nm. So, during the fabrication of radial junction by conventional semiconductor device processing technologies (like thermal diffusion, core–shell structure by VLS method, etc.) SiNWs get depleted and even for small diameter NWs polarity changes completely. To overcome this difficulty, Si microwires (SiMWs) or large-diameter Si nanorods (SiNRs) could play a vital role in achieving high PCE at low cost and simpler manner. The higher diameter microwire or nanorod array approach is therefore a draft for a system architecture that can decouple the optical absorption, electrical transport and ionic transport processes in a desirable fashion. SiMWs have been interrogated as a potential building block for wire-based radial junction solar cells [20–24]. Having the same 1D morphology as SiNWs, SiMWs using conventional doping methods could be doped easily to form radial p–n junctions. Seo group observed that 99% absorption of light and effective collection of photocarriers led to a PCE of 19% for radial junction SiMW arrays solar cell [21]. Additionally, the flexibility of the SiMWs adds another potential for their use in flexible devices because of their 1D structure with a high aspect ratio. Such property enables these SiMWs to release the stress developed during any fabrication process in the space among the MWs. Thus, the employment of SiMWs in flexible solar cells has received considerable attention. Compared to well-known organic flexible solar cells with low stability and PCE, the use of SiMWs in the fabrication of flexible solar cells widens the path for commercialization due to their stability and high efficiency. Thus, SiMWs are very imperative candidates for next-generation low-cost PV fabricated from the lower-grade Si wafers.

2 Advantage of Microwire Geometry

The performance of solar cells depends on the following steps: (i) photon absorption, (ii) generation of excitons, (iii) separation of photogenerated electron–hole pairs, i.e. the excitons, (iv) transport and collection of photogenerated free carriers in respective electrodes. Compared to the conventional micro-pyramid textured planar solar cells, SiMW array solar cells show enhanced light trapping and hence enhanced absorption properties. Due to the huge surface area compared to the same amount of Si material of the conventional solar cells, the generation of free carriers is also higher in the radial p-n junctions fabricated in SiMWs. SiMW radial p–n junction decouples the direction of light absorption with photogenerated charge carrier propagation direction, thus providing efficient transportation of the photogenerated charge carriers due to the short path to reach the respective electrodes. This property enables the fabrication of efficient solar cells by using low-grade Si materials which induces potential cost reduction in the production of Si solar cells.

2.1 Absorption and Exciton Formation

SiMW arrays show nearly 85% absorption of sunlight within visible range compared to only 60–70% absorption of the sunlight by the conventional micro-pyramid cells in the absence of any antireflection coatings. The preferable materials to be used for antireflection have a refractive index in between air and Si. In practice, multiple layers of antireflection coatings are employed to increase light absorption against the ideal need for a continuous layer of graded refractive indexed materials which in turn increases the cost of production of the solar cells. As the top nearly flat surface of the SiMWs occupies a reasonable area of the flat bottom surface, the reflection from SiMW arrays is higher compared to SiNW arrays. Depending on the size of the microwires, reflection increases. Several strategies have been used in order to reduce reflection more and more from SiMW arrays like use of Al_2O_3 nanoparticle light scatterer in between the space of microwires [22], the introduction of nanopores on the surface of the side walls of the microwires [25], fabrication of SiNW arrays on the top of the microwires [26], changing the shape from pillar type to needle or pencil type by tapering the microwires in such a way that their dimension gradually reduces from micron size at the bottom to nanosized at the top [27, 28] and so on. The use of Al_2O_3 nanoparticles of diameter 80 nm in between the space of microwires enhances the short-circuit current density (J_{sc}) from 11.2 to 16.6 mA/cm^2 of SiMW arrays solar cells [22]. On the other hand, the introduction of pores on the side wall of the microwires reduces the reflection down to 7%. Fabrication of SiNW arrays on the top of the microwires reduces the reflection by up to 2% within a range of 400–1000 nm compared to 10–20% reflection obtained from the MW arrays [26]. This type of nano- and micro-hybrid structure enhances J_{sc} from 34.1 to 39.5 mA/cm^2. The use of tapered SiMW arrays instead of cylindrical pillar type SiMW arrays enhances the absorption of incident photons up to 98% compared to 84.2% and 60.8% absorption for pillar type SiMW arrays and planar c-Si substrate, respectively [28].

2.2 Charge Carrier Separation and Transportation

In the conventional planar micro-pyramid textured solar cells separation of the photogenerated charge carriers occurs by the electric field at the junction and, after separation, travels a long path in the direction of light absorption. Such a long-distance journey in the same direction of light absorption generates huge recombination within the bulk material. For sufficient absorption of (nearly 90%) sunlight, the base of the solar cells needs to be thick enough ($\geq$ 125 μm) [29]. This requirement demands a high-quality Si material with a very high minority carrier lifetime (~ of the order of milliseconds) to avoid high recombination. In the case of radial junction, after the separation of the electron–hole pair at the electric field carriers propagates in a radial direction perpendicular to the light absorption direction. The path to travel by the carriers is less than the diameter of the microwires, and thus recombination of the carriers reduces much more than the planar junction. This advantage allows the use of very low-quality Si material where the radius of the microwire equal to the minority carrier diffusion length could give the efficient and optimal performance of the device [28]. This advantage of the separation of carriers in radial p-n junctions over the huge surface area and a short transportation of separated photogenerated charge carriers nearly doubles the J_{sc} in the SiMW array solar cells by suppressing carrier recombination during the transportation [30].

2.3 Carrier Collection and Cost

Radial p–n junction decouples the direction of light absorption from the direction of carrier collection orthogonally. In this kind of wire array geometry, light absorption occurs within the long wire arrays in the same direction as that of incident light with the carrier transportation after separation in the orthogonal direction. This short transportation of carriers enables effective collection by minimizing recombination during transport. After separation, the efficient carrier collection needs a separate electrode design. Conventional electrodes in finger-busbar design suffer from shadow loss owing to a good amount of metal coverage. In addition to that charge carriers moving between fingers suffer recombination due to the large separation between fingers that add to the electrical loss of the device. Moreover, large spacing between fingers also causes recombination loss. Thus, to efficiently collect the separated charge carriers from the huge entire surface of the microwires, electrodes should be in touch with the entire surface of all microwires (conformally) in the array in the ideal case. Considering this aspect of carrier collection transparent conducting oxides (TCOs) appear to be the best front electrode for the radial junction SiMW array solar cells [22]. However, TCOs suffer from parasitic absorption loss and high sheet resistance. To resolve such issue of carrier collection, other approaches such as edge electrodes and metal micro-grid mesh electrodes have been tried [31, 32]. The use of such kinds of electrode designs has been discussed in detail in Sect. 7.

Thus, a radial separation of charge carriers with a short transportation distance and effective light trapping management with a perfect design of front electrode not only reduces the necessity for high-quality Si material in the solar cell fabrication but also relaxes the need for multilayer and expensive antireflection coatings.

3 Fabrication Techniques

Several methods have been developed for vertical SiMWs fabrication. These methods are categorized into two major methods for the growth process: bottom-up and top-down. The bottom-up process is an assembly approach of joining Si atom by atom, whereas in the top-down approach reduction of bulk Si from the top surface of the wafer produces SiMWs. In the bottom-up approach, chemical vapour deposition (CVD) is a familiar and widely used technique. In the top-down approach, two types of etching techniques are used: dry etching and wet chemical etching in addition to conventional photolithographic or conventional lithography free mask patterning approach to control the diameter and pitch of the wires. Among conventional lithography free processes, use of the nanosphere lithography (NSL) technique is a relatively new approach and has become quite popular because of its simplicity, low-cost and effectiveness. In this section, an overview of these widely used techniques that have been employed to fabricate vertical Si MWs is presented.

3.1 Bottom-Up Fabrication

3.1.1 Chemical Vapour Deposition (CVD): Vapour–Liquid–Solid (VLS) and Vapour–Solid–Solid (VSS) Method

CVD is one of the popular choices for bottom-up growth of Si wires of various diameters depending on the use of metal films or metal nanoparticle suspension, which comes with the benefit of the growth of wires and other nanostructures on a variety of substrates. This technique could be scaled up for large-scale industrial production. However, this technique suffers from one major drawback of metal contamination during the growth process and control over growth orientation. For the bottom-up growth of the Si wire structures various types of Si gas precursors such as SiH_4, SiH_6 and $SiCl_4$ are introduced in the growth chamber at various temperatures and depending on growth temperatures growth rates vary. Choice of catalyst, growth temperature, precursor gases and flow rate gives the flexibility to tune wire diameter in a wide range from several micrometres to a few nanometers and control over doping of wires in both radial and axial directions. By controlling the reaction time, the wire length could be controlled. In the VLS growth mechanism, Si atoms come from gases in the vapour phase and go into the liquid droplets of metal-Si alloy that was formed above the eutectic temperature.

After becoming supersaturated, Si atoms precipitated out at the Si and droplet interface (Fig. 1a). In lesser temperatures instead of the above-mentioned VLS mechanism another growth mechanism works called vapour–solid–solid (VSS) growth mechanism. In the VSS mechanism, the metal catalyst instead of coming in the liquid state as occurs in the VLS method remains in the solid state. The conventional choice of metal catalyst for the VLS growth process is Au. However, Au introduces two sub-bandgap levels (one just below the band gap and another just above the valance band). These sub-levels act as deep-level trap states. Other metals such as Ag, Al, Cu and Ni have also been used as catalysts [33–36]. The Al catalyst comes with the benefit of doping of Al in Si that makes Si wires p-type [33]. The growth was VSS mediated as the growth temperature was maintained around 430–490 °C which was much lower than the eutectic temperature (~ 577 °C) of the Al-Si binary phase. Use of Al catalyst comes with caution of use of ultravacuum to avoid oxidation of Al.

Diameter control of the Si wires is achieved by using metal catalyst nanoparticles of various sizes and distributions. The VSS-mediated controlled diameter Si wires also were grown using Ag catalyst keeping the growth temperature below the Ag-Si eutectic (Fig. 1b and c) [34]. The low impact of the Cu catalyst was used to grow SiMWs of length 75–100 μm and diameter 2 μm and 1.5 μm having an average

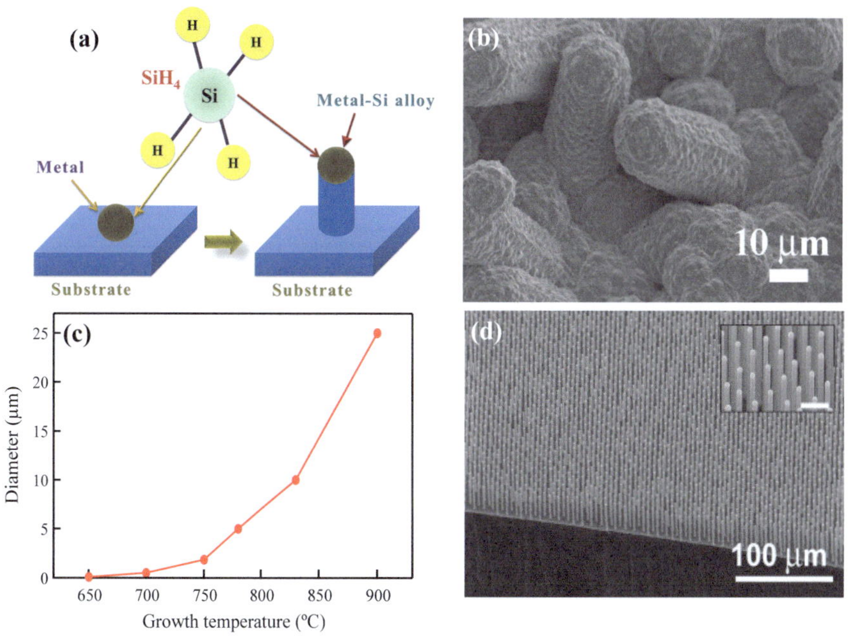

Fig. 1 **a** Schematic of the VLS growth process, **b** Ag catalyzed growth of SiMWs by VSS process, **c** variation in diameter of SiMWs depending on growth temperature in a Ag catalyzed VSS growth process, **d** Cu-catalyzed control growth of SiMW arrays by VLS process. (**b**) and (**c**) Reproduced with permission [34]. Copyright 2013, American Scientific Publishers. (**d**) Reproduced with permission [35]. Copyright 2007, American Institute of Physics

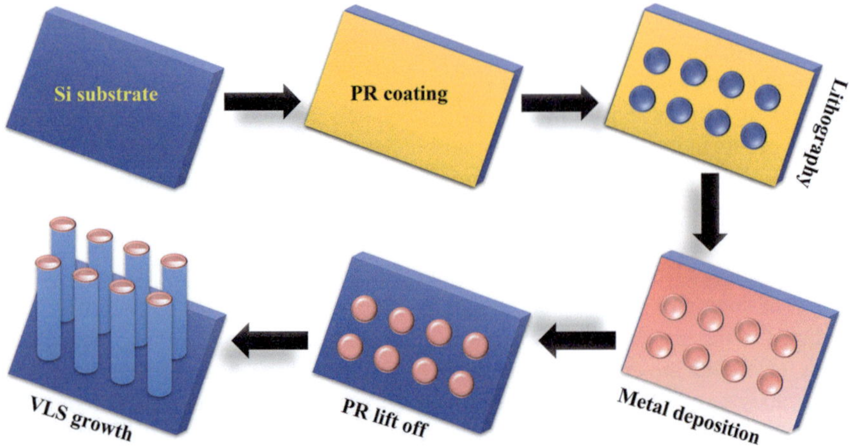

Fig. 2 Schematic of the photolithography-assisted fabrication of controlled diameter SiMW arrays by VLS process

minority carrier diffusion length of 10.5 μm (Fig. 1d) [35, 37]. As a precursor gas, a mixture of H_2:SiH_4 with BCl_3 in 5% H_2 was used. The resistivity of the SiMWs was estimated to be 0.19 Ω-cm corresponding to a concentration of acceptors as 1.05×10^{17} cm^{-3}. CVD grown Si wires suffer from the detrimental problem of the introduction of metal catalysts inside and on the surface of the nanostructures and microstructures during growth which affects the length, orientation and shape. To stop such surface migration of metal catalyst SiO_2 buffer layer was used [35], patterned using a conventional lithography process and then a metal catalyst was applied before introduction in the chamber as shown in Fig. 2.

3.2 Top-Down Fabrication

To control the diameter and the pitch of the Si wires various methods such as nanosphere lithography, block co-polymer and interference lithography besides conventional lithography has been used so far to deposit a pattern template on the Si wafer surface that acts as an etching mask for the top-down fabrication of size and space-controlled Si wire arrays by using either wet chemical etching methods or dry etching methods.

3.2.1 Nanosphere Lithography Method in Addition to Wet Chemical Etching Technique

To make periodic Si wire arrays with defined size and pitch, a self-assembled monolayer of either monosized polystyrene (PS) or SiO_2 (silica) spheres is formed on the Si wafer surface either by using spin coating, dip coating or Langmuir-Blodgett technique to transfer sphere pattern onto the Si wafer surface that allows the deposition of metal catalyst films (either by vacuum evaporation or coating of metal precursor solution) or to block designated areas equal to the size of wires [38–40]. Figure 3 shows the schematic of this process of patterning and growth of periodic size-controlled Si wire arrays. For coating purposes, a critical concentration of monosized nanoparticles or microparticle solution is coated onto the cleaned and hydrophilic surface of the Si wafer or substrate to allow the particles to be self-assembled and form a monolayer closed pack structure (Fig. 4) [41, 42]. This self-assembled closed pack monolayer structure is transferred to a non-closed pack monolayer by reducing the size of the particles either by RIE etching or by wet chemical process (Fig. 5). In the case of fabrication of a self-assembled monolayer closed pack structure of nano- or microspheres on Si substrate by Langmuir-Blodgett technique the self-assembled monolayer is separately transferred on the substrate surface and after that RIE is used to make a closed pack structure into non-closed pack structure by reducing the diameter of the spheres. Next, a noble metal film (e.g. Ag) is deposited over the non-closed pack structure. The metal goes into space among the spheres and touches the Si surface; however, in place of spheres the metal coating is blocked. After the metal coating, the sphere arrays were removed (in the case of silica spheres only) by sonication. Thus, a metal mask of hole arrays (area without metal) with continuous metal-coated area is formed on the substrate surface. Then, this patterned metal film is introduced in an etching solution consisting of HF and H_2O_2 for wet chemical etching. The vacant areas where metal is not present remain unetched whereas the areas under the metal coating are etched away. As a result, controlled diameter and pitch wire arrays are formed. The size of the silica or PS spheres after RIE or chemical reduction determines the size of the Si wires. By controlling the chemical etching time the length of Si wires could be controlled in a wide range as shown in Fig. 6. When bi-layer of two different metal (metal 1/metal 2) catalysts was used, different morphologies of wire array structures were formed [43]. Here metal 1 worked as a direct catalyst and metal 2 worked as mechanical support to metal 1. SiMW arrays with vertical walls and cone-shaped porous wires were formed when Ag/Au and Pt/Au bi-layer metals were used for wet chemical etching (Figs. 6 and 7). Before metal catalyst film deposition, a thin layer of Ti is usually deposited to increase the adhesion between Si surface and metal catalyst film. Chemically active Ag slowly dissolves in HF and H_2O_2 solution, forming isolated Ag particles during etching. As a result, nanowires formed among microwires. On the other hand, good chemical stability and continuity of Au film keep it intact during etching. However, Au film suffers from less penetration of etchant solution, making the etching slow and non-uniform. In bi-layer formation, it was found that etching becomes slow with a decrease in Ag/Au ratio, such as the etching rate of 0.3 μm/min reduces to 0.09 μm/

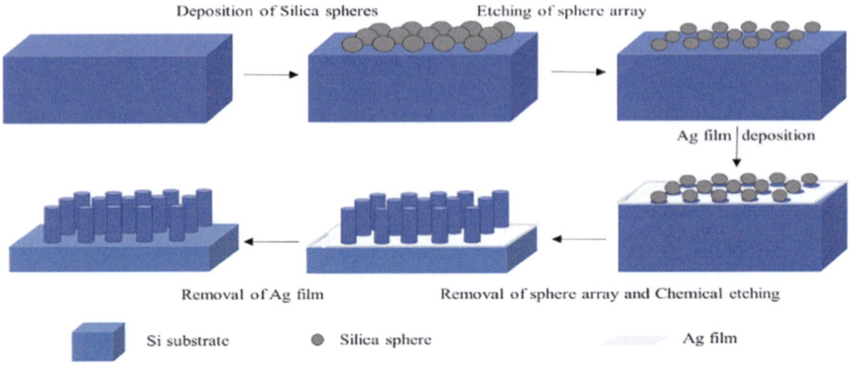

Fig. 3 Schematic of the SiMW arrays fabrication process by NSL technique in addition to wet chemical etching

min for 10 nm Ag/10 nm Au to 2.5 nm Ag/10 nm Au, respectively. By introducing co-solvents (such as methanol, 2-propanol, acetonitrile) in an aqueous solution of HF and H_2O_2 and controlling the amount of co-solvents curvature was introduced in Si wires, and moreover, slanting Si wire arrays were fabricated [44].

Nanopores with spongy nature are also introduced in SiMW array structure by simply varying the oxidant concentration in the presence of surfactant or co-solvent during the wet chemical etching after creating an Ag mesh pattern by NSL process as shown in Fig. 8a and b [25]. Figure 8c shows the schematic of the mechanism responsible for such bending in Si wire arrays due to the introduction of such co-solvents. In the course of the reaction, the co-solvent evaporates which creates an unbalance between surface pressure and surface tension and with time surface pressure increases which creates bending and pores in microwires.

3.2.2 Conventional Lithography, in Addition to Dry Etching and Wet Chemical Etching

Using photolithography to pattern Si wafer surface is a conventional way to make a mesh structure (as shown in Fig. 9a). In this method, the photoresist (PR) is spin coated on a Si wafer and dried. After certain UV exposure through an appropriate photomask and development by the use of a suitable developer the mask pattern of the photomask is transferred on Si wafer surface. So, a hole array pattern is developed on the Si surface. The diameter of the hole determines the diameter of the Si wire. After this process, oxygen plasma exposure is done to create sharp edges of photoresist and remove the leftover of the PR. In the next step, metal is deposited. In place of holes where PR exists metal does not touch the Si surface. The metal-coated Si wafer is then transferred to either the appropriate etching solution for wet etching or dry etching. This type of wet chemical etching has been tried to form longer SiMW arrays with longer etching time and higher temperature of etching solution [45]. However,

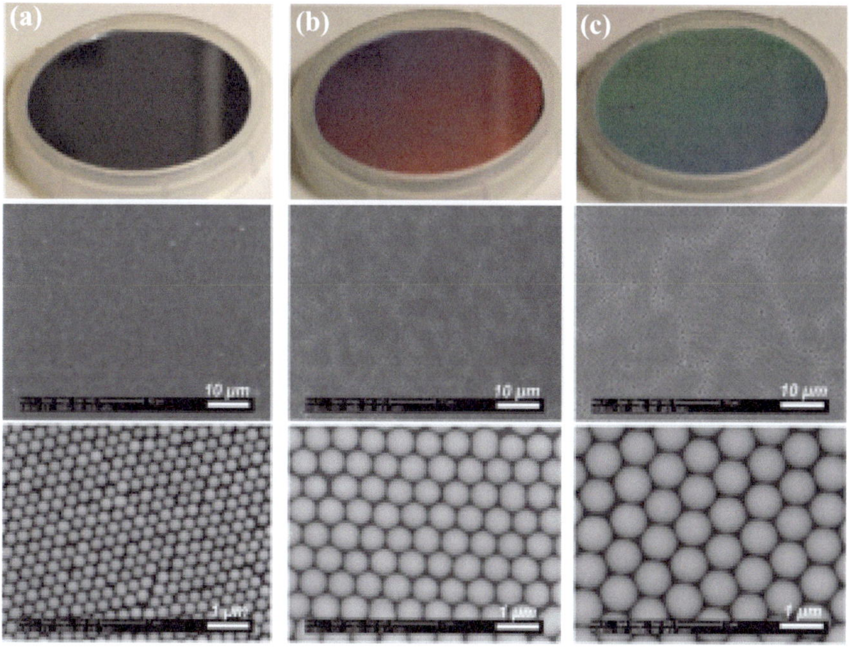

Fig. 4 Different size silica particles **a** 310 nm, **b** 600 nm and **c** 840 nm forming monolayer of closed pack structure on 2 inch Si wafer by spin coating and their corresponding FESEM images (middle and bottom panels represent lower and higher magnification images corresponding to top panel). Reproduced with permission [41]. Copyright 2015, American Chemical Society

horizontal etching with vertical etching affects the side walls of microwires and perfect vertical SiMWs were not formed. Later this lagging of the experimental arrangement was corrected by another group of researchers by controlling over the deposition rate and thickness of the metal catalyst [30]. For Au film of thickness less than 25 nm island type film formed during etching; thus, NWs were fabricated. For thicker Au films (> 50 nm), the etching solution cannot penetrate the Au film to reach the Si surface. For Au film of thickness in the range 30–40 nm, vertical SiMWs were fabricated. The same group found the relation between the deposition rate of Au film and the wet chemical etching rate. The higher deposition rate of the metal film forms a porous film. In contrast, a low deposition rate forms a dense Au film. It was observed that for Au film of thickness 30 nm the etching rate increased from 116 to 386 nm/min for increasing the deposition rate from 1 to 3 Å/sec.

In the top-down approach after etching mask patterning on the Si wafer surface apart from the wet chemical etching, dry etching most commonly deep reactive ion etching (DRIE) is used to etch the Si selectively. To fabricate an etching mask to carry out anisotropic etching of Si surface during dry etching most commonly prepared by photoresist (PR) as shown in Figs. 2 and 9 [46, 47]. Along with PR-based etching masks PS or silica spheres or metals have also been used to fabricate etching masks (Figs. 3 and 9b). The fabrication technique of PS or silica sphere-based etching masks

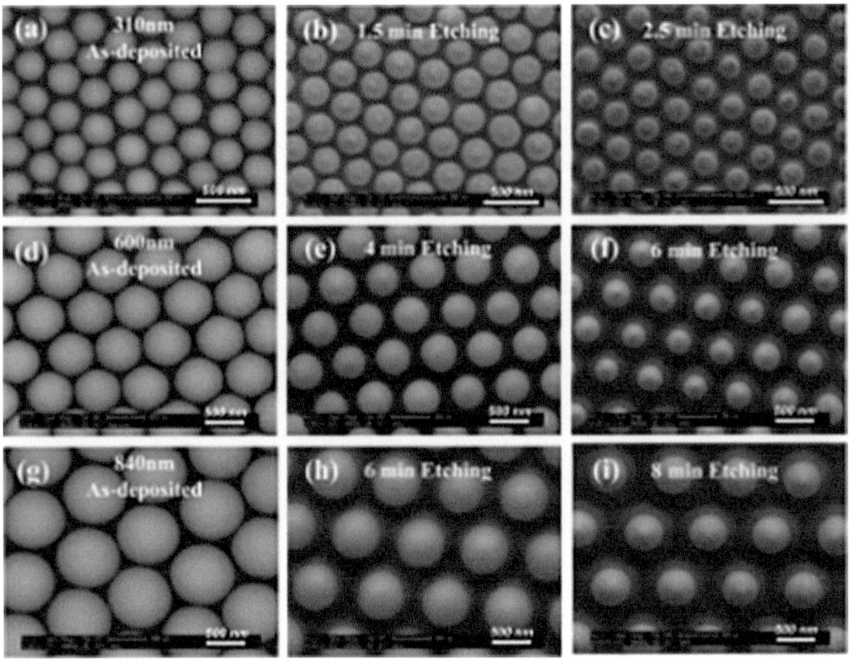

Fig. 5 FESEM images of monolayer of closed pack structure for different size silica particles **a** 310 nm, **d** 600 nm and **g** 840 nm. **b**, **e**, **h** and **c**, **f**, **i** are the FESEM images of non-closed pack structure formed after RIE etching for 1.5 min, 4 min and 6 min corresponding to closed pack structure FESEM images of (**a**), (**d**), (**g**), respectively. Reproduced with permission [41]. Copyright 2015, American Chemical Society

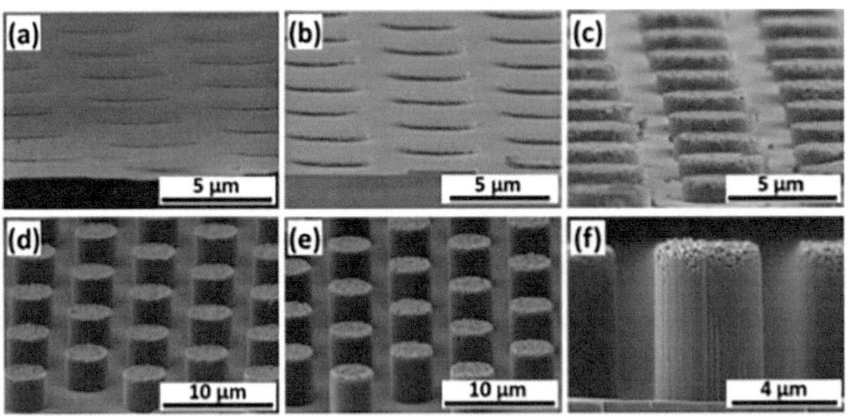

Fig. 6 Cross-sectional SEM images of SiMW arrays fabricated of using Ag/Au (10 nm/10 nm) bi-layer metal catalyst-assisted wet chemical etching for **a** 5 min, **b** 10 min, **c** 30 min, **d** 60 min and **e** 90 min, **f** higher magnification view of (**e**). Reproduced with permission [43]. Copyright 2016, American Institute of Physics

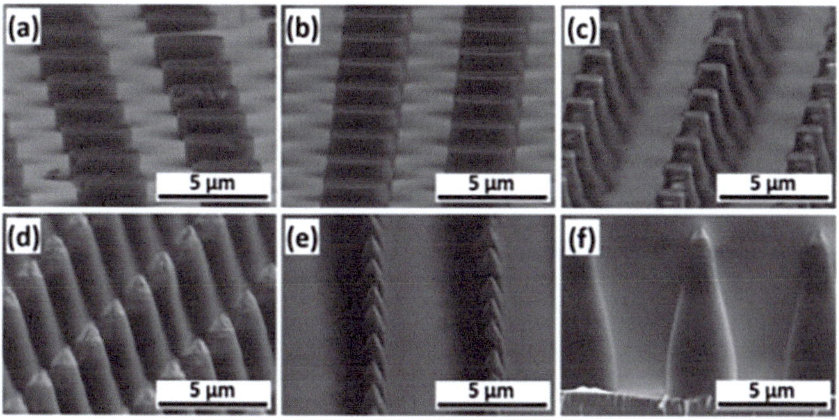

Fig. 7 Cross-sectional SEM images of cone shaped SiMW arrays fabricated of using Pt/Au (10 nm/ 10 nm) bi-layer metal catalyst-assisted wet chemical etching for **a** 5 min, **b** 10 min, **c** 30 min, **d** 60 min and **e** 90 min, **f** higher magnification view of (**e**). Reproduced with permission [43]. Copyright 2016, American Institute of Physics

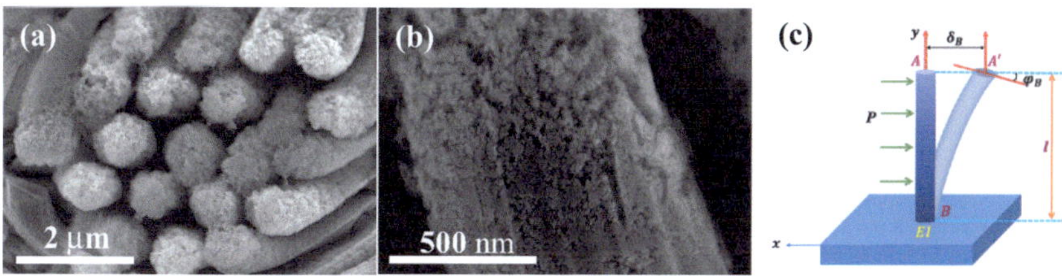

Fig. 8 FESEM **a** top and **b** side view images of spongy Si wires. **c** Schematic of the mechanism responsible for bending and pore formation in Si wires. Reproduced with permission [25]. Copyright 2022, Elsevier

is discussed in Sect. 3.2.1. However, these PR- or PS-based etching masks suffer a common problem of stability during longer periods of etching because of their fast etching rate and as a result, depleted quickly [48, 49]. So, in place of PR or PS or silica, some other hard materials like silicon dioxide, silicon nitride, Al_2O_3, metals, etc., are preferred for longer etching time due to their much higher resistance towards DRIE process compared to Si as shown in Fig. 9c and d [50, 51]. So, a designated structure could be formed without much damage to the etching mask in the long run. Using these harder materials such as Al_2O_3 as an etching mask vertical SiMWs with a length 80 μm have been fabricated by the DRIE process [49].

During the DRIE Bosch process composed of an alternative cycle of passivation of side walls and Si etching is used. Octafluorocyclobutane (C_4F_8) is the most commonly used fluorocarbon gas among the other gases such as C_2F_6 and CHF_3 for the passivation of side walls of SiMWs because of its fast deposition rate and SF_6 is used as Si etching gas. During the process after etching by SF_6, it is pumped out

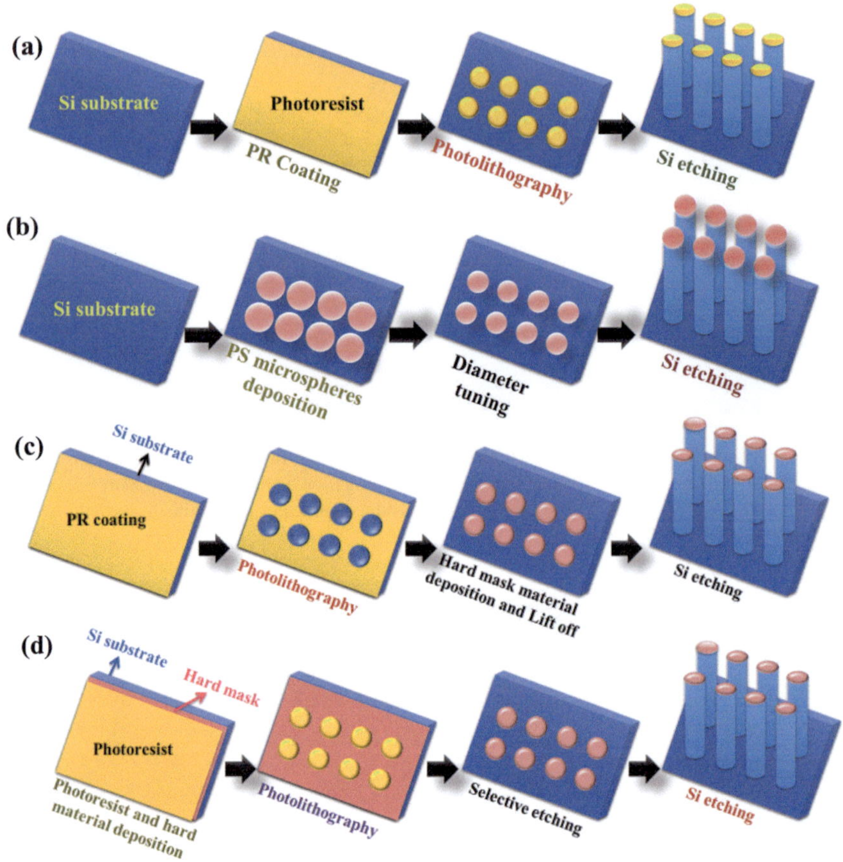

Fig. 9 Schematic of the **a** photolithography produced PR-based dot array-assisted dry etching, **b** NSL-assisted dry etching, **c** hard material (e.g. Al_2O_3, SiO_2, etc.) deposited on dot array pattern fabricated by photolithography process on PR coating followed by dry etching, **d** fabrication of dot array pattern on hard material deposited on Si prior PR coating followed by dry etching

from the chamber for the introduction of passivation gas. This fluorocarbon polymer of passivation gas coats entire exposed area including the side walls of the SiMWs. After this passivation, the fluorocarbon gas is pumped out and the etching gas SF_6 enters into the chamber. Because of ion bombardment, SF_6 etches anisotropically the bottom part of Si wafer. This process repeated in cycles until the desired structure is formed.

Depending on the gas flow ratio of etchant and passivation gases, shape of the SiMWs is determined as shown in Fig. 10. Three types of architectures are usually formed depending on the gas flow ratio. When the etching gas flow rate increases, the bottom part of the SiMWs etches more compared to the top part which makes the diameter of the top part to become higher than that of the lower part. This type of shape is called re-entrant. The opposite situation occurs when the flow rate of

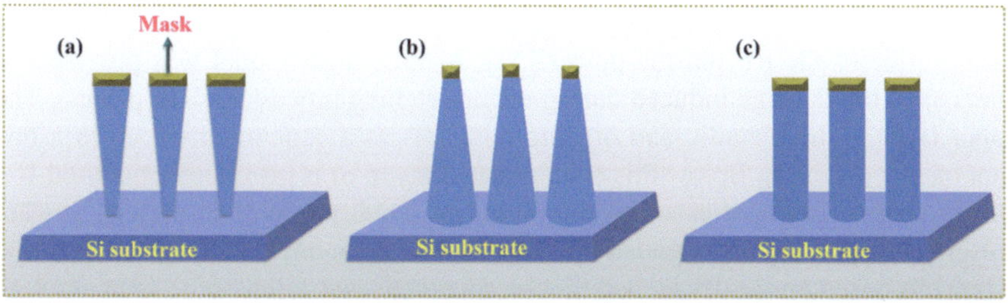

Fig. 10 Schematic of **a** re-entrant, **b** tapered and **c** vertical-shaped SiMWs fabricated by dry etching process

passivation gas increases. A tapered structure formed due to higher etching at the top side. Therefore, for obtaining vertical SiMWs a suitable ratio of flow rates of the etching and passivation gases is required. These gas flow rates were found to be 250 sccm for etching gas (e.g. SF_6) and 150 sccm for passivation gas (C_4F_8) for 2 s and 1.5 s, respectively [21]. This optimized condition was followed to fabricate 30 μm long SiMW arrays with vertical shape [26].

4 Surface Treatment to Remove Surface Damage and Reduce Dangling Bonds-Related Defects

Low performance of SiMW array solar cells occurs mainly due to the enhanced recombination that takes place at the large surface and also due to the damages created during the fabrication process. Thus, additional surface treatments are required to remove surface damage and it is also necessary to develop techniques to passivate the huge surface area of SiMWs compared to the conventional solar cell passivation.

For the SiMWs fabricated by wet chemical etching technique, the surface of the MWs becomes very rough and even porous. The roughness varies with the orientation of the crystal and depends on the etching solution concentration [52, 53]. Wires fabricated from c-Si with (100) orientation show much higher surface roughness. Depending on the crystal orientation and concentration and type of etching solution pores are also found to be introduced in the wall of the Si wires [31, 53]. The etching solution penetrates through the nanochannels formed between c-Si (wire wall) and the catalyst. As a result, pores formed throughout the wall of the wires. The rate of deposition of the catalyst and thickness is responsible for the formation of porous or compact metal catalyst films as discussed in earlier Sect. 3.2. The porous metal films help the etching solution to reach the nanochannels easily and form pores on the side wall surface and flat bottom part of the wafer. The variation in concentration of H_2O_2 and the presence of co-solvent also accelerate the process of formation of pores and even fractures on the side walls of SiMWs as discussed in earlier Sect. 3.2.

Compared to chemical etching, the use of RIE causes a larger amount of surface damage due to the RF plasma power [54]. A three-layer model was proposed to understand the damage induced during the dry etching process by RIE plasma. The upper layer is the metallic and organic remnants as a contamination up to a few nanometers (~ 5 nm). Below this contaminated layer, Si is profoundly polluted 10–20 nm deeper. Below this, up to 50–80 nm of depth particles of etching gas are present, which could be extended up to 1 µm. The open circuit voltage (V_{oc}) of the solar cells based on the dry etched Si wire solar cells was found to be decrease with increasing radio frequency plasma power during the dry etching process [51]. An increase in plasma power initially from 0 to 500 W damages the surface more and more by ion bombardment and V_{oc} decreased from ~ 608 to 590 mV. On the other hand in the same study, it was observed that removal of 20 nm of damaged c-Si after dry etching restored the minority carrier lifetime of the c-Si wafer before starting the dry etching process. As a whole, it could be concluded that the damage induced by the plasma power reduces the minority carrier lifetime, and as a result, V_{oc} of cells fabricated from such microstructures or nanostructures decreases.

To remove such a damaged or porous layer two types of methods could be used: first is thermal oxidation by which a small depth of wall of SiMWs could be oxidized and removed, the second one is chemical etching to etch away the side wall by a small amount. Thermal oxidation needs high temperatures in the range of 800–1100 °C. This process not only increases the thermal budget of production but also redistributes the dopant concentration at the interface of SiO_2 and Si as SiO_2 absorbs the dopant limited by the dopant solubility limit in SiO_2 [55]. Thus, redistribution of dopant atoms happens on the side walls as well as on the bottom surface of c-Si wafer which creates difficulties in achieving the desired dopant profile during diffusion to control junction depth. With this metal and other impurities reach deep into c-Si and degrade device performance by reducing V_{oc}. On the other hand, chemical etching of Si is free from the use of high temperature. The use of KOH as Si etchant to remove damages that occurred during the fabrication process improves the device performance by improving the J_{sc} by 18% [56]. For KOH etching times 0, 60, 180 and 270 s, the J_{sc} estimated from internal quantum efficiency measurements were found to be 27.1, 30.0, 31.1 and 31.9 mA cm^{-2}, respectively, indicating an 18% improvement in J_{sc} value for KOH etching time 270 s. However, this process of removal of surface damage is limited due to differences in the etching of KOH with different crystal orientations. Hence, the side wall of SiMWs which have different crystal orientations may not be equally etched. Using isotropic Si etchant mainly composed of HNA solution (HF/HNO_3/CH_3COOH) has been well accepted to remove surface damage. Longer time Si etching using such isotropic Si etchant results in a tapered shape from vertical SiMWs [57]. Different research groups have used a variety of Si etchants to improve device performance [21].

Treating the rough surface of the microwires by the above-mentioned methods could not be considered an ultimate solution to decrease recombination as surface area of microwire arrays is very high compared to commercial micro-pyramid textured solar cells. The huge number of dangling bonds present on the surface of the microwires need to be passivated to get high V_{oc} and J_{sc}. The conventional passivation materials used to passivate c-Si surface are SiO_2, Al_2O_3, SiN_x, hydrogenated a-Si and a-SiN_x [58]. Some of these materials activate negative charges (e.g. Al_2O_3) and some activate positive charges (e.g. SiO_2, SiN_x) at the interface to passivate p^+ and n^+ emitters of n-Si and p-Si PV cells, respectively. These activated negative and positive charges facilitate hole and electron collection at the emitter of n-Si and p-Si cells, respectively. Further, these activated negative and positive charges at the interface push the same polarity charge carriers towards the rear electrode of the solar cell. Thus reduces the recombination at the emitter surface to a greater extent. On the other hand, a-Si:H, a-SiN_x:H deposited by PECVD passivate both polarity Si surfaces as they formed a high-quality interface layer. The advantage of using such passivation materials lies behind the low-temperature process and scale-up possibility. However, the use of such passivation layers is constrained by thermal stability in addition to parasitic absorption.

5 Device Physics of Radial Junction

Low-quality Si material suffers from a high density of defects or impurities that result in low minority carrier diffusion lengths when used in solar cells. The use of such material results in the inferior performance of solar cells as the carrier collection after separation of photogenerated carriers is limited by minority carrier diffusion length in the base region. The solution of such a situation could be solved by using thicker material which increases light absorption at the same time subjected to less carrier collection or by use of efficient light trapping schemes. The use of radial junction instead of planar p–n junction solves this problem by orthogonalizing the light absorption and carrier transportation direction. Radial junctions use long-length nanorods/microwires to ensure effective light absorption with efficient carrier collection in short distances along with the photogenerated carrier separation and transportation in the radial direction.

The device physics [29] lies behind the efficient performance of radial p–n junction involving lower-grade material including the solution of the diffusion or drift equation for minority carriers, Poisson's equation and the current continuity equation for radial geometry. For establishing such a model two postulates with separate conditions need to be considered: (1) constant trap density throughout the material and (2) independent trap densities in quasi-neutral regions and depletion regions. Assuming the p–n junction to be abrupt and depletion approximation to be valid, the band diagram and geometry for the radial junction solar cell are presented in Fig. 11 by extending the planar cell geometry into a cylindrical geometry. The outer shell layer (emitter) is assumed to be *n*-type, and the inner core (base region) is assumed to be *p*-type. While estimating solar cell parameters only the Shockley–Read–Hall recombination from single trap level at midgap and surface recombination were considered neglecting the Auger recombination. The radial junction cell was divided into four parts: quasi-neutral *n*-type region of width x_1, depletion region in *n*-type

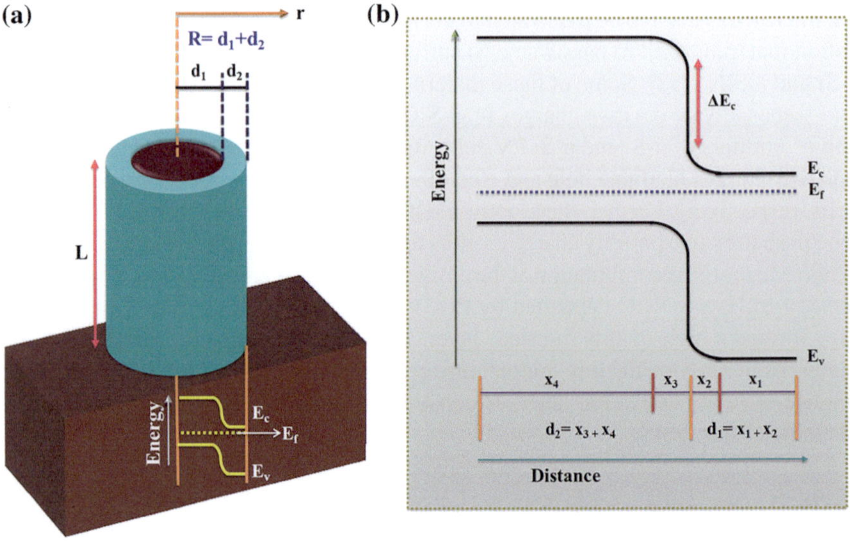

Fig. 11 Schematic of **a** radial p–n junction and corresponding **b** band diagram

side of width x_2, depletion region of width x_3 in *p*-type side and quasi-neutral part in p-type of width x_4.

Using the transport equation, the current density in these four regions could be given as follows: [29]

$$J^p = \frac{2x_4 \int_0^L J^p(z)dz}{R^2} \tag{1}$$

for the quasi-neutral part of the *p*-type region

where $J^p(z) = -qD_n \frac{\partial n'}{\partial r}$ at $r = x_4$ and n', R, L, D_n, q are the excess minority-electron concentration, radius, length of the rod, electron diffusion co-efficient and electronic charge in coulomb, respectively.

$$J^n(z) = \frac{2(R - x_1) \int_0^L J^n(z)dz}{R^2} \tag{2}$$

for the quasi-neutral part of n-type region

where $J^n(z) = qD_p \frac{\partial p'}{\partial r}$ at $r = R$–x_1 and p', D_p are the excess minority hole concentration and hole diffusion co-efficient.

The light-generated current density in the depletion region J_g^{dep} was estimated by considering the production of carriers through full absorption of incident light.

$$J_g^{dep,p}(V) = q\Gamma_o\left(1 - e^{-\alpha_2 L}\right)\frac{d_2^2 - x_4^2}{R^2} \tag{3}$$

for the p-type region of the depletion region

$$J_g^{dep,n}(V) = q\Gamma_o\left(1 - e^{-\alpha_1 L}\right)\frac{(d_2 + x_2)^2 - d_2^2}{R^2} \quad (4)$$

for the n-type region of the depletion region

where Γ_o is the incident photon flux, α_1 and α_2 are the absorption co-efficient of n-type and p-type material, respectively.

The solution of the above equations gives the current density of the radial junction as

$$J = \left(J_0^p + J_0^n\right)\left(e^{\frac{qv}{kT}} - 1\right) - J_l^p - J_l^n - J_g^{dep,p}(V) - J_g^{dep,n}(V) + J_r^{dep}(V) \quad (5)$$

where J_0^p, J_l^p and J_0^n, J_l^n are dark, light-generated current density in the quasi-neutral part of the p and n regions, $J_r^{dep}(V)$ is recombination current in the depletion region.

In radial junction, the efficiency reaches a maximum value for an optimal thickness L; however, in planar cell efficiency reaches a limiting value with increase of thickness as the two opposite effects work for determination of the light-generated current density and dark current density as follows:

$$J_l \propto \left(1 - e^{-\alpha L}\right) \quad (6)$$

and

$$J_l \propto \left(1 - e^{-\alpha L}\right) \quad (7)$$

where α is the absorption co-efficient and L is the cell thickness

The J_{sc} is independent of the trap density for radial junction solar cells whereas reduces with increase in trap density in case of the planar junction solar cells (Fig. 12). For example, J_{sc} of a nanowire-based radial junction solar cell with diameter 100 nm (L_n ~ 100 nm) and thickness 100 μm drops from 38 to 35 mA/cm^2 if the wire diameter increased above 1 μm. On the other hand, in the planar cells if L_n drops from 100 μm to 100 nm (which is equivalent to an increase in density of recombination centres (N_r) from 7×10^{12} to 7×10^{18} cm^{-3} in quasi-neutral region of base as L_n ~ $1/N_r$) the J_{sc} drops from 34 to 4 mA/cm^2. In contrast, for the radial junction V_{oc} decreases with an increase in the cell thickness as the junction area increases in contrast to V_{oc} of the planar junction which is independent of cell thickness. In both cases, V_{oc} decreases slowly if N_r is held fixed at relatively low values (or lifetime is fixed at a high value) in the depletion region and decreases rapidly if the trap densities are the same in quasi-neutral and depletion region. So, V_{oc} depends on the minority carrier diffusion length and decreases when diffusion length decreases. For 100 μm thick Si cell, the V_{oc} dropped from 0.580 to 0.380 V if the N_r is held fixed at 10^{14} cm^{-3} (for planar junction V_{oc} decreases from 0.590 to 0.490 V) and dropped from 0.580 to 0.010 V as N_r increased from 7×10^{12} to 7×10^{18} cm^{-3} (for planar junction V_{oc} decreases from 0.590 to 0.240 V). The same theoretical modelling shows that fill factor becomes nearly constant around 0.8 if the depletion region trap density N_r is maintained around 10^{14} cm^{-3} (equivalent to lifetime of 1 μs in the depletion region). However, for a constant trap density, $N_r > 3 \times 10^{15}$ cm^{-3} (equivalent to $L_n \leq 5$ μm) throughout the device fill factor drops sharply for radial junction but gradually for planar junction.

Thus, so far discussed device physics modelling indicates that the efficiency of a radial p–n junction solar cell in contrast to planar cells could be high despite high trap density in the quasi-neutral part, if the trap density is low at depletion region. In the planar cells, high trap density in quasi-neutral region makes the J_{sc} low regardless of maintaining low trap density at depletion region.

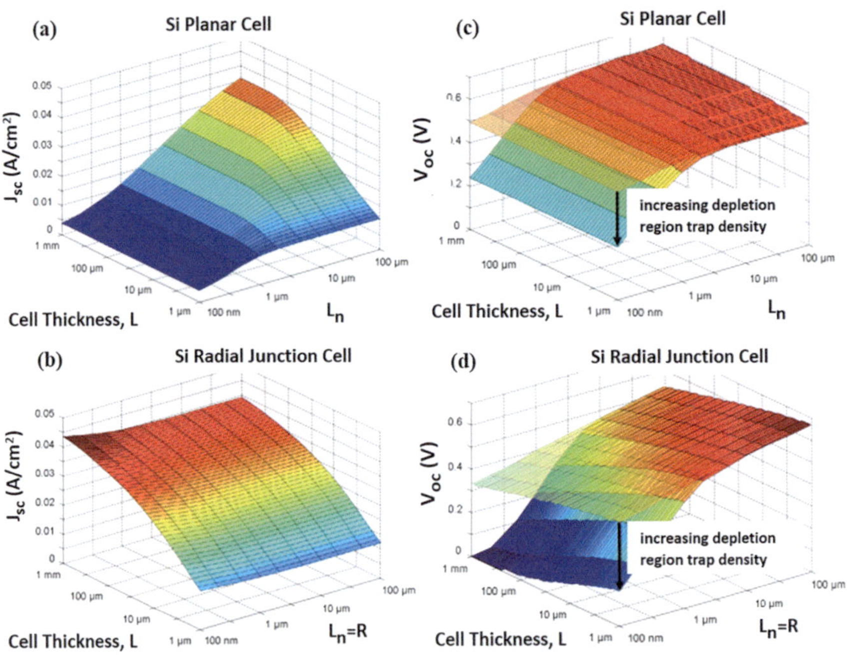

Fig. 12 J_{sc} vs cell thickness L and minority-electron diffusion length L_n for **a** conventional planar Si solar cell and **b** a radial p–n junction Si wire array solar cell. V_{oc} vs cell thickness L and minority-electron diffusion length L_n for **c** a conventional planar Si solar cell and **d** a radial p–n junction Si wire array solar cell. In the radial junction, the cell radius R is set equal to L_n (Reproduced with permission from Ref. [29]. Copyright 2005, American Institute of Physics)

6 Photovoltaics of Si Wire Solar Cells

6.1 Single Wire Radial Junction Solar Cell

To power nanoelectronics circuits such as NW-based logic circuits, AND logic gate single nanowire or microwire-based solar cells are the ideal choice and need to be developed to power future nanoelectronics devices that need power up to a few nanowatts. When the radial junction was fabricated on a *p*-type Si nanowire (core) surrounding an intermediate i-Si layer and a *n*-type Si shell layer, the single nanowire radial junction solar cell showed a conversion efficiency of 3.4% under 1 sun illumination and power output of 50–200 pW per nanowire device [59]. The co-axial formation of p-i-n junction in the same nanowire was confirmed by cross-field emission scanning electron microscopic (cross-FESEM) observations. This inferior device performance could be improved by providing good quality surface passivation as the large surface area of the Si wires is very high and becomes rough depending on the growth process. Passivation of such single nanowire solar cells by a-Si:H and SiN_x:H has improved the PCE up to 9% [60]. The V_{oc} reaches ~ 600 mV with an FF of 80%. Coating of nanostructure surface by a-Si:H and SiN_x:H comes with the additional benefits of production of heterojunction with very high V_{oc} (~ 0.743 V) [61] and further providing antireflection to the nanostructure surface [62]. Such a device carried selective formation of radial junction by phosphorus diffusion at one part of the wire and the contacts are made at both ends. The surface passivation by a-Si:H reduced the surface recombination velocity (450–600 cm/sec) and enhanced the L_n by more than 30 μm. The passivation by a-Si:H enhanced the conversion efficiency up to 7.4% whereas for the unpassivated device it was only 4.6%. Use of SiN_x:H coating as passivation layer produced best-performing single nanowire device of conversion efficiency 9.0%. The recombination velocity and L_n were estimated for SiN_x:H passivation to be << 70 cm/sec and 5–10 μm. In another work, a 40% enhancement in J_{sc} and efficiency has been predicted by changing the wire circular cross-section to a front-opening crescent shape [63].

6.2 Radial Junction Microwire Array Solar Cell

Fabrication of SiMW array solar cells by using the bottom-up or top-down approach with conventional gas phase diffusion or spin-on dopant technique could produce a path for the solar industry to use very low-grade Si material yet with reasonable efficiency. For the radial junction use of such low-grade Si comes with the advantage of very short charge carrier path (of the order of nm to μm) less than radius of the microwire after generation to reach the electrode. Thus, the minimum diffusion length of the Si material required to produce efficient solar cell in radial junction morphology is equivalent to the radius of the microwire, and this advantage allows use of very low-grade Si material. The last decade has witnessed a vast growth in the fabrication of SiMW array solar cell field [64–67]. Starting from the fabrication of initial low-performing p-n or p-i-n junction SiMW array solar cells to reasonable efficiency SiMW array solar cells up to date.

Atwater and co-workers [37] estimated the minority carrier diffusion length to be 10.5 ± 0.1 μm under illumination by scanning photocurrent microscopy of 2 μm diameter Si wires grown by Cu-catalyzed VLS growth process. Whereas for Au-assisted VLS grown Si wires [68, 69] the minority carrier diffusion length is estimated to be 4 μm due to the deep trap levels created by Au during the growth process as mentioned in Sect. 3.1.1. Atwater and co-workers reported the growth of a SiMW array solar cells of 2–3 μm diameter with pitch 7 μm grown on p^{++} Si (111) wafer using Cu-catalyzed VLS growth method [22, 35]. The VLS growth process involved the thermal growth of an oxide buffer layer of thickness 300 nm over Si substrate (111) and photolithographically patterned the buffer layer to confine the metal catalyst of thickness 500 nm deposited by thermal evaporation in the designated area followed by a lift-off process. After the growth of Cu-catalyzed Si microwires, Cu catalysts were removed and 200 nm thick oxide layer grown followed by poly-dimethyl siloxane (PDMS) deposition for selective etching of the oxide layer to protect the base of the microwires. During phosphorus diffusion (~ 80 nm junction depth) only the upper part of the microwires got converted in radial p–n junction not the base region. These diffused Si microwires were passivated by a SiN_x:H layer and filled with Al_2O_3 scatterer to scatter the light incident between microwires followed by filling of a transparent non-conducting thermoplastic polymer called mounting wax up to the tips and sputtering of ITO layer (120–150 nm thick, resistivity ~ 7×10^{-4} Ω-cm). The best-performing microwire radial junction solar cell showed a PCE of 7.9% under AM 1.5 G radiation with J_{sc} of 24.3 mA/cm^{-2}, V_{oc} of 0.498 V and FF of 65.4%. For Au catalyzed VLS grown Si microwire arrays of diameter 2.6 μm, pitch 7 μm and height 50 μm, the radial junction was produced by $POCl_3$ thermal diffusion at 1000 °C with an estimated junction depth of 350 nm, showed PCE of 2.3% (J_{sc} ~ 7.6 mA/cm^2, V_{oc} ~ 0.5 V and FF 57%) [66]. Epitaxial growth of radial p-n and p-i-n junction on wire array of diameter 0.3 μm and length 10 μm by cold wall low pressure chemical vapour deposition reactor, with 50% SiH_4 diluted in H_2 as a precursor, B_2H_6 and PH_3 as a *p*-type and *n*-type dopant gas, respectively, resulted in $V_{oc} = 0.440$ V, $J_{sc} =$ 40 mA/cm^2 and FF = 38% [64]. In this study the Si wire arrays were fabricated on (111) Si by Bosch etch process using a combination of DRIE and inductive coupled plasma etching with passivation as discussed in sec. 3.2.2. On the other hand, SiMW arrays of 7.5 μm diameter and 25 μm height fabricated from Si wafers with minority carrier diffusion length 10 μm and 1 μm by dry etching in combination with optical lithography for SiO_2 hard mask patterning, doped using gas phase diffusion process with junction depth of 300 nm and 200 nm respectively showed almost double PCE compared to the planar cells [30]. Again, simultaneous passivation of the junction and the surface of the microwire showed improvement in the conversion efficiency of the SiMW array solar cells by 55.7% compared to the cells without passivation [70]. In this study, CZ grade Si(100) wafer of thickness 500 μm was used for top-down fabrication of SiMW (p-core) arrays of diameter 3 ± 0.3 μm, pitch 7 ± 0.2 μm and length 9 ± 1 μm by DRIE. During the fabrication of the radial junction (core–shell structure) by CVD technique, a 50 nm thick intrinsic poly Si layer and a 40 nm thick *n*-type poly-Si layer were deposited simultaneously over the SiMWs. Passivation of p–n junction by the introduction of i-poly Si layer improved the ideality factor from 2.45 ± 0.11 to 1.89 ± 0.14 and the reduction in the saturation current density (J_0) by one order of magnitude reduced junction recombination which improved the conversion efficiency by 30%. To passivate the outer surface of shell layer a 80 nm thick a-SiN:H layer was deposited by plasma-enhanced chemical vapour deposition which increases the J_{sc} by 25% and light absorption from 74 to 92%. The a-SiN:H passivated SiMW array radial p-i-n junction solar cell showed a PCE of 10.6% with V_{oc}, J_{sc} and FF of 0.558 V, 30.6 mA/cm^2 and 0.625%, respectively.

In a recent study, a nano/micro-hybrid structure radial junction solar cell composed of nanowire arrays selectively fabricated on the top of microwires, proposed to demonstrate a PCE of 17.6% (effective cell area 1 cm^2) with J_{sc}, V_{oc} and FF of 39.5 mA/cm^2, 584 mV and 76.1%, respectively [26]. The addition of Si nanowire arrays on the top of microwires enables 97% absorption of sunlight and a 16.5% enhancement in efficiency compared to only SiMW array radial junction solar cell without such nanowires on the top. SiMWs of diameter 2 μm and spacing 2 μm were fabricated on *n*-type CZ grade 200 μm thick Si wafers of resistivity 1–3 Ω-cm using DRIE in addition to photolithography technique. Further, the surface damages caused by DRIE were removed from the side walls of the microwires by dipping in the HNA ($HF:HNO_3:CH_3COOH$ = 1:400:20) solution for 60 s. To fabricate selective nanowires on the top of the microwires, an additional passivation step was done during DRIE to protect the surface of the microwires during MCEE process. After removing the photoresist from the top of the microwires, 20 nm thick Ag film was selectively deposited on the top of SiMWs and Si nanowires were fabricated on the tops of microwires by MCEE method. The BSF and emitter were formed by spin-on-dopant technique, and a 10 nm thick Al_2O_3 layer was deposited by atomic layer deposition (ALD) as a surface passivation layer for the emitter. For electrode a 500 nm thick Al microgrid mesh structure was formed using the photolithography technique in addition to thermal evaporation of metal followed by lift-off. The effect of variation of the size of the microwires and the spacing (pitch) on the performance of SiMW array solar cells was also studied [71]. The microwire arrays were fabricated using RIE in addition to photolithography to design required dimension etching mask on a CZ grade 500 μm thick Si (100) wafer [72]. Radial junctions were formed by gas phase thermal diffusion of $POCl_3$ as a phosphorus dopant source at 800 °C and after deposition of SiN_x passivation layer, front and rear contacts were formed by screen printing. Four variations of the microwires diameter and their pitch were fabricated: (i) pillar-1 of diameter 1.58 μm, pitch 4.03 μm and height 5.2 μm, (ii) pillar-2 of diameter 1.26 μm, pitch 7.13 μm and height 5.43 μm, (iii) pillar-3 of diameter 5.11 μm, pitch 7.1 μm and height 5.71 μm and (iv) pillar-4 of diameter 4.12 μm, pitch 9.97 μm and height 6.2 μm. The solar cell based on pillar-1 showed the best performance of 16.92% with J_{sc}, V_{oc} and FF of 36.47 mA/cm^2, 608 mV and 76.29% respectively compared to 10.15% PCE of the planar cells formed under identical conditions.

Despite considerable worldwide efforts to enhance the PCE of SiMW-based solar cells the process of commercialization of these radial junction SiMW solar cells is yet to be achieved. So far most of the efforts in advancing the cell performance are only up to lab scale (active area of the cell around $1 cm^2$) and a minimal attempt has been made for scale-up. A very recent effort by researchers from the Ulsan National Institute of Science and Technology and the Kangwon National University of the Republic of Korea has reported 21.1% efficient (with 41.2 mA/cm^2, 0.628 V and FF 81.4%) tapered SiMW solar cells in wafer scale (4 inch ~ 25 cm^2) and for the lab-scale device (area ~ 1cm2) the PCE was observed to be 20.6% with 42.1 mA/ cm^2, 0.608 V and FF 80.3% [28]. The tapered SiMW arrays were fabricated on a CZ grade *n*-type Si wafer by DRIE in addition to photolithography followed by a short dip for 10 s in Si etchant (RSE-100). The diameters of the tapered microwires were gradually decreased from 2 μm at the bottom to nanometer dimension at the tip of the wires. The BSF and emitter (depth ~ 300 nm) were formed by spin-on dopant technique followed by thermal diffusion. Thereafter, a 10 nm thick Al_2O_3 layer by ALD and 10 nm thick SiO_2 layer were thermally grown above the emitter and rear side, respectively. In addition to these, a 200 nm thick SiO_2 layer was deposited on the rear side by PECVD process. The front metal grid and localized contact at the rear were patterned by using lithography. The masking of the edge electrode has a considerable effect on the degradation of device performance. The higher performance of 4-inch device compared to the lab-scale device in 1 cm^2 area was attributed to the reduced Si/edge electrode contact area in 4-inch device.

To avoid detrimental effects caused by the thermal diffusion process to establish a p–n junction, a dopant-free tapered SiMW array solar cell with a PCE of 20.1% was reported by establishing a field-induced radial junction [27]. The tapered SiMW arrays were fabricated on CZ grade *n*-type Si wafer by DRIE in addition to etching mask dot arrays pattern by photolithography and subsequent immersion in RSE-100 Si etchant for 10 s. A selective thermal SiO_2 layer was grown at the rear and Al_2O_3 layer (10 nm) deposited on the front surface by ALD. A localized back contact was formed by 1 nm/200 nm LiF/Al film deposition on a patterned SiO_2 layer and 10 nm/200 nm MoO_x/Al films were thermally evaporated as front electrodes on selective portions. The Al_2O_3 layer generated p^+ inversion layer on the surface of microwires which created radial junctions within the microwires with J_{sc}, V_{oc} and FF of 40.1 mA/cm^2, 0.628 V and 79.8%, respectively. Controlling the spacings among the microwires in the nanometer region (e.g. 200 nm) with a tapered structure consisting of a SiO_2 layer and 10 nm Al_2O_3 layer passivating in rear and front sides respectively with microgrid mesh electrode in front enables to enhance J_{sc} up to 42.5 mA/cm^2 and PCE of 20.5% (V_{oc} and FF of 0.608 V and 79.7%, respectively) [73].

6.3 *Flexible Microwire Array Solar Cell*

The need for flexible solar cells to integrate into wearable technologies, electric vehicles, foldable power sources, etc., has accelerated the research towards developing efficient and long-term flexible cells. However, long-life commercial c-Si solar could not solve the purpose because of mechanical rigidity. Nevertheless, c-Si solar cells with a thickness of less than 50 μm have shown the potential to be employed as flexible solar cells [74, 75]. These thin c-Si solar cells with micro-pyramids on the front side show crack initiation sites at the bending positions. Such problems could be overcome in flexible SiMW array-based solar cells where the mechanical stress of microwires could be released by movement towards the space among the microwires. Efficient and flexible thin c-Si solar cells with tapered SiMWs radial junctions have already been demonstrated [21]. The thin c-Si with SiMW arrays showed absorption up to 87.5% and in place of pillar type SiMWs the tapered SiMW arrays on thin c-Si showed reflection of less than 2%. The high J_{sc} shown by the tapered SiMW in flexible architecture was attributed to the efficient absorption of solar light and nearly 100% internal quantum efficiency. Another novel strategy to fabricate ultraflexible SiMW array solar cells is SiMW arrays embedded in PDMS [76]. To achieve such free-standing structures, PDMS is poured over the SiMW arrays and cured. These SiMWs embedded PDMS then detached mechanically from the Si substrate. These PDMS embedded SiMWs could be an excellent candidate for flexible solar cells. However, such kind of flexible solar cells suffers from the problems of parasitic absorption with high sheet resistance of ITO electrodes and ITO layer degradation due to bending [77]. To solve such issues metal nanowire-based flexible transparent electrodes with excellent flexibility, stretchability, high transmittance and low sheet resistance have been already demonstrated [78, 79].

7 Electrode Design

In order to collect photocarriers in respective electrodes, the design of the electrode should include minimum shading and resistive loss, Auger recombination loss, etc. The shadow loss depends on the transparency of the electrode. For the conventional metal grid electrode (finger-bush bar), the shading loss is estimated by the ratio of the area covered by the metal front electrode divided by the area of the cell as the places where metal is deposited are completely blocked from sunlight due to the height (~ few microns) of metal deposited by screen printing. Thus, in finger-bush bar design, the only way to reduce shading loss is to reduce the width of the fingers which reduces the total area covered by the metal. However, this reduction causes a decrease in FF as the series resistance increases and recombination of the photogenerated charge carrier increases because the carriers have to travel a longer distance with increased separation between fingers. The only advantage comes in the increase in J_{sc} which is again subjected to an increase in the two losses mentioned above. Metal grid electrodes have already been tried by some groups in SiMW solar cells [71]. However, the performance of these Si microwire solar cells was not as expected. The J_{sc} obtained from these cells is lower than the conventional solar cells caused by the covering of metal electrodes over the short-height microwires in addition to the higher reflectance of these short-height microwires and for longer microwires use of the screen-printing method damages the microwires. A shallow emitter configuration that could increase V_{oc} could not work efficiently in the case of microwire solar cells due to an increase in resistive loss faced during carrier collection in the finger-bush bar design. As a remedy to this problem edge electrodes were investigated by selectively patterning and depositing metal on the edge of the solar cells [31, 57]. However, this design suffers from the problem of inefficient collection of the charge carriers from places away from the electrode, especially from the middle of the solar cells. This incurs power loss for the shallow doped emitter. When the highly doped emitter layer formed, the FF improved to 71% but due to the enhanced Auger recombination J_{sc} and V_{oc} degraded, resulting in an overall low PCE of 13%.

The limitations of the conventional finger-bush bar and edge electrode design could be resolved by using a microgrid pattern designed by photolithography in addition to metal evaporation to deposit metals in the designated area [32]. The microgrid pattern can reduce shadow loss with effective carrier collection. In such a design, the width of the metal lines is kept at a few microns and the spacing between these lines is controlled to ensure efficient collection with minimal recombination of charge carriers flowing through the emitter layer. In one attempt, the width of the grid line was kept at 2 μm and the spacing was changed from 50 to 400 μm. It was found that with an increase of the spacing, the average transmittance and sheet resistance increase. For 400 μm spacing among the microgrid line, more than 97% transmittance with 25 Ω/sq sheet resistance was observed. The use of a microgrid as a front electrode in Si MW solar cells resulted in a high FF of around 81.2%. Wrap-around metal contact was also fabricated, where the electrode formed at the bottom of the wires, to reduce parasitic absorption [80]. Thus, as the electrode connects all the microwires the photogenerated charge carrier collection efficiency increases compared to conventional finger-bush bar electrodes and also this wrap-around metal contact acts as a back reflector. However, shorter metal electrode thickness and contact area increased the contact resistance and as a result, FF became low as 65.4%.

Thus, for Si MW solar cells where the surface area is too high, an increase in the J_{sc} and V_{oc} demand shallow emitter that will reduce Auger recombination. On the other hand, a shallow emitter reduces the conductivity of the emitter layer, so metal lines of microgrids or fingers need to be very close, increasing the shading loss. Thus, for efficient collection of charge carriers from each of the MWs total surface area without enhancing Auger recombination and shading loss, wrapping of all MWs surface by transparent conducting oxide (TCO) layer such as indium tin oxide (ITO), fluorine-doped tin oxide (FTO) or aluminium-doped zinc oxide (AZO) could provide a reasonable solution. However, applying such TCO as an effective front electrode raises mainly two concerns in addition to the increase in contact resistance: an increase in parasitic absorption in short wavelength and the requirement of metals such as indium for ITO. However, for the second one FTO and AZO could be a possible solution. Increasing the thickness of the TCO layer reduces the contact resistance at the same time increases the parasitic absorption.

8 Future Outlook and Challenges

Radial p–n junction SiMW solar cells offer the route to decouple the light absorption and photogenerated charge carrier transport direction where these carriers have to travel a very short path in the radial direction after separation. The path needed to travel the charge carrier is less than or equal to the radius of the microwires. This facility allows to use of low-grade Si material with a very short minority carrier diffusion length of the order of a few microns for the production of efficient solar cells and could reduce production costs. However, to obtain a reasonable power conversion efficiency for radial junction SiMW array solar cells many technological challenges need to be overcome. First of all the commercial fabrication process for the production of SiMW arrays on a wafer scale needs to be found out as the photolithography-assisted dry etching method is not suitable for commercial production. Another easy way to fabricate the SiMW arrays, NSL-assisted wet etching could be the one way. However, the technique to produce a monolayer closed pack structure of micro- or nanoparticles of SiO_2 or PS is not commercially viable as in most cases spin coating or dip coating is used. Langmuir–Blodgett method also could not be considered as a commercial option for this purpose. Only the chemical etching step involved in this process is already adapted or with minor modification could be used in commercial production. Again the light trapping of SiMW arrays is low in comparison with SiNW arrays. The SiMW arrays show 15% reflection of solar light in the visible region compared to only 2% reflection by SiNW arrays. The major cause of such reflection increase is the top flat surface of microwire tips. Many strategies such as selective fabrication of SiNW arrays at the top of microwires or tapering of the cylindrical pillar structure of microwires to produce pencil or cone-type tapered microwires have been tried. Among these efforts tapered microwires showed excellent results which have widened the use of tapered SiMW arrays for the production of commercial radial junction solar cells shortly. Another major drawback of these SiMWs is their huge surface area that enunciates huge surface recombination of charge carriers. The fabrication process of such SiMWs either by dry etching or by wet etching also damages the side walls of the microwires. To remove such fabrication damages use of KOH treatment or isotropic Si etching solutions is proven to be the best. A real challenge comes towards the uniform and conformal deposition of the passivation layer for long-length SiMWs. However, ALD-deposited Al_2O_3 has produced some good passivation but still, these efforts are not adequate. A major challenge that remains to be solved for such kind of radial junction SiMW array solar cells is the design, deposition and fabrication of the front electrode. As the p–n junction is spread over the full side wall of the microwires, the electrode design should be sufficient to collect charge carriers throughout the surface of the microwires. The conventional finger-bush bar electrode is not sufficient for collecting charge carriers for radial junction cells. The use of TCOs as collecting electrodes (the best design of front electrode for radial junction cells) was challenged by parasitic absorption and higher sheet resistance. In lab scale and wafer scale, the design and deposition of metal micro-grid mesh is proven to be the best front electrode so far for radial junction solar cells. However, this kind of electrode design involves lithography that is not suitable for commercial production. In this aspect, researchers should explore the technique of TCO deposition as used in the commercial production of HJT cells. So far most of the efforts are in lab-scale demonstration with a power conversion efficiency of around 20%. More efforts towards the fabrication of radial junction SiMW array solar cells in wafer scale (6 inches or 8 inches) are awaited. Nevertheless, a great number of efforts are being made worldwide and the full potential of the Si MW radial junction solar cells could be realized shortly with progressive technological advancements in the field.

References

1. Key World Energy Statistics (IEA, 2024), p 24. https://iea.blob.core.windows.net/assets/04f06925-a5f4-443d-8f1a-6daa31305aee/WorldEnergyOutlook2024.pdf
2. Renewables 2024 Global Status Report (REN21, 2024), p. 13. https://www.ren21.net/wp-content/uploads/2019/05/GSR2024_Demand_Full_Report.pdf
3. Y. Chu, P. Meisen, Review and comparison of different solar energy technologies. Report of Global Energy Network Institute (GENI), Diego (2011)
4. P.C. Choubey, A. Oudhia, R. Dewangan, A review: solar cell current scenario and future trends. Recent. Res. Sci. Technol. **4**, 99–101 (2012)
5. A. McEvoy, L. Castaner, T. Markvart, *Solar cells: materials, manufacture and operation*, 2nd edn. (Elsevier Ltd., Oxford, 2012), pp.3–25
6. A.L. Fahrenbruch, R.H. Bube, *Fundamentals of solar cells* (Academic Press Inc., New York, 1983)
7. A. Wall, Advantages and disadvantages of solar energy. Process Industry Forum, 7th Aug 2013. (Web. 2 Feb 2014) (2014)
8. A.M. Bagher, M.M.A. Vahid, M. Mohsen, Types of solar cells and application. Am. J. Opt. Photonics **3**, 94–113 (2015)
9. J.H. Zhao, A. Wang, M. Green, 24·5% Efficiency silicon PERT cells on MCZ substrates and 24·7% efficiency PERL cells on FZ substrates. Prog. Photovolt.: Res. Appl. **7**(6), 471–474 (1999). https://doi.org/10.1002/(SICI)1099-159X (199911/12)7:6<471: AID-PIP298>3.0.CO;2-7
10. M.A. Green, The path to 25% silicon solar cell efficiency: history of silicon cell evolution. Prog. Photovolt.: Res. Appl. **17**(3) 183–189 (2009). https://doi.org/10.1002/pip.892
11. K. Masuko, M. Shigematsu, T. Hashiguchi, D. Fujishima, M. Kai, N. Yoshimura, T. Yamaguchi, Y. Ichihashi, T. Mishima, N. Matsubara, T. Yamanishi, T. Takahama, M. Taguchi, E. Maruyama, S. Okamoto, Achievement of more than 25% conversion efficiency with crystalline silicon heterojunction solar cell. IEEE J. Photovoltaics **4**(6), 1433–1435 (2014). https://doi.org/10.1109/JPHOTOV.2014.2352151
12. K. Yoshikawa, H. Kawasaki, W. Yoshida, T. Irie, K. Konishi, K. Nakano, T. Uto, D. Adachi, M. Kanematsu, H. Uzu, K. Yamamoto, Silicon heterojunction solar cell with interdigitated back contacts for a photoconversion efficiency over 26%. Nat. Energy **2**, 17032 (2017). https://doi.org/10.1038/nenergy.2017.32

13. M.A. Green, Y. Hishikawa, E.D. Dunlop, D.H. Levi, J. Hohl-Ebinger, A.W.Y. Ho-Baillie, Solar cell efficiency tables (version 51). Prog. Photovolt.: Res. Appl. **26**(1), 3–12 (2018). https://doi.org/10.1002/pip.2978
14. E.A. Alsema, Energy pay-back time and CO_2 emissions of PV systems. Prog. Photovolt.: Res. Appl. **8**(1), 17–25 (2000). https://doi.org/10.1002/(SICI)1099-159X(200001/02)8:1<17::AID-PIP295>3.0.CO;2-C
15. T. Song, S.-T. Lee, B. Sunn, Silicon nanowires for photovoltaic applications: the progress and challenge. Nano Energy **1**(5), 654 (2012). https://doi.org/10.1016/j.nanoen.2012.07.023
16. S.W. Schmitt, F. Schechtel, D. Amkreutz, M. Bashouti, S.K. Srivastava, B. Hoffmann, C. Dieker, E. Spiecker, B. Rech, S.H. Christiansen, Nanowire arrays in multicrystalline silicon thin films on glass: a promising material for research and applications in nanotechnology. Nano Lett. **12**, 4050–4054 (2012). https://doi.org/10.1021/nl301419q
17. E. Garnett, P. Yang, Light trapping in silicon nanowire solar cells. Nano Lett. **10**(3), 1082 (2010). https://doi.org/10.1021/nl100161z
18. M. Dutta, N. Fukata, Effect of shell growth and doping conditions of core-shell homojunction Si nanowire solar cells. J. Nanosci. Nanotechnol. **15**(6), 4339–4346 (2015). https://doi.org/10.1166/jnn.2015.9766
19. D. Kumar, S.K. Srivastava, P.K. Singh, M. Hussain, V. Kumar, Fabrication of silicon nanowire arrays based solar cell with improved performance. Sol. Energy Mater. Sol. Cells **95**(1), 215–218 (2011)
20. J.Y. Jung, Z. Guo, S.W. Jee, H.D. Um, K.T. Park, M.S. Hyun, J.M. Yang, J.H. Lee, A wafer scale Si wire solar cell using radial and bulk p–n junctions. Nanotechnology **21**(44), 445303 (2010). https://doi.org/10.1088/0957-4484/21/44/445303
21. I. Hwang, H.D. Um, B.S. Kim, M. Wober, K. Seo, Flexible crystalline silicon radial junction photovoltaics with vertically aligned tapered microwires. Energy Environ. Sci. **11**, 641–647 (2018). https://doi.org/10.1039/c7ee03340k
22. M.C. Putnam, S.W. Boettcher, M.D. Kelzenberg, D.B. Turner-Evans, J.M. Spurgeon, E.L. Warren, R.M. Briggs, N.S. Lewis, H.A. Atwater, Si microwire-array solar cells. Energy Environ. Sci. **3**, 1037–1041 (2010). https://doi.org/10.1039/c0ee00014k
23. M.D. Kelzenberg, S.W. Boettcher, J.A. Petykiewicz, D.B. Turner-Evans, M.C. Putnam, E.L. Warren, J.M. Spurgeon, R.M. Briggs, N.S. Lewis, H.A. Atwater, Enhanced absorption and carrier collection in Si wire arrays for photovoltaic applications. Nat. Mater. **9**, 239–244 (2010). https://doi.org/10.1038/nmat2635
24. A.C. Tamboli, C.T. Chen, E.L. Warren, D.B. Turner-Evans, M.D. Kelzenberg, N.S. Lewis, H.A. Atwater, Wafer-scale growth of silicon microwire arrays for photovoltaics and solar fuel generation. IEEE J. Photovoltaics **2**, 294–297 (2012). https://doi.org/10.1109/JPHOTOV.2012.2191941
25. A. Saini, M. Abdelhameed, D. Rani, W. Jevasuwan, N. Fukata, P. Kumari, S.K. Srivastava, P. Pathi, A. Samanta, M. Dutta, Fabrication of periodic, flexible and porous silicon microwire arrays with controlled diameter and spacing: effects on optical properties. Opt. Mater. **134**, 113181 (2022). https://doi.org/10.1016/j.optmat.2022.113181
26. K. Lee, I. Hwang, N. Kim, D. Choi, H.-D. Um, S. Kim, K. Seo, 17.6%-efficient radial junction solar cells using silicon nano/micro hybrid structure. Nanoscale **8**, 14473–14479 (2016). https://doi.org/10.1039/C6NR04611H
27. D. Choi, K. Seo, Field-induced radial junction for dopant-free crystalline silicon microwire solar cells with an efficiency of over 20%. Adv. Energy Mater. **11**, 2003707 (2021). https://doi.org/10.1002/aenm.202003707
28. D. Choi, I. Hwang, Y. Lee, M. Lee, H.-D. Um, K. Seo, Wafer-scale radial junction solar cells with 21.1% efficiency using c-Si microwires. Adv. Funct. Mater 2208377 (2022). https://doi.org/10.1002/adfm.202208377
29. B.M. Kayes, H.A. Atwater, N.S. Lewis, Comparison of the device physics principles of planar and radial p-n junction nanorod solar cells. J. Appl. Phys. **97**, 114302 (2005). https://doi.org/10.1063/1.1901835

30. H.P. Yoon, Y.A. Yuwen, C.E. Kendrick, G.D. Barbaer, N.J. Podraza, J.M. Redwing, T.E. Mallouk, C.R. Wronski, T.S. Mayer, (2010) Enhanced conversion efficiencies for pillar array solar cells fabricated from crystalline silicon with short minority carrier diffusion length. Appl. Phys. Lett. **96**, 312503 (2010)
31. H.D. Um, N. Kim, K. Lee, I. Hwang, J.H. Seo, Y.J. Yu, P. Duane, M. Wober, K. Seo, Versatile control of metal-assisted chemical etching for vertical silicon microwire arrays and their photovoltaic applications. Sci. Rep. **5**, 1–11 (2015). https://doi.org/10.1038/srep11277
32. H.-D. Um, I. Hwang, N. Kim, Y.J. Yu, M. Wober, K.-H. Kim, K. Seo, Microgrid electrode for Si microwire solar cells with a fill factor of over 80%. Adv. Mater. Interfaces **2**, 1500347 (2015)
33. Y.W. Wang, V. Schmidt, S. Senz, U. Gosele, Epitaxial growth of silicon nanowires using an aluminium catalyst. Nat. Nanotechnol. **1**, 186 (2006)
34. M. Dutta, L. Thirugnanam, K. Sato, N. Fukata, Diameter-controlled growth and impurity doping of silver colloid-seeded silicon microwires to nanowires for the realization of solar cell materials. Mater. Express **3**, 85 (2013). https://doi.org/10.1166/mex.2013.1097
35. B.M. Kayes, M.A. Filler, M.C. Putnam, M.D. Kelzenberg, N.S. Lewis, H.A. Atwater, Growth of vertically aligned Si wire arrays over large areas (>1 cm^2) with Au and Cu catalysts. Appl. Phys. Lett. **91**, 103110 (2007). https://doi.org/10.1063/1.2779236
36. J.B. Hannon, S. Kodambaka, F.M. Ross, R.M. Tromp, The influence of the surface migration of gold on the growth of silicon nanowires. Nature **440**, 69–71 (2006). https://doi.org/10.1038/nature04574
37. M.C. Putnam, D.B. Turner-Evans, M.D. Kelzenberg, S.W. Boettcher, N.S. Lewis, H.A. Atwater, 10 μm minority-carrier diffusion lengths in Si wires synthesized by Cu-catalyzed vapor-liquid-solid growth. Appl. Phys. Lett. **95**, 163116 (2009). https://doi.org/10.1063/1.3247969
38. Z.P. Huang, H. Fang, J. Zhu, Fabrication of silicon nanowire arrays with controlled diameter, length, and density. Adv. Mater. **19**, 744 (2007). https://doi.org/10.1002/adma.200600892
39. K.Q. Peng, M.L. Zhang, A.J. Lu, N.B. Wong, R.Q. Zhang, S.T. Lee, Ordered silicon nanowire arrays via nanosphere lithography and metal-induced etching. Appl. Phys. Lett. **90**, 163123 (2007). https://doi.org/10.1063/1.2724897
40. F. Wu, H. Lin, Z. Yang, M. Liao, Z. Wang, Z. Li, P. Gao, J. Ye, W. Shen, Suppression of surface and Auger recombination by formation and control of radial junction in silicon microwire solar cells. Nano Energy **58**, 817–824 (2019). https://doi.org/10.1016/j.nanoen.2019.02.021
41. J.-Y. Choi, T.L. Alford, C.B. Honsberg, Fabrication of periodic silicon nanopillars in a two-dimensional hexagonal array with enhanced control on structural dimension and period. Langmuir **31**, 4018–4023 (2015). https://doi.org/10.1021/acs.langmuir.5b00128
42. A. Saini, H. Negi, A. Srivastava, S.K. Srivastava, P. Pathi, M. Dutta, Optical properties of one pot synthesized porous, monodisperse, size-controlled silica particles: effect of tetra alkoxysilane addition rate. SILICON **16**, 2985–2996 (2024). https://doi.org/10.1007/s12633-024-02899-9
43. R.W. Wu, G.D. Yuan, K.C. Wang, T.B. Wei, Z.Q. Liu, G.H. Wang, J.X. Wang, J.M. Li, Bilayer–metal assisted chemical etching of silicon microwire arrays for photovoltaic applications. AIP Adv. **6**, 025324 (2016). https://doi.org/10.1063/1.4943217
44. Y. Kim, A. Tsao, D.H. Lee, R. Maboudian, Solvent-induced formation of unidirectionally curved and tilted Si nanowires during metal-assisted chemical etching. J. Mater. Chem. C **1**, 220 (2013). https://doi.org/10.1039/C2TC00041E
45. H.J. Choi, S. Baek, H.S. Jang, S.B. Kim, B.Y. Oh, J.H. Kim, Optimization of metal-assisted chemical etching process in fabrication of p-type silicon wire arrays. Curr. Appl. Phys. **11**, S25–S29 (2011). https://doi.org/10.1016/j.cap.2010.11.047
46. J.O. De Beeck, W.D. Malsche, D.S. Tezcan, P.D. Moor, G. Desmet, Impact of the limitations of state-of-the-art micro-fabrication processes on the performance of pillar array columns for liquid chromatography. J. Chromatogr. A **1239**, 35–48 (2012). https://doi.org/10.1016/j.chroma.2012.03.054
47. H.-D. Um, K. Lee, I. Hwang, J. Park, D. Choi, N. Kim, H. Kim, K. Seo, Progress in silicon microwire solar cells. J. Mater. Chem. A **8**, 5395 (2020). https://doi.org/10.1039/C9TA12792E

48. B. Wu, A. Kumar, S. Pamarthy, High aspect ratio silicon etch: A review. J. Appl. Phys. **108**, 051101 (2010). https://doi.org/10.1063/1.3474652
49. M.D. Henry, S. Walavalkar, A. Homyk, A. Scherer, Alumina etch masks for fabrication of high-aspect-ratio silicon micropillars and nanopillars. Nanotechnology **20**, 255305 (2009). https://doi.org/10.1088/0957-4484/20/25/255305
50. H.P. Yoon, Y.A. Yuwen, C.E. Kendrick, G.D. Barber, N.J. Podraza, J.M. Redwing, T.E. Mallouk, C.R. Wronski, T.S. Mayer, Enhanced conversion efficiencies for pillar array solar cells fabricated from crystalline silicon with short minority carrier diffusion lengths. Appl. Phys. Lett. **96**, 213503 (2010). https://doi.org/10.1063/1.3432449
51. A. Solanki, H. Um, Top-down etching of Si nanowires. Semicond. Semimetals **98**, 71–149 (2018). https://doi.org/10.1016/bs.semsem.2018.04.001
52. W. Chern, K. Hsu, I.S. Chun, B.P.D. Azeredo, N. Ahmed, K.H. Kim, J.M. Zuo, N. Fang, P. Ferreira, X. Li, Nonlithographic patterning and metal-assisted chemical etching for manufacturing of tunable light-emitting silicon nanowire arrays. Nano Lett. **10**, 1582–1588 (2010). https://doi.org/10.1021/nl903841a
53. H. Han, Z. Huang, W. Lee, Metal-assisted chemical etching of silicon and nanotechnology applications. Nano Today **9**, 271–304 (2014). https://doi.org/10.1016/j.nantod.2014.04.013
54. S. Schaefer, R. Lüdemann, Low damage reactive ion etching for photovoltaic applications. J. Vac. Sci. Technol. A: Vac. Surf. Films **17**, 749–754 (1999). https://doi.org/10.1116/1.581644
55. X. Liu, P.R. Coxon, M. Peters, B. Hoex, J.M. Cole, D.J. Fray, Black silicon: fabrication methods, properties and solar energy applications. Energy Environ. Sci. **7**, 3223–3263 (2014). https://doi.org/10.1039/C4EE01152J
56. S.H. Zaidi, D.S. Ruby, J.M. Gee, Characterization of random reactive ion etched-textured silicon solar cells. IEEE Trans. Electron Devices **48**, 1200–1206 (2001). https://doi.org/10.1109/16.925248
57. K. Seo, Y.J. Yu, P. Duane, W. Zhu, H. Park, M. Wober, K.B. Crozier, Si microwire solar cells: improved efficiency with a conformal SiO_2 layer. ACS Nano **7**, 5539–5545 (2013). https://doi.org/10.1021/nn401776x
58. G. Dingemans, W.M.M. Kessels, Status and prospects of Al_2O_3-based surface passivation schemes for silicon solar cells. J. Vac. Sci. Technol. A **30**, 040802 (2012). https://doi.org/10.1116/1.4728205
59. B. Tian, X. Zheng, T.J. Kempa, Y. Fang, N. Yu, G. Yu, J. Huang, C.M. Lieber, Coaxial silicon nanowires as solar cells and nanoelectronic power sources. Nature **449**, 885 (2007). https://doi.org/10.1038/nature06181
60. M.D. Kelzenberg, D.B. Turner-Evans, M.C. Putnam, S.W. Boettcher, R.M. Briggs, J.Y. Baek, N.S. Lewis, H.A. Atwater, High-performance Si microwire photovoltaics. Energ. Environ. Sci. **4**, 866 (2011). https://doi.org/10.1039/c0ee00549e
61. E. Maruyama, A. Terakawa, M. Taguchi, Y. Yoshimine, D. Ide, T. Baba, M. Shima, H. Sakata, M. Tanaka, Sanyo's challenges to the development of high-efficiency HIT solar cells and the expansion of HIT business. in *Conference Record of the IEEE 4th World Conference on Photovoltaic Energy Conversion (WCPEC)*, Waikoloa, HI (2006)
62. P. Doshi, G.E. Jellison, A. Rohatgi, Characterization and optimization of absorbing plasma-enhanced chemical vapor deposited antireflection coatings for silicon photovoltaics. Appl. Optics **36**, 7826 (1997). https://doi.org/10.1364/AO.36.007826
63. Z. Yang, G. Cao, A. Shang, D.Y. Lei, C. Zhang, P. Gao, J. Ye, X. Li, Enhanced photo-electrical response of hydrogenated amorphous silicon single-nanowire solar cells by front-opening crescent design. Nanoscale Res. Lett. **11**, 233 (2016). https://doi.org/10.1186/s11671-016-1447-0
64. J. Yoo, S.A. Dayeh, W. Tang, S.T. Picaraux, Epitaxial growth of radial p-i-n junctions for photovoltaic applications. Appl. Phys. Lett. **102**, 093113 (2013). https://doi.org/10.1063/1.4794541
65. X. Wang, K.L. Pey, C.H. Yip, E.A. Fitzgerald, D.A. Antoniadis, Vertically arrayed Si nanowire/nanorod-based core-shell p-n junction solar cells. J. Appl. Phys. **108**, 124303 (2010). https://doi.org/10.1063/1.3520217

66. C.E. Kendrick, H.P. Yoon, Y.A. Yuwen, G.D. Barber, H. Shen, T.E. Mallouk, E.C. Dickey, T.S. Mayer, J.M. Redwing, Radial junction silicon wire array solar cells fabricated by gold-catalyzed vapor-liquid-solid growth. Appl. Phys. Lett. **97**, 143108 (2010). https://doi.org/10.1063/1.3496044
67. Y. Lu, A. Lal, High-efficiency ordered silicon nano-conical-frustum array solar cells by self-powered parallel electron lithography. Nano Lett. **10**, 4651–4656 (2010). https://doi.org/10.1021/nl102867a
68. M.D. Kelzenberg, D.B. Turner-Evans, B.M. Kayes, M.A. Filler, M.C. Putnam, N.S. Lewis, H.A. Atwater, Photovoltaic measurements in single-nanowire silicon solar cells. Nano Lett. **8**, 710 (2008). https://doi.org/10.1021/nl072622p
69. J.E. Allen, E.R. Hemesath, D.E. Perea, J.L. Lensch-Falk, Z.Y. Li, F. Yin, M.H. Gass, P. Wang, A.L. Bleloch, R.E. Palmer, L.J. Lauhon, High-resolution detection of Au catalyst atoms in Si nanowires. Nat. Nanotechnol. **3**, 168 (2008). https://doi.org/10.1038/nnano.2008.5
70. D.R. Kim, C.H. Lee, P.M. Rao, I.S. Cho, X. Zheng, Hybrid Si microwire and planar solar cells: passivation and characterization. Nano Lett. **11**, 2704–2708 (2011). https://doi.org/10.1021/nl2009636
71. H.-S. Kim, D.B. Patel, H. Kim, M. Patel, K.R. Chauhan, W. Park, J. Kim, Electrical and optical properties of Si microwire solar cells. Sol. Energy Mater. Sol. Cells **164**, 7–12 (2017). https://doi.org/10.1016/j.solmat.2017.01.046
72. H. Kim, J. Kim, E. Lee, D.-W. Kim, J.-H. Yun, J. Yi, Effect of the short collection length in silicon microscale wire solar cells. Appl. Phys. Lett. **102**, 193904 (2013). https://doi.org/10.1063/1.4804581
73. N. Kim, D. Choi, H. Kim, H.-D. Um, K. Seo, Silicon microwire arrays with nanoscale spacing for radial junction c-Si solar cells with an efficiency of 20.5%. ACS Nano **15**, 14756–14765 (2021). https://doi.org/10.1021/acsnano.1c04585
74. S. Jeong, M.D. McGehee, Y. Cui, All-back-contact ultra-thin silicon nanocone solar cells with 13.7% power conversion efficiency. Nat. Commun. **4**, 2950 (2013). https://doi.org/10.1038/ncomms3950
75. S. Zhou, Z. Yang, P. Gao, X. Li, X. Yang, D. Wang, J. He, Z. Ying, J. Ye, Wafer-scale integration of inverted nanopyramid arrays for advanced light trapping in crystalline silicon thin film solar cells. Nanoscale Res. Lett. **11**, 194 (2016). https://doi.org/10.1186/s11671-016-1397-6
76. K.E. Plass, M.A. Filler, J.M. Spurgeon, B.M. Kayes, S. Maldonado, B.S. Brunschwig, H.A. Atwater, N.S. Lewis, Flexible polymer-embedded Si wire arrays. Adv. Mater. **21**, 325–328 (2009). https://doi.org/10.1002/adma.200802006
77. E.H. Ko, H.J. Kim, S.J. Lee, J.H. Lee, H.K. Kim, Nano-sized Ag inserted into ITO films prepared by continuous roll-to-roll sputtering for high-performance, flexible, transparent film heaters. RSC Adv. **6**, 46634–46642 (2016). https://doi.org/10.1039/C6RA08704C
78. J.H. Seo, I. Hwang, H.-D. Um, S. Lee, K. Lee, J. Park, H. Shin, T.-H. Kwon, S.J. Kang, K. Seo, Cold isostatic-pressured silver nanowire electrodes for flexible organic solar cells via room-temperature processes. Adv. Mater. **29**, 1701479 (2017). https://doi.org/10.1002/adma.201701479
79. H. Hu, S. Wang, S. Wang, G. Liu, T. Cao, Y. Long, Aligned silver nanowires enabled highly stretchable and transparent electrodes with unusual conductive property. Adv. Funct. Mater. **29**, 1902922 (2019). https://doi.org/10.1002/adfm.201902922
80. S.M. Shin, J.Y. Jung, K.T. Park, H.D. Um, S.W. Jee, Y.H. Nam, J.H. Lee, A novel wrap-around metal contact optimized for radial p–n junction Si wire solar cells. Energy Environ. Sci. **6**, 1756–1760 (2013). https://doi.org/10.1039/C3EE24494F

MIX
Papier aus verantwortungsvollen Quellen
Paper from responsible sources
FSC® C105338

If you have any concerns about our products,
you can contact us on
ProductSafety@springernature.com

In case Publisher is established outside the EU,
the EU authorized representative is:
Springer Nature Customer Service Center GmbH
Europaplatz 3, 69115 Heidelberg, Germany

Printed by Libri Plureos GmbH
in Hamburg, Germany